T0184319

Lecture Notes in Computer Science 11436

Commenced Publication in 1973
Founding and Former Series Editors:
Gerhard Goos, Juris Hartmanis, and Jan van Leeuwen

More information about this series at http://www.springer.com/series/7407

T. V. Gopal · Junzo Watada (Eds.)

Theory and Applications of Models of Computation

15th Annual Conference, TAMC 2019
Kitakyushu, Japan, April 13–16, 2019
Proceedings

 Springer

Editors
T. V. Gopal
Anna University
Chennai, India

Junzo Watada
Waseda University
Kitakyushu, Japan

ISSN 0302-9743   ISSN 1611-3349   (electronic)
Lecture Notes in Computer Science
ISBN 978-3-030-14811-9   ISBN 978-3-030-14812-6   (eBook)
https://doi.org/10.1007/978-3-030-14812-6

Library of Congress Control Number: 2019932917

LNCS Sublibrary: SL1 – Theoretical Computer Science and General Issues

This Springer imprint is published by the registered company Springer Nature Switzerland AG
The registered company address is: Gewerbestrasse 11, 6330 Cham, Switzerland

Preface

Theory and Applications of Models of Computation (TAMC) is a series of annual conferences that aim to bring together a wide range of researchers with interest in computational theory and its applications.

There is a need for models. Most scientists begin with either explicit or implicit "word models" that describe their vision of how a system possibly works. The process of turning a word model into a formal mathematical model invariably forces the proponent to confront his or her hidden assumptions. The process ought to be supported by experimentation. Most of the experimental work tends to be preoccupied with the myriad and mundane details that are so crucial to doing experiments and analyzing data. Measurement is always an empirical procedure. Quantification is a kind of theorizing. The relationship between quantification and measurement is a "feedback loop." It is true that the patterns of data thus formulated may suggest theories and/or models. Ideally, models ought to be intermediate between theory and data. The distinction between theories and models is blurred. However, data do not impose themselves upon the research. "No data or missing data" are a widespread problem is several disciplines of research. There are some theories or models that may have no link with data. Theory then is nothing short of legitimized dreaming.

The solution for such problems is very difficult in computational approaches but is dominant in domains of research such as biology and cognitive sciences. The quest is for responsive and responsible theories and models of computation to address the evolving complex and interdisciplinary dynamics of the problem domains.

The TAMC series of conferences have a strong interdisciplinary character, bringing together researchers working in computer science, mathematics, and the physical sciences with predominantly computational and computability theoretic focus. The most important theoretical aspects of a model of computation are its *power, generality, simplicity, synthesizability, verifiability, and expressiveness*. The TAMC series of conferences explore the algorithmic foundations, computational methods, and computing devices to meet the rapidly emerging challenges of complexity, scalability, sustainability, and interoperability, with wide-ranging impacts on virtually every aspect of human endeavor. The TAMC series is distinguished by an appreciation for mathematical depth, scientific—rather than heuristic—approaches and the integration of theory and implementation. The quality of the conference has caught the attention of professionals all over the world, who eagerly look forward to the TAMC series.

The main themes of TAMC 2019 were computability, computer science logic, complexity, algorithms, models of computation, and systems theory. TAMC took place in Japan after a gap of 8 years with special sessions on "Soft Computing and AI Models." TAMC 2011 was held in Tokyo, Japan. TAMC 2019 was organized in association with the International Society of Management Engineers [ISME], Japan.

The review process was rigorous. There were at least two reviews for every paper. The authors of the papers and the reviewers are from 36 countries reflecting the international status of the TAMC series of conferences.

We are very grateful to the Program Committee of TAMC 2019 and the external reviewers they called on, for the hard work and expertise that they brought to the difficult selection process. We thank all those authors who submitted their work for our consideration. We thank the members of the Editorial Board of *Lecture Notes in Computer Science* and the editors at Springer for their encouragement and cooperation throughout the preparation of this conference.

January 2019

T. V. Gopal
Junzo Watada

Organization

TAMC 2019 was organized at the Kitakyushu International Conference Center, Kokura Station, Kitakyushu (City), Japan, in association with the International Society of Management Engineers [ISME], Japan.

Steering Committee

M. Agrawal	Indian Institute of Technology, Kanpur, India
Jin-Yi Cai	University of Wisconsin, USA
J. Hopcroft	Cornell University, USA
A. Li	Beihang University, Beijing, China
Z. Liu	Institute of Computing Technology, Chinese Academy of Sciences, China

Conference and Program Committee Chair

Junzo Watada   Waseda University, Japan

Program Committee Co-chair

T. V. Gopal   Anna University, Chennai, India

Finance Committee Chair

Yoshiyuki Matsumoto   Shimonoseki City University, Japan

Registration Committee Chair

Yoshiyuki Yabuuchi   Shimonoseki City University, Japan

Registration Committee Co-Chair

Hiroshi Sakai   Kyushu Institute of Technology, Japan

Program Committee

Manindra Agrawal	Indian Institute of Technology, Kanpur, India
Klaus Ambos-Spies	Heidelberg University, Germany
Abdel Monim Artoli	King Saud University, Saudi Arabia

Valentina Emilia Balas	Aurel Vlaicu University of Arad, Romania
Hans L. Bodlaender	Utrecht University/Eindhoven University of Technology, The Netherlands
Anthony Bonato	Ryerson University, Canada
Cristian S. Calude	The University of Auckland, New Zealand
Venkatesan Chakaravarthy	IBM Research, India
Partha Pratim Das	Indian Institute of Technology, Kharagpur, India
Maya Dimitrova	Institute of Robotics, Bulgarian Academy of Sciences, Bulgaria
Antonio Fernandez Anta	IMDEA Networks Institute, Spain
Toshihiro Fujito	Toyohashi University of Technology, Japan
Wu Guohua	Nanyang Technological University
Kouichi Hirata	Kyushu Institute of Technology, Japan
John Hopcroft	Cornell University, USA
Aaron D. Jaggard	U.S. Naval Research Laboratory, USA
Pushkar Joglekar	Vishwakarma Institute of Technology, Pune, India
Jan Kratochvil	Charles University, Czech Republic
Roman Kuznets	TU Wien, Austria
Steffen Lempp	University of Wisconsin-Madison, USA
Angsheng Li	Beihang University
Zhiyong Liu	Institute of Computing Technology, Chinese Academy of Science, China
Johann Makowsky	Technion Israel Institute of Technology, Israel
Klaus Meer	BTU Cottbus-Senftenberg, Germany
Philippe Moser	National University of Ireland, Maynooth, Ireland
Masaaki Nagahara	The University of Kitakyushu, Japan
Pan Peng	University of Sheffield, Sheffield, UK
Jose Rolim	University of Geneva, Switzerland
Hiroshi Sakai	Kyushu Institute of Technology, Japan
Yaroslav Sergeyev	University of Calabria, Italy
Rudrapatna Shyamasundar	Tata Institute of Fundamental Research and Indian Institute of Technology, Mumbai
Frank Stephan	National University of Singapore, Singapore
Gopal Tadepalli	Anna University, India
Junzo Watada	Waseda University, Japan
Yoshiyuki Yabuuchi	Shimonoseki City University, Japan
Thomas Zeugmann	Hokkaido University, Japan
Naijun Zhan	Institute of Software, Chinese Academy of Sciences, China

Additional Reviewers

Akshay, S.
B. Krishna Kumar
Fujiwara, Hiroshi
Ishii, Toshimasa
Jiao, Li
Kar, Purushottam
Kimura, Kei
Kucera, Antonin

Liu, Jiang
Meadows, Catherine
N. R., Aravind
Phawade, Ramachandra
R. S. Bhuvaneswaran
Rao, Raghavendra
Siromoney, Arul
Xue, Bai

In Association with:

Contents

Battery Scheduling Problem

Aakash Agrawal[1], Krunal Shah[1], Amit Kumar[1(✉)], and Ranveer Chandra[2]

[1] Indian Institute of Technology, Delhi, New Delhi, India
{ee1150421,ee1150476,amitk}@iitd.ac.in
[2] Microsoft Corporation, Redmond, USA
ranveer@microsoft.com
http://www.iitd.ac.in, https://www.microsoft.com

Abstract. Different batteries have different desirable properties like energy density, peak power, recharge time, longevity, efficiency, etc. So, it is beneficial if we multiplex different types of batteries in a single device. In this paper, we look at ways of scheduling workloads over the multiplexed batteries to maximize the overall efficiency. We consider two ways to model the efficiency and give efficient solutions to the same.

Keywords: Software defined batteries · Battery scheduling

1 Introduction

Software defined batteries seek to leverage the desirable properties of various types of batteries by multiplexing them. Different types of batteries have different desirable properties like energy density, cost, peak power, recharge time, longevity, and efficiency. Using a single kind of battery means compromising on other desirable features. Hence the hardware-software system, Software Defined Battery (SDB) presented in the paper by *Anirudh et al.* [1] which allows integrating the different desirable properties of different chemistries by dynamically scheduling the charge flowing in and out of each battery presents itself as an exceptional solution.

Software defined batteries can have a wide range of applications from laptops, mobile phones to drones, etc. [3]. One of the key applications of software defined batteries is battery driven vehicles [5]. In this paper, we initiate the study of the manner in which batteries should be multiplexed given a fixed workload at every point of time. In terms of battery-driven vehicles, this would mean that we are told the speed of the vehicle at every point of time, and then we want to figure out the optimal manner (in terms of energy) in which the batteries should be utilized. In a different setting, we consider the problem where the path followed by the vehicle is given to us, and we figure out the optimal usage of the batteries which trade-off time taken for travel and energy consumption. This formulation is motivated by the problem formulated in Energy-efficient algorithms for flow time minimization [6].

© Springer Nature Switzerland AG 2019
T. V. Gopal and J. Watada (Eds.): TAMC 2019, LNCS 11436, pp. 1–12, 2019.
https://doi.org/10.1007/978-3-030-14812-6_1

For both these problems, we design algorithms based on convex programming relaxations. We show that optimal solutions have nice properties, which allow us to solve these problems efficiently.

2   Preliminaries

We give the notation used in this paper, and then describe the problems considered. We are given n batteries $B_1, \ldots, B_n$. Each battery B_i has an internal resistance r_i which is assumed to be fixed throughout the lifespan of the battery. Further, it has an initial charge capacity $\mathtt{cap}_i$.

We consider two different problems. In the first problem, denoted EnergyMinimization, we are given *current* requirements $I_1, \ldots, I_T$ for each time $t \in \{1, \ldots, T\}$. A solution to such an instance has to specify the *current* from each of the batteries at each time step, i.e., quantities $i_{j,t}$ for all batteries $j = 1, \ldots, n$ and time $t = 1, \ldots, T$. These quantities should satisfy the following conditions: (i) for each time t, $\sum_j i_{j,t} = I_t$, and (ii) for all batteries B_j, the total *current* drawn from B_j, i.e., $\sum_t i_{j,t}$ should not exceed $\mathtt{cap}_j$. The goal is to minimize the total energy loss, i.e.,

$$\sum_{t=1}^{T}\sum_{j=1}^{n} i_{j,t}^2 \cdot r_i$$

In the second problem, denoted EnergyTimeTradeoff, we are given m ordered workload requirements[1] $W_1, \ldots, W_m$ and their corresponding positive weights $\alpha_1, \ldots, \alpha_m$. As in the previous problem, a solution to such an instance has to specify the *current* $i_{j,t}$ from each of the batteries B_j at each time step t. Let the times $t_1, \ldots, t_m$ denote the time taken to complete the corresponding workloads and let $T = \sum_{k=1}^{m} t_k$. These m tasks are performed in the given sequence. These quantities should satisfy conditions that for all batteries B_j, the total *current* drawn from B_j, i.e., $\sum_t i_{j,t}$ should not exceed $\mathtt{cap}_j$. The goal is to minimize a (weighted) sum of total energy consumed and the time taken to finish each of the m tasks, i.e.,

$$\sum_{t}\sum_{j=1}^{n} i_{jt}^2 \cdot r_j + \sum_{k=1}^{m} \alpha_k \cdot t_k$$

We shall use W to denote $\sum_{k=1}^{m} W_k$ and r_{eq} to denote the effective resistance of the batteries in parallel i.e. $\frac{1}{r_{eq}} = \sum_{j=1}^{n} \frac{1}{r_j}$.

In both EnergyMinimization and EnergyTimeTradeoff, we take the domain of *current* from $(-\infty, \infty)$ because we consider a negative value of current at a given instant for a battery to imply that the battery is being charged at that instant by the current drawn from other batteries. This assumption of ours is backed in the paper by *Anirudh et al.* [1] where they describe how one battery can be used to charge another battery in Software Defined Batteries [1].

---

[1] We will use the terms *workload* and *charge* interchangeably.

2.1 Our Results

For the two problems, we give efficient algorithms for finding the optimal solutions.

Theorem 1. *For the* `EnergyMinimization` *problem, there is an optimal algorithm with running time* $\mathcal{O}(n\log(n) + nm)$, *where* m *is the length of the time period.*

Theorem 2. *For the* `EnergyTimeTradeoff` *problem, there is an optimal algorithm with running time* $\mathcal{O}((n\log(n) + nm) \cdot \log T \cdot \log(\frac{mW^2 r_{eq}}{T^2}))$.

We observe that we could state the problems as a convex program, but solving a convex program in general can take much more time. We exploit the structure of an optimal solution to get a much simpler algorithm for both the problems.

In Sect. 3, we describe our algorithm for the `EnergyMinimization` problem. Subsequently in Sect. 4, we give the algorithm for the `EnergyTimeTradeoff` problem.

3 Algorithm for `EnergyMinimization`

In this section, we describe the algorithm for the `EnergyMinimization` problem, and subsequently show that it has the desired properties. Recall that an instance $\mathcal{I}$ consists of n batteries and workload requirement I_t at each time step t. The algorithm is given formally in Algorithm 1. We define W to be the total charge requirement, i.e., $\sum_t I_t$. The algorithm consists of two parts. In the first part, it figures out the total charge C_j which will be utilized by battery B_j, i.e., $\sum_t i_{j,t}$. We know that for a current flowing through a set of resistances in parallel, energy loss is minimized when the current through the resistors is distributed proportional to the inverse of their resistance values. This gives us the intuition for our greedy algorithm to assign C_j values to the batteries. We first try to distribute the charge contribution to batteries in their inverse resistance ratios but if a battery's capacity falls short of this assignment we assign it's contribution as it's capacity and then calculate the distribution of the remaining workload for the remaining batteries. Once we know the quantities C_j, the algorithm constructs $i_{j,t}$ in the second part of the algorithm. The formulation for the $i_{j,t}$ values is obtained by solving the corresponding Lagrangian optimization problem.

We now describe the first part of the algorithm in detail. We arrange the batteries in increasing order of $\mathtt{cap}_j \cdot r_j$ values in Step 2. It iteratively finds the C_j values of the batteries in this order. The algorithm maintains a quantity R which is the total charge required from the batteries which have not been considered so far. Initially, this quantity is W, the total charge requirement. It also maintains a variable r_{eq} which is the effective resistance (in parallel) of batteries which have yet to be assigned C_j values. In Step 6, we find the smallest indexed battery B_j in the set $\mathbf{B}$ for which the quantity $\mathtt{cap}_j \cdot r_j$ exceeds $R \cdot r_{eq}$.

Now if this battery happens to be the first indexed battery in the $\mathbf{B}$, we assign $C_{j'}$ values to all remaining batteries in Step 8; otherwise we assign $C_{j'}$ values to all batteries whose index j' is less than j in Step 11. We iterate after updating $R, r_{eq}, \mathbf{B}$ suitably. In Step 17, we define the corresponding $i_{j,t}$ values.

Algorithm 1. Assignment algorithm

1: $R \leftarrow W$
2: $\mathbf{B} = \{B_1, B_2, ...B_n\}$, where $\mathsf{cap}_1 \cdot r_1 \leq \mathsf{cap}_2 \cdot r_2 \ldots \mathsf{cap}_n \cdot r_n$.
3: $\frac{1}{r_{eq}} = \sum_i \frac{1}{r_i}$
4: **while $\mathbf{B}$ is not empty do**
5:     Let $j^\star$ be the smallest index of a battery in $\mathbf{B}$
6:     find the smallest index j such that $B_j \in \mathbf{B}$ and $\mathsf{cap}_j \geq \frac{R \cdot r_{eq}}{r_j}$
7:     **if** $j = j^\star$ **then**
8:         $C_{j'} = \frac{R \cdot r_{eq}}{r_{j'}}$ $\forall B_{j'} \in \mathbf{B}$
9:         $\mathbf{B} \leftarrow \emptyset$.
10:    **else**
11:        $C_{j'} = \mathsf{cap}_{j'}$ $\forall j' \in \{j^\star, j-1\}$
12:        $R = R - \sum_{j'=j^\star}^{j-1} C_{j'}$
13:        $\frac{1}{r_{eq}} = \frac{1}{r_{eq}} - \sum_{j=j^\star}^{j-1} \frac{1}{r_{j'}}$
14:        $\mathbf{B} = \mathbf{B} - \{B_{j^\star}, \ldots, B_{j-1}\}$
15: Define $\frac{1}{r_{eq}} = \sum_i \frac{1}{r_i}$, $\langle I_t \rangle = \frac{W}{T}$
16: **for** all batteries B_j and time $t = 1, \ldots T$ **do**
17:    $i_{j,t} \leftarrow \frac{C_j}{T} + \frac{r_{eq}}{r_j}(I_t - \langle I_t \rangle)$

3.1 Analysis

In this section, we analyze our algorithm. The analysis again proceeds along two steps. The first step shows that the computed C_j values are indeed same as the corresponding quantities for an optimal solution. In the second step, we show that the computed $i_{j,t}$ values minimize the total energy.

We consider an optimal solution $\mathcal{O}$. Let $i^{\mathcal{O}}_{j,t}$ be the *current* drawn from battery B_j at time t in this solution. Let $C^{\mathcal{O}}_j$ be the total charge from the battery B_j, i.e., $C^{\mathcal{O}}_j = \sum_t i^{\mathcal{O}}_{j,t}$. We begin with a characterization of the total charge used from each battery. From here on the index of a battery represents its index in the list of batteries sorted in increasing order of their $\mathsf{cap}_j \cdot r_j$ values.

Lemma 1. *If there exist indices k, j such that $C^{\mathcal{O}}_k \cdot r_k < C^{\mathcal{O}}_j \cdot r_j$, then $C^{\mathcal{O}}_k = \mathsf{cap}_k$.*

Proof. The proof is by contradiction. So assume that there exist indices k, j such that $C^{\mathcal{O}}_k \cdot r_k < C^{\mathcal{O}}_j \cdot r_j$, but $C^{\mathcal{O}}_k < \mathsf{cap}_k$. Since $C^{\mathcal{O}}_l = \sum_t i^{\mathcal{O}}_{l,t}$, it follows that there exists a time t for which $i^{\mathcal{O}}_{k,t} r_k < i^{\mathcal{O}}_{j,t} r_j$. At this time t, the total energy consumed by the these two batteries is given by

$$(i^{\mathcal{O}}_{k,t})^2 \cdot r_k + (i^{\mathcal{O}}_{j,t})^2 \cdot r_j$$

Now consider modifying this solution by increasing $i_{k,t}^{\mathcal{O}}$ by ε and decreasing $i_{j,t}^{\mathcal{O}}$ by ε, where ε is a small enough positive quantity. This can be done because we have not consumed the entire charge of battery B_k. Also, the total *current* requirement at time t does not change, and so, we still have a feasible solution. But now, the total energy consumed by these two batteries at time t is given by

$$(i_{k,t}^{\mathcal{O}} + \varepsilon)^2 r_k + (i_{j,t}^{\mathcal{O}} - \varepsilon)^2 r_j = (i_{k,t}^{\mathcal{O}})^2 r_k + (i_{j,t}^{\mathcal{O}})^2 r_j + 2\varepsilon \cdot (i_{k,t}^{\mathcal{O}} r_k - i_{j,t}^{\mathcal{O}} r_j)$$
$$+ \varepsilon^2 \cdot (r_k + r_j)$$
$$< (i_{k,t}^{\mathcal{O}})^2 r_k + (i_{j,t}^{\mathcal{O}})^2 r_j$$

where the last inequality follows because $i_{k,t}^{\mathcal{O}} r_k < i_{j,t}^{\mathcal{O}} r_j$ and we can always choose $0 < \varepsilon < \min\left(\mathsf{cap}_k - C_k^{\mathcal{O}}, 2 \cdot \frac{i_{j,t}^{\mathcal{O}} r_j - i_{k,t}^{\mathcal{O}} r_k}{r_k + r_j}\right)$. But this is a contradiction because we have a new solution which consumes less energy than $\mathcal{O}$. ∎

We now consider our algorithm. Recall that the jobs are ordered in ascending order of $\mathsf{cap}_j \cdot r_j$ values. Let C_j denote the value computed by the algorithm corresponding to battery B_j. Our first observation is that $C_j \cdot r_j$ values are also in ascending order.

Lemma 2. *If $k < j$, then $C_k \cdot r_k \leq C_j \cdot r_j$.*

Proof. If batteries B_j and B_k were assigned C_j and C_k values in the same iteration then we know that $C_k \cdot r_k \leq C_j \cdot r_j$ from Step 8 and Step 11 and the fact that the batteries are sorted in ascending order of their $\mathsf{cap} \cdot r$ values.

Let R_l, r_{eq}^l denote the value of R and r_{eq} at the beginning of iteration l, S^l be the set of batteries removed in iteration l and B^l be the set of batteries yet to be assigned C values before the start of iteration l. From the algorithm, we can see that for any battery B_j, removed in l^{th} iteration $C_j \leq \frac{R_l \cdot r_{eq}^l}{r_j}$ (from Step 6, 8 and 11). Using this we show that the value of $R_l \cdot r_{eq}^l$ keeps on increasing with iteration l.

$$R_{l+1} \cdot r_{eq}^{l+1} = \left(R_l - \sum_{B_p \in S^l} C_p\right) \cdot r_{eq}^{l+1} \geq \left(R_l - \sum_{B_p \in S^l} R_l \cdot \frac{r_{eq}^l}{r_p}\right) \cdot r_{eq}^{l+1}$$

$$= R_l \cdot r_{eq}^{l+1} \cdot \left(1 - \frac{\sum_{B_p \in S^l}\left(\frac{1}{r_p}\right)}{\sum_{B_q \in B^l}\left(\frac{1}{r_q}\right)}\right)$$

$$= R_l \cdot r_{eq}^{l+1} \cdot \left(\frac{\sum_{B_q \in B^l}\left(\frac{1}{r_q}\right) - \sum_{B_p \in S^l}\left(\frac{1}{r_p}\right)}{\sum_{B_q \in B^l}\left(\frac{1}{r_q}\right)}\right)$$

$$= R_l \cdot r_{eq}^{l+1} \cdot \left(\frac{\sum_{B_p \in B^{l+1}}\left(\frac{1}{r_p}\right)}{\sum_{B_p \in B^l}\left(\frac{1}{r_p}\right)}\right)$$

$$= R_l \cdot r_{eq}^{l+1} \cdot \left(\frac{r_{eq}^l}{r_{eq}^{l+1}}\right)$$

$$= R_l \cdot r_{eq}^l$$

Thus we know that the value of $R_l \cdot r_{eq}^l$ increases with iteration l. Now we observe that since $k < j$, if B_j is assigned C_j at Step 11 then B_k must also be assigned C_k at Step 11, hence $C_j = \mathsf{cap}_j \implies C_k = \mathsf{cap}_k$, in this case we know that the lemma holds since the batteries are sorted in increasing $\mathsf{cap}_j \cdot r_j$ values. In the other case, assume batteries B_j and B_k are removed from B in the jth and $k^{th}(< j)$ iteration respectively, then $C_j \cdot r_j = R_j \cdot r_{eq}^j \geq R_k \cdot r_{eq}^k \geq C_k \cdot r_k$. ■

We now show that the C_j values computed by our algorithm also satisfy the conditions of Lemma 1.

Lemma 3. *Let $C_1, \ldots, C_n$ be the quantities computed by Algorithm 1. If there exist indices k, j such that $C_k \cdot r_k < C_j \cdot r_j$, then $C_k = \mathsf{cap}_k$.*

Proof. We prove this by contradiction. Suppose B_k and B_j are two batteries for which $C_k \cdot r_k < C_j \cdot r_j$, but $C_k \neq \mathsf{cap}_k$. Lemma 2 implies that $k < j$ because if $j < k$, then from Lemma 2, $C_k \cdot r_k \geq C_j \cdot r_j$ which violates our assumption. It must be the case that battery C_k is assigned in Step 8 of Algorithm 1, otherwise this quantity is equal to cap_k. Since $j > k$, battery B_j will also get assigned C_j value in this step, and $C_j \cdot r_j$ will be same as $C_k \cdot r_k$, a contradiction. ■

We now show that the computed C_j values are feasible.

Lemma 4. *Assuming $\sum_j \mathsf{cap}_j \geq W$, the computed values C_j satisfy the property that $\sum_j C_j = W$. Further, $C_j \leq \mathsf{cap}_j$ for all batteries B_j.*

Proof. We first need to show that the algorithm will always find a job j in Step 6. This follows from the following invariant: if **B** is the set of batteries as defined in the algorithm at some time t, and R is the remaining workload at this time, then $\sum_{B_j \in \mathbf{B}} \mathsf{cap}_j \geq R$. This is true at the beginning by assumption. Suppose it is also true at some time, and the algorithm executes Step 11. Then both LHS and RHS in the above invariant decrease by the same quantity, and so the invariant continues to hold. When it executes Step 8, the total C_j values of batteries in **B** is equal to R, and so, the invariant now holds because both sides are 0.

So we can assume that the invariant holds at all times. Now, if no such battery B_j is found in Step 6, then it implies that for all $B_j \in \mathbf{B}$, $\mathsf{cap}_j < \frac{R \cdot r_{eq}}{r_j}$. Adding these, we see that $\sum_{B_j \in \mathbf{B}} \mathsf{cap}_j < R$, which is a violation of the invariant.

Having shown this, we now show that $\sum_j C_j$ is equal to W. The proof again uses the same arguments as above. We consider two quantities at any time t during the algorithm $\Delta(t)$ is defined as the remaining workload R at this time, and $\Lambda(t)$ is defined as $\sum_{j \in \mathbf{B}} C_j$, where **B** is the remaining set batteries at time t. Note that in each iteration of the while loop in Step 4, the algorithm either executes Step 11 or Step 8. When it executes Step 8, it comes out of the while loop. Whenever it executes Step 11, the quantities $\Delta(t)$ and $\Lambda(t)$ reduce by the same quantities. Now consider the final iteration when it executes Step 8. Let R be the value of $\Delta(t)$ at this time. Then the total charge assigned to all the batteries in **B** is equal to R as well. This proves the lemma.

The final statement follows similarly. We just need to check Step 8. In this step, the C_j assigned to any battery B_j is at most cap_j by the condition leading to this step. ■

Finally, we show that the C_j values match the C_j^O values.

Theorem 3. *For each battery B_j, $C_j^O = C_j$.*

Proof. Suppose the statement is not true. Since $\sum_j C_j^O = W$ and Lemma 4 shows that $\sum_j C_j = W$ as well, it follows that there are batteries B_j and B_k such that $C_j^O < C_j$, but $C_k^O > C_k$. Now consider several cases:

- $C_j \cdot r_j \leq C_k \cdot r_k$: Observe that $C_j^O \cdot r_j < C_j \cdot r_j \leq C_k \cdot r_k < C_k^O \cdot r_k$. Lemma 1 now implies that $C_j^O = \mathsf{cap}_j$. But then $C_j > \mathsf{cap}_j$, which is not possible (by Lemma 4).
- $C_j \cdot r_j > C_k \cdot r_k$: Lemma 3 implies that $C_k = \mathsf{cap}_k$, and so, $C_k^O > \mathsf{cap}_k$, again a contradiction.

This proves the desired result. ∎

This shows the correctness of the first part of algorithm. Now, we come to the assignment of $i_{j,t}$ values.

Note that $i_{j,t}$ values must be an optimal solution to the following convex program:

$$\min . \sum_{t=1}^{T} \sum_{j=1}^{n} i_{j,t}^2 \cdot r_j$$

$$\sum_{j=1}^{n} i_{j,t} = I_t \quad \forall t \qquad (1)$$

$$\sum_{t=1}^{T} i_{j,t} = C_j \quad \forall j \qquad (2)$$

We write the Lagrangian function for the above convex program as follows:

$$L(i_{j,t}, \lambda_t, \mu_j) = \sum_t \sum_j i_{j,t}^2 r_j - \sum_t [\lambda_t (\sum_j i_{j,t} - I_t)] - \sum_j [\mu_j (\sum_t i_{j,t} - C_j)]$$

For each battery B_j and time t, we get

$$\frac{\partial L}{\partial i_{j,t}} = 2 i_{jt} r_j - \lambda_t - \mu_j = 0,$$

which implies that

$$i_{j,t} = \frac{\lambda_t + \mu_j}{2 r_j} \qquad (3)$$

Combining the above with Eqs. (1) and (2), we see that

$$i_{j,t} = \frac{C_j}{T} + \frac{r_{eq}}{r_j}(I_t - \langle I_t \rangle),$$

where $\quad \langle I_t \rangle = \frac{\sum_t I_t}{T} = \frac{W}{T}$. Combining with Theorem 3, this proves the correctness of our algorithm.

We observe that the *current* $i_{j,t}$ above can be negative which means that the battery B_j should be charged with the *current* $i_{j,t}$ during time t. Further by our formulation we see that the battery can be charged from the *current* drawn from any of the remaining batteries which supply *current* in that time interval.

Finally, we consider the time complexity of the algorithm. Our algorithm takes the $\mathcal{O}(n \log(n))$ time for preprocessing and since we have a closed form solution for the *current* assignment for a given battery at a given time instant, the time complexity for calculating $m \cdot n$ *current* assignment values becomes $\mathcal{O}(n \log(n) + nm)$.

4   Algorithm for `EnergyTimeTradeoff`

In this section, we describe the algorithm for the `EnergyTimeTradeoff` problem, and subsequently show that it has the desired properties. Recall that an instance $\mathcal{I}$ consists of n batteries and m workloads W_k for $k \in \{1, \ldots m\}$ along with their corresponding weights α_k, and these have to be finished in this order. The algorithm is given formally in Algorithm 2.

The algorithm uses a parameter T, which is the total time needed to finish all the m tasks. We assume that the algorithm knows this parameter, because it can perform binary search like procedure on it – details are given later. In Step 1, it computes a parameter λ which satisfies the given equation – this parameter will turn out to be a Lagrange variable of a suitable convex program. Note that the LHS of this equation is monotone with respect to λ, and so given T, we can find λ (to desired accuracy) by binary search. In Step 3, we find the time t_i spent on each of the tasks W_i. For each of the tasks W_i, we shall show that the *current* requirement while performing it will remain constant, i.e, W_i/t_i. This gives us the total *current* requirement I_t at each time step. We now invoke Algorithm 1 to find the $i_{j,t}$ values.

Algorithm 2. Algorithm to find $i_{j,t}$ values for the `EnergyTimeTradeoff` problem, given parameter T.

1: Find λ which satisfies
2: $\quad \sum_k W_k \cdot \sqrt{\frac{r_{eq}}{\alpha_k + \lambda}} = T$
3: Initialize $t \leftarrow 0$.
4: **for** $k = 1, \ldots, m$ **do**
5: $\quad t_k \leftarrow W_k \cdot \sqrt{\frac{r_{eq}}{\alpha_k + \lambda}}$
6: $\quad$ **for** $t' = t, \ldots, t + t_k$ **do**
7: $\quad\quad I_{t'} \leftarrow \frac{W_k}{t_k}$.
8: $\quad t \leftarrow t + t_k$.
9: Invoke Algorithm 1 with the I_t values to compute $i_{j,t}$ values.

4.1   Analysis

In this section, we analyze our algorithm. We first show that the *current* supplied by a battery while performing a task W_k remains fixed.

Lemma 5. *In an optimal solution, the* current *supplied by a given battery during a workload W_k remains fixed.*

Proof. The proof is by contradiction. Consider an optimal solution which uses $i_{j,t}$ current from battery B_j at time t. By assumption, there exists a battery B_l, workload W_k and times t and t' during the processing of this workload such that $i_{l,t} \neq i_{l.t'}$. As in the proof of Lemma 1, one can check that incrementing one *current* by ϵ and decrementing another by ϵ strictly reduces the energy requirement, and the time taken for completing W_k remains unchanged. This leads to a contradiction.   ∎

Let t_k^O be the time spent by an optimal solution on workload W_k. The above Lemma shows that the total *current* requirement at any time during this workload would be $\frac{W_k}{t_k^O}$. Give this total *current* requirement at each time t, the optimal values of $i_{j,t}$ can be computed using Algorithm 1. Therefore, it suffices to show that the quantities t_k computed by Algorithm 2 are equal to t_k^O.

Lemma 6. *Let T denote the total time taken by an optimal solution. Using this value of T, the computed t_k values by Algorithm 2 are equal to the time taken by the optimal algorithm on the corresponding workload.*

Proof. Given the values t_k, the analysis of Algorithm 1 shows that the $i_{j,t}$ value for any time t during workload W_k is given by $\frac{C_j}{T} + \frac{r_{eq}}{r_j}\left(\frac{W_k}{t_k} - \frac{W}{T}\right)$, where C_j is the quantity computed by Algorithm 1 (note that C_j values do not depend on the t_k values, and are dependent on the total workload only).

Therefore, optimal t_k values are given by the solution to the following convex program:

$$\min \sum_{k=1}^{m}\sum_{j=1}^{n}\left(\frac{C_j}{T} + \frac{r_{eq}}{r_j}\left(\frac{W_k}{t_k} - \frac{W}{T}\right)\right)^2 r_j t_k + \sum_{k=1}^{m}\alpha_k t_k \qquad (4)$$

$$\text{subject to} \quad \sum_{k=1}^{m} t_k = T$$

Note that we have not explicitly added the constraint $t_k \geq 0$ – it will so happen that the optimal solution above will have this property.

Let λ be the Lagrange variable for the equality constraint above. Solving the above convex program yields the following solution:

$$t_k = \sqrt{\frac{r_{eq}}{\alpha_k + \lambda}} W_k \quad \forall k = 1,\ldots,m \qquad (5)$$

From the constraint $\sum_{k=1}^{m} t_k = T$, we get

$$T = \sum_{k=0}^{m} \left(\sqrt{\frac{r_{eq}}{\alpha_k + \lambda}} W_k \right) \tag{6}$$

We see that in the above equation, we can do a binary search to find the value of λ because the RHS is a monotonic function in λ (as in Step 1 of Algorithm 2). We find upper and lower bounds on λ to show that binary search needs to search in a small range only. Observe that

$$\lambda > - \min_k \alpha_k \tag{7}$$

Also by Cauchy-Schwarz,

$$\sqrt{\frac{\sum_k \frac{W_k^2 r_{eq}}{\alpha_k + \lambda}}{m}} \geq \sum_k \frac{W_k}{m} \sqrt{\frac{r_{eq}}{\alpha_k + \lambda}} = \frac{T}{m}$$

Therefore, we get

$$\lambda \leq r_{eq} \frac{m \sum_k W_k^2}{T^2} - \min_k \alpha_k \tag{8}$$

Thus, the time to perform binary search depends on $\log \left(r_{eq} \frac{m \sum_k W_k^2}{T^2} \right)$. ∎

Finally, we show how to search for T.

Lemma 7. *We can find the optimal value of T using Golden section search.*

Proof. The objective function of the convex program is given by

$$\sum_{k=1}^{m} \sum_{j=1}^{n} \left(\frac{C_j}{T} + \frac{r_{eq}}{r_j} \left(\frac{W_k}{t_k} - \frac{W}{T} \right) \right)^2 r_j t_k + \sum_{k=1}^{m} \alpha_k t_k \tag{9}$$

We shall show that the first term decreases as T increases, whereas the second term increases as T increases. Therefore the objective function is a unimodal function of T (i.e., in decreases with T, at some point reaches a minimum values, and then increases). The golden section search method can be used to find the optimal solution (to a desired accuracy) for any unimodal function [2].

From Eq. 6 we can see that on increasing T the value of λ decreases and from Eq. 5 on decreasing λ the t_k values increase. Hence the second term of the objective function increases with increasing T.

To show that the first term in the objective function decreases as T increases, we first observe that the first term of the objective function represents the total energy loss. Consider the optimal solution for some value of T, now when we increase the value of T to some T', let us construct a solution for T' such that the value of the second term for our solution is the same as the value of the second term for the optimal solution for T'. We know the t_k values for the optimal solution for T' from Eqs. 5 and 6 and we know that they increase on increasing T. Now, we construct a solution where we scale the *current* $i_{j,k}^T$ for any battery j in a given workload k by $t_k^T / t_k^{T'}$ (< 1) where t_k^T and $t_k^{T'}$ represent the t_k values for the optimal solutions to T and T' respectively. The energy loss for our constructed solution for battery B_j in workload k would be:

$$\left(\frac{i_{j,k}^T \cdot t_k^T}{t_k^{T'}} \right)^2 t_k^{T'} \cdot r_j = \left(i_{j,k}^T \right)^2 \cdot t_k^T \cdot r_j \left(\frac{t_k^T}{t_k^{T'}} \right) \leq \left(i_{j,k}^T \right)^2 \cdot t_k^T \cdot r_j$$

We can see that the energy loss during workload k of the battery B_j decreases without changing the charge supplied by the battery B_j in the workload, hence our constructed solution is a valid solution. Now, since the optimal solution for T' would only be better, we conclude that with increasing T, the value of the first term in the objective function decreases in the corresponding optimal solution. ∎

5   Conclusion

In this paper, we considered two different formulations for the problem of scheduling current from multiplexed batteries. In the first formulation, we considered the case where we are given the workload requirement at every point of time and the objective is to minimize the total energy loss. For this formulation we introduced an algorithm to calculate the overall contribution of every battery in any optimal solution and showed that using these values, the resulting convex program was solvable using Lagrange Optimization and resulted in a closed form solution.

In the second formulation, we considered the case when we are given a series of workloads and the objective is to minimize the weighted sum of the total energy loss and the time taken to complete each workload. We were able to reduce this problem to the first one by introducing new parameters whose value was search-able using search algorithms.

We have only looked at the case when the internal resistance of the batteries is constant, for future work - one can analyze the case where the internal resistance depends on the charge present inside the battery or consider different objective functions which deal with other factors like the longevity of the batteries.

References

1. Anirudh, B., et al.: Software defined batteries. In: Proceedings of the 25th Symposium on Operating Systems Principles, SOSP 2015, pp. 215–229. ACM, New York (2015). https://doi.org/10.1145/2815400.2815429
2. Heath, M.: Scientific Computing, 2nd edn. McGraw-Hill Inc., New York (2002)
3. Keshav, S.: Technical perspective: the chemistry of software-defined batteries. Commun. ACM **59**(12), 110 (2016). https://doi.org/10.1145/3007177

4. Nikhil, B., Tracy, K., Kirk, P.: Speed scaling to manage energy and temperature. J. ACM **54**(1), 3:1–3:39 (2007). https://doi.org/10.1145/1206035.1206038
5. Simonite, T.: Gadgets could get longer lives by combining batteries. https://www. technologyreview.com/s/541861/gadgets-could-get-longer-lives-by-combining-batteries/
6. Susanne, A., Hiroshi, F.: Energy-efficient algorithms for flow time minimization. ACM Trans. Algorithms **3**(4) (2007). https://doi.org/10.1145/1290672.1290686

The Volume of a Crosspolytope Truncated by a Halfspace

Ei Ando[✉] [ID] and Shoichi Tsuchiya [ID]

Senshu University, 2-1-1, Higashi-Mita, Tama-Ku, Kawasaki 214-8580, Japan
{ando.ei,s.tsuchiya}@isc.senshu-u.ac.jp

Abstract. In this paper, we consider the computation of the volume of an n-dimensional crosspolytope truncated by a halfspace. Since a crosspolytope has exponentially many facets, we cannot efficiently compute the volume by dividing the truncated crosspolytope into simplices. We show an $O(n^6)$ time algorithm for the computation of the volume. This makes a contrast to the $0-1$ knapsack polytope, whose volume is $\#P$-hard to compute. The paper is interested in the computation of the volume of the truncated crosspolytope because we conjecture the following question may have an affirmative answer: Does the existence of a polynomial time algorithm for the computation of the volume of a polytope K imply the same for K's geometric dual? We give one example where the answer is yes.

Keywords: Polynomial time algorithm · Volume computation · Geometric duality

1 Introduction

The computation of the volume of an n-dimensional polytope can be a hard problem, even for some simple cases. For example, the followings are some results about the hardness of computing the volume of polytopes.

- Dyer and Frieze [9] showed that computing the volume of a $0-1$ knapsack polytope is $\#P$-hard. A $0-1$ knapsack polytope is the intersection of a unit hypercube and a halfspace $\{x \in \mathbb{R}^n | a \cdot x \leq b\}$, where $a \in \mathbb{Z}_+^n$.
- Khachiyan [12,13] showed that computing the volume of the dual of a $0-1$ knapsack polytope is $\#P$-hard. The dual of a $0-1$ knapsack polytope is the convex hull of a crosspolytope $\mathrm{conv}(\{\pm e_1, \ldots, \pm e_n\})$ and a point $a \in \mathbb{Z}_+^n$.
- Dyer et al. [8] showed that computing the volume of a Zonotope is $\#P$-hard. Given m vectors $a_1, \ldots, a_m \in \mathbb{Z}^n$ ($m \geq n$), a *Zonotope* is a polytope given by the Minkowski sum of line segments given by $a_1, \ldots, a_m$.
- Ando and Kijima [4] showed that computing the volume of the intersection of two crosspolytopes is $\#P$-hard.

This research was supported by research grant of Information Sciences Institute of Senshu University.

© Springer Nature Switzerland AG 2019
T. V. Gopal and J. Watada (Eds.): TAMC 2019, LNCS 11436, pp. 13–27, 2019.
https://doi.org/10.1007/978-3-030-14812-6_2

If we consider the general convex body K in n-dimensional space, which can be accessed only by its membership oracle, any polynomial time deterministic algorithm cannot compute the volume of K. In fact, no polynomial time deterministic algorithm can achieve an exponentially large approximation ratio for $\mathrm{Vol}(K)$ [5,10].

Interestingly, these volumes can be approximated by some randomized algorithms. As for the problem of computing the volume of the general convex body K, there are fully polynomial time randomized approximation schemes (FPRAS) [6,7,15].

Recent results show that some of #P-hard volume actually yield *deterministic* FPTASes (Fully Polynomial Time Approximation Schemes). We here list up some results about the results.

- Lee and Shi [14] showed an FPTAS for the sum distribution of discrete random variables, which can be used to approximate the volume of the $0-1$ knapsack polytope. Their algorithm are based on the dynamic programming idea due to Štefankovič et al. [11,17].
- Ando and Kijima [3] showed an FPTAS for the volume of a truncated hypercube by a constant number of halfspace (see also [1]).
- Ando and Kijima [2] showed that there is an FPTAS for the volume of the dual of a $0-1$ knapsack polytope. In the proof, they also showed an FPTAS for the volume of the intersection of two crosspolytopes.

Since there are FPTASes for computing the volume of the knapsack polytope and its dual, the following question arises: If we know the existence of a deterministic polynomial time (approximation or exact) algorithm for computing the volume of a polytope K, is there always a deterministic polynomial time algorithm for computing the volume of the geometric dual of K?

In order to find out a way to answer this question, we consider how we can efficiently compute the volume of the geometric dual of polytopes.

The simplest example is a simplex. Let $S = \mathrm{conv}(\{a_0, a_1, \ldots, a_n\})$. Since the geometric dual of a simplex S is another simplex S^*, we can efficiently compute both $\mathrm{Vol}(S)$ and $\mathrm{Vol}(S^*)$.

As another example, we consider a convex hull of a hypercube and a point $a \in \mathbb{R}^n$. Then, its dual is a crosspolytope truncated by a halfspace, i.e., the intersection of a crosspolytope and a halfspace $\{x \in \mathbb{R}^n | a \cdot x \le b\}$ for $a \in \mathbb{R}^n$ and $b \in \mathbb{R}$. The volume of the former can easily be confirmed to have a polynomial time exact algorithm. That is, we can divide the polytope into a hypercube and at most n "pyramids" each of whose bottom is a hypercube facet visible from a and whose peak is a. This algorithm finishes in linear time. Thus, our interest is in whether there exists a polynomial time algorithm for the volume of the latter. We prove this in the main part of this paper.

In this paper, we show the following theorem.

Theorem 1. *Given $a \in \mathbb{R}^n$ and $b \in \mathbb{R}$, there exists an $O(n^6)$ time deterministic algorithm for computing the volume of the intersection of a crosspolytope and a halfspace $\{x \in \mathbb{R}^n | a \cdot x \le b\}$.*

The tricky part of the problem is that a crosspolytope has 2^n facets. We cannot efficiently compute the volume by dividing the truncated crosspolytope into simplices. In order to avoid exponential computational time, though our algorithm is deterministic, we use some probability arguments in the proof. We show that the volume can be given by a multiple integral using a kind of step function. Then, we compute the multiple integral symbolically.

The paper is organized as follows. In Sect. 2, we show that computing the volume of a convex hull of a hypercube and a point can be done in $O(n)$ time. In Sect. 3, we show the polynomial time algorithm for the volume of a truncated crosspolytope. We conclude the paper in Sect. 4.

2 The Volume of the Convex Hull of a Hypercube and a Point

In this section, we consider the running time for computing the volume of the convex hull of a hypercube $[-1/2, 1/2]^n$ and a point $a \in \mathbb{R}^n$.

Our algorithm is the following. For each facet F of the hypercube, we decide whether or not F is visible from a. To do this, let f be the normal of F directed to the outside of the hypercube. Then, we have the following observation.

Observation 1. F is visible from a if and only if $a \cdot f > 1/2$.

We compute the volume of the pyramid whose bottom is F and whose peak is a. This pyramid has bottom area (i.e., $n-1$ dimensional volume of F) 1 and height $a \cdot f - 1/2$. Thus, the volume of the pyramid is $(a \cdot f - 1/2)/n$. In the following algorithm, we keep in mind the following points: (1) If F is visible from a, then the opposite facet F' is not visible from a; (2) Since the normal of F is one of $\pm e_1, \ldots, \pm e_n$, we have that $a \cdot f$ is one of $\pm a_1, \ldots, \pm a_n$.

Algorithm 1. *Input: Vector $a = (a_1, \ldots, a_n) \in \mathbb{R}^n$.*
1. $V \leftarrow 0$;
2. *For $i = 1, \ldots, n$ do*
3.   *If $|a_i| \geq 1/2$ then*
4. $V \leftarrow V + (|a_i| - 1/2)/n$;
5. *done;*
6. *Output $V + 1$.*

Now, we have the following proposition.

Proposition 1. *Given $a \in \mathbb{R}^n$, the volume of $\mathrm{conv}([-1/2, 1/2]^n \cup \{a\})$ can be computed in $O(n)$ time.*

3 The Volume of a Crosspolytope Truncated by a Halfspace

In this section, we prove Theorem 1. We first show the basic definitions in Sect. 3.1. After that, we show our algorithm in Sect. 3.2.

3.1   Basic Definitions

Let $C(\mathbf{0}, 1)$ be a unit crosspolytope centered at $\mathbf{0}$. That is,

$$C(\mathbf{0}, 1) \overset{\text{def}}{=} \text{conv}(\{\pm e_1, \ldots, \pm e_n\}) = \left\{ \boldsymbol{x} \in \mathbb{R}^n \;\middle|\; ||\boldsymbol{x}||_1 = \sum_{i=1,\ldots,n} |x_i| \leq 1 \right\},$$

where e_i $(i = 1, \ldots, n)$ is the vector whose i-th component is 1 and the other components are all 0.

For $\boldsymbol{a} \in \mathbb{R}^n$ and $b \in \mathbb{R}$, we define $CT(\boldsymbol{a}, b)$ as a crosspolytope truncated by a halfspace

$$CT(\boldsymbol{a}, b) \overset{\text{def}}{=} C(\mathbf{0}, 1) \cap \{ \boldsymbol{x} \in \mathbb{R}^n | \boldsymbol{a} \cdot \boldsymbol{x} \leq b \}.$$

We are going to compute the volume of $CT(\boldsymbol{a}, b)$. Without loss of generality, we assume that $\boldsymbol{a} \in \mathbb{R}_+^n$.

The following definition of the geometric duality is in [16].

Definition 1. *For a point set $K \subseteq \mathbb{R}^n$, the dual set K^* of K is*

$$K^* \overset{\text{def}}{=} \{ \boldsymbol{y} \in \mathbb{R}^n | \boldsymbol{x} \cdot \boldsymbol{y} \leq 1 \text{ for all } \boldsymbol{x} \in K \}. \tag{1}$$

A vertex of K corresponds to a halfspace that gives a facet of K^*. For example, the hypercube $[-1, 1]^n$ and the crosspolytope $C(\mathbf{0}, 1)$ are geometric dual to each other. The vertex $\mathbf{1} = (1, 1, \ldots, 1)$ of the hypercube $[-1, 1]^n$ corresponds to a halfspace $\{ \boldsymbol{x} \in \mathbb{R}^n | \mathbf{1} \cdot \boldsymbol{x} \leq 1 \}$, which gives a facet of the crosspolytope $C(\mathbf{0}, 1)$. We have the following proposition.

Proposition 2. *For $\boldsymbol{a} \in \mathbb{R}_+^n$ and $b \in \mathbb{R}_+$, $CT(\boldsymbol{a}, b)$ is the geometric dual of* conv$([-1, 1]^n \cup \{\boldsymbol{a}/b\})$.

Since computing the volume of conv$([-1, 1]^n \cup \{\boldsymbol{a}/b\})$ can be finished in $O(n)$ time, we are interested in whether there is a polynomial time algorithm for computing the volume of $CT(\boldsymbol{a}, b)$. In the following, we assume that $b \in \mathbb{R}$ since our algorithm has no difficulty in dealing with the case where $b \leq 0$.

Let $f(x)$ be the density function of a uniform random variable in $[-1, 1]$. That is,

$$f(x) = \begin{cases} 1/2 & -1 \leq x \leq 1 \\ 0 & \text{otherwise.} \end{cases} \tag{2}$$

We use the following step function $H(x)$.

$$H(x) = \begin{cases} 0 & x < 0 \\ 1/2 & x = 0 \\ 1 & x > 0 \end{cases}$$

Intuitively, $H(x)$ is a probability distribution function of a normal distribution whose mean is 0 and whose variance is arbitrarily close to 0. We use $H(x)$ for writing the functions including breakpoints concisely. By ignoring the values at $x = \pm 1$, we use $H(x+1)H(1-x)/2$ instead of the above definition of $f(x)$.

We note that, like the example above, we may have some wrong values at the breakpoints (i.e., $x = \pm 1$ for $f(x)$) when we use the step function description (e.g., $H(x+1)H(1-x)/2$) instead of the casewise description (e.g., (2)). However, since the breakpoints are limited to the points where some of the step function's arguments are 0, we nevertheless use the step function description. We define the following 'equality almost everywhere'.

Definition 2. *Let $F(x)$ and $G(x)$ be two piecewise polynomial functions. If $F(x) = G(x)$ for $x \in \mathbb{R} \setminus I$ where $I \subseteq \mathbb{R}$ satisfies $\int_{x \in I} 1 \mathrm{d}x = 0$, then we write*

$$F(x) \overset{\text{a.e.}}{=} G(x). \tag{3}$$

If $F(x) \overset{\text{a.e.}}{=} G(x)$, we say $F(x)$ and $G(x)$ are equal almost everywhere.

Thus, for $f(x)$ defined by (2), we have $f(x) \overset{\text{a.e.}}{=} H(x+1)H(1-x)/2$.

Let $F(x) \overset{\text{a.e.}}{=} G(x)$. In case $c \in I$ and $F(x)$ is continuous, we have that $F(c) = \lim_{\epsilon \to 0} G(c - \epsilon)$. For the function of two variables, we similarly define the equality almost everywhere as follows:

Definition 3. *Let $F(x, y)$ and $G(x, y)$ be two piecewise polynomial functions. If $F(x, y) = G(x, y)$ for $(x, y) \in \mathbb{R} \setminus I$ where $I \subseteq \mathbb{R}^2$ satisfies $\iint_{(x,y) \in I} 1 \mathrm{d}x \mathrm{d}y = 0$, then we write*

$$F(x, y) \overset{\text{a.e.}}{=} G(x, y). \tag{4}$$

In case $(c, d) \in I$ and $F(x, y)$ is continuous, we have

$$F(x, y) = \lim_{\epsilon_1 \to 0} \lim_{\epsilon_2 \to 0} G(c - \epsilon_1 a, d - \epsilon_2 b) \tag{5}$$

for some $a, b \in \mathbb{R}$. In case we have the complete and finite description of $G(x, y)$, we can choose a and b where (5) converges to one value, and where there is no breakpoint in between (a, b) and (c, d). This is because the measure of I is 0 and I is given by a finite number of linear formulas.

In addition, the following definition allows us to argue how many memory bits we need to store a piecewise polynomial function.

Definition 4. *A piece I of a piecewise polynomial function $F(x)$ is a continuous interval where there exists a polynomial function $G(x)$ satisfying all the following conditions:*

1. *$\int_{x \in I} 1 \mathrm{d}x > 0$;*
2. *there exists $z \in I$ such that $F(z) = G(z)$;*
3. *$x \in I$ if and only if $F(y) = G(y)$ for all y such that $y \in [x, z] \cup [z, x]$.*

3.2 Our Algorithm

In the following, we show how we can compute $\text{Vol}(CT(a,b))$. Through the argument of probability, we first show that $\text{Vol}(CT(a,b))$ can be described by a multiple integral. Then, we prove that the running time of computing the multiple integral is bounded by a polynomial in n.

We note that a naive approach is not efficient for computing the volume of $CT(a,b)$. Since $C(0,1)$ has 2^n facets, $CT(a,b)$ may have exponentially many facets. If we divide $CT(a,b)$ into simplices, we get exponentially many simplices. Therefore, it takes exponentially long time to sum up the volume of those simplices.

Instead, we can compute the volume of $CT(a,b)$ efficiently by introducing probability. Let X be a uniform random vector in $[-1,1]^n$. Then,

$$\text{Vol}(CT(a,b)) = 2^n \Pr[\|X\|_1 \leq 1 \wedge a \cdot X \leq b].$$

To give the multiple integral expression of $\text{Vol}(CT(a,b))$, we define that

$$\Phi_0(u,v) = \Pr[0 \leq u \wedge 0 \leq t] \stackrel{\text{a.e.}}{=} H(u)H(v).$$

Let $X = (X_1, \ldots, X_n)$ and $a = (a_1, \ldots, a_n)$. We define $\Phi_i(u,v)$ as a function given by

$$\Phi_i(u,v) = 2^i \Pr[|X_1| + \cdots + |X_i| \leq u \wedge a_1 X_1 + \cdots + a_i X_i \leq v]$$

$$= 2^i \int_{x_i \in [-1,1]} \Pr\left[\sum_{j=1}^{i-1} |X_j| + |x_i| \leq u \wedge \sum_{j=1}^{i-1} a_j X_j + a_i x_i \leq v \,\middle|\, X_i = x_i\right] \mathrm{d}x_i$$

$$= 2^i \int_{x_i \in \mathbb{R}} \Pr\left[\sum_{j=1}^{i-1} |X_j| + |x_i| \leq u \wedge \sum_{j=1}^{i-1} a_j X_j + a_i x_i \leq v\right] f(x_i)\mathrm{d}x_i$$

$$= 2 \int_{x_i \in \mathbb{R}} \Phi_{i-1}(u - |x_i|, v - a_i x_i) f(x_i)\mathrm{d}x_i, \tag{6}$$

where $f(x) \stackrel{\text{a.e.}}{=} H(x+1)H(1-x)/2$. Then, we have that $\Phi_n(1,b) = \text{Vol}(CT(a,b))$. By the form of (6), we do not need the value of $\Phi_i(u,v)$ for $u > 1$ to compute the value of $\Phi_n(1,b)$. In our algorithm, we compute (6) for $u \leq 1$.

In the following, we prove Theorem 1, that is, $\Phi_n(1,b)$ can be computed in polynomial time. We first show that $\Phi_i(u,v)$ for any fixed $u \leq 1$ is a piecewise polynomial function with degree at most i, and that the number of pieces of $\Phi_i(u,v)$ is at most $2n + 1$. This allows us to process the integral in polynomial time, which implies Theorem 1. The following proposition is important for our proof.

Proposition 3. *Given* $a \in \mathbb{R}_+^n$, *let* K *be a convex polytope given by* $K = \text{conv}(\{v_1, \ldots, v_N\})$, *where* $v_1, \ldots, v_N \in \mathbb{R}^n$ *and* $a \cdot v_1 \leq a \cdot v_2 \leq \cdots \leq a \cdot v_N$. *Let* $F_K(a,b)$ *be given by*

$$F_K(a,b) = \text{Vol}(\{x \in K | a \cdot x \leq b\}). \tag{7}$$

Then, we have that

1. $F_K(a, b)$ *is a piecewise polynomial function of b with degree at most n;*
2. *there are at most $N + 1$ pieces of $F_K(a, b)$ for any fixed a;*
3. *the pieces of $F_K(a, b)$ for fixed a are given by $(-\infty, a \cdot v_1], [a \cdot v_N, +\infty)$ and $[a \cdot v_i, a \cdot v_{i+1}]$ for $i = 1, \ldots, N$.*

Proof. Without loss of generality, we assume that $a \cdot v_h \leq a \cdot v_{h+1}$ for $h = 1, \ldots, N-1$. We prove this proposition by showing how we can compute $F_K(a, b)$ as a polynomial function $G_K(a, b)$ of b when $a \cdot v_h \leq b \leq a \cdot v_{h+1}$. After that, for $b' \geq b$, we show that $G_K(a, b') = F_K(a, b')$ as long as $a \cdot v_h \leq b \leq b' < a \cdot v_{h+1}$. In case $a \cdot v_h \leq b < a \cdot v_{h+1} \leq b''$, we show that $G_K(a, b'')$ may not be equal to $F_K(a, b'')$, which shows the second claim of the proposition.

Let $K'(b)$ be the polytope given by $K'(b) = \{x \in K | a \cdot x \leq b\}$, so that $F_K(a, b) = \text{Vol}(K'(b))$. Fixing the value of b, consider dividing $K'(b)$ into M simplices $S_1, \ldots, S_M$.[1] This is called *triangulation* (see e.g., [16]). Especially, we assume a triangulation such that all vertices of $S_1, \ldots, S_M$ are the vertices of $K'(b)$. The existence of such triangulation can be proved by induction on n. See Proposition 4 in Appendix.

For $i = 1, \ldots, M$, let the vertices of S_i be given by $s_{i0}, s_{i1}, \ldots, s_{in} \in \mathbb{R}^n$. That is, $S_i = \text{conv}(\{s_{i0}, s_{i1}, \ldots, s_{in}\})$. Observe that there are two cases. In case (1), $s_{ij}(j = 0, 1, \ldots, n)$ is a vertex of K. In case (2), for s_{ij}, there exists a line segment $L_{ij} = \text{conv}(\{v, v'\})$, where v and v' are two vertices of K; and, L_{ij} satisfies that vertex s_{ij} of S_i is the intersection of L_{ij} and hyperplane $a \cdot x = b$.

Since, in case (1), it is clear that the coordinate of s_{ij} is constant with respect to b, we consider case (2) in the following.

In case (2), the coordinate of $s_{ij} = pv + (1 - p)v'$ is given by solving the equation $a \cdot (pv + (1-p)v') = b$ for p. Therefore, we can compute each component of s_{ij} $(j = 0, \ldots, n)$ as a constant or a linear function of b. Here, we write $s_{ij}(b)$ to show that the coordinate of s_{ij} is given depending on b.

We obtain the polynomial $G_K(a, b)$ as follows. Let $R_i(b)$ be a matrix, where j-th column of $R_i(b)$ is given by $s_{ij} - s_{i0}$ for $j = 1, \ldots, n$. The volume of S_i is given by $\frac{|\det R_i(b)|}{n!}$. Since each element of $R_i(b)$ is a constant or a linear function of b, the volume is given by a polynomial in b with degree at most n. We have that $G_K(b) = \sum_{i=1,\ldots,M} \frac{|\det R_i(b)|}{n!}$ as long as $s_{ij}(b) \in K$ for all $i = 1, \ldots, M$ and $j = 0, \ldots, n$.

Apparently, $G_K(a, b)$ is equal to $F_K(a, b)$ only in a certain interval of b, which coincides with a piece of $F_K(a, b)$. Suppose that the two vertices v and v' of K given in the above satisfy $a \cdot v \leq a \cdot v'$. Remember that $s_{ij}(b')$ is an interior point of L_{ij} if and only if $s_{ij}(b')$ satisfy

$$a \cdot v \leq a \cdot s_{ij}(b') \leq a \cdot v'. \tag{8}$$

---

[1] Though the number M of the simplices may be exponentially large with respect to n, it does not matter because, in the later part of the paper, we show another way to compute $F_K(a, b)$ efficiently for the case where K is a crosspolytope.

If b' does not satisfy (8), then $s_{ij}(b')$ is not an interior point of L_{ij}, which implies that $b' \notin K$. Thus, in case (8) is not satisfied for a combination of i, j, we have that $\text{Vol}(K'(b'))$ may not be equal to $G_K(a, b')$. The correct value of $\text{Vol}(K'(b'))$ is given by dividing $K'(b')$ into M' simplices, $S'_1, \ldots, S'_{M'}$ and then computing $\sum_{i=1,\ldots,M'} \text{Vol}(S')$, which is given by another polynomial in b'. Since there are N vertices of K, we have at most $N+1$ different polynomials in b for each number of vertices satisfying $a \cdot x \leq b$. We note that the third claim is obvious by the above arguments.   □

By Proposition 3, we have the following observation.

Observation 2. *Let $a_1 \geq a_2 \geq \cdots \geq a_i \geq 0$. Then we have,*

1. *$\Phi_i(u, v) = 0$ for $u \leq 1$ and $v \leq -a_1 u$; also, $\Phi_i(u, v) = u^i/i!$ for $0 \leq u \leq 1$ and $v \geq a_1 u$.*
2. *$\Phi_i(u, v)$ for any fixed $u \leq 1$ is a piecewise polynomial function. The number of pieces is at most $2i + 1$.*
3. *The pieces of $\Phi_i(u, v)$ are given by $(-\infty, \alpha_1 u]$, $[\alpha_{2i}u, +\infty)$ and $[\alpha_k u, \alpha_{k+1}u]$ for $k = 1, \ldots, 2i$, where*

$$
\alpha_k = \begin{cases} -a_k & \text{for } k = 1, \ldots, i, \\ a_{2i-k+1} & \text{for } k = i+1, \ldots, 2i. \end{cases} \tag{9}
$$

Then, we define the following.

Definition 5. *Let $V_k(u, v) = H(v - \alpha_k u)H(\alpha_{k+1}u - v)$ for $k = 1, \ldots, 2n - 1$. The normalized form of $\Phi_i(u, v)$ for $u \leq 1$ is*

$$
\Phi_i(u, v) \stackrel{\text{a.e.}}{=} H(u)H(1 - u) \sum_{k=0}^{2i} V_k(u, v)p_k(u, v)
$$
$$
+ H(u)H(1 - u)H(v + \alpha_{2i}u)u^i/i!,
$$

where $p_k(u, v)$ is a polynomial in u and v.

We do not compute the case where $u > 1$. Because, in that case, $\Phi_n(u, v)$ is equal to the volume of a polytope $C(\mathbf{0}, u) \cap [-1, 1]^n \cap \{x \in \mathbb{R}^n | a \cdot x \leq v\}$. This polytope may have exponentially many vertices.

For $u \leq 1$, our algorithm computes the normalized form of $\Phi_i(u, v)$ from the normalized form of $\Phi_{i-1}(u, v)$ for $i = 1, \ldots, n$. The normalized form of $\Phi_i(u, v)$ can be given by an array $A_i(k)$ and a three dimensional array $P_i(k, d_u, d_v)$ for $k = 0, \ldots, 2i$, $d_u = 0, \ldots, i$ and $d_v = 0, \ldots, i$. Here, k represents the index in the above definition of the normalized form; d_u and d_v represent the degree of u and v in $p_k(u, v)$. For $\Phi_i(u, v)$, the values of the arrays $A_i(k)$ and $P_i(k, d_u, d_v)$ represents the values α_k and the coefficient of a term in $p_k(u, v)$ with u'degree d_u and with v's degree d_v, respectively. Our algorithm is as follows.

Algorithm 2. *Input: $a \in \mathbb{R}^n_+$ and $b \in \mathbb{R}$.*
1. *Set $P_0(0, 0, 0) := 0$, $A_0(0) := 0$;*

2. For $i = 1, \ldots, n$ do
3.   For $k = 0, \ldots, 2i$ do
4.   Set $A_i(k) = \alpha_k$ as given in Observation 2;
5.   For all $(d_u, d_v) \in \{0, 1, \ldots, i\}^2$ do
6.   Compute $P_i(k, d_u, d_v)$;
7.   done;
8.   done;
9. done;
10. Find k such that $A_n(k-1) \le b \le A_n(k)$;
11. Output $\sum_{d_u=0}^{n} \sum_{d_v=0}^{n} P_n(k, d_u, d_v)b^{d_v}$.

We need to be careful when we consider the value of $\Phi_n(u, v)$ of a particular point where $(u, v) = (1, b)$. Since the normalized form of $\Phi_n(u, v)$ is equal to the original definition of $\Phi_n(u, v)$ only *almost everywhere* for $u \le 1$, the normalized form may give a wrong value when any step function argument is equal to zero. To avoid the wrong value, we use the value of $\Phi_n(1 - \epsilon, b - \epsilon')$, which tends to $\sum_{d_u=0}^{n} \sum_{d_v=0}^{n} P_n(k, d_u, d_v)b^{d_v}$ when $\epsilon, \epsilon' \to 0$.

We prove Theorem 1 using the following lemma.

Lemma 1. *For fixed $a \in \mathbb{R}_+^n$, the value of $\Phi_i(u, v)$ is given by the normalized form if all step function arguments in the normalized form are not zero. Moreover, $p_k(u, v)$ in the normalized form is a polynomial in u and v with degree at most i.*

Proof. As for the base case, we can compute $\Phi_1(u, v)$ by

$$\Phi_1(u, v) = 2 \int_{-\infty}^{\infty} \Phi_0(u - |x_1|, v - a_1 x_1) f(x_1) dx_1 \tag{10}$$

$$= \int_{-1}^{1} H(u - |x_1|) H(v - a_1 x_1) dx_1 \tag{11}$$

$$= \int_{-1}^{0} H(u) H(u + x_1) H(v - a_1 x_1) dx_1 + \int_{0}^{1} H(u - x_1) H(v - a_1 x_1) dx_1 \tag{12}$$

$$\stackrel{a.e.}{=} (\min\{0, v/a_1\} - \max\{-1, -u\}) H(\min\{0, v/a_1\} - \max\{-1, -u\})$$
$$+ \min\{1, u, v/a_1\} H(\min\{1, u, v/a_1\}). \tag{13}$$

These max and min can be rewritten by using step functions. That is,

$$\max_{i=1,\ldots,\ell} \{p_\ell\} \stackrel{a.e.}{=} \sum_{i=1,\ldots,\ell} \prod_{j \neq i} H(p_i - p_j) p_i \tag{14}$$

$$\min_{i=1,\ldots,\ell} \{q_\ell\} \stackrel{a.e.}{=} \sum_{i=1,\ldots,\ell} \prod_{j \neq i} H(q_j - q_i) q_i. \tag{15}$$

Therefore,

$$\min\{0, v/a_1\} \overset{\text{a.e.}}{=} H(-v/a_1)v/a_1,$$

$$\max\{-1, -u\} \overset{\text{a.e.}}{=} H(-1+u)(-1) + H(-u+1)(-u),$$

$$\min\{1, u, v/a_1\} \overset{\text{a.e.}}{=} H(u-1)H(v/a_1-1) + H(1-u)H(v/a_1-u)u$$
$$+ H(1-v/a_1)H(u-v/a_1)v/a_1.$$

Also, we have that

$$H\left(\min_{i=1,\ldots,\ell}\{p_i\} - \max_{j=1,\ldots,\ell'}\{q_j\}\right) \overset{\text{a.e.}}{=} \sum_{i=1,\ldots,\ell}\prod_{j=1,\ldots,\ell'} H(p_i - q_j). \qquad (16)$$

We have $H(\min\{0, v/a_1\} - \max\{-1, -u\}) \overset{\text{a.e.}}{=} H(u)H(v/a_1+1)H(v/a_1+u)$ and $H(\min\{1, u, v/a_1\}) \overset{\text{a.e.}}{=} H(u)H(v/a_1)$. This way, we have

$$\begin{aligned}\Phi_1(u,v) \overset{\text{a.e.}}{=} \ & H(u)H(v/a_1+1)H(v/a_1+u)H(-v/a_1)v/a_1 \\ &- H(u)H(v/a_1+1)H(v/a_1+u)H(-1+u)(-1) \\ &- H(u)H(v/a_1+1)H(v/a_1+u)H(-u+1)(-u) \\ &+ H(u)H(v/a_1)H(u-1)H(v/a_1-1) \\ &+ H(u)H(v/a_1)H(1-u)H(v/a_1-u)u \\ &+ H(u)H(v/a_1)H(1-v/a_1)H(u-v/a_1)v/a_1 \qquad (17)\end{aligned}$$

$$\overset{\text{a.e.}}{=} \begin{cases} 0 & u < 0 \text{ or } v < -a_1 u \\ u + v/a_1 & 0 < u < 1 \text{ and } -a_1 u < v < a_1 u \\ 2u & 0 < u < 1 \text{ and } a_1 u < v \\ 1 + v/a_1 & 1 < u \text{ and } -a_1 < v < a_1 \\ 2 & 1 < u \text{ and } a_1 < v. \end{cases} \qquad (18)$$

The normalized form is immediate from (18). That is,

$$\begin{aligned}\Phi_1(u,v) \overset{\text{a.e.}}{=} \ & H(u)H(1-u)H(v+a_1u)H(a_1u-v)(u+v/a_1) \\ &+ H(u)H(1-u)H(v-a_1u)2u, \qquad (19)\end{aligned}$$

for $u \leq 1$. Therefore, the claims of the lemma holds for the base case.

We proceed to the induction step. We assume that we have the normalized form of $\Phi_{i-1}(u,v)$, where $p_k(u,v)$ is a polynomial with degree at most $i-1$ for $k = 0, \ldots, 2i$. Since we already know how the pieces of $\Phi_i(u,v)$ are given by the third claim of Observation 2, we show how we can execute the multiple integral. For $0 \leq k \leq 2i$, we consider how to integrate each term given by $H(u - |x_i|)H(1 - (u - |x_i|))V_k(u - |x_i|, v - a_i x_i)p_k(u - |x_i|, v - a_i x_i)$. For the conciseness, we omit how the last term can be integrated since the execution of the integral is similar.

Here, we see what we get as the result of integrating the term. Let $q_k(u - x_i, v - a_i x_i)$ and $\tilde{q}_k(u + x_i, v - a_i x_i)$ satisfy $\frac{d}{dx_i}q_k(u - x_i, v - a_i x_i) = p_k(u -$

$x_i, v - a_i x_i)$ and $\frac{d}{dx_i}\tilde{q}_k(u + x_i, v - a_i x_i) = p_k(u + x_i, v - a_i x_i)$. In case $\alpha_k > 0$ and $\alpha_{k+1} > 0$, we have

$$\int_{-1}^{1} H(u - |x_i|)H(1 - (u - |x_i|))V_k(u - |x_i|, v - a_i x_i)p_k(u - |x_i|, v - a_i x_i)\mathrm{d}x_i$$

$$= \int_{0}^{1} H(u - x_i)H(1 + x_i - u)V_k(u - x_i, v - a_i x_i)p_k(u - x_i, v - a_i x_i)\mathrm{d}x_i$$

$$+ \int_{-1}^{0} H(u + x_i)H(1 - x_i - u)V_k(u + x_i, v - a_i x_i)p_k(u + x_i, v - a_i x_i)\mathrm{d}x_i \tag{20}$$

$$\stackrel{\text{a.e.}}{=} [q_k(u - x_i, v - a_i x_i)]_{S_k(u,v)}^{T_k(u,v)} H(T_k(u, v) - S_k(u, v))$$

$$+ [\tilde{q}_k(u + x_i, v - a_i x_i)]_{\tilde{S}_k(u,v)}^{\tilde{T}_k(u,v)} H(\tilde{T}_k(u, v) - \tilde{S}_k(u, v)), \tag{21}$$

where $T_k(u, v)$, $\tilde{T}_k(u, v)$, $S_k(u, v)$ and $\tilde{S}_k(u, v)$ are determined depending on the values of α_k and α_{k+1}. Since the integrands have step function factors $H(u \mp x_i)H(1 \pm x_i - u)$ and

$$V_k(u \mp x_i, v - a_i x_i) = H(v - a_i x_i - \alpha_k(u \mp x_i))H(\alpha_{k+1}(u \mp x_i) - v + a_i x_i),$$

these step functions give the upper or the lower limit of the definite integral, depending on the sign of the coefficient of x_i. In case $\alpha_k + a_i \geq \alpha_k - a_i > 0$ and $\alpha_{k+1} + a_i \geq \alpha_{k+1} - a_i > 0$, we have

$$T_k(u, v) = \min\left\{1, u, \frac{-v + \alpha_{k+1}u}{\alpha_{k+1} - a_i}\right\}$$

$$S_k(u, v) = \max\left\{0, u - 1, \frac{-v + \alpha_k u}{\alpha_k - a_i}\right\}$$

$$\tilde{T}_k(u, v) = \min\left\{0, 1 - u, \frac{v - \alpha_k u}{\alpha_k + a_i}\right\}$$

$$\tilde{S}_k(u, v) = \max\left\{-1, -u, \frac{v - \alpha_{k+1}u}{\alpha_{k+1} + a_i}\right\}.$$

In the above, $1, 0, -1$ are from the upper and the lower limit of the integral in (20). Then, (21) is equal to

$$q_k(u - T_k(u, v), v - T_k(u, v))H(T_k(u, v) - S_k)$$

$$- q_k(u - S_k(u, v), v - S_k(u, v))H(T_k(u, v) - S_k)$$

$$+ \tilde{q}_k(u + \tilde{T}_k(u, v), v - \tilde{T}_k(u, v))H(\tilde{T}_k(u, v) - \tilde{S}_k(u, v))$$

$$- \tilde{q}_k(u + \tilde{S}_k(u, v), v - \tilde{S}_k(u, v))H(\tilde{T}_k(u, v) - \tilde{S}_k(u, v)).$$

Since min and max can be rewritten by step functions, the resulting form is clearly a sum of polynomials in u and v with degree at most i, multiplied by some step functions. Then, by the way we replace the min and max by step function, the arguments of the step functions are the differences

$$T_k(u, v) - S_k(u, v)$$
$$= \min\left\{1, u, \frac{-v + \alpha_{k+1}u}{\alpha_{k+1} - a_i}\right\} - \max\left\{0, u - 1, \frac{-v + \alpha_k u}{\alpha_k - a_i}\right\},$$

and

$$\tilde{T}_k(u, v) - \tilde{S}_k(u, v)$$
$$= \min\left\{0, 1 - u, \frac{v - \alpha_k u}{\alpha_k + a_i}\right\} - \max\left\{-1, -u, \frac{v - \alpha_{k+1}u}{\alpha_{k+1} + a_i}\right\}.$$

Since the cases for the other combination of α_k and α_{k+1} are similar, we omit the other cases to keep the paper concise.

The normalized form of $\Phi_i(u, v)$ is given by the resulting form as follows. For $k = 1, \ldots, 2i$ and $\epsilon > 0$ arbitrarily close to 0, we check whether the step function factors are 0 or 1 if $u = \epsilon$ and $v = \alpha_k + \epsilon$. That is, we choose ϵ so that there is no step function argument equal to 0. We do not replace u and v in the polynomial factor by ϵ and $\alpha_k + \epsilon$. Then, $p_k(u, v)$ is given by the resulting form when we replace the step function factors by 0 or 1. Thus, the lemma is proved.   □

Then, we can prove Theorem 1 as follows.

Proof (of Theorem 1). We consider the memory space we need to store the description of $\Phi_i(u, v)$ (length of $\Phi_i(u, v)$) and the ongoing integral forms for $i = 1, \ldots, n$ and $u \leq 1$. Since we store $\Phi_i(u, v)$ by arrays $A_i(k)$ and $P_i(k, d_u, d_v)$ for $k = 0, \ldots, 2i$, $d_u = 0, \ldots, i$ and $d_v = 0, \ldots, i$, $O(i^3)$ space is sufficient for storing $\Phi_i(u, v)$ for $u \leq 1$.

Notice that the length of $\Phi_{i-1}(u - |x_i|, v - a_i x_i)$ may temporarily take i times the length of $\Phi_{i-1}(u, v)$ because we get sum of i terms by expanding each of $(u + x_i)^{i-1}$, $(u - x_i)^{i-1}$ and $(v - a_i x_i)^{i-1}$. Then, remember that executing the integral of a polynomial symbolically means rewriting the integrand into a resulting form that can be computed in a constant time for each term. It implies that executing the integral can be finished linear time with respect to the length of the integrand. Therefore, $O(i^4)$ time is sufficient for obtaining the resulting form of $\int_{-1}^{1} \Phi_{i-1}(u - |x_i|, v - a_i x_i) dx_i$.

Since this resulting form may not be the normalized form, we rewrite the resulting form into a normalized form in $O(i^4) \times 2i = O(i^5)$ time.

Therefore, we obtain a normalized form of $\Phi_i(u, v)$ from the normalized form of $\Phi_{i-1}(u, v)$ in $O(i^5)$ time. This implies the theorem.   □

4   Conclusion and Future Works

In this paper, we showed that there exists a polynomial time algorithm for computing the volume of the crosspolytope truncated by a halfspace. Unlike that computing the volume of a hypercube truncated by a halfspace is #*P*-hard, truncating a crosspolytope does not make the volume computation very hard. Together with the results [3, 4] and the result of this paper, the geometric duality seems to preserve the existence of the efficient volume computation algorithm.

As future works, we would like to prove or disprove the conjecture. Also, we conjecture that the volume of a polytope K with the following two conditions may be computed in polynomial time,

1. K has poly(n) vertices;
2. K is given by a constant number of linear constraints allowing the absolute value of the variables.

Since the running time of the algorithm for the truncated crosspolytope is $O(n^6)$ while the volume of the convex hull of a hypercube and a point can be computed in linear time, it is interesting to ask whether the difference of the running time is essential or not. Since we did not intend to optimize the running time, there may be much faster algorithm for computing the volume of the truncated crosspolytope.

Appendix   Supplemental Proof

To make this paper self-contained, we prove the following proposition which is used in the proof of Proposition 3.

Proposition 4. *Let $P \subseteq \mathbb{R}^n$ be a convex n-dimensional polytope. Then there exists a set of m simplices $S_1, \ldots, S_m$ satisfying the following three conditions:*

1. *$P = \bigcup_{i=1,\ldots,m} S_i$;*
2. *any vertex of the simplices $S_1, \ldots, S_m$ is a vertex of P;*
3. *$\mathrm{Vol}(S_i \cap S_j) = 0$ for any $1 \leq i < j \leq m$.*

Proof. The proof is the induction on n. As for the base case, we consider the case $n = 1$. In this case, P is always a bounded interval, which is a simplex. Therefore, the proposition holds for the base case.

We proceed to the induction step. We assume that we have the claims of the proposition in case $n = k$. Then, in case $n = k + 1$, we have that any facet of P can be divided into a set of k-dimensional simplices, satisfying the three conditions of the claim. Let $S'_1, \ldots, S'_M$ be the k-dimensional simplices obtained by dividing the P's facets satisfying the three conditions for each facet. Let v be one vertex of P. Then, we obtain the $(k + 1)$-dimensional simplices as the convex hulls $S_i = \mathrm{conv}(S'_i \cup \{v\})$ for $i = 1, \ldots, M$.

As for the first condition of the proposition, we show that for any internal point $p \in P$, there exists a point $q \in S'_i$ for some $1 \leq i \leq M$ such that $p \in \mathrm{conv}(\{v, q\})$. We consider a point given by $r(t) = t(p - v) + v$, where $t > 0$. Since P is bounded, we have that $r(t)$ is on a facet F of P for some $t > 1$. We have $t > 1$ since p is an internal point of P. Since each facet F can be divided into simplices, $q = r(t)$ is in one of these simplices.

Since the second condition of the claim clearly holds for $S_1, \ldots, S_M$ by definition, we proceed to the proof of the third condition. That is, $\text{Vol}(S_i \cap S_j) = 0$ for any $1 \leq i < j \leq M$. Let $\boldsymbol{p} \in S_i \cap S_j$. We consider the point $\boldsymbol{r}(t) = t(\boldsymbol{p} - \boldsymbol{v}) + \boldsymbol{v}$ as in the above for $t > 1$. Let t_0 be the value of t such that $\boldsymbol{r}(t_0)$ is on the surface of P. Since $S_i = \text{conv}(S_i' \cup \{\boldsymbol{v}\})$ and $S_j = \text{conv}(S_j' \cup \{\boldsymbol{v}\})$, we have that $\boldsymbol{r}(t_0) \in S_i' \cap S_j'$. Since the k-dimensional volume of $S_i' \cap S_j'$ is 0 by the assumption, we have that $\text{Vol}(\text{conv}(\{\boldsymbol{v}\} \cup (S_i' \cap S_j'))) = 0$, which shows the claim.   □

References

1. Ando, E.: An FPTAS for computing the distribution function of the longest path length in DAGs with uniformly distributed edge lengths. In: Poon, S.-H., Rahman, M.S., Yen, H.-C. (eds.) WALCOM 2017. LNCS, vol. 10167, pp. 421–432. Springer, Cham (2017). https://doi.org/10.1007/978-3-319-53925-6_33
2. Ando, E., Kijima, S.: An FPTAS for the volume of a $\mathcal{V}$-polytopes—it is hard to compute the volume of the intersection of two cross-polytopes. arXiv:1607.06173
3. Ando, E., Kijima, S.: An FPTAS for the volume computation of 0-1 knapsack polytopes based on approximate convolution. Algorithmica **76**(4), 1245–1263 (2016)
4. Ando, E., Kijima, S.: An FPTAS for the volume of some $\mathcal{V}$-polytopes—it is hard to compute the volume of the intersection of two cross-polytopes. In: Cao, Y., Chen, J. (eds.) COCOON 2017. LNCS, vol. 10392, pp. 13–24. Springer, Cham (2017). https://doi.org/10.1007/978-3-319-62389-4_2
5. Bárány, I., Füredi, Z.: Computing the volume is difficult. Discrete Comput. Geom. **2**, 319–326 (1987)
6. Cousins, B., Vempala, S.: Bypassing KLS: Gaussian cooling and an $O^*(n^3)$ volume algorithm. In: Proceedings of the STOC 2015, pp. 539–548 (2015)
7. Dyer, M., Frieze, A., Kannan, R.: A random polynomial-time algorithm for approximating the volume of convex bodies. J. Assoc. Comput. Mach. **38**(1), 1–17 (1991)
8. Dyer, M., Gritzmann, P., Hufnagel, A.: On the complexity of computing mixed volumes. SIAM J. Comput. **27**(2), 356–400 (1998)
9. Dyer, M., Frieze, A.: On the complexity of computing the volume of a polyhedron. SIAM J. Comput. **17**(5), 967–974 (1988)
10. Elekes, G.: A geometric inequality and the complexity of computing volume. Discrete Comput. Geom. **1**, 289–292 (1986)
11. Gopalan, P., Klivans, A., Meka, R., Štefankovič, D., Vempala, S., Vigoda, E.: An FPTAS for #knapsack and related counting problems. In: Proceedings of FOCS 2011, pp. 817–826 (2011)
12. Khachiyan, L.: The problem of computing the volume of polytopes is #P-hard. Uspekhi Mat. Nauk. **44**, 199–200 (1989)
13. Khachiyan, L.: Complexity of polytope volume computation. In: Pach, J. (ed.) New Trends in Discrete and Computational Geometry. AC, vol. 10, pp. 91–101. Springer, Heidelberg (1993). https://doi.org/10.1007/978-3-642-58043-7_5
14. Li, J., Shi, T.: A fully polynomial-time approximation scheme for approximating a sum of random variables. Oper. Res. Lett. **42**, 197–202 (2014)
15. Lovász, L., Vempala, S.: Simulated annealing in convex bodies and an $O^*(n^4)$ volume algorithm. J. Comput. Syst. Sci. **72**, 392–417 (2006)

16. Matoušek, J.: Lectures on Discrete Geometry. Graduate Texts in Mathematics. Springer, New York (2002). https://doi.org/10.1007/978-1-4613-0039-7
17. Štefankovič, D., Vempala, S., Vigoda, E.: A deterministic polynomial-time approximation scheme for counting knapsack solutions. SIAM J. Comput. **41**(2), 356–366 (2012)

Computable Isomorphisms of Distributive Lattices

Nikolay Bazhenov[1,2]($^{(\boxtimes)}$) , Manat Mustafa[3] , and Mars Yamaleev[4]

[1] Sobolev Institute of Mathematics,
4 Acad. Koptyug Ave., Novosibirsk 630090, Russia
[2] Novosibirsk State University,
2 Pirogova St., Novosibirsk 630090, Russia
bazhenov@math.nsc.ru
[3] Department of Mathematics, School of Science and Technology,
Nazarbayev University, 53 Qabanbaybatyr Avenue, Astana 010000, Kazakhstan
manat.mustafa@nu.edu.kz
[4] Kazan Federal University,
18 Kremlevskaya St., Kazan 420008, Russia
mars.yamaleev@kpfu.ru

Abstract. A standard tool for the classifying computability-theoretic complexity of equivalence relations is provided by computable reducibility. This gives rise to a rich degree-structure which has been extensively studied in the literature. In this paper, we show that equivalence relations, which are complete for computable reducibility in various levels of the hyperarithmetical hierarchy, arise in a natural way in computable structure theory. We prove that for any computable successor ordinal α, the relation of Δ^0_α isomorphism for computable distributive lattices is $\Sigma^0_{\alpha+2}$ complete. We obtain similar results for Heyting algebras, undirected graphs, and uniformly discrete metric spaces.

Keywords: Distributive lattice · Computable reducibility ·
Equivalence relation · Computable categoricity · Heyting algebra ·
Computable metric space

1 Introduction

We study computability-theoretic complexity of equivalence relations which arise in a natural way in computable structure theory. Our main working tool is *computable reducibility*.

The work was supported by Nazarbayev University Faculty Development Competitive Research Grants N090118FD5342. The first author was supported by Russian Science Foundation, project No. 18-11-00028. The last author was supported by the subsidy allocated to Kazan Federal University for the state assignment in the sphere of scientific activities, project № 1.13556.2019/13.1.

T. V. Gopal and J. Watada (Eds.): TAMC 2019, LNCS 11436, pp. 28–41, 2019.
https://doi.org/10.1007/978-3-030-14812-6_3

Definition 1.1. *Suppose that E and F are equivalence relations on the domain ω. The relation E is computably reducible to F (denoted by $E \leq_c F$) if there is a total computable function $f(x)$ such that for all $x, y \in \omega$, the following holds:*

$$(xEy) \iff (f(x)Ff(y)).$$

In what follows, we assume that every considered equivalence relation has domain ω.

The systematic study of c-degrees, i.e. degrees induced by computable reducibility, was initiated by Ershov [12,13]. His approach stems from the category-theoretic methods in the theory of numberings. In 1980s, the research in the area of c-degrees was concentrated on computably enumerable equivalence relations (or *ceers* for short): in particular, provable equivalence in formal systems was studied (see, e.g., [10,11,28]). Note that the acronym *ceer* was introduced in [17]. Recently, Andrews and Sorbi [1] provided a profound analysis of the structure of c-degrees of ceers. For the results and bibliographical references on ceers, the reader is referred to, e.g., the survey [2] and the articles [1,3,17].

Computable reducibility also proved to be useful for classifying equivalence relations having higher complexity than ceers. In particular, recent works [8,24] consider c-degrees of Δ_2^0 equivalence relations.

Definition 1.2. *Let Γ be a complexity class (e.g., Σ_1^0, d-Σ_1^0, Σ_2^0, or Π_1^1). An equivalence relation E is Γ complete (for computable reducibility) if $E \in \Gamma$ and for every equivalence relation $R \in \Gamma$, we have $R \leq_c E$.*

Examples of known Γ complete equivalence relations include:

- The relation of provable equivalence in Peano arithmetic is Σ_1^0 complete [11].
- 1-equivalence and m-equivalence on indices of c.e. sets are both Σ_3^0 complete [14].
- Turing equivalence on indices of c.e. sets is Σ_4^0 complete [21].
- For every $n \in \omega$, 1-equivalence on indices of $\emptyset^{(n+1)}$-c.e. sets is Σ_{n+4}^0 complete [21].

Furthermore, in [21], it was proved that for any computable ordinal α, there is no $\Pi_{\alpha+2}^0$ complete equivalence relation.

Some of Γ complete equivalence relations have origins in computable structure theory: Given a class of structures K, one can treat the *isomorphism relation* on (the set of computable members of) the class K as an equivalence relation on ω (to be formally explained in Sect. 2.1). In [15], it was proved that for each of the following classes K, the isomorphism relation on K is Σ_1^1 complete for computable reducibility: trees, graphs, torsion-free abelian groups, abelian p-groups, linear orders, fields (of arbitrary characteristic), 2-step nilpotent groups.

Fokina, Friedman, and Nies [14] investigated the relation of *computable isomorphism* on a given class. In particular, they showed that for predecessor trees, equivalence structures, and Boolean algebras, the computable isomorphism relation is Σ_3^0 complete.

In this paper, we study the relation of Δ_α^0 *isomorphism*, denoted by $\cong_{\Delta_\alpha^0}$, where α is a non-zero computable ordinal. Δ_α^0 isomorphisms and the closely related notion of Δ_α^0-*categoricity* have been extensively studied in the literature (see, e.g., [6,16] for a survey of results).

Following the approach of [14], our paper shows that the relation $\cong_{\Delta_\alpha^0}$ fits well in the setting of computable reducibility. The outline of the paper is as follows. Section 2 contains the necessary preliminaries. In Sect. 3, we prove our main result: For every computable successor ordinal α, the relation $\cong_{\Delta_\alpha^0}$ on computable distributive lattices is $\Sigma_{\alpha+2}^0$ complete for computable reducibility.

In Sect. 4, we prove consequences of the main theorem: similar results are obtained for Heyting algebras, undirected graphs, and uniformly discrete metric spaces. We also give a partial result for Boolean algebras with distinguished subalgebra. Section 5 discusses some open problems.

2   Preliminaries

We consider only computable languages. For any considered countable structure $\mathcal{S}$, its domain is contained in the set of natural numbers. By $D(\mathcal{S})$ we denote the atomic diagram of $\mathcal{S}$.

For a language L, *infinitary formulas* of L are formulas of the logic $L_{\omega_1,\omega}$. For a countable ordinal α, infinitary Σ_α and Π_α formulas are defined in a standard way (see, e.g., [6, Chap. 6]).

2.1   Isomorphism Relation

Suppose that L is a computable language. For a computable L-structure $\mathcal{S}$, its *computable index* is a number e such that the characteristic function $\chi_{D(\mathcal{S})}$ of the atomic diagram $D(\mathcal{S})$ is equal to φ_e, where $\{\varphi_e\}_{e\in\omega}$ is the standard enumeration of all unary partial computable functions.

For $e \in \omega$, by $\mathcal{M}_e$ we denote the structure with computable index e. Suppose that K is a class of L-structures. The *index set* of the class K is the set

$$I(K) = \{e : \mathcal{M}_e \in K\}.$$

Let $\sim$ be an equivalence relation on (computable members of) the class K. Then we will identify $\sim$ with the following equivalence relation $\sim_\#$ on the set of natural numbers:

$$(i \sim_\# j) \Leftrightarrow (i = j) \vee (i, j \in I(K)\,\&\,\mathcal{M}_i \sim \mathcal{M}_j).$$

Therefore, one can consider the relations of isomorphism and Δ_α^0 isomorphism in the setting of computable reducibility.

Lemma 2.1. *Let K be a class of structures, and α be a computable non-zero ordinal. If the index set $I(K)$ is $\Sigma_{\alpha+2}^0$, then the relation of Δ_α^0 isomorphism on computable members of K is also $\Sigma_{\alpha+2}^0$.*

Proof. Essentially follows from [19, Proposition 4.10].   □

It is not hard to establish the following result (e.g., compare [19, Proposition 4.1]).

Lemma 2.2. *For each of the following classes K (in an appropriate language, to be discussed in the corresponding sections), the index set $I(K)$ is Π_2^0:*

(a) distributive lattices,
(b) Heyting algebras,
(c) undirected graphs,
(d) Boolean algebras with distinguished subalgebra.

Lemmas 2.1 and 2.2 together show that on each of the classes K considered above, the relation $\cong_{\Delta_\alpha^0}$ is $\Sigma_{\alpha+2}^0$. Hence, in order to prove our results, it is sufficient to establish the $\Sigma_{\alpha+2}^0$ hardness of the relation $\cong_{\Delta_\alpha^0}$: Given an arbitrary $\Sigma_{\alpha+2}^0$ equivalence relation E, we produce a uniformly computable sequence $\{S_n\}_{n\in\omega}$ of structures from K such that:

$$(mEn) \ \Leftrightarrow \ (S_m \cong_{\Delta_\alpha^0} S_n).$$

We leave the discussion of metric spaces until Sect. 4.3.

2.2 Hyperarithmetical Equivalence Relations

In order to obtain our results on the relation of Δ_α^0 isomorphism, we will work with some special hyperarithmetical equivalence relations. Note that the exposition in this subsection mirrors the corresponding recursion-theoretical results from [14].

Consider an oracle $X \subseteq \omega$. For $e \in \omega$, by W_e^X we denote the X-c.e. set that has index e in the standard numbering of all X-c.e. sets.

Suppose that A and B are subsets of ω. We say that A is *1-X-reducible* to B, denoted by $A \leq_1^X B$, if there is a total X-computable, injective function $f(x)$ such that for every $x \in \omega$, we have $x \in A$ iff $f(x) \in B$. As usual, we write $A \equiv_1^X B$ if $A \leq_1^X B$ and $B \leq_1^X A$.

The sets A and B are *X-computably isomorphic* if there is an X-computable permutation σ of the set of natural numbers such that $\sigma(A) = B$. The following lemma is a relativization of Myhill Isomorphism Theorem [23].

Lemma 2.3. *Sets A and B are X-computably isomorphic iff $A \equiv_1^X B$.*

Now one can consider a relativized version of [14, Theorem 1]:

Theorem 2.1 (essentially [14]). *For any $\Sigma_3^0(X)$ equivalence relation E, there is a total computable function $g(x)$ such that:*

(a) If (yEz), then $W_{g(y)}^X \equiv_1^X W_{g(z)}^X$.
(b) If $\neg(yEz)$, then $W_{g(y)}^X \nleq_T W_{g(z)}^X \oplus X$ and $W_{g(z)}^X \nleq_T W_{g(y)}^X \oplus X$.

Proof (sketch). Proceed with a straightforward relativization of [14, Theorem 1]. Note that this gives only an X-*computable* function $g_0(x)$ with the desired properties. Nevertheless, there is a *computable* function $g(x)$ such that $W^X_{g(e)} = W^X_{g_0(e)}$ for all e. Indeed, the set $\{\langle k, e \rangle : k \in W^X_{g_0(e)}\}$ is c.e. in X and hence, the function g can be recovered by using s-m-n Theorem (see, e.g., Exercise 1.20 in [26, Chap. III] for more details). □

Suppose that α is a computable non-zero ordinal. For convenience, we use the following notation:

$$\emptyset_{(\alpha)} := \begin{cases} \emptyset^{(\alpha-1)}, \text{ if } \alpha < \omega, \\ \emptyset^{(\alpha)}, \quad \text{ if } \alpha \geq \omega. \end{cases}$$

Notice that for every α, we have $\Sigma^0_\alpha = \Sigma^0_1(\emptyset_{(\alpha)})$ and $\Delta^0_\alpha = \Delta^0_1(\emptyset_{(\alpha)})$. The theorem above implies the following.

Corollary 2.1. *Let α be a computable non-zero ordinal. Then the relation $\equiv^{\emptyset_{(\alpha)}}_1$ on the indices of $\emptyset_{(\alpha)}$-c.e. sets is $\Sigma^0_{\alpha+2}$ complete for computable reducibility.*

2.3 Pairs of Computable Structures

Our proofs heavily rely on the technique of pairs of computable structures developed by Ash and Knight [5,6]. Here we give necessary preliminaries on the technique.

Suppose that $\mathcal{A}$ and $\mathcal{B}$ are L-structures. We say that $\mathcal{B} \leq_\alpha \mathcal{A}$ if every infinitary Π_α sentence true in $\mathcal{B}$ is also true in $\mathcal{A}$.

Let α be a computable ordinal. A family $K = \{\mathcal{A}_i : i \in I\}$ of L-structures is α-*friendly* if the structures $\mathcal{A}_i$ are uniformly computable in $i \in I$, and the relations

$$B_\beta = \{(i, \bar{a}, j, \bar{b}) : i, j \in I, \bar{a} \in \mathcal{A}_i, \bar{b} \in \mathcal{A}_j, (\mathcal{A}_i, \bar{a}) \leq_\beta (\mathcal{A}_j, \bar{b})\}$$

are computably enumerable, uniformly in $\beta < \alpha$.

Theorem 2.2 ([5, Theorem 3.1]). *Suppose that α is a non-zero computable ordinal, $\mathcal{A}$ and $\mathcal{B}$ are L-structures. If $\mathcal{B} \leq_\alpha \mathcal{A}$ and the family $\{\mathcal{A}, \mathcal{B}\}$ is α-friendly, then for any Σ^0_α set X, there is a uniformly computable sequence of L-structures $\{\mathcal{C}_n\}_{n \in \omega}$ such that*

$$\mathcal{C}_n \cong \begin{cases} \mathcal{A}, \text{ if } n \notin X; \\ \mathcal{B}, \text{ if } n \in X. \end{cases}$$

Theorem 2.2 and the description of the relations $\leq_\alpha$ for countable well-orders [4,6] together imply the following:

Proposition 2.1. *Let β be a computable ordinal.*

(i) *For any $\Sigma^0_{2\beta+1}$ set S, there is a uniformly computable sequence of linear orders $\{C_n\}_{n\in\omega}$ such that*

$$C_n \cong \begin{cases} \omega^\beta, & \text{if } n \notin S; \\ \omega^\beta \cdot 2, & \text{if } n \in S. \end{cases}$$

(ii) *For any $\Sigma^0_{2\beta+2}$ set S, there is a uniformly computable sequence of linear orders $\{C_n\}_{n\in\omega}$ such that*

$$C_n \cong \begin{cases} \omega^{\beta+1}, & \text{if } n \notin S; \\ \omega^{\beta+1} + \omega^\beta, & \text{if } n \in S. \end{cases}$$

A sketch of the proof of Proposition 2.1 can be found, e.g., in [9, Theorem 4].

2.4   Distributive Lattices

Consider a language $L_{BL} := \{\vee, \wedge; 0, 1\}$. Recall that a lattice is *bounded* if it has the least element 0 and the greatest element 1. In this paper, we consider only bounded lattices. Thus, we treat lattices as L_{BL}-structures. The reader is referred to [20] for the background on lattice theory.

A partial order $\leq$ in a lattice $\mathcal{A}$ is recovered in a standard lattice-theoretical way: $x \leq y$ if and only if $x \vee y = y$. For elements $a, b \in \mathcal{A}$, by $[a; b]$ we denote the interval $\{c \in \mathcal{A} : a \leq c \leq b\}$.

Suppose that $\{\mathcal{A}_n\}_{n\in\omega}$ is a sequence of distributive lattices. The *direct sum* of the sequence $\{\mathcal{A}_n\}_{n\in\omega}$ (denoted by $\sum_{n\in\omega} \mathcal{A}_n$) is the substructure of the product $\prod_{n\in\omega} \mathcal{A}_n$ on the domain

$$\left\{ f \in \prod_{n\in\omega} \mathcal{A}_n : (\exists c \in \{0,1\})\exists m(\forall k \geq m)(f(k) = c^{\mathcal{A}_k}) \right\}.$$

It is not hard to show that $\sum_{n\in\omega} \mathcal{A}_n$ is a distributive lattice. Furthermore, if the sequence $\{\mathcal{A}_n\}_{n\in\omega}$ is computable, then one can build a computable copy of the sum $\sum_{n\in\omega} \mathcal{A}_n$, in a standard way (see, e.g., [9, §2.1] for details). Hence, in this case, we will identify the direct sum with its standard computable presentation.

If $a_i \in \mathcal{A}_i$, $i \leq n$, and $a_n \neq 0^{\mathcal{A}_n}$, then $(a_0, a_1, \ldots, a_n, \perp_{n+1})$ denotes the element $(a_0, a_1, \ldots, a_n, 0, 0, 0, \ldots)$ from $\sum_{n\in\omega} \mathcal{A}_n$. If $a_n \neq 1^{\mathcal{A}_n}$, then by $(a_0, a_1, \ldots, a_n, \top_{n+1})$ we denote the element $(a_0, a_1, \ldots, a_n, 1, 1, 1, \ldots)$.

If $\mathcal{L}$ is a linear order with the least and the greatest elements, then (as per usual) $\mathcal{L}$ can be treated as bounded distributive lattice $\mathcal{D}(\mathcal{L})$.

3 Δ^0_α Isomorphism for Distributive Lattices

Theorem 3.1. *Suppose that α is a computable successor ordinal. The relation of Δ^0_α isomorphism of computable distributive lattices is a complete $\Sigma^0_{\alpha+2}$ equivalence relation under computable reducibility.*

Proof. Here we give a detailed proof for the case when α is odd, i.e. $\alpha = 2\beta + 1$. At the end of the proof, we will briefly comment on how to deal with even α.

Suppose that E is a $\Sigma^0_{\alpha+2}$ equivalence relation on ω. Then by Corollary 2.1, there is a computable function $g(x)$ with the following property: for any $m, n \in \omega$,

$$(mEn) \;\Leftrightarrow\; W^{\emptyset(\alpha)}_{g(m)} \equiv^{\emptyset(\alpha)}_1 W^{\emptyset(\alpha)}_{g(n)}. \tag{1}$$

Since $\alpha = 2\beta + 1$, the first part of Proposition 2.1 gives a computable sequence $\{\mathcal{L}_{n,k}\}_{n,k\in\omega}$ of linear orders such that

$$\mathcal{L}_{n,k} \cong \begin{cases} \omega^\beta, & \text{if } k \notin W^{\emptyset(\alpha)}_{g(n)}; \\ \omega^\beta \cdot 2, & \text{if } k \in W^{\emptyset(\alpha)}_{g(n)}. \end{cases} \tag{2}$$

For a natural number n, we define a computable distributive lattice $\mathcal{S}_n$ as follows:

$$\mathcal{S}_n := \sum_{k\in\omega} \mathcal{D}(\mathcal{L}_{n,k} + 1).$$

Now it is sufficient to prove the following fact: For every $m, n \in \omega$,

$$(mEn) \;\Leftrightarrow\; (\mathcal{S}_m \text{ and } \mathcal{S}_n \text{ are } \Delta^0_\alpha\text{-computably isomorphic}).$$

For $n, k \in \omega$, consider the element $e_{n,k} := (0, 0, \ldots, 0, c_{n,k}, \perp_{k+1})$ from $\mathcal{S}_n$, where $c_{n,k}$ is the greatest element in the order $(\mathcal{L}_{n,k} + 1)$. Clearly, the sequence $\{e_{n,k}\}_{n,k\in\omega}$ is uniformly computable.

We define auxiliary finitary formulas

$$Lin(x) := \forall y \forall z [(y \leq x) \& (z \leq x) \to (y \leq z) \vee (z \leq y)],$$
$$MaxLin(x) := Lin(x) \& \forall y [(x \leq y) \& Lin(y) \to (y = x)].$$

The $\forall\exists$-formula $MaxLin(x)$ says that an element x is maximal such that the interval $[0; x]$ is linearly ordered. It is not hard to show that $MaxLin(\mathcal{S}_n) = \{e_{n,k} : k \in \omega\}$, see [9, Lemma 3] for details. Since the sequence $\{e_{n,k}\}_{n,k\in\omega}$ is computable, one may assume that the sets $MaxLin(\mathcal{S}_n)$ are computable, uniformly in n.

Lemma 3.1. *If $\mathcal{S}_m$ and $\mathcal{S}_n$ are Δ^0_α-computably isomorphic, then m and n are E-equivalent.*

Proof. Let F be a Δ^0_α isomorphism from $\mathcal{S}_m$ onto $\mathcal{S}_n$. Note that the map $F_1 := F \upharpoonright MaxLin(\mathcal{S}_m)$ is a Δ^0_α bijection from $MaxLin(\mathcal{S}_m)$ onto $MaxLin(\mathcal{S}_n)$. Define a map $\sigma \colon \omega \to \omega$ as follows:

$$\sigma(i) = j, \text{ if } F(e_{m,i}) = e_{n,j}.$$

It is easy to see that σ is well-defined. Moreover, σ is a Δ^0_α permutation of ω.

For every $i \in \omega$, the intervals $[0; e_{m,i}]_{\mathcal{S}_m}$ and $[0; e_{n,\sigma(i)}]_{\mathcal{S}_n}$ are isomorphic. Thus, for any i, the following conditions are equivalent:

$$i \in W^{\emptyset(\alpha)}_{g(m)} \;\Leftrightarrow\; \mathcal{L}_{m,i} \cong \omega^\beta \cdot 2 \;\Leftrightarrow\; \mathcal{L}_{n,\sigma(i)} \cong \omega^\beta \cdot 2 \;\Leftrightarrow\; \sigma(i) \in W^{\emptyset(\alpha)}_{g(n)}.$$

Therefore, the permutation σ witnesses that the sets $W_{g(m)}^{\emptyset(\alpha)}$ and $W_{g(n)}^{\emptyset(\alpha)}$ are $\emptyset_{(\alpha)}$-computably isomorphic. Equation (1) implies that the numbers m and n are E-equivalent. $\qquad\square$

Lemma 3.2. *If (mEn), then the lattices $\mathcal{S}_m$ and $\mathcal{S}_n$ are Δ_α^0-computably isomorphic.*

Proof. Assume that m and n are E-equivalent. By Eq. (1), there is a Δ_α^0 permutation σ such that $\sigma(W_{g(m)}^{\emptyset(\alpha)}) = W_{g(n)}^{\emptyset(\alpha)}$. Therefore, for every $i \in \omega$, the orders $\mathcal{L}_{m,i}$ and $\mathcal{L}_{n,\sigma(i)}$ are isomorphic. Recall that every $\mathcal{L}_{m,i}$ is isomorphic either to ω^β, or to $\omega^\beta \cdot 2$.

In [9, p. 609] (see also Proposition 2 in [9]), the following fact was proved: There is an effective procedure which, given computable indices of linear orders $\mathcal{M}$ and $\mathcal{N}$ such that $\mathcal{M}$ and $\mathcal{N}$ are both isomorphic to some $\mathcal{A} \in \{\omega^\beta, \omega^\beta \cdot 2\}$, computes a $\Delta_{2\beta+1}^0$ index of an isomorphism F from $\mathcal{M}$ onto $\mathcal{N}$.

Recall that $\alpha = 2\beta + 1$. Hence, using the fact above, one can produce a uniform sequence of Δ_α^0 isomorphisms $\{F_i\}_{i\in\omega}$ such that F_i maps $\mathcal{L}_{m,i}$ onto $\mathcal{L}_{n,\sigma(i)}$.

Now one can arrange a Δ_α^0 isomorphism G from $\mathcal{S}_m$ onto $\mathcal{S}_n$ in a pretty straightforward way. A typical example looks like follows: Consider an element $a = (p_0, p_1, p_2, \top_3)$ from $\mathcal{S}_m$, where $0 \le p_i < e_{m,i}$. Then

$$G(a) := F_0(p_0) \vee F_1(p_1) \vee F_2(p_2) \vee b,$$

where the jth coordinate of the element b (inside $\mathcal{S}_n$) is equal to

$$\begin{cases} 0, & \text{if } j \in \{\sigma(0), \sigma(1), \sigma(2)\}, \\ e_{n,j}, & \text{otherwise.} \end{cases}$$

Lemma 3.2 is proved. $\qquad\square$

The proof of Theorem 3.1 for the case $\alpha = 2\beta + 2$ is essentially the same, modulo the following key modification: one needs to use the ordinals $\omega^{\beta+1}$ and $\omega^{\beta+1} + \omega^\beta$ in place of ω^β and $\omega^\beta \cdot 2$, respectively. More details on this case can be recovered from the discussion in [9, p. 610]. Theorem 3.1 is proved. $\qquad\square$

4 Consequences of the Main Result

The (method of the) proof of Theorem 3.1 can be applied to obtain similar results for other familiar classes of structures.

4.1 Heyting Algebras

Heyting algebras are treated as structures in the language $L_{HA} = \{\vee, \wedge, \rightarrow; 0, 1\}$. An L_{HA}-structure $\mathcal{H}$ is a *Heyting algebra* if the $\{\vee, \wedge; 0, 1\}$-reduct of $\mathcal{H}$ is a bounded distributive lattice, and $\mathcal{H}$ satisfies the following three axioms:

(a) $\forall x \forall y [x \wedge (x \to y) = x \wedge y]$;
(b) $\forall x \forall y \forall z [x \wedge (y \to z) = x \wedge ((x \wedge y) \to (x \wedge z))]$;
(c) $\forall x \forall y \forall z [z \wedge ((x \wedge y) \to x) = z]$.

If $\mathcal{L}$ is a linear order with the least and the greatest elements, then it can be treated as Heyting algebra by introducing the operation:

$$x \to y := \begin{cases} 1, \text{ if } x \leq y; \\ y, \text{ if } x > y. \end{cases}$$

Therefore, essentially the same proof as for Theorem 3.1 provides us with the following result:

Corollary 4.1. *Let α be a computable successor ordinal. The relation of Δ_α^0 isomorphism of computable Heyting algebras is a $\Sigma_{\alpha+2}^0$ complete equivalence relation.*

More computability-theoretical results on Heyting algebras can be found in [7,9,27].

4.2   Undirected Graphs

Consider a linear order $\mathcal{L}$ on the domain $\{a_i : i \in \omega\}$. Assume that $\mathcal{L}$ has no greatest element. We define an undirected graph $G(\mathcal{L})$ as follows:

- $dom(G(\mathcal{L})) = dom(\mathcal{L}) \cup \{b_{i,j}, c_{i,j} : i < j\} \cup \{d, e, f\}$.
- We put (undirected) edges (d, e), (e, f), (f, d), $(a_i, b_{i,j})$, $(b_{i,j}, c_{i,j})$, $(c_{i,j}, a_j)$ for every $i < j$.
- Suppose that $i < j$. If $a_i <_\mathcal{L} a_j$, then add the edge $(c_{i,j}, d)$. Otherwise, put the edge $(b_{i,j}, d)$.

It is not hard to see that the set $dom(\mathcal{L})$ and the ordering $\leq_\mathcal{L}$ are definable by both $\exists$- and $\forall$-formulas inside $G(\mathcal{L})$.

The transformation $\mathcal{L} \mapsto G(\mathcal{L})$ allows us to obtain the following:

Proposition 4.1. *Let α be a computable successor ordinal. The relation of Δ_α^0 isomorphism of computable undirected graphs is a complete $\Sigma_{\alpha+2}^0$ equivalence relation under computable reducibility.*

Proof (sketch). We follow the lines of Theorem 3.1, and after obtaining the sequence $\{\mathcal{L}_{n,k}\}_{n,k\in\omega}$, we introduce a uniformly computable sequence of undirected graphs $\{\mathcal{G}_n\}_{n\in\omega}$ which is constructed as follows. Put into $\mathcal{G}_n$ the graphs $G(\mathcal{L}_{n,k})$, $k \in \omega$, on disjoint domains, i.e. $dom(G(\mathcal{L}_{n,k})) \cap dom(G(\mathcal{L}_{n,i})) = \emptyset$ for $k \neq i$. Suppose that $e_{n,k}$ is the element which "plays role" of the node e in the graph $G(\mathcal{L}_{n,k})$. Introduce a fresh cycle of size five, fix a node v_0 inside the cycle, and add an edge between every $e_{n,k}$ and v_0.

It is not difficult to prove that $\mathcal{G}_m$ and $\mathcal{G}_n$ are Δ_α^0 isomorphic if and only if $\mathcal{S}_m \cong_{\Delta_\alpha^0} \mathcal{S}_n$.   □

4.3 Metric Spaces

Consider a Polish metric space (M, d). Assume that $(q_i)_{i \in \omega}$ is a dense sequence in M without repetitions. A structure $\mathcal{M} = (M, d, (q_i)_{i \in \omega})$ is a *computable metric space* if the value $d(q_i, q_j)$ is a computable real, uniformly in i and j. The elements q_i are called *special points*. For the background on computable metric spaces, the reader is referred to [29].

Fix a (standard) effective enumeration $\{\psi_e\}_{e \in \omega}$ of all partial computable functions acting from ω^3 into the set $\{q \in \mathbb{Q} : q \geq 0\}$.

We say that a number $e \in \omega$ is a *computable index* of a computable metric space $\mathcal{M} = (M, d, (q_i)_{i \in \omega})$ if the function ψ_e is total and for all $i, j, t \in \omega$, the following holds:

$$|d(q_i, q_j) - \psi_e(i, j, t)| \leq 2^{-t}.$$

The notion of computable index allows us to introduce *index sets* in the same way as in Sect. 2.1 (for more details, we refer the reader to [22, 25]). Thus, one can treat the relation of surjective isometry on computable metric spaces as an equivalence relation on ω.

Recall that a computable metric space is *discrete* if every its point is isolated. Note that in such a space, every point is special. A computable metric space $\mathcal{M}$ is *uniformly discrete* if there is a real $\varepsilon > 0$ such that for any points $a \neq b$ from $\mathcal{M}$, we have $d(a, b) \geq \varepsilon$. It is easy to see that any uniformly discrete space is discrete.

Corollary 4.2. *Let α be a computable successor ordinal. The relation of Δ^0_α surjective isometry of computable, uniformly discrete metric spaces is a $\Sigma^0_{\alpha+2}$ complete equivalence relation.*

Proof. Note that the property "e is a computable index of a metric space" is equivalent to a Π^0_2 description (see, e.g., [22, p. 322]). A computable index e encodes a uniformly discrete space if and only if the following holds:

$$(\exists \varepsilon \in \mathbb{Q})[(\varepsilon > 0) \,\&\, \forall i \forall j (i \neq j \rightarrow \exists t (\psi_e(i, j, t) \geq \varepsilon + 2^{-t}))].$$

This is a Σ^0_3 description, hence the index set of uniformly discrete metric spaces is Σ^0_3. By (an analogue of) Lemma 2.1, we obtain that Δ^0_α surjective isometry for computable, uniformly discrete spaces is a $\Sigma^0_{\alpha+2}$ relation.

Given a countable undirected graph G on the domain $\{a_i : i \in \omega\}$, we introduce a discrete metric space $\mathcal{M}(G)$ as follows. The domain of $\mathcal{M}(G)$ is equal to $dom(G)$, and for every $i \neq j$, we set

$$d(a_i, a_j) = \begin{cases} 1, & \text{if } G \models Edge(a_i, a_j), \\ 3/2, & \text{if } G \models \neg Edge(a_i, a_j). \end{cases}$$

It is easy to see that there is a Δ^0_α surjective isometry from $\mathcal{M}(G)$ onto $\mathcal{M}(H)$ iff $G \cong_{\Delta^0_\alpha} H$. Thus, the desired result follows from Proposition 4.1. $\qquad\square$

4.4 Boolean Algebras with Distinguished Subalgebra

Consider a language $L_{BA} = \{\vee, \wedge, \overline{\cdot}\,; 0, 1\}$. A *Boolean algebra with a distinguished subalgebra* is a structure $\mathcal{S}$ in the language $L_{BA} \cup \{U\}$ such that:

– the L_{BA}-reduct of $\mathcal{S}$ (denoted by $\mathcal{S}_{BA}$) is a Boolean algebra, and
– the unary predicate U distinguishes a subalgebra of $\mathcal{S}_{BA}$.

Here we obtain a partial result on the relation of Δ^0_α isomorphism for this class of structures.

If $\mathcal{L}$ is a linear order with the least element, then $Int(\mathcal{L})$ denotes the corresponding *interval Boolean algebra*. The background on computable Boolean algebras can be found in [18].

Proposition 4.2. *Let β be a computable ordinal. The relation of $\Delta^0_{2\beta+1}$ isomorphism of computable Boolean algebras with distinguished subalgebra is a complete $\Sigma^0_{2\beta+3}$ equivalence relation under computable reducibility.*

Proof (sketch). Let $\alpha = 2\beta + 1$. As in Theorem 3.1, given a $\Sigma^0_{\alpha+2}$ equivalence relation E, we choose a computable function $g(x)$ which satisfies Eq. (1).

It is well-known that the transformation $\mathcal{L} \mapsto Int(\mathcal{L})$ is uniformly effective, i.e. given a computable index of a linear order $\mathcal{L}$ (with the least element), one can effectively find a computable index for the algebra $Int(\mathcal{L})$. Thus, using the sequence from Eq. (2), one can build a uniformly computable sequence of Boolean algebras

$$\mathcal{B}_{n,k} \cong \begin{cases} Int(\omega^\beta), & \text{if } k \notin W^{\emptyset(\alpha)}_{g(n)}, \\ Int(\omega^\beta \cdot 2), & \text{if } k \in W^{\emptyset(\alpha)}_{g(n)}. \end{cases}$$

Let $e_{n,k}$ be the greatest element in $\mathcal{B}_{n,k}$.

For a natural number n, we define the Boolean algebra $\mathcal{C}_n := \sum_{k\in\omega} \mathcal{B}_{n,k}$. Inside $\mathcal{C}_n$, we use a unary predicate U_n to distinguish the subalgebra generated by the elements $c_{n,k} := (0, 0, \dots, 0, e_{n,k}, \perp_{k+1})$, $k \in \omega$.

After that, one can show that

$$(mEn) \text{ iff } (\mathcal{C}_m, U_m) \text{ and } (\mathcal{C}_n, U_n) \text{ are } \Delta^0_\alpha\text{-computably isomorphic.}$$

First, note that the set $\{c_{n,k} : k \in \omega\}$ is precisely the set of atoms of the subalgebra U_n. This observation allows us to prove an analogue of Lemma 3.1.

In order to obtain an analogue of Lemma 3.2, we need the following fact: There is an effective procedure which, given computable indices of Boolean algebras $\mathcal{M}$ and $\mathcal{N}$ such that $\mathcal{M} \cong \mathcal{N} \cong \mathcal{A} \in \{Int(\omega^\beta), Int(\omega^\beta \cdot 2)\}$, computes a $\Delta^0_{2\beta+1}$ index of an isomorphism F from $\mathcal{M}$ onto $\mathcal{N}$. This is an easy consequence of the proofs of [6, Theorem 17.8] and [9, Proposition 2].   □

Note that in this setting, the proof of Theorem 3.1 for the case $\alpha = 2\beta + 2$ cannot be re-used in a direct way. Indeed, it is easy to see that the interval algebras $Int(\omega^{\beta+1})$ and $Int(\omega^{\beta+1} + \omega^\beta)$ are isomorphic, and hence, we cannot use these structures for encoding a $\Sigma^0_{\alpha+2}$ equivalence relation E.

5   Further Discussion

Note that in all our results, we consider only successor ordinals α. Therefore, the following is left open:

Question 5.1. Suppose that α is a computable limit ordinal. Is the relation of Δ^0_α isomorphism of computable structures $\Sigma^0_{\alpha+2}$ complete for computable reducibility?

Recall that in [14], it was shown that computable isomorphism of Boolean algebras is Σ^0_3 complete. We established $\Sigma^0_{\alpha+2}$ completeness of Δ^0_α isomorphism for Heyting algebras (Corollary 4.1). Since every Boolean algebra can be treated as Heyting algebra under the operation $x \to y := \overline{x} \vee y$, it is natural to ask the following:

Question 5.2. Suppose that α is a computable ordinal such that $\alpha \geq 2$. Is the relation of Δ^0_α isomorphism of computable Boolean algebras $\Sigma^0_{\alpha+2}$ complete for computable reducibility?

Acknowledgments. Part of the research contained in this paper was carried out while the first and the last authors were visiting the Department of Mathematics of Nazarbayev University, Astana. The authors wish to thank Nazarbayev University for its hospitality.

References

1. Andrews, U., Sorbi, A.: Joins and meets in the structure of ceers. Computability, Published online. https://doi.org/10.3233/COM-180098
2. Andrews, U., Badaev, S., Sorbi, A.: A survey on universal computably enumerable equivalence relations. In: Day, A., Fellows, M., Greenberg, N., Khoussainov, B., Melnikov, A., Rosamond, F. (eds.) Computability and Complexity. LNCS, vol. 10010, pp. 418–451. Springer, Cham (2017). https://doi.org/10.1007/978-3-319-50062-1_25
3. Andrews, U., Lempp, S., Miller, J.S., Ng, K.M., San Mauro, L., Sorbi, A.: Universal computably enumerable equivalence relations. J. Symb. Logic **79**(1), 60–88 (2014). https://doi.org/10.1017/jsl.2013.8
4. Ash, C.J.: Recursive labelling systems and stability of recursive structures in hyperarithmetical degrees. Trans. Am. Math. Soc. **298**(2), 497–514 (1986). https://doi.org/10.1090/S0002-9947-1986-0860377-7
5. Ash, C.J., Knight, J.F.: Pairs of recursive structures. Ann. Pure Appl. Log. **46**(3), 211–234 (1990). https://doi.org/10.1016/0168-0072(90)90004-L
6. Ash, C.J., Knight, J.F.: Computable Structures and the Hyperarithmetical Hierarchy. Studies in Logic and the Foundations of Mathematics, vol. 144. Elsevier Science B.V., Amsterdam (2000)
7. Bazhenov, N., Marchuk, M.: Degrees of categoricity for prime and homogeneous models. In: Manea, F., Miller, R.G., Nowotka, D. (eds.) CiE 2018. LNCS, vol. 10936, pp. 40–49. Springer, Cham (2018). https://doi.org/10.1007/978-3-319-94418-0_4

8. Bazhenov, N., Mustafa, M., San Mauro, L., Sorbi, A., Yamaleev, M.: Classifying equivalence relations in the Ershov hierarchy. arXiv:1810.03559
9. Bazhenov, N.A.: Effective categoricity for distributive lattices and Heyting algebras. Lobachevskii J. Math. **38**(4), 600–614 (2017). https://doi.org/10.1134/S1995080217040035
10. Bernardi, C.: On the relation provable equivalence and on partitions in effectively inseparable sets. Stud. Log. **40**(1), 29–37 (1981). https://doi.org/10.1007/BF01837553
11. Bernardi, C., Sorbi, A.: Classifying positive equivalence relations. J. Symb. Logic **48**(3), 529–538 (1983). https://doi.org/10.2307/2273443
12. Ershov, Y.L.: Positive equivalences. Algebra Log. **10**(6), 378–394 (1971). https://doi.org/10.1007/BF02218645
13. Ershov, Y.L.: Theory of Numberings. Nauka, Moscow (1977). (in Russian)
14. Fokina, E., Friedman, S., Nies, A.: Equivalence relations that are Σ_3^0 complete for computable reducibility. In: Ong, L., de Queiroz, R. (eds.) WoLLIC 2012. LNCS, vol. 7456, pp. 26–33. Springer, Heidelberg (2012). https://doi.org/10.1007/978-3-642-32621-9_2
15. Fokina, E.B., Friedman, S., Harizanov, V., Knight, J.F., McCoy, C., Montalbán, A.: Isomorphism relations on computable structures. J. Symb. Logic **77**(1), 122–132 (2012). https://doi.org/10.2178/jsl/1327068695
16. Fokina, E.B., Harizanov, V., Melnikov, A.: Computable model theory. In: Downey, R. (ed.) Turing's Legacy: Developments from Turing's Ideas in Logic. Lecture Notes in Logic, vol. 42, pp. 124–194. Cambridge University Press, Cambridge (2014)
17. Gao, S., Gerdes, P.: Computably enumerable equivalence relations. Stud. Log. **67**(1), 27–59 (2001). https://doi.org/10.1023/A:1010521410739
18. Goncharov, S.S.: Countable Boolean Algebras and Decidability. Siberian School of Algebra and Logic. Consultants Bureau, New York (1997)
19. Goncharov, S.S., Knight, J.F.: Computable structure and non-structure theorems. Algebra Log. **41**(6), 351–373 (2002). https://doi.org/10.1023/A:1021758312697
20. Grätzer, G.: Lattice Theory: Foundation. Birkhäuser/Springer, Basel (2011). https://doi.org/10.1007/978-3-0348-0018-1
21. Ianovski, E., Miller, R., Ng, K.M., Nies, A.: Complexity of equivalence relations and preorders from computability theory. J. Symb. Logic **79**(3), 859–881 (2014). https://doi.org/10.1017/jsl.2013.33
22. Melnikov, A.G., Nies, A.: The classification problem for compact computable metric spaces. In: Bonizzoni, P., Brattka, V., Löwe, B. (eds.) CiE 2013. LNCS, vol. 7921, pp. 320–328. Springer, Heidelberg (2013). https://doi.org/10.1007/978-3-642-39053-1_37
23. Myhill, J.: Creative sets. Z. Math. Logik Grundlagen Math. **1**, 97–108 (1955). https://doi.org/10.1002/malq.19550010205
24. Ng, K.M., Yu, H.: On the degree structure of equivalence relations under computable reducibility. Notre Dame J. Formal Log. (to appear)
25. Nies, A., Solecki, S.: Local compactness for computable polish metric spaces is Π_1^1-complete. In: Beckmann, A., Mitrana, V., Soskova, M. (eds.) CiE 2015. LNCS, vol. 9136, pp. 286–290. Springer, Cham (2015). https://doi.org/10.1007/978-3-319-20028-6_29
26. Soare, R.I.: Recursively Enumerable Sets and Degrees. Springer, Berlin (1987)
27. Turlington, A.: Computability of Heyting algebras and distributive lattices. Ph.D. thesis, University of Connecticut (2010)

28. Visser, A.: Numerations, λ-calculus and arithmetic. In: Seldin, J.P., Hindley, J.R. (eds.) To H.B. Curry: Essays on Combinatory Logic, Lambda Calculus and Formalism, pp. 259–284. Academic Press, London (1980)
29. Weihrauch, K.: Computable Analysis. Texts in Theoretical Computer Science. An EATCS Series. Springer, Berlin (2000). https://doi.org/10.1007/978-3-642-56999-9

Minmax-Regret Evacuation Planning
for Cycle Networks

Robert Benkoczi[1], Binay Bhattacharya[2], Yuya Higashikawa[3],
Tsunehiko Kameda[2(✉)], and Naoki Katoh[4]

[1] Department of Mathematics and Computer Science, University of Lethbridge,
Lethbridge, Canada
[2] School of Computing Science, Simon Fraser University, Burnaby, Canada
[3] School of Business Administration, University of Hyogo, Kobe, Japan
[4] School of Science and Technology, Kwansei Gakuin University, Sanda, Japan

Abstract. This paper considers the problem of evacuating people
located at vertices to a "sink" in a cycle network. In the "minmax-regret"
version of this problem, the exact number of evacuees at each vertex is
unknown, but only an interval for a possible number is given. We show
that a minmax-regret 1-sink in cycle networks with uniform edge capacities can be found in $O(n^2)$ time, where n is the number of vertices. No
correct algorithm was known before for this problem.

1 Introduction

Due to many recent disasters such as earthquakes, volcanic eruptions, typhoons,
and nuclear accidents, evacuation planning is getting increasing attention. The k-
sink problem is an attempt to model evacuation in such an emergency situation
by a network, where vertices represent the locations where the people to be
evacuated are located, and the edges represent the paths along which they can
evacuate [7]. Each edge has a capacity in terms of the number of people who can
enter it per unit time, and a transit time in terms of the time it takes to cross
the edge.

One of the useful objective functions is the evacuation completion time. Each
evacuee evacuates to one of the k sinks, and we want to place k sinks in a
network in such a way that the evacuation can be completed in minimum time.
One of the main features of evacuation problems, which are different from the
center location problem, for example, is that congestion may develop at non-sink
vertices, due to finite capacities of the outgoing edges.

Mamada *et al.* [16] solved the 1-sink problem for tree networks with non-
uniform edge capacities in $O(n \log^2 n)$ time, under the condition that only a
vertex can be a sink, where n is the number of vertices. When edge capacities
are uniform, Higashikawa et al. [11] and Bhattacharya and Kameda [4] presented
$O(n \log n)$ time algorithms with a more relaxed condition that the sink can be
on an edge, as well as at a vertex.

© Springer Nature Switzerland AG 2019
T. V. Gopal and J. Watada (Eds.): TAMC 2019, LNCS 11436, pp. 42–58, 2019.
https://doi.org/10.1007/978-3-030-14812-6_4

On path networks with uniform edge capacities, it is straightforward to find a 1-sink location in linear time, as shown by Cheng et al. [7]. Arumugam et al. [1] showed that the k-sink problem for the path networks can be solved in $O(kn\log^2 n)$ time when the edge capacities are non-uniform, and Higashikawa et al. [12] showed that it can be solved in $O(kn)$ time in the uniform capacity case. Bhattacharya et al. [3] then improved these results to $O(\min\{n\log^3 n, n\log n + k^2\log^4 n\})$ time in the non-uniform capacity case, and to $O(\min\{n\log n, n + k^2\log^2 n)$ time in the uniform capacity case, using a number of innovations, which include a new data structure, and a use of a sorted matrix [8], and parametric search [17].

Networks, which are more general than tree networks, contain cycles. A naïve algorithm for finding an optimal k-sink in a cycle network, which cuts one edge at a time and uses the path algorithm, takes at least $O(n^2)$ time. Practically nothing was known about cycle networks until quite recently. Benkoczi and Das [2] present an $O(n\log^3 n)$ (resp. $O(n\log n)$) time algorithm for the case where the edge capacities are non-uniform (resp. uniform). In the special case of $k = 1$ and uniform capacities, they can find a 1-sink in $O(n)$ time. Table 1 summarizes the most efficient algorithms that are currently known for path, tree, and cycle networks.

Table 1. Most efficient algorithms for computing sinks. [U] means Uniform edge capacities and [G] means General (non-uniform) edge capacities.

Topology	Problem	Time complexity
Path	1-sink [U]	$O(n)$ [12]
	2-sink [U]	$O(n)$ [12]
	2-sink [G]	$O(n\log n)$ [3]
	k-sink [U]	$O(kn)$ [12], $O(n + k^2\log^2 n)$ [3], $O(n\log n)$ [3]
	k-sink [G]	$O(n\log n + k^2\log^4 n)$ [3], $O(n\log^3 n)$ [3]
Tree	1-sink [U]	$O(n\log n)$ [4,11]
	1-sink [G]	$O(n\log^2 n)$ [15]
	k-sink [U]	$O(kn^2\log^4 n)$ [5], $O(\max\{k,\log n\}kn\log^3 n)$ [6]
	k-sink [G]	$O(kn^2\log^5 n)$ [5], $O(\max\{k,\log n\}kn\log^4 n)$ [6]
Cycle	1-sink [U]	$O(n)$ [2]
	1-sink [G]	$O(n\log n)$ [2]
	k-sink [U]	$O(n\log n)$ [2]
	k-sink [G]	$O(n\log^3 n)$ [2]

The concept of *regret* was introduced by Kouvelis and Yu [13], to model situations where optimization is required when the exact values (such as the number of evacuees at the vertices) are unknown, but are given by upper and lower bounds. A particular instance of the set of such numbers, one for each vertex, is called a *scenario*. The objective is to find a solution which is as good as any other solution in the worst case, where the actual scenario is the most unfavorable.

Cheng *et al.* [7] proposed an $O(n \log^2 n)$ time algorithm for finding a minmax-regret 1-sink in path networks with uniform edge capacities. This initial result was soon improved to $O(n \log n)$ [10,18], and further to $O(n)$ [4]. Bhattacharya and Kameda [4] proposed an $O(n \log^4 n)$ time algorithm to find a minmax-regret 2-sink on path networks. For the k-sink version of the problem, Arumugam *et al.* [1] give two algorithms, which run in $O(kn^3 \log n)$ and $O(kn^2 (\log n)^k)$ time, respectively. As for the tree networks with uniform edge capacities, Higashikawa *et al.* [11] propose an $O(n^2 \log^2 n)$ time algorithm for finding a minmax-regret 1-sink. Golin and Sandeep [9] recently proposed an $O(\max\{k^2, \log^2 n\} k^2 n^2 \log^5 n)$ time algorithm for finding a minmax-regret k-sink. Higashikawa *et al.* [11] and Bhattacharya and Kameda [4] proposed $O(n \log n)$ time algorithms to find a minmax-regret 1-sink in tree networks. Table 2 summarizes most efficient algorithms currently known for computing minmax-regret sinks.

Table 2. Most efficient algorithms for computing minmax-regret sinks.

Topology	Problem	Time complexity
Path	1-sink [U]	$O(n)$ [4]
	2-sink [U]	$O(n \log^4 n)$ [4]
	k-sink [U]	$O(kn^3 \log n)$ [1]
Tree	1-sink [U]	$O(n \log n)$ [4,11]
	k-sink [U]	$O(\max\{k^2, \log^2 n\} k^2 n^2 \log^5 n)$ [9]
Cycle	1-sink [U]	$O(n^2)$ [This paper]

This paper investigates the minmax-regret 1-sink problem in cycle networks. The only work on locating a minmax-regret sink in cycle networks that we are aware of is due to Xu and Li [19], who claim that a minmax-regret *vertex* 1-sink in cycle networks with uniform capacities can be found in $O(n^3 \log n)$ time.[1] Li et al. [14] proposed an $O(m^2 n^3)$ time algorithm for computing a minmax-regret 1-sink in general networks with the condition that the evacuees should evacuate to the nearest sink, where m is the number of edges. The main contribution of this paper is the following theorem.

Theorem 1. *We can find a minmax-regret 1-sink in dynamic flow cycle networks with uniform edge capacities in $O(n^2)$ time.*

As shown in Table 2, the minmax-regret 1-sink problem on path networks can be solved in $O(n)$ time. Cheng *et al.* [7] show that there are only $O(n)$ scenarios "of interest"[2] that need to be considered, and we can find a (conventional) 1-sink for a scenario in amortized constant time per such scenario [4]. On cycle networks, however, there are $O(n^2)$ scenarios that appear to be "of interest," and

---

[1] We thank Prof. M. Golin of Hong Kong University of Science and Technology for pointing out that their claim is incorrect.

[2] Formally, they are the *non-dominated* scenarios defined in Sect. 2.4.

we would need $O(n)$ time to find a (conventional) 1-sink for each of them [2]. So just finding the 1-sinks for all of them would take $O(n^3)$ time. Therefore, we cannot use our method in [4] in a straightforward way to achieve sub-cubic time complexity. One of our main contributions is to reduce the number of scenarios "of interest" down to $O(n)$.

The rest of this paper is organized as follows. In the next section, we introduce terms that are used throughout the paper, review some known facts, and describe our approach. Section 3 presents an algorithm that computes "critical vertices" that determine the completion time at different points. Then in Sect. 4, we first present a binary search based algorithm that runs in $O(n^2 \log n)$ time, and then improve it to $O(n^2)$. Finally Sect. 5 concludes the paper, mentioning some open problems. Our analysis is accurate for continuous supply or in the case where the edge capacity equals 1, we often substitute the term evacuees, which are discrete, for supply, and treat them as synonyms.

2 Preliminaries

2.1 Model

Let $\mathcal{C} = (V, E)$ be a cycle network consisting of n vertices, $v_1, v_2, \ldots, v_n \in V$, clockwise in this order.[3] The edges $e_i = (v_i, v_{i+1}) \in E$, $i = 1, \ldots, n - 1$ and $e_n = (v_n, v_1) \in E$ connect adjacent vertices. By $x \in \mathcal{C}$, we mean that point x lies on either an edge or at a vertex of $\mathcal{C}$. For $a, b \in \mathcal{C}$, $\mathcal{C}[a, b]$ denotes the cw section of $\mathcal{C}$ from a to b, and $V[a, b]$ denotes the set of vertices comprising $\mathcal{C}[a, b]$. If $c \in \mathcal{C}[a, b]$, we sometimes write $a \preceq c \preceq b$. Let $d(a, b)$ denote the length (sum of the edge lengths) of $\mathcal{C}[a, b]$. If a and/or b is on an edge, we use the prorated length on the edge. The transit time of an evacuee for a unit distance is denoted by τ, so that it takes $d(a, b)\tau$ time to travel from a to b cw, and τ is independent of the edge. Each vertex $v_i \in V$ has weight $w(v_i)$, which represents the number of evacuees located there before evacuation starts.

Each edge $e_i \in E$ has the same capacity c, which represents the number of evacuees who can enter it per unit time. It is assumed that once an evacuee enters an edge, he/she traverses the edge at constant speed to reach the other end vertex of the edge. The evacuees at all vertices start evacuation at the same time, and the evacuees who were originally at a vertex or who arrive there later move in the same direction (cw or ccw).[4] We define a weight array by

$$\boldsymbol{W}[v_i] \triangleq \sum_{j=1}^{i} w(v_j) \text{ for } 1 \leq i \leq n,$$

and a cw distance array

$$\boldsymbol{D}[v_i] \triangleq d(v_1, v_i) \text{ for } 1 \leq i \leq n.$$

---

[3] From now on clockwise and counterclockwise are abbreviated as *cw* and *ccw*, respectively.

[4] In the parlance of network flow theory, flow obeying the latter condition is called *confluent*.

It is easy to see that arrays $W[\cdot]$ and $D[\cdot]$ can be constructed in $O(n)$ time from the vertex weights and edge lengths, respectively. Once they are constructed, we can compute in constant time the sum of the weights of the vertices from v_h to v_i ($h \leq i$) by

$$W[v_i, v_j] \triangleq \sum_{h=i}^{j} w(v_h) = W[v_j] - W[v_{i-1}],$$

and can obtain $d(v_i, v_j)$ for any pair (v_i, v_j) in constant time as well.

2.2 Graphical Representation

If we remove an edge of $\mathcal{C}$, it becomes a path. Therefore, we can make use of some known results about the sink problem on path networks [3,7]. Assume that we remove edge (v_n, v_1). In Fig. 1(a), it is assumed that a sink is located to the right of v_4. The first evacuee from v_1 starts moving towards the sink from v_1 at time $t = 0$. He/she traverses edge $e_1 = (v_1, v_2)$ and arrives at v_2 at time $t_b = d(v_1, v_2)\tau$. It takes $w(v_1)/c$ time units for all the evacuees who were initially located at v_1 to vacate v_1, and all of them will have arrived at v_2 at time $t_c = d(v_1, v_2)\tau + w(v_1)/c$. Similarly, the evacuees who were initially located at v_2 leave v_2 completely at time $t_a = w(v_2)/c$. In the example in Fig. 1, inequality $t_a < t_b$ holds, so that when the first evacuee from v_1 arrives at v_2, finding no evacuee left there, he/she can immediately leave v_2 without any delay. The timing diagram in Fig. 1(b) shows the departure flow rate (c or 0) at v_2 as a function of time. Their arrival at v_3 is delayed by $d(v_2, v_3)\tau$, as shown in Fig. 1(c). Observe that $t_a < t_b$ if and only if

$$d(v_1, v_2)\tau > w(v_2)/c. \tag{1}$$

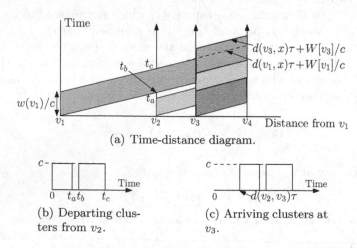

(a) Time-distance diagram.

(b) Departing clusters from v_2.

(c) Arriving clusters at v_3.

Fig. 1. Clockwise cluster sequences from v_1, v_2, and v_3.

Back to Fig. 1(a), when the first evacuee from v_2 arrives at v_3, there are still evacuees left there, waiting for departure from there. We call this *congestion*, and it causes the evacuees from v_2 to wait for departure from v_3, incurring delay for them. The evacuees from v_1, who encountered no delay at v_2, must now wait at v_3. Note that for $v_1 \prec x \preceq v_3$,

$$d(v_1, x)\tau + \boldsymbol{W}[v_1]/c$$

gives the time at which evacuation to x (if x were the sink) from all the vertices to the left of x completes, and for $v_3 \prec x \preceq v_4$,

$$d(v_1, x)\tau + \boldsymbol{W}[v_3]/c \tag{2}$$

gives the time at which evacuation to x from all the vertices to the left of x completes.

If we plot the arrival flow rate at, or departure flow rate from, a vertex as a function of time, as in Fig. 1(b) and (c), it consists of a sequence of *temporal clusters*, or just *clusters* for short, during which the flow rate is c. We call a cluster from the cw (resp. ccw) side of a point x in question a *cw-cluster* (resp. *ccw-cluster*).

2.3   Completion Time Functions

Since we consider only confluent flow to a sink, there is an edge, called the *split edge* $\hat{e}$, across which no evacuee travels. Clearly, removing the split edge converts $\mathcal{C}$ into a path. Let $v_{cw}(\hat{e})$ (resp. $v_{ccw}(\hat{e})$) denote the vertex at the cw (resp. ccw) end of $\hat{e}$. Then the evacuation completion time to a potential sink $x \notin \hat{e}$ is given by the maximum of the *cw-completion time* (*cw-time* for short) for all the evacuees on $\mathcal{C}[v_{cw}(\hat{e}), x]$ to move cw to x, and the *ccw completion time* (*ccw-time* for short) for all the evacuees on $\mathcal{C}[x, v_{ccw}(\hat{e})]$ to move ccw to x.

Let $f_{cw}((x, \hat{e}), v_h)$ (resp. $f_{ccw}((x, \hat{e}), v_h)$) denote the evacuation completion time to x with split edge $\hat{e}$, for the evacuees on $\mathcal{C}[v_{cw}(\hat{e}), v_h]$ (resp. $\mathcal{C}[v_h, v_{ccw}(\hat{e})]$) moving cw (resp. ccw), where $v_{cw}(\hat{e}) \preceq v_h \prec x$ (resp. $x \prec v_h \preceq v_{ccw}(\hat{e})$), under the assumption that the first evacuee from v_h encounters no delay on its way to x. By generalizing (2), we obtain

$$f_{cw}((x, \hat{e}), v_h) = d(v_h, x)\tau + \boldsymbol{W}[v_{cw}(\hat{e}), v_h]/c \text{ for } v_{cw}(\hat{e}) \preceq v_h \prec x. \tag{3}$$

It is just a lower bound on the actual completion time to x, because the actual flow may be intermittent (not continuous), or there may be congestion caused by other vertices on $\mathcal{C}[v_{h+1}, x]$. See the congestion at v_3 in Fig. 1(a). Similarly

$$f_{ccw}((x, \hat{e}), v_h) = d(x, v_h)\tau + \boldsymbol{W}[v_h, v_{ccw}(\hat{e})]/c \text{ for } x \prec v_h \preceq v_{ccw}(\hat{e}) \tag{4}$$

represents the evacuation completion time to x for all the evacuees on the vertices on $\mathcal{C}[v_h, v_{ccw}(\hat{e})]$, moving ccw.

To be precise, the second terms in (3) and (4) should be $\lceil \boldsymbol{W}[v_i, v_h]/c \rceil$ and $\lceil \boldsymbol{W}[v_h, v_j]/c \rceil$, respectively, if supply consists of discrete items, such as evacuees, but we adopt (3) and (4) as our cost functions for simplicity. It is accurate when $c = 1$. We refer to the first (resp. second) term in the righthand side of (3) and

(4) as the *distance cost* (resp. *weight cost*). Note that the distance cost is linear in the distance to x.

For two points a and b on $\mathcal{C}$, let $V[a, b)$ (resp. $V(a, b]$) denote the set of vertices that lie on $\mathcal{C}[a, b]$, excluding b (resp. a). For a given $\hat{e}$ and $x \notin \hat{e}$, we define[5] the *cw-cost* and *ccw-cost* at $(x, \hat{e})$ by

$$\Theta_{cw}(x, \hat{e}) \triangleq \max_{v_h \in V[v_{cw}(\hat{e}), x)} \{f_{cw}((x, \hat{e}), v_h)\}, \tag{5}$$

$$\Theta_{ccw}(x, \hat{e}) \triangleq \max_{v_h \in V(x, v_{ccw}(\hat{e})]} \{f_{ccw}((x, \hat{e}), v_h)\}, \tag{6}$$

respectively. Let

$$\Theta(x, \hat{e}) \triangleq \max \{\Theta_{cw}(x, \hat{e}), \Theta_{ccw}(x, \hat{e})\}. \tag{7}$$

A result for path networks in [7] implies the following lemma.

Lemma 1. *Given split edge $\hat{e}$, the evacuation completion time to $x \notin \hat{e}$ for all the supplies is given by $\Theta(x, \hat{e})$.*

The vertex $v_h \in V[v_{cw}(\hat{e}), x)$ that maximizes $f_{cw}((x, \hat{e}), v_h)$ is called the *cw-critical* vertex for x and is denoted by $c_{cw}(v_i, \hat{e})$ if $v_i \prec x \preceq v_{i+1}$. Similarly, the vertex $v_h \in V(x, v_{ccw}(\hat{e})]$ that maximizes $f_{ccw}((x, \hat{e}), v_h)$ is called the *ccw-critical* vertex for x, and is denoted by $c_{ccw}(v_i, \hat{e})$ if $v_{i-1} \preceq x \prec v_i$. Since the cw-critical vertex (resp. ccw-critical vertex) is the same for any point $v_i \prec x \preceq v_{i+1}$ (resp. $v_{i-1} \preceq x \prec v_i$), if we know $c_{cw}(v_i, \hat{e})$ and $c_{ccw}(v_i, \hat{e})$ for all i, then we know the critical vertices for all points x. The following lemma is easy to prove.

Lemma 2 [4]. *The cw-critical vertex (resp. ccw-critical vertex) is the first vertex of the last cw-cluster (resp. ccw-cluster).* □

By (3), we can compute the cw-cost at x ($v_i \prec x \preceq v_{i+1}$) as follows.

$$\Theta_{cw}(x, \hat{e}) = d(c_{cw}(v_i, \hat{e}), x)\tau + \boldsymbol{W}[v_{cw}(\hat{e}), c_{cw}(v_i, \hat{e})]/c. \tag{8}$$

We can similarly compute $\Theta_{ccw}(x, \hat{e})$. A point x that minimizes

$$\Theta(x) \triangleq \min_{\hat{e} \in E} \max\{\Theta_{cw}(x, \hat{e}), \Theta_{ccw}(x, \hat{e})\} \tag{9}$$

is a *1-sink*. The following lemma will be used in proving Lemma 10 later.

Lemma 3 [2]. *A 1-sink in cycle networks with uniform edge capacities can be found in $O(n)$ time.* □

Example 1. In Fig. 2, all edges have the same capacity and they all have a unit length, except edge (v_7, v_1) whose length is 2. Table 3 shows the costs on the edges for the split edge in the first column, as a function of the cw distance x from v_1. The 1-sink for each split edge is shown in bold face.

Table 4 reorganizes the costs at the vertices, so that the row and column unimodality can be more easily observed. However, the minimum elements in rows (the completion time at the 1-sinks, shown in bold face) are not unimodal.

□

---

[5] We have $\Theta_{cw}(x, \hat{e}) = 0$ (resp. $\Theta_{ccw}^s(x, \hat{e}) = 0$), if x and $\hat{e}$ are on the same edge, since $V[v_{cw}(\hat{e}), x) = \emptyset$ (resp. $V(x, v_{ccw}(\hat{e})] = \emptyset$).

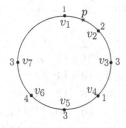

Fig. 2. Example cycle network.

Table 3. Completion time at point x.

$\hat{e}$	v_1	e_1	v_2	e_2	v_3	e_3	v_4	e_4	v_5	e_5	v_6	e_6	v_7	e_7
(v_7,v_1)	17	$17-x$	15	$16-x$	12	$14-x$	11	$14-x$	8	$x+6$	11	$x+9$	15	−
(v_1,v_2)	18	−	16	$17-x$	13	$15-x$	12	$15-x$	9	$x+5$	10	$x+8$	14	$x+10$
(v_2,v_3)	14	$16+x$	17	−	15	$17-x$	14	$17-x$	13	$15-x$	8	$x+6$	12	$x+8$
(v_3,v_4)	13	$x+13$	14	$x+13$	15	−	15	$18-x$	14	$18-x$	10	$15-x$	9	$x+5$
(v_4,v_5)	12	$x+12$	13	$x+12$	14	$x+14$	17	−	15	$19-x$	11	$16-x$	9	$x+4$
(v_5,v_6)	10	$\max\{10-x, x+9\}$	10	$x+9$	11	$x+11$	14	$x+11$	15	−	14	$19-x$	12	$18-x$
(v_6,v_7)	14	$14-x$	12	$13-x$	9	$x+7$	10	$x+7$	11	$x+9$	14	−	16	$21-x$

2.4 Regret

In the regret model [13], for each vertex $v_i \in V$, only the upper and lower bounds on its weight are known, which are denoted by $\overline{w}(v_i)$ and $\underline{w}(v_i)$, respectively. A *scenario s* assigns a particular weight $w^s(v_i)$, satisfying $\underline{w}(v_i) \leq w^s(v_i) \leq \overline{w}(v_i)$, to each vertex $v_i \in V$. Thus the set of all possible scenarios is a Cartesian product

$$\mathcal{S} \triangleq \prod_{v_i \in V} [\underline{w}(v_i), \overline{w}(v_i)].$$

The scenario under which all vertices have the minimum (resp. maximum) weights is denoted by s_0 (resp. s_M).

Since we need to express the dependency of various quantities on a scenario, under scenario s, we use $\Theta^s_{cw}(x, \hat{e})$, $c^s_{cw}(v_i, \hat{e})$, etc., from now on. Since removing $\hat{e}$ from $\mathcal{C}$ results in a path, the following lemma is implied by a result for path networks [4,7].

Lemma 4. *Let a scenario s and a split edge $\hat{e}$ be given, and assume that point x moves cw on the resulting path.*

(a) $\Theta^s_{cw}(x, \hat{e})$ is an increasing function of x.
(b) $\Theta^s_{ccw}(x, \hat{e})$ is a decreasing function of x.
(c) $\Theta^s(x, \hat{e})$ is unimodal.

Table 4. Completion times at the vertices and point p in Fig. 2.

$\hat{e}$	v_1	p	v_2	v_3	v_4	v_5	v_6	v_7	v_1	p	v_2	v_3	v_4	v_5	v_6	v_7
(v_7, v_1)	17	16.5	15	12	11	8	11	15	$\cdots$							
(v_1, v_2)		$\cdots$	16	13	12	9	10	14	18	$\cdots$						
(v_2, v_3)			$\cdots$	15	14	13	8	12	14	16.5	17	$\cdots$				
(v_3, v_4)				$\cdots$	15	14	10	9	13	13.5	14	15	$\cdots$			
(v_4, v_5)					$\cdots$	15	11	9	12	12.5	13	14	17	$\cdots$		
(v_5, v_6)						$\cdots$	14	12	10	9.5	10	11	14	15	$\cdots$	
(v_6, v_7)							$\cdots$	16	14	13.5	12	9	10	11	14	$\cdots$

Let $q^s \in \mathcal{C}$ denote a 1-sink under s. The *regret* [13] at 2-dimensional "point" $(x, \hat{e})$ under scenario s is defined by

$$R^s(x, \hat{e}) \triangleq \Theta^s(x, \hat{e}) - \Theta^s(q^s), \tag{10}$$

where $\Theta^s(q^s) \triangleq \min_e \Theta^s(q^s, e)$, i.e., the cost of the 1-sink under s. The following lemma is immediate from Lemma 4.

Lemma 5. *Under the assumption of Lemma 4, $R^s(x, \hat{e})$ is unimodal.*

Note that $\Theta^s(q^s)$ is independent of $\hat{e}$. Scenario s' is said to *dominate* another scenario s at $(x, \hat{e})$ if $R^{s'}(x, \hat{e}) \geq R^s(x, \hat{e})$ holds. A scenario under which all the max-weighted vertices are consecutive, including s_0 and s_M, is said to be *bipartite* [4]. Let $\tilde{S}$ denote the set of all bipartite scenarios. It is clear that $|\tilde{S}| = O(n^2)$.

The *maximum regret* at $(x, \hat{e})$ is defined by

$$R_{max}(x, \hat{e}) \triangleq \max_{s \in \tilde{S}} R^s(x, \hat{e}). \tag{11}$$

A scenario s that maximizes $R^s(x, \hat{e})$, namely, that dominates all others at $(x, \hat{e})$, is called a *worst case scenario* [13] for the pair $(x, \hat{e})$. The *minmax-regret 1-sink* solution is given by a pair $(x, \hat{e})$ that minimizes $R_{max}(x, \hat{e})$. This solution gives an optimal sink $q^* = x$ and the corresponding split edge e^*.

2.5   Road Map

Now that we have defined most of the necessary concepts and terms, we can describe our approach to finding a minmax-regret 1-sink on a cycle network. We call the last cw-cluster with respect to $(x, \hat{e})$ the *dominant cw-cluster* for $(x, \hat{e})$, where $v_i \prec x \preceq v_{i+1}$, and define the set of vertices it is comprised of.

$$D_{cw}^s(x, \hat{e}) \triangleq V[v_{cw}(\hat{e}), c_{cw}^s(v_i, \hat{e})].$$

Similarly, we define the vertex set of the *dominant ccw-cluster* for $(x, \hat{e})$, where $v_{i-1} \preceq x \prec v_i$, by

$$D^s_{ccw}(x, \hat{e}) \triangleq V[c^s_{ccw}(v_i, \hat{e}), v_{ccw}(\hat{e})].$$

See Fig. 3. If $w(v_j) = \overline{w}(v_j)$ (resp. $w(v_j) = \underline{w}(v_j)$), we say that v_j is *max-weighted* (resp. *min-weighted*). Let $s_i(\hat{e})$ denote the scenario such that $w^s(v_h) = w^{s_M}(v_h)$ for each $v_h \in D^{s_M}_{cw}(x, \hat{e})$, and all other vertices are min-weighted. Similarly, let $s'_i(\hat{e})$ denote the scenario such that $w^s(v_h) = w^{s_M}(v_h)$ for each $v_h \in D^{s_M}_{ccw}(x, \hat{e})$, and all other vertices are min-weighted. We proceed as follows:

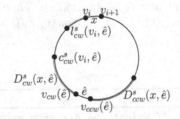

Fig. 3. Illustration of symbols.

1. For each split edge $\hat{e} \in E$, compute critical vertices

$$\{c^s_{cw}(v_i, \hat{e}) \mid i = 2, \ldots, n\} \cup \{c^s_{ccw}(v_i, \hat{e}) \mid i = 1, \ldots, n-1\} \qquad (12)$$

for $s = s_M$ and $s = s_0$. They help to compute $\Theta^s_{cw}(x, \hat{e})$ (resp. $\Theta^s_{ccw}(x, \hat{e})$) under $s = s_M$ and $s = s_0$, for x with $v_i \prec x \preceq v_{i+1}$ (resp. $v_{i-1} \preceq x \prec v_i$).

2. Based on (12), compute regret

$$R^{s_i(\hat{e})}_{cw}(x, \hat{e}) \triangleq \max\{\Theta^{s_i(\hat{e})}_{cw}(x, \hat{e}), \Theta^{s_0}_{ccw}(x, \hat{e})\} - \Theta^{s_i(\hat{e})}(q^{s_i(\hat{e})})$$

for $v_i \prec x \preceq v_{i+1}$, and similarly

$$R^{s'_i(\hat{e})}_{ccw}(x, \hat{e}) \triangleq \max\{\Theta^{s'_i(\hat{e})}_{ccw}(x, \hat{e}), \Theta^{s_0}_{cw}(x, \hat{e})\} - \Theta^{s'_i(\hat{e})}(q^{s'_i(\hat{e})})$$

for $v_i \preceq x \prec v_{i+1}$.

3. For each i, find upper envelope $R(x, \hat{e}) \triangleq \max\{R^{s_i(\hat{e})}_{cw}(x, \hat{e}), R^{s_{i+1}(\hat{e})}_{ccw}(x, \hat{e})\}$.

4. For each $\hat{e} \in E$, find $\overline{R}_{min}(\hat{e}) \triangleq \min_x R(x, \hat{e})$ by binary search on x.

5. Find $\min_{\hat{e}}\{\overline{R}_{min}(\hat{e})\}$, which is the cost of a minmax-regret 1-sink.

6. Output the minimizing $\hat{e}$ in Step 5, and the corresponding minimizing x in Step 4. This x is a minmax-regret 1-sink.

3   Critical Vertices and Dominant Clusters

Given a pair $(x, \hat{e})$, where $v_i \prec x \preceq v_{i+1}$, let $\sigma^s_{cw}(v_i, \hat{e})$ denote the cw-cluster sequence under scenario s with respect to v_i, starting in v_i and ending in $v_{cw}(\hat{e})$. The following lemma, which follows easily from an analogous lemma for path networks in [7], is needed to prove the correctness of **Algorithm 1**.

Lemma 6. *Let x satisfy $v_i \prec x \preceq v_{i+1}$.*

(a) *Let $v_{cw}(\hat{e}_1) \prec v_h \prec v_i$ and $\hat{e}_2 = (v_h, v_{h+1})$. Then the cw-critical vertex $c_{cw}(v_i, \hat{e}_1)$ is either $c_{cw}(v_h, \hat{e}_1)$ or $c_{cw}(v_i, \hat{e}_2)$, whichever has a higher cost at v_i.*

(b) *Let $v_i \prec v_h \prec v_{ccw}(\hat{e}_1)$ and $\hat{e}_2 = (v_h, v_{h+1})$. Then the ccw-critical vertex $c_{ccw}(v_i, \hat{e}_1)$ is either $c_{ccw}(v_{h+1}, \hat{e}_1)$ or $c_{ccw}(v_i, \hat{e}_2)$, whichever has a higher cost at v_i.* ☐

Algorithm 1. FIND-CRITICAL$_{cw}$

    Input : $W^s[\cdot]$, $D[\cdot]$, $\hat{e}$; // Weight and distance arrays, and split edge.
    Output: $\{(v_i, c^s_{cw}(v_i, \hat{e})) \mid 1 \leq i \leq n - 1\}$;
    // Assume $\hat{e} = (v_n, v_1)$, and write $\sigma^s_{cw}(v_i)$ for $\sigma^s_{cw}(v_i, \hat{e})$ for simplicity.
    Let v_0 be a dummy vertex with $d(v_0, v_1) = 0$.
1 $\sigma^s_{cw}(v_0) = \Lambda$;   // Null sequence.
2 **for** $i = 1, \ldots, n - 1$ **do**
3 | Shift $\sigma^s_{cw}(v_{i-1})$ in the positive direction of time by $d(v_{i-1}, v_i)\tau$;
4 | Let C be a single cluster of duration $w(v_i)/c$, starting at time 0 ;   // This
        | time is the local time at v_i.
5 | Set $l = i$;
6 | **if** $d(v_{i-1}, v_i)\tau \leq W^s[v_i, v_i]/c$ **then**
7 | | **while** $[d(v_{l-1}, v_i)\tau \leq W^s[v_l, v_i]/c] \wedge [\, v_l \neq v_1] $ **do**
8 | | | Enlarge C by merging it with the first cluster of $\sigma^s_{cw}(v_{i-1})$;
9 | | | Remove the first cluster from $\sigma^s_{cw}(v_{i-1})$;
10 | | | Update v_l to the last vertex of C ;
11 | | **end**
12 | | If $v_l = v_1$ then set $c^s_{cw}(v_i, \hat{e}) = v_i$;
13 | **end**
14 | Output $(v_i, c^s_{cw}(v_i, \hat{e}))$;
15 | Construct $\sigma^s_{cw}(v_i)$, concatenating C and $\sigma^s_{cw}(v_{i-1})$;
16 **end**

Lemma 7. Algorithm 1 *generates all the cw-critical vertices with respect to $(x, \hat{e})$ for split edge $\hat{e}$ under scenario s in $O(n)$ time.*

Proof. Step 1 initializes the cluster sequence. Step 3 adjusts the time origin of $\sigma^s_{cw}(v_{i-1})$ from the local time at v_{i-1} to the local time at v_i. The **if** clause in the

for loop tests if the new vertex, v_i, should be merged with the first cluster in $\sigma_{cw}^s(v_{i-1})$ (starting with v_{i-1}), and if so, the **while** loop merges all the clusters in $\sigma_{cw}^s(v_{i-1})$ that need to be merged by testing if

$$d(v_{l-1}, v_i)\tau \leq \boldsymbol{W}^s[v_l, v_i]/c, \qquad (13)$$

where v_l is the end vertex of the first cluster of $\sigma_{cw}^s(v_i)$ that is being formed. See $l_{cw}^s(v_i, \hat{e})$ in Fig. 3. Inequality (13) is a generalized version of the negation of (1). If the dominant cluster starting with $c_{cw}^s(v_{i-1}, \hat{e})$, where $v_{i-1} \prec x \preceq v_i$, is also merged, then $c_{cw}^s(v_i, \hat{e})$ is updated to v_i by Step 12. Steps 13 and 14 are self-explanatory. Note that a cw-cluster gets merged with C at most once and in $O(1)$ time. Thus the total execution time of **Algorithm 1** is $O(n)$. ☐

We can construct an algorithm, FIND-CRITICAL$_{ccw}$, which is symmetric to FIND-CRITICAL$_{cw}$, to generate all the ccw-critical vertices with respect to $(x, \hat{e})$, where $v_i \preceq x \prec v_{i+1}$, in $O(n)$ time.

4   Minmax-Regret 1-Sink

4.1   Worst-Case Scenarios

Given a split edge $\hat{e}$, we may assume without loss of generality that the evacuation completion time for the vertices on $\mathcal{C}[v_{cw}(\hat{e}), x]$ at x is not less than that for the vertices on $\mathcal{C}[x, v_{ccw}(\hat{e})]$. We are interested in the scenario that maximizes regret, as expressed by (11).

Clearly, reducing the weights of the vertices not in $D_{cw}^s(x, \hat{e})$ cannot decrease $\Theta^s(x, \hat{e})$, so assume that they are all min-weighted, so that $\Theta^s(q^s)$ in (10) is made small. In particular, the vertices on $\mathcal{C}[x, v_{ccw}(\hat{e})]$ are all min-weighted, because only vertices on $\mathcal{C}[v_{cw}(\hat{e}), x]$ may belong to $D_{cw}^s(x, \hat{e})$. Now, for any vertex $v \in D_{cw}^s(x, \hat{e})$, if we increase its weight by δ, i.e., $w^{s'}(v) = w^s(v) + \delta$, where s (resp. s') is the scenario before (resp. after) the increase, then we have $\Theta^{s'}(x, \hat{e}) = \Theta^s(x, \hat{e}) + \delta/c$. The cost of the sink, $\Theta^s(q^s)$, increases by the maximum amount δ/c, if $q^{s'} = q^s$ and v is in its dominant cw- or ccw-cluster. In all other cases, either the optimal split edge for the sink or the sink itself will move and its cost increase is less than δ/c. This implies that increasing the weights of all the vertices in $D_{cw}^s(x, \hat{e})$ to their maximum values cannot decrease regret. We thus end up with a bipartite scenario that dominates the original scenario at $(x, \hat{e})$.

Lemma 8. *In looking for a minmax-regret 1-sink in cycle networks with uniform edge capacities, we have the following.*

(a) Any scenario is dominated by a scenario in $\tilde{S}$.

(b) We may assume that, under a dominating scenario for $(x, \hat{e})$, all the vertices of the dominant cw-cluster or ccw-cluster, i.e., one ending at $\hat{e}$, are max-weighted and the rest are min-weighted.

Let us consider two special scenarios s_M and s_0, defined in Sect. 2.4. It is clear that for a given pair $(v_i, \hat{e})$, we generally have $v_{cw}(\hat{e}) \preceq c_{cw}^{s_0}(v_i, \hat{e}) \preceq c_{cw}^{s_M}(v_i, \hat{e})$. If $c_{cw}^{s_0}(v_i, \hat{e}) \prec c_{cw}^{s_M}(v_i, \hat{e})$, we can create a scenario satisfying the conditions of Lemma 8 by making all the vertices in $D_{cw}^{s_0}(x, \hat{e})$ max-weighted and the remaining vertices min-weighted. As observed earlier, these changes do not decrease the regret at $(x, \hat{e})$. Similarly, we can create a scenario satisfying the conditions of Lemma 8 by making all the vertices in the last two cw-clusters of $\sigma_{cw}^{s_0}(v_i, \hat{e})$ max-weighted and the remaining vertices min-weighted, and so forth, for all clusters of $\sigma_{cw}^{s_0}(v_i, \hat{e})$ whose vertices are in $D_{cw}^{s_M}(x, \hat{e})$. In the illustration in Fig. 4, there are three such cw-clusters under s_0, which are indicated by purple arcs.

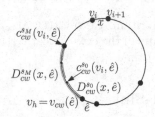

Fig. 4. $c_{cw}^{s_M}(v_i, \hat{e})$ and $c_{cw}^{s_0}(v_i, \hat{e})$.

The above argument proves the following lemma.

Lemma 9. *Let $s_i(\hat{e})$ denote the scenario such that $w^{s_i(\hat{e})}(v_h) = w^{s_M}(v_h)$ for each v_h such that $v_{cw}(\hat{e}) \preceq v_h \preceq c_{cw}^{s_M}(v_i, \hat{e})$, and all other vertices are min-weighted. Then $s_i(\hat{e})$ is the worst case scenario for $(x, \hat{e})$.*

Clearly, there is at most one worst-case scenario, $s_i(\hat{e})$, for any $(x, \hat{e})$ pair, where $v_i \prec x \preceq v_{i+1}$, i.e., as long as x lies on the same edge. Now that we have identified a unique worst case scenario, $s_i(\hat{e})$, for any given pair $(x, \hat{e})$, we shift gears to find the pair that *minimizes* the maximum regret $R_{max}(x, \hat{e})$ of (11).

4.2   Binary Search Based Algorithm

By Lemma 5, given the split edge $\hat{e}$, we can find the minimum value of the maximum regret using binary search on position x. For each probed x ($v_i \prec x \preceq v_{i+1}$), we first look up the precomputed $c_{cw}^{s_M}(v_i, \hat{e})$, make all the vertices in $D_{cw}^{s_M}(x, \hat{e})$ max-weighted, and then evaluate the regret of the 1-sink under the resulting scenario. After computing the lowest cost point of the regret function for each split edge, we identify the minimum among them, which is the minmax-regret 1-sink. Each binary probe costs $O(n)$ time by Lemma 3. Since a total of $O(n \log n)$ probes are needed for all the n split edges, they cost $O(n^2 \log n)$ time in total. **Algorithm 2**, presented below, formally states these steps, following the road map given in Sect. 2.5 fairly closely.

Lemma 10. Algorithm 2 *runs in $O(n^2 \log n)$ time.*

Algorithm 2. MINMAX-REGRET-SINK

> **Input** : $\boldsymbol{W}^{s_M}[\cdot], \boldsymbol{W}^{s_0}[\cdot], \boldsymbol{D}[\cdot]$;   // Weight and distance arrays.
> **Output:** Minmax-regret 1-sink and the corresponding split edge ;
> 1 **for** $j = 1, \ldots, n$ **do**
>> – Set $\hat{e} = (v_j, v_{j+1})$;
>> – Run FIND-CRITICAL$_{cw}$ with $s = s_M$ to compute $\{c^{s_M}_{cw}(v_k, \hat{e}) \mid v_k \in V\}$;
>> – Run FIND-CRITICAL$_{ccw}$ with $s = s_M$ to compute $\{c^{s_M}_{ccw}(v_k, \hat{e}) \mid v_k \in V\}$;
>> – Run FIND-CRITICAL$_{cw}$ with $s = s_0$ to compute $\{c^{s_0}_{cw}(v_k, \hat{e}) \mid v_k \in V\}$;
>> – Run FIND-CRITICAL$_{ccw}$ with $s = s_0$ to compute $\{c^{s_0}_{ccw}(v_k, \hat{e}) \mid v_k \in V\}$;
> 2 **end**
> 3 **for** *each* $\hat{e} \in E$ **do**
> 4 | For x selected by binary search, let $\overline{R}_{min}(\hat{e}) \triangleq \min_x R(x, \hat{e})$ by calling
> | **Procedure** R$(x, \hat{e})$;   // Given below.
> 5 **end**
> 6 Find $\min_{\hat{e}}\{\overline{R}_{min}(\hat{e})\}$;
> 7 For the minimizing split edge $\hat{e}$ in Step 6, find the corresponding minimizing x
> in Step 4 ;
> 8 Output them as a solution.

Procedure. R$(x, \hat{e})$

> **Input** : $(x, \hat{e})$
> **Output:** R$(x, \hat{e})$;
> 1 $R^{s_i(\hat{e})}_{cw}(x, \hat{e}) \triangleq \max\{\Theta^{s_i(\hat{e})}_{cw}(x, \hat{e}), \Theta^{s_i(\hat{e})}_{ccw}(x, \hat{e})\} - \Theta^{s_i(\hat{e})}(q^{s_i(\hat{e})})$;
> 2 $R^{s'_i(\hat{e})}_{ccw}(x, \hat{e}) \triangleq \max\{\Theta^{s'_i(\hat{e})}_{ccw}(x, \hat{e}), \Theta^{s_0}_{cw}(x, \hat{e})\} - \Theta^{s'_i(\hat{e})}(q^{s'_i(\hat{e})})$;
> 3 R$(x, \hat{e}) \triangleq \max\{R^{s_i(\hat{e})}_{cw}(x, \hat{e}), R^{s'_i(\hat{e})}_{ccw}(x, \hat{e})\}$;

Proof. Step 1 takes $O(n^2)$ time by Lemma 7. Step 4 makes $O(\log n)$ calls to **Procedure** R$(x, \hat{e})$ per split edge. **Procedure** R$(x, \hat{e})$ computes the upper envelope of four linear segments, one each from $\Theta^{s_i(\hat{e})}_{cw}(x, \hat{e})$, $\Theta^{s_0}_{ccw}(x, \hat{e})$, $\Theta^{s'_i(\hat{e})}_{ccw}(x, \hat{e})$, and $\Theta^{s_0}_{cw}(x, \hat{e})$, which is composed of at most two linear segments. This can be done constant time, using dominant clusters, precomputed in Step 1. The procedure also computes the cost of sinks, $\Theta^{s_i(\hat{e})}(q^{s_i(\hat{e})})$ and $\Theta^{s'_i(\hat{e})}(q^{s'_i(\hat{e})})$, which takes $O(n)$ time by Lemma 3. Since $O(n \log n)$ binary probes are made altogether, the total time required by all calls to **Procedure** R$(x, \hat{e})$ is $O(n^2 \log n)$ time. All other steps take less time.   □

4.3   Proof of Theorem 1

To explain our approach intuitively first, let us pretend the values in Table 4 in Example 1 were regret values. We can make this assumption, because the upper envelope of the regret functions is unimodal by Lemma 5. We start with scanning row 1 from left to right, until we find the smallest value (8 in this example). We then move to row 2 to look for the smallest value, which happens to be in the same column (9 in column v_5). We thus move to row 3 and look for the smallest value (8), moving right. This way, we can find the smallest value in every row by examining only a total of $O(n)$ entries of the table. We can then easily find the globally smallest value in $O(n)$ extra steps.

We perform similar operations for the upper envelope of regret functions. Just like we examined only $O(n)$ entries in the above paragraph, we identify a set of $O(n)$ worst-case scenarios. All others in $\tilde{S}$ can be ignored in computing maximum regret $R_{max}(x,\hat{e})$. We can do without binary search, once we find the lowest regret point for one split edge. After spending $O(n \log n)$ time for this, we can advance both x and $\hat{e}$ cw, one after another, so that the total number of $(x,\hat{e})$ pairs that we probe is reduced to $O(n)$ from $O(n \log n)$. Since each probe entails computing a 1-sink, which takes $O(n)$ time by Lemma 3, the total time is $O(n^2)$. We can move $\hat{e}$ cw within the cycle interval spanned by the vertices in $D_{cw}^{s_M}(x,\hat{e})$, until $v_{cw}(\hat{e})$ reaches $c_{cw}^{s_M}(v_i,\hat{e})$. Note that for any split edge $\hat{e}'$ between $\hat{e}$ and $c_{cw}^{s_M}(v_i,\hat{e})$, we have $c_{cw}^{s_M}(v_i,\hat{e}') = c_{cw}^{s_M}(v_i,\hat{e})$.

Let $R_{cw}^{s_i(\hat{e})}(x,\hat{e})$ and $R_{ccw}^{s_i'(\hat{e})}(x,\hat{e})$ be the regret functions defined in **Procedure** $R(x,\hat{e})$. Starting with $\hat{e} = (v_n, v_1)$ and $x \in (v_1, v_2)$, it is likely that $R_{cw}^{s_1(\hat{e})}(x,\hat{e}) < R_{ccw}^{s_1'(\hat{e})}(x,\hat{e})$ will hold for any x with $v_1 \prec x \preceq v_2$, so that $R(x,\hat{e}) = R_{ccw}^{s_n'(\hat{e})}(x,\hat{e})$. If so, we move x cw to the next edge, keeping $\hat{e}$ fixed, until $R_{cw}^{s_i(\hat{e})}(x,\hat{e}) \geq R_{ccw}^{s_i'(\hat{e})}(x,\hat{e})$ holds for some x with $v_i \prec x \preceq v_{i+1}$. (Of course $i = 1$ is possible.) See Fig. 5,[6] where they intersect at vertex v_{i+1}. At this point, we can determine the exact point x where the smallest max regret $R(x,\hat{e})$, i.e.,

$$\overline{R}_{min}(\hat{e}) = \min_x \max\{R_{cw}^{s_i(\hat{e})}(x,\hat{e}), R_{ccw}^{s_i'(\hat{e})}(x,\hat{e})\}, \tag{14}$$

is achieved and compute it. Note that moving x cw farther with $\hat{e}$ fixed makes regret larger by Lemma 5, and as observed in Fig. 5. We thus move $\hat{e}$ to the next cw edge, and repeat.

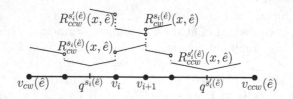

Fig. 5. $R_{cw}^{s_i(\hat{e})}(x,\hat{e})$ and $R_{ccw}^{s_i'(\hat{e})}(x,\hat{e})$.

Note that whenever $\hat{e}$ or x is moved to a new edge, $s_i(\hat{e})$ and $s_i'(\hat{e})$ may change and the 1-sinks under them must be computed, which takes $O(n)$ time by Lemma 3. Since $\hat{e}$ and x move at most $O(n)$ times, the total time required is $O(n^2)$. This proves Theorem 1.

---

[6] The tiny circle at an end of each linear segment means that that point is missing.

5   Conclusion

We have presented an $O(n^2)$ time algorithm that finds a minmax-regret 1-sink in cycle networks with uniform edge capacities. We are working on the extension of this work to compute the minmax-regret k-sink in a cycle network. Another problem of interest is to find a minmax-regret 1-sink when the edge capacities are non-uniform.

Acknowledgement. This work is supported in part by NSERC of Canada Discovery Grant, awarded to Robert Benkoczi and Binay Bhattacharya, in part by JST CREST (JPMJCR1402), granted to Naoki Kato and Yuya Higashikawa, and in part by JSPS KAKENHI Grant-in-Aid for Young Scientists (B) (17K12641), granted to Yuya Higashikawa.

References

1. Arumugam, G.P., Augustine, J., Golin, M., Srikanthan, P.: A polynomial time algorithm for minimax-regret evacuation on a dynamic path. arXiv:1404.5448 v1 [cs.DS] 22 April 2014
2. Benkoczi, R., Das, R.: The min-max sink location problem on dynamic cycle networks, October 2018. Submitted to a conference
3. Bhattacharya, B., Golin, M.J., Higashikawa, Y., Kameda, T., Katoh, N.: Improved algorithms for computing k-sink on dynamic flow path networks. Algorithms and Data Structures. LNCS, vol. 10389, pp. 133–144. Springer, Cham (2017). https://doi.org/10.1007/978-3-319-62127-2_12
4. Bhattacharya, B., Kameda, T.: Improved algorithms for computing minmax regret sinks on path and tree networks. Theor. Comput. Sci. **607**, 411–425 (2015)
5. Chen, D., Golin, M.: Sink evacuation on trees with dynamic confluent flows. In: Hong, S.-H. (ed.) Leibniz International Proceedings in Informatics, 27th International Symposium on Algorithms and Computation (ISAAC), pp. 25:1–25:13 (2016)
6. Chen, D., Golin, M.: Minmax centered k-partitioning of trees and applications to sink evacuation with dynamic confluent flows. CoRR abs/1803.09289 (2018)
7. Cheng, S.-W., Higashikawa, Y., Katoh, N., Ni, G., Su, B., Xu, Y.: Minimax regret 1-sink location problems in dynamic path networks. In: Chan, T.-H.H., Lau, L.C., Trevisan, L. (eds.) TAMC 2013. LNCS, vol. 7876, pp. 121–132. Springer, Heidelberg (2013). https://doi.org/10.1007/978-3-642-38236-9_12
8. Frederickson, G., Johnson, D.: Finding kth paths and p-centers by generating and searching good data structures. J. Algorithms **4**, 61–80 (1983)
9. Golin, M., Sandeep, S.: Minmax-regret k-sink location on a dynamic tree network with uniform capacities. arXiv:1806.03814v1 [cs.DS], pp. 1–32, 11 June 2018
10. Higashikawa, Y., et al.: Minimax regret 1-sink location problem in dynamic path networks. Theor. Comput. Sci. **588**(11), 24–36 (2015)
11. Higashikawa, Y., Golin, M.J., Katoh, N.: Minimax regret sink location problem in dynamic tree networks with uniform capacity. J. Graph Algorithms Appl. **18**(4), 539–555 (2014)
12. Higashikawa, Y., Golin, M.J., Katoh, N.: Multiple sink location problems in dynamic path networks. Theor. Comput. Sci. **607**(1), 2–15 (2015)

13. Kouvelis, P., Yu, G.: Robust Discrete Optimization and its Applications. Kluwer Academic Publishers, London (1997)
14. Li, H., Xu, Y.: Minimax regret 1-sink location problem with accessibility in dynamic general networks. Eur. J. Oper. Res. **250**, 360–366 (2016)
15. Mamada, S., Uno, T., Makino, K., Fujishige, S.: An $O(n \log^2 n)$ algorithm for a sink location problem in dynamic tree networks. In: Levy, J.-J., Mayr, E.W., Mitchell, J.C. (eds.) TCS 2004. IIFIP, vol. 155, pp. 251–264. Springer, Boston, MA (2004). https://doi.org/10.1007/1-4020-8141-3_21
16. Mamada, S., Uno, T., Makino, K., Fujishige, S.: An $O(n \log^2 n)$ algorithm for a sink location problem in dynamic tree networks. Discret. Appl. Math. **154**, 2387–2401 (2006)
17. Megiddo, N.: Combinatorial optimization with rational objective functions. Math. Oper. Res. **4**, 414–424 (1979)
18. Wang, H.: Minmax regret 1-facility location on uncertain path networks. Eur. J. Oper. Res. **239**(3), 636–643 (2014)
19. Xu, Y., Li, H.: Minimax regret 1-sink location problem in dynamic cycle networks. Inf. Process. Lett. **115**(2), 163–169 (2015)

Planar Digraphs for Automatic Complexity

Achilles A. Beros, Bjørn Kjos-Hanssen[✉][iD], and Daylan Kaui Yogi

University of Hawai'i at Mānoa, Honolulu, HI 96822, USA
{beros,bjoernkh,dkyogi64}@hawaii.edu

Abstract. We show that the digraph of a nondeterministic finite automaton witnessing the automatic complexity of a word can always be taken to be planar. In the case of total transition functions studied by Shallit and Wang, planarity can fail.

Let $s_q(n)$ be the number of binary words x of length n having nondeterministic automatic complexity $A_N(x) = q$. We show that s_q is eventually constant for each q and that the eventual constant value of s_q is computable.

Keywords: Automatic complexity · Planar graph · Möbius function · Nondeterministic finite automata

1 Introduction

Automatic complexity, introduced by Shallit and Wang [7], is an automata-based and length-conditional analogue of Sipser's CD complexity [8] which is in turn a computable analogue of the noncomputable Kolmogorov complexity. The nondeterministic case was taken up by Hyde and Kjos-Hanssen [3], who gave a table of the number of words of length n of a given complexity q for $n \leq 23$. The numbers in the table suggested (see Table 2) that the number may be eventually constant for each fixed q. Here we establish that that is the case (Theorem 9), and show that the limit is computable (in exponential time). Moreover, we narrow down the possible automata that are needed to witness nondeterministic automatic complexity: they must have planar digraphs, in fact their digraphs are trees of cycles in a certain sense.

We recall our basic notion.

Definition 1 ([7]). *The **nondeterministic automatic complexity** $A_N(x)$ of a word x is the minimal number of states of a nondeterministic finite automaton M (without ϵ-transitions) accepting x such that there is only one accepting path in M of length $|x|$.*

This work was partially supported by a grant from the Simons Foundation (#315188 to Bjørn Kjos-Hanssen). We are indebted to Jeff Shallit and Malik Younsi for helpful comments.

T. V. Gopal and J. Watada (Eds.): TAMC 2019, LNCS 11436, pp. 59–73, 2019.
https://doi.org/10.1007/978-3-030-14812-6_5

2   Automatic Complexity as Chains of Trees of Lumps

Consider the version of automatic complexity where the transition functions are not required to be total.[1] Then we claim that the digraphs representing the witnessing automata are planar, in fact they are "trees of cycles". As an example, for the word $0^5 10^5 1^6 010^3$, we have the following witnessing automaton:

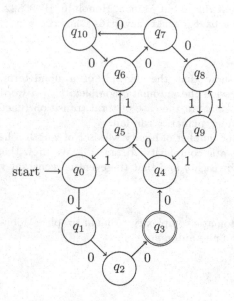

To explain this, first let us say that a cycle is a sequence of states that starts and ends with the same state. Let us say that a lump is the automaton whose transitions come from a given cycle. So if a cycle is repetitive, like 3456734567345673, then it generates the same lump as just 345673.

Consider the sequence of states visited during processing of a unique accepted word x of length n. Let us call the first visited state 0, the next distinct state 1, and so on. (So for example the permitted state sequences of length 3 are only 000, 001, 010, 011, 012.)

Then the state sequence starts $0, 1, \ldots, q, q+1, \ldots, q$ where q is the first state that is visited twice. Now the claim is that there will never, at a later point in the state sequence, be a transition (an edge) q_1, q_2 such that q_2 occurs within the lump generated by the cycle $q, q+1, \ldots, q$ and such that the transition q_1, q_2 does not occur in that lump. Indeed, otherwise our state sequence would start

$$0, 1, \ldots, \underbrace{q, \ldots, q_2}_{\text{first}}, \ldots, \underbrace{q, \ldots, q_1, q_2}_{\text{second}}$$

---

[1] Whether determinism is required is not important in the following, but in the non-deterministic case we assume we require there to be only one accepting path, as usual.

and then there is a second accepting path of the same length where the first and second segments are switched.

Consequently, the path can only return to states that are not yet in any lumps. This leaves only two choices whenever we decide to create a new edge leading to a previously visited state:

Case 1. Go back to a state that was first visited after the last completed lump so far seen, or Case 2. Go back to a state that was first visited at some earlier time, before some of the lumps so far seen started (and in general after some of them were complete).

This gives a tree of lumps where each new lump either (Case 1) creates a new sibling for the previous lump, or (Case 2) creates a new parent for a final segment of the so far seen top-level siblings. In this tree of lumps, only the leaves (the lumps that are not anybody's parents) can be traversed more than once by the uniquely accepted path of length n.

So if the first lump created is l_1 then next we can have two cases:

$$(l_1, \quad l_2) \hspace{4cm} \text{(Case 1)}$$

$$l_1 \rightarrow l_2 \hspace{4cm} \text{(Case 2)}$$

In Case 1, l_1 and l_2 are siblings ordered from first to second. In Case 2, $\rightarrow$ denotes *is a child of*, which by definition is the same as *sub-digraph*. Now for the third lump l_3, we have only the following possibilities:

$$(l_1, \quad l_2, \quad l_3) \hspace{3cm} \text{(Subcase 1.1)}$$

$$(l_1, \quad l_2 \rightarrow l_3) \hspace{3cm} \text{(Subcase 1.2)}$$

$$(l_1, l_2) \rightarrow l_3 \hspace{3cm} \text{(Subcase 1.3)}$$

$$(l_1 \rightarrow l_2, \quad l_3) \hspace{3cm} \text{(Subcase 2.1)}$$

$$l_1 \rightarrow l_2 \rightarrow l_3 \hspace{3cm} \text{(Subcase 2.2)}$$

In Subcase 1.2, l_1 and l_3 are siblings and l_2 is a child of l_3. In Subcase 1.3, l_3 is a common parent of l_1 and l_2. In Subcase 2.1, l_3 is a new sibling for l_2, and l_2 still has l_1 as its child. In Subcase 2.2, l_3 is a parent of l_2.

For instance, the state sequence 01234567345673456720 has the structure of Subcase 2.2, with l_1 being the lump generated from 345673, l_2 being generated from 23456734567345672, and l_3 being generated from the whole sequence 01234567345673456720. The corresponding automaton is shown in an online tool.[2] Using this planarity result, we are able to increase the speed of our algorithm for calculating $A_N(x)$. Consequently, we have been able to extend the string length in our computations from $n = 23$ to $n = 25$. The number of maximally complex binary words of a given length are shown in Table 1. A similar table for $n \leq 23$ was given in [3].

---

[2] http://math.hawaii.edu/wordpress/bjoern/complexity-of-00011110111110111111/.

Table 1. Lengths n, number of words of length n of maximal $A_N(x)$, 2^n, percentage of maximally complex words, number of non-maximally complex words.

n	#	2^n	%complex	2^n-#
0	1	1	100.00%	0
1	2	2	100.00%	0
2	2	4	50.00%	2
3	6	8	75.00%	2
4	8	16	50.00%	8
5	24	32	75.00%	8
6	30	64	46.88%	34
7	98	128	76.56%	30
8	98	256	38.28%	158
9	406	512	79.30%	106
10	344	1,024	33.59%	680
11	1,398	2,048	68.26%	650
12	1,638	4,096	39.99%	2,458
13	5,774	8,192	70.48%	2,418
14	5,116	16,384	31.23%	11,268
15	23,018	32,768	70.25%	9,750
16	22,476	65,536	34.30%	43,060
17	86,128	131,072	65.71%	44,944
18	89,566	262,144	34.17%	172,578
19	351,250	524,288	67.00%	173,038
20	375,710	1,048,576	35.83%	672,866
21	1,461,670	2,097,152	69.70%	635,482
22	1,539,164	4,194,304	36.70%	2,655,140
23	5,687,234	8,388,608	67.80%	2,701,374
24	6,814,782	16,777,216	40.62%	9,962,434
25	24,031,676	33,554,432	71.62%	9,522,756
26	27,782,964	67,108,864	41.40%	39,325,900
27	97,974,668	134,217,728	73.00%	36,243,060

3   The Asymptotic Number of Words of Given Complexity

In this section, we examine the asymptotic behavior of the number of words with automatic complexity q for a fixed $q \in \mathbb{N}$.

Definition 2. *A binary word x is* right inextendible *if $A_N(x) < A_N(x0)$ and $A_N(x) < A_N(x1)$.*

Inextendibility is closely related to volatility of the automatic complexity, as examined in the Complexity Option Game [5]. The number and proportion of right-inextendible words of length n and complexity q can be examined using an online database [4] and is shown in Table 2 for small q and n.

A basic procedure in our results will be the counting of periodic words, since a cycle containing a periodic word can be shortened and an automaton containing such a cycle will not be optimal.

Definition 3. *A word x is* periodic *if there exists a subword $y \neq x$ and an integer n such that*

$$\underbrace{yyy \cdots y}_{n} = x.$$

A non-periodic word [2] is also called a primitive word and one starting with 0, in our setting, is called a *Lyndon word* [6].

Definition 4 ([1]). *Let n be a positive integer with $\omega(n)$ denoting the number of distinct prime factors of n and $\Omega(n)$ denoting the total number of prime factors (i.e., with repetition) of n. The Möbius function μ is defined as*

$$\mu(n) := \begin{cases} (-1)^{\omega(n) \bmod 2} & \text{if } \Omega(n) = \omega(n), \\ 0 & \text{if } \Omega(n) > \omega(n). \end{cases}$$

Theorem 5 ([2]). *The number of unique periodic binary words of length n is given by $Z(0) = 0$ and for $n \geq 1$,*

$$Z(n) = 2^n - \sum_{d \mid n} \mu\left(\frac{n}{d}\right) \cdot 2^d.$$

Recall that a *necklace* is an equivalence class of non-periodic words under cyclic rotation. Thus, for instance, $\{0011, 0110, 1100, 1001\}$ is a necklace. Theorem 5 is a restatement of the following classical result.

Theorem 6 (Witt's Formula [9]**).** *The number of necklaces of binary words of length n is*

$$\frac{1}{n} \sum_{d \mid n} \mu\left(\frac{n}{d}\right) \cdot 2^d.$$

Definition 7. *We define the set $S_q(n) = \{x \in \{0,1\}^n : A(x) = q\}$ and $s_q(n) = |S_q(n)|$.*

Definition 8. *Given an automaton, G, whose set of states is Q, we define a* detour *to be a pair of finite non-trivial sequences of states, $\alpha, \beta \in Q^*$, such that $\alpha(0) = \beta(0)$, $\alpha(|\alpha| - 1) = \beta(|\beta| - 1)$ and $\alpha \neq \beta$. We call a detour* minimal *if $\{\alpha(i) : 0 < i < |\alpha| - 1\} \cap \{\beta(i) : 0 < i < |\beta| - 1\} = \emptyset$.*

Table 2. Proportions $r_q(n)/s_q(n)$ of right-inextendible binary words of automatic complexity q and length n.

n	q							
	3	4	5	6	7	8	9	10
22	8/20	28/58	86/164	322/502	1288/2846	6594/16024	44922/94732	220544/451368
21	8/20	28/58	98/176	292/496	1318/3168	8472/18720	52178/108042	266760/504794
20	8/20	28/58	86/164	238/430	1478/3814	11670/23328	54990/115896	278696/529148
19	8/20	28/58	86/164	402/582	2380/4996	12312/26542	78892/410668	134578/351250
18	8/20	28/58	110/188	356/598	2070/5692	14456/29990	68288/36024	0/0
17	8/20	28/58	104/200	262/514	2850/7102	20516/37042	30486/86128	
16	8/20	28/58	80/164	536/752	2908/7738	14230/34320	0/22476	
15	8/20	28/58	148/226	578/908	3338/8530	7524/23018		
14	8/20	28/58	112/244	774/1270	4442/9868	0/5116		
13	8/20	28/58	120/250	1396/2076	1736/5774			
12	8/20	28/58	158/282	1048/2090	0/1638			
11	8/20	28/58	384/564	576/1398				
10	8/20	34/64	244/588	0/344				
9	8/20	48/78	112/406					
8	8/20	82/130	0/98					
7	10/22	38/98						
6	14/26	0/30						
5	8/24							
4	0/8							

Consider an automaton with a single cycle (Fig. 5). Suppose the automaton has i states before the cycle and ℓ states after the cycle (which implies that there are $q - (i + \ell)$ states within the cycle). We now obtain a formula for the limit of the number of binary words of given complexity q.

Theorem 9. s_q is eventually constant, with limiting value

$$\sum_{\substack{i,\ell \geq 0 \\ i+\ell < q}} 2^{(i-1)^+} \cdot [2^{q-(i+\ell)} - Z(q - (i + \ell))] \cdot 2^{(\ell-1)^+},$$

where Z was defined in Theorem 5 and

$$x^+ = \max\{x, 0\}.$$

Proof. Consider an arbitrary automaton G with q states. There are a finite number of such automata. We will prove that unless G has at most one minimal detour, there is an N such that, for all $n \geq N$, G cannot accept a unique word of length n.

We begin with the observation that we may assume that G has a unique initial state and a unique accepting state.

If G has at most one detour, then G has one of the following forms.

If G is of the type on the right and G accepts a unique word σ of length n, then any accepting path for σ either uses the k states that comprise the top path of the detour, or uses the j states that comprise the bottom path, but no both. Thus, if both k and j are non-zero, there is an automaton with fewer states that accepts only σ among all words of length n. We conclude that in the case of automata with at most one minimal detour, we need only consider ones of the form on the left.

Now, we consider the possibilities for automata with at least two distinct minimal detours. Each of the twelve cases in Fig. 1 falls into one of three cases.

1. On any accepting path, each detour can be used at most once ((1), (2) and (3)).
2. On any accepting path, one of the detours can be used at most once ((7), (8), (10), (11) and (12)).
3. There are accepting paths that use each of the detours an arbitrary number of times ((4), (5), (6) and (9)).

These further break down as follows:

– (1), (4), (7), (10) represent two separated cycles;
– (2), (5), (8), (11) represent overlapping cycles.
– (3), (6), (9), (12) represent nested cycles; and

If G falls into the first case, then σ is also uniquely accepted among words of length n by an automaton with at most q states and no detours. If G falls into the second case, then σ is uniquely accepted by an automaton with at most q states and at most one detour. If G falls into the third category, then there are two cycles (although they may have common transitions) which can each be traversed and independent and arbitrary number of times on an accepting path. Thus, for large enough n, the cycles can be traversed in different orders or different numbers of times and still reach an accepting state, thereby violating the requirement that G accept exactly one word of length n.

As an example of the third case, suppose that G is of the type shown in (9). G has two independent cycles, one of length $p+j+k+\ell$ and the other of length $p+j+n+\ell$. Let $N = i + a(p+j+k+\ell) + m = i + b(p+j+n+\ell) + m$, where $a, b \in \mathbb{N}$. There are at least two words of length N that G accepts, and for any $M \geq N$ such that G accepts a word of length M, G must accept at least two words of length M.

In conclusion, we may assume our automata have at most one detour. Thus they consist of a chain of states, followed by a single (in general multi-state) cycle, followed by another chain. Let i be the number of states before the cycle, ℓ the number of states after the cycle, so that $q - (i + \ell)$ if the number of states

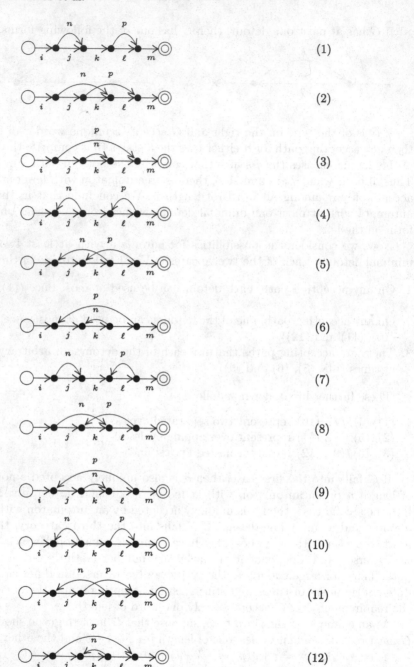

Fig. 1. The possibilities for automata with at least two distinct minimal detours.

within the cycle. If the bits read within the cycle do not form a necklace, we can reduce the number of states. Thus there are $[2^{q-(i+\ell)} - Z(q-(i+\ell))]$ states within the cycle. The an upper bound for the total number of binary words with $A_N(x) = q$ is

$$2^i \cdot 2^\ell \cdot [2^{q-(i+\ell)} - Z(q-(i+\ell))].$$

Let ξ be the bit that advances the automaton from the ith state to the $(i+1)$th state (i.e. the transition that takes the automaton into the cycle) and η be the bit that advances that automaton from the $q-(i+\ell)$th state to the $(i+1)$th state (i.e, the transition that completes the cycle). If $\xi = \eta$, then it is possible to create an automaton with fewer states that accepts the same word and no other of length n. A similar consideration applies upon leaving the cycle. Thus, we have

$$2^{(i-1)^+} \cdot [2^{q-(i+\ell)} - Z(q-(i+\ell))] \cdot 2^{(\ell-1)^+}$$

possible words.

Finally, to conclude that $s_q(n)$ is eventually constant, note that while the single cycle will have to be exited at different points depending on $n \bmod k$, where k is the length of the main cycle, there will always be exactly one value of $n \bmod k$ and hence exactly one automaton contributed from the cycle and the given "head" and "tail" words. See Figs. 2, 3, and 4 for illustrations of the cases $q = 2, 3, 4$, respectively.

Remark 10. *Here is perhaps a simpler view of the classification of detours in Fig. 1. Suppose A is an NFA that uniquely accepts some word. Now consider some shortest directed path P from q_0 to the unique final state q_f. Let us say*

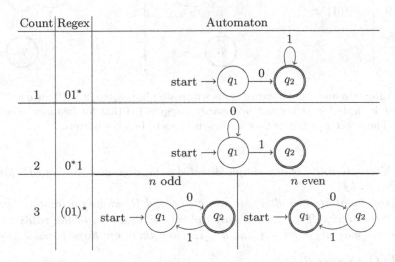

Count	Regex	Automaton
1	01*	
2	0*1	
3	(01)*	

Fig. 2. The witnessing automata for $\lim_n s_q(n)/2 = 3$, $q = 2$. The first two are used at any length n, whereas the bottom two are each used only for one value of $n \bmod 2$, illustrating Theorem 11.

Count	Regex	Automata
1–5	001* (shown) 010*, 01*0 0*10, 0*11	
6–8	(001)* (shown) (010)*, (011)*	
9	0(01)*	
10	(01)*x	

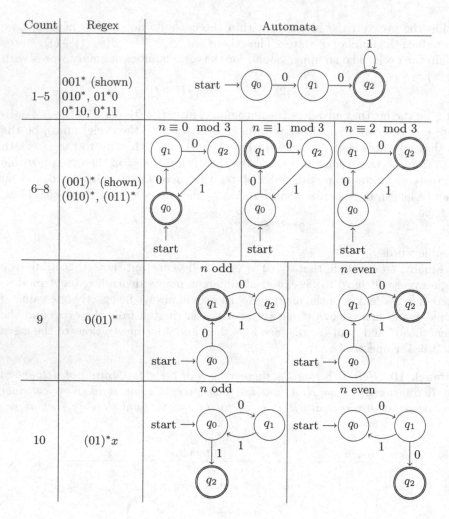

Fig. 3. Automata and regular expressions witnessing $\lim_n s_q(n)/2 = 10$ for $q = 3$. The exponents indicated by $*$ are not necessarily integers (so that for instance $abcd^{1.5} = abcdab$). The letter x indicates 0 or 1, chosen so as to break a pattern.

that an alternate route *is any simple directed path, edge-disjoint from* P, *joining two vertices of* P.

Suppose there are two alternate routes, Q and R, joining q_i and q_j, and q_k and q_l, respectively. If we do not worry about the direction of the paths for the moment, we may assume $i \leq j$ and $k \leq l$. Then there are three possibilities:

1. $j \leq k$: Q precedes R;
2. $k \leq i$ and $j \leq l$: Q encompasses R;
3. $i \leq k \leq j \leq l$: Q and R overlap.

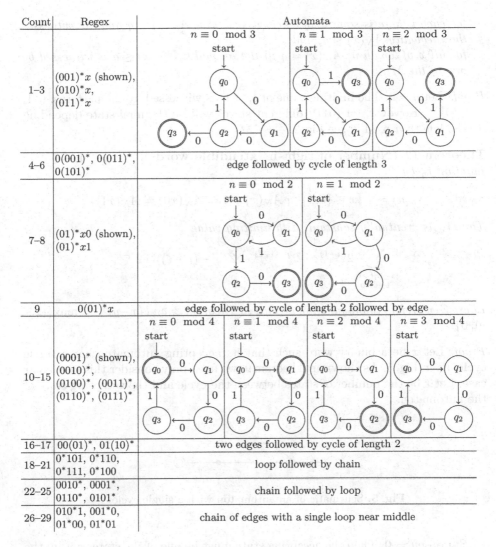

Count	Regex	Automata

Fig. 4. Witnessing automata for $\lim_n s_q(n)/2 = 29$, $q = 4$. The exponents indicated by $*$ are not necessarily integers, and the letter x indicates 0 or 1, chosen so as to break a pattern.

Furthermore, for Q and R one can choose the direction of the edges independently. This gives $3 \cdot 4 = 12$ possibilities to consider.

The main proviso to Theorem 9 may be that while *the number of words with given complexity* reaches a limit, *the set of witnessing automata* does not quite. To wit:

Theorem 11. *There is a q such that there is no set of automata $M_1, \ldots, M_s$ such that for all sufficiently large n,*

– *for each i there is some x of length n such that $A_N(x) = q$ and M_i witnesses the inequality $A_N(x) \leq q$, and*
– *for all x of length n, $A_N(x) = q$ iff the inequality $A_N(x) \leq q$ is witnessed by one of the M_i.*

Proof. Let $q = 2$. The limiting value of s_q is 6 as witnessed by the patterns: 0^*1, 01^*, $(01)^*$. However, for $(01)^*$, different states will be the final state depending on the length $n \bmod 2$; see Fig. 2.

Theorem 12 (Number of right-inextendible words). *For $q \geq 1$, define a function r_q by*

$$r_q(n) = \#\{x \in \{0,1\}^n \mid A_N(x) + 1 = A_N(x0) = A_N(x1)\}.$$

Then r_q is eventually constant, with limiting value

$$\sum_{\substack{i \geq 0, \ell > 0 \\ i+\ell < q}} 2^{(a-1)^+} \cdot [2^{q-(i+\ell)} - Z(q - (i+\ell))] \cdot 2^{\ell-1},$$

where $Z(n)$ refers to the function defined in Theorem 2, and $(x-y)^+ := \max\{(x-y), 0\}$.

Proof. Let x be a binary word such that its accepting automaton has a single cycle, as in Fig. 5. As shown in Theorem 9, we need only consider this particular case. Let ℓ be the number of states between the cycle and the accepting state of the automaton.

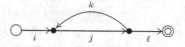

Fig. 5. Schematic of an automaton with a single cycle.

Suppose $\ell = 0$. Then the accepting state must be one of the states within the cycle. Without loss of generality, suppose the path out of the accepting state is triggered by a 0 input. Then $x0$ must have the same automatic complexity as x, as appending 0 to x does not require the addition of any additional states, and x is thus not inextendible. Thus, for a word to be inextendible, it is necessary that $\ell > 0$.

Theorem 13. *$s_q(n)$ is eventually bounded by $2^{q-2}\left(\frac{q(q+5)}{2} + 1\right)$.*

Proof. By Theorem 9, we can upper bound the sum by

$$\sum_{i,\ell \geq 0, i+\ell < q} 2^q = \binom{q+1}{2} 2^q.$$

```
oeisValues = [
        0, 0, 2, 2, 4, 2, 10, 2, 16, 8, 34, 2, 76, 2, 130, 38, 256, 2,
        568, 2, 1036, 134, 2050, 2, 4336, 32, 8194, 512, 16396, 2, 33814,
        2, 65536, 2054, 131074, 158, 266176, 2, 524290, 8198, 1048816, 2,
        2113462, 2, 4194316, 33272, 8388610, 2, 16842496, 128, 33555424
]# from http://oeis.org/A152061
def Z(n): # number of periodic binary strings of length n
        return oeisValues[n]
def plus(k):
        if k<0:
                return 0
        return k
def limS(q): #limitingNumberOfStringsWithNFAComplexity(q):
        num = 0
        print "."
        for i in range(0, q):
                for l in range(0, q):
                        if i+l<q:
                                left = 2**(plus(i-1))
                                right = 2**(plus(l-1))
                                middle = (2**(q-(i+l))-Z(q-(i+l)))
                                num += left*middle*right
        return num
def answer(q):
        bound = 2**(q-2)*(1+q*(q+5)/2)
        print "q=" + str(q) + ",_" + str(limS(q)),
        print ",_bound_=_" + str(bound) + ",_",
        print str(limS(q)/float(bound))
for q in range(3, len(oeisValues)):
        answer(q)
```

Fig. 6. Python code which when run hints at the sharpness of Theorem 13.

In fact, by considering the four possible truth values for the cases $i = 0$, $\ell = 0$, we get the upper bound

$$\sum_{i=\ell=0} 2^q + \sum_{i\ell=0, i+\ell>0} 2^{q-1} + \sum_{i>0, \ell>0} 2^{q-2} = 2^q + 2(q-1)2^{q-1} + \binom{q-1}{2} 2^{q-2}$$

$$= 2^{q-2}\left(4q + \binom{q-1}{2}\right) = 2^{q-2}\left(\frac{q(q+5)}{2} + 1\right).$$

Remark 14. *A comparison of s_q with the bound in Theorem 13 can be done using the computer code in Fig. 6. The number in the title of this section was calculated using that Python script and using a table of values of Z from the OEIS database. Table 3 shows an initial segment of the resulting sequence. There we count only words starting with 0, so that the full number would be twice that, matching the impression that $\lim_n s_3(n) = 20$ given by Table 2.*

Table 3. The number of binary words $0x$ of length n with $A_N(0x) = q$, for sufficiently large n. The value for $q = 7$ is surprisingly small when comparing with Table 2.

q	$\lim_n s_q(n)/2$	q	$\lim_n s_q(n)/2$
1	1	21	64 594 576
2	3	22	141 046 655
3	10	23	306 858 874
4	29	24	665 342 837
5	82	25	1 438 134 475
6	215	26	3 099 548 927
7	556	27	6 662 442 946
8	1 385	28	14 285 118 725
9	3 391	29	30 557 828 119
10	8 135	30	65 225 030 201
11	19 261	31	138 937 277 596
12	44 963	32	295 385 810 819
13	103 906	33	626 867 939 224
14	237 719	34	1 328 075 901 017
15	539 458	35	2 809 126 944 436
16	1 214 993	36	5 932 793 909 801
17	2 718 760	37	12 511 847 996 740
18	6 047 426	38	26 350 575 690 893
19	13 380 766	39	55 423 630 773 538
20	29 463 632	40	116 429 658 505 697

References

1. Bender, E.A., Goldman, J.R.: On the applications of Möbius inversion incombinatorial analysis. Am. Math. Mon. **82**(8), 789–803 (1975). https://doi.org/10.2307/2319793
2. Choi, J.S.: Counts of unique periodic binary strings of length n, September 2011. http://oeis.org/A152061
3. Hyde, K.K., Kjos-Hanssen, B.: Nondeterministic automatic complexity of overlap-free and almost square-free words. Electron. J. Combin. **22**(3), 18 (2015). paper 3.22
4. Kjos-Hanssen, B.: Complexity lookup. http://math.hawaii.edu/wordpress/bjoern/complexity-of-0110100110010110/
5. Kjos-Hanssen, B.: Complexity option game. http://math.hawaii.edu/wordpress/bjoern/complexity-option-game/
6. Lyndon, R.C.: On Burnside's problem. Trans. Am. Math. Soc. **77**, 202–215 (1954). https://doi.org/10.2307/1990868
7. Shallit, J., Wang, M.W.: Automatic complexity of strings. J. Autom. Lang. Comb. **6**(4), 537–554 (2001). 2nd Workshop on Descriptional Complexity of Automata, Grammars and Related Structures, London, ON (2000)

8. Sipser, M.: A complexity theoretic approach to randomness. In: Proceedings of the Fifteenth Annual ACM Symposium on Theory of Computing, STOC 1983, pp. 330–335. ACM, New York (1983). https://doi.org/10.1145/800061.808762
9. Witt, E.: Treue Darstellung Liescher Ringe. J. Reine Angew. Math. **177**, 152–160 (1937). https://doi.org/10.1515/crll.1937.177.152

Approximation Algorithms for Graph Burning

Anthony Bonato[1(✉)] and Shahin Kamali[2]

[1] Ryerson University, Toronto, ON, Canada
abonato@ryerson.ca
[2] University of Manitoba, Winnipeg, MB, Canada
shahin.kamali@umanitoba.ca

Abstract. Numerous approaches study the vulnerability of networks against social contagion. Graph burning studies how fast a contagion, modeled as a set of fires, spreads in a graph. The burning process takes place in synchronous, discrete rounds. In each round, a fire breaks out at a vertex, and the fire spreads to all vertices that are adjacent to a burning vertex. The selection of vertices where fires start defines a schedule that indicates the number of rounds required to burn all vertices. Given a graph, the objective of an algorithm is to find a schedule that minimizes the number of rounds to burn graph. Finding the optimal schedule is known to be NP-hard, and the problem remains NP-hard when the graph is a tree or a set of disjoint paths. The only known algorithm is an approximation algorithm for disjoint paths, which has an approximation ratio of 1.5.

We present approximation algorithms for graph burning. For general graphs, we introduce an algorithm with an approximation ratio of 3. When the graph is a tree, we present another algorithm with approximation ratio 2. Moreover, we consider a setting where the graph is a forest of disjoint paths. In this setting, when the number of paths is constant, we provide an optimal algorithm which runs in polynomial time. When the number of paths is more than a constant, we provide two approximation schemes: first, under a regularity condition where paths have asymptotically equal lengths, we show the problem admits an approximation scheme which is fully polynomial. Second, for a general setting where the regularity condition does not necessarily hold, we provide another approximation scheme which runs in time polynomial in the size of the graph.

Keywords: Approximation algorithms · Graph algorithms ·
Graph burning problem · Information dissemination · Social contagion

1 Introduction

Numerous efforts were initiated to characterize and analyze social contagion or social influence in networks; see, for example, [8,16,29,30]. These studies

© Springer Nature Switzerland AG 2019
T. V. Gopal and J. Watada (Eds.): TAMC 2019, LNCS 11436, pp. 74–92, 2019.
https://doi.org/10.1007/978-3-030-14812-6_6

investigate the vulnerabilities and strengths of these networks against the spread of an emotional state or other data, such as a meme or gossip. For example, there are studies that suggest emotional states can be transferred to others via emotional contagion on Facebook; such emotional contagion is known to occur without direct interaction between people and in the complete absence of nonverbal cues [30].

The *burning number* [5,6] measures how prone a network is to fast social contagion. In the burning protocol, like many other network protocols, data is communicated between nodes in discrete rounds. The input is an undirected, unweighted, finite simple graph. We say a node is burning if it has received data. Initially, no vertex is burning. In each round, a burning vertex sends data to all its neighbors, and all neighbors will be on fire at the end of the round; this is consistent with the fact that a user in the network can expose all its neighbours to a posted piece of data. In addition, in each given round, a new fire starts at a non-burning vertex called an *activator*; this can be interpreted as a way to target additional users that initiate the contagion. Note that the burning protocol does not provide a specified algorithm of how the fire spreads. However, the algorithm can choose where to initiate the fire. The decisions of the algorithm for the location of activators define a *schedule* that can be described by a *burning sequence*: the ith member of the burning sequence indicates the vertex at which a fire is started in round i. We say the graph is *burned* when all vertices are on fire; that is, all members of the network have received the data. Figure 1 provides an illustration of the burning process.

To understand how prone a graph is to the spread of data, we are interested in schedules that minimize the number of rounds required to burn the whole graph. The burning number of a given graph is the minimum such number; hence, an optimal algorithm burns the graph in a number of rounds that is equal to the burning number. Unfortunately, finding optimal solutions is NP-hard even for elementary graph families [2]. The focus of this paper is to provide approximation algorithms for burning graphs.

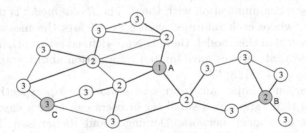

Fig. 1. Burning a graph in three rounds using a schedule defined by burning sequence $\langle A, B, C \rangle$. The number on each vertex indicates the rounds at which the vertex becomes a burning vertex. At round 1, a fire starts at A. At round 2, another fire starts at B while the fire at A spreads to all neighbors of A. At round 3, the fire spreads to all vertices except for C, where a new fire is started.

Previous Work

Bonato et al. [5,6] first introduced the burning process as a way to model spread of contagion in a social network; they characterized the burning number for some graph classes and proved some properties for the burning number. The results of [32,33] extended these results for additional graph families and also studied a variant of burning number in which the burning sequence is selected according to some probabilistic rule. Bessy et al. [3] further studied the burning number and proved that for a connected graph of size n the burning number is at most $2\lceil\sqrt{n}\rceil - 1$ and conjectured that this number is indeed at most $\lceil\sqrt{n}\rceil$. They proved better bounds for the burning number of trees. Land and Lu [31] slightly improved the upper bound to $\frac{\sqrt{6}}{2}\sqrt{n}$. In [4], burning densities were considered for the infinite Cartesian grid. Sim et al. [39] provided tight bounds for the burning number of generalized Petersen graphs. Bonato et al. [2] proved that it is NP-hard to find a schedule that completes burning in the minimum number of rounds (in time equal to the burning number). Interestingly, their hardness result holds for basic graph families such as acyclic graphs with maximum degree three, spider graphs, and path forests (that is, a disjoint union of paths).

There are numerous gossiping and broadcasting protocols that aim to model the amount of time it takes to spread information throughout a given network. For example, in the *telephone model* for gossiping, there is a distinguished originator that starts spreading the gossip. In a given round, each node that has received a piece of data (gossip) can inform one of its neighbors via a phone call. A gossip schedule defines the order in which each node informs its neighbors. The goal of a schedule is to minimize the number of rounds required to inform all vertices. This problem is known to be NP-hard [20,40] (in fact, APX-hard [38]) and there is an approximation algorithm completes within a sublogarithmic factor of optimal schedule [15] (whether a constant approximation algorithm exists is an open problem). We refer the reader to [22,34,36] for more results on telephone broadcasting. It is evident that the telephone model is not suitable for situations where a user can expose all its neighbors by posting a gossip and without in-person communication with them. The *Radio model* is more relevant in this context, where each informed vertex broadcasts the message to all its neighbors; however, in this model, there is a pre-defined set of originators and it is often assumed that vertices have limited information about graph structures (see, for example, [13,21,28,35]).

Social contagion is important from a *viral marketing* perspective, based on the observation that targeting a small set of users can have a cascading word-of-mouth effect in a social network. Domingos and Richardson [14,37] define influence maximization problems that aim to define a set of initially activated user that can eventually influence a maximum members of the network. This problem is known to be NP-hard. Kempe et al. provide several approximation algorithms for several simple diffusion models [24,26] as well as a more general decreasing cascade model, where a behaviour spreads in a cascading fashion according to a probabilistic rule [25]. These results were followed by more approximation algorithms and inapproximability results for these models (see, for example, [10–12]). We refer the reader to Kleinberg [27] for the economic aspects of cascading behaviour on social networks. Note that besides the diffusion model, the influence maximization problem is different from burning in the sense that initial informed users start spreading data at the same time (while in burning they start one at a time).

Another problem related to graph burning is the *Firefighter Problem*, which also assumes discrete, synchronous rounds. Given a graph G, at round 1, a fire starts at a given node r of G. In each subsequent round, a firefighter can defend one non-burning vertex while the fire spreads to all undefended neighbours of each burning vertex. Once burning or defended, a vertex remains so for all subsequent rounds. The process ends when the fire can no longer spread. The goal of an algorithm (which we identify with a firefighter) is to defend a maximum number of vertices that can be saved; that is, that are not burning at the end of the process. Despite similarities in the underlying model, the objective in the Firefighter problem is quite different from the burning problem. As expected, the Firefighter problem is NP-hard [18], and it is known that no approximation algorithms can achieve a factor of n^α for any $\alpha < 1$, assuming $P \neq NP$ [1]. The problem remains NP-hard for the trees [18]; however, there are constant-factor approximation algorithms for trees (see, for example, [9, 19]).

Contributions

The burning problem is NP-hard, which is not surprising as many related problems are NP-hard. However, the fact that the problem remains NP-hard for elementary graph families such as path forests (that is, disjoint unions of paths) raises questions about its computational complexity. In particular, we may ask whether there is a polynomial algorithm that has a constant approximation ratio. Bonato and Lidbetter [7] answered this question for path forests in the affirmative by introducing a 3/2-approximation algorithm. The problem remained open for other graph families. This question is particularly interesting because it has different answers for similar problems (as described in the previous section): for telephone broadcasting, it remains open whether there is a constant approximation algorithm. For influence maximization, there is a constant approximation algorithm, while for the Firefighter problem, it is NP-hard to achieve a sublinear approximation ratio.

In this paper, we show that there is indeed a simple polynomial algorithm with constant approximation ratio of at most 3 for *any* graph. Our algorithm is intuitive and runs in time $O(m \log n)$ for a graph with n vertices and m edges. When the graph is a tree, we present another algorithm with improved approximation ratio of 2. Finally, we consider the problem when the graph is a path forest. In case the graph is formed by a constant number of paths, we present a dynamic programming algorithm that creates an optimal solution in polynomial time. When the number of paths is not a constant, we provide two approximation schemes. The first scheme works under a regularity condition which implies the lengths of paths are asymptotically equal. For this scheme, we reduce the problem to the bin covering problem to achieve a fully polynomial time approximation scheme (FPTAS) for the problem. For the general setting, when there is no assumption on the length of the paths, we use a different approach to present a polynomial time approximation scheme (PTAS) which runs in time polynomial in the size of the graph.

2   Approximation Algorithm for Burning Graphs

In this section, we devise an approximation algorithm with approximation factor of 3 for the burning problem. Throughout the section, we use $G = (V, E)$ to denote an input graph and OPT to denote the optimal algorithm for the problem. We use $\text{OPT}(G)$ to denote the burning number of G. We begin with the following lemma.

Lemma 1. *For a positive integer r, if there are r vertices at pairwise distance at least $2r - 1$, then any burning schedule requires at least r rounds to complete.*

Proof. Let $x_1, x_2, \ldots, x_r$ be r vertices of pairwise distance at least $2r - 1$. For each x_i, consider the ball of radius $r - 1$ formed by vertices of distance at most $r - 1$ from x_i. Since the distance of x_i and any x_j is at least $2r - 1$, their balls do not intersect. Assume that there is a schedule that completes in at most $r - 1$ rounds. That schedule should have a fire started inside each ball (a fire started at a distance r or more reaches x_i after at least r rounds). Hence, at least r fires must be started, which implies the burning completes in at least r rounds. This contradicts the initial assumption that the schedule completes within $r - 1$ rounds.

We devise a procedure Burn-Guess(G, g) that receives a 'guess' value g for the number of rounds required to burn graph G. The output of Burn-Guess is one of the following.

1. A schedule that completes burning in at most $3g - 3$ rounds.
2. 'Bad-Guess' that guarantees any schedule requires at least g rounds to complete.

To devise an approximation algorithm, it suffices to find the smallest guess value g^* so that Burn-Guess(g^*) returns a schedule (which implies Burn-Guess$(g^* - 1)$ returns Bad-Guess). In this way, the returned schedule completes in at most $3g^* - 3$ rounds while OPT requires at least $g^* - 1$ rounds to complete. This results in an algorithm with approximation ratio of at most 3.

Burn-Guess processes vertices one-by-one in an arbitrary order and maintains a set of 'centers' that is initially empty. When processing a vertex v, the algorithm checks the distance of v to its closest center. If such distance is at most $2g - 2$, then v is marked as 'non-center'; otherwise, v is added to the set of centers. In this way, all centers are at pairwise distance of at least $2g - 1$. After processing any vertex, if the number of centers becomes equal to g, then Burn-Guess returns Bad-Guess. When all vertices are processed, the algorithm returns a schedule defined by a burning sequence formed by an arbitrary ordering of centers.

Lemma 2. *If Burn-Guess(G, g) returns Bad-Guess, then there is no burning schedule for G that completes in less than g number of rounds.*

Proof. Burn-Guess returns Bad-Guess if the number of centers becomes equal to g. Since all centers are at pairwise distance of at least $2g - 1$, we have that there are g vertices at pairwise distance of $2g - 1$ or more. Applying Lemma 1, we conclude that any burning schedule requires at least g rounds to burn the graph.

Lemma 3. *If Burn-Guess(G, g) returns a burning sequence, then the burning of that sequence completes in at most $3g - 3$ rounds.*

Proof. All non-center vertices are at distance at most $2g - 2$ of at least one center. Recall that the burning schedule uses centers as activators. The fire starts at the last center at round $g - 1$; all vertices within distance $2g - 2$ of that center burn by round $3g - 3$. We conclude that all non-center vertices burn by round $3g - 3$.

We now arrive at our main result.

Theorem 1. *There is a polynomial algorithm with approximation ratio of at most 3 for burning any graph $G = (V, E)$.*

Proof. Let $n = |V|$. The algorithm finds the smallest value g^* for which Burn-Guess returns a schedule ($g^* \leq n$). By Lemma 3, the schedule returned by Burn-Guess completes in at most $3g^* - 3$ rounds. Meanwhile, since Burn-Guess returns Bad-Guess for $g^* - 1$, be Lemma 2, no schedule completes in $g^* - 1$ rounds.

It is not hard to see the upper bound in Theorem 1 is tight. Consider the graph in Fig. 2, where $c_1, \ldots, c_k$ are the centers selected by the algorithm. Note the pairwise distance between any two centers is $2k$ and the distance of a non-center and a center is at most $2k - 2$. Thus, Burn-Guess returns a schedule when its parameter is $g = k$ while it returns Bad-Guess when $g = k - 1$. Assuming centers are burned in the same order, the cost of the algorithm is $3k - 2$ (a fire starts at c_k at round k and reaches b at round $3k - 2$). On the other hand, there is a better scheme that burns vertex a at round 1 and burns the middle point of the path p between c_k and b at round 2. This scheme burns all vertices by round $k + 1$. Consequently, the approximation ratio of the algorithm is at least $\frac{3k-2}{k+1}$ which converges to 3 for large values of g.

A straightforward implementation of the Burn-Guess uses breadth-first traversal of the graph. Starting with an unvisited node v, we add v to the set of centers and apply breath first to visit all vertices within distance $2g - 2$ of v. After reaching 'depth' of $2g - 2$, we stop the breath search and pick another unvisited vertex as the next center and start another breath first traversal. This process continues until all vertices are visited or the number of centers exceeds g. Clearly, any edge is visited at most once and hence Burn-Guess runs in time $O(m)$. Since Burn-Guess is called $O(\log n)$ times (via a binary search in the space of g), we conclude that our algorithm for burning graphs runs in time $O(m \log n)$.

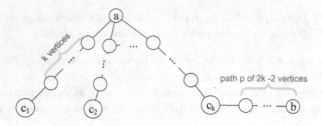

Fig. 2. An instance for which the scheme by the burning algorithm takes three times more rounds than the optimal algorithm to burn the graph.

The above implementation is useful when the order in which vertices are processed is not defined by the algorithm. This is particularly handy when Burn-Guess has to work based on partial information; for example, a parallel setting where only a partition of the input graph is available to each processor. When there is no such restriction, we can apply optimizations like selecting the point located at the maximum distance to all current centers as the next center (this is similar to farthest-first algorithm for the metric k-center problem; see, for example, [41]). While this optimization is likely to improve the approximation ratio (albeit with analysis techniques that would be more involved) it degrades the running time: an efficient implementation of requires pre-computing all-pair shortest-path distances in $O(mn + n^2 \log n)$ using Dijkstra's algorithm. Provided with these distances, running an instance of Burn-Guess(G, g) takes $O(ng)$, which is $O(n^2)$ for general graphs (and $O(n^{3/2})$ for connected graphs since $g \in O(\sqrt{n})$ when the graph is connected). Burn-Guess is called $O(\log n)$ times, which gives a total time complexity of $O(n^2 \log n)$. This complexity is dominated by the $O(mn + n^2 \log n)$ of pre-computing pair-wise distances.

3   Approximation Algorithm for Trees

In this section, we show that there is an algorithm with an approximation ratio of at most 2 for burning a tree T. In a way analogous to general graphs, the algorithm is based on a procedure Burn-Guess-Tree(T, g) that guarantees the following for a given guess value g.

1. If the algorithm returns Bad-Guess, then any schedule for burning T requires at least g rounds to complete.
2. If the algorithm returns a schedule for burning T, then that schedule completes in at most $2g$ rounds.

It is evident that, provided with the above guarantees, the schedule returned for the smallest value of g completes within twice the optimal schedule.

Given an input tree T, Burn-Guess-Tree(T, g) selects an arbitrary node s as the *root* of the tree. The *level* of a node v is the distance of v to s and the *k-ancestor* of v is the vertex at distance k from v on its path to the root. The procedure works in a number of steps and maintains a set of centers as well as a set of marked vertices. Initially, the set of centers is empty, and all vertices are unmarked. At the beginning of a step i (where $i \leq g$), the algorithm finds an unmarked vertex v with the highest level. If the level of v is at least g, then the g-ancestor of v is added to the set of centers; otherwise, the root of T is added to the set of centers. Meanwhile, all vertices within distance g of the added center are marked. The procedure continues until all vertices are marked. In this case, the algorithm returns a burning sequence defined by an arbitrary ordering of centers as activators. If the number of centers becomes larger than g before all vertices are marked, then the algorithm returns Bad-Guess. Figure 3 illustrates the Burn-Guess-Tree procedure.

Lemma 4. *If Burn-Guess-Tree(T, g) returns a burning sequence, then the burning of that sequence completes in at most $2g$ rounds.*

Proof. Since all vertices are marked, they are all within distance g of a center. In the returned schedule, a fire is activated at all centers by round g, and consequently, all vertices are burned by round $2g$.

Define a 'g-site partition' as a set of at most g vertices, called g 'sites', so that every vertex is within distance g of its closest site. We say that the tree admits the 'g-site condition' if it has a g-site partition. Clearly, in order to burn a tree in less than g rounds, the tree should pass the g-site condition; otherwise, any set of at most g activators leaves a vertex outside of combination of the spheres of all the activators and hence, the burning process cannot complete within g rounds.

Lemma 5. *If Burn-Guess-Tree(T, g) returns Bad-Guess, then T does not admit g-site condition.*

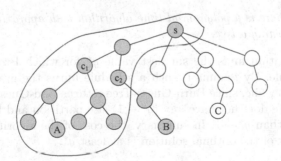

Fig. 3. An illustration of Burn-Guess-Tree with parameter $g = 2$. The tree is rooted at s. The first selected vertex is A and the first center is c_1. The next unmarked vertex with maximum level is B and c_2 is selected as the next center. At this point, since there are still unmarked vertices (nodes that are not highlighted), the algorithm returns Bad-Guess. In the next iteration with $g = 3$, the algorithm returns a schedule formed by the parent of c_1 and s.

Proof. Consider otherwise; that is, there is a g-site partitioning P defined by at most g sites so that every vertex in the tree is within distance g of one of the sites in P. We show that it is possible to update P so that it is still a g-site partitioning while having the set of g activators selected by Burn-Guess-Tree(T, g) as its set of sites. If that it is true, then every vertex is within distance g of the g centers selected by the algorithm. However, we know at the time Bad-Guess was returned, there was an unmarked node at distance more than g of the closest center (contradiction).

Let $c_1, c_2, \ldots, c_g$ be the centers selected by Burn-Guess-Tree (in the same order they are selected). We iteratively update P by including these centers in its set of sites. At the beginning of iteration i, P has $c_1, \ldots, c_{i-1}$ in it set of sites. In Burn-Guess-Tree, vertices at distance g of these centers are marked. Let v be the unmarked vertex with maximum distance from a marked node; hence, following the definition of the Burn-Guess-Tree, c_i is the g-ancestor of v. Since P is a g-site partitioning, it should have a site c' within distance g of v. Note that such site cannot be any of c_j with $j < i$ since v is unmarked. We argue that if c' is replaced with c_i in S, the partitioning still remains a g-site partitioning. For that, we show unmarked vertices within distance g of c' form a subset of unmarked vertices within distance g of c_i. Consider otherwise; that is, there is an unmarked vertex w at distance more than g of c_i and within distance g of c'. If w is in the tree rooted at c_i, then level of w will be more than v that contradicts v being the unmarked vertex with the highest level. Next assume w is outside of the tree rooted at c_i. Since w has distance more than g to c_i and is within distance g of c', we conclude that c' should be outside of the tree rooted at c_i; this contradicts v being within distance g of c' (since c_i is at distance g of v). To summarize, after replacing c' with c_i in P, the partitioning remains a g-site partitioning. After repeating this process g times, P will be a g-site partitioning formed by the g centers selected by Burn-Guess-Tree that is a contradiction as mentioned above.

Theorem 2. *There is a polynomial time algorithm with approximation ratio of at most 2 for burning a tree.*

Proof. The algorithm finds the smallest value g^* for which Broad-Guess-Tree returns a schedule. By Lemma 4, such a schedule burns the graph in at most $2g^*$ rounds. Meanwhile, since Burn-Guess-Tree returns Bad-Guess for $g^* - 1$, by Lemma 5, the tree does not have any $(g^* - 1)$-site partition and hence, the cost of OPT is more than $g^* - 1$. In summary, the cost of the algorithm is at most $2g^*$, and the cost of the optimal solution is at least g^*.

4   Algorithms for Disjoint Paths

Consider the burning problem when the input is a disjoint forest of paths. This problem is NP-hard and a 1.5 approximation exists for it [7]. In this section, we present exact and approximation algorithms for this problem. When the graph is formed by a constant number of paths, we provide an exact algorithm that runs in polynomial time. For the more interesting case when the graph is formed by a non-constant number of paths, we provide two approximation schemes for the problem. Throughout the section, we assume the input is a graph of size n formed by b paths of length $n_1, n_2, \ldots, n_b$, where $n_i \leq n_{i+1}$.

Constant Number of Disjoint Paths. We show that when the number of disjoint paths is constant, there is a polynomial time algorithm which provides optimal solution. We call a graph G an r-*subset of graph* G' if both G and G' are formed by b disjoint paths of the same lengths, except for one path p which has length x in G and length $x + i$ in G' for some i in the range $[0, 2r + 1]$.

Lemma 6. *A graph G' formed by a forest of disjoint paths can be burned in t rounds if and only if it has a t-subset G which can be burned in $t - 1$ rounds.*

Proof. Assume that G' can be burned in t rounds. Remove all vertices burned through the fire started at round 1. There will be at most $2t + 1$ such vertices. Removing them will form a t-subset of G and the same schedule can be used to burn that subgraph in $t - 1$ rounds. Next, assume G' has a t-subset G which can be burned in $t - 1$ rounds. To burn G', we use the same schedule for burning G except that at round 1 a fire is started at a node at distance t of one endpoint of path p which differentiates the two graphs. By round t, $2t + 1$ vertices at distance t of that node are burned. The remaining vertices that form G can be burned in t rounds following the same burning schedule for G.

The above lemma helps us devise a straightforward dynamic programming solution. We fill a table of size polynomial to n (size of the graph) which has a boolean entry for each graph formed by b paths of total size at most n and for each deadline value τ (which is at most n). Such entry indicates whether the graph can be burned in τ rounds. Using Lemma 6, we can fill the table in a bottom-up approach. Additional bookkeeping when filling the table leads us to the optimal burning scheme.

Theorem 3. *Given a graph of size n formed by a forest of $b = \Theta(1)$ disjoint paths, there is an algorithm that generates an optimal burning scheme. The time complexity of the algorithm is polynomial in n.*

Proof. Consider a dynamic-programming table A of dimension $b + 1$. Here, $A[t][x_1][x_2] \ldots [x_b]$ is a boolean value that indicates whether it is possible to burn a forest of b paths within t rounds, where the first path of the forest has length x_1, the second path has length x_2, and so on. In other words, an entry in the table is associated with a *deadline* time t (first dimension) and a graph formed by b disjoint paths (subsequent b dimensions). Note that the first dimension takes values between 1 and n (the upper bound for burning time), while any other dimension takes values between 1 and n_b, where n_b is the maximum length of any path. Consequently, the size of the table is $O(n^{b+1})$, which is polynomial in n for constant values of b. To find the optimal burning time, after filling the table, find the smallest t^* for which $A[t^*][n_1][n_2] \ldots [n_b]$ is True; recall that

$n_1, \ldots, n_b$ are the lengths of paths in the input forest. Additional bookkeeping when filling the table leads us to the optimal burning scheme which completes in t^* rounds.

Next, we describe how to fill the table. Assume that the table is processed (filled) for values up to $t-1$ for the first dimension; that is, the entries for graphs which can be burned within deadline $t - 1$ are set to True. By Lemma 6, graphs with True entries for deadline t have a t-subset with True entry for deadline $t - 1$. Hence, for any entry with True value associated with deadline $t - 1$ and graph G', we set all entries associated with deadline t and graphs having G' as a t-subset to be True. In other words, if entry $A[t - 1][x_1][x_2] \ldots [x_b]$ is true, then for any $j \le b$, the entry $A[t][x_1][x_2] \ldots [x_j + i] \ldots x[b]$ will be set to True for $i \le 2t - 1$. In doing so, we also record the value of j. In this fashion, we can retrieve a burning schedule by looking at the index of the path at which a fire is started in each given round.

FPTAS for Non-constant Number of Regular Disjoint Paths. In this section, we use a reduction to the *bin covering problem* to show the burning problem admits a *fully* polynomial time approximation scheme (FPTAS) when the input graph is formed by a non-constant number of 'regular' disjoint paths. Here, we assume the paths are regular in the sense that the lengths of all paths are asymptotically equal. The bin covering problem is the dual to the classic bin packing problem and can be defined as follows.

Definition 1. *The input to the bin covering problem is a multi-set of items with sizes in the range (0,1]. The goal is to 'cover' a maximum number of size 1 with these items. By covering a bin, we mean assigning a multiset of items with total size at least 1 to the bin.*

Bin covering is NP-hard [17]; but there is an FPTAS for the problem:

Lemma 7 [23]. *There is an algorithm A that, given a multiset L of n items with sizes $s(a_i) \in (0, 1]$ and a positive number $\epsilon_0 > 0$, produces a bin covering of L such that $A(L) \ge (1 - \epsilon_0)OPT(L)$, assuming $OPT(L)$ is sufficiently large. The time complexity of A is polynomial in n and $1/\epsilon_0$.*

To provide an FPTAS for the burning problem on regular disjoint paths, we reduce the problem to the bin covering problem. Before presenting the reduction, we state two lemmas with respect to the bin covering problem:

Lemma 8. *Assume that two bins B_1 and B_2 are covered with a multiset of items so that B_1 only includes items of sizes at most 1/3 and B_2 includes two items of size at least 2/3. It is possible to modify the covering so that each bin has an item of size at least 2/3 and both bins are still covered.*

Proof. Since all items in B_1 have size at most 1/3, it is possible to select a subset S of these items which has total size in $(1/3, 2/3]$ (start with an empty S and repeatedly add items until the total size is in the desired range). Move items of S from B_1 to B_2 and move an item of size at least 2/3 from B_2 to B_1. Both bins will be covered in the result and each contain an item of size at least 2/3.

Lemma 9. *If we remove a multiset of total size $x \geq 2$ from an instance of bin covering, then the number of covered bins in the optimal packing reduces by at least $\lfloor x/2 \rfloor$.*

Proof. If we remove a multiset of items with total size at least 1, then the number of covered bins decreases by at least 1. Otherwise, if removing a set of items with total size at least 1 does not reduce the number of covered bins, then these items can cover a new bin without impacting coverage of other bins. This contradicts the optimality assumption for the covering. Given a multiset of total size x, we can partition it into $\lfloor x/2 \rfloor$ multisets, all having size at least 1; this is possible because all items have size at most 1. Repeating the above argument $\lfloor x/2 \rfloor$ times completes the proof.

Consider an instance I of the graph burning problem formed by b paths $P_1, \ldots, P_b$ of lengths $n_1, n_2, \ldots, n_b$ such that $n_i \leq n_{i+1}$. Let $m_i = \lceil (n_i + 1)/2 \rceil$ and $C = 3m_b$. We define the k-*instance of bin covering associated with I* as an instance of bin covering formed by b 'large' items $\{p_1, \ldots, p_b\}$, where p_i has size $1 - m_i/C$ for $1 \leq i \leq b$. We also define k 'small' items $\{q_1, \ldots, q_k\}$, where q_j has size $\min\{j/C, 1/3\}$ for $1 \leq j \leq k$. Note that all large items have size at least $2/3$ and small items have size at most $1/3$. Also note that large items appear in same way for any value of k in the k-instances of the bin covering problem. Figure 4 illustrates this construction. Since paths are regular, the size large items is upper-bounded by a constant $c^* = 1 - m_1/(3m_b)$ which we refer to as the *canonical constant* of the graph burning instance. Intuitively, burning the b disjoint paths is translated to covering b bins. By Lemma 8, the b large items can be placed in distinct bins without changing the number of covered bins. The remaining space of bins (to be covered) translates to paths of different length that should be burned. Small items are associated with the radii of the fires started at different rounds. These intuitions are formalized in the following two lemmas:

Lemma 10. *Given a solution for the k-instance of bin covering that covers at least b bins, one can find, in polynomial time, a burning scheme that completes in at most k rounds.*

Proof. Given the solution for bin covering, we apply Lemma 8 to ensure that there are b bins that each include exactly one large item (this is possible because large items have size at least $2/3$ and small items have size at most $1/3$). Call the resulting bins $B_1, \ldots, B_b$, where B_i is the bin that includes the large item p_i. Let S_i be the set of small items in B_i. We associate items in S_i with activators in a burning schedule. Assume that initially all vertices are unmarked. We process small items in the solution for bin covering in the following manner. If q_j ($1 \leq j \leq k$) appears in set S_i in the covering solution, then at time $k - j$ we start a fire at distance j of the left-most marked node in path P_i and mark any node at distance j of it. In this way, by the end of round k all marked nodes will be burned. Since the total size of items in S_i is at least m_i/C, the number of marked vertices by the time k would be $2m_i + 1 \geq n_i$; that is, all vertices will be burned by the end of round k.

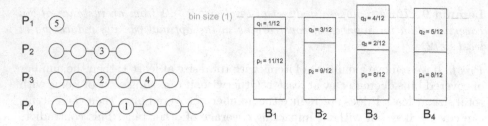

Fig. 4. A burning scheme for an instance of the burning problem on disjoint paths (left) and the equivalent covering for the 5-instance of the covering problem (right). Here, we have $m_1 = 1, m_2 = 3, m_3 = 4$, and $m_4 = 4$.

Lemma 11. *Given a burning scheme that completes within $k - 1$ rounds, it is possible to create a solution for the k-instance of bin covering, in polynomial time, so that at least b bins are covered in the solution.*

Proof. We say an edge is burned if its both endpoints are burned. Since the burning scheme completes in at most $k - 1$ rounds, we can burn all edges within k rounds even if the lengths of all paths is increased by 2.

Consider a path P_i of length n_i. Assume that the burning schedule starts fires at rounds $k - y_1, k - y_2, \ldots, k - y_t$ in P_i. Note that a fire started at round x burns at most $2(k - x)$ edges within k rounds. Hence, if Y denotes the total sum of y_j's, then at most $2Y$ edges are burned by round k. Since all edges can be burned within k rounds even in a longer path of $n_i + 2$ vertices, we have that $2Y \geq n_i + 1$; that is $Y \geq m_i$.

We create a solution for the covering problem as follows. Place the large items in separate bins, and let B_i be the large bin at which p_i is placed. Recall that fires at the path P_i are started at rounds $k - y_1, \ldots, k - y_t$. Consider the set $Q_i = \{\min\{y_1/C, 1/3\}, \ldots, \min\{y_t/C, 1/3\}\}$, which is a subset of small items in the covering instance. We place items in Q_i in the bin that contains large item p_i. Next, we show the total size of items in the bin B_i is at least 1. First, note that if any item in Q_i has size 1/3, since p_i has size at least 2/3, the total size of these two items will be 1 and we are done. Next, assume all items in S are smaller than 1/3; that is, $Q_i = \{y_1/C, \ldots, y_t/C\}$. The total size of items in Q_i is equal to Y/C which is at least m_i/C. Hence, the total size of items in the bin will be at least m_i/C (for small items) plus $1 - m_i/C$ (for the large item b_i) which sums to at least 1. In summary, for any $i \leq k$, if we can burn edges in path P_i within k rounds, we can cover the bin B_i with small items associated with the rounds at which P_i is burned.

We repeatedly apply the FPTAS of Lemma 7 (with a carefully chosen value of ϵ_0) to find the smallest k such that, for the k-instance of bin covering, the FPTAS returns a solution that covers at least b bins. By Lemma 10, such solution can be converted to a burning scheme. Using Lemmas 9 and 11, we can show that this solution achieves approximation ratio of $1 + \epsilon$ while running in time polynomial in both n and $1/\epsilon$. More formally, we can prove the following theorem:

Theorem 4. *Given a graph of size n formed by a forest of $b = \omega(1)$ regular disjoint paths and a positive value ϵ, there is an algorithm that generates a burning scheme that completes within a factor $1 + \epsilon$ of an optimal scheme. The time complexity of the algorithm is polynomial in both n and $1/\epsilon$.*

Proof. Define $\epsilon_0 = \frac{(1-c^*)\epsilon}{4+(5-c^*)\epsilon}$ (recall that c^* is the canonical constant of the regular instance of the burning problem). Find the smallest k such that for the k-instance of bin covering, the FPTAS of Lemma 7 with parameter ϵ_0 returns a solution that covers at least b bins. Let that value of k be k^*. By Lemma 10, that solution can be converted, in polynomial time, to a burning scheme that completes in k^* rounds. Note that the total size of small items in the k^*-instance is $k^*(k^*+1)/(2C)$, and the total size of large items is at most bc^*. Since b bins are covered, we conclude that $k^*(k^*+1)/(2C) \geq b - bc^*$; that is $\frac{bC}{k^*(k^*+1)} \leq \frac{1}{2(1-c^*)}$. This implies that for large values of k we have $\frac{bC}{k'^2} < \frac{1}{1-c^*}$ where $k' = k^* - 1$ (we refer to this fact later).

Next, we provide a lower bound for the cost of OPT. Since k^* is the smallest value for which the FPTAS failed to cover b bins in the k'-instance of bin covering, by Lemma 7, an optimal covering algorithm OPT cannot cover more than $b/(1 - \epsilon_0)$ bins in the k'-instance. Let $\epsilon_1 = \epsilon_0/(1 - \epsilon_0)$. Thus, OPT cannot cover more than $b(1 + \epsilon_1)$ bins in the k'-instance. Let $\alpha = (1 - \epsilon_2)k'$, where $\epsilon_2 = \frac{4}{1-c^*}\epsilon_1$. We claim that OPT cannot cover b bins in the α-instance of the bin covering problem. If this claim is true, then there is no burning scheme that completes within $\alpha - 1$ rounds; otherwise, by Lemma 11 that burning scheme yields to a covering solution that covers b bins of the α-instance of the bin covering problem. In summary, we will have a burning scheme that completes in k^* rounds while an optimal burning algorithm requires $\alpha - 1$ rounds to burn the graph. This gives an approximate ratio of $k^*/(\alpha - 1)$ which approaches to $\frac{1}{1-\epsilon_2} = 1 + \frac{4\epsilon_0}{(1 - \epsilon_0)(1 - c^*) - 4\epsilon_0} = 1 + \epsilon$ for large values of k^*. Note that, since k^* is lower-bounded by the number of paths, we have $k^* \in \omega(1)$.

It remains to show that an optimal covering algorithm cannot cover b bins in the α-instance of bin covering. Note that the α-instance is similar to the k'-instance except that, among the small items, the $\epsilon_2 k'$ largest items are missing. Call these items *critical items*. We claim that the total size of critical items, denoted by X, is more than $2\epsilon_1 b$. For now assume it is true; by Lemma 9, removing items with total size at least X decreases the number of covered bins in an optimal solution of the k'-instance by at least $\lfloor X/2 \rfloor$. Thus, if it is possible to cover b bins in the α-instance then it is possible to cover at least $b + \epsilon_1 b$ bins in the k'-instance, which we know is not possible. We conclude that we cannot cover b bins in the α-instance. We are just left to show $X > 2\epsilon_1 b$. We have $X = k'^2(2\epsilon_2 - \epsilon_2^2)/2C + k'\epsilon_2/2C > k'^2\epsilon_2/2C$. Therefore, it suffices to have $k'^2\epsilon_2/2C > 2\epsilon_1 b$; that is, $\epsilon_2 > \frac{4bc}{k'^2}\epsilon_1$. We previously observed that $\frac{bc}{k'^2} < \frac{1}{1-c^*}$. Therefore, the inequality holds as long as $\epsilon_2 \geq \frac{4}{1-c^*}\epsilon_1$.

PTAS for the General Case of Non-constant Disjoint Paths. In this section, we use a direct approach to provide a PTAS for graphs formed by non-constant number of disjoint paths. Unlike previous section, we do not make any assumption on the length of the paths, in particular, the length of paths can be asymptotically larger than the number of paths. We note that when regularity condition holds, the result of the previous section is stronger as the provided algorithm is *fully* polynomial.

Assume that the graph is formed by b disjoint paths, each of length at most n. An instance of the decision variant of the burning problems has a parameter g and asks whether it is possible to burn the graph with fires started at times $1, 2, \ldots, g$. We define the *radius* of a fire started at round t as $g - t + 1$; so an instance $I(g)$ of the decision problem asks whether it is possible to burn the graph with fires of radii $1, 2, \ldots, g$. Given a constant integer k, we form at most $k + 1$ groups of fires, each containing fires of close radii such that the difference in the radii of any two fire in a group is at most $\beta = \lfloor g/k \rfloor$ (there will be β fires in each group, except potentially the last one). Based on this grouping, we define two new instances of the decision problem: in the *weak instance* $I'(g, k)$, the fires of the first group (with smallest radii) are removed and the radii of fires of other groups is rounded to the smallest radius in the group. In the *strong instance* $I''(g, k)$ the radii are rounded up to the largest radius in the group. Note that in both weak and strong instances, there are $k + 1$ radius sizes, and each fire has radius at least $\lfloor g/k \rfloor + 1$. In addition, note that if we remove the β fires of largest radii from the strong instance $I''(g, k)$, the result will be the weak instance $I'(g, k)$. We prove the following lemma, which will be later applied on the weak instances of the problem.

Lemma 12. *Consider an instance of the burning problem on disjoint paths in which there are g fires each having a radius among $k + 1$ possible radii for some constant k so that each radius is in the range $(\lfloor g/k \rfloor, g]$. There is an algorithm that answers the problem with the following guarantees. If the answer is 'yes', then it is possible to burn the graph with the fires in the instances. If the answer is 'no', then there is a number $p \in o(g)$ so that it is not possible to burn the graph when the p fires of largest radii are removed.*

Proof. Assume that there are n vertices in the graph. We divide the paths in the graph into short paths with length $O(g)$ and long paths with length $\omega(g)$. If the number of long paths is $\Omega(g)$, the algorithms sends 'no': there are $\omega(g^2)$ vertices in the graph while the maximum number of vertices that can be burned with the instance is $O(g^2)$. Next, assume the number of long paths is p where $p \in o(g)$. Further, assume the number of paths is at most g; otherwise, the algorithm returns 'no' as there is no way to burn more than g disjoint paths with g fires. In order to achieve the desired guarantees, we exhaustively check all possible burning schemes for short paths and use a simple strategy to burn long paths. Assume that all short paths have length at most αg for some constant α. Since all fires have radius more than g/k, it suffices to use at most $\lceil \alpha k/2 \rceil$ fires to burn each path. Hence, for each path, we have at most $\lceil \alpha k/2 \rceil$ fires each having one of the k possible radii. There are $\tau = \binom{\lceil \alpha k/2 \rceil + k}{k}$ ways to assign fires to each path; define each such assignment a 'fire schedule' for a path. Note that τ is a constant. There are at most g short paths each taking one of the possible τ fire schedules. It follows that there are $\binom{g+\tau}{\tau}$ ways to assign fire schedules to these paths; this value is polynomial in g as τ is constant. We conclude that there is a polynomial number of possible burning schedules for short paths. For each such schedule for the short paths, we complete the burning by using the fires absent in the schedule to burn long paths. We process these fires in an arbitrary order and assign them one by one to the long paths. A fire of radius r burns up to $2r - 1$ vertices. When this fire is assigned to a path, $2r - 1$ vertices in the path are declared 'burned' and the process continues until all vertices in the path are burned, after which the fires are assigned to burn the next path. This process continues until all paths are burned, in which case the algorithm returns 'yes'. If we run out of fires and not all paths are burned, the burning schedule for the short paths is not useful and the process continues by checking the next schedule for short paths. If all schedules for short paths are checked and for all of them we fail to burn long paths with the remaining fires, the algorithm returns 'no'.

Next, we show the algorithm provides the desired guarantees. First, if the algorithm returns 'yes', then there has been a schedule to burn short paths and the remaining fires have successfully burned the long paths. Hence, there is a schedule for fires in the instance that burns the whole graph. Next, assume the algorithm returns 'no'; we claim no algorithm can burn the graph using the same fires when the p fires with the largest radii are removed (recall that $p \in o(g)$ is the number of long paths). Consider otherwise, that is, assume it is possible to burn the graphs with the mentioned fires. The burning schedule for assigning fires to short paths in such solution S is checked also by the algorithm. The difference is that the algorithm assigns fires to long paths differently from S. Since the algorithm returns 'no', it fails to cover all paths with fires. Hence, if we remove the last fire assigned to each path by the algorithm, the number of vertices that can be burned by the remaining fires will be less than the total size of long paths. Consequently, if we remove the p fires with the largest radii, the remaining fires do not suffice to burn the long paths. In summary, if we assign fires to short paths in the same way that S does and remove the largest p fires from the rest of fires, the remaining fires cannot burn long paths. This contradicts our assumption that S can burn all graphs with the same fires.

Theorem 5. *Given a graph of size n formed by a forest of $b = \omega(1)$ disjoint paths and a positive value ϵ, there is an algorithm that generates a burning scheme that completes within a factor $1 + \epsilon$ of an optimal scheme. The time complexity of the algorithm is polynomial in n.*

Proof. Let $k = \lceil 1/\epsilon \rceil + 1$. We exhaustively apply Lemma 12 to find the smallest value of g so that the algorithm of the lemma returns 'yes' for the weak instance $I'(g, k)$ of the problem. Since the graph can be burned with such weak instance, it can be burned with the actual instance formed by fires of radii $(1, 2, \ldots, g)$ (this only involves increasing the radii of fires in the solution provided by the weak instance). Hence, we can burn the graph in g rounds. Next we provide a lower bound for Opt.

Since the algorithm returns 'no' for the weak instance $I'(g - 1, k)$, by Lemma 12, it is not possible to burn the graph with fires in the weak-instance in which $p \in o(g)$ largest fires are removed for some value of p. Recall that the weak instance $I'(g - 1, k)$ is similar to the strong instance $I''(g - 1, k)$ in which $\beta = \lfloor (g - 1)/k \rfloor$ fires of largest radii are removed. We conclude that, the strong instance $I''(g-1, k)$ in which $\beta - o(g)$ fires with largest radii are removed cannot burn the graph. Meanwhile, such strong instance is similar to the regular instance formed by fires of radii $1, 2, \ldots, g - 1 - \beta - o(g)$ in which some fires radii is increased. We conclude that it is not possible to burn the graph in $g - 1 - \beta - o(g)$ rounds. This implies $\mathrm{OPT} \geq g(1 - 1/k) - o(g)$. The ratio between the cost of the algorithm and OPT approaches to $\frac{g}{g(1-1/k)} = 1 + 1/(k - 1)$, which is at most $1 + \epsilon$.

5   Concluding Remarks

For general graphs, we provided an approximation algorithm with constant factor of 3. This result shows the burning problem is different from problems such as the Firefighter problem that do not admit constant approximations. The approximation factor is likely to be improved. However, such improvement requires a different (and more involved) argument that improves the lower bounds of Lemma 1 for the cost of OPT.

It is not clear whether the burning problem admits a PTAS or is APX-hard for general graphs. A potential APX-hardness proof requires an approach different from the current reductions which are confined to input graphs that are forests of paths. Recall that we showed there is a PTAS for these instances. As the existing negative results are confined to sparse, disconnected graphs, and since a PTAS exists for disjoint forests of paths, it might be possible that a PTAS exists for general graphs. We note that the hardness results concerning similar problems such as k-center and dominating set problems cannot be applied to show APX-hardness of the burning problem.

References

1. Anshelevich, E., Chakrabarty, D., Hate, A., Swamy, C.: Approximability of the firefighter problem - computing cuts over time. Algorithmica **62**(1–2), 520–536 (2012)
2. Bessy, S., Bonato, A., Janssen, J.C.M., Rautenbach, D., Roshanbin, E.: Burning a graph is hard. Discret. Appl. Math. **232**, 73–87 (2017)
3. Bessy, S., Bonato, A., Janssen, J.C.M., Rautenbach, D., Roshanbin, E.: Bounds on the burning number. Discret. Appl. Math. **235**, 16–22 (2018)
4. Bonato, A., Gunderson, K., Shaw, A.: Burning the plane: densities of the infinite cartesian grid. Preprint (2019)
5. Bonato, A., Janssen, J., Roshanbin, E.: Burning a graph as a model of social contagion. In: Bonato, A., Graham, F.C., Prałat, P. (eds.) WAW 2014. LNCS, vol. 8882, pp. 13–22. Springer, Cham (2014). https://doi.org/10.1007/978-3-319-13123-8_2

6. Bonato, A., Janssen, J., Roshanbin, E.: How to burn a graph. Internet Math. **12**(1–2), 85–100 (2016)
7. Bonato, A., Lidbetter, T.: Bounds on the burning numbers of spiders and path-forests. ArXiv e-prints, July 2017
8. Bond, R.M., et al.: A 61-million-person experiment in social influence and political mobilization. Nature **489**(7415), 295–298 (2012)
9. Cai, L., Verbin, E., Yang, L.: Firefighting on trees: $(1-1/e)$-approximation, fixed parameter tractability and a subexponential algorithm. In: Hong, S.-H., Nagamochi, H., Fukunaga, T. (eds.) ISAAC 2008. LNCS, vol. 5369, pp. 258–269. Springer, Heidelberg (2008). https://doi.org/10.1007/978-3-540-92182-0_25
10. Chen, N., Gravin, N., Lu, P.: On the approximability of budget feasible mechanisms. In: Proceedings of Annual ACM-SIAM Symposium on Discrete Algorithms SODA, pp. 685–699 (2011)
11. Chen, W., et al.: Influence maximization in social networks when negative opinions may emerge and propagate. In: Proceedings of SIAM International Conference on Data Mining, SDM, pp. 379–390 (2011)
12. Chen, W., Wang, Y., Yang, S.: Efficient influence maximization in social networks. In: Proceedings of ACM International Conference on Knowledge Discovery and Data Mining (SIGKDD), pp. 199–208 (2009)
13. Czumaj, A., Rytter, W.: Broadcasting algorithms in radio networks with unknown topology. J. Algorithms **60**(2), 115–143 (2006)
14. Domingos, P.M., Richardson, M.: Mining the network value of customers. In: Proceedings of ACM International Conference on Knowledge Discovery and Data Mining (SIGKDD), pp. 57–66 (2001)
15. Elkin, M., Kortsarz, G.: Sublogarithmic approximation for telephone multicast. J. Comput. Syst. Sci. **72**(4), 648–659 (2006)
16. Fajardo, D., Gardner, L.M.: Inferring contagion patterns in social contact networks with limited infection data. Netw. Spat. Econ. **13**(4), 399–426 (2013)
17. Assmann, S.F.: Problems in discrete applied mathematics. Ph.D. thesis, MIT (1983)
18. Finbow, S., King, A.D., MacGillivray, G., Rizzi, R.: The firefighter problem for graphs of maximum degree three. Discret. Math. **307**(16), 2094–2105 (2007)
19. Fitzpatrick, S.L., Li, Q.: Firefighting on trees: how bad is the greedy algorithm? Congr. Numer. **145**, 187–192 (2000)
20. Garey, M.R., Johnson, D.S.: Computers and Intractability: A Guide to the Theory of NP-Completeness. W. H. Freeman, Stuttgart (1979)
21. Ghaffari, M., Haeupler, B., Khabbazian, M.: Randomized broadcast in radio networks with collision detection. Distrib. Comput. **28**(6), 407–422 (2015)
22. Hedetniemi, S.M., Hedetniemi, S.T., Liestman, A.L.: A survey of gossiping and broadcasting in communication networks. Networks **18**(4), 319–349 (1988)
23. Jansen, K., Solis-Oba, R.: An asymptotic fully polynomial time approximation scheme for bin covering. Theor. Comput. Sci. **306**(1–3), 543–551 (2003)
24. Kempe, D., Kleinberg, J.M., Tardos, E: Maximizing the spread of influence through a social network. In: Proceedings of the International Conference on Knowledge Discovery and Data Mining (SIGKDD), pp. 137–146 (2003)
25. Kempe, D., Kleinberg, J., Tardos, É.: Influential nodes in a diffusion model for social networks. In: Caires, L., Italiano, G.F., Monteiro, L., Palamidessi, C., Yung, M. (eds.) ICALP 2005. LNCS, vol. 3580, pp. 1127–1138. Springer, Heidelberg (2005). https://doi.org/10.1007/11523468_91
26. Kempe, D., Kleinberg, J.M., Tardos, É.: Maximizing the spread of influence through a social network. Theory Comput. **11**, 105–147 (2015)

27. Kleinberg, J.M.: Cascading behavior in social and economic networks. In: Proceedings of ACM Conference on Electronic Commerce (EC), pp. 1–4 (2013)
28. Kowalski, D.R., Pelc, A.: Optimal deterministic broadcasting in known topology radio networks. Distrib. Comput. **19**(3), 185–195 (2007)
29. Kramer, A.D.I.: The spread of emotion via Facebook. In: CHI Conference on Human Factors in Computing Systems, (CHI), pp. 767–770 (2012)
30. Kramer, A.D.I., Guillory, J.E., Hancock, J.T.: Experimental evidence of massive-scale emotional contagion through social networks. In: Proceedings of the National Academy of Sciences, pp. 8788–8790 (2014)
31. Land, M.R., Lu, L.: An upper bound on the burning number of graphs. In: Bonato, A., Graham, F.C., Prałat, P. (eds.) WAW 2016. LNCS, vol. 10088, pp. 1–8. Springer, Cham (2016). https://doi.org/10.1007/978-3-319-49787-7_1
32. Mitsche, D., Pralat, P., Roshanbin, E.: Burning graphs: a probabilistic perspective. Graphs Comb. **33**(2), 449–471 (2017)
33. Mitsche, D., Pralat, P., Roshanbin, E.: Burning number of graph products. Theor. Comput. Sci. **746**, 124–135 (2018)
34. Nikzad, A., Ravi, R.: Sending secrets swiftly: approximation algorithms for generalized multicast problems. In: Esparza, J., Fraigniaud, P., Husfeldt, T., Koutsoupias, E. (eds.) ICALP 2014. LNCS, vol. 8573, pp. 568–607. Springer, Heidelberg (2014). https://doi.org/10.1007/978-3-662-43951-7_48
35. Peleg, D.: Time-efficient broadcasting in radio networks: a review. In: Janowski, T., Mohanty, H. (eds.) ICDCIT 2007. LNCS, vol. 4882, pp. 1–18. Springer, Heidelberg (2007). https://doi.org/10.1007/978-3-540-77115-9_1
36. Ravi, R.: Rapid rumor ramification: approximating the minimum broadcast time (extended abstract). In: Proceedings of Symposium on Foundations of Computer Science (FOCS), pp. 202–213 (1994)
37. Richardson, M., Domingos, P.M.: Mining knowledge-sharing sites for viral marketing. In: Proceedings of the ACM International Conference on Knowledge Discovery and Data Mining (SIGKDD), pp. 61–70 (2002)
38. Schindelhauer, C.: On the inapproximability of broadcasting time. In: Jansen, K., Khuller, S. (eds.) APPROX 2000. LNCS, vol. 1913, pp. 226–237. Springer, Heidelberg (2000). https://doi.org/10.1007/3-540-44436-X_23
39. Sim, K.A., Tan, T.S., Wong, K.B.: On the burning number of generalized petersen graphs. Bull. Malays. Math. Sci. Soc. **6**, 1–14 (2017)
40. Slater, P.J., Cockayne, E.J., Hedetniemi, S.T.: Information dissemination in trees. SIAM J. Comput. **10**(4), 692–701 (1981)
41. Vazirani, V.V.: Approximation Algorithms. Springer, Heidelberg (2001). https://doi.org/10.1007/978-3-662-04565-7

Sublinear Decoding Schemes
for Non-adaptive Group Testing
with Inhibitors

Thach V. Bui[1(✉)], Minoru Kuribayashi[3], Tetsuya Kojima[4], and Isao Echizen[1,2]

[1] SOKENDAI (The Graduate University for Advanced Studies),
Hayama, Kanagawa, Japan
[2] National Institute of Informatics, Tokyo, Japan
{bvthach,iechizen}@nii.ac.jp
[3] Okayama University, Okayama, Japan
kminoru@okayama-u.ac.jp
[4] National Institute of Technology, Tokyo College, Hachioji, Japan
kojt@tokyo-ct.ac.jp

Abstract. Identification of up to d defective items and up to h inhibitors in a set of n items is the main task of non-adaptive group testing with inhibitors. To reduce the cost of this Herculean task, a subset of the n items is formed and then tested. This is called *group testing*. A test outcome on a subset of items is positive if the subset contains at least one defective item and no inhibitors, and negative otherwise. We present two decoding schemes for efficiently identifying the defective items and the inhibitors in the presence of e erroneous outcomes in time $\mathsf{poly}(d, h, e, \log_2 n)$, which is sublinear to the number of items. This decoding complexity significantly improves the state-of-the-art schemes in which the decoding time is linear to the number of items, i.e., $\mathsf{poly}(d, h, e, n)$. Moreover, each column of the measurement matrices associated with the proposed schemes can be nonrandomly generated in polynomial order of the number of rows. As a result, one can save space for storing them. Simulation results confirm our theoretical analysis. When the number of items is sufficiently large, the decoding time in our proposed scheme is smallest in comparison with existing work. In addition, when some erroneous outcomes are allowed, the number of tests in the proposed scheme is often smaller than the number of tests in existing work.

Keywords: Non-adaptive group testing · Sublinear algorithm · Sparse recovery

1 Introduction

Group testing was proposed by an economist, Robert Dorfman, who tried to solve the problem of identifying which draftees had syphilis [1] in WWII. Nowaday, it is known as a problem of finding up to d defective items in a colossal number

© Springer Nature Switzerland AG 2019
T. V. Gopal and J. Watada (Eds.): TAMC 2019, LNCS 11436, pp. 93–113, 2019.
https://doi.org/10.1007/978-3-030-14812-6_7

of items n by testing t subsets of n items. It can also be translated into the classification of up to d defective items and at least $n - d$ negative items in a set of n items. The meanings of "items", "defective items", and "tests" depend on the context. Normally, a test on a subset of items (a test for short) is positive if the subset has at least one defective item, and negative otherwise. For testing design, there are two main approaches: adaptive and non-adaptive designs. In *adaptive group testing*, the design of a test depends on the earlier tests. With this approach, the number of tests can be theoretically optimized [2]. However, it would take a long time to proceed such sequential tests. Therefore, *non-adaptive group testing* (NAGT) [2,3] is preferable to be used: all tests are designed in prior and tested in parallel. The proliferation of applying NAGT in various fields such as DNA library screening [4], multiple-access channels [5], data streaming [6], neuroscience [7], has made it become more attractive recently. We thus focus on NAGT in this work.

The development of NAGT applications in the field of molecular biology led to the introduction of another type of item: *inhibitor*. An item is considered to be an inhibitor if it interferes with the identification of defective items in a test, i.e., a test containing at least one inhibitor item returns negative outcome. In this "Group Testing with Inhibitors (GTI)" model, the outcome of a test on a subset of items is positive iff the subset has at least one defective item and no inhibitors. Due to great potential for use in applications, the GTI model has been intensively studied for the last two decades [8–11].

In NAGT using the GTI model (NAGTI), if t tests are needed to identify up to d defective items and up to h inhibitors among n items, it can be seen that they comprise a $t \times n$ measurement matrix. The procedure for obtaining the matrix is called the *construction procedure*. The procedure for obtaining the outcome of t tests using the matrix is called *encoding procedure*, and the procedure for obtaining the defective items and the inhibitor items from t outcomes is called the *decoding procedure*. Since noise typically occurs in biology experiments, we assume that there are up to e erroneous outcomes in the test outcomes. The objective of NAGTI is to *efficiently* classify all items from the encoding procedure and from the decoding procedure in the presence of noise.

There are two approaches when using NAGTI. One is to identify defective items only. Chang et al. [12] proposed a scheme using $O((d + h + e)^2 \log_2 n)$ tests to identify all defective items in time $O((d + h + e)^2 n \log_2 n)$. Using a probabilistic scheme, Ganesan et al. [13] reduced the number of tests to $O((d + h) \log_2 n)$ and the decoding time to $O((d + h)n \log_2 n)$. However, this scheme proposed is applicable only in a noise-free setting, which is restricted in practice. The second approach is to identify both defective items and inhibitors. Chang et al. [12] proposed a scheme using $O(e(d+h)^3 \log_2 n)$ tests to classify n items in time $O(e(d + h)^3 n \log_2 n)$. Without considering the presence of noise in the test outcome, Ganesan et al. [13] used $O((d + h^2) \log_2 n)$ tests to identify at most d defective items and at most h inhibitor items in time $O((d + h^2)n \log_2 n)$.

1.1 Problem Definition

We address two problems. The first is how to efficiently identify defective items in the test outcomes in the presence of noise. The second is how to efficiently identify both defective items and inhibitor items in the test outcome in the presence of noise. Let z be an odd integer and $e = \frac{z-1}{2}$ be the maximum number of errors in the test outcomes.

Problem 1. There are n items including up to d defective items and up to h inhibitor items. Is there a measurement matrix such that

- All defective items can be identified in time $\mathsf{poly}(d, h, e, \log_2 n)$ in the presence of up to e erroneous outcomes, where the number of rows in the measurement matrix is much smaller than n?
- Each column of the matrix can be nonrandomly generated in polynomial time of the number of rows?

Problem 2. There are n items including up to d defective items and up to h inhibitor items. Is there a measurement matrix such that

- All defective items and inhibitors items can be identified in time $\mathsf{poly}(d, h, e, \log_2 n)$ in the presence of up to e erroneous outcomes, where the number of rows in the measurement matrix is much smaller than n?
- Each column of the matrix can be nonrandomly generated in polynomial time of the number of rows?

We note that some previous works such as [14,15] do not consider inhibitor items. In these works, Problems 1 and 2 can be reduced to the same problem by eliminating all terms related to "inhibitor items."

1.2 Problem Model

We model NAGTI as follows. Suppose that there are up to $1 \leq d$ defectives and up to $0 \leq h$ inhibitors in n items. Let $\mathbf{x} = (x_1, \ldots, x_n)^T \in \{0, 1, -\infty\}^n$ be the vector representation of n items. Note that the number of defective items must be at least one. Otherwise, the outcomes of the tests designed would yield negative. Item j is defective iff $x_j = 1$, is an inhibitor iff $x_j = -\infty$, and is negative iff $x_j = 0$. Suppose that there are at most d 1's in $\mathbf{x}$, i.e., $|D = \{j \mid x_j = 1, \text{ for } j = 1, \ldots, n\}| \leq d$, and at most h $-\infty$'s in $\mathbf{x}$, i.e., $|H = \{j \mid x_j = -\infty, \text{ for } j = 1, \ldots, n\}| \leq h$.

Let $\mathcal{Q} = (q_{ij})$ be a $q \times n$ binary measurement matrix which is used to identify defectives and inhibitors in n items. Item j is represented by column j of $\mathcal{Q}$ ($\mathcal{Q}_j$) for $j = 1, \ldots, n$. Test i is represented by row i in which $q_{ij} = 1$ iff the item j belongs to test i, and $q_{ij} = 0$ otherwise, where $i = 1, \ldots, q$. Then the outcome vector using the measurement matrix $\mathcal{Q}$ is

$$\mathbf{r} = \mathcal{Q} \otimes \mathbf{x} = \begin{bmatrix} r_1 \\ \vdots \\ r_q \end{bmatrix}, \tag{1}$$

where $\otimes$ is called the NAGTI operator, test outcome $r_i = 1$ iff $\sum_{j=1}^{n} q_{ij}x_j \geq 1$, and $r_i = 0$ otherwise for $i = 1, \ldots, q$. Note that we assume $0 \times (-\infty) = 0$ and there may be at most e erroneous outcomes in $\mathbf{r}$.

Given l binary vectors $\mathbf{y}_w = (y_{1w}, y_{2w}, \ldots, y_{Bw})^T$ for $w = 1, \ldots, l$ and some integer $B \geq 1$. The union of $\mathbf{y}_1, \ldots, \mathbf{y}_l$ is defined as vector $\mathbf{y} = \vee_{i=1}^{l} \mathbf{y}_i = (\vee_{i=1}^{l} y_{1i}, \ldots, \vee_{i=1}^{l} y_{Bi})^T$, where $\vee$ is the OR operator. Then when vector $\mathbf{x}$ is binary, i.e., there are no inhibitors in n items, (1) can be represented as

$$\mathbf{r} = \mathcal{Q} \otimes \mathbf{x} = \bigvee_{j=1}^{n} x_j \mathcal{Q}_j = \bigvee_{j \in D} \mathcal{Q}_j. \tag{2}$$

Our objective is to design the matrix $\mathcal{Q}$ such that vector $\mathbf{x}$ can be recovered when having $\mathbf{r}$ in time $\mathsf{poly}(q) = \mathsf{poly}(d, h, e, \log_2 n)$.

1.3   Our Contributions

Overview: Our objective is to reduce the decoding complexity for identifying up to d defectives and/or up to h inhibitors in the presence of up to e erroneous test outcomes. We present two deterministic schemes that can efficiently solve both Problems 1 and 2 with the probability 1. These schemes use two basic ideas: each column of a $t_1 \times n$ $(d+h, r; z]$-disjunct matrix (defined later) must be generated in time $\mathsf{poly}(t_1)$ and the tensor product (defined later) between it and a special signature matrix. These ideas reduce decoding complexity to $\mathsf{poly}(t_1)$. Moreover, the measurement matrices used in our proposed schemes are nonrandom, i.e., their columns can be nonrandomly generated in time polynomial of the number of rows. As a result, one can save space for storing the measurement matrices. Simulation results confirm our theoretical analysis. When the number of items is sufficiently large, the decoding time in our proposed scheme is smallest in comparison with existing work.

Comparison: We compare our proposed schemes with existing schemes in Table 1. There are six criteria to be considered here. The first one is construction type, which defines how to achieve a measurement matrix. It also affects how defectives and inhibitors are identified. The most common construction type is random; i.e., a measurement matrix is generated randomly. The six schemes evaluated here use random construction except for our proposed schemes.

The second criterion is decoding type: "Deterministic" means the decoding objectives are always achieved with probability 1, while "Randomized" means the decoding objectives are achieved with some high probability. Ganesan et al. [13] used randomized decoding schemes to identify defectives and inhibitors. The schemes in [12] and our proposed schemes use deterministic decoding.

The remaining criteria are: identification of defective items only, identification of both defective items and inhibitor items, error tolerance, the number of tests, and the decoding complexity. The only advantage of the schemes proposed by Ganesan et al. [13] is that the number of tests is less than ours. Our schemes outperformed the existing schemes in other criteria such as error-tolerance, the

Table 1. Comparison with existing schemes. "Deterministic" and "Randomized" are abbreviated as "Det. and "Rnd.". The $\sqrt{}$ sign means that the criterion holds for that scheme, while the $\times$ sign means that it does not. We set $e = \frac{z-1}{2}$, $\lambda = \frac{(d+h)\ln n}{W((d+h)\ln n)} + z$, and $\alpha = \max\left\{\frac{\lambda}{(d+h)^2}, 1\right\}$, where $W(x) = \Theta\left(\ln x - \ln\ln x\right)$.

Scheme	Construction type	Decoding type	Max. no. of # errors	Defectives only	Defectives and inhibitors	Number of tests (t)	Decoding complexity
Chang et al. [12]	Random	Det.	e	$\sqrt{}$	$\times$	$O((d+h+e)^2 \ln n)$	$O(tn)$
Ganesan et al. [13]	Random	Rnd.	0	$\sqrt{}$	$\times$	$O((d+h)\ln n)$	$O(tn)$
Proposed (Theorem 4)	Nonrandom	Det.	e	$\sqrt{}$	$\times$	$\Theta\left(\lambda^2 \ln n\right)$	$O\left(\frac{\lambda^5 \ln n}{(d+h)^2}\right)$
Chang et al. [12]	Random	Det.	e	$\sqrt{}$	$\sqrt{}$	$O(e(d+h)^3 \ln n)$	$O(tn)$
Ganesan et al. [13]	Random	Rnd.	0	$\sqrt{}$	$\sqrt{}$	$O((d+h^2)\ln n)$	$O(tn)$
Proposed (Theorem 5)	Nonrandom	Det.	e	$\sqrt{}$	$\sqrt{}$	$\Theta\left(\lambda^3 \ln n\right)$	$O\left(d\lambda^6 \times \alpha\right)$

decoding type, and the decoding complexity. The number of tests with our proposed schemes for identifying defective items only (both defective items and inhibitor items, resp.) is smaller (larger, resp.) than that with the scheme proposed by Chang et al. [12]. The decoding complexity in our proposed scheme is much less than theirs when the number of items is sufficiently large.

2 Preliminaries

Notation is defined here for consistency. We use capital calligraphic letters for matrices, non-capital letters for scalars, bold letters for vectors, and capital letters for sets. Capital letters with asterisk is denoted for multisets in which elements may appear multiple times. For example, $S = \{1, 2, 3\}$ is a set and $S^* = \{1, 1, 2, 3\}$ is a multiset. Here we assume $0 \times (-\infty) = 0$.

Some frequent notations are listed as follows:

- $n; d$: number of items; maximum number of defective items. For simplicity, we suppose that n is the power of 2.
- $|\cdot|$: the weight, i.e., the number of non-zero entries in the input vector or the cardinality of the input set.
- $\otimes, \odot$: operator for NAGTI and tensor product, respectively.
- $[n]$: $\{1, 2, \ldots, n\}$.
- $\mathcal{S}$: $s \times n$ measurement matrix used to identify at most one defective item or one inhibitor item, where $s = 2\log_2 n$.
- $\mathcal{M} = (m_{ij})$: $m \times n$ disjunct matrix, where integer $m \geq 1$ is number of tests.
- $\mathcal{T} = (t_{ij})$: $t \times n$ measurement matrix used to identify at most d defective items, where integer $t \geq 1$ is number of tests.
- $\mathbf{x}; \mathbf{y}$: representation of n items; binary representation of the test outcomes.

- $S_j, M_j, M_{i,*}$: column j of matrix S, column j of matrix M, and row i of matrix M.
- $D; H$: index set of defective items; index set of inhibitor items.
- $\mathsf{supp}(\mathbf{c})$: support set of vector $\mathbf{c} = (c_1, \ldots, c_k)$; i.e., $\mathsf{supp}(\mathbf{c}) = \{j \mid c_j \neq 0\}$. For example, the support vector for $\mathbf{v} = (1, 0, 0, -\infty)$ is $\mathsf{supp}(\mathbf{v}) = \{1, 4\}$.
- $\mathsf{diag}(M_{i,*}) = \mathsf{diag}(m_{i1}, m_{i2}, \ldots, m_{in})$: diagonal matrix constructed from input vector $M_{i,*} = (m_{i1}, m_{i2}, \ldots, m_{in})$.
- $e; \log; \ln$: base of natural logarithm; logarithm of base 2; natural logarithm.
- $\lceil x \rceil; \lfloor x \rfloor$: ceiling function of x; floor function of x.
- $\mathsf{W}(x)$: the Lambert W function in which $\mathsf{W}(x)e^{\mathsf{W}(x)} = x$.

2.1 Tensor Product

Let $\odot$ be the tensor product notation. Note that the tensor product defined here is not the usual tensor product used in linear algebra. Given an $a \times n$ matrix $\mathcal{A} = (a_{ij})$ and an $s \times n$ matrix $\mathcal{S} = (s_{ij})$, their tensor product is defined as

$$\mathcal{R} = \mathcal{A} \odot \mathcal{S} := \begin{bmatrix} \mathcal{S} \times \mathsf{diag}(\mathcal{A}_{1,*}) \\ \vdots \\ \mathcal{S} \times \mathsf{diag}(\mathcal{A}_{f,*}) \end{bmatrix} = \begin{bmatrix} a_{11}\mathcal{S}_1 \ldots a_{1n}\mathcal{S}_n \\ \vdots \quad \ddots \quad \vdots \\ a_{a1}\mathcal{S}_1 \ldots a_{an}\mathcal{S}_n \end{bmatrix}, \tag{3}$$

where $\mathsf{diag}(.)$ is the diagonal matrix constructed from the input vector, and $\mathcal{A}_{h,*} = (a_{h1}, \ldots, a_{hn})$ is the hth row of $\mathcal{A}$ for $h = 1, \ldots, a$. The size of $\mathcal{R}$ is $r \times n$, where $r = a \times s$.

2.2 Reed-Solomon Codes

Let n_1, r_1, Λ, q be positive integers. Let Σ be a finite field and $|\Sigma| = q$. From now, we set $\Sigma = \mathbb{F}_q$. Each codeword is considered as a vector of $\mathbb{F}_q^{n_1 \times 1}$. Let C be a subset of Σ^{n_1}. Assume that for any $\mathbf{y} \in C$, there exists a message $\mathbf{x} \in \mathbb{F}_q^{r_1}$ such that $\mathbf{y} = \mathcal{G}\mathbf{x}$, where matrix $\mathcal{G}$ is a full-rank $n_1 \times r_1$ matrix in $\mathbb{F}_q$. Then C is called a linear code with minimum distance $\Lambda = \min_{\mathbf{y} \in C} |\mathsf{supp}(\mathbf{y})|$ and denoted as $[n_1, r_1, \Lambda]_q$. The cardinality of C is q^{r_1}. Let $\mathcal{M}_C$ denote the $n_1 \times q^{r_1}$ matrix whose columns are the codewords in C.

An $[n_1, r_1, \Lambda]_q$-Reed-Solomon (RS) code [16] is an $[n_1, r_1, \Lambda]_q$ code with $\Lambda = n_1 - r_1 + 1$. Since the parameter Λ can be obtained from n_1 and r_1, we usually refer to an $[n_1, r_1, \Lambda]_q$-RS code as $[n_1, r_1]_q$-RS code.

2.3 Disjunct Matrix

Superimposed code was introduced by Kautz and Singleton [17] and then generalized by D'yachkov et al. [18] and Stinson and Wei [19]. A superimposed code is defined as follows.

Definition 1. *An* $m \times n$ *binary matrix* $\mathcal{M}$ *is called an* $(d, r; z]$-*superimposed code if for any two disjoint subsets* $S_1, S_2 \subset [n]$ *such that* $|S_1| = d$ *and* $|S_2| = r$, *there exists at least* z *rows in which there are all* 1's *among the columns in* S_2 *while all the columns in* S_1 *have* 0's, *i.e.,* $\left| \bigcap_{j \in S_2} \mathsf{supp}\,(\mathcal{M}_j) \setminus \bigcup_{j \in S_1} \mathsf{supp}\,(\mathcal{M}_j) \right| \geq z$.

Matrix $\mathcal{M}$ is usually referred to as an $(d, r; z]$-disjunct matrix. Parameter $e = \lfloor (z-1)/2 \rfloor$ is referred to as the *error tolerance* of a disjunct matrix. It is clear that for any $d' \leq d$, $r' \leq r$, and $z' \leq z$, an $(d, r; z]$-disjunct matrix is also an $(d', r'; z']$-disjunct matrix.

Let $\mathbf{x} = (x_1, \ldots, x_n)^T \in \{0, 1\}^n$ be the binary representation vector of n items, where $|\mathbf{x}| \leq d$. From (2), the outcome vector of m tests by using $\mathcal{M}$ and $\mathbf{x}$ is defined as follows:

$$\mathbf{y} = \mathcal{M} \otimes \mathbf{x} = \bigvee_{j=1}^{n} x_j \mathcal{M}_j = \bigvee_{j \in D} \mathcal{M}_j, \tag{4}$$

where $D = \mathsf{supp}(\mathbf{x}) = \{j \mid x_j = 1\}$. The procedure to get $\mathbf{y}$ is called *encoding procedure*. It includes the construction procedure, which is to get a measurement matrix $\mathcal{M}$. The procedure to recover $\mathbf{x}$ from $\mathbf{y}$ and $\mathcal{M}$ is called *decoding procedure*. Our objective is to recover $\mathbf{x}$ when the outcome vector $\mathbf{y}$ and the matrix $\mathcal{M}$ are given.

The number of rows in an $m \times n$ $(d, r; z]$-disjunct matrix is usually exponential to d [15, 20]. Cheraghchi [21] proposed a nonrandom construction for $(d, r; z]$-disjunct matrices in which the number of tests is larger than the existing works as d or r increases.

Theorem 1 (Lemma 29 [21]). *For any positive integers* d, r, z *and* n *with* $d + r \leq n$, *there exists an* $m \times n$ *nonrandom* $(d, r; z]$-*disjunct matrix where* $m = O\left((rd \ln n + z)^{r+1}\right)$. *Moreover, each column of the matrix can be generated in time* $\mathsf{poly}(m)$.

An $(d, r; z]$-disjunct matrix is called an $(d; z]$-disjunct matrix when $r = 1$, and a d-disjunct matrix when $r = z = 1$. For efficient decoding in the NAGTI model, we pay attention only to an $m \times n$ binary $(d, r; z]$-disjunct matrix in which each column can be generated in time $\mathsf{poly}(m)$.

2.4 Bui et al.'s Scheme

In this section, the scheme proposed by Bui et al. [14] is described. Its main contribution is that, given any $m \times n$ $(d-1)$-disjunct matrix, a bigger $t \times n$ measurement matrix can be generated such that up to d defective items (in a set of n items having only defective and negative items) can be identified in time $O(t) = O(m \log n)$, where $t = 2m \log n$.

Encoding procedure: Let $\mathcal{S}$ be an $s \times n$ measurement matrix:

$$\mathcal{S} := \begin{bmatrix} \mathbf{b}_1 \ \mathbf{b}_2 \ \ldots \ \mathbf{b}_n \\ \overline{\mathbf{b}_1} \ \overline{\mathbf{b}_2} \ \ldots \ \overline{\mathbf{b}_n} \end{bmatrix} = \begin{bmatrix} \mathcal{S}_1 \ \ldots \ \mathcal{S}_n \end{bmatrix}, \tag{5}$$

where $s = 2\log n$, $\mathbf{b}_j$ is the $\log n$-bit binary representation of integer $j - 1$, $\overline{\mathbf{b}}_j$ is the complement of $\mathbf{b}_j$, and $\mathcal{S}_j := \begin{bmatrix} \mathbf{b}_j \\ \overline{\mathbf{b}}_j \end{bmatrix}$ for $j = 1, 2, \ldots, n$. Item j is characterized by column $\mathcal{S}_j$ and that the weight of every column in $\mathcal{S}$ is $s/2 = \log n$. Furthermore, the index j is uniquely identified by $\mathbf{b}_j$.

Given an $m \times n$ $(d-1)$-disjunct matrix $\mathcal{M}$, the new measurement $t \times n$ matrix is constructed as follows:

$$\mathcal{T} = \mathcal{M} \odot \mathcal{S}, \tag{6}$$

where $\odot$ is the tensor product defined in Sect. 2.1 and $t = ms$. For any binary input vector $\mathbf{x}$, its outcome using measurement matrix $\mathcal{T}$ is

$$\mathbf{y} = \mathcal{T} \otimes \mathbf{x} = \begin{bmatrix} \mathbf{y}_1 \\ \vdots \\ \mathbf{y}_m \end{bmatrix}, \tag{7}$$

where $\mathbf{y}_i = (\mathcal{S} \times \mathrm{diag}(\mathcal{M}_{i,*})) \otimes \mathbf{x} = \bigvee_{j=1}^{n} x_j m_{ij} \mathcal{S}_j$ for $i = 1, \ldots, m$.

Decoding Procedure: The decoding procedure is quite simple. We can scan all $\mathbf{y}_i$ for $i = 1, \ldots, m$. If $\mathrm{wt}(\mathbf{y}_i) = \log n$, the defective item can be identified by calculating the first half of $\mathbf{y}_i$. Otherwise, no defective item is identified. The procedure is described in Algorithm 1.

Algorithm 1. GetDefectives($\mathbf{y}, n$): detection of up to d defective items.

Input: number of items n; outcome vector $\mathbf{y}$
Output: defective items
1:   $s = 2\log n$.
2:   $S = \emptyset$.
3:   Divide $\mathbf{y}$ into $m = t/s$ smaller vectors $\mathbf{y}_1, \ldots, \mathbf{y}_m$ such that $\mathbf{y} = (\mathbf{y}_1, \ldots, \mathbf{y}_m)^T$ and their size are equal to s, where t is the number of entries in $\mathbf{y}$.
4:   **for** $i = 1$ to m **do**
5:      **if** $\mathrm{wt}(\mathbf{y}_i) = \log n$ **then**
6:         Get defective item d_0 by checking first half of $\mathbf{y}$.
7:         $S = S \cup \{d_0\}$.
8:      **end if**
9:   **end for**
10: **return** S.

This scheme can be summarized as the following theorem:

Theorem 2. *Let an $m \times n$ matrix $\mathcal{M}$ be $(d-1)$-disjunct. Suppose that a set of n items has up to d defective and no inhibitors. Then there exists a $t \times n$ matrix $\mathcal{T}$ constructed from $\mathcal{M}$ that can be used to identify up to d defective items in time $t = m \times 2\log n$. Further, suppose that each column of $\mathcal{M}$ can be computed in time β. Then every column of $\mathcal{T}$ can be computed in time $2\log n \times \beta = O(\beta \log n)$.*

Algorithm 1 is modified and denoted as GetDefectives$^*(\mathbf{y}, n)$ if we substitute S by multiset S^*; i.e., the output of GetDefectives$^*(\cdot)$ may have duplicated items which are used to handle the presence of erroneous outcomes in Sects. 4 and 5. Line 7 is interpreted as "Add d_0 to set S^*".

3   Improved Instantiation of Nonrandom $(d, r; z]$-Disjunct Matrices

We first state the useful nonrandom construction of $(d, r; z]$-disjunct matrices, which is an instance of Theorem 1:

Theorem 3 (Lemma 29 [21]). *Let $1 \leq d, r, z < n$ be integers and C be a $[n_1 = q - 1, k_1]_q$-RS code. For any $d < \frac{n_1 - z}{r(k_1 - 1)} = \frac{q - 1 - z}{r(k_1 - 1)}$ and $n \leq q^{k_1}$, there exists a $t \times n$ nonrandom $(d, r; z]$-disjunct matrix where $t = O\left(q^{r+1}\right)$. Moreover, each column of the matrix can be constructed in time $O\left(q^{r+2}/(r^2 d^2)\right)$.*

An approximation of a Lambert W function $\mathsf{W}(x)$ [22] is $\ln x - \ln \ln x \leq \mathsf{W}(x) \leq \ln x - \frac{1}{2} \ln \ln x$ for any $x \geq e$. Then an improved instantiation of nonrandom $(d, r; z]$-disjunct matrix is stated as follows:

Corollary 1. *For any positive integers d, r, z, and n with $d + r \leq n$, there exists a $t \times n$ nonrandom $(d, r; z]$-disjunct matrix with $t = \Theta\left(\lambda^{r+1}\right)$, where $\lambda = (rd \ln n)/(\mathsf{W}(d \ln n)) + z$. Moreover, each column of the matrix can be constructed in time $O\left(\lambda^{r+2}/(r^2 d^2)\right)$.*

Proof. From Theorem 3, we only need to find an $[n_1 = q - 1, k_1]_q$-RS code such that $d < \frac{n_1 - z}{r(k_1 - 1)} = \frac{q - 1 - z}{r(k_1 - 1)}$ and $q^{k_1} \geq n$. One chooses

$$q = \begin{cases} \frac{rd \ln n}{\mathsf{W}(d \ln n)} + z + 1 & \text{if } \frac{rd \ln n}{\mathsf{W}(d \ln n)} + z + 1 \text{ is the power of 2.} \\ 2^{\eta+1}, & \text{otherwise.} \end{cases} \tag{8}$$

where η is an integer satisfying $2^\eta < \frac{rd \ln n}{\mathsf{W}(d \ln n)} + z + 1 < 2^{\eta+1}$. We have $q = \Theta\left(\frac{rd \ln n}{\mathsf{W}(d \ln n)} + z\right)$ in both cases because $\frac{rd \ln n}{\mathsf{W}(d \ln n)} + z + 1 \leq q < 2\left(\frac{rd \ln n}{\mathsf{W}(d \ln n)} + z + 1\right)$.

Set $k_1 = \left\lceil \frac{q - z - 1}{rd} \right\rceil \geq \frac{\ln n}{\mathsf{W}(d \ln n)}$. Note that the condition on d in Theorem 3 always holds because:

$$k_1 = \left\lceil \frac{q - z - 1}{rd} \right\rceil \implies k_1 < \frac{q - z - 1}{rd} + 1 \implies d < \frac{q - 1 - z}{r(k_1 - 1)} = \frac{n_1 - z}{r(k_1 - 1)}.$$

Finally, our task is to prove that $n \leq q^{k_1}$. Indeed, we have:

$$q^{k_1} \geq \left(\frac{rd \ln n}{\mathsf{W}(d \ln n)} + z + 1\right)^{\frac{\ln n}{\mathsf{W}(d \ln n)}} \geq \left(\frac{d \ln n}{\mathsf{W}(d \ln n)}\right)^{\frac{\ln n}{\mathsf{W}(d \ln n)}}$$

$$\geq \left(e^{\mathsf{W}(d \ln n)} e^{\mathsf{W}(d \ln n)}\right)^{1/d} = \left(e^{d \ln n}\right)^{1/d} = n.$$

This completes our proof.

The number of tests in our construction is better than the one in Theorem 1. Furthermore, there is no decoding scheme associated with matrices in this corollary. However, when $r = z = 1$, the scheme in [14] achieves the same number of tests and has an efficient decoding algorithm.

4   Identification of Defective Items

In this section, we answer Problem 1 that there exists a $t \times n$ measurement matrix such that: it can handle at most e errors in the test outcome; each column can be nonrandomly generated in time $\mathsf{poly}(t)$; and all defective items can be identified in time $\mathsf{poly}(d, h, e, \log n)$, where there are up to d defective items and up to h inhibitor items in n items. The main idea is to use the modified version of Algorithm 1 to identify all potential defective items. Then a sanitary procedure is proceeded to remove all false defective items.

Theorem 4. *Let* $1 \leq d, h, d + h \leq n$ *be integers,* z *be odd, and* $\lambda = \frac{(d+h)\ln n}{W((d+h)\ln n)} + z$. *A set of* n *items includes up to* d *defective items and up to* h *inhibitors. Then there exists a* $t \times n$ *nonrandom matrix such that up to* d *defective items can be identified in time* $O\left(\frac{\lambda^5 \log n}{(d+h)^2}\right)$ *with up to* $e = \frac{z-1}{2}$ *errors in the test outcomes, where* $t = \Theta\left(\lambda^2 \log n\right)$. *Moreover, each column of the matrix can be generated in time* $\mathsf{poly}(t)$.

The proof is given in the following sections.

4.1   Encoding Procedure

We set $e = \frac{z-1}{2}$ and $\lambda = \frac{(d+h)\ln n}{W((d+h)\ln n)} + z$. Let an $m \times n$ matrix $\mathcal{M}$ be an $(d + h; z]$-disjunct matrix in Corollary 1 $(r = 1)$, where

$$m = \Theta\left(\left(\frac{(d+h)\ln n}{W((d+h)\ln n)} + z\right)^2\right) = O(\lambda^2).$$

Each column in $\mathcal{M}$ can be generated in time $t_1 = O\left(\frac{\lambda^3}{(d+h)^2}\right)$. Then the final $t \times n$ measurement matrix $\mathcal{T}$ is

$$\mathcal{T} = \mathcal{M} \circledcirc \mathcal{S}, \tag{9}$$

where the $s \times n$ matrix $\mathcal{S}$ is defined in (5) and $t = ms = \Theta\left(\lambda^2 \log n\right)$. Then it is easy to see that each column of $\mathcal{T}$ can be generated in time $t_1 \times s = \mathsf{poly}(t)$.

Any input vector $\mathbf{x} = (x_1, \ldots, x_n)^T \in \{0, 1, -\infty\}^n$ contains at most d 1's and at most h $-\infty$'s as described in Sect. 1.2. Note that D and H are the index sets of the defective items and the inhibitor items, respectively. Then the binary outcome vector using the measurement matrix $\mathcal{T}$ is $\mathbf{y} = \mathcal{T} \otimes \mathbf{x} = \begin{bmatrix} \mathbf{y}_1 \\ \vdots \\ \mathbf{y}_m \end{bmatrix}$, where

$$\mathbf{y}_i = (\mathcal{S} \times \text{diag}(\mathcal{M}_{i,*})) \otimes \mathbf{x} = \begin{bmatrix} y_{(i-1)s+1} \\ \cdots \\ y_{is} \end{bmatrix}, \text{ and } y_{(i-1)s+l} = 1 \text{ iff } \sum_{j=1}^{n} m_{ij} s_{lj} x_j \geq$$

1, and $y_{(i-1)s+l} = 0$ otherwise, for $i = 1, \ldots, m$, and $l = 1, \ldots, s$. We assume that there are at most e incorrect outcomes in the outcome vector $\mathbf{y}$.

4.2   Decoding Procedure

Given outcome vector $\mathbf{y} = (\mathbf{y}_1, \ldots, \mathbf{y}_m)^T$, we can identify all defective items by using Algorithm 2. Step 1 is to identify all potential defectives and put them in the set S^*. Then Steps 3 to 8 are to remove duplicate items in the new potential defective set S_0. After that, Steps 9 to 16 are to remove all false defectives. Finally, Step 17 returns the defective set.

Algorithm 2. GetDefectivesWOInhibitors($\mathbf{y}, n, e$): detection of up to d defective items without identifying inhibitors.

Input: a function to generate $t \times n$ measurement matrix $\mathcal{T}$; outcome vector $\mathbf{y}$; maximum number of errors e
Output: defective items

1: $S^* = \text{GetDefectives}^*(\mathbf{y}, n)$.   ▷ Identify all potential defectives.
2: $S_0 = \emptyset$.   ▷ Defective set.
3: **foreach** $x \in S^*$ **do**
4:       **if** x appears in S^* at least $e + 1$ times **then**
5:           $S_0 = S_0 \cup \{x\}$.
6:           Remove all elements that equal x in S^*.
7:       **end if**
8: **end foreach**
9: **for all** $x \in S_0$ **do**   ▷ Remove false defectives.
10:      ▷ Get column corresponding to defective item x.
11:      Generate column $\mathcal{T}_x = \mathcal{M}_x \odot \mathcal{S}_x$.
12:      **if** $\exists i_0 \in [t] : t_{i_0 x} = 1$ and $y_{i_0} = 0$ **then**   ▷ Condition for a false defective.
13:          $S_0 = S_0 \setminus \{x\}$.   ▷ Remove false defectives.
14:          break;
15:      **end if**
16: **end for**
17: **return** S_0.   ▷ Return set of defective item.

4.3   Correctness of Decoding Procedure

Since matrix $\mathcal{M}$ is an $(d+h; z]$-disjunct matrix, there are at least z rows i_0 such that $m_{i_0 j} = 1$ and $m_{i_0 j'} = 0$ for any $j \in D$ and $j' \notin D \cup H \setminus \{j\}$. Since up to $e = (z-1)/2$ errors may appear in test outcome $\mathbf{y}$, there are at least $e+1$ vectors $\mathbf{y}_{i_0}$ such that the condition in Step 5 of Algorithm 1 holds. Consequently, each value $j \in D$ appears at least $e+1$ times. Therefore, Steps 1 to 8 return a set S_0 containing all defective items and some false defectives.

Steps 9 to 16 are to remove false defectives. For any index $j \notin D$, since there are at most $e = (z-1)/2$ erroneous outcomes, there is at least 1 row i_0 such that $t_{i_0 j} = 1$ and $t_{i_0 j'} = 0$ for all $j' \in D \cup H$. Because item $j \notin D$, the outcome of that row (test) is negative (0). Therefore, Step 12 is to check whether an item in S_0 is non-defective. Finally, Step 17 returns the set of defective items.

4.4   Decoding Complexity

The time to run Step 1 is $O(t)$. Since $|S^*| \leq m$, it takes m time to run Steps 3 to 8. Because $|S^*| \leq m$, the cardinality of S_0 is up to m. The loop at Step 9 runs at most m times. Steps 11 and 12 take time $s \times \frac{m^{1.5}}{(d+h)^2}$ and t, respectively. The total decoding time is:

$$O(t) + m + m \times \left(s \times \frac{m^{1.5}}{(d+h)^2} + t \right) = O\left(\frac{sm^{2.5}}{(d+h)^2} \right) = O\left(\frac{\lambda^5 \log n}{(d+h)^2} \right).$$

5   Identification of Defectives and Inhibitors

In this section, we answer Problem 2 that there exists a $v \times n$ measurement matrix such that: it can handle at most e errors in the test outcome; each column can be nonrandomly generated in time $\mathsf{poly}(v)$; and all defective items and inhibitor items can be identified in time $\mathsf{poly}(d, h, e, \log n)$, where there are up to d defective items and up to h inhibitor items in n items.

Theorem 5. *Let $1 \leq d, h, d+h \leq n$ be integers, z be odd, and $\lambda = \frac{(d+h)\ln n}{W((d+h)\ln n)} + z$. A set of n items includes up to d defective items and up to h inhibitors. Then there exists a $v \times n$ nonrandom matrix such that up to d defective items and up to h inhibitor items can be identified in time $O\left(d\lambda^6 \times \max\left\{ \frac{\lambda}{(d+h)^2}, 1 \right\} \right)$, with up to $e = \frac{z-1}{2}$ errors in the test outcomes, where $v = \Theta\left(\lambda^3 \log n \right)$. Moreover, each column of the matrix can be generated in time $\mathsf{poly}(v)$.*

To detect both up to h inhibitors and d defectives, we have to use two types of matrices: an $(d+h; z]$-disjunct matrix and an $(d+h-2, 2; z]$-disjunct matrix. The main idea is as follows. We first identify all defective items. Then all potential inhibitors are located by using an $(d+h-2, 2; z]$-disjunct matrix. The final procedure is to remove all false inhibitor items.

5.1   Identification of an Inhibitor

Let $\underline{\vee}$ be the notation for the union of the column corresponding to the defective item and the column corresponding to the inhibitor item. We suppose that there is an outcome $\mathbf{o} := (o_1, \ldots, o_s)^T = \mathcal{S}_a \underline{\vee} \mathcal{S}_b$, where the defective item is a and the inhibitor item is b, and that $\mathcal{S}_a$ and $\mathcal{S}_b$ are two columns in the $s \times n$ matrix $\mathcal{S}$ in (5). Note that $o_i = 1$ iff $s_{ia} = 1$ and $s_{ib} = 0$, and $o_i = 0$ otherwise, for

---

Algorithm 3. GetInhibitorFromADefective($\mathbf{o}, \mathcal{S}_a, n$): identification of an inhibitor when defective item and union of corresponding columns are known.

Input: outcome vector $\mathbf{o} := (o_1, \ldots, o_s) = \mathcal{S}_a \vee \mathcal{S}_b$; number of items n; vector $\mathcal{S}_a$ corresponding to defective item a

Output: inhibitor item b

1: $s = 2 \log n$.
2: Set $\mathcal{S}_b = (s_{1b}, \ldots, s_{sb})^T = (-1, -1, \ldots, -1)^T$.
3: **for** $i = 1$ to s **do**   ▷ Obtain $s/2$ entries of $\mathcal{S}_b$.
4:     If $s_{ia} = 1$ and $o_i = 1$ then $s_{ib} = 0$. **end if**
5:     If $s_{ia} = 1$ and $o_i = 0$ then $s_{ib} = 1$. **end if**
6: **end for**
7: **for** $i = 1$ to $s/2$ **do**   ▷ Obtain $s/2$ remaining entries of $\mathcal{S}_b$.
8:     If $s_{ib} = -1$ then $s_{ib} = 1 - s_{i+s/2,b}$. **end if**
9:     If $s_{ib} = 0$ then $s_{i+s/2,b} = 1$. **end if**
10:     If $s_{ib} = 1$ then $s_{i+s/2,b} = 0$. **end if**
11: **end for**
12: Get index b by checking first half of $\mathcal{S}_b$.
13: **return** b.   ▷ Return the inhibitor item.

$i = 1, \ldots, s$. Assume that the defective item a is already known. The inhibitor item b is identified as in Algorithm 3.

The correctness of the algorithm is described here. Step 2 initializes the corresponding column of inhibitor b in $\mathcal{S}$. Since column $\mathcal{S}_a$ has exactly $s/2$ 1's, Steps 3 to 6 are to obtain $s/2$ positions of $\mathcal{S}_b$. Since the first half of $\mathcal{S}_a$ is the complement of its second half, it does not exist two indexes i_0 and i_1 such that $s_{i_0 a} = s_{i_1 a} = 1$, where $|i_0 - i_1| = \log n$. As a result, it does not exist two indexes i_0 and i_1 such that $s_{i_0 b} = s_{i_1 b} = -1$, where $|i_0 - i_1| = \log n$. Moreover, the first half of $\mathcal{S}_b$ is the complement of its second half. Therefore, the remaining $s/2$ entries of $\mathcal{S}_b$ can be obtained by using Steps 7 to 11. The index of inhibitor b can be identified by checking the first half of $\mathcal{S}_b$, which is done in Step 12. Finally, Step 13 returns the index of the inhibitor.

It is easy to verify that the decoding complexity of Algorithm 3 is $O(s)$.

Example: Let $\mathcal{S}$ be the matrix in (5), where $n = 8$ and $s = 2 \log n = 6$. Given item 1 is the unknown inhibitor and that item 3 is the known defective item, assume that the observed vector is $\mathbf{o} = (0, 1, 0, 0, 0, 0)^T$. The corresponding column of the defective item is $\mathcal{S}_3$. We set $\mathcal{S}_b = (-1, -1, -1, -1, -1, -1)^T$. We get $\mathcal{S}_b = (-1, 0, -1, 1, -1, 1)^T$ from Steps 3 to 6 and the complete column $\mathcal{S}_b = (0, 0, 0, 1, 1, 1)^T$ from Steps 7 to 11. Because the first half of $\mathcal{S}_b$ is $(0, 0, 0)^T$, the index of the inhibitor is 1.

5.2   Encoding Procedure

We set $e = \frac{z-1}{2}$ and $\lambda = \frac{(d+h) \ln n}{W((d+h) \ln n)} + z$. Let an $m \times n$ matrix $\mathcal{M}$ and a $g \times n$ matrix $\mathcal{G}$ be an $(d + h; z]$-disjunct matrix and an $(d + h - 2, 2; z]$-disjunct matrix in Corollary 1, respectively, where

$$m = \Theta \left(\left(\frac{(d+h) \ln n}{\mathrm{W}((d+h) \ln n)} + z \right)^2 \right) = \Theta \left(\lambda^2 \right),$$

$$g = \Theta \left(\left(\frac{(d+h) \ln n}{\mathrm{W}((d+h) \ln n)} + z \right)^3 \right) = \Theta \left(\lambda^3 \right).$$

Each column in $\mathcal{M}$ and $\mathcal{G}$ can be generated in time t_1 and t_2, respectively, where

$$t_1 = O \left(\frac{\lambda^3}{(d+h)^2} \right), t_2 = O \left(\frac{\lambda^4}{(d+h)^2} \right). \tag{10}$$

The final $v \times n$ measurement matrix $\mathcal{V}$ is

$$\mathcal{V} = \begin{bmatrix} \mathcal{M} \odot \mathcal{S} \\ \mathcal{G} \odot \mathcal{S} \\ \mathcal{G} \end{bmatrix} = \begin{bmatrix} \mathcal{T} \\ \mathcal{H} \\ \mathcal{G} \end{bmatrix}, \tag{11}$$

where $\mathcal{T} = \mathcal{M} \odot \mathcal{S}$ and $\mathcal{H} = \mathcal{G} \odot \mathcal{S}$. The sizes of matrices $\mathcal{T}$ and $\mathcal{H}$ are $t \times n$ and $h \times n$, respectively. Then we have $t = ms = 2m \log n$ and $h = gs = 2g \log n$. Note that the matrix $\mathcal{T}$ is the same as the one in (9). The number of tests of the measurement matrix $\mathcal{V}$ is

$$v = t + h + g = ms + gs + g = O((m+g)s) = \Theta \left(\lambda^3 \log n \right).$$

Then it is easy to see that each column of matrix $\mathcal{V}$ can be generated in time $(t_1 + t_2) \times s + t_2 = \mathrm{poly}(v)$.

Any input vector $\mathbf{x} = (x_1, \ldots, x_n)^T \in \{0, 1, -\infty\}^n$ contains at most d 1's and at most h $-\infty$'s as described in Sect. 1.2. The outcome vector using measurement matrix $\mathcal{T}$, i.e., $\mathbf{y} = \mathcal{T} \otimes \mathbf{x}$, is the same as the one in Sect. 4.1. The binary outcome vector using the measurement matrix $\mathcal{H}$ is

$$\mathbf{h} = \mathcal{H} \otimes \mathbf{x} = \begin{bmatrix} \mathbf{h}_1 \\ \vdots \\ \mathbf{h}_g \end{bmatrix}, \tag{12}$$

where $\mathbf{h}_i = (\mathcal{S} \times \mathrm{diag}(\mathcal{G}_{i,*})) \otimes \mathbf{x} = \begin{bmatrix} h_{(i-1)s+1} \\ \cdots \\ h_{is} \end{bmatrix}$, $h_{(i-1)s+l} = 1$ iff $\sum_{j=1}^{n} g_{ij} s_{lj} x_j$
≥ 1, and $h_{(i-1)s+l} = 0$ otherwise, for $i = 1, \ldots, g$, and $l = 1, \ldots, s$. Therefore, the outcome vector using the measurement matrix $\mathcal{V}$ in (11) is:

$$\mathbf{v} = \mathcal{V} \otimes \mathbf{x} = \begin{bmatrix} \mathcal{T} \\ \mathcal{H} \\ \mathcal{G} \end{bmatrix} \otimes \mathbf{x} = \begin{bmatrix} \mathcal{T} \otimes \mathbf{x} \\ \mathcal{H} \otimes \mathbf{x} \\ \mathcal{G} \otimes \mathbf{x} \end{bmatrix} = \begin{bmatrix} \mathbf{y} \\ \mathbf{h} \\ \mathbf{g} \end{bmatrix}, \tag{13}$$

where $\mathbf{y}$ is as same as the one in Sect. 4.1, $\mathbf{h}$ is defined in (12), and $\mathbf{g} = \mathcal{G} \otimes \mathbf{x} = (r_1, \ldots, r_g)^T$. We assume that $0 \times (-\infty) = 0$ and there are at most $e = (z-1)/2$ incorrect outcomes in the outcome vector $\mathbf{v}$.

5.3 Decoding Procedure

Given outcome vector $\mathbf{v}$, number of items n, number of tests in matrix $\mathcal{M}$, number of tests in matrix $\mathcal{G}$, maximum number of errors e, and functions to generate matrix $\mathcal{V}, \mathcal{G}, \mathcal{M}$, and $\mathcal{S}$. The details of the proposed scheme is described in Algorithm 4. Steps 1 to 2 are to divide the outcome vector $\mathbf{v}$ into three smaller vectors $\mathbf{y}, \mathbf{h}$, and $\mathbf{g}$ as (13). Then Step 3 is to get the defective set. All potential inhibitors would be identified in Steps 5 to 12. Then Steps 14 to 23 are to remove most of false inhibitors. Since there may be some duplicate inhibitors and some remaining false inhibitors in the inhibitor set, Step 25 to 31 are to remove the remaining false inhibitors and make each element in the inhibitor set unique. Finally, Step 32 is to return the defective set and the inhibitor set.

5.4 Correctness of the Decoding Procedure

Because of the construction of $\mathcal{V}$, the three vectors split from the outcome vector $\mathbf{v}$ in Step 2 are $\mathbf{y} = \mathcal{T} \otimes \mathbf{x}, \mathbf{h} = \mathcal{H} \otimes \mathbf{x}$, and $\mathbf{g} = \mathcal{G} \otimes \mathbf{x}$. Therefore, the set D achieved in Step 3 is the defective set as analyzed in Sect. 4.

Let H be the true inhibitor set which we will identify. Since $\mathcal{G}$ is an $(d + h - 2, 2; z]$-disjunct matrix $\mathcal{G}$, for any $j_1 \in H$ (we have not known H yet) and $j_2 \in D$, there exists at least z rows i_0's such that $g_{i_0 j_1} = g_{i_0 j_2} = 1$ and $g_{i_0 j'} = 0$, for all $j' \in D \cup H \setminus \{j_1, j_2\}$. Then, since there are at most $e = (z-1)/2$ errors in $\mathbf{v}$, there exists at least $e+1 = (z-1)/2+1$ index i_0's such that $\mathbf{h}_{i_0} = \mathcal{S}_{j_1} \vee \mathcal{S}_{j_2}$. As analyzed in Sect. 5.1, for any vector which is the union of the column corresponding to the defective item and the column corresponding to the inhibitor item, the inhibitor item is always identified if the defective item is known. Therefore, the set H_0^* obtained from Steps 7 to 12 contains all inhibitors and may contain some false inhibitors. Our next goal is to remove false inhibitors.

To remove the false inhibitors, we first remove all defective items in the set H_0^* as Step 16. Therefore, there are only inhibitors and negative items in the set H_0^* after implementing Step 16. One needs to exploit the property of the inhibitor that it will make the test outcome negative if there are at least one inhibitor and at least one defective in the same test. We pick an arbitrary defective item $y \in D$ and generate its corresponding column $\mathcal{G}_y$ in the matrix $\mathcal{G}$. Since $\mathcal{G}$ is an $(d + h - 2, 2; z]$-disjunct matrix $\mathcal{G}$ and there are at most $e = (z - 1)/2$ errors in $\mathbf{v}$, for any $j_1 \in H$ (we have not known H yet) and $y \in D$, there exists at least $z - e = e + 1$ rows i_0's such that $g_{i_0 j_1} = g_{i_0 y} = 1$ and $g_{i_0 j'} = 0$, for all $j' \in D \cup H \setminus \{j_1, y\}$. The outcome of these tests would be negative. Therefore, Steps 14 to 23 removes most of false inhibitors. Note that since there are at most e errors, the are at most e false inhibitors and each of them appears at most e times in the set H_0^*. Then Step 25 to 31 are to completely remove false inhibitors and make each element in the inhibitor set unique. Finally, Step 32 returns the sets of defective items and inhibitor items.

Algorithm 4. GetInhibitors($\mathbf{v}, n, e, m, g$): identification of up to d defectives and up to h inhibitors.

Input: outcome vector $\mathbf{v}$; number of items n; number of tests in matrix $\mathcal{M}$; number of tests in matrix $\mathcal{G}$; maximum number of errors e; and functions to generate matrix $\mathcal{V}, \mathcal{G}, \mathcal{M}$, and $\mathcal{S}$

Output: defective items and inhibitor items

1: $s = 2 \log n$.   ▷ number of rows in the matrix $\mathcal{S}$.
2: Divide vector $\mathbf{v}$ into three smaller vectors $\mathbf{y}, \mathbf{h}$, and $\mathbf{g}$ such that $\mathbf{v} = (\mathbf{y}^T, \mathbf{h}^T, \mathbf{g}^T)^T$
    and number of entries in $\mathbf{y}, \mathbf{h}$, and $\mathbf{g}$ are ms, gs, and g, respectively.
3: $D = $ GetDefectivesWOInhibitors($\mathbf{y}, n, e$).   ▷ defective set.
4: ▷ Find all potential inhibitors.
5: Divide vector $\mathbf{h}$ into g smaller vectors $\mathbf{h}_1, \ldots, \mathbf{h}_g$ such that $\mathbf{h} = (\mathbf{h}_1^T, \ldots, \mathbf{h}_g^T)^T$ and
    their size are equal to s.
6: $H_0^* = \emptyset$.   ▷ Initialize inhibitor multiset.
7: **for** $i = 1$ to g **do**   ▷ Scan all outcomes in $\mathbf{h}$.
8:    **foreach** $x \in D$ **do**
9:       $i_0 = $ GetInhibitorFromADefective($\mathbf{h}_i, \mathcal{S}_x, n$).
10:       Add item i_0 to multiset H_0^*.
11:    **end foreach**
12: **end for**
13: ▷ Remove most of false inhibitors.
14: Assign $(r_1, \ldots, r_g)^T = \mathbf{g}$.
15: Generate a column $\mathcal{G}_y$ for any $y \in D$.   ▷ Get the column of a defective.
16: $H_0^* = H_0^* \setminus D$.
17: **foreach** $x \in H_0^*$ **do**   ▷ Scan all potential inhibitors.
18:    Generate column $\mathcal{G}_x$
19:    **if** $\exists i_0 \in [g] : g_{i_0 x} = g_{i_0 y} = 1$ and $r_{i_0} = 1$ **then**
20:       Remove all elements that equal x in H_0^*.   ▷ Remove the false inhibitor.
21:       break;
22:    **end if**
23: **end foreach**
24: ▷ Completely remove false inhibitors and duplicate inhibitors.
25: $H = \emptyset$.
26: **foreach** $x \in H_0^*$ **do**
27:    **if** x appears in H_0^* at least $e + 1$ times **then**
28:       $H = H \cup \{x\}$.
29:       Remove all elements that equal x in H_0^*.
30:    **end if**
31: **end foreach**
32: **return** D and H.   ▷ Return set of defective items.

5.5   Decoding Complexity

First, we find all potential inhibitors. It takes time $O(v)$ for Step 2. The time to get the defective set D is $O\left(\frac{sm^{2.5}}{(d+h)^2}\right) = O\left(\frac{\lambda^5 \log n}{(d+h)^2}\right)$ as analyzed in Theorem 4. Steps 7 and 8 have up to g and $|D| \leq d$ loops, respectively. Since Step 9 takes time $O(s)$, the running time from Steps 7 to 12 is $O(gds)$ and the cardinality of H_0^* is up to gd.

Second, we analyze the complexity of removing false inhibitors. Step 15 takes time t_1 as in (10). Since $|H_0^*| \leq gd$, the number of loops at Step 17 is at most gd. For the next step, it takes time t_2 for Step 18 as in (10). And it takes time $O(g)$ from Steps 19 to 22. As a result, it takes time $O(t_1 + gd(t_2 + g))$ for Steps 14 to 23.

Finally, Steps 25 to 31 are to remove duplicate inhibitors in the new defective set H. It takes time $O(gd)$ to do that because we know $|H_0^*| \leq gd$.

In summary, the decoding complexity is:

$$O\left(\frac{sm^{2.5}}{(d+h)^2}\right) + O(gds) + O(t_1 + gd \times (t_2 + g)) + O(gd)$$

$$= O\left(\frac{sm^{2.5}}{(d+h)^2}\right) + O(gd(t_2 + g)) = O\left(\frac{\lambda^5 \log n}{(d+h)^2}\right) + O\left(d\lambda^3 \times \left(\frac{\lambda^4}{(d+h)^2} + \lambda^3\right)\right)$$

$$= O\left(d\lambda^6 \times \max\left\{\frac{\lambda}{(d+h)^2}, 1\right\}\right).$$

6 Simulation

In this section, we visualize the number of tests and decoding times in Table 1. We evaluated variations of our proposed scheme by simulation using $d = 2, 4, \ldots, 2^{10}$, $h = 0.2d$, and $n = 2^{32}$ in Matlab R2015a on an HP Compaq Pro 8300SF desktop PC with a 3.4-GHz Intel Core i7-3770 processor and 16-GB memory. Two scenarios are considered here: identification of defective items (corresponding to Sect. 4) and identification of defectives and inhibitors (corresponding to Sect. 5). For each scenario, two models of noise are considered in test outcomes: noiseless setting and noisy setting. In the noisy setting, the number of errors is set to be as 100 times the summation of the number of defective items and the number of inhibitor items. Moreover, in some special cases, the number of items and the number of errors may be reconsidered.

All figures are plotted in 3 dimensions in which the x-axis (on the right of figures), y-axis (in the middle of figures), z-axis (the vertical line) represent number of defectives, number of inhibitors, and number of tests. Our proposed scheme, Ganesan et al.'s scheme, and Chang et al.'s scheme are visualized with red color with marker of circle, green color with marker of pentagram, and blue color with marker of asterisk. In the noisy setting, Ganesan et al.'s scheme is not plotted because the authors of that scheme did not consider the noisy setting.

For decoding time, when the number of items is sufficiently large, the decoding time in our proposed scheme is smaller than that of Chang et al.'s scheme and Ganesan et al.'s scheme.

6.1 Identification of Defective Items

We illustrate decoding time when defective items are the only items that we want to recover here. When there are no errors in test outcomes, as shown in Fig. 1, the

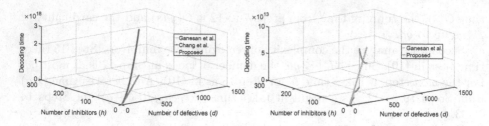

Fig. 1. Decoding time vs. number of defectives and number of inhibitors for identifying only defective items when there are no errors in test outcomes.

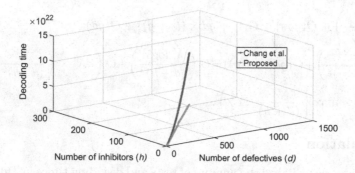

Fig. 2. Decoding time vs. number of defectives and number of inhibitors for identifying only defective items with presence of erroneous outcomes.

decoding time in our proposed scheme is lowest. Since the decoding times in our proposed scheme and Ganesan et al.'s scheme are relatively equal, only one line is visible in the left subfigure of Fig. 1. Therefore, we zoomed in on those lines to see how close these two decoding times are. As plotted in the right subfigure of Fig. 1, when the number of defective items and the number of inhibitor items are small, the decoding time in our proposed scheme is always smaller the one in Ganesan et al.'s scheme. As the number of defective items and the number of inhibitor items increase, the decoding time in our proposed scheme first becomes larger the one in Ganesan et al.'s scheme, though it becomes smaller after the number of items reaches some threshold. We note that if the number of defective items and inhibitor items are fixed while the number of total items is sufficiently large, the decoding time in our proposed scheme is always smaller than the ones in Chang et al.'s scheme and Ganesan et al.'s scheme.

When some erroneous outcomes are allowed, the decoding time in our proposed scheme is always smaller than the one in Chang et al.'s scheme as shown in Fig. 2.

6.2   Identification of Defectives and Inhibitors

We illustrate decoding time for classifying all items. In principle, the complexity of the decoding time in our proposed scheme is smallest in comparison with the

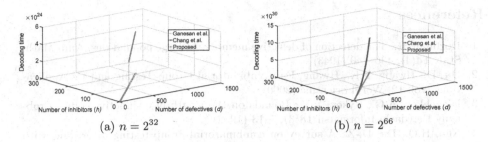

(a) $n = 2^{32}$ (b) $n = 2^{66}$

Fig. 3. Decoding time vs. number of defectives and number of inhibitors for classifying items when there are no errors in test outcomes.

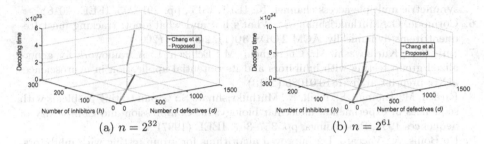

(a) $n = 2^{32}$ (b) $n = 2^{61}$

Fig. 4. Decoding time vs. number of defectives and number of inhibitors for classifying items when there are some erroneous outcomes.

ones in Chang et al.'s scheme and Ganesan et al.'s scheme when the number of items is sufficiently large. When there are no errors in test outcomes, the decoding time of the proposed scheme is smallest when the number of items is at least 2^{66}, as shown in subfigure (b) of Fig. 3. When some erroneous outcomes are allowed, the decoding time in our proposed scheme is always smaller than the one in Chang et al.'s scheme when the number of items is at least 2^{61}, as shown in subfigure (b) of Fig. 4.

7   Conclusion

We have presented two schemes efficiently identifying up to d defective items and up to h inhibitors in the presence of e erroneous outcomes in time $\text{poly}(d, h, e, \log n)$. This decoding complexity is substantially less than that of state-of-the-art systems in which the decoding complexity is $\text{poly}(d, h, e, n)$. However, the number of tests with our proposed schemes is slightly higher. Moreover, we have not considered an inhibitor complex model [12] in which each inhibitor in this work would be transferred to a bundle of inhibitors. Such a model would be much more complicated and is left for future work.

References

1. Dorfman, R.: The detection of defective members of large populations. Ann. Math. Stat. **14**(4), 436–440 (1943)
2. Du, D., Hwang, F.K., Hwang, F.: Combinatorial Group Testing and Its Applications, vol. 12. World Scientific (2000)
3. D'yachkov, A.G., Rykov, V.V.: Bounds on the length of disjunctive codes. Problemy Peredachi Informatsii **18**(3), 7–13 (1982)
4. Ngo, H.Q., Du, D.-Z.: A survey on combinatorial group testing algorithms with applications to DNA library screening. Discret. Math. Probl. Med. Appl. **55**, 171–182 (2000)
5. D'yachkov, A., Polyanskii, N., Shchukin, V., Vorobyev, I.: Separable codes for the symmetric multiple-access channel. In: IEEE ISIT, pp. 291–295. IEEE (2018)
6. Cormode, G., Muthukrishnan, S.: What's hot and what's not: tracking most frequent items dynamically. ACM TODS **30**(1), 249–278 (2005)
7. Bui, T.V., Kuribayashi, M., Cheraghchi, M., Echizen, I.: A framework for generalized group testing with inhibitors and its potential application in neuroscience. arXiv preprint arXiv:1810.01086 (2018)
8. Farach, M., Kannan, S., Knill, E., Muthukrishnan, S.: Group testing problems with sequences in experimental molecular biology. In: Compression and Complexity of Sequences 1997, Proceedings, pp. 357–367. IEEE (1997)
9. De Bonis, A., Vaccaro, U.: Improved algorithms for group testing with inhibitors. Inf. Process. Lett. **67**(2), 57–64 (1998)
10. De Bonis, A., Gasieniec, L., Vaccaro, U.: Optimal two-stage algorithms for group testing problems. SIAM J. Comput. **34**(5), 1253–1270 (2005)
11. Hwang, F.K., Liu, Y.: Error-tolerant pooling designs with inhibitors. J. Comput. Biol. **10**(2), 231–236 (2003)
12. Chang, H., Chen, H.-B., Fu, H.-L.: Identification and classification problems on pooling designs for inhibitor models. J. Comput. Biol. **17**(7), 927–941 (2010)
13. Ganesan, A., Jaggi, S., Saligrama, V.: Non-adaptive group testing with inhibitors. In: ITW, pp. 1–5. IEEE (2015)
14. Bui, T.V., Kojima, T., Kuribayashi, M., Haghvirdinezhad, R., Echizen, I.: Efficient (nonrandom) construction and decoding for non-adaptive group testing. arXiv preprint arXiv:1804.03819. J. Inf. Process. (to appear)
15. Bui, T.V., Kuribayashil, M., Cheraghchi, M., Echizen, I.: Efficiently decodable non-adaptive threshold group testing. In: ISIT, pp. 2584–2588. IEEE (2018)
16. Reed, I.S., Solomon, G.: Polynomial codes over certain finite fields. JSIAM **8**(2), 300–304 (1960)
17. Kautz, W., Singleton, R.: Nonrandom binary superimposed codes. IEEE Trans. Inf. Theory **10**(4), 363–377 (1964)
18. D'yachkov, A., Vilenkin, P., Torney, D., Macula, A.: Families of finite sets in which no intersection of ℓ sets is covered by the union of s others. J. Combin. Theory Ser. A **99**(2), 195–218 (2002)
19. Stinson, D.R., Wei, R.: Generalized cover-free families. Discret. Math. **279**(1–3), 463–477 (2004)
20. Chen, H.-B., Fu, H.-L., Hwang, F.K.: An upper bound of the number of tests in pooling designs for the error-tolerant complex model. Optim. Lett. **2**(3), 425–431 (2008)

21. Cheraghchi, M.: Improved constructions for non-adaptive threshold group testing. Algorithmica **67**(3), 384–417 (2013)
22. Hoorfar, A., Hassani, M.: Inequalities on the lambert w function and hyperpower function. J. Inequal. Pure Appl. Math **9**(2), 5–9 (2008)

Compacting and Grouping Mobile Agents on Dynamic Rings

Shantanu Das[1]([✉]), Giuseppe Di Luna[2], Linda Pagli[3], and Giuseppe Prencipe[3]

[1] Aix-Marseille University, CNRS, LIS, Marseille, France
shantanu.das@lis-lab.fr
[2] CINI, Rome, Italy
[3] Department of Computer Science, Università di Pisa, Pisa, Italy

Abstract. We consider computations by a distributed team of autonomous mobile agents that move on an unoriented dynamic ring network. In particular, we consider 1-interval connected dynamic rings (i.e. at any time, at most one of the edges might be missing). The agents move according to a Look-Compute-Move life cycle, under a synchronous scheduler. The agents may be homogenous (thus identical and monochromatic) or they may be heterogenous (distinct agents have distinct colors from a set of $c \geq 1$ colors). For monochromatic agents starting from any dispersed configuration we want the agents to form a compact segment, where agents occupy a continuous part of the ring and no two agents are on the same node – we call this the Compact Configuration Problem. In the case of multiple colors ($c > 1$), agents of the same color are required to occupy continuous segments, such that agents having the same color are all grouped together, while agents of distinct colors are separated. These formation problems are different from the classical and well studied problem of *Gathering* all agents at a node, since unlike the gathering problem, we do not allow collisions (each node may host at most one agent of a color).

We study these two problems and determine the necessary conditions for solving the problems. For all solvable cases, we provide algorithms for both the monochromatic and the colored version of the compact configuration problem, allowing for at most one intersection between the colored segments (which cannot be avoided in a dynamic ring). All our algorithms work even for the simplest model where agents have no persistent memory, no communication capabilities and do not agree on a common orientation. To the best of our knowledge this is the first work on the compaction problem in any type of dynamic network.

1 Introduction

Research in the field of distributed computing has always considered fault tolerance as an important aspect of algorithm design and there are many

G. Di Luna—This work was performed while the author was affiliated with Aix-Marseille University, France.

T. V. Gopal and J. Watada (Eds.): TAMC 2019, LNCS 11436, pp. 114–133, 2019.
https://doi.org/10.1007/978-3-030-14812-6_8

studies on algorithms tolerating e.g. failures of nodes or links in a network. However, in recent years researchers started to investigate so called *dynamic graphs*, that is graphs where the topological changes are not localized and sporadic; on the contrary, the topology changes continuously and at unpredictable locations, and these changes are not anomalies (e.g., faults) but rather an integral part of the nature of the system [4,9,10,16]. The study of distributed computations in such dynamic graphs has concentrated on problems of information diffusion, reachability, agreement, and other communication problems (see e.g., [2,6,11,14,15,19]). These studies are on message passing networks under various different models of dynamic changes of topology. A general theoretical model for dynamic networks is the evolving graph model, where the network is modelled as a sequence of graphs each of which is a subgraph of the so-called footprint graph which represents the underlying topology. In order to allow useful tasks to be performed on such a network, we need to make some assumptions on the connectivity of the network. One natural way of modelling this is the k-interval connected dynamic graph model (See e.g. [16]). The most restricted of these models is the 1-interval connected dynamic graph model where the only assumption is that at each round, the instance of the graph is connected.

One of the ways of dealing with a highly dynamic environment is the use of mobile code, allowing processes to migrate from node to node on a network during the process of computation. This initiated research in algorithms for mobile agents, where an agent is an autonomous process that moves along the edges of the a network and can perform computations at the nodes of the network, using its own memory and state information, as well as the information stored in the nodes. Mobile agents can also represent agents moving in a dynamic environment. In this case, the agents may have some vision allowing them to see parts of the network and take decisions based on this knowledge. There are many different models for mobile agents depending on their capabilities of remembering (memory), visualizing (vision range), communication and computation abilities.

There has been a lot of research on mobile agents moving in static graphs. The fundamental problems studied are exploration and patrolling, where a team of agents has to visit all nodes of the graph, either once or periodically. A related problem is information dissemination or data collection from the nodes. Several coordination problems for teams of agents have been studied where the agents need to form a particular configuration. One of the most studied problem is rendezvous or gathering where all agents need to meet at a single node of the graph. This requires mechanisms for symmetry breaking as in the leader election problem in distributed computing. The problem has been studied both for agents with identities or anonymous (and thus identical) agents. For homonymous agents (where multiple agents share the same name or color), the problem of grouping the agents into teams with specific colors, is called the team assembling problem, and has been proposed and studied in [17] for agents moving freely in a plane. In the above problems, all agents of the same team must be at the same point or at the same node of the graph. However, it may not always be possible for a single node to host many agents at the same time. In this paper, we avoid multiple agents in the same node, but we want the agents in a team to be close to each other (e.g. to be able to exchange information and coordinate with each other). Motivated by this requirement, we define and study the Compact Configuration Problem (CCP) problem: starting from any configuration of mobile agents scattered in a graph G, the objective is to reach a configuration where each node contains at most one agent and the nodes occupied by agents of the same color induce a connected subgraph of G. To the best of our knowledge, this is the first time compaction problems have been studied at least for distributed teams of autonomous agents. As a preliminary investigation in this paper we consider one of the simplest topology - the ring network. In a ring, solving the CCP problem requires agents of the same team to occupy the nodes of a continuous segment of the ring, without any multiplicities. Although conceptually simple, a ring is highly symmetrical, and it is challenging to solve problems in the ring that require symmetry breaking. We assume that neither the nodes or the agents possess any unique identifiers, which makes the problem much harder. Moreover we consider the network to be dynamic where at any stage of the algorithm, some edges may be unavailable. In this paper the network is a 1-interval ring network, at most one edge of the ring may be missing at any round of the algorithm.

Previous studies of mobile agents in dynamic graphs has focussed on the fundamental problems of exploration, patrolling, gathering and dispersion [1,5, 7,8,12,13]. All these results consider t-interval connected graphs. Under weaker models of dynamicity, only weaker versions of gathering may be solved [3]. The problem of compaction is loosely related to the problem of *near-gathering* that has been studied recently in [18].

Our Contribution. In this paper we investigate the problem of compacting groups of mobile agents initially scattered on *dynamic rings*. We study the problem in two different scenarios: the agents either have all the same color ($c = 1$) or, there are $c > 1$ colors. We show that only local visibility is not sufficient for solving the problem even if the agents have unbounded memory. On the other hand, under global visibility, even oblivious agents (agents with no persistent memory) can solve the problem in all solvable instances. However, due to the dynamicity of the graph, we cannot always avoid intersections between the compacted segments. Our algorithms solve the CCP problem for many colors, with at most one intersection between two colored compact segments, while all other segments are separated. The results of this paper provide the full characterization of solvable instances for the above problems. Due the space limitations, some of the proofs have been omitted.

2   Preliminaries

Interval Connected Ring. A dynamic graph $\mathcal{G}$ is an infinite sequence of static graphs $(G_0, G_1, \ldots)$. For each round $r \in \mathbb{N}$ we have a graph $G_r : (V, E(r))$ where $V : \{v_0, \ldots, v_{n-1}\}$ is a set of nodes and $E : \mathbb{N} \to V \times V$ is a function mapping a round r to a set of undirected edges. Given a dynamic graph $\mathcal{G}$, its footprint G is the graph obtained by the union of all graph instances $G = (V, E_\infty) = (V, \cup_{i=0}^{+\infty} E(i))$. A dynamic graph $\mathcal{G}$ is a 1-interval connected ring if its footprint is a ring and G_r is connected, for each round r. In this paper, we assume 1-interval connected ring such that at most one edge of the ring can be missing at any time; such an edge is arbitrarily chosen by an adversary. Throughout the paper we use the term *dynamic ring* to always mean such a network. The graph $\mathcal{G}$ is anonymous, i.e. all nodes are identical to the agents, the endpoints of each edge are unlabelled, and we do not assume any common orientation (i.e the ring is not oriented).

The Agents. We consider a set of oblivious agents, $A = \{a_1, \ldots, a_k\}$ that are initially located on distinct nodes of a dynamic ring. The agents have no persistent memory, and each agent has an initial color in $[1, c]$ (when $c = 1$, all agents have the same color). When $c > 1$, we assume that the sets of agents having the same color all have the same size h, with $h > 2$. Also, we assume that the size of the ring is at least $2hc + c$. Also, we assume a total ordering on the colors; we call *max_color* the first color in this ordering. Note that the color of the agents is fixed at the beginning and it cannot be changed.

Agents follow the same algorithm executing a sequence of Look, Compute, Move cycles. In the Look phase of each cycle, the agent gets a *snapshot* of the environment. In the Compute phase the agent uses the information from the snapshot and the contents to compute the next destination, which may be the current node or one of its neighbours. During the Move phase an agent traverses an edge to reach the destination node. Given a direction of movement, we say that an agent a is *blocked* by the missing edge, if the edge adjacent to a, in the chosen direction of movement, is missing. We say that two agents *collide* if they occupy the same node at the same round. When two (or more) agents with distinct colors occupy the same node, we say that the collision is *admissible*.

The visibility of the agents may be either global or local:

- Global Snapshot: The snapshot obtained by an agent in round r contains the graph G_r (with the current location of the agent marked), and $\forall v \in G_r$, the colors of the agents (if any) that are located in node v.
- Local Snapshot: The snapshot obtained by an agent at a node v in round r contains the same information as in the Global snapshot, but restricted to a distance of R hops from node v.

Synchronous System. The system is *synchronous*, agents perform each (Look, Compute, Move) cycle in a discrete time unit called round. Rounds are univocally mapped to numbers in $\mathbb{N}$, starting from 0. All agents start the execution at round 0. In each round, each agent in A executes exactly one entire (Look, Compute, Move) cycle.

Configurations and Other Definitions. The configuration of the set of agents A at round r, is a function $C_r : A \to V$ that maps agents in A to nodes of V where agents are located. The term initial configuration indicates the configuration of agents at round 0, when it is clear from the context we omit the round and

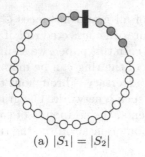
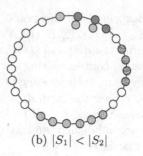

(a) $|S_1| = |S_2|$ (b) $|S_1| < |S_2|$

Fig. 1. (a) Impossibility with no overlap. (b) A Solution for ColoredCCP problem.

we use C to indicate the current configuration. We use the notation $C_r(A)$ to indicate the set of nodes where agents in A are located at round r, and we use $G[C_r(A)]$ to indicate the subgraph induced by the locations of agents in A in graph G.

A *segment* indicates a set of nodes of $\mathcal{G}$ that have connected footprint and that do not form a cycle. Given a node $v \in \mathcal{G}$ we say that the node is *empty* at round r, if in C_r there is no agent on v. Similarly, we say that a segment of nodes is *empty* at round r if all nodes of the segment are empty. We say that a segment is *full* if each node of the segment contain agents of the same color. Given two full segments S_1 and S_2, let a be any agent in S_1 and b any agent in S_2; we define the *distance* between S_1 and S_2 as the smallest number of consecutive empty nodes between a and b. We say that a full segment S is *blocked* by the missing edge if the first agent in S is blocked according to the chosen direction of movement. Also, the "full segment" is said to move when all agents in the segment do a move in a given direction. Given two disjoint segments the distance between them is the minimum number of nodes between two endpoints of the segments.

Any given configuration at a round r can be represented by a sequence of n sets, representing the contents of the n nodes of the ring, starting from any given node. The configuration is said to be: (1) **periodic** if this sequence is periodic, (2) **Palindrome** if some cyclic rotation of this sequence is a palindrome, and (3) **Asymmetric** if it is neither Periodic nor Palindrome.

The Compact Configuration Problem. We introduce the problems we will investigate in the following. The first definition is for monochromatic agents.

Definition 1 (Compact Configuration Problem). *Given a dynamic graph $\mathcal{G}$ with footprint G and a set of agents A, we say that an algorithm solves the distributed* Compact Configuration Problem *(CCP) if and only if there exists a round r, when $G[C_r(A)]$ is connected and each agent occupies a distinct node.*

For multi-colored agents, we want agents of the same color to occupy continuous segments, while agents of distinct colors should be separated. Consider the example configuration in Fig. 1a where agents of two colors are interleaved.

Suppose the adversary blocks the edge marked with a dash, forever. In this case the only way to solve the problem would require two pairs of agents of two distinct colors to cross each other along the other continuous segment of the ring. Now during this process the agents would form an interleaved configuration in another part of the ring and now the adversary can block a new edge (releasing the previously blocked edge) in such a way that a similar configuration as in Fig. 1a is created again. Thus it is not possible to completely segregate the agents of different colors under such an adversary. We therefore allow at most one overlap between two segments of different colors as in Fig. 1b.

Definition 2 (Colored Compact Configuration Problem). *Given a dynamic graph $\mathcal{G}$ with footprint G and set of agents A_i having color $i \in [1, c]$, where $c \geq 2$, we say that an algorithm solves the distributed* Colored Compact Configuration Problem *(ColoredCCP) if and only if there exists a round r where, for each $i \in [1, c]$ except two colors j, p, each agent in A_i occupies a different node and $G[C_r(S_i)]$ is connected. Moreover, if $p \neq j$ it holds that $G[C_r(S_p)]$ and $G[C_r(S_j)]$ intersect.*

Intuitively, in the CCP problem we ask all agents, initially arbitrarily placed, to move so to form one full segment (i.e., with no empty nodes). While in the ColoredCCP problem, we require that all agents having the same color form one full segment, and that at most two of these full segments intersect. All the algorithms presented here allow only admissible collisions: i.e., at any point in time no two agents having the same color occupy the same node of the ring.

Table 1. Results for the CCP and ColoredCCP problems.

	$c = 1$, Global	$c = 2$, Global	$c > 2$, Global	Local
Asymmetric	✓ (Sect. 3.1)	✓ (Sect. 5)	✓ (Sects. 4.1, 4.2)	✗ (Theorem 2)
Palindrome	✓ (Sect. 3.2)	✓ (Sect. 5.2)	✓ (Sect. 4.3)	✗ (Theorem 2)

A summary of the results that we show in this paper is reported in Table 1. The first thing to notice is that solving CCP is impossible when the initial configuration is periodic as shown below (Theorem 1).

Theorem 1. *Given a dynamic ring $\mathcal{G}$, and a set of agents A initially placed on G_0 in a configuration that is periodic and disconnected, it is impossible to solve the* CCP *or the* ColoredCCP *problem, even if the agents have global visibility.*

Proof. In a periodic configuration, the ring can be partitioned into identical segments and none of these are full segments. In case no edge is ever missing, the symmetry between the agents in the two consecutive segments cannot be broken deterministically, thus agents in equivalent positions take the same action in each step and the resulting configuration remains periodic. Since any compacted configuration (with $k < n$) is not periodic, the theorem follows.

Therefore, in the following we assume that the initial configuration is either asymmetric or palindrome. These two cases are handled separately. However even for aperiodic configurations, the compaction problem cannot be solved in the local visibility model. In the following, the visibility graph of a configuration C is the defined as the graph $G_{vis} = (A, E)$, where A is the set of agents and there is an edge $(a, b) \in E$ whenever agent b is within distance R from a.

Theorem 2. *In the local snapshot model, starting from a configuration C such that C is asymmetric and has a connected visibility graph, there is no correct algorithm that solves CCP, avoiding collisions. The result holds even if the agents have unbounded memory.*

Thus, in the following, we will consider the global snapshot model. We assume that the initial configuration is aperiodic (i.e. it is either asymmetric or palindrome). We will also assume that there are more than two agents in total (the special case of exactly $k = 2$ agents is handled separately in Sect. 6).

3   CCP with Global Snapshot

3.1   The Asymmetric Case

First, let us consider the case when the initial configuration is asymmetric. We denote by $\mathcal{E}_r$ the empty segment of maximum size in the configuration at round r. If initially there is only one empty segment of maximum size, we call this segment D. Otherwise, if there is more than one empty segment of maximum size, we can deterministically select one of these as segment D (since the initial configuration is asymmetric). Let S_1 and S_2 be the full segments of length at least 1 on the two sides of segment D (see Fig. 2a). In case $|S_1| \neq |S_2|$, without loss of generality let $|S_1| < |S_2|$; we define the *augmented* S_1, denoted by S_1^+, as the block of nodes constituted by the nodes in S_1 (all non empty), plus the empty node v close to S_1 and not in D, plus, if any, all agents between v and the next empty node (moving away from S_1, see Fig. 3a).

The algorithm for solving CCP tries to increase the length of the empty segment D in each step, while preserving the asymmetric configuration. This is done by moving either S_1 or S_2 or both. The details are explained in Algorithm 1 (Fig. 4).

Lemma 1. *Starting from an asymmetric configuration, by executing Algorithm* ONE COLOR CONNECTED FORMATION, *at any round $r \geq 0$:*

(i) $|\mathcal{E}_r| > |\mathcal{E}_{r-1}|$, and
(ii) The configuration is either asymmetric or solves CCP.

By previous lemma, since the size of $\mathcal{E}_r$ strictly increases at each round, we can state the following:

Theorem 3. *If the initial configuration is asymmetric, the agents executing Algorithm* ONE COLOR CONNECTED FORMATION, *solve CCP within at most n rounds.*

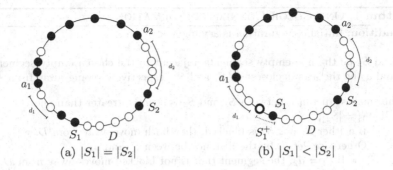

(a) $|S_1| = |S_2|$ (b) $|S_1| < |S_2|$

Fig. 2. Asymmetric initial configuration.

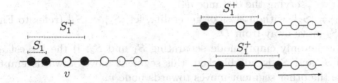

Fig. 3. (a) Definition of S_1^+ (b) Movement of S_1^+ (The arrows denotes the direction of movement)

3.2   The Palindrome Case

Let us now consider the case where the initial configuration C is palindrome i.e., there exists an axis of symmetry.

Theorem 4. *Let the initial configuration be palindrome, aperiodic, and not compact. Then, if the axis of symmetry passes through two empty nodes, then* CCP *is not solvable.*

Proof. Let us assume that the problem is solvable, and that, by contradiction, the axis of symmetry of the initial configuration passes through two empty nodes (see Fig. 5(a)). If no edge is missing during the algorithm, the agents in both sides

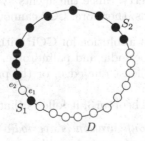

Fig. 4. Case 2 of Algorithm 1: the distance between S_1 and S_2 is 1.

Algorithm 1. ONE COLOR CONNECTED FORMATION

Pre-condition: Initial configuration is asymmetric.

Let S_1 and S_2 be the non-empty segments adjacent to the chosen empty segment D. Let a_1 and a_2 be the agents closest to S_1 and S_2 respectively (going away from D).

1. If the smallest distance between S_1 and S_2 is strictly greater than one:
   (a) If $|S_1| = |S_2|$,
       – If neither S_1 nor S_2 is blocked, they both move away from D.
       – Otherwise, let d_i be the distance between S_i and a_k,
         • If $d_1 = d_2$, the segment that is not blocked moves away from D.
         • Otherwise, without loss of generality, let $d_1 < d_2$.
           * If S_1 is not blocked, then S_1 moves away from D.
           * If S_1 is blocked, then all agents not in S_1 move towards S_1 (preserving the distance d_2).
   (b) If $|S_1| \neq |S_2|$, without loss of generality, let $|S_1| < |S_2|$ (refer to Fig. 2b). S_1^+ and S_2 move away from D.
2. Else: let v the only empty node separating S_1 and S_2. If the largest among the segments S_1 and S_2 is not blocked, this segment moves towards empty node v. Otherwise the other segment moves towards node v.

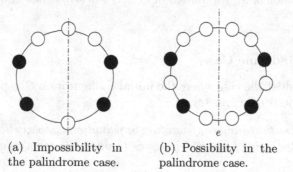

(a) Impossibility in the palindrome case.   (b) Possibility in the palindrome case.

Fig. 5. Example configurations for CCP in the palindrome case.

of the axis perform symmetric actions and the configuration stays palindrome with the same axis of symmetry. Since the agents avoid collision, no agent can move to the nodes on the axis; therefore, CCP cannot be solved in this case.

In Algorithm 2, we present a solution for CCP with more than 2 agents, when the initial configuration is aperiodic and palindrome, and the axis of symmetry either (a) passes through at least one edge, or (b) passes through at least one non empty node.

By Algorithm 2, and by Theorem 3, it follows that:

Theorem 5. *If the initial configuration is aperiodic and palindrome with more than two agents, and the axis of symmetry either (a) passes through at least one edge, or (b) passes through at least one non empty node, then CCP is solvable.*

Algorithm 2. ONE COLOR PALINDROME

Pre-condition: Initial configuration is aperiodic and palindrome, with more than two agents. The axis of symmetry does not pass through two empty nodes.

(a) **If the axis of symmetry passes through at least one edge.** Since the configuration is aperiodic, we can elect a unique edge e that is crossed by the axis of symmetry ax. Once e has been elected, the two agents nearest to e that do not belong to a full segment containing e, are selected to move towards e. If none of these agents are blocked by a missing edge, the symmetry axis is preserved after the moves of the agents. Otherwise, if an agent cannot move because of a missing edge, the next configuration becomes asymmetric, and Algorithm 1 can be applied.

(b) **If the axis of symmetry passes through at least one non empty node.** In aperiodic configurations, it is always possible to elect one of the agents (agent a) among those that occupy the nodes crossed by the unique axis of symmetry.

    1. If the neighbors nodes of a are empty, a moves to one of the neighbors (chosen arbitrarily when both incident edges are available); After the move, the configuration becomes asymmetric and Algorithm 1 can be applied.

    2. If the two neighbors nodes of a are both occupied, and the axis of symmetry passes through another node occupied by agent b, and the two neighbors nodes of b are both empty, then a moves to one of the neighbors (chosen arbitrarily when both incident edges are available); After the move, the configuration becomes asymmetric.

    3. If no agent on the symmetry axis can move, since the configuration is palindrome, there must be two (full) segments of equal size to both the left and the right of a. These two segments move away from a by one position. Now, either the configuration becomes asymmetric (if one of the two segments cannot completely move because of a missing edge), or previous Case b.1 applies.

---

4   COLOREDCCP with Global Snapshot and $c > 2$

In this section, we investigate the compaction problem for heterogenous agents having $c > 2$ distinct colors. Recall that $h = k/c$ is the number of agents of each color. Obviously there is nothing to solve when $h = 1$.

4.1   Asymmetric Initial Configuration and $h \geq 3$

The algorithm for this case builds segments around some specific points of the ring, called *rally points*. These points are identified during the execution of the algorithm, and to each color is assigned a specific rally point.

Definition 3. *We say that agents are forming a* compact line *if they are forming a full segment of size h around the rally point of their color. We say that agents are forming an* almost compact line *if they are forming a full segment of size $h - 1$ around the rally point of their color; the only agent that is not part of the almost compact line is called a* dangling *agent.*

FC denotes the set of agents colored with *max_color*. We say that the current configuration is *correctly placed* if and only if both the following conditions hold:

(i) There are at least $c - 2$ compact lines that do not overlap;
(ii) There is at most one almost compact line.

Algorithm 3. MULTI COLOR CONNECTED SEGMENT (First Step)

Pre-condition: Current configuration is not correctly placed and FC is symmetric. Let a be the first agent in FC, according to the total ordering; a will move of one step to make FC asymmetric.

The algorithm is split into three main steps, described in Algorithms 3, 4, and 6, respectively. Let us first describe the intuition behind each step.

- **First Step** (Algorithm 3). The main idea of the first step is to make an agent with color FC move in such a way that all agents with color FC become asymmetrically placed (this step is skipped if FC is already asymmetric). Once FC is asymmetric, we keep still the agents in FC until the last phase of the algorithm: these agents are used as reference points to univocally identify both the rally points and a unique orientation of the ring.

- **Second Step** (Algorithm 4). In the second step, the algorithm proceeds by making each color but FC to form a full segment around the respective rally point. This step lasts until the configuration becomes correctly placed. Note that it is not possible to wait until all agents not in FC form compact lines (i.e., with no dangling agents): in fact, one of the agents not in FC might become blocked by a missing edge, and the whole system become blocked forever.

- **Third Step** (Algorithm 6). Once the configuration is correctly placed, the only agents still to fix in order to solve the problem, are the agents in FC (that are still asymmetrically placed), and the only dangling agent (that has a color different from FC), if any. Note that, if there is no dangling agents, then there are $c - 1$ compact lines, and no almost compact line.

  The idea is to use the compact lines formed so far to establish a global chirality of the ring, and a rally point for FC. In particular, the already formed compact lines do not move, hence the computed chirality can be kept; the other agents (i.e., those in FC and the dangling agent) move as done in the second step. The movements go on until either ColoredCCP is solved, or there are $c - 1$ compact lines and one almost compact line. In this second case, the only dangling agent and the almost compact line (by construction, all these agents have the same color) move one towards each other until they form a compact line.

Correctness: Since the initial configuration is asymmetric, we have the following:

Lemma 2. *If in the initial configuration FC is not asymmetric, by executing Algorithm 3 (Step One), within finite time agents in FC are placed asymmetrically on the ring.*

Once the agents in FC occupy asymmetric positions on the ring, it is possible to elect one of them as a leader, which provides a global orientation to the ring. Once a global orientation has been computed, the positions of agents in FC allow also to compute the rally points where all other agents will form their respective compact lines (Algorithms 4). Let us denote these points by rp_i, $0 \leq i \leq c-1$. To each rally point rp_i, $0 \leq i \leq c-1$, is assigned a color, c_i (color c_0 is max_color, and is assigned to FC): all agents of color c_i will gather around rp_i, as described in Routine RALLY POINTS CONNECTED FORMATION, reported in Algorithm 5. Given a rally point rp_i, let us call the *rally line* of color c_i a full segment of color c_i that is formed around rp_i. Extending Definition 3, we will call dangling any agent that is not part of a rally line.

Algorithm 4. MULTI COLOR CONNECTED SEGMENT (Second Step)

Precondition: Current configuration is not correctly placed, and FC is asymmetric.

During this step, FC never moves until current configuration is correctly placed. Since FC is asymmetric, it can be used to establish an orientation of the ring; also let v_f the first node in FC according to this orientation.

1. **Rally Points Computation.** FC is now used to compute $c - 1$ *rally points*, as follows: v_f is the first rally point, rp_0. The $i - th$ rally point rp_i is the node of the ring at distance $i * (2 \cdot h + 1)$ from rp_0 (in the clockwise direction; we assume the ring size is at least $2 \cdot h \cdot c + c$).
2. **Formation using Rally Points.** The rally points are now used by the other colored classes to form a line, by executing routine RALLY POINTS CONNECTED FORMATION in Algorithm 5.

Lemma 3. *Within finite time, by executing Routine RALLY POINTS CONNECTED FORMATION in Algorithm 5, the system reaches a configuration with $c - 1$ almost compact rally lines.*

Proof. If $c - 1$ rally lines are almost compact, the lemma trivially follows. Thus, let us assume that there exists at least one rally line, rl_i, that has at least two dangling agents. By construction, only Pattern 1 of RALLY POINTS CONNECTED FORMATION can be executed. Let us consider only agents having color c_i. Let a be the closest agent in the counter-clockwise direction to rp_i that has not

---

Algorithm 5. MULTI COLOR CONNECTED SEGMENT (Auxiliary routine)

There are c rally points, sorted according to the ring orientation.

Case:(Pattern 1) There exists a rally line rl_i of color different from *max_color* that is being formed around rally point rp_i that has at least two dangling agents. Given a dangling agent a, let p be the counter-clockwise path that connects a with its own rally line.

Movement (see Fig. 6):

  - If a is not the farthest agent from its rally line (according to the counter-clockwise direction), and on p there is a missing edge, then a does not move.
  - If on p there is no missing edge, then a moves counterclockwise.
  - If on p there is a missing edge, and a is the farthest agent from it rally line (according to the counter-clockwise direction), then a moves clockwise.

Case:(Pattern 2) For all rally lines of color different from *max_color*, there is at most one dangling agent; let m be the number of rally lines with exactly $h-1$ agents (i.e., only one dangling agent). Given a dangling agent a, let p be the counter-clockwise path that connects a with its own rally line.

Movement (see Fig. 7):

  - If a does not have the shortest distance to its own rally line among all distances of all other dangling agents from their own rally lines (according to clockwise direction), then a does not move.
  - If the first edge on p is not missing, then a moves counter-clockwise.
  - If there are $m-1$ dangling agents that are blocked by a missing edge, and a has the shortest distance to its own rally line among all distances of all other dangling agents from their own rally lines, then a moves clockwise.

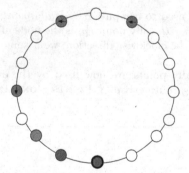 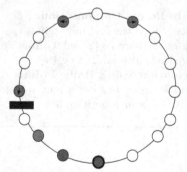

(a) The dangling agents are not blocked. They move counterclockwise towards their rally line.

(b) The dangling agents are blocked. The last agent changes direction and move clockwise towards its rally line.

Fig. 6. Pattern 1 of Algorithm 5.

reached rl_i yet. We will show that, within finite time, the size of rl_i increases. Note that, as long as a is not blocked, it will always move towards its own rally line, even if other agents are blocked.

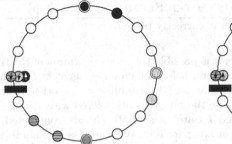
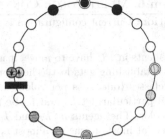

(a) The black agent switches direction.

(b) The vertical striped agent switches direction.

Fig. 7. Pattern 2 of Algorithm 5.

Therefore, if a is never blocked by the missing edge, the statement trivially follows. Otherwise, let r be the furthest agent from rl_i: by Pattern 1, r switches direction, and starts moving towards rl_i. As long as a is blocked, r keeps approaching rl_i. If r becomes blocked, a does at least one step towards rl_i decreasing its distance from rl_i. Thus, within finite time, either a or r will join rl_i.

In conclusion, within finite time, rl_i becomes almost compact, and the lemma follows.

Lemma 4. *Let us assume that in the current configuration there exist $m > 2$ rally lines with exactly one dangling agent each, and $c - 1 - m$ compact lines. Within finite time, by executing Routine* RALLY POINTS CONNECTED FORMATION *in Algorithm 5, m decreases.*

Thus, by previous Lemmas 3 and 4, the following holds:

Lemma 5. *Within finite time, by executing Algorithm 4, the configuration becomes correctly placed.*

Finally, by executing Algorithm 6, agents are able to solve the problem. In particular, at the beginning of this step, there are at least $c - 2$ compact lines, at most one line with just one dangling agent, and finally the agents in FC, that still needs to be compacted. Thus, the agents that still need to be placed to correctly solve the problem, are those in FC and the dangling agent.

Lemma 6. *If there are 3 or more colors, then, within finite time, by executing Algorithm 6 (Third Step), the* ColoredCCP *is solved.*

Combining all the results from this section, we have the following result:

Theorem 6. *Starting from an asymmetric initial configuration, with $c \geq 3$ and $h \geq 3$, algorithm* MULTI COLOR CONNECTED SEGMENT *correctly solves the* ColoredCCP *problem.*

Algorithm 6. MULTI COLOR CONNECTED SEGMENT (Third Step)

Precondition: Current configuration is correctly placed.

- Since agents in FC have to move, it is possible that the orientation of the ring that FC is establishing gets lost. Therefore, before moving any agent in FC, the other $c - 1$ classes (one class per color) are used to establish a new orientation of the ring: in particular, let L_2 and L_3 be the set of agents colored with the second and third color. The agents in L_2 and L_3 are either both already compacted, or one of them (at most) forms an almost compact line. Without loss of generality, let us assume that L_2 forms a compact line. The new orientation of the ring follows the smallest distance from L_2 to L_3 (note that, by the definition of rally points, this distance is unique).
  Now, the rally point for FC, call it rp^*, is computed by taking the middle point of the largest segment between the lines that are not colored FC.
- The agents in FC and the dangling agent starts compacting, using rules described in RALLY POINTS CONNECTED FORMATION, as follows: agents in FC use rp^* as rally point, and the dangling agent uses as rally point the middle point of the almost compact line having its own color.
- If all lines are formed, and at most one has a dangling agent, the two portion of the last line to be compacted move towards each other.

4.2 Asymmetric Initial Configuration and $h = 2$

In this section we focus on the case of agents with many colors ($c > 2$) but only two agents of each color ($h = 2$). In this case, agents execute again the three steps of previous section with a slight modification; The agents of the two maximum colors act as a single team having just one color. Thus, FC is the union of the agents having these two colors. This ensures that there are at least 3 agents in FC, such that the previous algorithm can still be executed.

At the end of the algorithm, agents of $c - 2$ colors have formed compact lines and only the agents in FC form a segment where two colors are interleaved. More specifically, Configuration A in Fig. 8 is, up to symmetries, the only possible interleaved configuration. At this point we run a simple separation procedure that separates the agents of distinct colors and forms the remaining two compact lines. As shown in Fig. 8 from the configuration A, we can reach either configuration B or configuration C by swapping the agents on either edge e_1 or edge e_2 (at least one of these edges must be available). Thus we reach a configuration where $c - 1$ compact lines are already formed. For the two agents of the last color that is not compacted yet, these two agents can simply move towards each-other. Since there are at least 2 compact lines of other colors already formed, the configuration remains asymmetric after any movement of these two agents. Thus, eventually these two agents will reach adjacent nodes and thus, the ColoredCCP problem would be solved.

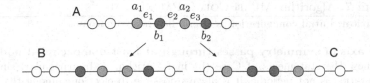

Fig. 8. Separating an interleaved line with $h = 2$ and two colors.

Theorem 7. *Starting from an asymmetric initial configuration, with $c \geq 3$ and $h = 2$, the modified algorithm* MULTI COLOR CONNECTED SEGMENT *in this section correctly solves the* ColoredCCP *problem.*

4.3 Palindrome Initial Configuration for $c > 2$ Colors

We now consider the only remaining case for ColoredCCP with $c > 2$ colors.

Theorem 8. *Starting from an initial configuration that is palindrome, aperiodic, and not compact, the* ColoredCCP *problem for $c > 2$ is not solvable if*

1. *the axis of symmetry passes through two empty nodes, or,*
2. *the axis of symmetry passes through one edge and one empty node, or,*
3. *the axis of symmetry passes through two edges and $c > 3$.*

Proof. We prove each of the statements independently.

1. The proof comes directly from Theorem 4.
2. By hypothesis, the configuration is palindrome; moreover, the symmetry axis intersects the ring on a node v and an edge e. Therefore, the agents can form the compact lines either around v or around e. If the lines are formed around v, since the ring not oriented, two agents with the same color would move to v, thus violating the no collision requirement of the problem. If the line would be formed around e, then there would be three compact lines of three different colors around e, intersecting, and thus violating the ColoredCCP specification.
3. Since the configuration is palindrome, then compact lines formable by agents have to be centred around the symmetry axis. By construction, it is only possible to form two disjoint compact lines. Since there are more than 3 colors, by the pigeonhole principle, these three compact lines will intersect, thus violating the specification of ColoredCCP.

Algorithm 7 solves the remaining cases when (a) the axis of symmetry passes through at least one occupied node, or, (b) there is an axis of symmetry passing through two edges, and $c = 3$. We can thus conclude that:

Theorem 9. *If the initial configuration is aperiodic and palindrome and either (a) the axis of symmetry passes through at least one occupied node, or (b) there is an axis of symmetry passing through two edges, and $c = 3$, then* ColoredCCP *is solvable.*

Algorithm 7. Algorithm MULTI COLOR PALINDROME

Pre-condition: Initial configuration is aperiodic and palindrome.

(a) **If the axis of symmetry passes through at least one occupied node.**

We follow the statements of Case (b) in Algorithm 2. In particular, since the configuration is not periodic, it is always possible to elect one among the agents that are on the axis of symmetry, let this agent be a. We distinguish the three possible cases:

1. If the neighbors nodes of a are empty, a moves of one position, and the configuration becomes asymmetric. Now, Algorithm of Sect. 4.1 can be run.

2. If the neighbors nodes of a are occupied, and the axis of symmetry passes through another node b, and the neighbors nodes of b are empty, then b moves of one position, and the configuration becomes asymmetric. Now, Algorithm of Sect. 4.1 can be run.

3. Finally, no node on the symmetry axis can move. In this case, since the configuration is palindrome, there must be two block of nodes of equal size to the left and to the right of a. These two block of nodes move away from a of one position. Now, either the configuration becomes asymmetric (one of the two block does not move because of a missing edge), or previous Case a.1 applies.

(b) **If the axis of symmetry passes through two edges, and $c = 3$.**

Let e be one of the edges intersected by the symmetry axis, elected as in Case (a) of Algorithm 2. The agents proceed as follows: at each round, only agents with maximum color are allowed to move. In particular, the two agents nearest to e that do not belong to a full segment containing e, move towards e. If no agent is blocked by an edge removal, the symmetry axis is preserved and eventually all agents with maximum color form a full segment around e. Otherwise, if an agent is blocked, the next configuration becomes asymmetric; thus we can apply the algorithm of Sect. 3.1.

Once we have a compact segment of the first color, following the same strategy, the second color in the order will form a full segment around the antipodal edge e' of e. Finally, the agents of the third color form a full segment around edge e, solving ColoredCCP.

---

5   COLORED CCP with Global Snapshot and $c = 2$

5.1   Asymmetric Initial Configuration

If $h = 2$, the agents act like they have the same a color, and form one full segment using the algorithm presented in Sect. 3.1. Once this full segment is formed, they separate using the technique described in Sect. 4.2.

If $h \geq 3$, algorithm TWO COLOR CONNECTED SEGMENT (Algorithm 8) solves the problem. The algorithm is based on a modification of the strategy in Sect. 4.1: eventually two compact lines are formed, with possible overlap.

By the discussion presented in previous Sect. 4.1, we have:

Theorem 10. *If $c = 2$, and the initial configuration is asymmetric, within finite time, Algorithm 8 solves* ColoredCCP.

Algorithm 8. TWO COLOR CONNECTED SEGMENT

1. **First Step** is the same as in Algorithm 3.
2. **Second Step** is the same as in Algorithm 4. Please note that, at the end of this step, agents in FC are not forming a full segment, while the agents with the other color form an almost compact line.
3. **Third Step**: at this point the dangling agent and the almost compact line will form a unique line by moving towards each other. Once this is done, the agents in FC form a compact line, by executing Algorithm 1.

5.2 Palindrome Initial Configuration

First of all, by Theorem 4, we can state that:

Theorem 11. *Let the initial configuration be palindrome, aperiodic, and not compact. If the axis of symmetry passes through two empty nodes, then* Colored-CCP *is not solvable.*

Finally, following the lines of previous Algorithm 7, it is easy to show that (a) if the axis of symmetry passes through at least one occupied node, or (b) if $h \geq 2$, $c = 2$ and there is an axis of symmetry passing through at least one edge, then ColoredCCP is solvable.

6 Special Case: Compaction of $k = 2$ Agents

For the CCP problem, we assumed that there are $k > 2$ agents throughout this paper and we now consider the remaining case. For the case of $k = 2$ agents of the same color, the CCP cannot be solved in dynamic rings using oblivious agents. Any configuration with two agents is a palindrome configuration. Thus, if the axis of symmetry passes through two nodes (i.e the distance between the agents is even on both sides), then the problem in not solvable due to previous results. On the other hand if the symmetry axis passes through an edge, when the agents try to approach this edge, one agent may be blocked, so the resulting configuration would have the axis of symmetry passing through a node and the agents would not be able to solve the problem. Thus, some form of persistent memory is needed to solve the problem. If the agents have some persistent memory of at least one bit, then during the execution of the algorithm when one of the agents is blocked, then this agent can be elected, by setting a flag in the memory of this agent; during the rest of the algorithm, the agent can simply approach each-other until they are compacted or there is exactly one empty node between them. If there is only one empty node between the agents and none of the agents are blocked then the leader agent can move to the empty node to solve the problem. On the other hand, if none of the agents are blocked during the execution of the algorithm, then both agents can move in synchronous steps (maintaining the same symmetry axis) and eventually reaching the two end-points of the edge through which the axis passes. Thus we can state the following result:

Theorem 12. *For exactly two agents,* ColoredCCP *is solvable if and only if (i) the initial configuration has the axis of symmetry passing through at least one edge and (ii) the agents have persistent memory.*

7   Conclusions

In this paper we introduced and studied the Compact Configuration Problem and the Colored Compact Configuration Problem for a set of autonomous mobile agents on a dynamic ring networks. We showed that both the problems can be solved only if the initial configuration is aperiodic. The results of this paper provides the exact characterization of the solvable initial configurations for the CCP and ColoredCCP problems. We also showed that having persistent memory is not necessary for solving the problem (except in the special case of two agents). It would be interesting to determine what additional capabilities of the agents would allow them to the solve the ColoredCCP problem without any overlaps. Future investigations on this problem could also consider other graph topologies under either the same or a more relaxed model for dynamicity. Another interesting issue is to consider less synchronous models where all agents may not start at the same time and they may not be active at the same time.

References

1. Agarwalla, A., Augustine, J., Moses, W.K., Madhav, S., Sridhar, A.K.: Deterministic dispersion of mobile robots in dynamic rings. In: Proceedings of 19th International Conference on Distributed Computing and Networking (ICDCN), pp. 19:1–19:4 (2018)
2. Biely, M., Robinson, P., Schmid, U., Schwarz, M., Winkler, K.: Gracefully degrading consensus and k-set agreement in directed dynamic networks. In: 2nd International Conference on Networked Systems (NETSYS), pp. 109–124 (2015)
3. Bournat, M., Dubois, S., Petit, F.: Gracefully degrading gathering in dynamic rings. In: Proceedings of the 20th International Symposium on Stabilization, Safety, and Security of Distributed Systems (SSS), pp. 349–364 (2018)
4. Casteigts, A., Flocchini, P., Quattrociocchi, W., Santoro, N.: Time-varying graphs and dynamic networks. Int. J. Parallel Emergent Distrib. Syst. **27**(5), 387–408 (2012)
5. Das, S., Di Luna, G.A., Gasieniec, L.: Patrolling on dynamic ring networks. In: 45th International Conference on Current Trends in Theory and Practice of Computer Science (SOFSEM), pp. 150–163 (2019)
6. Di Luna, G., Baldoni, R.: Brief announcement: investigating the cost of anonymity on dynamic networks. In: 34th Symposium on Principles of Distributed Computing (PODC), pp. 339–341 (2015)
7. Di Luna, G., Dobrev, S., Flocchini, P., Santoro, N.: Live exploration of dynamic rings. In: 36th IEEE International Conference on Distributed Computing Systems (ICDCS), pp. 570–579 (2016)
8. Di Luna, G., Flocchini, P., Pagli, L., Prencipe, G., Santoro, N., Viglietta, G.: Gathering in dynamic rings. Theor. Comput. Sci. (2018, in press)

9. Flocchini, P., Enriques, A.M., Pagli, L., Prencipe, G., Santoro, N.: Point-of-failure shortest-path rerouting: computing the optimal swap edges distributively. IEICE Trans. Inf. Syst. **E89–D**(6), 700–708 (2006)
10. Flocchini, P., Pagli, L., Prencipe, G., Santoro, N., Widmayer, P.: Computing all the best swap edges distributively. J. Parallel Distrib. Comput. **68**(7), 976–983 (2008)
11. Haeupler, B., Kuhn, F.: Lower bounds on information dissemination in dynamic networks. In: Aguilera, M.K. (ed.) DISC 2012. LNCS, vol. 7611, pp. 166–180. Springer, Heidelberg (2012). https://doi.org/10.1007/978-3-642-33651-5_12
12. Ilcinkas, D., Klasing, R., Wade, A.M.: Exploration of constantly connected dynamic graphs based on cactuses. In: Halldórsson, M.M. (ed.) SIROCCO 2014. LNCS, vol. 8576, pp. 250–262. Springer, Cham (2014). https://doi.org/10.1007/978-3-319-09620-9_20
13. Ilcinkas, D., Wade, A.M.: Exploration of the T-interval-connected dynamic graphs: the case of the ring. In: Moscibroda, T., Rescigno, A.A. (eds.) SIROCCO 2013. LNCS, vol. 8179, pp. 13–23. Springer, Cham (2013). https://doi.org/10.1007/978-3-319-03578-9_2
14. Jadbabaie, A., Lin, J., Morse, A.S.: Coordination of groups of mobile autonomous agents using nearest neighbor rules. IEEE Trans. Autom. Control **48**(6), 988–1001 (2003)
15. Kuhn, F., Lynch, N., Oshman, R.: Distributed computation in dynamic networks. In: 42th Symposium on Theory of Computing (STOC), pp. 513–522 (2010)
16. Kuhn, F., Oshman, R.: Dynamic networks: models and algorithms. SIGACT News **42**(1), 82–96 (2011)
17. Liu, Z., Yamauchi, Y., Kijima, S., Yamashita, M.: Team assembling problem for asynchronous heterogeneous mobile robots. Theor. Comput. Sci. **721**, 27–41 (2018)
18. Pagli, L., Prencipe, G., Viglietta, G.: Getting close without touching: near-gathering for autonomous mobile robots. Distrib. Comput. **28**(5), 333–349 (2015)
19. Ren, W., Beard, R.W.: Consensus seeking in multi-agent systems under dynamically changing interaction topologies. IEEE. Trans. Autom. Control **50**(5), 655–661 (2005)

Maximum Independent and Disjoint Coverage

Amit Kumar Dhar[1], Raghunath Reddy Madireddy[2], Supantha Pandit[3(✉)], and Jagpreet Singh[4]

[1] Department of Electrical Engineering and Computer Science,
Indian Institute of Technology Bhilai, Datrenga, Chhattisgarh, India
amitkdhar@iitbhilai.ac.in
[2] Department of Computer Science and Engineering,
Indian Institute of Technology Ropar, Rupnagar, Punjab, India
raghunath.reddy@iitrpr.ac.in
[3] Stony Brook University, Stony Brook, NY, USA
pantha.pandit@gmail.com
[4] Department of Information Technology,
Indian Institute of Information Technology Allahabad,
Allahabad, Uttar Pradesh, India
jagpreets@iiita.ac.in

Abstract. Set Cover is one of the most studied optimization problems in Computer Science. In this paper, we target two interesting variations of this problem in a geometric setting: (*i*) *Maximum Disjoint Coverage (MDC)*, and (*ii*) *Maximum Independent Coverage (MIC)* problems. In both problems, the input consists of a set P of points and a set O of geometric objects in the plane. The objective is to maximize the number of points covered by a set O' of selected objects from O. In the *MDC* problem we restrict the objects in O' are pairwise disjoint (non-intersecting). Whereas, in the *MIC* problem any pair of objects in O' should not share a point from P (however, they may intersect each other). We consider various geometric objects as covering objects such as axis-parallel infinite lines, axis-parallel line segments, unit disks, axis-parallel unit squares, and intervals on a real line. For axis-parallel infinite lines both *MDC* and *MIC* problems admit polynomial time algorithms. On the other hand, we prove that the *MIC* problem is NP-complete when the objects are horizontal infinite lines and vertical segments. We also prove that both *MDC* and *MIC* problems are NP-complete for axis-parallel unit segments in the plane. For unit disks and axis-parallel unit squares, we prove that both these problems are NP-complete. Further, we present PTASes for the *MDC* problem for unit disks as well as unit squares using Hochbaum and Maass's "shifting strategy". For unit squares, we design a PTAS for the *MIC* problem using

S. Pandit—The author is partially supported by the Indo-US Science & Technology Forum (IUSSTF) under the SERB Indo-US Postdoctoral Fellowship scheme with grant number 2017/94, Department of Science and Technology, Government of India.

T. V. Gopal and J. Watada (Eds.): TAMC 2019, LNCS 11436, pp. 134–153, 2019.
https://doi.org/10.1007/978-3-030-14812-6_9

Chan and Hu's "mod-one transformation" technique. In addition to that, we give polynomial time algorithms for both *MDC* and *MIC* problems with intervals on the real line.

Keywords: Set cover · Maximum coverage · Independent set · NP-hard · PTAS · Line · Segment · Disk · Square

1 Introduction

The *Set Cover* problem along with its geometric variations are fundamental problems in Computer Science with numerous applications in different fields. In the geometric set cover problem, we are given a set of points P and a set of objects O, the objective is to cover all the points in P by choosing the minimum number of objects from O. A variation of the geometric set cover problem is the *Maximum Coverage* problem, where in addition to P and O, an integer k is given as a part of the input. The objective is to select at most k objects from O that cover the maximum number of points from P. In this paper, we consider two interesting variations of the maximum coverage problem. In the following, we give the formal definitions of the problems.

Maximum Disjoint Coverage (*MDC*) Problem: Given a set P of points and a set O of objects in the plane. The objective is to find a set of disjoint (pairwise non-intersecting) objects $O' \subseteq O$ that covers maximum number of points from P (see Fig. 1(a)).

Maximum Independent Coverage (*MIC*) Problem: Given a set P of points and a set O of objects in the plane. The objective is to find a set $O' \subseteq O$ of objects that covers maximum number of points from P such that no two objects in O' share a point from P (see Fig. 1(b)).

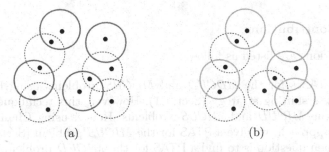

(a)   (b)

Fig. 1. (a) An example of the *MDC* problem. (b) An example of the *MIC* problem.

These problems have applications in wireless communication networks, where the objective is to service each receiver from only a single base station from a set of given base stations and also to maximize the receivers serviced. In case a receiver receives signals from more than one base stations then it may not be able to communicate at all because of signal interference. While in the case of static receivers, the regions covered by base stations can intersect, the same is not favorable for moving receivers as the receivers can eventually reach the intersection of two base stations.

The *MDC* problem is closely related to the *Maximum Weighted Independent Set (MWIS)* problem. In the *MWIS* problem, we are given a set of weighted objects O, and the objective is to find a set of pairwise non-intersecting objects from O whose total weight is maximized. We can interpret the *MDC* problem as the *MWIS* problem where the set of objects in the *MDC* problem is same as the set of objects in the *MWIS* problem and the weight of an object is the number of points it covers. Hence, the *MDC* problem is same as the *MWIS* problem with a special weight function (the number of points covered by objects). By a similar argument, the *MIC* problem is also closely related to the *Maximum Weighted Discrete Independent Set (MWDIS)* problem [4]. In the *MWDIS* problem, we are given a set P of points and a set O of weighted objects, the objective is to select a subset $O' \subseteq O$ of the maximum total weight such that no pair of objects in O' share a point in P.

In this paper, we consider the following problems.

- **MICL**: The *MIC* problem with axis-parallel lines.
- **MICHLVSeg**: The *MIC* problem with horizontal infinite lines and vertical Segments.
- **MICUSeg**: The *MIC* problem with axis-parallel unit segments.
- **MICUD**: The *MIC* problem with unit disks.
- **MICUS**: *MIC* problem with axis-parallel unit squares.
- **MICI**: The *MIC* problem with intervals on a real line.

In a similar fashion, we consider the *MDC* problem with axis-parallel lines (**MDCL**), axis-parallel unit segments (**MDCUSeg**), unit disks (**MDCUD**), axis-parallel unit squares (**MDCUS**), and intervals on a real line (**MDCI**).

1.1   Our Contributions

Our contributions are listed as follows:

- We give PTASes for the *MDCUD* and *MDCUS* problems using Hochbaum and Maass's shifting strategy (Sect. 2.1). However, this technique does not work for the *MICUD* and *MICUS* problems. Hence, using Chan and Hu's mod-one approach, we give a PTAS for the *MICUS* problem (Sect. 2.2). The natural open question is to find a PTAS for the *MICUD* problem.
- We prove that the *MICHLVSeg* problem is NP-complete (Sect. 3.1). We also prove that the *MDCUSeg* and *MICUSeg* problems are NP-complete

(Sect. 3.2). Similar reduction shows that the *MDCUD* and *MICUD* problems are NP-complete (Sect. 3.3). Finally, we prove that both *MDCUS* and *MICUS* problems are also NP-complete (Sect. 3.4).

➤ We present a polynomial time algorithm for the *MICL* problem by reducing it to the *MWIS* problem in vertex weighted bipartite graph (Sect. 4.1). We also note that the *MDCL* problem is easy to solve in polynomial time. Further, we provide polynomial time algorithms for both the *MDCI* and *MICI* problems (Sect. 4.2).

1.2   Related Work

The set cover problem is NP-complete and a greedy algorithm achieves a $\ln n$ factor approximation algorithm [9] and this bound is essentially tight unless P=NP [9]. Similarly, for the maximum coverage problem, there is a well-known greedy algorithm which produces an approximation factor of $1 - \frac{1}{e}$, which is essentially optimal (unless P=NP) [9]. Most of the geometric versions of these problems are NP-hard as well. Another variation of the set cover problem is the **Maximum Unique Coverage** problem [8]. In this problem, we are given a set of points P and a set of objects O in the plane and one has to find a subset $O' \subseteq O$ which uniquely covers the maximum number of points in P. The authors have shown that the problem is NP-hard for unit disks and provided a 18 factor approximation algorithm. Later, Ito et al. [11], improved this factor to 4.31. For the case of unit squares, a PTAS is known [12]. Recently, Mehrabi [15], studied a variation of this problem, called the **Maximum Unique Set Cover**, in which one needs to cover all the points in P while maximizing the number of uniquely covered points in P. Further, he proved that for unit disks and unit squares, the problem is NP-hard [15] and gave a PTAS for unit squares. However, no approximation algorithm is known for the unit disks. Note that the NP-hardness of the *MICUD* and *MICUS* problems can also be obtained from the NP-hardness result of [15]. In [17], the authors show that the problem of **Min-max-coverage-for-unit-square** with depth 1 is NP-hard. The same proof essentially shows that *MDCUS* is NP-hard.

The *MWIS* problem is known to be NP-hard for unit disks graphs [6]. Further, PTASes are also known for disks and squares [7,18]. For the case of axis-parallel rectangles, a $(1 + \epsilon)$-approximation algorithm which runs in quasi-polynomial time is also known [1]. For pseudo-disks, Chan and Har-Peled [4] gave an $O(1)$-approximation algorithm for the *MWIS* problem using linear programming. Their algorithm can be extended to the *MWDIS* problem for pseudo-disks in the plane. Chan and Grant [3] considered the unweighted version of the discrete independent set problem with the downward shadows of horizontal segments in the plane. They gave a polynomial time algorithm for this problem. For the case of *MWDIS* problem, Chan and Har-Peled [4] gave a $O(1)$-factor approximation algorithm for pseudo-disks. On a related note, PTASes are known for *MWDIS* with disks and axis-parallel squares when all objects have the same weight [14].

2   PTASes

2.1   PTASes: The *MDCUD* and *MDCUS* Problems

We first design a PTAS for the *MDCUD* problem. The algorithm is based on Hochbaum and Maass's [10] shifting strategy and follows the outline of the algorithm developed in [7] for providing a similar approximation guarantee. Let P be a set of n points, D be a set of m unit disks in the plane. We first enclose P and D inside a rectangular box B. Next, we partition B into vertical strips of unit width. Let us fix a constant k, called the *shifting parameter*. We define a *fat-strip* as the collection of at most k consecutive unit strips. In the i-th shift, $shift_i$, the first fat-strip consists of the first i consecutive unit vertical strips and the subsequent fat-strips, except possibly the last one, each contains exactly k consecutive unit vertical strips. We apply k shifts, $shift_i$ for $1 \le i \le k$ in the horizontal direction.

The idea of the algorithm is to find a solution to cover the maximum number of points for each shift, $shift_i$ for $1 \le i \le k$, and then select the solution among them which covers the maximum points. A solution for a particular $shift_i$ is obtained by finding solutions in each fat-strip S_j, for $j = 1, 2, \ldots$, during that shift and then taking the union of all such solutions. To obtain a solution for each fat-strip, the shifting strategy is reapplied to each fat-strip in the vertical direction. As a result each fat-strip is partitioned into "squares" of size at most $k \times k$. Later in this section, we will design an exact algorithm that covers a maximum number of points in each such square.

Let X be an approximation algorithm applied to each fat-strip $S_j : j = 1, 2, \ldots$ which return a solution W_j^X (maximum number of points) and α_X be the approximation factor. Further, let sh be the shifting algorithm that applies X in each fat-strip of a particular shift and α_{sh} be its approximation factor. Now, we prove the following lemma:

Lemma 1 (Shifting Lemma [10]). $\alpha_{sh} \ge \alpha_X(1 - \frac{1}{k})$, *where* k, X, α_{sh}, *and* α_X *are defined above.*

Proof. Let S_j be a fat-strip of width k during $shift_i$. Let opt_j be the optimal number of points covered by disjoint disks in S_j and W_j^X be the number of points covered by disjoint disks while algorithm X applied to S_j. Then, by the definition of α_X we have,

$$W_j^X \ge \alpha_X \cdot opt_j$$

Let $W^{X(shift_i)}$ denote the number of points return by algorithm X for $shift_i$. Now summing the solutions over all the fat-strips during $shift_i$ we have,

$$W^{X(shift_i)} = \sum_{j \in shift_i} W_j^X \ge \alpha_X \sum_{j \in shift_i} opt_j$$

Let opt be the number of points covered by disjoint disks in an optimal solution and $opt^{(i)}$ be the number of points in an optimal solution which are covered by disjoint disks in optimal solution covering two adjacent fat-strips in

i-th shift. Let W_{sh} be the number of points returned by the shifting algorithm sh. Then, we have:

$$\sum_{j \in shift_i} opt_j \geq opt - opt^{(i)}$$

and now:

$$
\begin{aligned}
W_{sh} &= \max_{j=1 \text{ to } k} W^{X(shift_i)} \\
&\geq \tfrac{1}{k} \sum_{j=1}^{k} W^{X(shift_i)} \\
&= \tfrac{1}{k} \sum_{j=1}^{k} \left(\sum_{j \in shift_i} W_j^X \right) \\
&\geq \tfrac{1}{k} \alpha_X \sum_{j=1}^{k} \left(\sum_{j \in shift_i} opt_j \right) \\
&\geq \tfrac{1}{k} \alpha_X \sum_{j=1}^{k} \left(opt - opt^{(i)} \right)
\end{aligned}
$$

There can be no disk which covers points from the optimal solution that cover points in two adjacent strips in more than one shift partition. Therefore, the sets $opt^{(1)}, \ldots, opt^{(k)}$ are disjoint and can add up to at most opt. Hence, $\sum_{j=1}^{k} \left(opt - opt^{(i)} \right) \geq (k-1)opt$. Finally we have, $W_{sh} \geq \alpha_X (1 - \tfrac{1}{k})opt$. Hence the lemma. □

Lemma 2. *Let T be a square of size $k \times k$ and L_v be a vertical line which bisects T vertically into two equal rectangles. Then at most $\lceil \sqrt{2}k \rceil$ pairwise disjoint unit disks can intersect L_v.*

Proof. Consider a vertical line segment L_v of length k. We want to find a maximum cardinality pairwise non-intersecting unit disks that intersects L_v. Consider a rectangle R of width 2 and height k such that L_v vertically partition into two equal parts. Observe that, any unit disk which intersects L_v has its center inside R. We take $\lceil \sqrt{2}k \rceil$ squares each of length $\sqrt{2}$. We arrange these squares into two columns on both sides of L_v with $\lceil \tfrac{k}{\sqrt{2}} \rceil$ in each column (the two columns share the common boundary with L_v). Then, this arrangement completely covers the rectangle R. Consider a single square s of length $\sqrt{2}$. Observe that all unit disks whose centers are inside s, are pairwise intersected. Hence, at most one of them can be part of a maximum disjoint set. □

We now describe an algorithm which finds an optimal solution in a $k \times k$ square T. Let $P_T \subseteq P$ and $D_T \subseteq D$ be the set of points and disks inside T respectively. Consider a vertical line L_v and a horizontal line L_h that partition T into four squares T_1, T_2, T_3, and T_4 of size $k/2 \times k/2$ each. Let $D_{vh} \subseteq D_T$ be the set of unit disks which intersect either L_v or L_h, or both. Let D_1, D_2, D_3, and D_4 be the set of unit disks in T_1, T_2, T_3, and T_4 respectively such that they do not intersect L_v and L_h. Now we have the following observations. We can find the set of non-intersecting disks in optimal solution for D_{vh}, since the size of maximum disjoint set is at most $2\lceil \sqrt{2}k \rceil$ (by Lemma 2). Any two disk that belongs to two different D_i's are disjoint. Moreover, any disk from any of the D_i's in the optimal solution cannot intersect L_v and L_h.

Now our algorithm is as follows. Consider all possible subsets $D'_{vh} \subseteq D_{vh}$ of size at most $2\lceil \sqrt{2}k \rceil$. For each of these choices, do the following in each T_i. Remove all the points which are covered by D'_{vh} and remove all the disks from

D_i which have an intersection with D'_{vh}. We now apply the same algorithm recursively on each T_i on the modified points and disks. Thus, the number of combinations of points to be chosen for testing for an optimum solution follows the recursion relation $T(n, k) = 4 * T(n, k/2) \times n^{2\lceil \sqrt{2k} \rceil} = n^{O(k)}$.

Theorem 1. *There exists an algorithm which yields a* PTAS *for the MDCUD problem with performance ratio at least* $(1 - \frac{1}{k})^2$.

Proof. We use two nested applications of the shifting strategy to solve the problem. First, we apply shifting strategy on vertical strips of width at most k. Then by Lemma 1, we get $\alpha_{sh} \geq \alpha_X (1 - \frac{1}{k})$. Further, to solve each vertical strip of width k, we again apply the shifting strategy on horizontal strips of height at most k. However, we can solve the *MDCUD* problem optimally inside $k \times k$ square. Thus, we get $\alpha_X \geq (1 - \frac{1}{k})$. Hence, the theorem follows. $\square$

Corollary 1. *By similar analysis as above, we can prove that, there exists an algorithm which yields a* PTAS *for the MDCUS problem with performance ratio at least* $(1 - \frac{1}{k})^2$.

2.2 PTAS: The *MICUS* Problem

In this section, we give a PTAS for the *MICUS* problem. Our PTAS is on the same lines of the PTAS-es designed by Chan and Hu [5] and Mehrabi [15]. Our main contribution is an exact algorithm for the *MICUS* problem when the points and unit squares are inside a $k \times k$ square, where k is a fixed constant. With the help of the shifting strategy of Hochbaum and Maass [10], we obtain a PTAS for the *MICUS* problem.

Let $s_1, s_2, \ldots, s_\ell$ be a sequence of unit squares containing a common point such that their centers are increasing x-coordinate. If the centers of $s_1, s_2, \ldots, s_\ell$ are either in increasing or in decreasing y-coordinate, then we say that the set $\{s_1, s_2, \ldots, s_\ell\}$ forms a monotone set. The boundaries of the union of these squares form two monotone chains (staircases), called complementary chains (see Fig. 2).

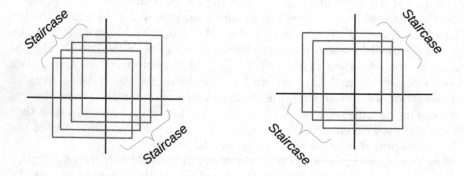

Fig. 2. Two sets of staircases.

We consider the following lemma from [15] (Lemma 4 in [15]). One can look [5] also for a similar result.

Lemma 3. *Let (P, S) be an instance of the MICUS problem such that all the points in P are inside a $k \times k$ square. Further, let $OPT \subseteq S$ be the optimal set of squares for the instance (P, S). Then, OPT can be decomposed into $O(k^2)$ (disjoint) monotone sets.*

We now define the *mod-one* transformation given by Chan and Hu [5]. Let (x, y) be a point in the plane. Then (x, y) mod-one is defined as (x', y') where x' and y' are the fractional parts of x and y respectively.

Theorem 2. *There exists a polynomial time algorithm for the MICUS problem where the given points and squares are inside a $k \times k$ square, for some constant $k > 0$.*

Proof. Note that the squares in an optimal solution can be decomposed into $O(k^2)$ monotone sets (Lemma 3). Every monotone set forms two staircases. Under mod-one transformation, both staircases map to two monotone chains which join at the corners after the mod-one. Thus, at a point (after mod-one) where a square disappears from the boundary of a staircase, the square starts appearing on the boundary of another staircase. Our dynamic programming algorithm is based on the above facts.

We now discuss the sweep-line based dynamic programming in the form of a state-transition diagram. Every state stores $O(k^2)$ 6-tuples of unit squares. More specifically, the following defines a state in the state-transition diagram.

1. A vertical sweep-line l, which is always placed at a corner of any one of the given unit squares, and
2. $O(k^2)$ 6-tuples of unit squares and each 6-tuple forms a monotone set i.e., every 6-tuple is $(s_{start}, s_{prev'}, s_{prev}, s_{curr}, s_{curr'}, s_{end})$ such that
   (a) the sequence of squares $s_{start}, s_{prev'}, s_{prev}, s_{curr}, s_{curr'},$ and s_{end} are in the increasing x-coordinate of their centers and hence, they form a monotone set, and
   (b) sweep-line l lies between the corners of squares s_{prev} and s_{curr} after mod 1 transformation.

In a 6-tuple $(s_{start}, s_{prev'}, s_{prev}, s_{curr}, s_{curr'}, s_{end})$, the squares s_{start} and s_{end} are the start and end squares of the corresponding monotone set, $s_{prev'}$ is the square which is the immediate predecessor of s_{prev} and $s_{curr'}$ is the square which is the immediate successor of s_{curr} in the monotone set. Further, $s_{prev'}$ and $s_{curr'}$ are stored to verify a point is uniquely covered or not.

We now describe the transitions from a state A to a state B. Assume that the current position of the sweep-line is l. Let $(s_{start}, s_{prev'}, s_{prev}, s_{curr}, s_{curr'}, s_{end})$ be the 6-tuple in A such that the x-coordinate, after mod 1, of the corner of s_{curr} is the smallest among all other tuples and which is to the right of the position of the sweep-line l. Then the next possible position of the sweep-line is, l', the x-coordinate of the corner (after mod 1) of s_{curr}. Hence, there can be a transition from the state A to a state B that has all other tuples as in the state A except the 6-tuple $(s_{start}, s_{prev'}, s_{prev}, s_{curr}, s_{curr'}, s_{end})$ that is changed to $(s_{start}, s_{prev}, s_{curr}, s_{curr'}, s_{new}, s_{end})$ for some unit squares s_{new} only if no point in P between l and l', after taking mod 1, are covered by more than one square in the $O(k^2)$ 6-tuples in state B, before taking mod 1. Further, the cost of the transition is the number of points in P between l and l', after taking mod 1, which are uniquely covered by the squares in the $O(k^2)$ 6-tuples in the state B, before taking mod 1.

Note that there are only $O(n^{O(k^2)})$ states and transitions in the diagram. However, to check the existence of a transition and finding the cost of a transition requires $O(m)$ time. Hence, the state-transition diagram can be constructed in $O(m\dot{n}^{O(k^2)})$ time. Since the sweep-line always moves to the right, the state-transition diagram is a directed acyclic graph (DAG). Further, one can suitably add a source X and sink Y to this DAG. It is easy to observe that the cost of an optimal solution to an instance of the $MICUS$ problem is nothing but the cost of the longest path from X to Y in the corresponding DAG, which can be computed in the time polynomial with respect to the size of the DAG.   □

We now apply the shifting strategy of Hochbaum and Maass [10] to obtain a PTAS for the $MICUS$ problem (see Sect. 2.1 above for the explanation).

Theorem 3. *There exists a* PTAS *for the MICUS problem.*

3   NP-Completeness Results

In this section we prove the NP-completeness results for the MDC and MIC problems with various geometric objects. To proceed further, we require the following definitions and results. We first define the *Positive Exactly 1-in-3SAT (P1-in-3SAT)* problem [19]. We are given a 3SAT formula ϕ with n variables $x_1, x_2, \ldots, x_n$ and m clauses $C_1, C_2, \ldots, C_m$ such that each clause contains exactly 3 positive literals. The objective is to find an assignment to the variables of ϕ such that exactly one literal is true in every clause. Schaefer [19] proved that this problem is NP-complete. A planar variation of this problem is the *rectilinear-positive-planar-one-in-3-SAT (RPP1-in-3SAT)* problem [16]. In this case, the variables are placed horizontally on a line. Each clause is connected with exactly three variables either from the top or from the bottom such that the clause-variable connection graph is planar. The objective is to find a truth assignments to the variables of ϕ such that exactly one literal in each clause of ϕ is true. Refer Fig. 3 for an instance of the *RPP1-in-3SAT* problem. Mulzer and Rote [16] proved that this problem is also NP-complete.

3.1   NP-Completeness: The *MICHLVSeg* Problem

In this section, we prove that the MIC problem with infinite horizontal lines and vertical segments (*MICHLVSeg* problem) is NP-complete. We give a reduction

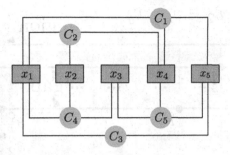

Fig. 3. An instance of the *RPP1-in-3SAT* problem.

from the *P1-in-3SAT* problem. Let ϕ be an instance of the *P1-in-3SAT* problem. We create an instance $\mathcal{I}$ of the *MICHLVSeg* problem as follows.

Variable Gadget: For each variable x_i, we take a horizontal infinite line h_i, a vertical line v_i, and a point p_i. The point p_i is placed in the intersection between h_i and v_i (see Fig. 4). In order to cover p_i, either one of these two lines needs to be picked. Note that between any pair of consecutive horizontal lines there is a horizontal *gap*. In the later stage, we place some points corresponding to clauses in these gaps.

Clause Gadget: For each clause we take a vertical infinite strip say *region* of that clause. We place the regions side by side to the right of all the vertical lines for variables such that no two regions intersect. Let C_c be a clause containing variables x_i, x_j and x_k. For this clause we take 5 points $\{p_1^c, p_2^c, p_3^c, p_4^c, p_5^c\}$ and 4 vertical segments $\{s_1^c, s_2^c, s_3^c, s_4^c\}$. All the points and segments are on a vertical line and placed inside the region of C_c. The points p_1^c, p_3^c, and p_5^c are on h_i, h_j and h_k respectively. The point p_2^c are inside a gap between h_i, and h_j and p_4^c are inside a gap between h_j, and h_k. The segment s_ℓ^c covers only the points $\{p_\ell^c, p_{\ell+1}^c\}$, for $1 \le \ell \le 4$.

This completes the construction. See Fig. 4 for the complete construction. Clearly, the construction can be done in polynomial time with respect to the number of variables and clauses in ϕ. We now prove the following theorem.

Theorem 4. *The MICHLVSeg problem is* NP-*Complete.*

Proof. It is easy to prove that the *MICHLVSeg* problem is in NP. We now show that exactly one literal is true in every clause of ϕ if and only if $\mathcal{I}$ has a solution that covers all the points.

Assume that ϕ has a satisfying assignment such that exactly one literal is true in every clause of ϕ. For the gadget of x_i, select h_i in the solution if x_i is true. Otherwise select v_i. Now consider a clause $C_c = (x_i \lor x_j \lor x_k)$. Note that exactly one literal is true for C_c i.e., exactly one of h_i, h_j, or h_k is selected in the solution. So exactly one of p_1^c, p_3^c, or p_5^c is covered by the variable gadgets. Hence the remaining 4 points are covered by exactly two segments from the gadget of C_c. Thus we have a solution for $\mathcal{I}$ covering all the points.

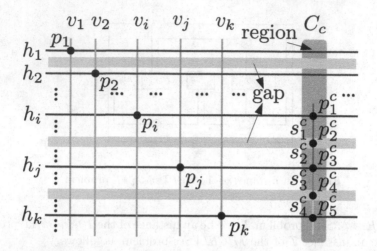

Fig. 4. Variable and clause gadgets and their interaction.

On the other hand, assume that $\mathcal{I}$ has a solution covering all the points. Now to cover p_i in the gadget of x_i either h_i or v_i is in the solution. So we set x_i to be true if h_i is in the solution. Otherwise, x_i is false. Now we show that this is a satisfying assignment of ϕ. Let $C_c = (x_i \vee x_j \vee x_k)$ be a clause. To cover all the points corresponding to C_c, exactly one of p_1^c, p_3^c, or p_5^c is covered from variable gadgets, and we set that variable to be true.   □

3.2   NP-Completeness: The *MDCUSeg* and *MICUSeg* Problems

In this section, we first prove that the *MDCUSeg* problem is NP-complete by a reduction from the *RPP1-in-3SAT* problem. The reduction is in the same line of the reduction provided for the Min-max-coverage-for-unit-square problem in [17]. We present a variation of the proof for vertical and horizontal unit segments in detail here to be self-complete. This reduction also directly implies that the *MICUSeg* problem is NP-complete.

We now describe the construction to convert an instance ϕ of the *RPP1-in-3SAT* problem to an instance Γ of the *MDCUSeg* problem in polynomial time. As shown in Fig. 3, a clause can connect to exactly three positive literals, either from top or from bottom. We represent these clause-literal connections as *loops* and more specifically as *left*, *middle*, and *right* loops. For example, Fig. 5 shows some of these loops for the instance ϕ in Fig. 3.

Variable Gadgets: The variable gadget for x_i is represented as a *rectangular loop* as is shown in Fig. 6. This loop has 2α points (value of α is established later), which are covered by 2α unit horizontal and vertical segments (see Fig. 6). Each segment covers exactly two consecutive points along the loop. A segment t_i, $1 \le i \le 2\alpha - 1$ covers i-th and $(i+1)$-th points. The segment $t_{2\alpha}$ covers 1-st and the 2α-th point.

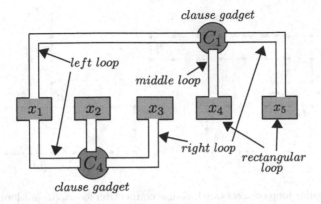

Fig. 5. A schematic diagram of the construction of a *MDCUSeg* problem instance.

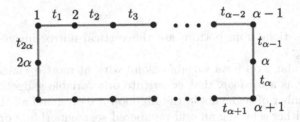

Fig. 6. Rectangular variable loop

Note that, a clause C_c can connect to x_i from top or bottom. Hence, apart from a rectangular loop, every variable gadget has multiple connections through its left, middle and right loops from top or bottom. Let σ represents the maximum number of clause connections to the rectangular loop of any variable from either top or bottom. For example, in Fig. 3 the value of σ is two since in each variable at most two clauses are connected either from the top or from the bottom. Figure 7 describes a left and a middle loop which connect to the rectangular loop of x_i from the top. The right loop is a horizontal mirror image of the left loop. In the design of every connection loop, there are three special segments namely t^*, t^{**} and t_i^c. The segments t^* and t^{**} connects the vertical arrangement of every loop to the rectangular loops. Whereas, the segment t_i^c is a *clause segment*, which is used to connect with the clause gadget.

Figure 7 also shows an example of connecting left/middle loop with the variable rectangular loop. All other loops are similarly connected with the variable rectangular loop. If clause connections to a variable rectangular loop from top are numbered as $1, 2, ..., l, ..., \delta$ ($\delta \leq \sigma$) then the l-th connection is made through the t_{4l}-th segment on the rectangular loop. The segment t_{4l} is removed from the variable rectangular loop and the points $4l$ and $4l + 1$ covered by t_{4l} are now covered by the segments t^* and t^{**} respectively (see Fig. 7). To accommodate this arrangement, we set the value of α to be $4\sigma + 4$. Similarly, different loops

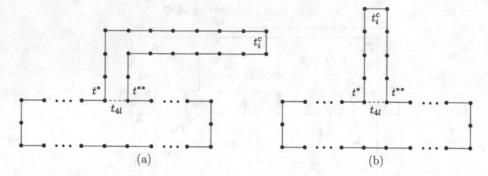

Fig. 7. Rectangular loop connects with clause connection loops, (a) left loop (b) middle loop, segment t_{4l} is removed and points $4l$ and $4l + 1$ are covered by segments t^* and t^{**} respectively. (Color figure online)

for clause connections from bottom are the vertical mirror image of the loops described in Fig. 7.

The rectangular loop for a variable along with at most σ clause connection loops are called as a *big-loop*, that constitute our variable gadget. It is easy to see that the number of segments in the big-loop are even and all the points can be covered by either selecting all odd numbered segments (blue) or by selecting all even numbered segments (orange). These two sets are disjoint and represent the truth value of the corresponding variable.

Clause Gadgets: Let $C_c = (x_i \lor x_j \lor x_k)$ be a clause. For this clause, we take four points and three unit segments as shown in Fig. 8(a). Figure 8(b) shows the interaction of clause segments s_i^c, s_j^c and s_k^c from variables x_i, x_j, and x_k with the clause gadget. The clause point p^c is covered by all three clause segments s_i^c, s_j^c, and s_k^c corresponding to the variables in C_c. Hence, to have a valid solution (maximum coverage) for the *MDCUSeg* problem, the clause point p^c has to be covered by exactly one of these three segments.

Since the number of points and segments in Γ is a polynomial function on the number of clauses and variables in ϕ. Hence, the construction can be done in polynomial time. This completes the construction.

Theorem 5. *The MDCUSeg problem is* NP-*complete.*

Proof. Clearly, the problem is in NP. We now prove that exactly one literal is true in every clause of ϕ if and only if Γ has a solution of size $|P|$, where P is the point-set in Γ. Observe that, in a variable gadget (big-loop) there are only two ways to cover all points using disjoint segments, either by selecting all even numbered (orange colored) or all odd numbered (blue colored) segments.

Assume that exactly one literal is true in every clause by a truth assignment to the variables of ϕ. If x_i is true then in the gadget of x_i, blue segments are chosen. Otherwise, orange segments are chosen. From the above observation, the chosen segments will cover all the points in a variable gadget. Since, exactly one

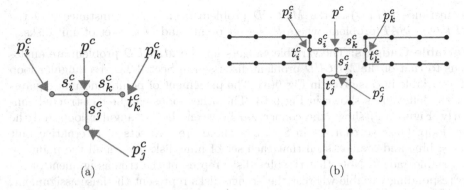

Fig. 8. (a) Structure of a clause gadget (b) Connection between clause and variable gadgets. (Color figure online)

literal is true for each clause C_c, exactly two of the three points p_i^c, p_j^c, and p_k^c are covered by the false variable. So the remaining two uncovered points (p^c and one of the uncovered point p_i^c, p_j^c, and p_k^c) is covered by the corresponding clause segment. Hence, we have a solution of size $|P|$.

On the other hand, assume that there exists a covering of all the points in the instance Γ of *MDCUSeg* problem using disjoint segments. Recall that, all points in a variable gadget is covered by either all blue or orange segments. Thus, we can construct a truth assignment as follows. For the gadget of x_i, if all blue segments are chosen then set x_i to be 1, otherwise set x_i to be 0. Since every point is covered by all the segments, in any clause gadget C_c, the point p^c can be covered by exactly one of the three segments s_i^c, s_j^c, and s_k^c. Without loss of generality assume that s_i^c is selected in the solution. Then the point p_i^c on segment t_i^c (orange colored) in the variable gadget of x_i is also covered. Therefore in the gadget of x_i we can not select the orange segments. Further, in the gadgets of x_j and x_k, we need to select the orange segments since the points p_j^c and p_k^c are on orange segments in their corresponding variable gadgets. As a result, x_i becomes true and both x_j and x_k become false. Therefore, exactly one of the three literals is true in every clause of ϕ.   □

We note that the above reduction also works for *MICUSeg* problem. Hence, we have the following theorem.

Theorem 6. *The MICUSeg problem is* NP-*complete.*

3.3   NP-Completeness: The *MDCUD* and *MICUD* Problems

We prove that both *MDCUD* and *MICUD* problems are NP-complete. Here, we give a brief outline for the *MDCUD* problem. A similar reduction can be done for the *MICUD* problem. We give a reduction from the *RPP1-in-3SAT* problem to the *MDCUD* problem. Similar to the construction in Sect. 3.2, we construct an instance $\Gamma(P, D)$ of the $MDCUD$ problem from a given instance ϕ of the $RPP1$-in-$3SAT$ problem, where P is a set points and D is a set of unit disks.

Variable Gadgets: The variable gadgets for the $MDCUD$ problem are analogous to that of the $MDCUD$ problem discussed in Sect. 3.2. A rectangular loop of a variable in ϕ is given in Fig. 9(a). The placement of points and unit squares for the left loop is shown in Fig. 9(b). The other loops can be constructed similarly. Figure 9(c) shows the connection between the rectangular loop and the left loop. Observe that, as in Sect. 3.3, there are two sets of alternative unit disks, blue and orange, such that each set of unit disks covers all the points in a variable gadget. Note that the blue disks represent the true assignment of the corresponding variable whereas the orange disks represent the false assignment.

Clause Gadgets: The clause gadget is similar to the clause gadget of $MDCUSeg$ problem (Sect. 3.2) with the following differences. Here the gadget for clause C_c consists of a single point p^c only. Let C_c be a clause with variables $x_i, x_j,$ and x_k. Then, the three squares $d_i^c, d_j^c,$ and d_k^c corresponding to the variables $x_i, x_j,$ and x_k respectively contain the clause point p^c as shown in Fig. 9(d). Now to cover p^c, exactly one of the three literals $x_i, x_j,$ and x_k in the clause C_c is true.

Since the rest of the argument is almost similar to that in Sect. 3.2, we conclude the following theorem.

Theorem 7. *The MDCUD and MICUD problems are* NP-*complete.*

3.4   NP-Completeness: The $MDCUS$ and $MICUS$ problems

We prove that both $MDCUS$ and $MICUS$ problems are NP-complete. We give a polynomial time reduction from the $RPP1$-in-$3SAT$ problem [16] to $MDCUS$ (and hence $MICUS$) problem. Similar to the construction in Sect. 3.3, we construct an instance $\beta(P, O)$ of $MDCUS$ problem from a given instance ϕ of $RPP1$-in-$3SAT$ problem, where O is the set of unit squares.

Variable Gadgets: Variable gadgets for the $MDCUS$ problem are analogous to that of the $MDCUD$ problem discussed in Sect. 3.3. A rectangular loop of a variable in ϕ is given in Fig. 10(a). The placement of points and unit squares for the left loop is shown in Fig. 10(b). The other loops can be constructed similarly. Figure 10(c) shows the connection between the rectangular loop and the left loop. Observe that, as in Sect. 3.3, there are two sets of alternative unit squares, blue and orange, such that each set of unit squares covers all the points in a variable gadget.

Clause Gadgets: This is similar to the clause gadget of $MDCUD$ problem in Sect. 3.3. Let C_a be a clause with variables $x_i, x_j,$ and x_k. Then, the three squares $s_i^a, s_j^a,$ and s_k^a will contain the clause point p^a as shown in Fig. 10(d).

Since the rest of the argument is similar to the that in Sect. 3.3, we conclude the following theorem.

Theorem 8. *The MDCUS and MICUS problems are* NP-*complete.*

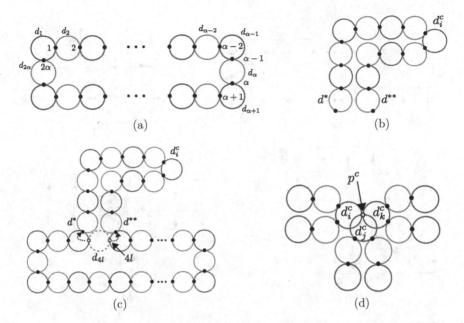

Fig. 9. (a) Rectangular variable loop (b) A left loop (c) Connection of rectangular loop with the vertical arrangement of any other loop (d) Clause gadget and variable clause interaction.

4   Polynomial Time Algorithms

4.1   The *MDCL* and *MICL* Problems

In this section, we show that the *MDCL* and *MICL* problems can be solved in polynomial-time. We recall that the input for both problems is a set L of axis-parallel lines and a set P of points in the plane.

The *MDCL* Problem: The polynomial-time algorithm for the *MDCL* problem is straightforward. We need to find a set of non-intersecting axis-parallel lines covering the maximum number of points. Thus the optimal solution selects the set of either all horizontal lines or all vertical lines based on the set which covers the maximum number of points.

The *MICL* Problem: We show that the *MICL* problem can be solved in poly-nomial time. To do so, we first reduce this problem to an equivalent problem, the maximum weight independent set problem in bipartite graphs. Let L be a set of lines and P be a set of points. Also let $L_h \subseteq L$ and $L_v \subseteq L$ be sets of vertical and horizontal lines respectively. We generate the vertex weighted bipartite graph $G(U, W, E)$ as follows. For each vertical line $v_i \in L_v$, we take a vertex u_i in U and for each horizontal line $h_j \in L_h$, we take a vertex w_j in W. For each point $p \in P$, if p is on the intersection between a vertical line v_i and a horizontal line h_j, take an edge e_{ij} between vertices u_i and w_j. Finally, we assign the weight of

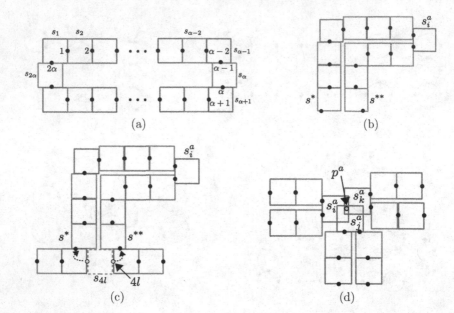

Fig. 10. (a) A rectangular loop (b) A left loop (c) Connection of rectangular loop with the vertical arrangement of any other loop (d) Clause gadget and variable clause interaction.

a vertex as the total number of the points contained by its corresponding lines. See the construction in Fig. 11.

It is now observed that finding a maximum weight independent set in G is equivalent to finding a solution to its corresponding *MICL* problem instance. Since the maximum weight independent set problem on a bipartite graph can be solved in polynomial time[1] [2], and so the *MICL* problem.

4.2   The *MDCI* and *MICI* Problems

We give polynomial time algorithms for both *MDCI* and *MICI* problems. The algorithm is very similar to that of [17], and hence we would only provide outlines for the same. Let I be a set of n intervals $\{i_1, \ldots, i_n\}$ and P be a set of m points $\{p_1, \ldots, p_m\}$ on the real line.

The *MDCI* Problem: We formulate the *MDCI* problem as a *MWIS* problem with weighted intervals. Let us define a weight function $w : I \to \mathbb{N}$ as follows: the weight of an interval $i_t \in I$ is the total number of points in P covered by i_t, i.e., $w(i_t) = \{|p_j|$ such that $p_j \in P, p_j \in i_t\}$. We first build a binary search tree $\mathcal{T}$ on P. Next for each interval $i_j \in I$, we make a counting query on $\mathcal{T}$. Hence, the total time will be taken as $O(m \log m + n \log m)$.

---

[1] Let $G(V, E)$ be a bipartite graph. Finding a minimum weight vertex cover $V^* \subset V$ in G can be solved by a minimum cut computation or a maximum flow computation in a related graph. Then the maximum weight independent set of G is $V \setminus V^*$.

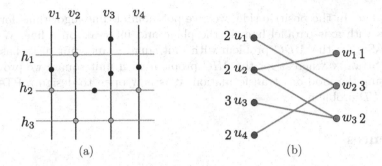

Fig. 11. (a) An instance of the *MICL* problem. (b) A vertex weighted bipartite graph instance constructed from the instance of the *MICL* problem in (a).

We now have a set I_w of n weighted intervals, and the objective is to compute a subset $I'_w \subseteq I_w$ of pairwise non-intersecting intervals of maximum weight. Note that *MWIS* problem in weighted intervals can be solved in $O(n \log n)$-time [13]. Let $I_w^* = \{i_{w1}^*, i_{w2}^*, \ldots, i_{wk}^*\}$ be an optimal solution to the *MWIS* problem on I_w. Then, clearly the set I^* corresponding to the intervals in I_w^* is also an optimal solution for *MDCI* problem which covers $w(I_w^*) = \sum_{j=1}^{k} w(i_{wj}^*)$ points. Hence, we can find the optimal solution to *MDCI* problem in $O(m \log m + n \log m + n \log n)$-time.

Theorem 9. *The MDCI problem can be solved in $O(m \log m + n \log m + n \log n)$ time.*

The *MICI* Problem: Here we apply the same algorithm presented for the *MDCI* problem in the previous section. To apply the algorithm, we first truncate every interval $i_j \in I$ such that it starts with the leftmost and ends with the rightmost point covered by i_j. Next, we assign a weight to each interval as the number of points in P that are covered by it. Finally, we find maximum weight non-intersecting intervals in the generated weighted intervals. Thus, we can conclude the following theorem.

Theorem 10. *There exists an $O(m \log m + n \log m + n \log n)$ time algorithm for the MICI problem.*

5   Conclusion

In this paper, we consider the Maximum Disjoint Coverage (*MDC*) and Maximum Independent coverage (*MIC*) problems. For both problems, we present some positive and negative results on various geometric objects. In the negative side, we show that when the objects are horizontal lines and vertical segments, the *MIC* problem is NP-hard. Further, with unit axis-parallel segments, unit squares, and unit disks in the plane both the *MIC* and *MDC* problems are

NP-complete. In the positive side, we give polynomial-time algorithms for both problems with axis-parallel lines in the plane and intervals on a line. We provide PTASes for the *MDC* problem with unit squares and unit disks based on Shifting Strategy whereas for the *MIC* problem with unit square we provide a PTAS using the mod-one transformation. It is now open to design a PTAS for the *MICUD* problem.

References

1. Adamaszek, A., Wiese, A.: Approximation schemes for maximum weight independent set of rectangles. In: Proceedings of the 2013 IEEE 54th Annual Symposium on Foundations of Computer Science, FOCS 2013, pp. 400–409 (2013)
2. Ahuja, R.K., Magnanti, T.L., Orlin, J.B.: Network Flows: Theory, Algorithms, and Applications. Prentice-Hall Inc., Upper Saddle River (1993)
3. Chan, T.M., Grant, E.: Exact algorithms and APX-hardness results for geometric packing and covering problems. Comput. Geom. **47**(2), 112–124 (2014)
4. Chan, T.M., Har-Peled, S.: Approximation algorithms for maximum independent set of pseudo-disks. Discrete Comput. Geom. **48**(2), 373–392 (2012)
5. Chan, T.M., Hu, N.: Geometric red-blue set cover for unit squares and related problems. Comput. Geom. **48**(5), 380–385 (2015)
6. Clark, B.N., Colbourn, C.J., Johnson, D.S.: Unit disk graphs. Discrete Math. **86**(1), 165–177 (1990)
7. Erlebach, T., Jansen, K., Seidel, E.: Polynomial-time approximation schemes for geometric intersection graphs. SIAM J. Comput. **34**(6), 1302–1323 (2005)
8. Erlebach, T., van Leeuwen, E.J.: Approximating geometric coverage problems. In: Proceedings of the Nineteenth Annual ACM-SIAM Symposium on Discrete Algorithms, SODA 2008, pp. 1267–1276 (2008)
9. Feige, U.: A threshold of $\ln n$ for approximating set cover. J. ACM **45**(4), 634–652 (1998)
10. Hochbaum, D.S., Maass, W.: Approximation schemes for covering and packing problems in image processing and VLSI. J. ACM **32**(1), 130–136 (1985)
11. Ito, T., et al.: A 4.31-approximation for the geometric unique coverage problem on unit disks. Theor. Comput. Sci. **544**, 14–31 (2014)
12. Ito, T., et al.: A polynomial-time approximation scheme for the geometric unique coverage problem on unit squares. Comput. Geom. **51**, 25–39 (2016)
13. Kleinberg, J., Tardos, E.: Algorithm Design. Addison Wesley, Boston (2006)
14. Madireddy, R.R., Mudgal, A., Pandit, S.: Hardness results and approximation schemes for discrete packing and domination problems. In: Kim, D., Uma, R.N., Zelikovsky, A. (eds.) COCOA 2018. LNCS, vol. 11346, pp. 421–435. Springer, Cham (2018). https://doi.org/10.1007/978-3-030-04651-4_28
15. Mehrabi, S.: Geometric unique set cover on unit disks and unit squares. In: Proceedings of the 28th Canadian Conference on Computational Geometry, CCCG 2016, pp. 195–200 (2016)
16. Mulzer, W., Rote, G.: Minimum-weight triangulation is NP-hard. J. ACM **55**(2), 11:1–11:29 (2008)
17. Nandy, S.C., Pandit, S., Roy, S.: Covering points: Minimizing the maximum depth. In: Proceedings of the 29th Canadian Conference on Computational Geometry, CCCG 2017, pp. 37–42 (2017)

18. Nieberg, T., Hurink, J., Kern, W.: A robust PTAS for maximum weight independent sets in unit disk graphs. In: Hromkovič, J., Nagl, M., Westfechtel, B. (eds.) WG 2004. LNCS, vol. 3353, pp. 214–221. Springer, Heidelberg (2004). https://doi.org/10.1007/978-3-540-30559-0_18

19. Schaefer, T.J.: The complexity of satisfiability problems. In: Proceedings of the Tenth Annual ACM Symposium on Theory of Computing, STOC 1978, pp. 216–226. ACM, New York (1978)

Algorithms for Closest and Farthest String Problems via Rank Distance

Liviu P. Dinu[1,2], Bogdan C. Dumitru[1,2], and Alexandru Popa[1,3(✉)]

[1] Faculty of Mathematics and Computer Science, University of Bucharest,
Bucharest, Romania
{ldinu,alexandru.popa}@fmi.unibuc.ro, bogdan27182@gmail.com
[2] Human Language Technologies Research Center, University of Bucharest,
Bucharest, Romania
[3] National Institute for Research and Development in Informatics,
Bucharest, Romania

Abstract. A new distance between strings, termed *rank distance*, was introduced by Dinu (Fundamenta Informaticae, 2003). Since then, the properties of rank distance were studied in several papers. In this article, we continue the study of rank distance. More precisely we tackle three problems that concern the distance between strings.

1. The first problem that we study is *String with Fixed Rank Distance (SFRD)*: given a set of strings S and an integer d decide if there exists a string that is at distance d from every string in S. For this problem we provide a polynomial time exact algorithm.
2. The second problem that we study is named is the *Closest String Problem under Rank Distance (CSRD)*. The input consists of a set of strings S, asks to find the minimum integer d and a string that is at distance at most d from all strings in S. Since this problem is NP-hard (Dinu and Popa, CPM 2012) it is likely that no polynomial time algorithm exists. Thus, we propose three different approaches: a heuristic approach and two integer linear programming formulations, one of them using geometric interpretation of the problem.
3. Finally, we approach the *Farthest String Problem via Rank Distance (FSRD)* that asks to find two strings with the same frequency of characters (i.e. the same Parikh vector) that have the largest possible rank distance. We provide a polynomial time exact algorithm for this problem.

1 Introduction

In many important problems from various research fields (e.g., computational biology, image processing, computational linguistics, information retrieval), when a new objects is given, the first step taken by a specialist is to compare the new object with objects that are already well studied and annotated. Objects that are similar probably have the same function and behavior. For example, in bioinformatics, a common task is to design a new sequence that is similar to the

© Springer Nature Switzerland AG 2019
T. V. Gopal and J. Watada (Eds.): TAMC 2019, LNCS 11436, pp. 154–171, 2019.
https://doi.org/10.1007/978-3-030-14812-6_10

input given sequences: the design of genetic drugs with a structure similar to a set of existing sequences of RNA [22], PCR primer design [16,22], genetic probe design [22], antisense drug design [6], finding unbiased consensus of a protein family [4], motif finding [24,34]. Language classification, authorship identification, historical linguistics, etc., are few of the computational linguistics' topics which deal with similarity problems. In information retrieval, similarity is at the basis of most techniques that seek an optimal match between query and document, and the vision domain is full of applications regarding the similarity of images or video objects.

Even if a lot of measures are already available and well studied in mathematics [7] alternative approaches are explored in daily applications. In a workshop dedicated to Linguistics distances [27], the organizers assume that "there is always a *hidden variable* in the similarity relation, so that we should always speak of similarity with respect to some property", and they suggest that the researchers are often unclear on this point.

In computational biology the standard method for sequence comparison is by sequence alignment. Sequence alignment is the procedure of comparing two sequences (pairwise alignment) or more sequences (multiple alignment) by searching for a series of individual characters or character patterns that are in the same order in the sequences. The standard pairwise alignment method is based on dynamic programming: the algorithm compares every pair of characters of the two sequences and generates an alignment and a score (edit distance is based on a scoring scheme for insertion or deletion penalties). The sequence alignment procedure is by far too slow to compare a large number of sequences and therefore alternative approaches might be explored in bioinformatics if we can answer the following question: is it possible to design a sequence distance which is at the same time easily computable and non-trivial? This important problem, known also as DNA sequence comparison, is a major open problem in bioinformatics [35], being ranked on the first position in two list of major open problems in bioinformatics [21,35].

The standard distances with respect to the alignment principle are edit (Levenshtein) distance or ad-hoc variants. However, as previously mentioned, Levenstein distance is too slow in some computational biology tasks which imply a large number of sequences. In other fields, like natural language classification, the accuracy of results obtained with Levensthein distance is too low, and new methods are required by a linguist. Moreover, notice the surprising conclusion through the severity of tone *"there is no reason to continue using the Levenshtein distance to classify languages"* [17].

Rank Distance is a metric introduced with linguistics motivation in 2003 by Dinu [9] (it is also called *Dinu rank distance* by Gagolewski [15]) in order to measure the similarity between strings, and begin to be very attractive for various research topics in the last decade: sequence aligner and phylogenetic analysis [10], predicting synergistic drugs for cancer [33], DNA similarity [13], natural language identification [19], authorship identification [30], language similarity [8], image processing [18], etc.

To measure the distance between two strings, Rank distance uses the following strategy: we scan (from left to right) both strings and for each letter from the first string we count the number of elements between its position in first string and the position of its first occurrence in the second string. Finally, we sum all these scores and obtain the rank distance. In other words, rank distance measures the *gap* between the positions of a letter in the two given strings, and then sums up these values. Intuitively, rank distance gives us the total non-alignment score between two sequences. Clearly, the rank distance gives a score zero only to letters which are in the same position in both strings, as Hamming distance does. Rank distance can be computed fast and benefits from some features of the edit distance. On the other hand, an important aspect is that the reduced sensitivity of the rank distance w.r. to deletions and insertions is of paramount importance, since it allows us to make use of ad hoc extensions to arbitrary strings, such as do not affect its low computational complexity.

1.1   Our Results

In this paper we study three problems related to string similarity and for each we propose one or more solutions, aiming for polynomial time algorithms when possible.

String with Fixed Rank Distance (SFRD). The input of the SFRD problem is a set of n strings $S = \{s_1, s_2, \dots, s_n\}$ and an integer d decide if there exists a string that is at distance d from every string in S. For this problem we provide a polynomial time exact algorithm.

We start from a geometrical interpretation of the problem and we consider the input strings as vectors in the n-dimensional space. Then, we observe that the vector of interest is the center of a $(n-1)$-sphere and sits on another $(n-1)$-sphere centered in the origin. We also show how to transform our problems from the rank distance to the l_2 norm and then we construct an equation system whose solution is equivalent to the solution of the SFRD problem. This equation system is polynomial time solvable.

Closest String Problem Under Rank Distance (CSRD). The input of the second problem consists of a set of strings S and the goal is to find the minimum integer d and a string that is at distance at most d from all strings in S.

We begin by introducing the notion of *degenerate strings*, that are strings where two characters might have the same index. Then, based on the properties of degenerate strings, we present a heuristic algorithm. Next, we propose two integer programming (IP) formulations, the second one being an improved version of the first by adding additional constraints based on the properties of heuristic approach. At last, we gave a geometrical interpretation of CSRD and using the bounding sphere center projection, we propose another IP formulation which has the least number of variables and constraints.

Farthest String Problem via Rank Distance (FSRD). Finally, we aim to find two strings with the same frequency of characters (i.e., the same Parikh vector) that have the largest possible rank distance. For this problem, we provide a polynomial time exact algorithm.

1.2 Previous Work

The above mentioned problems were previously studied in relation with other distances. For example, the first similarity measure used in the closest string problem is the Hamming distance and emerged from a coding theory application [14]. The closest string problem via Hamming distance is known to be NP-complete [14] and there exist a number of approximation algorithms and heuristics (see, for example, [2,22,24,25]).

As the closest string problem is used in many contexts, alternative distance measures were introduced. The most studied alternative approach is the edit distance. The closest string problem via edit distance is NP-hard even for binary alphabets [28,29]. The existence of fast exact algorithms when the number of input strings is fixed is investigated in [28].

The study of genome rearrangement introduces new problems related to closest string via new distances. Popov [31] shows that the closest string via swap distance (or Kendal distance) and element duplication distance (i.e. the minimum number of element duplications needed to transform a string into another) is also NP-complete.

Another remarkable fact is that the median string in the rank case is a tractable problem [11], while in the edit case it is NP-hard [5]. In terms of the closest string problem, both in the case of edit [28] and rank [12] both approaches are NP-hard. An IP based technique has been proposed for the closest string via rank distance [1] and via Hamming distance [26].

2 Preliminaries

In this section we introduce notation and preliminaries. We first introduce the rank distance and then we define the problems that we study in this paper.

Let $\Sigma = \{a_1, a_2, \ldots, a_n\}$ be an alphabet. The *Parikh vector* $p : \Sigma^* \to \mathbb{N}^n$ of a word $w \in \Sigma^*$ is defined as $p(w) = (|w|_{a_1}, \ldots, |w|_{a_n})$, where $|w|_{a_i}$ is the number of occurrences of character a_i in w. Given a string $s \in \Sigma^*$, we denote by s_i the i-th character of the string.

Definition 1. *Given two permutations σ and τ over the same universe $\mathcal{U}$, we define the rank distance between them as:*

$$\Delta(\sigma, \tau) = \sum_{x \in \mathcal{U}} |\sigma(x) - \tau(x)|.$$

In [9] Dinu proves that Δ is a distance function. The rank distance is naturally extended to strings. The next definition [11] formalizes the transformation of strings into rankings.

Definition 2. *Let n be an integer and let $w = a_1 \ldots a_n$ be a finite word of length n over an alphabet Σ. We define the extension of w, $\bar{w} = a_{1(1)}a_{1(2)} \cdots a_{1(j)} \cdots a_{n(i)}$, where $i(j)$ is the j occurrence of a_i in the string $a_1 a_2 \ldots a_n$.*

Example 1. If $w = aaababbbac$ then $\bar{w} = a_1a_2a_3b_1a_4b_2b_3b_4a_5c_1$.

Observe that given $\bar{w}$ we can obtain w by simply deleting all the indexes. Note that the transformation of a string into a permutation can be done in linear time (by storing for each symbol, in an array, the number of times it appears in the string). We extend the rank distance to arbitrary strings with the same Parikh verctor[1] as follows:

Definition 3. *Given $w_1, w_2 \in \Sigma^*$, we define $\Delta(w_1, w_2) = \Delta(\bar{w}_1, \bar{w}_2)$.*

Example 2. Consider the following two strings $x = abca$ and $y = baac$. Then, $\bar{x} = a_1b_1c_1a_2$ and $\bar{y} = b_1a_1a_2c_1$. The order of the characters in $\bar{x}$ and $\bar{y}$ is the following:

- a_1: 1 and 2;
- a_2: 4 and 3;
- b_1: 2 and 1;
- c_1: 3 and 4;

Thus, the rank distance between x and y is the sum of the absolute differences between the orders of the characters in $\bar{x}$ and $\bar{y}$

$$\Delta(x, y) = |1 - 2| + |4 - 3| + |2 - 1| + |3 - 4| = 4$$

The computation of the RD between two strings can be done in linear time in the length of the strings.

Let χ_n be the space of all strings of length n over an alphabet Σ and let $p_1, p_2, \ldots, p_k$ be k strings with the same Parikh vector from χ_n. The first problem that we study in this paper, namely the *String with Fixed Rank Distance Problem (SFRD)*, is defined as follows.

Problem 1 (String with Fixed Rank Distance (SFRD)). Given a set of strings $S = \{s_1, s_2, \ldots, s_k\}$, the goal is to find a string t of size n such the rank distance between t and all strings from S is equal with a given d. Notice that t may not be unique. The set of strings t is defined formally as follows:

$$T = \{t \in \Sigma^n \mid \Delta(t, s) = d, \forall s \in S\} \tag{1}$$

The SFRD problem is illustrated in the following example.

---

[1] Rank distance can be defined for strings that do not necessarily have the same Parikh vector (see, e.g., [12]). However, these strings can be transformed into strings with the same Parikh vector without affecting the rank distance. Thus, for the sake of simplicity, we do not consider such strings in our paper.

Example 3. Consider a set of three permutations $S = \{(6,7,2,4,1,5,0,2),$ $(4,6,7,3,0,2,1,5),\ (1,3,0,5,4,2,6,7)\}$. In this case $T = \emptyset, \forall d = 1,\ldots,k$. If $S = \{(6,4,2,7,5,0,3,1),\ (6,0,2,7,4,3,5,1)\}$, then $T = \{(6,2,4,7,0,3,5,1),$ $(6,4,2,0,7,3,5,1)\}$ for $d = 6$ and T has cardinality 14 for $d = 8$.

Then, we study the closest string problem under the metric defined by the rank distance.

Problem 2 (Closest string via rank distance (CSRD)). Let $P = \{p_1, p_2, \ldots, p_k\}$ be a set of k length n strings over an alphabet Σ. The *closest string problem via rank distance (CSRD)* is to find a minimal integer d (and a corresponding string t of length n) such that the maximum rank distance from t to any string in P is at most d. We say that t is the closest string to P and we name d the radius. Formally, the goal is to compute:

$$\min_{x \in \chi_n} \max_{i=1..k} \Delta(x, p_i) \tag{2}$$

Remark 1. The string t that minimizes (2) is not necessary unique. For example, if $P = \{(1,2,3),\ (3,1,2),\ (2,3,1)\}$ is a set of three length 3 permutations, then every length 3 permutation is a closest string to P.

We introduce now the concept of partition, necessary to define the third problem that we study, namely, the *Furthest String Problem under Rank Distance (FSRD)*

Definition 4. *Given an alphabet Σ and a number k where $k \leq |\Sigma|$, we define a partition, denoted $part(k,n)$ as a set of k multisets $M_1, \ldots, M_k$ such that $\sum_{i=1}^{k} |M_i| = n$ and each multiset M_i has only one distinct element.*

Example 4. If $\Sigma = \{a,b,c\}$ and $n = 5$, then a partition $part(2,5)$ could be: $M_1 = \{a,a\}, M_2 = \{b,b,b\}$. A partition $part(3,5)$ could be: $M_1 = \{a\}, M_2 = \{b,b\}, M_3 = \{c,c\}$.

We say that a string x is *generated* by a partition $part(k,n)$ if and only if $|x| = n$ and x contains all M_i elements $\forall i = \overline{1,n}$. For example, let $\Sigma = \{a,b,c\}$, $n = 5$ and partition $part(2,5)$. Thus, the partition has two multisests M_1 and M_2 with $|M_1| + |M_2| = 5$. We can choose $M_1 = \{a,a\}$ and $M_2 = \{b,b,b\}$. The string $x = abbab$ is generated from partition $part$, while $x = aaabb$ is not since it has three a.

We introduce now the notion of a diameter of a partition.

Definition 5. *The diameter of partition $part(x,n)$ is defined as:*

$$D_{(\Sigma,part(k,n))} = max(\Delta(x_1, x_2))$$

$\forall x_1, x_2$ are generated from partition $part(k,n)$.

We are now ready to define the third problem.

Problem 3 (Furthest Strings under Rank Distance (FSRD)). The input of the problem consists of an alphabet Σ with $|\Sigma| > 2$ and two integers k and n, $k, n \in \mathbb{N}$, $k \le n$. The goal is to find two strings x_1 and x_2 generated from a partition $part(k,n)$ such that $D_{(\Sigma, part(k,n))}$ is maximized.

3   An Exact Algorithm for the SFRD Problem

In this section we present a polynomial time algorithm for the SFRD problem defined in Sect. 2. As described in Sect. 2, we can transform the input strings into permutations in $O(n)$ time. Thus, to ease the presentation, throughout this section we consider that the input strings are permutations, and more precisely, n-dimensional vectors.

Our approach is based on the following geometric observation. Since every string $s \in S$ represents an n-dimensional vector it follows that all strings $s \in S$ reside on a $(n-1)$ sphere. Even more, since we are looking for a string c situated at a fixed distance d, then c is the center of another sphere in $(n-1)$ dimensions.

Our geometrical intuition uses l_2 2-norm distance and not the rank distance Δ. In fact, in our current configuration, Δ is actually the l_1 1-norm distance. Hence, the key to obtain a polynomial time algorithm for SFRD is to exploit the relation between l_1 and l_2 distances. The lemma below establishes the connection between the two aforementioned norms.

Lemma 1. *For any $x \in R^n$ the following inequality holds:*

$$\|x\|_1 \le \sqrt{n}\|x\|_2$$

Proof. Applying Cauchy-Schwarz inequality we have:

$$\|x\|_1 = \sum_{i=1}^{n} |x_i|$$

$$= \sum_{i=1}^{n} |x_i| \cdot 1$$

$$\le (\sum_{i=1}^{n} |x_i|^2)^{1/2} \cdot (\sum_{i=1}^{n} 1)^{1/2}$$

$$= \sqrt{n}\|x\|_2$$

Based on Lemma 1 we can reformulate the SFRD problem in relation with l_2 by observing that all string s and c resides on the surface of the $(n-1)$-sphere with the center at the origin.

Theorem 1. *The SFRD problem can be reduced in polynomial time to solving a system of inequalities with $n+1$ variables and k constraints.*

Proof. Formalizing all the observation above we have that SFRD is equivalent to the following systems of equations:

$$\sum_{j=1}^{n} s_{ij}^2 = \gamma^2, \; \forall i = 1, \ldots, k$$

$$\sum_{j=1}^{n} c_j^2 = \gamma^2$$

$$\sum_{j=1}^{n} |s_{ij} - c_j| = d, \; \forall i = 1, \ldots, k \tag{3}$$

where $c_1, \ldots, c_j, d, \gamma$ are variables. Using Eq. 3 we expand l_2:

$$\sum_{j=1}^{n} (s_{ij} - c_j)^2 = \sum_{j=1}^{n} s_{ij}^2 - 2 \sum_{j=1}^{n} s_{ij} c_j + \sum_{j=1}^{n} c_j^2$$

$$= 2\gamma^2 - 2 \sum_{j=1}^{n} s_{ij} c_j, \; \forall i = 1, \ldots, k \tag{4}$$

Based on out initial assumption that $s_i \in S$ are permutations over Σ with distinct consecutive elements, we have that:

$$\gamma^2 = \|c\|_2^2 = \sum_{i=1}^{n} c_i^2 = \frac{1}{6}(n-1)n(2n-1)$$

Next, based on Lemma 1 we have:

$$\sum_{j=1}^{n} |s_{ij} - c_j| \le \sqrt{n} \cdot (\sum_{j=1}^{n} (s_{ij} - c_j)^2)^{1/2}$$

$$d \le \sqrt{n} \cdot (\sum_{j=1}^{n} (s_{ij} - c_j)^2)^{1/2}$$

$$d^2 \le n \cdot \sum_{j=1}^{n} (s_{ij} - c_j)^2, \; \forall i = 1, \ldots, k \tag{5}$$

From Eqs. 5 and 4

$$d^2 \le n \cdot (2\gamma^2 - 2\sum_{j=1}^{n} s_{ij} c_j)$$

$$\sum_{j=1}^{n} s_{ij} c_j \le \gamma^2 - \frac{d^2}{2n}, \; \forall i = 1, \ldots, k \tag{6}$$

Equation 6 is an inequality system with $n+1$ variables (c_j and d) and k constraints and can be solved in polynomial time since the number of variables c_i is constant in given n-dimensional space where c and s_i resides, while the number of points s may vary. This is a special case of integer programming problem: *integer programming with a fixed number of variables* and algorithms proposed by [23] and [20] can offer a polynomial time solution. If Lenstra's algorithm will produce a solution, then it guaranties it will have integer coefficients. The inequality system can be extended to include constraints of $s \in V$ like distinctness of elements and fact that they are permutations and d can be regarded as constant, chosen form a heuristically determined interval.   □

Observation 1. *Equation system 6 can be extended for strings that do not have distinct elements and the only thing that needs to be changed is the computation of γ^2 which is a simple sum of k squared values, where k is the size of string s from S.*

4   Heuristic Algorithm for the CSRD Problem

In this section we introduce a heuristic algorithm for the CSRD problem. As in the previous section, without loss of generality, we consider that the input strings are permutations and more precisely, vectors from an n-dimensional space. To present our algorithm (formally defined in Algorithm 1) we introduce some preliminary notions. Since all the strings are permutations, for a string s and a character $x \in s$ we define $s(x) = i$ if and only if $s_i = x$. This is the *position* function that for a given character x returns the index in the string.

Let $min(x, S)$ be the lowest position of symbol x found in all $s \in S$. In a similar manner we define $max(x, S)$ as being the highest position of symbol x. For example, if $S = \{3142, 3241, 1423\}$, then we have $min(1, S) = 1$, $max(1, S) = 4$ or $min(4, S) = 2$, $max(4, S) = 3$.

Intuitively, if we plan to construct an optimal string $\hat{t}$, then we have to place every symbol somewhere between the positions given by min and max of those symbols in relation with S. To be more precise we need to place them in the middle of the $[min, max]$ interval.

We now formalize this idea by introducing the concept of *degenerated string* which is a string that accepts none or multiple symbols on the same position (index). The position of a character in the case of degenerated string is a real value. For example, consider the string $s = 1234$ with all four symbols on position 2 while the other indexes are empty. In the case of degenerated strings the positions of two distinct characters might be the same. For example, $s(1) = s(2) = s(3) = s(4)$.

Our goal is to find a degenerated string $\hat{t}_*$ in which every symbol c has the index given by formula:

$$s(c) = \frac{min(c, S) + max(c, S)}{2} \tag{7}$$

where c is a symbol from Σ.

The string $\hat{t}_*$ has several interesting properties and one of them in particular because it gives a lower bound about minimal Δ achievable by t.

Property 1. The string $\hat{t}_*$ is the closest degenerated-string for S.

Proof. Assume there exists another $\hat{t}_{**}$ as closest degenerated-string for S. We have $\Delta(\hat{t}_{**}) < \Delta(\hat{t}_*)$ meaning

$$\sum_{i=1}^{n} |\hat{t}_{**}(i) - s(i)| < \sum_{i=1}^{n} |\hat{t}_*(i) - s(i))| \tag{8}$$

It follows that there exists at least one i with $|\hat{t}_{**}(i) - s(i)| < |\hat{t}_*(i) - s(i)|$. Since $\hat{t}_*$ elements are in the middle of lower and upper bound in relation with S, it follows there is at least one string s' from S that gets added to Δ the same amount that is subtracted from $\Delta(\hat{t}_{**}, s')$. By the definition of Problem 2 we have a contradiction. $\square$

Now we have $\hat{t}_*$ and we want to design an algorithm to construct $\hat{t}$. Before we present the algorithm, we introduce another useful property of $\hat{t}$ that will help us to lower the complexity for both heuristic and IP formulation.

Observation 2. *In the string $\hat{t}$ the index of every symbol c is in the interval $[min(c, S), max(c, S)]$.*

Proof. If at least one symbol c is outside of his $[min(c, S), max(c, S)]$ with a value x then $\Delta(t, S)$ is increased with at least x. $\square$

Next we define *cluster* C as being the set of symbols c having the same integer part of $s(c)$. Formally:

$$C_\alpha = \{c \in \Sigma \mid [s(c)] = \alpha\} \tag{9}$$

α is called *cluster center*.

We are now in the position to formulate a simple algorithm to compute $\hat{t}$. Starting from $\hat{t}_*$ we align the symbols to an integer index called *cluster center* and the we start moving symbols from the same index to nearby empty indexes. We alternate between left and right part when searching for an empty index near the cluster center. An outline of the algorithm is presented in Algorithm 1.

Next, we analyze the running time of Algorithm 1.

Theorem 2. *The complexity of the algorithm is $O(|\Sigma|)$.*

Proof. The efficiency of our algorithm in terms of how close is $\hat{t}$ from t (CSRD string) is highly impacted by the method of selecting symbols from clusters. A naive way is to selected them based on list order. But minimization of $\Delta(\hat{t}, S)$ is influenced by how items are positioned on empty indexes around cluster center. For example, if we place a symbol to the left that increases the distance from $\hat{t}$ to a string s' and we place a symbol to the right that also increases s' distance, then we diverge from t. To control this outcome we maintain a weight vector w where we count the impact, the increase caused by placing previous symbols for every vector s. Is this manner we can approach a greedy method to always select the symbol that minimize the difference between $max(w)$ and $min(w)$. $\square$

Data: Set S
Result: $\hat{t}$ string
for *every symbol c of Σ* **do**
  | compute $[min(c,S), max(c,S)]$;
end
compute $\hat{t}*$;
compute clusters center;
for *every symbol c of Σ* **do**
  | alternate direction (starting from cluster center) ;
  | select non-allocated cluster item ;
  | allocate it to the next empty index in the selected direction ;
end

Algorithm 1. Heuristic algorithm for the construction of a string $\hat{t}$.

5 Integer Programming Formulations for the CSRD Problem

In this section we present integer programming (IP) formulations for the CSRD problem. As before, we consider that the input strings are n-dimensional vectors. Initially, we start with a simple formulation which we improve based on the properties discussed in the previous sections.

5.1 The Standard IP Formulation

Recall that the goal of the CSRD problem is to minimize distance d between t all s from S. An IP sketch is the following:

$$
\begin{aligned}
\min \quad & d \\
\text{s.t.} \quad & \Delta(t, s_i) \leq d \ i = 1, \ldots, k \\
& d \in \mathbf{Z}_+ \\
& t_i \in \mathbf{Z}_+ \ i = 1, \ldots, n
\end{aligned}
\tag{10}
$$

We expand Δ in Eq. 10 and we obtain:

$$
\begin{aligned}
\min \quad & d \\
\text{s.t.} \quad & \sum_{c \in \Sigma} |t(c) - s_i(c)| \leq d \ i = 1, \ldots, k \\
& d \in \mathbf{Z}_+ \\
& t_i \in \mathbf{Z}_+ \ i = 1, \ldots, n
\end{aligned}
\tag{11}
$$

In Eq. (11) and the following ones, $t(c)$ and $s_i(c)$ represent *position functions* as defined at the beginning of Sect. 4.

The inequalities are not linear so we need to define additional variables for inequalities to have a proper IP formulation. We use that $|x| \leq y$ can be replaced with the following two constraints $-x \leq y$ and $x \leq t$. First we replace $\sum_{c \in \Sigma} |t(c) - s_i(c)|$ by adding new variables $ts_ic = |t(c) - s_i(c)|$ and reformulating constraints. We obtain the following formulation which is now correct but not complete for our problem:

$$
\begin{aligned}
\min \quad & d \\
\text{s.t.} \quad & \sum_i^n ts_ic \leq d && i = 1, \ldots, k \\
& ts_ic \geq t(c) - s_i(c) \; i = 1, \ldots, k; \; \forall c \in \Sigma \\
& ts_ic \geq s_i(c) - t(c) \; i = 1, \ldots, k; \; \forall c \in \Sigma \\
& d \in \mathbf{Z}_+ \\
& t_i \in \mathbf{Z}_+ && i = 1, \ldots, n
\end{aligned} \tag{12}
$$

Next, we want to make sure that t is properly defined, meaning that is a string of size n with symbols appearing only once, meaning that is a permutation. We introduce another set of variables as follows: for every symbol t_i of t we introduce n variables $b_{i1}, b_{i2}, \ldots, b_{in}$ to control what values t_i can have. We constrain all $0 \leq b_{ij} \leq 1$ and $\sum_{j=1}^{n} b_{ij} = 1$. Therefore, exactly one variable from $b_{i1}, b_{i2}, \ldots, b_{in}$ can be 1. If we set $t_i = 1b_{i1} + 2b_{i2} + \cdots + nb_{in}$ then t_i can have a value from 1 to n. To control distinctiveness between t_i symbols, we need to add an additional constrain as $\sum_j b_{ji} \leq 1$. Rewriting Eq. (12) with new constraints we obtain:

$$
\begin{aligned}
\min \quad & d \\
\text{s.t.} \quad & \sum_i^n ts_ic \leq d && i = 1, \ldots, k \\
& ts_ic \geq t(c) - s_i(c) && i = 1, \ldots, k; \; \forall c \in \Sigma \\
& ts_ic \geq s_i(c) - t(c) && i = 1, \ldots, k; \; \forall c \in \Sigma \\
& 0 \leq b_{ij} \leq 1 \; i = 1, \ldots, n; \; j = 1, \ldots, n \\
& \sum_{j=1}^{n} b_{ij} = 1 && i = 1, \ldots, n \\
& \sum_{j=1}^{n} b_{ji} \leq 1 && i = 1, \ldots, n \\
& t_i = 1b_{i1} + 2b_{i2} + \cdots + kb_{in} && i = 1, \ldots, n \\
& t(i) \in \mathbf{Z}_+ && i = 1, \ldots, n \\
& d \in \mathbf{Z}_+
\end{aligned} \tag{13}
$$

The IP formulation is now complete. Next, by using Observation 2 we can improve on IP formulation (13), since we can introduce additional constrains for t_i: $min(c, S) \leq t_i \leq max(c, S)$, where $c = t_i$.

$$
\begin{aligned}
&\text{min} && d \\
&\text{s.t.} && \sum_{i}^{n} ts_i c \leq d && i = 1, \ldots, k \\
& && ts_i c \geq t(c) - s_i(c) && i = 1, \ldots, k;\ \forall c \in \Sigma \\
& && ts_i c \geq s_i(c) - t(c) && i = 1, \ldots, k;\ \forall c \in \Sigma \\
& && 0 \leq b_{ij} \leq 1 && i = 1, \ldots, n;\ j = 1, \ldots, n \\
& && \sum_{j=1}^{k} b_{ij} = 1 && i = 1, \ldots, k \\
& && \sum_{j=1}^{n} b_{ji} \leq 1 && i = 1, \ldots, n \\
& && t_i = 1 b_{i1} + 2 b_{i2} + \cdots + n b_{in} && i = 1, \ldots, n \\
& && d \in \mathbf{Z}_+ \\
& && min(t_i, S) \leq t_i \leq max(t_i, S) && i = 1, \ldots, k \\
& && t_i \in \mathbf{Z}_+ && i = 1, \ldots, n \\
& && d \in \mathbf{Z}_+
\end{aligned}
\tag{14}
$$

Condition $t_i \in \mathbf{Z}_+$ can be droped in both formulations since $t_i = 1 b_{i1} + 2 b_{i2} + \cdots + n b_{in}$ force t_i to be an integer between $1, \ldots, n$.

5.2   Geometric Formulation

Since we consider that all the strings are permutations of the alphabet Σ, we treat strings s as vectors in a space $\mathbb{R}^n$. Then, we observe that all vectors s from S reside on the $(n-1)$ sphere.

Now we are interested to find a vector t with the property that the distance between t and all vectors s is minimized.

$$
t = min_t(max\|t - s_i\|_2)
\tag{15}
$$

Equation (15) can be transformed into the *bounding sphere problem* and can be approximated by using various algorithms like *Ritters bounding sphere* [32], *Linear programming* or *Core-set based 1+ϵ approximation* [3].

Given a good approximation of t we attempt to find an approximation of $\hat{t}$ that resides on the $(n-1)$ sphere and the vector binds to the restrictions described above. Let $Proj(t)$ be the projection of t on the unit n-sphere. We now can use another IP formulation to minimize $\|\hat{t} - Proj(t)\|_1$ which is exactly $\Delta(\hat{t}, Proj(t))$.

$$
\begin{aligned}
&\text{min} && \sum_{i=1}^{n} |\hat{t}_i - Proj(t)_i| \\
&\text{s.t.} && \hat{t} = \sigma(s_e) \quad s_e = \{1, 2, \ldots, n\}
\end{aligned}
\tag{16}
$$

Now we add new variables $pt_i = \hat{t}_i - Proj(t)_i$ and new constraints using the same method as in equation system 12 and we obtain:

$$
\begin{aligned}
\min \quad & \sum_{i=1}^{k} pt_i \\
\text{s.t.} \quad & \hat{t} = \sigma(s_e) \quad s_e = \{1, 2, \ldots, n\} \\
& pt_i \geq \hat{t}_i - Proj(t)_i \quad i = 1, \ldots, n \\
& pt_i \geq Proj(t)_i - \hat{t}_i \quad i = 1, \ldots, n
\end{aligned}
\tag{17}
$$

Next step is to transform $\hat{t} = \sigma(s_e)$ in a propper IP formulation. We use the same method as in Eq. (13) by adding variables n_{ij} along with necessary constraints:

$$
\begin{aligned}
\min \quad & \sum_{i=1}^{n} pt_i \\
\text{s.t.} \quad & pt_i \geq \hat{t}_i - Proj(t)_i \quad i = 1, \ldots, n \\
& pt_i \geq Proj(t)_i - \hat{t}_i \quad i = 1, \ldots, n \\
& 0 \leq b_{ij} \leq 1 \quad i = 1, \ldots, n;\, j = 1, \ldots, n \\
& \sum_{j=1}^{n} b_{ij} = 1 \quad i = 1, \ldots, n \\
& \sum_{j=1}^{n} b_{ji} \leq 1 \quad i = 1, \ldots, n \\
& t_i = 1b_{i1} + 2b_{i2} + \cdots + nb_{in} \quad i = 1, \ldots, n \\
& t_i \in \mathbf{Z}_+ \quad i = 1, \ldots, n
\end{aligned}
\tag{18}
$$

The IP formulation is now complete. We can drop the last constrain since $t_i = 1b_{i1} + 2b_{i2} + \cdots + nb_{in}$ forces t_i to be a integer between $1, \ldots, n$.

6   An Exact Algorithm for the FSRD Problem

In this section, we study the FSRD problem defined in Sect. 2. More precisely, we describe an algorithm that finds two strings generated from the same partition that have the maximum rank distance.

Our target is to find a formula for $D_{(\Sigma, part(k,n))}$ where $part(k, n)$ is a partition and Σ is an alphabet with $|\Sigma| > 2$, $k, n \in \mathbb{N}$, $k \leq n$. From now on, without loss of generality, we assume n is even. Observe that all the strings x generated by a partition $part(k, n)$ have the same Parikh vector. For a string x, we denote by $\bar{x}$ the reversed string. Observe that, given a string x, the string with the same Parikh vector that is farthest apart from x is $\bar{x}$.

To design our algorithm we start with a simple lemma.

Lemma 2. *To maximize $\Delta(x, \bar{x})$, x must have identical elements in consecutive order and near one of the two ends of the vector.*

Proof. Let k and l be the positions of the identical elements in x. Assume $k < l$. The two identical elements contribute to $\Delta(x, \bar{x})$ with $|k - n + l| + |l - n + k| = 2 * (n - l - k)$. To maximize $n - l - k$ we need to have $k = 1$ and $l = 2$.

For a given string x we call the pair (x_i, x_j) a *canceling pair* iff $x_i = x_j$ and $j = n - i + 1$. A canceling pair does not contribute with anything to $\Delta(x, \overline{x})$. Based on previous work we know that:

$$\Delta(x, \overline{x}) = \frac{n^2}{2} - \lceil \frac{n}{2} \rceil - \lfloor \frac{n}{2} \rfloor$$

if and only if x does not have overlapping elements with $\overline{x}$ (i.e. $\forall i = \overline{1, n}, x_i \neq \overline{x}_i$)

We continue with another lemma.

Lemma 3. *To maximize $\Delta(x, \overline{x})$, x must have the canceling elements in the middle of the vector.*

Proof. Let k be the position of one of the canceling elements. The direct contribution of those two identical elements to $\Delta(x, \overline{x})$ is 0. Assume $k = 0$. Compute $\Delta(x, \overline{x})$ and let $2m = \Delta(x, \overline{x})$. Now let $k = 1$. It follows by simple comutation that $\Delta(x, \overline{x}) = 2(m+2)$. Continuing this approach we obtain $max(\Delta(x, \overline{x}))$ when $k = \lfloor \frac{n}{2} \rfloor$.

Based on above properties we can derive an algorithm to generate a vector x from a given partition p, such that $D_{(V, part(k,n))} = \Delta(x, \overline{x})$, as described in Algorithm 2. Running time is $O(n)$.

Data: given partition
Result: build string x to maximize Δ
Initialize vector x of size n;
  for *every multiset of partition* **do**
    read next multiset;
    for *every element of multiset* **do**
      write element at first empty slot of x;
      alternate the first empty slot between left and right of the vector;
    end
  end
$D_{(V, part(k,n))} = \Delta(x, \overline{x})$,;

Algorithm 2. The computation of the diameter of a partition.

Next, we present an example that illustrates our algorithm.

Example 5. Consider the partition $p(3, 5)$ with $M_1 = \{a\}, M_2 = \{b, b\}, M_3 = \{c, c\}$. We initialize x of size 5. We write M_1 starting from left, then write M_2 to the right. Finally starting from the left first empty slot, we fill x with M_3 elements. We obtain $x = accbb$.

In the next lemma we determine the number of canceling pairs for a given partition $part(k, n)$.

Lemma 4. *Given a partition $part(k,n)$ the number of canceling pairs is*

$$C(part(k,n)) = \left|\left| \sum_{i=1}^{\lfloor \frac{k}{2} \rfloor} |M_{2i-1}| - |M_{2i}| \right| - \left(\left\lceil \frac{k}{2} \right\rceil - \left\lfloor \frac{k}{2} \right\rfloor \right) |M_n| \right|$$

Proof. We follow the steps of the algorithm presented above and at each step we count the number of the remaining elements that need to be further paired to get a non-canceling pairs. This number is M_{2i-1} - M_{2i}. Using the equality $\sum_{i=1}^{k} |M_i| = n$ we conclude that the sum of the differences for every $\frac{n}{2}$ steps needs to 0 to have no canceling pairs.

As an example, for the previous lemma, consider the partition $part(3,5)$ with $M_1 = \{a\}, M_2 = \{b,b\}, M_3 = \{c,c\}$. We have $|M_1| - |M_2| = -1$. Then $|1-2| = 1$. Thus, we have one canceling pair.

Now we prove the main result of this section, namely we give a formula for D given a partition $part(k,n)$.

Theorem 3. *Given a partition $part(k,n)$ then*

$$D_{(V,part(k,n))} = \frac{n^2 - C(part(k,n))^2}{2}$$

Moreover, $D_{(V,part(k,n))}$ can be computer in $O(n)$ time.

Proof. The x is built using the Algorithm 2. It follows that the canceling pairs are located in the middle of x. If we have l canceling pairs then the amount not added to the $\Delta(x,\overline{x})$ is $\Delta(y,\overline{y})$ where y is a vector with distinct elements and $|y| = l$. Then $\Delta(x,\overline{x}) = \Delta(x',\overline{x'})$ - $\Delta(y,\overline{y})$, where x' is a vector with distinct elements and $|x'| = n$. It follows $\Delta(x,\overline{x}) = \frac{n^2}{2} - \Delta(y,\overline{y})$. But $|y| = C(p_{k,n})$. We have already showed that $\Delta(x,\overline{x})$ is maximal. The algorithm runs in linear time since we have one step for each character of the string x.

Example 6. If $part(3,5)$ is a partition with $M_1 = \{a\}, M_2 = \{b,b\}, M_3 = \{c,c\}$ then we have $C(p_{3,5}) = 1$ meaning that $max(\Delta(p_{k,n})) = \frac{5^2-1}{2} = 12$.

7 Conclusions and Future Work

In this paper we enhanced the study of the rank distance and we tackled three problems, String with Fixed Rank Distance, Closest String Problem under Rank Distance and Farthest String via Rank Distance. We provided several exact algorithms, IP formulation and heuristic algorithms.

There are still many interesting problems. For example, it is interesting to provide an experimental study of our heuristic algorithm for the CSRD problem. From our tests, Algorithm 1 produces a very good approximation of $\hat{t}$. For example in our experiments with 10–20 strings of size 7–9 the average distance between t and $\hat{t}$ was 2.

References

1. Arbib, C., Felici, G., Servilio, M., Ventura, P.: Optimum solution of the closest string problem via rank distance. In: Cerulli, R., Fujishige, S., Mahjoub, A.R. (eds.) ISCO 2016. LNCS, vol. 9849, pp. 297–307. Springer, Cham (2016). https://doi.org/10.1007/978-3-319-45587-7_26
2. Babaie, M., Mousavi, S.R.: A memetic algorithm for closest string problem and farthest string problem. In: 2010 18th Iranian Conference on Electrical Engineering. IEEE, May 2010
3. Bădoiu, M., Har-Peled, S., Indyk, P.: Approximate clustering via core-sets. In: Proceedings of the Thiry-Fourth Annual ACM Symposium on Theory of Computing, STOC 2002, pp. 250–257. ACM, New York (2002)
4. Ben-Dor, A., Lancia, G., Ravi, R., Perone, J.: Banishing bias from consensus sequences. In: Apostolico, A., Hein, J. (eds.) CPM 1997. LNCS, vol. 1264, pp. 247–261. Springer, Heidelberg (1997). https://doi.org/10.1007/3-540-63220-4_63
5. de la Higuera, C., Casacuberta, F.: Topology of strings: median string is NP-complete. Theor. Comput. Sci. **230**(1–2), 39–48 (2000)
6. Deng, X., Li, G., Li, Z., Ma, B., Wang, L.: Genetic design of drugs without side-effects. SIAM J. Comput. **32**(4), 1073–1090 (2003)
7. Deza, E., Deza, M.: Dictionary of Distances. North-Holland, Amsterdam (2006)
8. Dinu, A., Dinu, L.P.: On the syllabic similarities of romance languages. In: Gelbukh, A. (ed.) CICLing 2005. LNCS, vol. 3406, pp. 785–788. Springer, Heidelberg (2005). https://doi.org/10.1007/978-3-540-30586-6_88
9. Dinu, L.P.: On the classification and aggregation of hierarchies with different constitutive elements. Fundam. Inform. **55**(1), 39–50 (2003)
10. Dinu, L.P., Ionescu, R., Tomescu, A.: A rank-based sequence aligner with applications in phylogenetic analysis. PLoS ONE **9**(8), e104006 (2014)
11. Dinu, L.P., Manea, F.: An efficient approach for the rank aggregation problem. Theor. Comput. Sci. **359**(1–3), 455–461 (2006)
12. Dinu, L.P., Popa, A.: On the closest string via rank distance. In: Kärkkäinen, J., Stoye, J. (eds.) CPM 2012. LNCS, vol. 7354, pp. 413–426. Springer, Heidelberg (2012). https://doi.org/10.1007/978-3-642-31265-6_33
13. Dinu, L.P., Sgarro, A.: A low-complexity distance for DNA strings. Fundam. Inform. **73**(3), 361–372 (2006)
14. Frances, M., Litman, A.: On covering problems of codes. Theory Comput. Syst. **30**(2), 113–119 (1997)
15. Gagolewski, M.: Data Fusion: Theory, Methods, and Applications. Institute of Computer Science, Polish Academy of Sciences, Warsaw, Poland (2015)
16. Gramm, J., Huffner, F., Niedermeier, R.: Closest strings, primer design, and motif search. In: Currents in Computational Molecular Biology. RECOMB, pp. 74–75 (2002)
17. Greenhill, S.J.: Levenshtein distances fail to identify language relationships accurately. Comput. Linguist. **37**(4), 689–698 (2011)
18. Ionescu, R.T., Popescu, M.: Knowledge Transfer between Computer Vision and Text Mining - Similarity-Based Learning Approaches. Advances in Computer Vision and Pattern Recognition. Springer, Cham (2016). https://doi.org/10.1007/978-3-319-30367-3
19. Ionescu, R.T., Popescu, M., Cahill, A.: String kernels for native language identification: insights from behind the curtains. Comput. Linguist. **42**(3), 491–525 (2016)

20. Kannan, R.: Minkowski's convex body theorem and integer programming. Math. Oper. Res. **12**(3), 415–440 (1987)
21. Koonin, E.V.: The emerging paradigm and open problems in comparative genomics. Bioinformatics **15**(4), 265–266 (1999)
22. Lanctot, J.K., Li, M., Ma, B., Wang, S., Zhang, L.: Distinguishing string selection problems. Inf. Comput. **185**(1), 41–55 (2003)
23. Lenstra, H.W.: Integer programming with a fixed number of variables. Math. Oper. Res. **8**(4), 538–548 (1983)
24. Li, M., Ma, B., Wang, L.: Finding similar regions in many sequences. J. Comput. Syst. Sci. **65**(1), 73–96 (2002)
25. Liu, X., He, H., Sýkora, O.: Parallel genetic algorithm and parallel simulated annealing algorithm for the closest string problem. In: Li, X., Wang, S., Dong, Z.Y. (eds.) ADMA 2005. LNCS (LNAI), vol. 3584, pp. 591–597. Springer, Heidelberg (2005). https://doi.org/10.1007/11527503_70
26. Meneses, C.N., Lu, Z., Oliveira, C.A.S., Pardalos, P.M.: Optimal solutions for the closest-string problem via integer programming. INFORMS J. Comput. **16**(4), 419–429 (2004)
27. Nerbonne, J., Hinrichs, E.W.: Linguistic distances. In: Proceedings of the Workshop on Linguistic Distances, Sydney, July 2006, pp. 1–6 (2006)
28. Nicolas, F., Rivals, E.: Complexities of the centre and median string problems. In: Baeza-Yates, R., Chávez, E., Crochemore, M. (eds.) CPM 2003. LNCS, vol. 2676, pp. 315–327. Springer, Heidelberg (2003). https://doi.org/10.1007/3-540-44888-8_23
29. Nicolas, F., Rivals, E.: Hardness results for the center and median string problems under the weighted and unweighted edit distances. J. Discrete Algorithms **3**(2–4), 390–415 (2005)
30. Popescu, M., Dinu, L.P.: Rank distance as a stylistic similarity. In: 22nd International Conference on Computational Linguistics, Posters Proceedings, COLING 2008, 18–22 August 2008, Manchester, UK, pp. 91–94 (2008)
31. Popov, V.Y.: Multiple genome rearrangement by swaps and by element duplications. Theor. Comput. Sci. **385**(1–3), 115–126 (2007)
32. Ritter, J.: An efficient bounding sphere. In: Graphics Gems, pp. 301–303. Elsevier (1990)
33. Sun, Y., et al.: Combining genomic and network characteristics for extended capability in predicting synergistic drugs for cancer. Nat. Commun. **6**, 8481 (2015)
34. Wang, L., Dong, L.: Randomized algorithms for motif detection. J. Bioinf. Comput. Biol. **3**(5), 1039–1052 (2005)
35. Wooley, J.C.: Trends in computational biology: a summary based on a RECOMB plenary lecture. J. Comput. Biol. **6**(3/4), 459–474 (1999)

Computable Analysis of Linear Rearrangement Optimization

Amin Farjudian[(✉)] [iD]

School of Computer Science, University of Nottingham Ningbo China, Ningbo, China
Amin.Farjudian@nottingham.edu.cn
http://www.nottingham.edu.cn/en/science-engineering/staffprofile/amin-farjudian.aspx

Abstract. Optimization problems over rearrangement classes arise in various areas such as mathematics, fluid mechanics, biology, and finance. When the generator of the rearrangement class is two-valued, they reduce to shape optimization and free boundary problems which can exhibit intriguing symmetry breaking phenomena. A robust framework is required for computable analysis of these problems. In this paper, as a first step towards such a robust framework, we provide oracle Turing machines that compute the distribution function, decreasing rearrangement, and linear rearrangement optimizers, with respect to functions that are *continuous and have no significant flat zones*. This assumption on the reference function is necessary, as otherwise, the aforementioned operations may not be computable. We prove that the results can be computed to within any degree of accuracy, conforming to the framework of Type-II Theory of Effectivity.

Keywords: Computable analysis · Rearrangements of functions · Optimization

1 Introduction

The aim of the current paper is to lay the foundation for computability and complexity analysis of optimization problems over rearrangement classes, in the framework of Type-II Theory of Effectivity (TTE) [17].

The theory of rearrangements of functions may be traced back to 1899, when it was introduced as a framework for the study of a problem in hydrostatics [16]. In the following decades, even though it attracted attention from some of the most prominent mathematicians of the twentieth century, it remained a peripheral tool in mathematical analysis, until it re-emerged in the 1970s in the work of Benjamin [1]. Specifically, a problem in fluid mechanics related to steady vortices was formulated by Benjamin as an optimization problem over a rearrangement class.

In response, G. R. Burton laid out a theory for optimization of convex functionals over rearrangement classes [3,5]. Although PDE-constrained rearrangement optimization problems have their origins in fluid mechanics, the abstract

© Springer Nature Switzerland AG 2019
T. V. Gopal and J. Watada (Eds.): TAMC 2019, LNCS 11436, pp. 172–187, 2019.
https://doi.org/10.1007/978-3-030-14812-6_11

formulation of these problems has shown greater potential, and Burton's theory has been used in the study of PDE-constrained rearrangement optimization problems in several areas, e. g., finance [12], free boundary problems [8], non-local problems [11], population biology [10], and eigenvalue problems [9], to name a few. Of special interest is the case when the generator of the rearrangement class is a two-valued function. Rearrangement optimization problems with two-valued generators form some important examples of shape optimization and free boundary problems.

In virtually all but exceptional cases, analytic solutions do not exist for these problems. Even when existence of (not necessarily analytic) solutions is guaranteed by the theory, there are no rigorous computational frameworks for computability and complexity analysis of the solutions. Apart from rare cases, even qualitative accounts of the optimal shapes are missing from the literature. In summary, assuming the existence of a solution is guaranteed, in the majority of cases, the answers to the following questions are still unknown:

(1) Is the optimal solution computable?
(2) Does there exist a Type-II Turing machine, which takes the input parameters of the problem, and returns the optimal solution? Note that this is the uniform version of Question (1).
(3) What are the robustness properties of the optimal solutions?

The following question arises in the related shape optimization problems:

(4) If symmetry breaking occurs, how does it occur? If symmetry is preserved for the n-dimensional ball B, but breaks for the given reference domain Ω, and if we obtain Ω from B through a smooth deformation, at what point does symmetry break?

1.1 Contributions of the Paper

On the computational side of PDE-constrained rearrangement optimization problems, only sporadic attempts have been made, with the main focus on numerical algorithms based on floating-point arithmetic [7,10,11,13]. The current paper is meant to serve as a starting point towards answering questions of the type just listed. *We aim to lay the foundation for a computational framework that allows us to study rearrangement optimization problems in a validated setting.*

Specifically, we provide oracle Turing machines that compute the distribution function, decreasing rearrangement, and linear rearrangement optimizers, with respect to functions that are *continuous and have no significant flat zones*. This assumption on the reference function is necessary, as otherwise, the aforementioned operations may not be computable. Furthermore, the reference functions are solutions of the constraint partial differential equations (PDEs), which, in virtually all applications, satisfy the assumption on continuity and flat zones. For this reason, many of the results in the literature on rearrangement optimization problems—including Burton's seminal papers [3,5]—are formulated with respect to this assumption.

We prove that the results can be computed to within any degree of accuracy, conforming to the framework of Type-II Theory of Effectivity.

Remark 1. It should be noted that the existence of linear rearrangement optimizers is a fundamental result in the theory of rearrangements of functions (Lemma 1 (iv)). Our aim here is to investigate its computability.

1.2 Our Approach

Throughout the paper, we make sure that the arguments adhere to Type-II Theory of Effectivity, as presented in [17]. Yet, instead of using the concepts and notations of [17], we work directly with the discretizations of the domains and maps involved. We hope that this makes the content accessible to a broader audience, as we believe that this direct approach allows us to express our algorithms in a way that is more intuitive.

1.3 Structure of the Paper

- In Sect. 2, we present the background concepts and results from the theory of rearrangements, together with some basic notations that we will be adopting for the remainder of the paper.
- Section 3 contains our main results regarding computable analysis of the distribution function and the decreasing rearrangements.
- In Sect. 4, we present our main results regarding linear rearrangement optimization.
- Section 5 discusses some generalizations of the main results to other domains and dimensions.
- We conclude the paper in Sect. 6, where we also discuss some future work on computable analysis of rearrangement optimization problems.

2 Preliminaries

The material in this section includes some basic concepts and results from the theory of rearrangements of functions. We will also establish some notations that we will be adopting throughout the paper.

2.1 Rearrangement Theory

Let (Ω, Σ, μ) be a measure space, in which Ω is a non-empty set, Σ is a σ-algebra on Ω, and μ is a positive measure on Ω satisfying $\mu(\Omega) < \infty$.

Definition 1 (distribution function λ_f). *For a real measurable $f : \Omega \to \mathbb{R}$, the distribution function $\lambda_f : \mathbb{R} \to \mathbb{R}$ is defined by:* $\forall s \in \mathbb{R} : \lambda_f(s) := \mu\left(f^{-1}[s, \infty)\right)$.

Definition 2 (rearrangement class $\mathcal{R}_\Omega(f_0)$). *Let $(\Omega_0, \Sigma_0, \mu_0)$ and (Ω, Σ, μ) be two measure spaces, such that $\mu_0(\Omega_0) = \mu(\Omega)$.*

(a) We say that $f_0 : \Omega_0 \to \mathbb{R}$ and $f : \Omega \to \mathbb{R}$ are rearrangements of each other if and only if $\forall s \in \mathbb{R} : \mu_0\left(f_0^{-1}[s, \infty)\right) = \mu\left(f^{-1}[s, \infty)\right)$.

(b) The rearrangement class $\mathcal{R}_\Omega(f_0)$ generated by f_0 is defined as follows:

$$\mathcal{R}_\Omega(f_0) := \{f : \Omega \to \mathbb{R} \mid f \text{ is a rearrangement of } f_0\}.$$

Whenever $(\Omega_0, \Sigma_0, \mu_0) = (\Omega, \Sigma, \mu)$, we may simply write $\mathcal{R}(f_0)$. If both f_0 and Ω are clear from the context, we may just use the symbol $\mathcal{R}$ to denote the rearrangement class.

For any Lebesgue-measurable $\Omega \subseteq \mathbb{R}^n$ and $f : \Omega \to \mathbb{R}$, we let $\| f \|_p$ denote the usual L^p norm:

$$\| f \|_p := \begin{cases} \left(\int_\Omega | f(x) |^p \, dx\right)^{\frac{1}{p}}, & \text{if } p \in [1, \infty), \\ \text{ess sup}\{| f(x) | \mid x \in \Omega\}, & \text{if } p = \infty. \end{cases}$$

For every $p \geq 1$, we let q denote its conjugate exponent satisfying $1/p + 1/q = 1$ when $p > 1$, and $q = \infty$ when $p = 1$.

Henceforth, we make the following assumptions:

- Ω denotes a bounded, open, and connected domain in $\mathbb{R}^n$;
- Σ denotes the Lebesgue σ-algebra over Ω, with μ denoting the Lebesgue measure. Indeed, for simplicity, we denote the n-dimensional Lebesgue measure of any Lebesgue-measurable $E \subseteq \mathbb{R}^n$ by $| E |$.
- Ω_0 denotes the open interval $(0, | \Omega |)$.

Definition 3 ($f^\triangle$, $f_\triangle$: non-increasing and non-decreasing rearrangements). *For a real measurable $f : \Omega \to \mathbb{R}$:*

(i) The (essentially unique) non-increasing rearrangement $f^\triangle$ of f is defined on Ω_0 by $f^\triangle(s) := \sup\{\alpha \in \mathbb{R} \mid \lambda_f(\alpha) \geq s\}$. In case f can be extended to $\overline{\Omega}$, with an essential infimum a and an essential supremum b, we extend $f^\triangle$ to $\overline{\Omega_0}$ by letting $f^\triangle(0) := b$ and $f^\triangle(| \Omega |) := a$.

(ii) The (essentially unique) non-decreasing rearrangement $f_\triangle$ of f is defined on Ω_0 by $f_\triangle(s) := f^\triangle(| \Omega | - s)$.

Definition 4 (significant flat zones). *A measurable function $f : \Omega \to \mathbb{R}$ is said to have no significant flat zones on Ω if $\forall c \in \mathbb{R} : | f^{-1}(c) | = 0$.*

The following is very easy to establish:

Proposition 1. *(i) If f is continuous, then λ_f has no significant flat zones.*

(ii) If f has no significant flat zones, then $f^\triangle$ is decreasing, and $f_\triangle$ is increasing.

(iii) If f is continuous and has no significant flat zones, then $f^\triangle$ and λ_f are both continuous, decreasing, and are the inverses of each other.

We will consider linear rearrangement optimization against functions that have no significant flat zones. This condition guarantees uniqueness of solutions, and provides a convenient optimality condition, as summarized in the following lemma:

Lemma 1. *Assume that $f_0 \in L^p(\Omega_0)$, and let $\mathcal{R} := \mathcal{R}_\Omega(f_0)$ be its rearrangement class over Ω. Then:*

(i) *$\mathcal{R} \subseteq L^p(\Omega)$.*

(ii) *$\forall f \in \mathcal{R} : \| f \|_p = \| f_0 \|_p$.*

(iii) *The weak closure of $\mathcal{R}$ in $L^p(\Omega)$, denoted by $\overline{\mathcal{R}}$, is weakly compact and convex.*

(iv) *For every $h \in L^q(\Omega)$, the linear functional $L_h : L^p(\Omega) \to \mathbb{R}$ defined by:*

$$L_h(f) := \int_\Omega f(x)h(x) \, \mathrm{d}x \qquad (1)$$

has a maximizer $\hat{f}$ over $\mathcal{R}$.

(v) *If $\hat{f}$ is the unique maximizer of the linear functional L_h over $\mathcal{R}$, then it is the unique maximizer of L_h over all of $\overline{\mathcal{R}}$. Moreover, $\hat{f} = \hat{\psi}(h)$, almost everywhere in Ω, for some non-decreasing function $\hat{\psi}$.*

(vi) *For any $h \in L^q(\Omega)$ with no significant flat zones, there exists a non-decreasing function $\hat{\psi}$ such that $\hat{\psi}(h) \in \mathcal{R}$, and $\hat{f} := \hat{\psi}(h)$ is the unique maximizer of the linear functional L_h defined in (1) over $\overline{\mathcal{R}}$. Furthermore:*

$$\hat{\psi} = f_0^\Delta \circ \lambda_h. \qquad (2)$$

(vii) *Items (iv), (v), and (vi) remain valid if one replaces 'maximizer' with 'minimizer', and 'non-decreasing function $\hat{\psi}$' with 'non-increasing function $\check{\psi}$', in which case, Eq. (2) becomes $\check{\psi} = f_{0\Delta} \circ \lambda_h$.*

Proof. See [3] and [4]. □

We will also refer to the following results related to non-increasing rearrangements from the literature:

Lemma 2. *Assume that $1 \le p < \infty$. Then:*

(i) *For any given $f \in L^p(\Omega)$, there exists a measure-preserving map $\rho : \Omega \to [0, |\Omega|]$ such that $f = f^\Delta \circ \rho$.*

(ii) *$\forall f, g \in L^p(\Omega) : \| f^\Delta - g^\Delta \|_p \le \| f - g \|_p$.*

Proof. (i) See [4, Lemma 2.4].
(ii) See [6], or [4, Lemma 2.7]. □

2.2   Further Definitions and Notations

The set of dyadic numbers will be denoted by $\mathbb{D}$, i.e., $\mathbb{D} := \{p/2^n \mid p \in \mathbb{Z}, n \in \mathbb{N}\}$. For any $n, k \in \mathbb{N}$, let M_k^n be the meshgrid with granularity k over $[0,1]^n$ whose vertices are:

$$\left\{ \left(\frac{p_1}{2^k}, \frac{p_2}{2^k}, \ldots, \frac{p_n}{2^k} \right) \Big| p_1, p_2, \ldots, p_n \in \{0, 1, \ldots, 2^k\} \right\}.$$

By an element (or a cell) in a meshgrid M_k^n, we mean a compact box of the form:

$$\left[\frac{p_1}{2^k}, \frac{p_1+1}{2^k} \right] \times \left[\frac{p_2}{2^k}, \frac{p_2+1}{2^k} \right] \times \cdots \times \left[\frac{p_n}{2^k}, \frac{p_n+1}{2^k} \right], \quad p_1, p_2, \ldots, p_n \in \{0, 1, \ldots, 2^k - 1\}.$$

Clearly, every meshgrid M_k^n is the union of 2^{kn} such cells, and for each such box S:

$$\forall x, y \in S: \quad \| x - y \|_\infty \leq 2^{-k},$$

in which $\| . \|_\infty$ is the sup norm on $\mathbb{R}^n$. When n is clear from the context, we may just write M_k.

A box $\prod_{i=1}^n [a_i, b_i]$ with rational vertices will be represented by the following element of $\mathbb{Q}^{2n}$:

$$(a_1, b_1, a_2, b_2, \ldots, a_n, b_n).$$

For any set T, we denote the set of finite subsets of T as $\mathcal{P}_{fin}(T)$. For instance, a finite set of two-dimensional rational boxes is an element of $\mathcal{P}_{fin}(\mathbb{Q}^4)$.

Definition 5 (simple step function). *Let $q > 0$ be a rational number. A function $f : [0, q] \to \mathbb{R}$ is said to be a* simple step function *if for some $n \in \mathbb{N}$, there is a set $\{x_0, x_1, \ldots, x_n, y_1, \ldots, y_n\} \subseteq \mathbb{Q}$ such that:*

(a) $0 = x_0 < x_1 < \cdots < x_n = q$.
(b) $f(0) = y_1$ and $\forall x \in (x_{i-1}, x_i) : f(x) = y_i$ for $1 \leq i \leq n$.

We denote the closure and interior of a set A by $\overline{A}$ and A°, respectively. The characteristic function of a subset A of a reference set Y will be denoted as χ_A, which is defined as:

$$\forall y \in Y: \quad \chi_A(y) := \begin{cases} 0, & \text{if } y \notin A, \\ 1, & \text{if } y \in A. \end{cases}$$

Another concept that we will refer to quite frequently is that of a modulus of continuity of a function:

Definition 6 (modulus of continuity). *Let (X, d_X) and (Y, d_Y) be two metric spaces, and assume that $f : X \to Y$ is continuous. Then, a function $\phi : \mathbb{N} \to \mathbb{N}$ is said to be a* modulus function *for f (on X) iff:*

$$\forall n \in \mathbb{N}, \forall x, y \in X: \quad d_X(x, y) \leq 2^{-\phi(n)} \Rightarrow d_Y(f(x), f(y)) \leq 2^{-n}. \tag{3}$$

A fundamental property of computable real functions is that they are continuous, and over compact domains, they have a recursive modulus of continuity:

Theorem 1. *Let* $f : [0,1]^n \to \mathbb{R}$ *be computable. Then,* f *is continuous, and has a recursive modulus of continuity.*

Proof. See, e.g., [14, Theorem 2.13]. $\qquad\qquad\square$

3 Distribution Function and Non-increasing Rearrangement

In this section, we discuss computable analysis of the distribution function λ_u of a given $u : [0,1]^n \to \mathbb{R}$, and its non-increasing rearrangement u^Δ. For every $d \in \mathbb{R}$, define $A_u(d) := \{x \in [0,1]^n \mid u(x) > d\}$, and $B_u(d) := \{x \in [0,1]^n \mid u(x) < d\}$. In [14, Theorems 5.14 and 5.15], it has been shown that:

- u is computable iff the classes of sets $A_u(d)$ and $B_u(d)$ (when d ranges over $\mathbb{D}$) are uniformly recursively open.
- u is recursively approximable iff the classes of sets $A_u(d)$ and $B_u(d)$ (when d ranges over $\mathbb{D}$) are uniformly recursively G_δ.

In both cases, by uniform, we mean uniform in d.

Important as they are, these results do not address the question of computability of λ_u. In fact, it is not difficult to see that:

Proposition 2. *If* u *has a significant flat zone, then* λ_u *is not computable.*

Proof. Let us assume that u has a significant flat zone with value c. Then λ_u is not continuous at c. By Theorem 1, λ_u is not computable. $\qquad\qquad\square$

Here, we consider the case where u and one of its moduli of continuity are provided, respectively, as oracles O_u and ϕ_u—hence, their computability is not assumed—but we have to demand u not to have any significant flat zones. For simplicity, we focus on the case $n = 2$, though the results may be generalized in a straightforward way to any finite dimension.

First, we consider the machine $\mathrm{DistFun}_{\mathbb{Q}}$, which operates under Algorithm 1. This machine takes a height $c \in \mathbb{Q}$ together with an accuracy parameter $k \in \mathbb{N}$, and then, through querying the oracles O_u and ϕ_u, provides an approximation of $\{x \in [0,1]^2 \mid u(x) \geq c\}$ to within 2^{-k} accuracy. This can then be used to obtain a rational approximation of $\lambda_u(c)$ to within 2^{-k} accuracy. The way the oracle O_u operates is as expected:

- On any input query $((x,y), m) \in (\mathbb{Q} \cap [0,1])^2 \times \mathbb{N}$ received on its input channel, O_u outputs a rational $\hat{u} := O_u((x,y), m) \in \mathbb{Q}$ such that $|\hat{u} - u(x,y)| \leq 2^{-m}$.

The correctness of the algorithm for rational input values hinges on the following:

Algorithm 1. Pseudocode for DistFun$_\mathbb{Q}$

Input: Received on four channels:
  $c \in \mathbb{Q}$: height of the level set;
  $k \in \mathbb{N}$: accuracy;
  $u \in C([0,1]^2)$: target function, queried through oracle O_u;
  $\phi_u : \mathbb{N} \to \mathbb{N}$: a modulus of continuity of u.
Output: Approximation of $\{x \in [0,1]^2 \mid u(x) \geq c\}$ to within 2^{-k} accuracy.
  $n \leftarrow 0$
  error $\leftarrow 1 + 2^{-k}$ // anything larger than 2^{-k} would do
  while error $\geq 2^{-k}$ **do**
    Query ϕ_u with $n + 1$
    Create meshgrid $M_{\phi_u(n+1)}$
    for $S \in M_{\phi_u(n+1)}$ **do**
      $S_\sigma \leftarrow$ centroid of S
      $u_S \leftarrow O_u(S_\sigma, n + 1)$ // u_S is assigned the reply to query $(S_\sigma, n + 1)$ sent to O_u
    end for
    $A \leftarrow \{S \in M_{\phi_u(n+1)} \mid u_S > c + 2^{-n}\}$
    $B \leftarrow \{S \in M_{\phi_u(n+1)} \mid u_S < c - 2^{-n}\}$
    $C \leftarrow \{S \in M_{\phi_u(n+1)} \mid c - 2^{-n} \leq u_S \leq c + 2^{-n}\}$
    error $\leftarrow \Sigma_{S \in C} |S|$
    $n \leftarrow n + 1$
  end while
  return A

Lemma 3. *Assume that* $u : \Omega \to \mathbb{R}$ *is continuous with no significant flat zones, where* $\Omega \subseteq \mathbb{R}^n$ *is a bounded domain. Then:*

$$\forall c \in \mathbb{R}, \forall \epsilon > 0, \exists \delta > 0 : |\{x \in \Omega \mid |u(x) - c| < \delta\}| < \epsilon. \tag{4}$$

Proof. Assume that c and ϵ are given. As u is continuous with no significant flat zones, the level set $u_c := \{x \in \Omega \mid u(x) = c\}$ is closed and has Lebesgue measure zero. Hence, there exists an open set $O \subseteq \mathbb{R}^n$ such that $|O| < \epsilon$ and $u_c \subseteq O$. The function $v(x) := |u(x) - c|$ is continuous over the compact domain $\overline{\Omega} \setminus O$. Hence, it attains its minimum at some point (say) x_0, for which, we have $v(x_0) \neq 0$. The value $\delta := |u(x_0) - c|$ satisfies (4). $\square$

Theorem 2. *Assume that* $u : [0,1]^2 \to \mathbb{R}$ *is a continuous function with no significant flat zones, and let* $\phi_u : \mathbb{N} \to \mathbb{N}$ *be a modulus of continuity for* u. *Then, the machine* DistFun$_\mathbb{Q}$, *operating under Algorithm 1, halts on any input* $c \in \mathbb{Q}$ *and* $k \in \mathbb{N}$, *and returns a finite set* A *of two dimensional rational boxes, such that:*

  – $\cup A \subseteq \alpha_c := \{x \in \Omega \mid u(x) \geq c\}$, *and* $|\alpha_c \setminus \cup A| \leq 2^{-k}$.
  – $q := |\cup A| \in \mathbb{Q}$ *and satisfies:* $|q - \lambda_u(c)| \leq 2^{-k}$.

Proof. A careful inspection of Algorithm 1 reveals that, at every iteration of the **while** loop, we have:

$$\cup A \subseteq \alpha_c \subseteq (\cup A) \cup (\cup C) \tag{5}$$

To see this, assume that $S \in A$, and S_σ is its centroid. For all $y \in S$, we have $\| y - S_\sigma \|_\infty < 2^{-\phi_u(n+1)}$, which, by (3), entails that $u(y) \geq u(S_\sigma) - 2^{-(n+1)}$. As the oracle O_u has been queried with accuracy $n+1$, we have $u(S_\sigma) \geq u_S - 2^{-(n+1)}$. Therefore $u(y) \geq u_S - 2^{-n} > c$. This proves that $S \subseteq \alpha_c$, hence $\cup A \subseteq \alpha_c$. A similar argument shows that $\forall S \in B : \forall y \in S : u(y) < c$, which proves that $\alpha_c \subseteq (\cup A) \cup (\cup C)$.

From (5), we obtain:

$$| \cup A | \leq | \alpha_c | \leq | \cup A | + | \cup C |. \tag{6}$$

Now, at the n-th iteration of the while loop, we have:

$$\forall x \in \cup C : | u(x) - c | \leq 2^{-n} + 2^{-(n+1)} < 2^{-(n-1)}. \tag{7}$$

According to (4), for the given $c \in \mathbb{Q}$ and $k \in \mathbb{N}$, there exists an $n_0 \in \mathbb{N}$ such that:

$$\forall n \geq n_0 : | \{x \in \Omega \mid | u(x) - c | < 2^{-(n-1)}\} | < 2^{-k}. \tag{8}$$

From (8) and (7), we infer that at iterations $n > n_0$, we have $| \cup C | < 2^{-k}$. But $| \cup C |$ is exactly the value of error in Algorithm 1, and the output of the algorithm is A. This, together with (6), proves the result.   □

Although Theorem 2 is sufficient for our purposes, we discuss the general case where $c \in \mathbb{R}$ for completeness. As rational numbers are finitely representable, in Algorithm 1, the value of c is provided to the machine DistFun$_\mathbb{Q}$ in one transaction. When $c \in \mathbb{R}$, the value should be provided to the respective machine DistFun$_\mathbb{R}$ (Algorithm 2) through an oracle O_c, which, on any given input $n \in \mathbb{N}$, supplies DistFun$_\mathbb{R}$ with a rational $c_n \in \mathbb{Q}$ satisfying:

$$| c - c_n | \leq 2^{-n}. \tag{9}$$

Theorem 3. *Assume that $u : [0,1]^2 \to \mathbb{R}$ is a continuous function with no significant flat zones, and let $\phi_u : \mathbb{N} \to \mathbb{N}$ be a modulus of continuity for u. Then, the machine DistFun$_\mathbb{R}$, operating under Algorithm 2, halts on any input $c \in \mathbb{R}$ and $k \in \mathbb{N}$, and returns a finite set A of two-dimensional rational boxes, such that:*

- *$\cup A \subseteq \alpha_c := \{x \in \Omega \mid u(x) \geq c\}$, and $| \alpha_c \setminus \cup A | \leq 2^{-k}$.*
- *$q := | \cup A | \in \mathbb{Q}$ and satisfies: $| q - \lambda_u(c) | \leq 2^{-k}$.*

Proof. By (9) we have $\forall n : c_n - 2^{-n} \leq c \leq c_n + 2^{-n}$. As λ_u is decreasing (Proposition 1), we have:

$$\lambda_u(c_n + 2^{-n}) < \lambda_u(c) < \lambda_u(c_n - 2^{-n}). \tag{10}$$

Let us define $q_n := | \cup A_n |$ and $q_n' := | \cup A_n' |$. By Theorem 2, we have:

$$\begin{cases} | q_n - \lambda_u(c_n - 2^{-n}) | \leq 2^{-n}, \\ | q_n' - \lambda_u(c_n + 2^{-n}) | \leq 2^{-n}. \end{cases} \tag{11}$$

Algorithm 2. Pseudocode for DistFun$_\mathbb{R}$

Input: Corresponding to the four channels:
  $c \in \mathbb{R}$: height of the level set, queried through oracle O_c;
  $k \in \mathbb{N}$: accuracy;
  $u \in C([0,1]^2)$: target function, queried through oracle O_u;
  $\phi_u : \mathbb{N} \to \mathbb{N}$: a modulus of continuity of u.
Output: Approximation of $\{x \in [0,1]^2 \mid u(x) \geq c\}$ to within 2^{-k} accuracy.
  $n \leftarrow 0$
  error $\leftarrow 1 + 2^{-k}$ // anything larger than 2^{-k} would do
  while error $\geq 2^{-k}$ **do**
    $c_n \leftarrow O_c(n)$ // Note that $|c_n - c| < 2^{-n}$
    $A_n \leftarrow \text{DistFun}_\mathbb{Q}(c_n - 2^{-n}, n, u, \phi_u)$
    $A'_n \leftarrow \text{DistFun}_\mathbb{Q}(c_n + 2^{-n}, n, u, \phi_u)$
    error $\leftarrow |\cup A_n| - |\cup A'_n| + 2^{-(n-1)}$
    $n \leftarrow n + 1$
  end while
  return A_n

From (10) and (11), we get:

$$q'_n - 2^{-n} \leq \lambda_u(c) \leq q_n + 2^{-n},$$

which explains why error is defined as the value of $q_n - q'_n + 2^{-(n-1)}$.

Note that λ_u is continuous (Proposition 1). Therefore:

$$\lim_{n \to \infty} \lambda_u(c_n - 2^{-n}) = \lim_{n \to \infty} \lambda_u(c_n + 2^{-n}) = \lambda_u(c).$$

In particular $\lim_{n \to \infty} q_n = \lim_{n \to \infty} q'_n = \lambda_u(c)$. This, together with (11), ensure that the aforementioned error goes below 2^{-k} for sufficiently large n. This proves halting of the algorithm. □

By Proposition 1, when u is continuous and has no significant flat zones, then u^Δ and λ_u are inverses of each other. This, together with the fact that both functions are one-to-one, continuous, and decreasing, provides a simple way of obtaining the decreasing rearrangement of u from its distribution function:

Corollary 1. *There exists an oracle machine* InvDistFun$_\mathbb{R}$ *which, given:*

- *a continuous $u \in C([0,1]^2)$ with no significant flat zones and a modulus of continuity $\phi_u : \mathbb{N} \to \mathbb{N}$ for u;*
- *a real number $r \in [0,1]$ and an accuracy $k \in \mathbb{N}$,*

returns a rational number $c \in \mathbb{Q}$, together with a finite set of cells $A_r \subseteq M_{n_0}$, for some meshgrid of granularity n_0, such that $|c - u^\Delta(r)| \leq 2^{-(k+1)}$, $\cup A_r \subseteq \{x \in [0,1]^2 \mid u(x) \geq c\}$, and $|\Sigma_{S \in A_r} |S| - r| \leq 2^{-(k+1)}$.

Furthermore, if $c_r = u^\Delta(r)$ and $\alpha_{c_r} := \{x \in [0,1]^2 \mid u(x) \geq c_r\}$, then:

$$|\cup A_r \Delta \alpha_{c_r}| \leq 2^{-k}, \tag{12}$$

in which Δ denotes symmetric difference of sets defined as $X \Delta Y := (X \setminus Y) \cup (Y \setminus X)$.

Proof. As a modulus of continuity is provided, a lower bound a and an upper bound b for u can be easily obtained. All that remains to do for InvDistFun$_\mathbb{R}$ is to perform a binary search using the machine DistFun$_\mathbb{Q}$ of Theorem 2.   □

So far, we have only demanded the level sets of u to have measure zero. By Theorem 3, these level sets become *computably* measure zero provided that u is computable:

Corollary 2. *Assume that $u : [0,1]^2 \to \mathbb{R}$ is a computable function with no significant flat zones. Then for any computable $c \in \mathbb{R}$, the level set $u^{-1}(c)$ is computably measure zero.*

4 Linear Rearrangement Optimization

Assume that, for some $p \in [1,\infty)$, we are given a generator $f_0 \in L^p([0,1])$ and a function $u \in C([0,1]^2)$ which has no significant flat zones. Our task is to compute the necessarily unique $\hat{f} \in \mathcal{R}_{[0,1]^2}(f_0)$ which maximizes the functional L_u as defined in (1). Equation (2) provides the basis for the results of this section. Nonetheless, as f_0 is in $L^p([0,1])$—hence might have discontinuities—we need to go through some careful computable analysis to make sure that error estimates are accurately accounted for.

Requiring u to be continuous might seem like a strong condition, but in practice, solutions of the PDE constraints to rearrangement optimization problems that we have in mind invariably are 'continuous', i.e., the solutions lie in $C(\overline{\Omega})$, where Ω is the domain over which the PDE is stated. Furthermore, requiring u not to have any significant flat zones ensures uniqueness of solutions, and again, it is a condition that is satisfied in the vast majority of PDE-constrained rearrangement optimization problems in the literature.

For computational purposes, the function u and one of its moduli of continuity ϕ_u will be provided as oracles, in the same manner as in Sect. 3. As for the generator f_0, we first note that for any given $p \in [1,\infty)$, the set of simple step functions is dense in $L^p([0,1])$. Thus, we represent $f_0 \in L^p([0,1])$ as the limit of a Cauchy sequence of simple step functions converging to it. Using the results of Sect. 3, we obtain tight approximations of linear rearrangement maximizers for the approximant simple step functions, and then prove that these approximations, in turn, converge to the true maximizer for the given (potentially not finitely representable) function f_0.

Remark 2. Although we focus on maximization, corresponding results for linear rearrangement minimization may be obtained with straightforward tweaking of the arguments and the proofs.

4.1 Simple Step Function Generator

For some $p \in [1,\infty)$, assume that $f \in L^p([0,1])$ is a simple step function, represented by the set:

$$\{x_0, x_1, \ldots, x_n, y_1, \ldots, y_n\} \subseteq \mathbb{Q},$$

Without loss of generality, we assume that f is non-increasing and $\forall i \neq j : y_i \neq y_j$. If this is not the case, a simple sorting and then gluing of subintervals can ensure these two conditions. Let $\hat{f}$ be the unique maximizer of L_u over $\mathcal{R}_{[0,1]^2}(f)$ as in Lemma 1 (vi), and let $\gamma := 2\max\{|y_i| \mid 1 \leq i \leq n\}$.

Now, assume that we are given an $\epsilon > 0$, and our aim is to find some $\tilde{f}$ which approximates $\hat{f}$ to within ϵ accuracy. Let $k \in \mathbb{N}$ be large enough such that:

$$2^{-k} < \frac{\epsilon^p}{n\gamma^p}. \tag{13}$$

Together with u, we provide this value k, and successive values of x_i, to the machine InvDistFun$_{\mathbb{R}}$, and for each $i \in \{1, \ldots, n\}$, let A_{x_i} be as in Corollary 1. In particular, if $c_i := \lambda_u^{-1}(x_i)$ and $\alpha_{c_i} := \{x \in [0,1]^2 \mid u(x) \geq c_i\}$, then by (12) we know that $\cup A_i$ approximates α_{c_i} to within 2^{-k} accuracy. Next, we define:

$$\begin{cases} \hat{A}_{x_1} := \cup A_{x_1}, \\ \hat{A}_{x_k} := \cup A_{x_k} \setminus \cup A_{x_{k-1}}, \quad (2 \leq k \leq n), \end{cases}$$

and note that $\hat{A}_{x_i}$'s partition $[0,1]^2$. Hence, we can define a piecewise constant function $\tilde{f} : [0,1]^2 \to \mathbb{R}$ as follows:

$$\forall x \in [0,1]^2 : \tilde{f}(x) := \sum_{i=1}^{n} y_i \chi_{\hat{A}_{x_i}}(x).$$

From Corollary 1, we deduce that $\tilde{f}$ and $\hat{f}$ coincide on all of $[0,1]^2$ except perhaps on a set of Lebesgue measure at most $n2^{-k}$. Thus:

$$\left(\int_{[0,1]^2} |\tilde{f}(x) - \hat{f}(x)|^p \, dx\right)^{1/p} < \left(n2^{-k}\gamma^p\right)^{1/p}$$

$$\text{(by (13))} < \epsilon.$$

Putting all of the above together, we obtain:

Lemma 4. *There exists an oracle machine M_1 which, given:*

- *a continuous $u \in C([0,1]^2)$ with no significant flat zones, and a modulus of continuity $\phi_u : \mathbb{N} \to \mathbb{N}$ for u;*
- *a real number $p \in [1, \infty)$, and a simple step function $f \in L^p([0,1])$;*
- *an accuracy parameter $n \in \mathbb{N}$;*

returns a piecewise constant function $\tilde{f} \in L^p([0,1]^2)$ such that $\|\tilde{f} - \hat{f}\|_p < 2^{-n}$, in which $\hat{f}$ is the unique maximizer of L_u over $\mathcal{R}_{[0,1]^2}(f)$ as in Lemma 1 (vi).

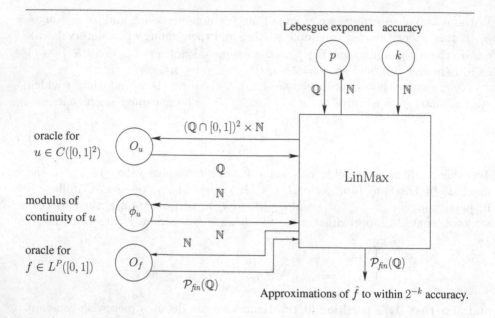

Fig. 1. An oracle machine representation of LinMax, which computes the linear rearrangement maximizer $\hat{f}$ of L_u over $\mathcal{R}_{[0,1]^2}(f)$.

4.2   General Case

For an arbitrary $f \in L^P([0,1])$, we consider the oracle machine LinMax of Fig. 1. On receiving the accuracy demand $k \in \mathbb{N}$, the machine sends $k+1$ as a query to the oracle O_f, which in turn returns a simple step function $f_{k+1} : [0,1] \to \mathbb{R}$ satisfying:

$$\| f_{k+1} - f \|_p \leq 2^{-(k+1)}. \tag{14}$$

Subsequently, LinMax uses the machine M_1 from Lemma 4, providing it with p, u, ϕ_u, f_{k+1}, and accuracy parameter $k+1$. The machine M_1, in turn, returns a piecewise constant $\tilde{f}_{k+1} \in L^P([0,1]^2)$ which satisfies:

$$\| \tilde{f}_{k+1} - \hat{f}_{k+1} \|_p \leq 2^{-(k+1)}. \tag{15}$$

Finally, LinMax returns $\tilde{f}_{k+1}$ as output.

To prove that LinMax is working correctly, we need to prove that:

$$\| \tilde{f}_{k+1} - \hat{f} \|_p \leq 2^{-k}. \tag{16}$$

By Lemma 2 (i), there exists a measure-preserving map $\rho : [0,1]^2 \to [0,1]$ satisfying:

$$u = u^{\Delta} \circ \rho, \tag{17}$$

where u^Δ is the decreasing rearrangement of u. As $\hat{f}$ and $\hat{f}_{k+1}$ are maximizers of L_u over $\mathcal{R}_{[0,1]^2}(f)$ and $\mathcal{R}_{[0,1]^2}(f_{k+1})$, respectively, by Lemma 1 (vi), there are non-decreasing functions ψ_1 and ψ_2 such that:

$$\hat{f} = \psi_1 \circ u \quad \text{and} \quad \hat{f}_{k+1} = \psi_2 \circ u. \tag{18}$$

As ψ_1 and ψ_2 are non-decreasing, it is straightforward to show that:

$$f^\Delta = \psi_1 \circ u^\Delta \quad \text{and} \quad f^\Delta_{k+1} = \psi_2 \circ u^\Delta. \tag{19}$$

From (17), (18), and (19), we deduce $\hat{f} = f^\Delta \circ \rho$ and $\hat{f}_{k+1} = f^\Delta_{k+1} \circ \rho$, which implies that:

$$\| \hat{f} - \hat{f}_{k+1} \|_p = \| f^\Delta - f^\Delta_{k+1} \|_p$$
$$\text{(by Lemma 2 (ii))} \leq \| f_{k+1} - f \|_p$$
$$\text{(by (14))} \leq 2^{-(k+1)}. \tag{20}$$

By combining (15) and (20), we obtain (16). Hence, we have:

Theorem 4. *There exists an oracle machine* LinMax *which, given:*

- *a continuous $u \in C([0,1]^2)$ with no significant flat zones, and a modulus of continuity $\phi_u : \mathbb{N} \to \mathbb{N}$ for u;*
- *a real number $p \in [1, \infty)$, and a function $f \in L^p([0,1])$;*
- *an accuracy parameter $k \in \mathbb{N}$;*

returns a piecewise constant function $\tilde{f} \in L^p([0,1]^2)$ such that $\| \tilde{f} - \hat{f} \|_p \leq 2^{-k}$, in which $\hat{f}$ is the unique maximizer of L_u over $\mathcal{R}_{[0,1]^2}(f)$ as in Lemma 1 (vi).

Essentially, we have fleshed out the algorithm for computing $\hat{f}$ which is suggested by Eq. (2). If we define $\psi := f^\Delta \circ \lambda_u$, then we will have $\hat{f} = \psi \circ u$. Now it should be clear that the measure-preserving transformation ρ of (17) is just $\lambda_u \circ u$. Note that λ_u can be obtained by Theorem 3, and f^Δ may be approximated using the simple step function approximations $\{f_n \mid n \in \mathbb{N}\}$ of f.

5   Generalizations

To stay focused on the essence of rearrangements, we presented our results for the simple two dimensional cube $[0,1] \times [0,1]$. Generalizations to the following, however, are straightforward:

- Linear rearrangement *minimization*;
- n-dimensional cube $[0,1]^n$, for all $n \in \mathbb{N}$;
- Open, bounded, and connected domains $\Omega \subseteq \mathbb{R}^n$, for which $\overline{\Omega}$ is a union of n-dimensional cubes with rational coordinates.

Some careful error analysis, together with (say) Delaunay triangulation, may provide a further generalization to polygonal domains with rational coordinates. Indeed, domains Ω that can be approximated *from within* via rational polygonal domains, whose boundary vertices lie on the boundary of Ω, may also be treated, using more careful error analysis.

Going further to general domains might need substantial change in approach, especially if the approximants of the domain Ω have to cover locations out of Ω. This is reminiscent of the finite element methods for numerical solutions of PDEs, where care is taken to have the finite element space as a subspace of the reference Sobolev space.

6   Conclusions and Future Work

We have taken some steps towards computable analysis of rearrangement optimization problems. We provided oracle Turing machines that compute the distribution function, decreasing rearrangement, and linear rearrangement optimizers, with respect to functions that are continuous and have no significant flat zones.

The next step will be the computable analysis of a complete PDE-constrained rearrangement optimization problem. Note that linear rearrangement optimization is one of the two main components of some numerical methods for solving PDE-constrained rearrangement optimization problems [7,10], the other being PDE solving. Apart from some isolated work (e. g., [2,15]) computable analysis of PDE solving is largely an unexplored area.

In longer term, we aim to develop validated methods for shape optimization and free boundary problems arising as PDE-constrained rearrangement optimization problems.

References

1. Benjamin, T.B.: The alliance of practical and analytical insights into the nonlinear problems of fluid mechanics. In: Germain, P., Nayroles, B. (eds.) Applications of Methods of Functional Analysis to Problems in Mechanics. LNM, vol. 503, pp. 8–29. Springer, Heidelberg (1976). https://doi.org/10.1007/BFb0088744
2. Brattka, V., Yoshikawa, A.: Towards computability of elliptic boundary value problems in variational formulation. J. Complex. **22**(6), 858–880 (2006)
3. Burton, G.R.: Rearrangements of functions, maximization of convex functionals, and vortex rings. Math. Ann. **276**(2), 225–253 (1987)
4. Burton, G.R.: Variational problems on classes of rearrangements and multiple configurations for steady vortices. Ann. Inst. H. Poincaré Anal. Non Linéaire **6**(4), 295–319 (1989)
5. Burton, G.R., McLeod, J.B.: Maximisation and minimisation on classes of rearrangements. Proc. R. Soc. Edinb. Sect. A **119**(3–4), 287–300 (1991)
6. Crowe, J.A., Zweibel, J.A., Rosenbloom, P.C.: Rearrangements of functions. J. Funct. Anal. **66**(3), 432–438 (1986)
7. Elcrat, A., Nicolio, O.: An iteration for steady vortices in rearrangement classes. Nonlinear Anal. **24**(3), 419–432 (1995)

8. Emamizadeh, B., Marras, M.: Rearrangement optimization problems with free boundary. Numer. Funct. Anal. Optim. **35**(4), 404–422 (2014)
9. Emamizadeh, B., Zivari-Rezapour, M.: Rearrangements and minimization of the principal eigenvalue of a nonlinear Steklov problem. Nonlinear Anal. **74**(16), 5697–5704 (2011)
10. Emamizadeh, B., Farjudian, A., Liu, Y.: Optimal harvesting strategy based on rearrangements of functions. Appl. Math. Comput. **320**, 677–690 (2018)
11. Emamizadeh, B., Farjudian, A., Zivari-Rezapour, M.: Optimization related to some nonlocal problems of Kirchhoff type. Canad. J. Math. **68**(3), 521–540 (2016)
12. Emamizadeh, B., Hanai, M.A.: Rearrangements in real estate investments. Numer. Funct. Anal. Optim. **30**(5–6), 478–485 (2009)
13. Kao, C.Y., Su, S.: Efficient rearrangement algorithms for shape optimization on elliptic eigenvalue problems. J. Sci. Comput. **54**(2), 492–512 (2013)
14. Ko, K.I.: Complexity Theory of Real Functions. Birkhäuser, Boston (1991)
15. Selivanova, S., Selivanov, V.: Computing the solution operators of symmetric hyperbolic systems of PDE. J. Univers. Comput. Sci. **15**(6), 1337–1364 (2009)
16. Talenti, G.: The art of rearranging. Milan J. Math. **84**(1), 105–157 (2016)
17. Weihrauch, K.: Computable Analysis, An Introduction. Springer, Heidelberg (2000). https://doi.org/10.1007/978-3-642-56999-9

On the Power of Oritatami Cotranscriptional Folding with Unary Bead Sequence

Szilárd Zsolt Fazekas[1], Kohei Maruyama[2(✉)], Reoto Morita[2],
and Shinnosuke Seki[2(✉)]

[1] Graduate School of Engineering Science, Akita University,
1-1 Tegate Gakuen-machi, Akita 0108502, Japan
[2] The University of Electro-Communications,
1-5-1 Chofugaoka, Chofu, Tokyo 1828585, Japan
{k.maruyama,s.seki}@uec.ac.jp

Abstract. An oritatami system is a novel mathematical model of RNA cotranscriptional folding, which has recently proven extremely significant in information processing in organisms and also controllable artificially in a test tube to construct an artificial structure by folding an RNA sequence. This model has turned out to be Turing universal. One next step is to simplify the Turing universal oritatami system and another is to characterize weaker oritatami systems as we may not need Turing universality for applications. In this paper, we look at oritatami systems that folds a unary sequence, and show that under reasonable assumptions, these systems are not universal.

1 Introduction

Transcription is the first essential step of gene expression, in which a DNA template sequence is copied into a single stranded RNA sequence of nucleotides A, C, G, and U (letter of RNA alphabet) by a 'molecular Xerox' called RNA polymerase, nucleotide by nucleotide according to the complimentarity relation $A \rightarrow U$, $G \rightarrow C$, $C \rightarrow G$, and $T \rightarrow A$. The copied RNA sequence is called *transcript*. The transcript does NOT remain single-stranded until it is fully synthesized. It rather starts folding upon itself into intricate stable conformations (structures) primarily via hydrogen bonds, immediately after it emerges from the polymerase, as illustrated in Fig. 1 (Left).

In a recent breakthrough in molecular engineering by Geary, Rothemund and Andersen [8] the co-transcriptional folding of RNA is controlled by careful design of the DNA template. As demonstrated in laboratory, this method, called RNA Origami, makes it possible to cotranscriptionally self-assemble a unique RNA

This work is supported in part by KAKENHI Grant-in-Aid for Challenging Research (Exploratory) No. 18K19779 granted to S. Z. F. and S. S. and JST Program to Disseminate Tenure Tracking System No. 6F36 granted to S. S.

T. V. Gopal and J. Watada (Eds.): TAMC 2019, LNCS 11436, pp. 188–207, 2019.
https://doi.org/10.1007/978-3-030-14812-6_12

Fig. 1. (Left) RNA origami. (Right) an abstraction of its product, i.e., an RNA tile, as a configuration of an oritatami system. A dot • in the figure on the right represents a sequence of 3-4 nucleotides (oligonucleotides). The solid arrow and dashed lines represent respectively its RNA transcript and interactions based on hydrogen bonds between nucleotides.

rectangular tile highly probably (see Fig. 1 (Left)). This breakthrough and the design of RNA tile has encouraged the research on the nano-scale RNA structure self-assembly so that several successful attempts to self-assemble artificial structures by folding a single-stranded RNA sequence [4,10]. Geary et al. [6] proposed a mathematical model for this process, called oritatami system. In this model, an oligonucleotide (a short sequence of RNA nucleotides) is considered to be a bead and an RNA structure is abstracted as a directed path with information on hydrogen-based interaction (bonds) between beads over the triangular grid graph $\mathbb{T}$ as illustrated in Fig. 1. An oritatami system folds a transcript of abstract molecules (beads) of finite number of types over $\mathbb{T}$. This model has been just proved efficiently Turing universal in [7] by simulating cyclic tag systems introduced by Cook [2]. The simulation involves a very large and complex oritatami system. This system is deterministic in the sense that every bead is stabilized uniquely point-wise as well as interaction-wise (for the formal definition, see Sect. 2). One future direction of research is to quest for a smaller Turing-universal oritatami system.

Closely related is the question of where not to look for universal systems, i.e., what are the limitations of simple oritatami system. In search for simple oritatami systems, there are a number of restrictions one can pose on them:

- bounds on the relative speed of transcription to folding (delay), the number of bead types, or the number of hydrogen bonds per bead (arity);
- bounds on the length of the transcript or on the complexity of rules to decide what types of beads interact with each other (attraction rules);
- structural conditions on the transcript or the attraction rules.

In [3], Demaine et al. proved that at delay 1 and arity 1, upon an initial structure of n beads, a deterministic oritatami system cannot fold into any conformation of more than $10n$ beads, no matter how many bead types are available. We consider this finiteness problem for unary oritatami systems under various settings of the values of delay and arity, which is formalized as follows (Table 1).

Problem 1. Give an upper bound on the length of a transcript of a delay-δ, arity-α deterministic unary oritatami system whose seed is of length n by a function in δ, α, and n.

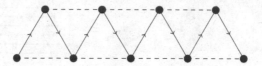

Fig. 2. The zig-zag conformation. This is the only one infinite conformation foldable deterministically by an unary oritatami system at delay 1 and arity 2.

Table 1. Upper bounds on the length of a conformation foldable by a deterministic unary oritatami system at delay δ and arity α. At any combination of delay and arity without anything written, no upper bound is known yet.

$\alpha \backslash \delta$	1	2 and 3	4 and larger
1	$10n$ [3]	$3n^2+4n+1$ (Theorem 6)	
2	∞ but zigzag after $(27n^2+9n+1)$-th bead (Theorem 5)		
3	$4n+14$ (Theorem 4)		
4	$3n(n+1)+1$ (Theorem 3)		

In this paper, we will solve this problem completely at delay 1 and partially at arity 1 in Sect. 4. At delay 1, we will provide a quadratic upper bound $3n(n+1)+1$ for the case of arity being 4, while a linear upper bound $4n+14$ for arity 3. At the delay 1 and arity 2, one infinite structure turns out to be foldable deterministically, which is the zigzag conformation shown in Fig. 2. At arity 1, we will prove that at delay 2 or 3, the upper bound is $(3n+1)(n+1)$. Upper bounds for longer delays remain open. These results as well as known upper bounds are summarized in Fig. 1. As shown at the end of Sect. 2, the stabilization of the first t beads of transcript by a deterministic oritatami system can be simulated by a deterministic Turing machine within t^3 steps (Corollary 1). Thus, the above mentioned upper bounds show that at delay 1 and arity 1, 3, or 4, or at delay 2 or 3 and arity 1, the class of deterministic unary oritatami systems is not Turing universal. The Turing universal oritatami system by Geary et al. [7] employs more than 500 types of beads. Thus, this weakness result is not surprising at all. The unary oritatami system might not be practical very much, though one bead may abstract an oligonucleotide (a short sequence of nucleotides), and in that case, unary transcript can be a repetitive but nonunary sequence, which is not so unrealistic in experiments (see [5]). Nevertheless, this paper makes a first considerable step towards the characterization of non-Turing-universal oritatami systems.

As a result of independent significance, in Sect. 3, we show that increasing the delay from 1 to 2 enables an oritatami system to yield a conformation of quadratic length in n as long as 9 types of beads are available.

2   Preliminaries

Let Σ be a finite set of types of abstract molecules, or *beads*. A bead of type $a \in \Sigma$ is called an a-bead. By Σ^* and Σ^ω, we denote the set of finite sequences of beads and that of one-way infinite sequences of beads, respectively. The empty sequence is denoted by λ. Let $w = b_1 b_2 \cdots b_n \in \Sigma^*$ be a sequence of length n for some integer n and bead types $b_1, \ldots, b_n \in \Sigma$. The *length* of w is denoted by $|w|$, that is, $|w| = n$. For two indices i, j with $1 \le i \le j \le n$, we let $w[i..j]$ refer to the subsequence $b_i b_{i+1} \cdots b_{j-1} b_j$; if $i = j$, then $w[i..i]$ is simplified as $w[i]$. For $k \ge 1$, $w[1..k]$ is called a *prefix* of w.

Oritatami systems fold their transcript, which is a sequence of beads, over the triangular grid graph $\mathbb{T} = (V, E)$ cotranscriptionally. We designate one point in V as the origin O of $\mathbb{T}$. For a point $p \in V$, let $\bigcirc_p^d$ denote the set of points which lie in the regular hexagon of radius d centered at the point p. Note that $\bigcirc_p^d$ consists of $3d(d+1) + 1$ points. A directed path $P = p_1 p_2 \cdots p_n$ in $\mathbb{T}$ is a sequence of *pairwise-distinct* points $p_1, p_2, \ldots, p_n \in V$ such that $\{p_i, p_{i+1}\} \in E$ for all $1 \le i < n$. Its i-th point is referred to as $P[i]$. Now we are ready to abstract RNA single-stranded structures in the name of conformation. A *conformation C* (over Σ) is a triple (P, w, H) of a directed path P in $\mathbb{T}$, $w \in \Sigma^*$ of the same length as P, and a set of h-interactions $H \subseteq \{\{i, j\} \mid 1 \le i, i+2 \le j, \{P[i], P[j]\} \in E\}$. This is to be interpreted as the sequence w being folded along the path P in such a manner that its i-th bead $w[i]$ is placed at the i-th point $P[i]$ and the i-th and j-th beads are bound (by a hydrogen-bond-based interaction) if and only if $\{i, j\} \in H$. The condition $i + 2 \le j$ represents the topological restriction that two consecutive beads along the path cannot be bound. The *length* of C is defined to be the length of its transcript w (that is, equal to the length of the path P). A *rule set* $R \subseteq \Sigma \times \Sigma$ is a symmetric relation over Σ, that is, for all bead types $a, b \in \Sigma$, $(a, b) \in R$ implies $(b, a) \in R$. A bond $\{i, j\} \in H$ is *valid with respect to R*, or simply R-valid, if $(w[i], w[j]) \in R$. This conformation C is R-valid if all of its bonds are R-valid. For an integer $\alpha \ge 1$, C is *of arity α* if it contains a bead that forms α bonds but none of its beads forms more. By $\mathcal{C}_{\le \alpha}(\Sigma)$, we denote the set of all conformations over Σ whose arity is at most α; its argument Σ is omitted whenever Σ is clear from the context.

The oritatami system grows conformations by an operation called elongation. Given a rule set R and an R-valid conformation $C_1 = (P, w, H)$, we say that another conformation C_2 is an elongation of C_1 by a bead $b \in \Sigma$, written as $C_1 \xrightarrow{R}_b C_2$, if $C_2 = (Pp, wb, H \cup H')$ for some point $p \in V$ not along the path P and set $H' \subseteq \{\{i, |w|+1\} \mid 1 \le i < |w|, \{P[i], p\} \in E, (w[i], b) \in R\}$ of bonds formed by the b-bead; this set H' can be empty. Note that C_2 is also R-valid. This operation is recursively extended to the elongation by a finite sequence of beads as: for any conformation C, $C \xrightarrow{R}_\lambda^* C$; and for a finite sequence of beads $w \in \Sigma^*$ and a bead $b \in \Sigma$, a conformation C_1 is elongated to a conformation C_2 by wb, written as $C_1 \xrightarrow{R}_{wb}^* C_2$, if there is a conformation C' that satisfies $C_1 \xrightarrow{R}_w^* C'$ and $C' \xrightarrow{R}_b C_2$.

An *oritatami system* (OS) $\Xi = (\Sigma, R, \delta, \alpha, \sigma, w)$ is composed of

- a set Σ of bead types,
- a rule set $R \subseteq \Sigma \times \Sigma$,
- a positive integer δ called the *delay*,
- a positive integer α called the *arity*,
- an initial R-valid conformation $\sigma \in \mathcal{C}_{\leq\alpha}(\Sigma)$ called the *seed*, whose first bead is assumed to be at the origin O without loss of generality,
- a (possibly infinite) *transcript* $w \in \Sigma^* \cup \Sigma^\omega$, which is to be folded upon the seed by stabilizing beads of w one at a time so as to minimize energy collaboratively with the succeeding $\delta-1$ nascent beads.

The energy of a conformation $C = (P, w, H)$, denoted by $\Delta G(C)$, is defined to be $-|H|$; the more bonds a conformation has, the more stable it gets. The set $\mathcal{F}(\Xi)$ of conformations *foldable* by the system Ξ is recursively defined as: the seed σ is in $\mathcal{F}(\Xi)$; and provided that an elongation C_i of σ by the prefix $w[1..i]$ be foldable (i.e., $C_0 = \sigma$), its further elongation C_{i+1} by the next bead $w[i+1]$ is foldable if

$$C_{i+1} \in \underset{\substack{C \in \mathcal{C}_{\leq\alpha} \, s.t. \\ C_i \xrightarrow{R}_{w[i+1]} C}}{\arg\min} \ \min\left\{ \Delta G(C') \ \Big| \ C \xrightarrow{R}{}^*_{w[i+2...i+k]} C', k \leq \delta, C' \in \mathcal{C}_{\leq\alpha} \right\}. \quad (1)$$

Then we say that the bead $w[i+1]$ and the bonds it forms are *stabilized* according to C_{i+1}. The easiest way to understand this stabilization process should be the video available at https://www.dailymotion.com/video/x3cdj35, in which the Turing universal oritatami system by Geary et al. [7], whose delay is 3, is running. Note that an arity-α oritatami system cannot fold any conformation of arity larger than α. A conformation foldable by Ξ is *terminal* if none of its elongations is foldable by Ξ. The oritatami system Ξ is *deterministic* if for all $i \geq 0$, there exists at most one C_{i+1} that satisfies (1). A deterministic oritatami system folds into a unique terminal conformation. An oritatami system with the empty rule set just folds into an arbitrary elongation of its seed nondeterministically. Thus, the rule set is reasonably assumed non-empty.

In this paper, we considerably focus on the unary oritatami system. An oritatami system is *unary* if it involves only one type of bead, say a, that is, $\Sigma = \{a\}$. Its rule set is $R = \{(a, a)\}$. Its transcript is a sequence of a-beads so that nothing can be hardcoded on it.

Proposition 1. *For any rule set R, arity α and conformation $C = (P, w, H)$ it is possible to check whether C is R-valid and whether $C \in \mathcal{C}_{\leq\alpha}$ in time $\mathcal{O}(|H| \cdot |w| \cdot |R|)$.*

Proof. To check whether C is R-valid:

1. FOR each $(i, j) \in H$:
2.   IF $(w[i], w[j]) \notin R$ THEN answer NO and HALT
3. answer YES and HALT.

Checking the condition in 2. can be done in $\mathcal{O}(|w| \cdot |R|)$ time for any reasonable representation of w and R, hence the whole process takes $\mathcal{O}(|H| \cdot |w| \cdot |R|)$ time. To check the arity constraint $C \in \mathcal{C}_{\leq \alpha}$:

1. FOR each $i \in \{1, \ldots, |w|\}$:
2.   IF degree$(i) = |\{j|(i,j) \in H\}| > \alpha$ THEN answer NO and HALT
3. answer YES and HALT.

Checking the condition in 2. can be done in $\mathcal{O}(|H|)$ time for any reasonable representation of H, hence the whole process takes $\mathcal{O}(|w| \cdot |H|)$ time. $\qquad\square$

Theorem 1. *There is an algorithm that simulates any deterministic oritatami system* $\varXi = (\Sigma, R, \delta, \alpha, \sigma, w)$ *in time* $2^{\mathcal{O}(\delta)} \cdot |R| \cdot |w|$.

Proof. Take any step in the computation, up to which some $i \geq 0$ first beads of w have been stabilized, with the last bead at a point p. The number of all possible elongations of the current conformation by the next δ-beads is $(6 \times 5^{\delta-1}) \times ((2^4)^{\delta-1} \times 2^5) \in 2^{\mathcal{O}(\delta)}$. By Proposition 1, we can check for each of these elongations whether its arity is at most α or not and whether it is R-valid or not in time $\mathcal{O}((2^4)^{\delta-1} \cdot 2^5 \cdot \delta \cdot |R|) = 2^{\mathcal{O}(\delta)} \cdot |R|$. Therefore, the total running time is $2^{\mathcal{O}(\delta)} \cdot |R| \cdot |w|$. $\qquad\square$

Corollary 1. *For fixed* δ, *the class of problems solvable by deterministic oritatami systems* $(\Sigma, R, \delta, \alpha, \sigma, w)$ *is included in* DTIME$(|w|^3)$.

Proof. The claim follows from Theorem 1 and the fact that $|R|$ is implicitly bounded by $|w|^2$. $\qquad\square$

Considering the following decision problem: given an oritatami system, integer i, and a point p, decides whether the bead $w[i]$ is stabilized at p. By Corollary 1, this problem is in P for a fixed delay δ. Because of the time hierarchy theorems, we know that P $\subsetneq$ EXP (see, e.g., [1]), so we can conclude that OS which cannot deterministically fold transcripts of length exponential in the length of the seed are not computationally universal.

3 Quadratic Lower Bound for Delay 2, Arity 1

First we present a lower bound construction for arity 1 systems. At $\alpha = 1$, having delay $\delta = 2$ allows the deterministic folding of quadratic length transcripts compared with $\delta = 1$, where, as stated before, the maximum length is linear in the length of the seed. We demonstrate this with an infinite family of OS, which fold deterministically a transcript of length $\frac{(n-1)^2}{4}$ starting from a given seed of length n.

Consider the following $\delta = 2$, $\alpha = 1$ system with bead types $\{0, 1, \ldots, 8\}$ and attraction rules $\{(i,i) \mid 1 \leq i \leq 8\}$. Let the seed σ be a conformation of a $4k + 1$ long bead sequence of the form $(10205060)^{k/2}0$ and $(10205060)^{(k-1)/2}(1020)0$, for k even and odd, respectively. Bead $\sigma[i]$ of the seed is stabilized at point $(i, 0)$, for all $1 \leq i \leq 4k - 1$. Bead $4k$ is at $(4k - 1, -1)$ and bead $4k + 1$ is at $(4k, 0)$.

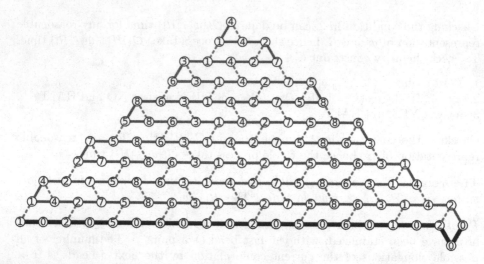

Fig. 3. Quadratic length transcript folding deterministically into pyramid shape. Seed: thick black path. Transcript: thin blue path. Bonds: dashed red lines. (Color figure online)

The transcript is $w = \text{row}_1 \cdots \text{row}_{2k}$, where

- $\text{row}_1 = (24136857)^{(k-1)/2}241$ if k is odd, and $\text{row}_1 = (68572413)^{k/2-1}6857241$ if k is even;
- $\text{row}_{i+1} = (\text{row}_i[2..|\text{row}_i| - 1])^r$ for $i \in \{1, \ldots, 2k-1\}$, where w^r is the reverse of w. In other words, each row is the reverse of the previous without its first and last bead.

The transcript above is written in rows which correspond to beads in the conformation stabilized along the same row on the grid. To simplify the argument we will use *row* both for the transcript above and for the conformation it stabilizes in (note that in the figure the row index grows from bottom to top).

Row 1 is of length $4k - 1$ and row $\ell + 1$ is two beads shorter than row ℓ, so the length of the whole transcript is $|w| = 4k^2 = \frac{(4k+1-1)^2}{4} = \frac{(|\sigma|-1)^2}{4}$. As an example, see Fig. 3, where $k = 5$, so the length of the seed is $4k + 1 = 21$ and the transcript is $4k^2 = 100$ beads long.

Stabilizing the first bead of a row goes by binding to the penultimate bead of the previous row, as they are of the same type according to how we constructed the transcript, and they have free hands (see Fig. 5).

As for the other beads, in rows $j \equiv 1, 3 \mod 4$, beads of type $1, 2, 5, 6$ bind to a bead in row $j - 1$. In rows $j \equiv 2, 4 \mod 4$ beads of type $3, 4, 7, 8$ can bind to a bead in row $j - 1$. This is true for row 1, because beads of type $1, 2, 5, 6$ from row 1 can only bind to every second bead of the seed, whereas the other beads of row 1 cannot bind to anything (Fig. 4, (c)). Once this dynamic holds for a row, it holds inductively for the next, as a bead that binds to another loses its only free hand at arity 1.

Within one row of the transcript, no bead i can bind to a preceding bead, because if there is a previous bead of the same type in that row, it is stabilized at a distance of at least 6 from any point where i could be placed.

By the arguments above, the beads in row i of the transcript are stabilized along row i on the grid, forming the pyramid-like conformation from Fig. 3.

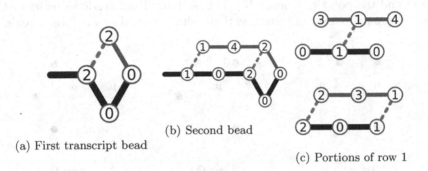

(a) First transcript bead

(b) Second bead

(c) Portions of row 1

Fig. 4. Fixing transcript beads in first row, when k is odd

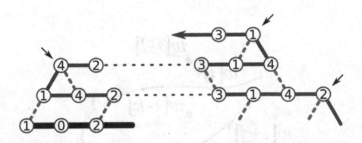

Fig. 5. Beads $4k, 8k-2, \ldots$ stabilize at turning points because the bead two positions before is the same type and has a free hand.

4   Upper Bounds for Determinisitc Unary Oritatami Systems at Delay 1 or Arity 1

In this section, we consider Problem 1 at delay 1 first and then at arity 1. Let $\varXi = (\varSigma, R, \delta, \alpha, \sigma, w)$ be a deterministic oritatami system of delay 1. For $i \geq 0$ let C_i be the unique elongation of σ by $w[1..i]$, that is, foldable by $\varXi$. Hence $C_0 = \sigma$.

At delay 1, a bead cannot collaborate with its successors in order to stabilize itself. In fact, there are just two ways for a bead to get stabilized at delay 1 (or the bead has no place to be stabilized around so that the system halts), as observed in [3]. One is to be bound to another bead and the other is through a 1-in-1-out structure called the tunnel section. See Fig. 6. A *tunnel section* consists of one free point p_c and four beads that occupy four neighbors of p_c. In order for an oritatami system to stabilize the bead $w[i]$ at the central point p_c, its predecessor $w[i-1]$ must be put at one of the two free neighbors of p. Thus, at the stabilization of $w[i]$, only one neighbor of p is left free so that the successor $w[i+1]$ is to be stabilized there, even without being bound. In this case, the point p_e where $w[i-1]$ is stabilized is considered to be an *entrance of the tunnel section* and the point p_s where $w[i+1]$ is stabilized is considered as its *exit*. A *tunnel* is a maximal set of tunnel sections whose central points form a path.

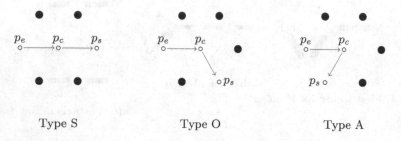

Type S   Type O   Type A

Fig. 6. Tunnel sections of all possible three types: straight (Type S), obtuse turn (Type O), and acute turn (Type A).

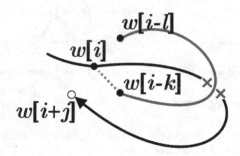

Fig. 7. A tunnel divides the world into two.

The behavior of an oritatami system at delay 1 can be described by a sequence of S of b (bound), t_s (straight tunnel section), t_o (obtuse-turn tunnel section), and t_a (acute-turn tunnel section); priority is given to tunnel, that is, $S[i]$ is t_s (resp. t_o, t_a) if the i-th bead of the system is stabilized not only by being bonded but also through a straight (resp. obtuse-turn, acute-turn) tunnel section. Let us introduce t as a wildcard for t_s, t_o, and t_a. We also let S take the value ■ for halt (due to the lack of free neighbors).

We say that a neighbor of a point p is *reachable* from a conformation C if there exists an elongation of C in which a bead occupies the neighbor and it binds with a bead at p. For example, in Fig. 7, the transcript w is about to step into a tunnel. The wall beads $w[i-\ell]$ and $w[i-k]$ are both older than the bead at the entrance, $w[i]$. Even if $w[i]$ leaves a free neighbor, the neighbor is not reachable because the path of a conformation must be non-self-intersecting, and its subpath between these two wall beads divides the plane into two regions, one of which includes the entrance of the tunnel and the other of which includes the exit (Jordan curve theorem [9]). Taking this reachability into account, we define the *binding capability* of a conformation as the number of free bonds of its beads available geometrically for elongations of C. It is defined formally as follows:

Definition 1 (Binding Capability). *Let α be an arity and $C = (P, w, H)$ be a conformation of arity at most α. Let $H_k = H \cap \{(i, j) \mid i = k \text{ or } j = k\}$. Moreover, let R_k be a set of neighbors of the point $P[k]$ that are free and reachable from C. The binding capability of C at arity α, denoted by $\#bc_\alpha(C)$, is defined by $\sum_{k=1}^{|w|} \min\{\alpha - |H_k|, |R_k|\}$. The subscript α is omitted whenever it is clear from context.*

Observe one almost-trivial but important fact that a bead inside a tunnel does not increase the binding capability. This is because for such a bead $w[k]$, $|R_k| = 0$.

We now prove that in "almost all" tunnels is a troll domiciled and robs the transcript of binding capability (the original story is from [11]). Originally, we tried to find a troll in every tunnel but failed; a troll seems to dislike the very first bead $w[1]$, or its property that only $\alpha+1$ beads around may take all hands of $w[1]$ thanks to the absence of its predecessor; any other bead must be surrounded by at least $\alpha + 2$ beads in order to be free from free hand because a bead cannot bind with its predecessor or successor. We call a bead *singular* if it is surrounded by only $\alpha+1$ beads but forms α bonds. No bead but $w[1]$ can be singular because of their predecessor and successor. A tunnel is *singular* if its entrance or exit is next to $w[1]$ that is singular. There can be at most 3 tunnels around one bead so that no more than 3 tunnels can be singular. A singular tunnel will be denoted with the superscript $\times$ like $t_s^\times$ or $t^\times$. In contrast, the notation without $\times$ such as t_s and t shall imply their non-singularity.

Theorem 2 (Tunnel Troll Theorem). *Let Ξ be a deterministic unary oritatami system of delay $\delta = 1$. The following statements hold.*

1. *At arity $\alpha \geq 3$, if $S[i] = t$ (i.e. the tunnel that stabilizes $w[i]$ is not singular) and $S[i + 1] \neq \blacksquare$, then $\#bc(C_{i-1}) > \#bc(C_i)$.*
2. *At arity $\alpha = 2$, if m is the number of occurrences of bt as a factor in $S[1..k]$ for an index k, then $\#bc(C_0) - m \geq \#bc(C_k)$.*

In order to prove this theorem, we use the following three lemmas.

Lemma 1. *Let Ξ be a deterministic unary oritatami system at delay $\delta = 1$ and arity $\alpha = 2$. Assume Ξ stabilizes the transcript until $w[i - 1]$. If $S[i + 1] = b$ and $S[i + 2] \in \{t_s, t_o\}$, then $\#bc(C_{i-1}) > \#bc(C_i)$.*

Proof. See Fig. 8. $S[i + 2] \in \{t_s, t_o\}$ means that $w[i + 2]$ is stabilized by a tunnel section of type S or O. Thus, its predecessor $w[i + 1]$ must be inside the tunnel section, that is, n_1 and n_2 must be occupied. Free bonds of $w[i]$, if any, cannot be used in future by another bead $w[j]$ because otherwise the part of transcript $w[i..j]$ and the bond between $w[i]$ and $w[j]$ would form a closed curve and the curve would cross the path of C_{i-1} between n_1 and n_2, contradiction. Therefore, if $w[i]$ forms a bond at its stabilization $\#bc(C_{i-1}) > \#bc(C_i)$ holds. We now prove that $w[i]$ must form a bond.

Suppose $w[i]$ were stabilized without any bond, that is, by a tunnel. For that the two points that are a neighbor of both $w[i-1]$ and $w[i]$ must be occupied already. In addition, at least one of the neighbors of $w[i]$ must be free because $S[i+1] = b$. Thus, only the case to be considered is Fig. 8 (middle) with n_5 being occupied (that is, n_4 is free). In this case, before $w[i]$ is stabilized, at lest three neighbors of n_2 were free and hence, a bead at n_2 was provided with one free bond and could form a bond with $w[i]$. □

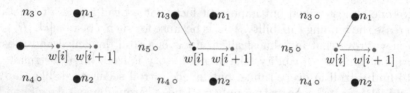

Fig. 8. The three ways to enter a tunnel: (Left) straight, (Middle) obtuse, (Right) acute. The bead $w[i]$ is stabilized at the entrance and $w[i+1]$ is stabilized inside.

Fig. 9. Two kinds of exit of a tunnel: (Left) Both n_1 and n_2 are free, (Right) One of n_1 and n_2 is occupied.

Lemma 2. *Let Ξ be a deterministic unary oritatami system of delay $\delta = 1$ and arity $\alpha = 2$. If $S[i+1..j+1] = bt^{(j-i-1)}b$ for some i, j with $i \leq j - 2$ and $S[j] \in \{t_s, t_o\}$, then $\#bc(C_{j-2}) \geq \#bc(C_j)$, and hence, $\#bc(C_i) \geq \#bc(C_j)$. If $i \leq j - 3$, then the second inequality is strengthened as $\#bc(C_i) > \#bc(C_j)$.*

Proof. Since the binding capability never increases inside a tunnel, we just need to consider the exit of a tunnel. See Fig. 9. At least one of points n_1 or n_2 must be free because otherwise $w[j]$ would be inside of a tunnel, that is, $S[j+1]$ would not be b.

Let m be the number of bonds $w[j-1]$ forms, that is, $\#bc(C_{j-2}) - \#bc(C_{j-1}) = m$. We claim $\#bc(C_j) - \#bc(C_{j-1}) \leq m$. Indeed, if both n_1 and n_2 are free (see Fig. 9), the predecessor $w[j-1]$ must be bound to both beads at

n_3 and at n_4 because both of them still have a free hand. Hence, $m \geq 2$. Since $\#bc(C_j) - \#bc(C_{j-1})$ is less than the arity, this difference is at most m.

If n_1 is occupied, then n_2 is free. The predecessor $w[j-1]$ must be bound n_4. Hence, $m \geq 1$. The bead $w[j]$ can increase the binding capability at most by 1 because one of its free neighbors would, n_0 or n_2, is to be occupied by the successor $w[j+1]$. Therefore, $\#bc(C_j) - \#bc(C_{j-1}) \leq m$.

Thus, $\#bc(C_{j-2}) \geq \#bc(C_j)$, and hence, $\#bc(C_i) \geq \#bc(C_j)$. If $i \leq j-3$, then the second inequality is strengthened as $\#bc(C_i) > \#bc(C_j)$ because $S[i+1] = b$, that is, $\#bc(C_i) > \#bc(C_{i+1})$ and $\#bc(C_{i+1}) \geq \#bc(C_j)$. □

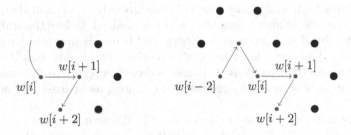

Fig. 10. The bead $w[i+2]$ is stabilized by a tunnel of type A. (Right) Moreover $S[i] = t_a$.

Lemma 3. *Let Ξ be a deterministic unary oritatami system of delay $\delta = 1$ and arity $\alpha = 2$. If $S[i+2] = t_a$, the following statements hold.*

1. *If $w[i]$ forms at least one bond, $\#bc(C_{i-1}) > \#bc(C_{i+2})$.*
2. *If $w[i]$ does not consume any bond and $S[i] \in \{t_s, t_o\}$, $\#bc(C_{i-2}) > \#bc(C_{i+2})$.*
3. *If $w[i]$ does not consume any bond and $S[i] = t_a$, $\#bc(C_{i-3}) - 2 \geq \#bc(C_{i+2})$.*

Proof. We consider each statement. First we prove Statement 1. The bead $w[i+1]$ consumes one hand and provides nothing. If $w[i]$ forms two bonds, then even if $w[i+2]$ provides two free hands, $\#bc(C_{i-1}) > \#bc(C_{i+2})$. On the other hand, if it leaves a free hand, it will be used by $w[i+2]$, and hence, $w[i+2]$ does not increase the binding capability. Thus, $\#bc(C_{i-1}) > \#bc(C_{i+2})$. This argument actually work also for the case when $w[i]$ is stabilized rather by binding.

Let us proceed to Statement 2. See Fig. 10. Consider the case when $w[i]$ is stabilized by a tunnel section of type S or O. As prove in Lemma 2, $\#bc(C_{i-2}) \geq \#bc(C_i)$. The bead $w[i]$ leaves two free hand, it will be used by $w[i+2]$, and hence, $w[i+2]$ does not increase the binding capability. Thus, $\#bc(C_{i-2}) > \#bc(C_{i+2})$. This argument actually work also for the case when $w[i]$ is stabilized rather by binding.

We finalize this proof by showing Statement 3. In order for $w[i]$ not to bind, $w[i-2]$ must have already used up its hands. The bead $w[i-1]$ consumes one hand and provides nothing. Thus, $\#bc(C_{i-3}) - 1 \geq \#bc(C_i)$. The bead

$w[i + 1]$ consumes one hand and provides nothing. Finally, $w[i + 2]$ uses one hand of $w[i]$, and hence, does not increase the binding capability. Therefore, $\#bc(C_{i-3}) - 2 \geq \#bc(C_{i+2})$. This argument actually works also for the case when $w[i]$ is stabilized rather by binding. □

Now we are ready to prove the Tunnel Troll Theorem.

Proof. Let us first consider cases of $\delta \geq 3, \alpha = 1$. See Fig. 9. Consider the stabilization of $w[i]$. This bead $w[i]$, once stabilized, shares two neighbors with its predecessor $w[i - 1]$, which are denoted by n_3, n_4. Both of them have been already occupied because $S[i] = t$.

Since $S[i + 1] \neq \blacksquare$, at least one of the other three neighbors, denoted by n_0, n_1, n_2, must be free. Assume that in the neighborhood of $w[i]$, there are two beads with one free neighbor even after $w[i]$ is stabilized. Before the stabilization of $w[i]$, such a bead had two free neighbors, and hence, is provided with at least one free bond. Thus, $w[i]$ is to be bonded to these two beads, and it decreases the binding capability by at least 1. It now suffices to check that this assumption holds no matter how n_0, n_1, n_2 are occupied as long as at least one of them is left free.

Next, we consider the case of $\delta = 2, \alpha = 1$. We assume there are indices i, j such that $S[i + 1..j + 1] = bt^{(j-i-1)}b$. If $S[i + 2]$ is t_s or t_o, then Lemma 1 implies $\#bc(C_{i-1}) > \#bc(C_i)$ and Lemma 2 implies $\#bc(C_i) \geq \#bc(C_j)$. Thus, binding capability decreases by 1 per a factor bt_s or bt_o.

Now, we assume $S[i + 2] = t_a$. Then, we have to make sure that one troll is not double-counted. If $w[i]$ forms a bond, Lemmas 2 and 3 imply that binding capability decreases through this tunnel.

Assume $w[i]$ forms no bond. If $S[i] \in \{t_s, t_o\}$, Lemmas 2 and 3 imply $\#bc(C_{i-2}) > \#bc(C_{i+2})$. Observe that the bead $w[i - 1]$ is at the entrance of the previous tunnel or inside. It is when a bead is stabilized at the entrance of a tunnel that the troll of the tunnel decreases binding capability. Thus, the inequality does not rely on the troll of previous tunnel. If $S[i] = t_a$, Lemma 3 implies $\#bc(C_{i-3}) - 2 \geq \#bc(C_{i+2})$. This inequality involves two tunnels but its difference 2 enables us to consider that binding capability decreases by 1 through this tunnel. □

4.1   Upper Bounds on the Length of Conformation Foldable Deterministically at Delay $\delta = 1$

Theorem 3 ($\delta = 1, \alpha = 4$). *The terminal conformation of a deterministic unary oritatami system of $\delta = 1, \alpha = 4$ is of length at most $3n^2 + 3n + 1$.*

Proof. Consider the moment when a bead, say b, is stabilized outside $\bigcirc_O^n$ for the first time. The bead must be bound a bead at the periphery of $\bigcirc_O^n$ as depicted in Fig. 11. In order to avoid nondeterminism, the bead b must not be attracted anyhow else by beads around.

The point p_1 must be empty because a bead there would have at least two free neighbors and hence is provided with a free hand. If there is a bead at p_2,

n must be at least 2 so that the bead is not singular. Since p_1 is empty, this bead has at least one free hand, a contradiction. Thus, p_2 must be also empty. In the same way, we can easily show that the point p_3 must not be occupied by a non-singular bead. Suppose $p_3 = O$. The point p_4 must not be empty; otherwise the singular bead, at O, would have a free hand. However, then a bead at p_4 would be provided with a free hand, a contradiction. □

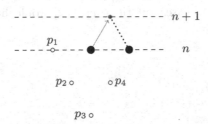

Fig. 11. The first bead out of $\bigcirc_O^n$

Theorem 4 ($\delta = 1, \alpha = 3$). *The terminal conformation of a deterministic unary oritatami system of $\delta = 1, \alpha = 3$ is of length at most $4n + 14$.*

Proof. In this proof, we shall verify the claim that when the bead $w[i]$ is stabilized with $S[i] = b$ and $S[i+1] \neq \blacksquare$, if the circle of radius 2 centered at its predecessor $w[i-1]$ is free from the singular point, then $w[i]$ must form at least 2 bonds. Recall that the circle of radius 2 centered at the origin O, where the only candidate of singularity is, consists of 19 points including O. In order for the bead at O to be singular, its $\alpha + 1 = 4$ neighbors must be occupied. This means that there are at most 14 points where a bead find a singular bead within 2 points. Therefore, the claim, once proved, and the Tunnel Troll Theorem imply that all but at most 14 beads strictly decrease the binding capability. The binding capability of the seed is at most $3n$. Consequently, this theorem holds.

Now let us verify the claim. Suppose $w[i]$ were stabilized by just one bond. There are three cases to be considered as depicted in Fig. 12, depending on the relative position of $w[i]$ to $w[i-2]$ and $w[i-1]$. Since $S[i] = b$, at least one of the four neighbors of $w[i-1]$ must be empty. If n_3 is free, then $w[i-2]$ must have used up all of its hands; otherwise, $w[i]$ would be stabilized also at n_3 nondeterministically, a contradiction. Thus, $\alpha + 2 = 5$ neighbors of $w[i-2]$, that is, all of its neighbors, must be occupied. Hence, n_5 is occupied. (All the remaining arguments are based on this "merry-go-round" occupation. This works only if the circle of radius 2 around $w[i-1]$ is free from the singular bead). In the same way, all the neighbors of n_3 turned out to be occupied in the clockwise order, but eventually, we would encounter a neighbor that is adjacent to also the point where $w[i]$ is supposed to go. Thus, n_3 must be occupied (in the left and middle cases). In the left case, n_4 is symmetric to n_3, and hence, it must be occupied, too. Since $S[i] = b$, n_1 or n_2 must be free; assume n_1 is. Going clockwise around n_1 implies that n_{-1} is occupied, but the bead at n_1 has a free hand and would cause a nondeterminism.

Let us focus on the remaining cases: middle and right. Suppose that among the 5 neighbors of $w[i]$ at which $w[i-1]$ is not, only one can be occupied. Otherwise, a bead without any hand is found at one of them, and staring from the point, merry-go-round occupies all the neighbors, but then $w[i+1]$ would lose its way to go. In the middle case, this means n_0 is free. Since n_{-1} is free, so must be n_2. Repeating this, we get that both n_4 and n_5 are free, but then $w[i-2]$ would have a free hand and a free neighbor, attract $w[i]$, and cause a nondeterminism. Even in the right case, the points n_1, n_0, n_2, n_4 turn out to be free one after another likewise, but then $w[i-2]$ would have a free hand and neighbor, a contradiction.   □

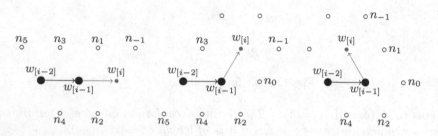

Fig. 12. All possible directions of $w[i]$: straight, obtuse, acute.

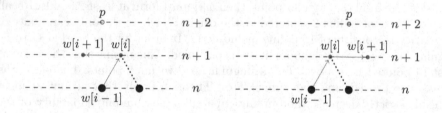

Fig. 13. The moment when the transcript steps outside $\bigcirc_O^n$.

Theorem 5 ($\delta = 1, \alpha = 2$). *A deterministic unary oritatami system of $\delta = 1, \alpha = 2$ can fold into an infinite conformation, but its transcript folds into the zig-zag conformation (Fig. 2) after its $(27n^2 + 9n + 1)$-th bead.*

Proof. We assume $w[i]$ is the first bead stabilized outside $\bigcirc_O^n$. See Fig. 13. The next bead $w[i+1]$ is to be bound for stabilization. Hence, it goes to the west or to the east (Fig. 13). Once $w[i+2]$ is stabilized at p, the remaining transcript folds into the zig-zag conformation. In order to avoid this or nondeterminism, $w[i+2]$ must form two bonds; it thus decreases binding capability by 1. Until when a bead is stabilized outside $\bigcirc_O^{n+1}$, binding capability never increases because of arity being 2 and of the Tunnel Troll Theorem. This means that, only at most $\#bc(C_0) \leq 2n$ times we can thus expand that hexagonal region. In other words, outside $\bigcirc_O^{3n}$ the transcript cannot help but fold zig-zag.   □

4.2   Quadratic Upper Bounds for Arity 1 and Delay 2 or 3

In this section we will argue that unary systems at arity 1 and delay 2 and 3, respectively, cannot fold infinite transcripts deterministically. As we will see, in fact, the length of transcripts deterministically foldable by these systems has an upper bound quadratic in the length n of the seed. The main result is the following theorem, which is a direct consequence of Lemmas 4 and 5 which follow.

Theorem 6 ($\delta \in \{2,3\}, \alpha = 1$). *The terminal conformation of a deterministic unary oritatami system at arity $\alpha = 1$ and delay $\delta \in \{2,3\}$ is of length at most $3n^2 + 4n + 2$.*

Let us fix some common starting points for Lemmas 4 and 5. Let the point where the first transcript bead was fixed be p. We will argue about the situation when the first bead is stabilized outside $\bigcirc_p^n$ (a hexagon of radius n). Let this be the ith bead of the transcript. Without loss of generality, we can translate the origin $(0,0)$ to the coordinates of bead $i - 1$ (which is still in $\bigcirc_p^n$), and we can assume that bead i is fixed at $(1,1)$ (see Fig. 14).

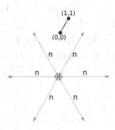

Fig. 14. $\bigcirc_p^n$ and the position $(1,1)$ of the first bead fixed outside of it.

Lemma 4 ($\delta = 2, \alpha = 1$). *The terminal conformation of a deterministic unary oritatami system of $\delta = 2$ and $\alpha = 1$ is of length at most $3n^2 + 4n + 2$.*

Proof. In the elongation that places bead i at $(1,1)$ there are two possibilities.

- i forms a bond with a bead at $(1,0)$.
- i does not bond to anything and $i + 1$ is at $(2,1)$ bonding with a bead at $(2,0)$. If there is no bead at $(1,0)$, then placing i at $(1,0)$ instead of $(1,1)$ results in the same number of bonds, leading to nondeterminism. Therefore, there has to be a bead at $(1,0)$ and it is inactive, otherwise it would bond to i. This is analogous to case 1. below, with the only difference being that the bond between $(1,1)$ and $(1,0)$ is missing.

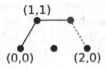

Because of the above, we need only consider the case when i binds to a bead at $(1,0)$. The next bead, $i + 1$, can be fixed at $(2,1)$ or at $(0,1)$ as all other possibilities result in nondeterministic behavior immediately, so we have two cases.

Case 1. bead $i + 1$ is fixed at $(2,1)$ and can bond with a bead at $(2,0)$ (see Fig. 15). Now consider bead $i+2$. For $i+1$ to be fixed at $(2,1)$, $i+2$ needs to form a bond somewhere, otherwise $i + 2$ could go to $(2, 1)$ forming the bond with the bead at $(2, 0)$ and there would be two conformations with the maximal 1 bond. The only possibility is that there is a bead at $(3, 0)$ and $i + 2$ can bond with it when placed at $(3, 1)$. We can apply the same argument inductively: there is some $m \geq 0$ such that grid points $(\ell, 0)$ are occupied by beads with free hands, for all $\ell \in \{2, \ldots, 2 + m\}$, and there is no bead at $(3 + m, 0)$. Such an m exists, and it is not greater than n, because those beads are all stabilized along the same side of $\bigcirc_p^n$. Then, bead $i + \ell$ is fixed at $(\ell + 1, 1)$ and bonds with $(\ell + 1, 0)$. However, bead $i + 2 + m$ cannot be fixed anywhere, because $i + 2 + m$ and $i + 3 + m$ can only add one bond to the conformation, and that is possible either with $i + 2 + m \rightarrow (2 + m, 1)$, $i + 3 + m \rightarrow (3 + m, 1)$ or with $i + 2 + m \rightarrow (2 + m, 2)$, $i + 3 + m \rightarrow (2 + m, 1)$. Intuitively, when we reach a corner of the hexagon $\bigcirc_p^n$, the next bead of the transcript cannot deterministically stabilize, as depicted in Fig. 15. In this case, the size of the transcript which was deterministically stabilized is bounded by the size of $\bigcirc_p^n$ plus the length of one side of $\bigcirc_p^{n+1}$, so by $(3n^2 + 3n + 1) + (n + 1) = 3n^2 + 4n + 2$.

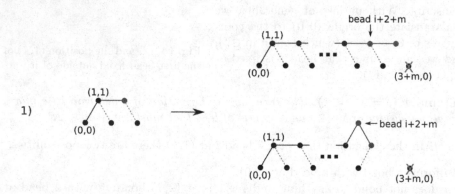

Fig. 15. When bead $i + 2 + m$ is fixed

Case 2. bead $i + 1$ is fixed at $(0, 1)$. This is only possible if
(a) there is an inactive bead at $(-1, 0)$ and one with a free hand one at $(-2, 0)$. This case is symmetrical to (1). there is no bead at $(-1, 0)$, bead $i + 1$ can bond with bead $i - 1$ at $(0, 0)$ and the bead $i + 2$ can be placed at $(-1, 0)$ where it can bond with $(-2, 0)$, $(-2, -1)$ or $(-1, -1)$. This leads to nondeterminism, because placing bead i at $(-1, 0)$ and bead $i + 1$ at $(0, 1)$ would yield two bonds, just as the original conformation.
(b) there is a bead at $(-1, 0)$ and bead $i + 1$ can bond with that or with bead $i - 1$ at $(0, 0)$. However, this means that placing bead i at $(0, 1)$ at bead $i + 1$ at $(1, 1)$ creates the same number of hydrogen bonds, thus resulting in bead i not being placed deterministically.

   Case 2.(a) gives the same upper bound as case 1. Cases 2.(b)-(c) give the smaller upper bound $|\bigcirc_p^n| + 1$, thereby concluding the proof.   □

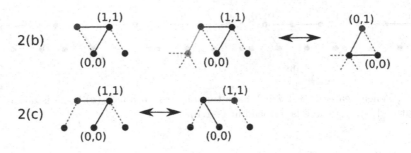

Fig. 16. When bead $i + 1$ is fixed at $(0,1)$

Lemma 5 ($\delta = 3, \alpha = 1$). *The terminal conformation of a deterministic unary oritatami system of $\delta = 3$ and $\alpha = 1$ is of length at most $3n^2 + 4n + 2$.*

Proof. We will argue similarly to the $\delta = 2$ case: when we stabilize beads outside $\bigcirc_p^n$, we can only do so at points right next to one of the sides and even there only until we reach a corner. This yields the upper bound immediately.

As before, let us assume that the last bead stabilized within $\bigcirc_p^n$ is at point $(0,0)$ and the next bead is the first to be stabilized outside $\bigcirc_p^n$ at point $(1,1)$. Depending on whether bead $i - 1$ at point $(0,0)$ has a free hand to form a bond or not, we distinguish two cases (similarly to Fig. 16).

Case 1. Bead $i - 1$ at $(0,0)$ has a free hand. If the most stable conformation formed by beads $i, i + 1, i + 2$ adds only two bonds, it will be nondeterministic because there are at least two possibilities (see Fig. 19, except it starts from $(0,0)$ not $(1,1)$). Therefore, it needs to make three new bonds to deterministically stabilize i. There are five possible cases in which beads $i, i + 1, i + 2$ can add three bonds, see Fig. 17. In the cases (b) and (e) in Fig. 17, there are beads having a free hand at $(1,0)$ and $(-1,0)$ already stabilized before bead $i - 1$ is fixed at $(0,0)$. One of these two beads may be a predecessor of bead $i - 1$, but at least one of them is not. When bead $i - 1$ is fixed at $(0,0)$, it makes a bond with the one of these two beads which is not its predecessor. This means that it is impossible to have three beads at $(-1,0)$, $(0,0)$ and $(1,0)$, each with a free hand, when bead i is fixed, and consequently, cases (b) and (e) in Fig. 17 cannot occur. Case (d) in Fig. 17 becomes nondeterministic when bead i is fixed because bead i can be fixed at $(-1,0)$ and bond with $(-2,0)$, bead $i + 1$ can be placed $(0,1)$ and bond with $(0,0)$ and bead $i + 2$ can be placed $(1,1)$ and bond with $(0,1)$. If it makes three bonds once, such as (a) and (c) in Fig. 17, it will need to make three bonds forever to be deterministic. Similarly to case 1. in the previous section, cases (a) and (c) in Fig. 17 lead to nondeterministism eventually, when the transcript reaches a first corner of $\bigcirc_p^n$.

Case 2. Bead $i - 1$ at $(0,0)$ does not have a free hand.

(i) First, let us assume that there is a bead at $(1,0)$ which has a free hand. If there is a bead at $(1,0)$ and beads $i, i + 1, i + 2$ add only two bonds, we

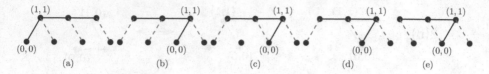

Fig. 17. When beads $i, i+1, i+2$ make three hydrogen bonds (c) there is an inactive bead at $(-1, 0)$. (d) there is no bead at $(-1, 0)$.

instantly get nondeterministism as in Fig. 18. Hence, beads $i, i+1, i+2$ need to make three bonds to stabilize i, but if they make three hydrogen bonds, the situation is analogous to (a) and (c) in Fig. 17.

(ii) Now consider when there is no bead at $(1, 0)$ or there is one, but with no free hand. Let us discuss the moment after bead i is fixed outside $\bigcirc_p^n$. Now bead i has a free hand because it cannot bind to $(1, 0)$. If beads $i+1, i+2, i+3$ can form only two bonds, it will be nondeterministic because there are at least two possible such conformations, as in Fig. 19. Hence, they need to form three bonds to deterministically stabilize, such as in Fig. 20. Case (a) in Fig. 20 becomes the same as case (a) in Fig. 17. Case (b) in Fig. 20 becomes nondeterministic already when bead i is fixed because bead i could also be fixed at $(1, 0)$.

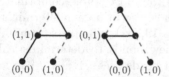

Fig. 18. Two conformations when a bead at $(1, 0)$ has a free hand.

Fig. 19. Two conformations when a bead i at $(1, 1)$ has a free hand

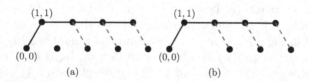

Fig. 20. When beads $i+1, i+2, i+3$ make three hydrogen bonds (a) there is an inactive bead at $(1, 0)$. (b) there is no bead at $(1, 0)$.

We have shown that at $\alpha = 1$ and $\delta \in \{2, 3\}$, unary oritatami systems can only fold finite length transcripts deterministically, and the length of that transcript is bounded by the size of the regular hexagon of radius n plus the length of one side of the surrounding hexagon. This gives us the upper bound

$|\bigcirc_p^n| + (n + 1) = 3n^2 + 4n + 2$, where n is the length of the seed, concluding the proof of Theorem 6.

5 Conclusion

In this work, we have considered unary oritatami systems at $\delta = 1$ or $\alpha = 1$. As a result, we found that a unary oritatami system does not have Turing universality at $\delta = 1$ and at $\delta \leq 3$, $\alpha = 1$. This non-Turing universality was obtained by the following results. One is that unary oritatami systems are not able to make any infinite structures at $\delta = 2, 3$, $\alpha = 1$ and at $\delta = 1$, $\alpha = 1, 3, 4$ (Theorems 3, 4, 6 and due to [3]). The other is that a unary oritatami system can only produce a single type of simple infinite structures, which is zigzag at $\delta = 1$, $\alpha = 2$ (Theorem 5).

The case of $\delta \geq 4$ and $\alpha = 1$ remains an open problem. Our results should be extended to non-unary oritatami systems, that is, characterization Turing universality of oritatami systems with respect to delay, arity or some other parameters such as the number of bead types.

Acknowledgements. We thank Yo-Sub Han for his valuable comments on the contents of this paper.

References

1. Arora, S., Barak, B.: Computational Complexity: A Modern Approach. Cambridge University Press, Cambridge (2009)
2. Cook, M.: Universality in elementary cellular automata. Complex Syst. **15**, 1–40 (2004)
3. Demaine, E., et al.: Know when to fold 'Em: self-assembly of shapes by folding in oritatami. In: Doty, D., Dietz, H. (eds.) DNA 2018. LNCS, vol. 11145, pp. 19–36. Springer, Cham (2018). https://doi.org/10.1007/978-3-030-00030-1_2
4. Elonen, A., et al.: Algorithmic design of 3D wireframes RNA polyhedral. DNA 24 poster (2018)
5. Geary, C.W., Andersen, E.S.: Design principles for single-stranded RNA origami structures. In: Murata, S., Kobayashi, S. (eds.) DNA 2014. LNCS, vol. 8727, pp. 1–19. Springer, Cham (2014). https://doi.org/10.1007/978-3-319-11295-4_1
6. Geary, C., Meunier, P.E., Schabanel, N., Seki, S.: Programming biomolecules that fold greedily during transcription. In: Proceedings of MFCS 2016, LIPIcs, vol. 58, pp. 43:1–43:14 (2016)
7. Geary, C., Meunier, P.E., Schabanel, N., Seki, S.: Proving the turing universality of oritatami cotranscriptional folding. In: Proceedings of ISAAC 2018 (2018, in press)
8. Geary, C., Rothemund, P.W.K., Andersen, E.S.: A single-stranded architecture for cotranscriptional folding of RNA nanostructures. Science **345**(6198), 799–804 (2014)
9. Hales, T.C.: The Jordan curve theorem, formally and informally. Am. Math. Mon. **114**(10), 882–894 (2007)
10. Han, D., et al.: Single-stranded DNA and RNA origami. Science **358**(6369), 1402 (2017)
11. Pratchett, T.: Troll Bridge. Pan Books, London (1992)

Stochastic Programming for Energy Plant Operation

Tomoki Fukuba[1], Takayuki Shiina[1(✉)], and Ken-ichi Tokoro[2]

[1] Waseda University, 3-4-1 Ohkubo, Shinjuku, Tokyo 169-855, Japan
tshiina@waseda.jp
[2] Central Research Institute of Electric Power Industry,
2-6-1 Nagasaka, Yokosuka, Kanagawa 240-0196, Japan

Abstract. A stochastic programming model of the operation of energy plants with the introduction of photovoltaic generation and a storage battery is developed. The uncertainty of the output of the photovoltaic generation is represented by a set of discrete scenarios, and the expected value of the operation cost is minimized. The effectiveness of the stochastic programming model by comparing it with the deterministic model is shown. As an economic evaluation, the recovery period for the initial investment of photovoltaic generation and storage battery is also shown.

Keywords: Stochastic programming · Optimization · Energy plant ·
Operational planning · Photovoltaic generation · Unit commitment problem

1 Introduction

Because of the prevalent environmental problems, the need to spread awareness about the use of renewable energy is an urgent concern worldwide. In the Paris Agreement issued in 2016, the long-term goal was to keep the global average temperature increase within 2 °C as a result of the industrial revolution. To achieve this, efforts are under way to adapt smart community all over the world. Furthermore, in Japan, the reexamination of energy costs has been dealt with as a problem of management engineering due to the liberalization of electricity. Based on these, introduction of renewable energy as a new type of energy supply in large-scale facilities such as factories and shopping centers is being studied. However, as the output of renewable energy is unstable, decision making under uncertain conditions is required at the time of introduction.

In this research, an optimization model for operation planning by stochastic programming by introducing photovoltaic power generation as renewable energy into factory energy plant was developed. We showed that the modeling by stochastic programming is more suitable as a realistic operation plan than conventional deterministic mixed integer programming method and economic evaluation on implementation plan was carried out.

Figure 1 shows the outline of the basic model which can quantitatively evaluate the energy cost etc. of the smart community. An industrial model of energy consumption at the factory in the basic model is modeled based on a benchmark problem seeking an optimum operation plan of an energy plant presented by Suzuki and Okamoto [1], Inui and Tokoro [2].

© Springer Nature Switzerland AG 2019
T. V. Gopal and J. Watada (Eds.): TAMC 2019, LNCS 11436, pp. 208–221, 2019.
https://doi.org/10.1007/978-3-030-14812-6_13

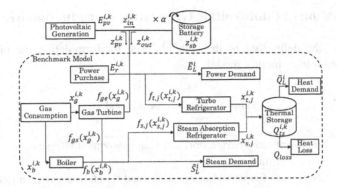

Fig. 1. Energy plant with photovoltaic generation.

2   Introduction Model for Photovoltaic Generation

The optimization model for photovoltaic generation extends the benchmark problem by including photovoltaic generation and a storage battery. To consider the uncertainty of photovoltaic power generation, stochastic programming is applied.

The benchmark problem is a model of an energy plant that purchases electricity and gas as shown in the dotted frame in Fig. 1, and generates electricity, heat, and steam to meet the demand. As equipment, there are a gas turbine, a boiler, two kinds of refrigerators, and a thermal storage tank. The objective of the benchmark problem is to establish an operation plan that minimizes the cost of purchasing electricity and gas while satisfying the constraints on equipment and energy balance. Decision variables consists of variables related to the amount of purchase and generation of energy, and variables concerning the start and stop of each device.

The photovoltaic generation introduction model includes MW-class photovoltaic generation equipment so that the generated electricity flows to the demand and turbo refrigerators. The photovoltaic generation of power is for in-house power consumption. We assumed that the storage battery can store only the electric power of photovoltaic power generation, and that some of the charged electricity will be lost by the time of discharge. Figure 1 shows the energy flow of the energy plant when photovoltaic generation and storage batteries are introduced.

The uncertainty of photovoltaic generation output is expressed using a set of deterministic scenarios. Because the uncertainty of the output of photovoltaic generation also affects the entire energy flow, decision variables related to the purchase amount of energy and generation amount are also defined for each scenario. As a result, the number of decision variables increases according to the number of scenarios. When the number of scenarios is 30, the number of decision variables increases from 192 to 5826, considering the introduction of photovoltaic power generation.

The objective of the photovoltaic generation introduction model is to establish an operation plan that minimizes the expected value of electricity and gas purchase cost while satisfying the restrictions on equipment and storage battery and energy balance.

3 Formulation of Photovoltaic Generation Introduction Model

Table 1 shows the definitions of the symbols used for formulating the photovoltaic power generation introduction model:

Table 1. Notation for the model.

Parameter	
N_t, N_s:	Number of turbo refrigerators and steam absorption refrigerators
I:	Number of time zones
a_{ge}:	Coefficient of input and output of gas and the electric power in gas turbine
a_{gs}:	Coefficient of input and output of gas and steam in gas turbine
a_b:	Coefficient of input and output of gas and steam in boiler
$a_{t,j}$:	Coefficient of input and output of power and heat in the turbo refrigerator j
$a_{s,j}$: $b_{s,j}$: $c_{s,j}$:	Coefficient of input and output relational expression of quantity of steam and heat at steam absorption refrigerator j
$Q_{t,j}^{\min}, Q_{t,j}^{\max}$	Lower and upper limits of heat production of turbo refrigerator j
$Q_{s,j}^{\min}, Q_{s,j}^{\max}$	Lower and upper limits of heat production amount of steam absorption refrigerator j
$E_g^{\min}, E_g^{\max}$	Lower and upper limits of power generation amount of gas turbine
$S_b^{\min}, S_b^{\max}$	Lower and upper limits of boiler steam production
$Q_{ts}^{\min}$:	Lower limit of heat storage amount of thermal storage tank
$Q_{ts}^{\max 1}$:	Upper limit of the thermal storage amount of the thermal storage tank in the first time zone to the $(I-1)$ time zone
$Q_{ts}^{\max 2}$:	Upper limit of the thermal storage amount of the thermal storage tank in the Ith time zone
Q_{ts}^{init}:	Initial thermal storage of the thermal storage tank
Q_{loss}:	The amount of heat loss in the thermal storage tank
$L_{t,j}, L_{s,j}$:	Minimum startup/stop time of the turbo refrigerator j and steam absorption refrigerator j
L_g, L_b:	Minimum startup/stop time of the gas turbine and boiler
C_{Er}^i, C_{Fr}^i:	Purchase cost of electricity and gas in time i
$\bar{E}_L^i, \bar{Q}_L^i, \bar{S}_L^i$:	Demand of electricity, heat, and steam in time i
$E_{rm}^i, \tilde{S}_{rm}^i$:	Remaining amount of electric energy and steam in time i
K:	Number of scenarios
Pr_k:	Probability of scenario k
$E_{pv}^{i,k}$:	Electric energy of photovoltaic power generation in time i under scenario k
z_{sb}^{init}:	Initial storage of storage battery
$E_{sb}^{\max}$:	Capacity of storage battery
α:	Charge and discharge efficiency

(continued)

Table 1. (*continued*)

Parameter	
Decision variable	
$x_{t,j}^{i,k}, x_{s,j}^{i,k}$:	Heat production of the turbo refrigerator j and steam absorption refrigerator j in time i under scenario k
$x_g^{i,k}, x_b^{i,k}$:	Gas consumption of gas turbine and the boiler in time i under scenario k
$y_{t,j}^i, y_{s,j}^i$:	State of the turbo refrigerator j and the steam absorption refrigerator j in time i(1 for running, 0 for stopped)
y_g^i, y_b^i:	State of the gas turbine and the boiler in time i (1 for running, 0 for stopped)
$z_{in}^{i,k}$:	Electric energy to be charged from the photovoltaic generation to the storage battery in time i under scenario k
$z_{out}^{i,k}$:	Discharge of storage battery in time i under scenario k
$z_{pv}^{i,k}$:	Electric energy directly consumed from photovoltaic power generation in time i under scenario k
$z_{sb}^{i,k}$:	The storage of the storage battery in time i under scenario k
Function	
$Q_{ts}^{i,k}\left(x_t^{i,k}, x_s^{i,k}; Q_{ts}^{i-1,k}, \bar{Q}_L^i\right)$:	
	The heat storage of the thermal storage tank in time i under scenario k
$E_r^{i,k}\left(x_g^{i,k}, x_t^{i,k}, z_{pv}^{i,k}, z_{out}^{i,k}; \bar{E}_L^i, E_{rm}^i\right)$:	
	Purchase of electric power in time i under scenario k
$S_{rm}^{i,k}\left(x_g^{i,k}, x_b^{i,k}, x_s^{i,k}; \bar{S}_L^i\right)$:	
	Remaining amount of steam in time i under scenario k
$f_{ge}\left(x_g^{i,k}\right)$:	Power generation of the gas turbine in time i under scenario k
$f_{gs}\left(x_g^{i,k}\right)$:	The steam generation of the gas turbine in time i under scenario k
$f_b\left(x_b^{i,k}\right)$:	Steam generation in the boiler in time i under scenario k
$f_{t,j}\left(x_{t,j}^{i,k}\right)$:	Power input of the turbo refrigerator j in time i under scenario k
$f_{s,j}\left(x_{s,j}^{i,k}\right)$:	Steam input of steam absorption refrigerator j in time i under scenario k

In this model, $x_t^{i,k}, x_s^{i,k} \forall i(i = 1, \cdots, I), \forall k(k = 1, \cdots, K)$ is defined as $x_t^{i,k} = \left(x_{t,1}^{i,k}, \cdots, x_{t,N_t}^{i,k}\right)^{\mathrm{T}}, x_s^{i,k} = \left(x_{s,1}^{i,k}, \cdots, x_{s,N_s}^{i,k}\right)^{\mathrm{T}}$.

The formulation of the photovoltaic generation introduction model is shown below:

$$\min \sum_{k=1}^K \left[Pr_k \sum_{i=1}^I \left\{ C_{Er}^i E_r^{i,k}\left(x_g^{i,k}, x_t^{i,k}, z_{pv}^{i,k}, z_{out}^{i,k}; \bar{E}_L^i, E_{rm}^i\right) + C_{Fr}^i\left(x_g^{i,k} + x_b^{i,k}\right) \right\} \right] \quad (1)$$

s.t.
$$Q_{ts}^{\min} \le Q_{ts}^{i,k}\left(x_t^{i,k}, x_s^{i,k}; Q_{ts}^{i-1,k}, \bar{Q}_L^i\right) \le Q_{ts}^{\max 1}, i = 1, \cdots, I - 1; k = 1, \cdots, K \quad (2)$$

$$Q_{ts}^{\min} \le Q_{ts}^{i,k}\left(\boldsymbol{x}_t^{i,k}, \boldsymbol{x}_s^{i,k}; Q_{ts}^{i-1,k}, \bar{Q}_L^i\right) \le Q_{ts}^{\max 2}, i = I; k = 1, \cdots, K \tag{3}$$

$$S_{rm}^{i,k}\left(\boldsymbol{x}_g^{i,k}, \boldsymbol{x}_b^{i,k}, \boldsymbol{x}_s^{i,k}; \bar{S}_L^i\right) = \tilde{S}_{rm}^i, i = 1, \cdots, I; k = 1, \cdots, K \tag{4}$$

$$Q_{t,j}^{\min}y_{t,j}^i \le x_{t,j}^{i,k} \le Q_{t,j}^{\max}y_{t,j}^i, i = 1, \cdots, I; j = 1, \cdots, N_t; k = 1, \cdots, K \tag{5}$$

$$Q_{s,j}^{\min}y_{s,j}^i \le x_{s,j}^{i,k} \le Q_{s,j}^{\max}y_{s,j}^i, i = 1, \cdots, I; j = 1, \cdots, N_s; k = 1, \cdots, K \tag{6}$$

$$E_g^{\min}y_g^i \le f_{ge}\left(x_g^{i,k}\right) \le E_g^{\max}y_g^i, i = 1, \cdots, I; k = 1, \cdots, K \tag{7}$$

$$S_b^{\min}y_b^i \le f_b\left(x_b^{i,k}\right) \le S_b^{\max}y_b^i, i = 1, \cdots, I; k = 1, \cdots, K \tag{8}$$

$$y_{t,j}^i - y_{t,j}^{i-1} \le y_{t,j}^\tau, \tau = i+1, \cdots, \min\{i + L_{t,j} - 1, I\}; i = 2, \cdots, I; j = 1, \cdots, N_t \tag{9}$$

$$y_{t,j}^{i-1} - y_{t,j}^i \le 1 - y_{t,j}^\tau, \tau = i+1, \cdots, \min\{i + L_{t,j} - 1, I\};$$
$$i = 2, \cdots, I; j = 1, \cdots, N_t \tag{10}$$

$$y_{s,j}^i - y_{s,j}^{i-1} \le y_{s,j}^\tau, \tau = i+1, \cdots, \min\{i + L_{s,j} - 1, I\};$$
$$i = 2, \cdots, I; j = 1, \cdots, N_s \tag{11}$$

$$y_{s,j}^{i-1} - y_{s,j}^i \le 1 - y_{s,j}^\tau, \tau = i+1, \cdots, \min\{i + L_{s,j} - 1, I\};$$
$$i = 2, \cdots, I; j = 1, \cdots, N_s \tag{12}$$

$$y_g^i - y_g^{i-1} \le y_g^\tau, \tau = i+1, \cdots, \min\{i + L_g - 1, I\}; i = 2, \cdots, I \tag{13}$$

$$y_g^{i-1} - y_g^i \le 1 - y_g^\tau, \tau = i+1, \cdots, \min\{i + L_g - 1, I\}; i = 2, \cdots, I \tag{14}$$

$$y_b^i - y_b^{i-1} \le y_b^\tau, \tau = i+1, \cdots, \min\{i + L_b - 1, I\}; i = 2, \cdots, I \tag{15}$$

$$y_b^{i-1} - y_b^i \le 1 - y_b^\tau, \tau = i+1, \cdots, \min\{i + L_b - 1, I\}; i = 2, \cdots, I \tag{16}$$

$$z_{sb}^{i-1,k} + \alpha z_{in}^{i,k} = z_{sb}^{i,k} + z_{out}^{i,k}, i = 1, \cdots, I; k = 1, \cdots, K; z_{sb}^{0,k} = z_{sb}^{I,k} = z_{sb}^{init} \tag{17}$$

$$z_{sb}^{i,k} \le E_{sb}^{\max}, i = 1, \cdots, I; k = 1, \cdots, K \tag{18}$$

$$E_{pv}^{i,k} = z_{pv}^{i,k} + z_{in}^{i,k}, i = 1, \cdots, I; k = 1, \cdots, K \tag{19}$$

$$x_{t,j}^{i,k} \ge 0, i = 1, \cdots, I; j = 1, \cdots, N_t; k = 1, \cdots, K \tag{20}$$

$$x_{s,j}^{i,k} \ge 0, i = 1, \cdots, I; j = 1, \cdots, N_s; k = 1, \cdots, K \tag{21}$$

$$x_g^{i,k}, x_b^{i,k}, z_{in}^{i,k}, z_{out}^{i,k}, z_{pv}^{i,k} \geq 0, i = 1, \cdots, I; k = 1, \cdots, K \tag{22}$$

$$z_{sb}^{i,k} \geq 0, i = 1, \cdots, I - 1; k = 1, \cdots, K \tag{23}$$

$$y_{t,j}^{i} \in \{0, 1\}, i = 1, \cdots, I; j = 1, \cdots, N_t \tag{24}$$

$$y_{s,j}^{i} \in \{0, 1\}, i = 1, \cdots, I; j = 1, \cdots, N_s \tag{25}$$

$$y_g^{i}, y_b^{i} \in \{0, 1\}, i = 1, \cdots, I \tag{26}$$

Where

$$Q_{ts}^{i,k}\left(\boldsymbol{x}_t^{i,k}, \boldsymbol{x}_s^{i,k}; Q_{ts}^{i-1,k}, \bar{Q}_L^i\right) = -\sum_{j=1}^{N_t}\left\{x_{t,j}^{i,k}\right\} - \sum_{j=1}^{N_s}\left\{x_{s,j}^{i,k}\right\} + Q_{ts}^{i-1,k} + \bar{Q}_L^i + Q_{loss}, \tag{27}$$
$$i = 1, \cdots, I; k = 1, \cdots, K; Q_{ts}^{0,k} = Q_{ts}^{init}$$

$$E_r^{i,k}\left(x_g^{i,k}, \boldsymbol{x}_t^{i,k}, z_{pv}^{i,k}, z_{out}^{i,k}; \bar{E}_L^i, E_{rm}^i\right) =$$
$$\sum_{j=1}^{N_t}\left\{f_{t,j}\left(x_{t,j}^{i,k}\right)\right\} + \bar{E}_L^i - f_{ge}\left(x_g^{i,k}\right) + E_{rm}^i - z_{pv}^{i,k} - z_{out}^{i,k} \geq 0, \tag{28}$$
$$i = 1, \cdots, I; k = 1, \cdots, K$$

$$S_{rm}^{i,k}\left(x_g^{i,k}, x_b^{i,k}, \boldsymbol{x}_s^{i,k}; \bar{S}_L^i\right) = f_{gs}\left(x_g^{i,k}\right) + f_b\left(x_b^{i,k}\right) - \sum_{j=1}^{N_s}\left\{f_{s,j}\left(x_{s,j}^{i,k}\right)\right\} - \bar{S}_L^i, \tag{29}$$
$$i = 1, \cdots, I; k = 1, \cdots, K$$

$$f_{ge}\left(x_g^{i,k}\right) = a_{ge}x_g^{i,k}, i = 1, \cdots, I; k = 1, \cdots, K \tag{30}$$

$$f_{gs}\left(x_g^{i,k}\right) = a_{gs}x_g^{i,k}, i = 1, \cdots, I; k = 1, \cdots, K \tag{31}$$

$$f_b\left(x_b^{i,k}\right) = a_b x_b^{i,k}, i = 1, \cdots, I; k = 1, \cdots, K \tag{32}$$

$$f_{t,j}\left(x_{t,j}^{i,k}\right) = a_{t,j}x_{t,j}^{i,k}, i = 1, \cdots, I; j = 1, \cdots, N_t; k = 1, \cdots, K \tag{33}$$

$$f_{s,j}\left(x_{s,j}^{i,k}\right) = \frac{x_{s,j}^{i,k}}{-a_{s,j}\left\{x_{s,j}^{i,k}\right\}^2 + b_{s,j}x_{s,j}^{i,k} + c_{s,j}}, i = 1, \cdots, I; j = 1, \cdots, N_s; k = 1, \cdots, K$$
$$\tag{34}$$

Objective function (1) represents the minimization of the expected value of the purchase cost of electricity and gas. Constraints (2) and (3) represent the capacity constraints of the thermal storage tank. Inequality (2) represents the case of the first time zone to the $(I - 1)$ time zone, and (3) represents the case of the I time zone. Equation (4) is a constraint on the remaining amount of steam. Inequalities (5)–(8) represent the capacity constraints of the energy production amounts of two types of refrigerators, gas turbines, and boilers. Inequalities (9)–(16) are constraints on on/off decision of each device, and they are represented by a linear inequality based on the unit commitment problem [3]. For these restrictions, two constraints are used for each unit, and the first one means that once the unit starts, it must keep its operating state for a certain period of time. The second one represents the case of stop. Constraint (17) is related to the storage amount of the storage battery. To consider that a certain amount of electric power is lost at the time of charge and discharge, $z_{in}^{i,k}$ is multiplied by efficiency α. The initial charge amount and the storage amount in time zone I are the same. Constraint (18) represents the capacity of the storage battery.

Equation (19) shows that the power generated by the photovoltaic generation is divided into the power to satisfy the demand and the power charged in the storage battery. Constraints (20)–(26) represent non-negative constraints and 0-1 constraints of decision variables.

Function (27) represents the heat storage amount of the thermal storage tank. Function (28) represents the purchase amount of the electricity. However, as it does not consider power selling, it imposes non-negativity constraints. Function (29) represents the remaining amount of steam. Function (30)–(34) represent the relational expressions of the input and output quantities of the gas turbine, the boiler, and the two types of refrigerators. Function (34) represents the relationship between the input and output amounts of the steam absorption refrigerator. This function is expressed in the form of non-convex nonlinear constraints to account for a practical operating plan. Because these types of constraints are non-convex, they are difficult to deal with.

4 Piecewise Linear Approximation of Nonlinear Constraint Equation

This research model includes the nonlinear constraint Eq. (34) expressing the relationship between the input and output quantities of the steam absorption refrigerator. To treat the model as a mixed integer programming problem, it was linearized by piecewise linear approximation [4] on (34).

Let $\bar{f}_{s,j}\left(x_{s,j}^{i,k}\right)$ be the approximation of $f_{s,j}\left(x_{s,j}^{i,k}\right)$ and $\left(x_{s,j,l}, f_{s,j}\left(x_{s,j,l}\right)\right) \forall i, j,$ $k(i = 1, \cdots, I; j = 1, \cdots, N_s; k = 1, \cdots, K)$ be the split points of the function. The approximation $\bar{f}_{s,j}\left(x_{s,j}^{i,k}\right)$ is given by Eqs. (36) and (37).

Constraints (38) and (39) represent the SOS 2 constraint that requires only two adjacent $\lambda_{j,l}^{i,k}$ is at most positive. Because of adding the decision variables to the piecewise linear approximation, the number of decision variables further increases, resulting in a large-scale mixed integer programming problem.

$$x_{s,j}^{i,k} = \sum_{l=1}^{p_j} \lambda_{j,l}^{i,k} x_{s,j,l} \tag{35}$$

$$\bar{f}_{s,j}\left(x_{s,j}^{i,k}\right) = \sum_{l=1}^{p_j} \lambda_{j,l}^{i,k} f_{s,j}\left(x_{s,j,l}\right) \tag{36}$$

$$\sum_{l=1}^{p_j} \lambda_{j,l}^{i,k} = 1, \lambda_{j,l}^{i,k} \geq 0, l = 1, \cdots, p_j \tag{37}$$

$$\left.\begin{aligned} \lambda_{j,1}^{i,k} &\leq \mu_{j,1}^{i,k} \\ \lambda_{j,2}^{i,k} &\leq \mu_{j,1}^{i,k} + \mu_{j,2}^{i,k} \\ &\vdots \\ \lambda_{j,p_j}^{i,k} &\leq \mu_{j,p_j-1}^{i,k} + \mu_{j,p_j}^{i,k} \end{aligned}\right\} \tag{38}$$

$$\sum_{l=1}^{p_j} \mu_{j,l}^{i,k} = 1, 0 \leq \mu_{j,l}^{i,k} \in Z, l = 1, \cdots, p_j \tag{39}$$

The exact algorithm using piecewise linear approximation is as follows: For each iteration, we increase the number of split points and improve the accuracy of piece-wise linear approximation. This made it possible to solve a large-scale mixed integer programming problem.

Piecewise Linear Approximation Algorithm

Step 0:	Given initial number of split points, and tolerance ε
Step 1:	Set the initial split points
Step 2:	Solve the problem of piecewise linear approximation of constraint Eq. (34)
Step 3:	If $\left|f_{s,j}\left(\hat{x}_{s,j}^{i,k}\right) - \bar{f}_{s,j}\left(\hat{x}_{s,j}^{i,k}\right)\right| > \varepsilon$ at the optimal solution $\hat{x}_{s,j}^{i,k}$, add $\left(\hat{x}_{s,j}^{i,k}, f_{s,j}\left(\hat{x}_{s,j}^{i,k}\right)\right)$ to the set of split points
Step 4:	If there are additional split points, the process returns to Step 2. If there are no points added, stop

5   Evaluation of Solution by Stochastic Programming

For evaluating the solution of the stochastic programming method, we use the value VSS (value of stochastic solution) [5] of the solution of the stochastic programming problem.

If the random variable is defined as ξ, we define the optimization problem on the realization ξ of the random variable ξ as follows:

$$\min_x z(x, \xi) = c^T x + \min_y \{q^T y | Wy = h - Tx, y \geq 0\} \tag{40}$$

$$\text{s.t. } Ax = b, x \geq 0 \tag{41}$$

The optimum objective function value RP (recourse problem) of the stochastic programming problem is defined as follows:

$$RP = \min_x E_\xi z(x, \xi) \tag{42}$$

We define the optimal objective function value ADP (average deterministic problem) of a deterministic problem in which the random variable ξ is replaced by its mean value $\bar{\xi}$, and let the optimal solution of the problem be $\bar{x}(\bar{\xi})$.

$$ADP = \min_x z(x, \bar{\xi}) \tag{43}$$

The optimal objective function value $RP(\bar{\xi})$ when $\bar{x}(\bar{\xi})$ is applied to the stochastic programming problem is defined as follows.

$$RP(\bar{\xi}) = E_\xi \left(z(\bar{x}(\bar{\xi}), \xi) \right) \tag{44}$$

VSS is defined as follows.

$$VSS = RP(\bar{\xi}) - RP \tag{45}$$

Because the optimal solution obtained in the problem of finding $RP(\bar{\xi})$ is a feasible solution to the problem for obtaining RP, the following relation holds.

$$RP \leq RP(\bar{\xi}), VSS \geq 0 \tag{46}$$

The problem for obtaining RP is a stochastic programming model that is, formulated as (1)–(34). On the other hand, the problem of finding ADP is a deterministic model that considers the output of the photovoltaic generation fixed at the average value. $RP(\bar{\xi})$ becomes a problem to find out how much expense the deterministic solution in the problem for ADP will be under uncertainty.

Table 2 lists the definitions of symbols in the problem of ADP.

Table 2. Notation for ADP problem.

Parameter	
$\bar{E}^i_{pv}$:	Average value of photovoltaic generation amount in time i
Decision variable	
$x^i_{t,j}$:	Heat production of the turbo refrigerator j in time i
$x^i_{s,j}$:	Heat production of the steam absorption refrigerator j in time i
x^i_g:	Gas consumption of gas turbine in time i
x^i_b:	Gas consumption of boiler in time i
z^i_{in}:	The amount of electricity stored in the storage battery
z^i_{out}:	Discharge amount of storage battery in time i
z^i_{pv}:	The amount of electricity directly consumed from photovoltaic power generation in time i
z^i_{sb}:	The storage amount of the storage battery in time i
Function	
$Q^i_{ts}\left(x^i_t, x^i_s; Q^{i-1}_{ts}, \bar{Q}^i_L\right)$:	
	The heat storage of the thermal storage tank in time i
$E^i_r\left(x^i_g, x^i_t, z^i_{pv}, z^i_{out}; \bar{E}^i_L, E^i_{rm}\right)$:	
	Purchase amount of electricity in time i
$S^i_{rm}\left(x^i_g, x^i_b, x^i_s; \bar{S}^i_L\right)$:	
	Remaining amount of steam in time i
$f_{ge}\left(x^i_g\right)$:	Power generation of gas turbine in time i
$f_{gs}\left(x^i_g\right)$:	Steam generation of gas turbine in time i
$f_b\left(x^i_b\right)$:	Steam generation in boiler in time i
$f_{t,j}\left(x^i_{t,j}\right)$:	Power input of the turbo refrigerator j in time i
$f_{s,j}\left(x^i_{s,j}\right)$:	Steam input of steam absorption refrigerator j in time i

The formulation of the problem for solving the problem for ADP, using $x^i_t = \left(x^i_{t,1}, \cdots, x^i_{t,N_t}\right)^T, x^i_s = \left(x^i_{s,1}, \cdots, x^i_{s,N_s}\right)^T$, is as shown below:

$$\text{ADP} = \min \sum_{i=1}^{I} \left\{ C^i_{Er} E^i_r\left(x^i_g, x^i_t, z^i_{pv}, z^i_{out}; \bar{E}^i_L, E^i_{rm}\right) + C^i_{Fr}\left(x^i_g + x^i_b\right) \right\} \quad (47)$$

s.t.

$$Q^{min}_{ts} \leq Q^i_{ts}\left(x^i_t, x^i_s; Q^{i-1}_{ts}, \bar{Q}^i_L\right) \leq Q^{max1}_{ts}, i = 1, \cdots, I - 1 \quad (48)$$

$$Q^{min}_{ts} \leq Q^i_{ts}\left(x^i_t, x^i_s; Q^{i-1}_{ts}, \bar{Q}^i_L\right) \leq Q^{max2}_{ts}, i = I \quad (49)$$

$$S_{rm}^i\left(x_g^i, x_b^i, x_s^i; \bar{S}_L^i\right) = \tilde{S}_{rm}^i, i = 1, \cdots, I \tag{50}$$

$$Q_{t,j}^{\min} y_{t,j}^i \leq x_{t,j}^i \leq Q_{t,j}^{\max} y_{t,j}^i, i = 1, \cdots, I; j = 1, \cdots, N_t \tag{51}$$

$$Q_{s,j}^{\min} y_{s,j}^i \leq x_{s,j}^i \leq Q_{s,j}^{\max} y_{s,j}^i, i = 1, \cdots, I; j = 1, \cdots, N_s \tag{52}$$

$$E_g^{\min} y_g^i \leq f_{ge}\left(x_g^i\right) \leq E_g^{\max} y_g^i, i = 1, \cdots, I \tag{53}$$

$$S_b^{\min} y_b^i \leq f_b\left(x_b^i\right) \leq S_b^{\max} y_b^i, i = 1, \cdots, I \tag{54}$$

(9)–(16)

$$z_{sb}^{i-1} + \alpha z_{in}^i = z_{sb}^i + z_{out}^i, i = 1, \cdots, I; z_{sb}^0 = z_{sb}^I = z_{sb}^{init} \tag{55}$$

$$z_{sb}^i \leq E_{sb}^{\max}, i = 1, \cdots, I \tag{56}$$

$$\bar{E}_{pv}^i = z_{pv}^i + z_{in}^i, i = 1, \cdots, I \tag{57}$$

$$x_{t,j}^i \geq 0, i = 1, \cdots, I; j = 1, \cdots, N_t \tag{58}$$

$$x_{s,j}^i \geq 0, i = 1, \cdots, I; j = 1, \cdots, N_s \tag{59}$$

$$x_g^i, x_b^i, z_{in}^i, z_{out}^i, z_{pv}^i \geq 0, i = 1, \cdots, I \tag{60}$$

$$z_{sb}^i \geq 0, i = 1, \cdots, I - 1 \tag{61}$$

(24)–(26)
where

$$Q_{ts}^i\left(x_t^i, x_s^i; Q_{ts}^{i-1}, \bar{Q}_L^i\right) = -\sum_{j=1}^{N_t}\left\{x_{t,j}^i\right\} - \sum_{j=1}^{N_s}\left\{x_{s,j}^i\right\} + Q_{ts}^{i-1} + \bar{Q}_L^i + Q_{loss},$$
$$i = 1, \cdots, I; Q_{ts}^0 = Q_{ts}^{init} \tag{62}$$

$$E_r^i\left(x_g^i, x_t^i, z_{pv}^i, z_{out}^i; \bar{E}_L^i, E_{rm}^i\right) = \sum_{j=1}^{N_t}\left\{f_{t,j}\left(x_{t,j}^i\right)\right\} + \bar{E}_L^i - f_{ge}\left(x_g^i\right) + E_{rm}^i - z_{pv}^i - z_{out}^i$$
$$\geq 0, i = 1, \cdots, I \tag{63}$$

$$S_{rm}^i\left(x_g^i, x_b^i, x_s^i; \bar{S}_L^i\right) = f_{gs}\left(x_g^i\right) + f_b\left(x_b^i\right) - \sum_{j=1}^{N_s}\left\{f_{s,j}\left(x_{s,j}^i\right)\right\} - \bar{S}_L^i, i = 1, \cdots, I \tag{64}$$

$$f_{ge}\left(x_g^i\right) = a_{ge}x_g^i, i = 1, \cdots, I \tag{65}$$

$$f_{gs}\left(x_g^i\right) = a_{gs}x_g^i, i = 1, \cdots, I \tag{66}$$

$$f_b\left(x_b^i\right) = a_b x_b^i, i = 1, \cdots, I \tag{67}$$

$$f_{t,j}\left(x_{t,j}^i\right) = a_{t,j} x_{t,j}^i, i = 1, \cdots, I; j = 1, \cdots, N_t \tag{68}$$

$$f_{s,j}\left(x_{s,j}^i\right) = \frac{x_{s,j}^i}{-a_{s,j}\left\{x_{s,j}^i\right\}^2 + b_{s,j} x_{s,j}^i + c_{s,j}}, i = 1, \cdots, I; j = 1, \cdots, N_s \tag{69}$$

The optimal solution concerning the on/off scheduling for ADP is given by $\hat{y}_{t,j}^i(i = 1, \cdots, I; j = 1, \cdots, N_t)$, $\hat{y}_{s,j}^i(i = 1, \cdots, I; j = 1, \cdots, N_s)$, $\hat{y}_g^i(i = 1, \cdots, I)$, $\hat{y}_b^i(i = 1, \cdots, I)$. The problem of finding $\mathrm{RP}(\bar{\xi})$ is shown below:

The difference from the original photovoltaic power generation introduction model is that the decision variable regarding on/off is fixed and becomes a constant, and the restriction on on/off is removed.

$$\mathrm{RP}(\bar{\xi}) = \min \sum_{k=1}^{K}\left[Pr_k \sum_{i=1}^{I}\left\{C_{Er}^{i,k} E_r^{i,k}\left(x_g^{i,k}, x_t^{i,k}, z_{pv}^{i,k}, z_{out}^{i,k}; \bar{E}_L^i, E_{rm}^i\right) + C_{Fr}^i\left(x_g^{i,k} + x_b^{i,k}\right)\right\}\right] \tag{70}$$

s. t.

(2)–(4)

$$Q_{t,j}^{\min}\hat{y}_{t,j}^i \leq x_{t,j}^{i,k} \leq Q_{t,j}^{\max}\hat{y}_{t,j}^i, i = 1, \cdots, I; j = 1, \cdots, N_t; k = 1, \cdots, K \tag{71}$$

$$Q_{s,j}^{\min}\hat{y}_{s,j}^i \leq x_{s,j}^{i,k} \leq Q_{s,j}^{\max}\hat{y}_{s,j}^i, i = 1, \cdots, I; j = 1, \cdots, N_s; k = 1, \cdots, K \tag{72}$$

$$E_g^{\min}\hat{y}_g^i \leq f_{ge}\left(x_g^{i,k}\right) \leq E_g^{\max}\hat{y}_g^i, i = 1, \cdots, I; k = 1, \cdots, K \tag{73}$$

$$S_b^{\min}\hat{y}_b^i \leq f_b\left(x_b^{i,k}\right) \leq S_b^{\max}\hat{y}_b^i, i = 1, \cdots, I; k = 1, \cdots, K \tag{74}$$

(17)–(23)
Where
(27)–(34)

6 Numerical Experiments

Based on the benchmark problem data, we conducted numerical experiments to obtain a daily operation plan. However, only one steam absorption refrigerator is used. Demand for power, heat, and steam during the daytime is approximately about 20 [MWh], 20 [GJ], 10 [t], respectively. The capacity of photovoltaic power generation is set to 4 [MW], and the capacity of storage battery is compared in 4 ways for of 0.5, 1, 1.5, 2 [MWh].

The initial storage amount z_{sb}^{init} of the storage battery was 30% of the storage battery capacity and the charge/discharge efficiency α was set to 0.8.

A scenario representing the uncertainty of photovoltaic power generation is created on a monthly basis based on the horizontal level total solar insolation in Tokyo in the average year of NEDO (New Energy and Industrial Technology Development Organization) database (METPV - 11) [6].

According to the Agency for Natural Resources and Energy of Ministry of Economy, Trade and Industry [7, 8], the annual power generation of photovoltaic power is approximately 1100 [kWh] per 1 [kW] capacity. To satisfy the relationship between this capacity and annual power generation amount, the scenario was generated by multiplying the horizontal plane total solar insolation by a constant. The number of scenarios is the number of days of the month, and the probability of each scenario is 1/(the number of days of the month). We used AMPL as the modeling language and Gurobi 7.5.0 as the solver. The parameter of Gurobi was MIPGAP = 10^{-7}. The piecewise linear approximation parameter was set as 256 for the initial number of split points and $\varepsilon = 10^{-6}$.

First, we compare RP and $\mathrm{RP}(\bar{\xi})$ for evaluating the model by stochastic programming. Table 3 shows a comparison of RP and $\mathrm{RP}(\bar{\xi})$ in the case the storage battery capacity is set to 1 [MWh]. The maximum value of VSS is 2,007 [yen] in November, equivalent to 2.3 [%] of cost reduction due to the introduction of photovoltaic power. The operating cost per day before the introduction of photovoltaic generation was 4,042,763 [yen].

Next, the recovery period for the initial investment cost is calculated. According to the Ministry of Economy, Trade and Industry [7, 8], the initial investment cost of photovoltaic generation of capacity 1 [MW] or more is 27.5 [yen/kW], the cost of NAS battery used as a large storage battery is 4 [10^4 yen/kWh]. Based on the internal rate of return [9], the recovery period for the initial investment cost is calculated by finding n as shown in (75) given the discount rate r. In this formula, the difference in cost is the difference in the annual cost before and after installing photovoltaic power generation.

$$\text{initial investment cost} = \sum_{i=1}^{n} \frac{\text{difference of cost}}{(1+r)^i} \qquad (75)$$

The calculation result of the discount rate r = 1 [%] is listed in Table 4. Comparing the recovery years with the present model by using the stochastic programming method and deterministic model, the initial investment cost recovery period is smaller in this

Table 3. RP and $\mathrm{RP}(\bar{\xi})$ $(E_{sb}^{max} = 1[\mathrm{MWh}])$.

Month	RP [yen]	$\mathrm{RP}(\bar{\xi})$ [yen]	Month	RP [yen]	$\mathrm{RP}(\bar{\xi})$ [yen]
1	3,946,209	3,947,316	7	3,872,237	3,873,733
2	3,921,607	3,921,607	8	3,880,117	3,881,212
3	3,906,488	3,908,086	9	3,925,601	3,925,601
4	3,881,487	3,882,983	10	3,938,757	3,938,757
5	3,868,814	3,870,060	11	3,953,877	3,955,884
6	3,896,726	3,898,624	12	3,956,763	3,956,763

Table 4. Recovery period for initial investment ($r = 1[\%]$).

Storage battery capacity	Stochastic programming	Deterministic model
0.5 [MWh]	27.02	27.19
1 [MWh]	27.50	27.74
1.5 [MWh]	28.00	28.30
2 [MWh]	28.52	28.78

research model. Compared with the capacity of the storage battery, the larger the capacity, the larger the recovery period.

7 Concluding Remarks

In this research, we extended the benchmark problem on energy plant operation of large-scale facilities to the problem including photovoltaic power generation and storage battery and showed that modeling by stochastic programming method is useful. The new model for the introduction of large capacity photovoltaic power generation was developed. And the evaluation for the introduction of photovoltaic power generation for the purpose of self-power generation at large facilities has been made possible.

References

1. Suzuki, R., Okamoto, T.: An introduction of the energy plant operational planning problem: a formulation and solutions. In: IEEJ International Workshop on Sensing, Actuation, and Motion Control (2015)
2. Inui, N., Tokoro, K.: Finding the feasible solution and lower bound of the energy plant operational planning problem by an MILP formulation. In: IEEJ International Workshop on Sensing, Actuation, and Motion Control (2015)
3. Shiina, T., Birge, J.R.: Stochastic unit commitment problem. Int. Trans. Inoper. Res. **11**(1), 19–32 (2004)
4. Nemhauser, G., Wolsey, L.A.: Integer and Combinatorial Optimization. Wiley, London (1989)
5. Birge, J.R., Louveaux, F.: Introduction to Stochastic Programming. Springer, New York (1997). https://doi.org/10.1007/978-1-4614-0237-4
6. NEDO New Energy and Industrial Technology Development Organization.http://www.nedo. go.jp/library/nissharyou.html. Accessed 03 Nov 2017
7. Ministry of Economy, Trade and Industry. http://www.enecho.meti.go.jp/category/saving_ and_new/ohisama_power/common/pdf/guideline-2013.pdf Accessed 18 Dec 2017
8. Ministry of Economy, Trade and Industry. http://www.enecho.meti.go.jp/committee/council/ basic_problem_committee/028/pdf/28sankou2-2.pdf Accessed 18 Dec 2017
9. Luenberger, D.G.: Investment Science. Oxford University Press, Oxford (1998)

Compact I/O-Efficient Representation of Separable Graphs and Optimal Tree Layouts

Tomáš Gavenčiak and Jakub Tětek(✉)

Department of Applied Mathematics, Charles University,
Malostranské nám. 25, 11800 Prague, Czech Republic
{gavento,jtetek}@kam.mff.cuni.cz

Abstract. Compact and I/O-efficient data representations play an important role in efficient algorithm design, as memory bandwidth and latency can present a significant performance bottleneck, slowing the computation by orders of magnitude. While this problem is very well explored in e.g. uniform numerical data processing, structural data applications (e.g. on huge graphs) require different algorithm-dependent approaches. Separable graph classes (i.e. graph classes with balanced separators of size $\mathcal{O}(n^c)$ with $c < 1$) include planar graphs, bounded genus graphs, and minor-free graphs.

In this article we present two generalizations of the separator theorem, to partitions with small regions *only on average* and to weighted graphs. Then we propose I/O-efficient succinct representation and memory layout for random walks in (weighted) separable graphs in the pointer machine model, including an efficient algorithm to compute them. Finally, we present a worst-case I/O-optimal tree layout algorithm for root-leaf path traversal, show an additive $(+1)$-approximation of optimal compact layout and contrast this with NP-completeness proof of finding an optimal compact layout.

1 Introduction

Modern computer memory consists of several memory layers that together constitute a memory hierarchy with every level further from the CPU being larger and slower [2], usually by more than an order of magnitude, e.g. CPU registers, L1–L3 caches, main memory, disk drives etc. In order to simplify the model, commonly only two levels are considered at once, called *main memory* and *cache* of size M. There, the main memory access is block-oriented, assuming unit time for reading and writing of a block of size B, making random byte access very inefficient. While some I/O-efficient algorithms need to know the values of B and M (generally called *cache-aware*)[3], *cache-oblivious algorithms*[13] operate efficiently without this knowledge.

The work was supported by the Czech Science Foundation (GACR) project 17-10090Y "Network optimization".

T. V. Gopal and J. Watada (Eds.): TAMC 2019, LNCS 11436, pp. 222–241, 2019.
https://doi.org/10.1007/978-3-030-14812-6_14

Computations that process medium to large volumes of data therefore call for space-efficient data representations (to utilize the memory capacity and bandwidth) and strongly benefit from optimized memory access patterns and layouts (to utilize the data in fast caches and read-ahead mechanisms). While this area is very well explored in e.g. numerical data processing and analysis (e.g. [24]), structural data applications (e.g. huge graphs) require different and application-dependent approaches. We describe a representations to address these issues in separable graphs and trees.

Separable graphs satisfy the n^c-*separator theorem* for some $c < 1$, shown for planar graphs in 1979 by Lipton and Tarjan [29] (with $c = 1/2$), where every such graph on n vertices has a vertex subset of size $\mathcal{O}(n^c)$ that is a 2/3-balanced separator (i.e. it separates the graph into two subgraphs each having at most 2/3-fraction of vertices). These graphs not only include planar graphs [29] but also bounded genus graphs [17] and minor-free graph classes in general [22]. Small separators are also found in random graph models of small-world networks (e.g. geometric inhomogeneous random graphs by Bringmann et al. [7] have sublinear separators w.h.p. for all subgraphs of size $\Omega(\sqrt{\log n})$). Some graphs which come from real-world applications are also separable, such as the road network graphs [33,35]. Separable graph classes have linear information entropy (i.e. a separable class can contain only $2^{\mathcal{O}(n)}$ graphs of size n) and have efficient representations using only $\mathcal{O}(1)$ bits per vertex on average [4] and therefore utilize the memory capacity and bandwidth very efficiently.

This paper is organized as follows: Sects. 1.1 and 1.2 give an overview of the prior work and our contribution. Section 2 recalls used concepts and notation. Section 3 contains our results on random walks in separable graphs. Section 4 generalizes the separator theorem. Section 5 discusses the layout of trees.

1.1 Related Work

Turán [34] introduced a succinct representation[1] of planar graphs, Blandford et al. [4] introduced compact representations for separable graphs and Blelloch and Farzan [5] presented a succinct representation of separable graphs. However, none of those representations is cache-efficient (or can be easily made so). Analogous representations for general graphs suffer similar drawbacks [12,32].

Agarwal et al. [1] developed a representation of planar graphs allowing I/O-efficient path traversal, requiring $\mathcal{O}(K/\log B)$ block accesses[2] for arbitrary path of length K. This has been extended to a succinct planar graph representation by Dillabaugh et al. [11] with the same result for arbitrary path traversal. It appears unlikely that the representation of [11] could be easily modified to match the I/O complexity $\mathcal{O}(K/B)$ of our random-walk algorithm due to their use of a global indexing structure.

---

[1] A succinct (resp. compact) data representation uses $H + o(H)$ (resp. $\mathcal{O}(H)$) bits where H is the class information entropy.

[2] Note that $\Omega(K/\log B)$ blocks may be required even for trees. Standard graph representation would access $\mathcal{O}(K)$ blocks.

Dillabaugh et al. [10] describes a succinct data structure for trees that uses $\mathcal{O}(K/B)$ I/O operations for leaf-to-root path traversal. For root-to-leaf traversal, they offer a similar but only compact structure.

Among other notable I/O-efficient algorithms, Maheshwari and Zeh [30] develop I/O-efficient algorithms for computing vertex separators, shortest paths and several other problems in planar and separable graphs. Jampala and Zeh [20] extends this to a cache-oblivious algorithm for planar shortest paths. While there are representations even more efficient than succinct (e.g. implicit representations, which use only $\mathcal{O}(1)$ bits more than the class information entropy, see Kannan et al. [21] for an implicit graph representation), these do not seem to admit I/O-efficient access.

Random walks on graphs are commonly used in Monte Carlo sampling methods, among others in Markov Chain Monte Carlo methods for inference on graphical models [14], Markov decision process (MDP) inference and even in partial-information game theory algorithms [25].

1.2   Our Contribution

Random Walks on Separable Graphs. We present a compact cache-oblivious representation of graphs satisfying the n^c edge separator theorem. We also present a cache-oblivious representation of weighted graphs satisfying weighted n^c edge separator theorem, where the transition probabilities depend on the weights. The representations are I/O-efficient when performing random walks of any length on the graph, starting from a vertex selected according to the stationary distribution and with transition probabilities at each step proportional to the weights on the incident edges, respectively choosing a neighbor uniformly at random for the unweighted compact representation.

Namely, if every vertex contains q bits of extra (user) information, the representation uses $\mathcal{O}(n \log(q+2)) + qn$ bits and a random path of length K (sampled w.r.t. edge weights) uses $\mathcal{O}(K/(\frac{Bw}{(1+q)})^{1-c})$ I/O operations with high probability.

The graph representation is compact (as the structure entropy including the extra bits is $\Theta((q+1)n)$. The amount of memory used for the representation of the graph is asymptotically strictly smaller than the memory used by the user data already for the common case of $q = \Theta(w)$, in which case only $\mathcal{O}(K/B^{1-c})$ I/O operations are used. For $q = \mathcal{O}(1)$, the representation uses $\mathcal{O}(n)$ bits.

In contrast with previous I/O-efficient results for planar graphs, our representation is only compact (and not succinct) but works for all separable graph classes, is cache-oblivious (in contrast to only cache-aware in prior work), and, most importantly, comes with a much better bound on the number of I/O operations for randomly sampled paths (order of $\mathcal{O}(K/B^{1-c})$ rather than $\mathcal{O}(K/\log B)$).

Fast tree path traversal is a ubiquitous requirement for tree-based structures used in external storage systems, database indexes and many other applications. With Theorem 9, we present a linear time algorithm to compute a layout of the vertices in memory minimizing the worst-case number of I/O operations for leaf-to-root paths in general trees and root-to-leaf paths in trees with unit vertex size. We show an additive $(+1)$-approximation of an optimal *compact* layout (i.e. one that fully uses a consecutive block of memory) and show that finding an optimal compact layout is NP-hard.

The above layout optimality is well defined assuming unit vertex size, an assumption often assumed and satisfied in practice. Using techniques from Sect. 3 we can turn the layout into a compact representation using $\mathcal{O}(n)$ bits of memory, requiring at most OPT_L I/O operations for leaf-to-root paths in general trees and root-to-leaf paths in trees of fixed degree where OPT_L is the I/O complexity of the optimal layout, i.e. I/O-optimal layout with the vertices using any conventional vertex representation with $\Theta(w)$ bits for inter-vertex pointers. See Theorem 10.

Compared to previous results [10], our representation is compact and we present the exact optimum over all layouts while they provide the asymptotic optimum $\mathcal{O}(K/B)$. However, this does not guarantee that our representation has lower I/O complexity, since our notion of optimality only considers different layouts with each vertex stored by a structure of unit size.

Separable Graph Theorems. We prove two natural generalizations of the separator theorem (Theorem 7) and show that their natural joint generalization does not hold by providing a counterexample (Theorem 8). The Recursive Separator Theorem involves graph partitions coming from recursive applications of the Separator Theorem. Let r and $\bar{r}$ denote the maximum and average size of a region in the partition, respectively. We prove stronger bound on number of edges going between regions – $\mathcal{O}(\frac{n}{\bar{r}^{1-c}})$ instead of $\mathcal{O}(\frac{n}{r^{1-c}})$. The second generalization is for weighted graphs, showing that n in the bound $\mathcal{O}(\frac{n}{r^{1-c}})$ can be replaced by the total weight W to get $\mathcal{O}(\frac{W}{r^{1-c}})$. We show that the bound $\mathcal{O}(\frac{W}{\bar{r}^{1-c}})$ does not hold in general by providing a counterexample.

2 Preliminaries

Throughout this paper, we use standard graph theory notation and terminology as in Bollobas [6]. We denote the subtree of T rooted in vertex v by T_v, the root of tree T by r_T and the set of children of a vertex v as $\delta(v)$. All the logarithms are binary unless noted otherwise.

We use standard notation and results for Markov chains as introduced in the book by Grinstead and Snell [19] (Chapter 11) and mixing in Markov chains, as introduced in the chapter on mixing times in a book by Levin and Peres [27].

2.1 Separators

Let S be a class of graphs closed under the subgraph relation. We say that S satisfies the vertex (edge) $f(n)$-separator theorem iff there exist constants $\alpha < 1$ and $\beta > 0$ such that any graph in S has a vertex (edge) cut of size at most

$\beta f(n)$ that separates the graph into components of size at most αn. We define a weighted version of vertex (edge) separator theorem, which requires that there is a balanced vertex (edge) separator of total weight at most $\beta \frac{f(n)}{n} W$, where W is the sum of weights of all the edges. Note that these definitions make sense even for directed graphs. $f(n)$-separator theorem without explicit statement whether it is edge or vertex separator, means $f(n)$ vertex separator theorem.

Many graphs that arise in real-world applications satisfy n^c vertex or edge separator theorem.

It has been extensively studied how to find balanced separators in graphs. In planar graphs, a separator of size $\sqrt{n}$ can be found in linear time [29]. Separators of the same size can be found in minor-closed families in time $\mathcal{O}(n^{1+\epsilon})$ for any $\epsilon > 0$ [22]. A balanced separator of size $n^{1-1/d}$ can be found in finite-element mesh in expected linear time [31]. Good heuristics are known for some graphs which arise in real-world applications, such as the road network [33]. A poly-logarithmic approximation which works on any graph class is known [26]. A poly-logarithmic approximation of the separators will be sufficient to achieve almost the same bounds in our representation (differing by a factor at most poly-logarithmic in B).

We define a *recursive separator partition* to be a partition of vertex set of a graph, obtained by the following recursive process. Given a graph G, we either set the whole $V(G)$ to be one set of the partition or do the following:

1. Apply separator theorem. This gives us partition of $V(G)$ into two sets A, B from the separator theorem.
2. Recursively obtain recursive separator partitions of A and B.
3. Return the union of the partitions of A and B as the partition of $V(G)$.

We call the sets in a recursive separator partition regions.

If there is an algorithm that computes balanced separator in time $\mathcal{O}(f(n))$, there is an algorithm that computes recursive separator partition with region size $\Theta(r)$ in time $\mathcal{O}(f(n) \log n)$ for any r. A stronger version called r-division can be computed in linear time on planar graphs [18].

2.2 I/O Complexity

For definitions related to I/O complexity, refer to Demaine [8]. We use the standard notation with B being the block size and M the cache size. Both B and M is counted in words. Each word has w bits and it is assumed that $w \in \Omega(\log n)$.

3 Representation for Random Walks

In this section, we present our cache-oblivious representation of separable graphs optimized for random walks and related results.

Theorem 1. *Let G be a graph from a graph class satisfying the n^c edge separator theorem where every vertex contains q extra bits of information. Then there* *is a cache-oblivious representation of G using $\mathcal{O}\left(n\log(q+2)\right)+qn$ bits in which a random walk of length k starting in a vertex sampled from the stationary distribution uses in expectation $\mathcal{O}\left(k/\left(\frac{Bw}{(1+q)}\right)^{1-c}\right)$ I/O operations. Moreover, such representation can be computed in time $\mathcal{O}(n^{1+\epsilon})$ for any $\epsilon > 0$.*

For other random walks and weighted graphs where the transition probabilities are proportional to the random walk stationary distribution, we can show a weaker result. Namely, we can no longer guarantee a compact representation.

Theorem 2. *Let M be any Markov chain of random walks on a graph G and assume M has a unique stationary distribution π. Assume G satisfies the n^c edge separator theorem with respect to the edges-traversal probabilities in π. Let M' be a Markov chain of random walks on G with transition probabilities proportional to M, e.g. $\pi'(e) = \Theta(\pi(e))$. Then there is a layout of vertices of G into blocks with $\Theta(B)$ vertices each such that a random walk in M' of length k crosses memory block boundary in expectation $\mathcal{O}(k/B^{1-c})$ times.*

Note that this gives an efficient memory representation when $N_G(v)$ and the probabilities on incident edges can be represented by (or computed from) $\mathcal{O}(1)$ words, which is the case for bounded degree graphs with some chains M'. We also note that such partially-implicit graph representations are present in the state graphs of some MCMC probabilistic graphical model inference algorithms.

Additionally, we present a result on the concentration of the number of I/O operations which applies to both Theorems 1 and 2.

Theorem 3. *Let G be a fixed graph, t_{mix} the mixing time of G and X the number of edges going between blocks crossed during the random walk. Then the probability that $(1 - \delta)E(X) \leq X \leq (1 + \delta)E(x)$ does not hold is $\mathcal{O}\left(me^{-c'\frac{\delta^2 nB^{c-1}}{m}}\right)$ for some value c' and $m = t_{mix}\log(n^2/E(X_1))$, where the variable X_i indicates if the walk crossed an edge between two different blocks in step i.*

The following lemma is implicit in [4], as the authors use the same layout to get compact representation of separable graphs and they use the following property.

Lemma 1 (Blandford et al. [4]). *If π in Theorem 2 gives the same traversal probability to all edges, the representation induces a vertex order $l : V \to 1 \ldots n$ such that $\sum_{e=uv \in E} \log |l(u) - l(v)| = \mathcal{O}(n)$.*

3.1   Proofs of Theorems 1–3

Proof (Proof of Theorem 1).
Since the stationary distribution on an undirected graph assigns equal probability to every edge, we can apply Lemma 1 on G to obtain vertex ordering $r : V \to 1 \ldots n$ such that $\sum_{e=uv \in E_G} \log |r(u) - r(v)| = \mathcal{O}(n)$. We could therefore compactly store the edges as variable-width vertex order differences (offsets). However, it is not straightforward to find the memory location of a given vertex

when a variable-width encoding is used. To avoid an external (and I/O inefficient) index used in some other approaches, we replace the edge offset information with relative bit-offsets, directly pointing to the start of the target vertex,
using Theorem 4 on the edge offsets. We expand the representation by inserting
the q bits of extra information to every vertex, adjusting the pointers and thus
widening each by $\mathcal{O}(\log q)$ bits.

To prove the bound on I/O complexity, we use the same argument as in the
proof of Theorem 2. Average of $\mathcal{O}(1 + q)$ bits is used for representation of single
vertex and, therefore, average of $\Theta(\frac{Bw}{q+1})$ vertices fit into one cache line. By
Theorem 7, part i, the total probability on edges going between memory blocks
is $\mathcal{O}(1/\frac{Bw}{q+1})$. Again, by linearity of expected value, this proves the claimed I/O
complexity.

Compact representation as in Theorem 4 can be computed in the claimed
bound, as is shown in Theorem 5.   □

Proof (Proof of Theorem 2). We use the following recursive layout. Let S be an
edge separator with respect to edge-traversal probabilities in π. Then S partitions G into two subgraphs X and Y. We recursively lay out X and Y and
concatenate the layouts. Note that X and Y are stored in memory contiguously. At some level of recursion, we get partition into subgraphs represented
by between ϵB and B words for $\epsilon > 0$ constant. We call these subgraphs block
regions. Since the average degree in graphs satisfying n^c edge separator theorem
is $\mathcal{O}(1)$ [28], the average vertex representation size is also $\mathcal{O}(1)$ and the average
number of vertices in a block region is, therefore, $\Theta(B)$. It follows from Theorem 7, part ii, that the total probability on edges going between block regions
is $\mathcal{O}(1/B^{1-c})$. From linearity of expectation, $\mathcal{O}(1/B^{c-1})$-fraction of steps in the
random walk cross between block regions in expectation. Moreover, each of the
block regions in the partition is stored in $\mathcal{O}(1)$ memory blocks, which proves the
claimed bound on I/O complexity.   □

Proof (Proof of Theorem 3). Let X be the number of edges crossed during the
random walk that go between blocks. We are assuming that there is at least one
edge going between two blocks in the graph.

We choose $\delta' = \sqrt{\frac{3}{4}\delta}$ (arbitrary constant $c'' < 1$ would work). Note that m
is a number of steps, after which the probabilities on edges differ from those in
stationary distribution by at most $E(X_1)/n^2$, regardless from what distribution
we started the random walk since $t_{mix}(\epsilon) \leq \lceil \log \epsilon^{-1} \rceil t_{mix}$ [27]. This means that
the probability that an edge going between two blocks is crossed after m steps
differs by at most $\frac{1}{n}$-fraction from the probability in stationary distribution.

Let X_i be indicator random variable that is 1 iff the random walk crosses
edge going between blocks in step i. We consider the following sets of random
variables $S_i = \{X_j | X_{j-m} : j \mod m\} = i\}$ for $1 \leq i \leq m$ (not conditioning on
variables with nonpositive indices). Note that the random variables in each of
sets S_i are independent and $(1 - \frac{1}{n})E(X_j) \leq E(X_j | X_{j-m}) \leq (1 + \frac{1}{n})E(X_j)$, as
mentioned above. Let μ_i be $E(\sum_{X \in S_i} X)$ and $\mu = E(\sum_i \sum_{X \in S_i} X)$. Note that
$\mu_i \in \Theta(nB^{c-1}/m)$ for each i. By applying the Chernoff inequality, we get that

the following bounds hold for all $n \geq n_0$ for some n_0 for each i:

$$P\left(\sum_{X \in S_i} X \geq (1 + \delta')\mu_i\right) \leq e^{-\frac{\delta'^2 \mu_i}{3}} = e^{-\frac{\delta^2 \mu_i}{4}}$$

$$P\left(\sum_{X \in S_i} X \leq (1 - \delta')\mu_i\right) \leq e^{-\frac{\delta'^2 \mu_i}{2}} \leq e^{-\frac{\delta^2 \mu_i}{4}}$$

The probability that there exists i such that either $\sum_{X \in S_i} X \geq (1 + \delta')\mu_i$ or $\sum_{X \in S_i} X \leq (1 - \delta)\mu_i$ is by the union bound for some value of c' at most the following:

$$2\lceil \log(n/E(X_1))\rceil t_{mix} e^{-\frac{\delta^2 \mu}{4m}} \in \mathcal{O}(me^{-c'\frac{\delta^2 nB^{c-1}}{m}})$$

Note that μ_i converges to $|S_i|E(X_1)$, which is the value that we are showing concentration of $\sum_{X \in S_i} X$ around. The asymptotic bound on the probability follows. □

3.2   Expanding Relative Offsets to Relative Bit-Offsets

Having the edges of a graph encoded as relative offsets to the target vertex and having these numbers encoded by a variable-length encoding, we need a way to find the exact location of the encoded vertex. Others have used a global index for this purpose but this is generally not I/O-efficient.

Our approach encodes the relative offsets as slightly wider numbers that directly give the relative bit-index of the target. However, this is not straightforward as expanding just one relative offset to a relative bit-offset can make other bit-offsets (spanning over this value) larger and even requiring more space, potentially cascading the effect.

Note that one simple solution would be to widen every offset representation by $\Theta(\log \log N)$ bits where N is the total number of bits required to encode all the n offsets, yielding $N + n * \mathcal{O}(\log \log N)$ encoding. $\log n$ bits are sufficient to store each offset. Therefore, by expanding the offsets, they increase at most $\log n$ times. By adding $\log(2 \log n)$ bits, we can encode increase of offsets by factor of up to $2 \log n \geq \log n + \log(2 \log n)$.

However, we propose more efficient encoding with the following theorem. We interpret the numbers a_i as relative pointers, i-th number pointing to the location of the $(i + a_i)$-th value. In the proof, we use a dynamic width *gamma number encoding* in the form $[(\text{sign})B_0 0 B_1 0 B_2 0 \ldots B_i 1]$, where $2i + 1$-th bit encodes whether B_i is the last bit encoded.

Theorem 4. *Let $a_1 \ldots a_n$ be a sequence of numbers such that $-i \leq a_i \leq n - i$ and $\sum_{i=0}^n \log |a_n| = m$. Then there are n-element sequences $\{w_i\}$ (the encoded bit-widths) and $\{b_i\}$ (the bit-offsets) of numbers such that for all $1 \leq i \leq n$, $w_i \geq 2 \log |b_i| + 1$ (i.e. b_i can be gamma-encoded in w_i bits), $P(i) + w_i = P(i + a_i)$ where $P(j) := \sum_{i=1}^{j-1} w_i$ (so w_i is a relative bit-offset of encoded position $i + a_i$) and $\sum_{i=1}^n w_i = \mathcal{O}(m + n)$.*

Proof. There are certainly *some* non-optimal valid choices for w_i's and b_i's, and we can improve upon them iteratively by shrinking w_i's to fit gamma-encoded b_i with sign (i.e. $w_i = 1 + 2\log|b_i|$), which may, in turn, decrease some b_i's. Being monotonic, this process certainly has a fixpoint $\{b_i\}_i$ and $\{w_i\}_i$ and we assume arbitrary such fixpoint.

Let $C < 1$ and $D > 1$ be constants to be fixed below. Denote $v_i = \log|a_i|$ and $R_i = \{i \ldots i + a_i - 1\}$ (resp. $\{i + a_i \ldots i - 1\}$ when $a_i < 0$). Intuitively, when expanding offsets a_x to bit offsets b_x, it may happen that R_x contains y with $w_y \gg a_x$, forcing $w_x \gg v_x$. We amortize such cases by distributing "extra bits" to such "smaller" offsets.

Let $x \prec y \iff y \in R_x \wedge v_x \leq C \log w_y \wedge v_x > D$ and let $x^\uparrow = \arg\max_{y \succ x} w_y$ (or undefined if there is no such y) and let $y^\downarrow = \{x | y \in x^\uparrow\}$. Observe that $|y^\downarrow| \leq 2 \cdot 2^{C \log w_y} = 2w_y^C$ since all $x \in y^\downarrow$ have $|a_x| \leq 2^{v_x} \leq w_y^C$. We also note that $y = x^\uparrow$ implies $w_x < w_y$ since $w_y \leq w_x$ would imply $b_x \leq |a_x| w_x$ and $w_x > 2^{v_x/C}$ leading to $w_x \leq v_x + \log w_x$ and $2^{v_x/C} < w_x \leq 2v_x$, which gives the desired contradiction with D large enough (depending only on C).

We will distribute the extra bits starting from the largest w_i's. Every y uses w_y bits for its encoding and distributes another w_y bits to $y^\downarrow$. Let $r_x = w_{x^\uparrow} / |(x^\uparrow)^\downarrow| \geq \frac{1}{2} w_{x^\uparrow}^{1-C}$ be the number of extra bits received from $x^\uparrow$ in this way.

For every offset x we use $10v_x + 2D$ bits and the received bits r_x. Since the received bits are accounted for in other offsets, this uses $\sum_{i=1}^{n} 10v_x + D = 10m + \mathcal{O}(n)$ bits in total. Therefore we only need to show that the number of bits thus available at x is sufficient, i.e. that $2w_x \leq r_x + 10v_x + 2D$ (one w_x to represent b_x, one to distribute to $x^\downarrow$).

Now either there is $y = x^\uparrow$ and we have $b_x \leq |a_x| w_y$ so $w_x \leq 1 + 2v_x + 2\log w_y$ and noting that for large enough D only depending on C: $2\log w_y \leq \frac{1}{4} w_y^{1-C} + D \leq \frac{1}{2} r_x + D$, so we obtain $w_x \leq \frac{1}{2} r_x + 5v_x + 2D$ as desired.

On the other hand, undefined $x^\uparrow$ implies that $\forall y \in R_x : w_y \leq 2^{v_x/C}$. Therefore $b_x \leq |a_x| 2^{v_x/C}$ and $w_x \leq 1 + 2v_x + 2v_x/C = 1 + (2 + 2/c)v_x$. Now we may fix $C = 2/3$, obtaining $w_x \leq 5v_x + D$ as required for $D \geq 1$. This finishes the proof for any fixpoint $\{b_i\}_i$ and $\{w_i\}_i$. $\square$

The algorithm from the beginning of the proof can be shown to run in polynomial time. We start with e.g. $w_i = w_0 = 1 + 4\log n$ and $b_i = \text{sign}(a_i) \sum_{j \in R_i} w_j$. Then we iteratively update $w_i := 1 + 2\lceil \log b_i \rceil$ and recompute b_i as above. Since every iteration takes $\mathcal{O}(n^2)$ time and in every iteration at least one w_i decreases, the total time is at most $\mathcal{O}(n^3 \log n)$. In the following section, we show an algorithm that computes a representation with the same asymptotic bounds, running in time $\mathcal{O}(n^{1+\epsilon})$ for any $\epsilon > 0$.

Constructing the Compact Representation. In this section, we use notation defined in Sect. 3.2, specifically R_e and b_e. Recall that R_e is the set of edges of G spanned by the edge e in the representation and b_e is the relative offset of edge e in the (expanded) representation). Let G be the graph we want to represent. We assume that G satisfies the n^c edge separator theorem.

We find a representation using $\mathcal{O}(n \log \log n)$ bits, as mentioned above by expanding all pointers and then modify it to make it compact.

We define a directed graph H on the set $E(G)$ with arc going from v to u iff $v \in R_u$. Let us fix a recursive separator hierarchy of G. We call $l(e)$ the level of recursion on which the edge e is part of the separator. We define a graph $H_{\leq k}$ to be the subgraph of H induced by vertices corresponding to edges of G which appear in the recursive separator hierarchy in a separator of subgraph of size at most k.

The following lemma will be used to bound the running time of the algorithm:

Lemma 2. *The maximum out-degree of $H_{\leq n^{c'}}$ is $n^{c*c'}$. For any fixed $c' > 0$, $|H \setminus H_{\leq n^{c'}}| \in n^{1-\epsilon'}$ where $\epsilon' > 0$ is some constant depending only on c and c'.*

Proof. We first prove that maximum out-degree of H is $\mathcal{O}(n^c)$.

There are $\mathcal{O}(n^c)$ edges $e \in G$ with $l(e) = 1$ spanning any single vertex. The number of edges e spanning some vertex with $l(e) = k$ decreases exponentially with k, resulting in a geometric sequence summing to $\mathcal{O}(n^c)$.

The maximum out-degree of $H_{\leq n^{c'}}$ is the same as that of graph H' corresponding to a subgraph of G of size at most $n^{c'}$. Maximum out-degree of $H_{\leq n^{c'}}$ is, therefore, $\mathcal{O}(n^{c*c'})$.

The number of vertices in $H \setminus H_{\leq n^{c'}}$ is equal to the number of edges in G going between blocks of size $\Theta(n^{c'})$. This number is, by Theorem 7, equal to $n/n^{c'(1-c)}$, which is $\mathcal{O}(n^{1-\epsilon'})$ for some $\epsilon' > 0$. □

Theorem 5. *Given a separator hierarchy, the representation from Theorem 1 can be computed in time $\mathcal{O}(n^{1+\epsilon})$ for any $\epsilon > 0$.*

Proof. We first describe an algorithm running in time $\mathcal{O}(n^{1+c} \log \log n)$, where c is the constant from the separator theorem, and then improve it.

Just as in the proof of Theorem 4, b_v denotes the relative offset of edge v in the representation. We store a counter c_v for each vertex $v \in H$ equal to the decrease of b_v required to shrink its representation by at least one bit. That is, $c_v = b_v - \lfloor b_v \rfloor_{2^k} + 1$, where $\lfloor i \rfloor_{2^k}$ is i rounded down to closest power of two. When we shrink the representation of edge corresponding to vertex $v \in H$, we have to update counters c_u for all u, such that $vu \in E(H)$. Since the out-degree of H is $\mathcal{O}(n^c)$, the updates take $\mathcal{O}(n^c)$ time. We start with representation with $\mathcal{O}(n \log \log n)$ bits and at each step, we shorten the representation by at least one bit. This gives the running time of $\mathcal{O}(n^{1+c} \log \log n)$.

To get the running time of $\mathcal{O}(n^{1+\epsilon} \log \log n)$, we consider the graph $H_{\leq n^{\epsilon'}}$ for some sufficiently small epsilon. Note that the maximum out-degree of $H_{\leq n^{\epsilon'}}$ is $\mathcal{O}(n^{c\epsilon'})$. We can fix ϵ' small enough to decrease the maximum out-degree to n^{ϵ}. Therefore, by using the same algorithm as above on graph $H_{\leq n^{\epsilon'}}$ for ϵ' sufficiently small, we can get a running time of $\mathcal{O}(n^{1+\epsilon} \log \log n)$ for any fixed $\epsilon > 0$. The representations of edges corresponding to vertices not in the graph $H_{\leq n^{\epsilon'}}$ are not shrunk.

Note that the presumptions of Theorem 4 are fulfilled by the edges corresponding to vertices in $H_{\leq n^\epsilon}$ and the obtained representation of graph $G' = (V(G), V(H_{\leq n^\epsilon}))$, is therefore compact. The edges not in $H_{\leq n^\epsilon}$ are then added, increasing some offsets. The representation of an offset of length at least $n^{\epsilon''}$ for $\epsilon'' > 0$ is never increased asymptotically by inserting edges since it already has $\Theta(\log n)$ bits. There are at most $\mathcal{O}(n^{\epsilon''})$ edges of G' shorter than $n^{\epsilon''}$ that span any single inserted edge. Lengthening of offsets shorter than $n^{\epsilon''}$, therefore, contributes at most $\mathcal{O}(n^{1-\epsilon'} n^{\epsilon''} \log\log n) \in o(n)$ for some ϵ'' sufficiently small. The inserted edges themselves have representations of total length $\mathcal{O}(n^{1-\epsilon'} \log n) \in o(n)$. Additional $o(n)$ bits are used after the insertion of edges and the representation, therefore, remains compact.   □

4   Separator Hierarchy

In this section, we prove two generalizations of the separator hierarchy theorem. Our proof is based on the proof from [23]. Most importantly, we show that the recursive separator theorem also holds if we want the regions to have small size on average and not in the worst case. We also prove the theorem for weighted separator theorem with weights on edges. We show that the natural generalization of our two generalizations does not hold by presenting a counterexample.

Since the two theorems are very similar and their proofs only differ in one step, we present them as one theorem with two variants and show only one proof proving both variants. The difference lies in the reason why the Inequality 1 holds. The following lemma and observation prove the inequality under some assumptions and they will be used in the proof of the theorem.

$$\frac{c'\gamma_w W}{r_1^{1-c}} + \frac{c'(1-\gamma_w)W}{r_2^{1-c}} \leq \frac{c'W_n}{r^{1-c}} \tag{1}$$

Observation 6. *The Inequality 1 holds for $r_1 = r_2 = r$.*

Lemma 3. *The Inequality 1 holds for $\gamma_w = \gamma_n$ and r_1, r_2 and r satisfying the following.*

$$r = \frac{1}{\frac{\gamma_n}{r_1} + \frac{1-\gamma_n}{r_2}} = \frac{r_1 r_2}{\gamma_n r_2 + (1-\gamma_n)r_2}. \tag{2}$$

Proof. Let $\gamma = \gamma_w = \gamma_n$. We simplify the inequality

$$\frac{\gamma}{r_1^{1-c}} + \frac{1-\gamma}{r_2^{1-c}} \leq \frac{1}{r^{1-c}}$$

for r_1, r_2 and r satisfying the equality (2). By substituting for r and rearranging the inequality, we get

$$\gamma r_1^{1-c} + (1-\gamma)r_2^{1-c} \leq (\gamma r_1 + (1-\gamma)r_2)^{1-c}.$$

We substitute $r_2 = \lambda r_1$. Note that this holds for $\lambda = 1$ and that we may assume $r_1 \leq r_2$ by symmetry. Since the inequality holds for $\lambda = 1$, it is sufficient to show the inequality for $\lambda \geq 1$ with both sides differentiated with respect to λ. By differentiating both sides and simplifying the inequality, we get

$$(x - (\lambda - 1)\gamma)^{-c} \geq x^{-c}$$

which obviously holds, since $\lambda \geq 1$ and $\gamma > 0$.

Now we proceed to prove the two generalizations of the recursive separator theorem. Note that in the following, r is the average or maximum region size, depending on whether the graph is weighted or not.

Theorem 7. *Let G be a (possibly weighted) graph satisfying the n^c separator theorem with respect to its weights and let P be its recursive balanced separator partition. Then if either*

(i) the graph in not weighted and r is the average size of a region in the partition P, or

(ii) the graph is weighted and r is the maximum size of a region in the partition P.

Then the total weight of edges not contained inside a region of P is $\mathcal{O}(W/r^{1-c})$, where W is the total weight (resp. number if unweighted) of all edges of G.

In this proof, let $w(S)$ be the total weight of the edges in S with $w(e)$ denoting the weight of the single edge e.

Proof. We use induction on the number of vertices to prove the following claim.

Claim. Let us have a recursive separator partition P of n-vertex graph G of average region size r. Then $w(E(G) \setminus \bigcup_{p \in P} p) < \frac{c'W}{r^{1-c}} - \frac{c''W}{n^{1-c}}$ for some c' and c''.

Before the actual proof of this claim, let us define some notation. Let c, α and β be the constants from the separator theorem (recall that separator theorem ensures existence of a partition of $V(G)$ into two sets of size at least $\alpha V(G)$ with edges of total weight at most $\beta \frac{W}{n^{1-c}}$ going across). Let $B(W, n, r)$ be the maximum value of $w(E(G) \setminus \bigcup_{p \in P} p)$ over all n-vertex graphs of total weight W and all their recursive separator partitions with average region size r. We use γ_n to denote a fraction of the number of vertices and γ_w to denote a fraction of the total weight.

Proof (Proof of the claim). We defer the proof of the base case until we fix the constant c'.

By the separator theorem, $B(W, n, r)$ satisfies the following recurrence.

$$B(W, n, r) = 0 \text{ for } n \leq r$$

$$B(W, n, r) \leq \beta \frac{W}{n^{1-c}} + \max_{\substack{\alpha \leq \gamma_n \leq 1-\alpha \\ \gamma_w \in [0,1]}} B(\gamma_w W, \gamma_n n, r_1) + B((1 - \gamma_w)W, (1 - \gamma_n)n, r_2)$$

where r_1, r_2 are the respective average region sizes in the two subgraphs. It, therefore, holds that $r = \frac{1}{\frac{\gamma_n}{r_1} + \frac{1-\gamma_n}{r_2}} = \frac{r_1 r_2}{\gamma_n r_2 + (1-\gamma_n) r_2}$.

From the inductive hypothesis, we get the first inequality of the following. The second inequality follows from the Observation 6 for the case i and from the Lemma 3 for the case ii.

$$B(W, n, r) \leq \beta \frac{W}{n^{1-c}} + \frac{c' \gamma_w W}{r_1^{1-c}} + \frac{c'(1-\gamma_w)W}{r_2^{1-c}} - c'' \frac{W}{n^{1-c}} (\gamma_n^c + (1-\gamma_n)^c) \leq \quad (3)$$

$$\leq \beta \frac{W}{n^{1-c}} + \frac{c' W_n}{r^{1-c}} - c'' \frac{W}{n^{1-c}} (\gamma_n^c + (1-\gamma_n)^c)$$

It holds that $\gamma_n^c + (1-\gamma_n)^c \geq 1 + \epsilon_\alpha$, where $\epsilon_\alpha > 0$ is a constant depending only on α, since $\gamma_n \in [\alpha, 1-\alpha]$ for $\alpha > 0$. We can therefore set c'' such that

$$c'' \frac{W}{n^{1-c}} (\gamma_n^c + (1-\gamma_n)^c) - \beta \frac{W}{n^{1-c}} \geq c'' \frac{W}{n^{1-c}}$$

This completes the induction step.

For c' large enough, the claimed bound in the base case is negative and it, therefore, holds. $\qquad \square$

We conclude this section by showing that the following natural generalization of Theorem 7 does not hold:

Theorem 8. *The following generalization does* not *hold: Let G be a weighted graph satisfying the n^c separator theorem with respect to its weights and let P be its recursive separator partition. Let r be the average size of a region in the partition P. Then the total weight of edges not contained in a region of P is $\mathcal{O}(W/r^{1-c})$, where W is the total weight of all edges of G.*

Proof. We show that there is a weighted graph satisfying the n^c-separator theorem with respect to its weight and a recursive partition P of G with edges going between partition regions of P that have total weight $\Theta(W)$, where W is the total weight of all edges, and with average region size of $\Theta(n/\log n)$.

Let G be an unweighted graph of bounded degree satisfying the n^c-separator theorem. We set weights of all its edges to be 1, except for one arbitrary edge e with weight $m-1$, where m is the number of edges of G. Note that $w(e) = W/2$. We denote this weighted graph by G_w.

Let S be a separator in G from the separator theorem. We modify S in order to obtain a balanced separator S_w in G_w of weight $\mathcal{O}(W/n^{1-c})$. If $e \notin S$, we set $S_w = S$. Otherwise, we remove e from S and add all other edges incident to its endpoints. This gives us S_w which is a separator and its weight differs from the weight of S only by an additive constant, since the graph G has bounded degree. It follows that G_w satisfies the n^c-separator theorem with respect to its weights.

We consider a partition P constructed by the following process. Let S be a separator from the separator theorem on G_w, partitioning $V(G_w)$ into vertex sets A and B. If $e \in S$, we stop and set A and B as the regions of P. Otherwise, without loss of generality, $e \in A$. We set B as a region of P and recursively partition A.

At the end of this process, we get P with edges of total weight at least $W/2$ between regions (as e is not contained within any region). The partition P has $\Theta(\log n)$ regions, so the average region size is $\Theta(n/\log n)$. □

5   Representation for Paths in Trees

In this section, we show a linear algorithm that computes a cache-optimal layout of a given tree. We are assuming that the vertices have unit size and B is the number of vertices that fit into a memory block. The same assumption has been used previously by Gil and Itai [16]. This is a reasonable assumption for trees of fixed degree and for trees in which each vertex only has a pointer to its parent. It does not matter in which direction the paths are traversed and we may, therefore, assume that the paths are root-to-leaf.

We also show that it is *NP*-hard to find an optimal compact layout of a tree and show an algorithm which gives a compact layout with I/O complexity at most $OPT + 1$.

Definition 1. *Laid out tree: A laid out tree is an ordered triplet $T = (V, E, L)$, where (V, E) is a rooted tree and $L : V \to \{0, 1, 2, \cdots, |V|\}$ assigns to each vertex the memory block that it is in. We require that at most B vertices are assigned to any block. We treat the block 0 specially as the block already in the cache.*

We define $c'_L(P) = |\{L(v)$ for $v \in P\} \setminus \{0\}|$ to be the cost of path P in a given layout L. We define $c(T, k)$, the worst-case I/O complexity given k free slots, as

$$c(T, k) = \min_L (\max_P (c(P)))$$

where P ranges over all root-to-leaf paths and L over all layouts that assign at most k vertices to block 0. Since block 0 is assumed to be already in cache, accessing these vertices does not count towards the I/O complexity. We define $c(T)$, the worst-case I/O complexity of laid out tree T, to be $c(T, 0)$. This means $c(T)$ is the maximum number of blocks on a root-to-leaf path. We define a worst-case optimal layout of a tree T given k free memory slots as a layout attaining $c(T, k)$.

We can observe that $c(T) \leq 1 + \max_{u \in \delta(r_T)}(c(T_u))$. From the lemmas below follows that $c(T)$ only depends on the subtrees rooted in children of r_T with the maximum value of $c(T_u)$.

Lemma 4. *For any $k_1, k_2 \in [B]$, $|c(T, k_1) - c(T, k_2)| \leq 1$ and $c(T, k)$ is non-increasing in k.*

Proof. The function $c(T, k)$ is monotonous in k since a layout given k_1 free slots is a valid layout given k_2 slots for $k_2 \geq k_1$. Moreover $c(T, 0) = c(T, B) - 1$, since we can map vertices in the root's block to block 0 instead. From this and the monotonicity, the lemma follows. □

We define *deficit* of a tree $k(T) = min\{k$, such that $c(T,k) < c(T,0)\}$. Note that $k(T) \leq B$. It follows from Lemma 4 that $c(T,k') = c(T,0) = c(T,B)+1$ for all $k' < k(T)$ and $c(T,k') = c(T,0) - 1 = c(T,B)$ for $k' \geq k(T)$.

Lemma 5. *For $k \geq 1$, there is a worst-case optimal layout attaining $c(T,k)$ such that root is in block* 0.

Proof. Let L be a layout that does not assign block 0 to the root. If no vertex is mapped to block 0, we can move root to block 0. Since block 0 does not count towards I/O complexity, doing this can only improve the layout. Otherwise, let v be vertex, which is mapped to block 0. We construct layout L' such that $L'(v) = L(r)$, $L'(r) = L(v)$ and $L'(u) = L(u)$ for all other vertices u. For any path P, $c'_L(P) \geq c'_{L'}(P)$, since any path which contains v in layout L' already contained it in L and block 0 does not count towards the I/O complexity. □

It is natural to consider layouts in which blocks form connected subgraphs. This motivates the following definition

Definition 2. *A partition of a rooted tree is convex if the intersection of any root-to-leaf path with any set of the partition is a (possibly empty) path.*

Let M_v be the set of successors u of vertex v with maximum value of $c(T_u)$.

Lemma 6. *The function $c(T,k)$ satisfies the following recursive formula for $k \geq 1$.*

$$c(T,k) = \min_{\{k_u\}} \max_{u \in M_v} c(T_u, k_u)$$

where the min is over all sequences $\{k_u\}$ such that $\sum_{u \in \delta(v)} k_u = k - 1$.

Proof. By Lemma 5, we may assume that an optimal layout attaining $c(T,k)$ for $k \geq 1$ puts the root to block 0 and allocates the remaining $k - 1$ slots of block 0 to root's subtrees, k_u slots to the subtree T_u. On the other hand, from values of k_u, we can construct a layout with cost $\max_{u \in M_v}(c(T_u, k_u))$. □

Problem 1.
Input: Rooted tree T
Output: Worst-case optimal memory layout of T.

Theorem 9. *There is an algorithm which computes a worst-case optimal layout in time $\mathcal{O}(n)$. Moreover, this algorithm always outputs a convex layout.*

Proof. We solve the problem using a recursive algorithm. For each vertex, we compute $k(T_v)$ and $c(T_v)$. First, we define $d(T)$ and $c_{max}(v)$.

$$d(T_v) = 1 + \sum_{u \in M_v} k(u), \qquad c_{max}(v) = \max_{u \in \delta(v)} (c(T_u))$$

If $d(T) < B$, we let $k(T_v) = d(T)$ and $c(T_v) = c_{max}(v)$. Otherwise $k(T_v) = 1$ and $c(T_v) = c_{max}(v)+1$. As a base case, we use that $c(T,k) = 0$ when $|V(T)| \leq k$. For $k = 0$, we use that $c(T,0) = c(T,B) + 1$.

Using the values $k(T_u)$ and $c(T_u)$ calculated using the above recurrence, we reconstruct the worst-case optimal layout in a recursive manner. When laying out a subtree given k free slots, we check whether $k \geq d(T)$. If it is, we distribute the $k - 1$ empty slots (one is used for the root) in a way that subtrees T_v for $v \in M(r_T)$ get at least $k(T_v)$ empty slots. Otherwise, distribute them arbitrarily. We put the root of a subtree into a newly created block if the subtree gets 0 free slots. Otherwise, we put the root into the same block as its parent. It follows from the way we construct the solution that it is convex.

It follows from Lemmas 4 and 6 that $c(T, k) = c(T, 0) - 1$ if and only if $k - 1$ free slots can be allocated among the subtrees $T_u, u \in \delta(r_T)$ such that subtree T_u gets at least $k(T_u)$ of them. It can be easily proven by induction that the algorithm finds for each vertex the smallest number of free slots required to make the allocation possible and calculates the correct value of $c(T_v)$. $\square$

If the subtree sizes are computed beforehand, we spend $deg(v)$ time in vertex v. By charging this time to the children, we show that the algorithm runs in linear time.

This algorithm can be easily modified to give a compact layout which ensures I/O complexity of walking on a root-to-leaf path to be at most $c(T) + 1$. This is especially relevant since finding the worst-case optimal layout is **NP**-hard, as we show in Sect. 5.1. The algorithm can be modified to give a compact layout by changing the reconstruction phase such that we never give more than $|V(T_v)|$ free slots to the subtree of T rooted in v unless $k > |V(T)|$. Note that only the last block on a path can have unused slots. We can put blocks which are not full consecutively in memory, ignoring the block boundaries. Any path goes through at most $c(T)$ blocks out of which at most one is not aligned, which gives total I/O complexity of $c(T) + 1$.

The following has been proven before in [9] and follows directly from Theorem 9.

Corollary 1. *For any tree T, there is a convex partition of T which is worst-case optimal.*

Proof. The corollary follows from Theorem 9, since the algorithm given in the proof is correct and always gives a convex solution. $\square$

Since the layout computed by the algorithm is always convex, we never re-enter a block after leaving it. This means that $c(T)$ really is the worst-case I/O complexity.

Finally, we show how to construct a compact representation with similar properties. Note that we do not claim I/O optimality among all compact representations but only relative to the tree layout optimality as in Theorem 9.

Theorem 10. *For a given tree T with q bits of extra data per vertex, there is a compact memory representation of T using $\mathcal{O}(nq)$ bits of memory requiring at most OPT_L I/O operations for leaf-to-root paths in general trees and root-to-leaf paths in bounded degree d trees. Here OPT_L is the I/O complexity of the optimal* *layout from Theorem 9 when we set the vertex size to be $q+2\log n$ for leaf-to-root paths, or to $q + 2d\log n$ for root-to-leaf paths.*

Proof. The theorem is an indirect corollary of Theorems 9 and 4. We set the vertex size as indicated in the theorem statement (depending on the desired direction of paths) and obtain an assignment of vertices to blocks by Theorem 9. We call the set of the blocks D. Note that for $q = \Omega(\log n)$, this is already a compact representation.

For smaller q, we construct an auxiliary tree T' on the blocks D representing their adjacency in T. We can assume that T' is a tree due to the convexity of the blocks of D. We apply the separator decomposition to obtain an ordering R of $V_{T'}$ with short representation of offset edge representation (Lemma 1). Similarly, we can get an ordering for each block in D. We order the vertices of T' according to R, ordering the vertices within blocks according to orderings of the individual blocks. We obtain an ordering having offset edge representation of total length $\mathcal{O}(n \log q)$, as there is $\mathcal{O}(n/B)$ edges going between blocks with offset edge representations of total length $\mathcal{O}(n \log B \log q/B)$ and edges within blocks with offset edge representations of total length $\mathcal{O}(n \log q)$.

We now apply Theorem 4 on the edge offsets still split in memory blocks according to D, obtaining a bit-offset edge representation where the vertex representation of every block of D still fits within one memory block, as we have previously reserved $2\log n + \Theta(1)$ memory for every pointer and $w_i \leq 1 + 2\log n$. We merge consecutive blocks whose vertices fit together into one block. This ensures that every block has at least $B/2$ vertices.   □

5.1   Hardness of Worst-Case Optimal Compact Layouts

In this section, we prove that it is **NP**-hard to find a worst-case optimal compact layout (that is, the packing with minimum I/O complexity out of all compact layouts). We show this by reduction from the 3-partition problem, which is strongly **NP**-hard [15] (i.e. it is **NP**-hard even if all input numbers are written in unary). This result is in contrast with Theorem 9 which shows how to find worst-case optimal non-compact layout.

Problem 2 (3-partition).
Input: Natural numbers $x_1, \cdots, x_n$.
Output: Partition of $\{x_i\}_1^n$ into sets $Y_1, \cdots, Y_{n/3}$ such that $\sum_{x \in Y_i} x = 3(\sum_1^n x_i)/n = S$ for each i.

Theorem 11. *It is **NP**-hard to find a worst-case optimal compact layout of a given tree T.*

Proof. We let $B = S$. We construct the following tree. It consists of a path $P = p_1 p_2 \cdots p_B$ of length B rooted in p_1. For each number x_i from the 3-partition instance, we create a path of length x_i. We connect one of the end vertices of each of these paths to p_B.

Next, we prove the following claim. There is a layout of I/O complexity 2 iff the instance of 3-partition is a yes instance. We can get such layout from a valid partition easily by putting in a memory block exactly the paths corresponding to x_i's that are in the same partition set. For the other implication, we first prove that P is stored in one memory block. If it were not, we would visit at least two different memory block while traversing P and there would be a root-to-leaf path that would visit three memory blocks. If P is stored in one memory block, the I/O complexity of the tree is 2 iff the paths p_i can be partitioned such that ever no part is stored in multiple memory blocks. There is such partition iff the instance of 3-partition is a yes instance.   □

6 Further Research

Finally, we propose several open problems and future research directions.

Experimental comparison of traditional graph layouts with the layouts presented in our work and layouts proposed in prior work could both direct and motivate further research in this area.

While we optimize the separable graph layout for random walks it is conceivable that a minor modification would also match the worst-case performance of the previous results.

The worst-case performance of the algorithm for finding the bit-offsets in Sect. 3.2 is most likely not optimal, and we suspect that the practical performance would be much better.

For the sake of simplicity, both our and prior representations of trees assume fixed vertex size (e.g. implicitly in the results on layouts) or allow $q = \mathcal{O}(1)$ extra bits per vertex in the compact separable graph representation. This could be generalized for vertices of different sizes and unbounded degrees.

References

1. Agarwal, P.K., Arge, L., Murali, T., Varadarajan, K.R., Vitter, J.S.: I/O-efficient algorithms for contour-line extraction and planar graph blocking. In: SODA, pp. 117–126 (1998)
2. Aggarwal, A., Chandra, A.K., Snir, M.: Hierarchical memory with block transfer. In: Foundations of Computer Science, pp. 204–216. IEEE (1987)
3. Aggarwal, A., Vitter, J.S.: The input/output complexity of sorting and related problems. Commun. ACM **31**(9), 1116–1127 (1988). https://doi.org/10.1145/48529.48535
4. Blandford, D.K., Blelloch, G.E., Kash, I.A.: Compact representations of separable graphs. In: Proceedings of the Fourteenth Annual ACM-SIAM Symposium on Discrete Algorithms, pp. 679–688. SIAM (2003)
5. Blelloch, G.E., Farzan, A.: Succinct representations of separable graphs. In: Amir, A., Parida, L. (eds.) CPM 2010. LNCS, vol. 6129, pp. 138–150. Springer, Heidelberg (2010). https://doi.org/10.1007/978-3-642-13509-5_13
6. Bollobas, B.: Modern Graph Theory. Springer, Heidelberg (2010). https://doi.org/10.1007/978-1-4612-0619-4

7. Bringmann, K., Keusch, R., Lengler, J.: Sampling geometric inhomogeneous random graphs in linear time. In: 25th Annual European Symposium on Algorithms, ESA 2017, 4–6 September 2017, Vienna, Austria, pp. 20:1–20:15 (2017)
8. Demaine, E.: Cache-oblivious algorithms and data structures. Lect. Notes EEF Summer Sch. Massive Data Sets **8**, 1–249 (2002)
9. Demaine, E.D., Iacono, J., Langerman, S.: Worst-case optimal tree layout in external memory. Algorithmica **72**(2), 369–378 (2015)
10. Dillabaugh, C., He, M., Maheshwari, A.: Succinct and I/O efficient data structures for traversal in trees. Algorithmica **63**(1), 201–223 (2012). https://doi.org/10.1007/s00453-011-9528-z
11. Dillabaugh, C., He, M., Maheshwari, A., Zeh, N.: I/O-efficient path traversal in succinct planar graphs. Algorithmica **77**(3), 714–755 (2017)
12. Farzan, A., Munro, J.I.: Succinct encoding of arbitrary graphs. Theor. Comput. Sci. **513**, 38–52 (2013). http://www.sciencedirect.com/science/article/pii/S0304397513007238
13. Frigo, M., Leiserson, C.E., Prokop, H., Ramachandran, S.: Cache-oblivious algorithms. In: Proceedings of the 40th Annual Symposium on Foundations of Computer Science, FOCS 1999, p. 285. IEEE Computer Society, Washington, DC (1999). http://dl.acm.org/citation.cfm?id=795665.796479
14. Gamerman, D., Lopes, H.F.: Markov Chain Monte Carlo: Stochastic Simulation for Bayesian Inference. Chapman and Hall/CRC, Boca Raton (2006)
15. Garey, M.R., Johnson, D.S.: Computers and Intractability; A Guide to the Theory of NP-Completeness. W. H. Freeman & Co., New York (1979)
16. Gil, J., Itai, A.: How to pack trees. J. Algorithms **32**(2), 108–132 (1999)
17. Gilbert, J.R., Hutchinson, J.P., Tarjan, R.E.: A separator theorem for graphs of bounded genus. J. Algorithms **5**(3), 391–407 (1984)
18. Goodrich, M.: Planar separators and parallel polygon triangulation. J. Comput. Syst. Sci. **51**(3), 374–389 (1995)
19. Grinstead, C.M., Snell, J.L.: Introduction to Probability. American Mathematical Society, Providence (2006)
20. Jampala, H., Zeh, N.: Cache-oblivious planar shortest paths. In: Caires, L., Italiano, G.F., Monteiro, L., Palamidessi, C., Yung, M. (eds.) ICALP 2005. LNCS, vol. 3580, pp. 563–575. Springer, Heidelberg (2005). https://doi.org/10.1007/11523468_46
21. Kannan, S., Naor, M., Rudich, S.: Implicit representation of graphs. SIAM J. Discrete Math. **5**(4), 596–603 (1992)
22. Kawarabayashi, K.I., Reed, B.: A separator theorem in minor-closed classes. In: 2010 IEEE 51st Annual Symposium on Foundations of Computer Science, pp. 153–162, October 2010
23. Klein, P., Mozes, S.: Optimization algorithms for planar graphs (no date). http://planarity.org/
24. Kowarschik, M., Weiß, C.: An overview of cache optimization techniques and cache-aware numerical algorithms. In: Meyer, U., Sanders, P., Sibeyn, J. (eds.) Algorithms for Memory Hierarchies. LNCS, vol. 2625, pp. 213–232. Springer, Heidelberg (2003). https://doi.org/10.1007/3-540-36574-5_10
25. Lanctot, M., Waugh, K., Zinkevich, M., Bowling, M.: Monte Carlo sampling for regret minimization in extensive games. In: Advances in Neural Information Processing Systems, pp. 1078–1086 (2009)
26. Leighton, T., Rao, S.: An approximate max-flow min-cut theorem for uniform multicommodity flow problems with applications to approximation algorithms. In: 29th Annual Symposium on Foundations of Computer Science, pp. 422–431, October 1988

27. Levin, D.A., Peres, Y.: Markov Chains and Mixing Times, 2nd edn. American Mathematical Society, Providence, Rhode Island (2017)
28. Lipton, R.J., Rose, D.J., Tarjan, R.E.: Generalized nested dissection. SIAM J. Numer. Anal. **16**(2), 346–358 (1979). http://www.jstor.org/stable/2156840
29. Lipton, R.J., Tarjan, R.E.: A separator theorem for planar graphs. SIAM J. Appl. Math. **36**(2), 177–189 (1979)
30. Maheshwari, A., Zeh, N.: I/O-optimal algorithms for planar graphs using separators. In: Proceedings of the Thirteenth Annual ACM-SIAM Symposium on Discrete Algorithms, pp. 372–381. Society for Industrial and Applied Mathematics (2002)
31. Miller, G.L., Teng, S.H., Thurston, W., Vavasis, S.A.: Geometric separators for finite-element meshes. SIAM J. Sci. Comput. **19**(2), 364–386 (1998)
32. Naor, M.: Succinct representation of general unlabeled graphs. Discrete Appl. Math. **28**(3), 303–307 (1990)
33. Schild, A., Sommer, C.: On balanced separators in road networks. In: Bampis, E. (ed.) Experimental Algorithms, pp. 286–297. Springer, Cham (2015). https://doi.org/10.1007/978-3-319-20086-6_22
34. Turán, G.: On the succinct representation of graphs. Discrete Appl. Math. **8**(3), 289–294 (1984)
35. Wang, J., Zheng, K., Jeung, H., Wang, H., Zheng, B., Zhou, X.: Cost-efficient spatial network partitioning for distance-based query processing. In: 2014 IEEE 15th International Conference on Mobile Data Management, vol. 1, pp. 13–22, July 2014

Unshuffling Permutations: Trivial Bijections and Compositions

Guillaume Fertin[1], Samuele Giraudo[2], Sylvie Hamel[3],
and Stéphane Vialette[2(✉)]

[1] Université de Nantes, LS2N (UMR 6004), CNRS, Nantes, France
guillaume.fertin@univ-nantes.fr
[2] Université Paris-Est, LIGM (UMR 8049), CNRS, ENPC, ESIEE Paris, UPEM,
77454 Marne-la-Vallée, France
{samuele.giraudo,stephane.vialette}@u-pem.fr
[3] Université de Montréal, DIRO, Montréal, QC, Canada
sylvie.hamel@umontreal.ca

Abstract. Given permutations π, σ_1 and σ_2, the permutation π (viewed as a string) is said to be a *shuffle* of σ_1 and σ_2, in symbols $\pi \in \sigma_1 \sqcup\!\sqcup \sigma_2$, if π can be formed by interleaving the letters of two strings p_1 and p_2 that are order-isomorphic to σ_1 and σ_2, respectively. Given a permutation $\pi \in S_{2n}$ and a bijective mapping $f : S_n \to S_n$, the f-UNSHUFFLE-PERMUTATION problem is to decide whether there exists a permutation $\sigma \in S_n$ such that $\pi \in \sigma \sqcup\!\sqcup f(\sigma)$. We consider here this problem for the following bijective mappings: inversion, reverse, complementation, and all their possible compositions. In particular, we present combinatorial results about the permutations accepted by this problem. As main results, we obtain that this problem is NP-complete when f is the reverse, the complementation, or the composition of the reverse with the complementation.

Keywords: Permutation · Shuffle product ·
Computational complexity

1 Introduction

Given permutations π, σ_1 and σ_2, π is said to be a *shuffle* of σ_1 and σ_2, in symbols $\pi \in \sigma_1 \sqcup\!\sqcup \sigma_2$, if π is the disjoint union of two patterns p_1 and p_2 (*i.e.*, if, viewed as a string, it can be formed by interleaving the letters of p_1 and p_2) that are order-isomorphic to σ_1 and σ_2, respectively. This shuffling operation $\sqcup\!\sqcup$ was introduced by Vargas [12] in an algebraic context and was called *supershuffle*. In case $\sigma = \sigma_1 = \sigma_2$, the permutation π is said to be a *square w.r.t. the shuffle product* (or simply a *square* when clear from the context), and σ is said to

Partially supported by Laboratoire International Franco-Québécois de Recherche en Combinatoire (LIRCO) (G. Fertin and S. Vialette), supported by the Individual Discovery Grant RGPIN-2016-04576 from Natural Sciences and Engineering Research Council of Canada (NSERC) (S. Hamel).

T. V. Gopal and J. Watada (Eds.): TAMC 2019, LNCS 11436, pp. 242–261, 2019.
https://doi.org/10.1007/978-3-030-14812-6_15

be a *square root* of π *w.r.t. the shuffle product* (or simply a *square root* of π when clear from the context). Note that a permutation may have several square roots. For example, $\pi = 18346752 \in 1234 \, \sqcup\!\sqcup \, 4321$ since π is a shuffle of the patterns $p_1 = 1347$ and $p_2 = 8652$ that are order-isomorphic to $\sigma_1 = 1234$ and $\sigma_2 = 4321$, respectively (as shown in $1_8 3 4 6 7_5 2$). However, π is not a square. Besides, $\pi' = 24317856$ is a square as it is a shuffle of the patterns 2175 and 4386 that are both order-isomorphic to 2143 (as shown in $2_4 3 1 7_8 5 6$). Note that 2143 is not the unique square root of π' since π' is also a shuffle of patterns 2156 and 4378 that are both order-isomorphic to 2134 (as shown in $2_4 3 1 7 8 5 6$).

The above definitions are of course intended to be natural counterparts to the ordinary shuffle of words and languages. Given words u, v_1 and v_2, u is said to be a *shuffle* of v_1 and v_2 (denoted $u \in v_1 \, \sqcup\!\sqcup \, v_2$), if u can be formed by interleaving the letters of v_1 and v_2 in a way that maintains the left-to-right ordering of the letter from each word. For example, $u = abccbabacbb \in abcab \, \sqcup\!\sqcup \, cbabcb$ (as shown in $ab_c c_{bab} a_c^{bb}$). Similarly, in case $v = v_1 = v_2$, the word u is said to be a *square w.r.t. the shuffle product*, and v is said to be a *square root of u w.r.t. the shuffle product*. For example, $abaaabaabb$ is a square w.r.t. the shuffle product since it belongs to the shuffle of $abaab$ with itself (as shown in $ab_a aa_{baa} b_b$). Given words u, v_1 and v_2, deciding whether $u \in v_1 \, \sqcup\!\sqcup \, v_2$ is $O(|u|^2/\log(|u|))$ time solvable [6]. To the best of our knowledge, the first $O(|u|^2)$ time algorithm for this problem appeared in [7]. However, deciding whether a given word u is a square w.r.t. the shuffle product is NP-complete [2,9]. Finally, for a given word u, deciding whether there exists a word v such that u is in the shuffle of v with its reverse v^R (*i.e.*, $u \in v \, \sqcup\!\sqcup \, v^R$) is NP-complete as well [9]. However, the problem is polynomial-time solvable for words built from a binary alphabet.

Coming back to permutations, by using a pattern avoidance criterion on directed perfect matchings, it is proved in [4] that recognizing permutations that are squares is NP-complete, and a bijection between $(213, 231)$-avoiding square permutations and square binary words is presented. Given $(3142, 2413)$-avoiding (*a.k.a. separable*) permutations π, σ_1 and σ_2, it can be decided in polynomial time whether $\pi \in \sigma_1 \, \sqcup\!\sqcup \, \sigma_2$ [8].

We finally observe that deciding whether a permutation is a shuffle of two monotone permutations is in P:

(i) *merge permutations* are the union of two increasing subsequences and are characterized by the fact that they contain no decreasing subsequence of length 3 [5], whereas (ii) *skew-merged permutations* are the union of an increasing subsequence with a decreasing subsequence and are characterized by the fact that they contain no subsequence ordered in the same way as 2143 or 3412 [11].

In this paper, we are interested in a generalization of the problem consisting in recognizing square permutations. We call this new problem the f-UNSHUFFLE-PERMUTATION problem, which is defined as follows. Given a permutation $\pi \in S_{2n}$ and a bijective mapping $f : S_n \rightarrow S_n$, the f-UNSHUFFLE-PERMUTATION problem asks whether there exists a permutation $\sigma \in S_n$ such that $\pi \in \sigma \, \sqcup\!\sqcup \, f(\sigma)$. We say in this case that π is a *generalized square permutation*. In case $f = \text{id}$, the id-UNSHUFFLE-PERMUTATION problem reduces to recognizing square permutations, and hence is NP-complete. This paper is devoted to studying the f-UNSHUFFLE-PERMUTATION problem in case f is either a trivial bijection (identity, complement, reverse or inverse) or obtained by composing trivial bijections. These bijections act on permutations by performing a transformation on their permutation matrices and can hence be seen as elements of the Dihedral group D_4, as discussed in the next section. The paper is organized as follows. In Sect. 2, we provide the needed definitions. Section 3 is devoted to presenting generalized square permutations and some associated enumerative properties. Finally, in Sect. 4, we show hardness of recognizing some generalized square permutations.

2 Definitions

For any nonnegative integer n, $[n]$ is the set $\{1, \ldots, n\}$. We follow the usual terminology on words [3]. Let us recall here the most important ones. Let u be a word. The length of u is denoted by $|u|$. The *empty word*, the only word of null length, is denoted by ϵ. For any $i \in [|u|]$, the i-th letter of u is denoted by $u(i)$. If I is a subset of $[|u|]$, $u_{|I}$ is the subword of u consisting in the letters of u at the positions specified by the elements of I. A *permutation* of size n is a word π of length n on the alphabet $[n]$ such that each letter admits exactly one occurrence. The set of all permutations of size n is denoted by S_n.

Definition 1 (Reduced form). *If λ is a list of distinct integers, the* reduced form *of λ, denoted* reduce(λ), *is the permutation obtained from λ by replacing its i-th smallest entry with i. For instance,* reduce(31845) = 21534.

Definition 2 (Order-isomorphism). *Let λ_1 and λ_2 be two words of distinct integers such that* reduce(λ_1) = reduce(λ_2). *We say that λ_1 and λ_2 are order-isomorphic and we denote it by $\lambda_1 \simeq \lambda_2$.*

Definition 3 (Pattern containment). *A permutation σ is said to be contained in, or to be a* subpermutation *of, another permutation π, written $\sigma \preceq \pi$, if π has a (not necessarily contiguous) subsequence whose terms are order-isomorphic to σ. We also say that π admits an occurrence of the pattern σ.*

Thus, $\sigma \preceq \pi$ if there is a set I of positions in π such that reduce($\pi_{|I}$) = σ. For example, $1423 \preceq 149362785$ since reduce $(149362785_{|\{2,3,5,8\}})$ = reduce (4968) = 1423.

Definition 4 (Trivial bijections). *Let $\pi = \pi(1)\pi(2)\ldots\pi(n)$ be a permutation of size n. The* reverse *of π is the permutation* $r(\pi) = \pi(n)\pi(n-1)\ldots\pi(1)$. *The* complement *of π is the permutation* $c(\pi) = \pi'(1)\pi'(2)\cdots\pi'(n)$, *where $\pi'(i) = n - \pi(i) + 1$. The* inverse *is the regular group theoretical inverse on permutations, that is the $\pi(i)$-th position of the inverse $i(\pi)$ is occupied by i. The reverse, complement and inverse are called the* trivial bijections *from S_n to itself.*

Abusing notations, for any $S \subseteq S_n$, we let c(S), i(S) and r(S) stand respectively for $\{c(\pi) : \pi \in S\}$, $\{i(\pi) : \pi \in S\}$ and $\{r(\pi) : \pi \in S\}$.

Definition 5 (Shuffle). *Let $\sigma_1 \in S_{k_1}$ and $\sigma_2 \in S_{k_2}$ be two permutations. A permutation $\pi = \pi(1)\pi(2)\ldots\pi(k_1 + k_2)$ is a* shuffle *of σ_1 and σ_2 if there is a subset I of $[n]$ such that $\pi_{|I} \simeq \sigma_1$ and $\pi_{|\bar{I}} \simeq \sigma_2$, where $\bar{I} = [n] \setminus I$ and $n = k_1 + k_2$. In other terms, π is obtained by interleaving the letters of two words respectively order-isomorphic with σ_1 and σ_2. We denote by $\sigma_1 \sqcup\!\sqcup \sigma_2$ the set of all shuffles of σ_1 and σ_2.*

For example,

$$12 \sqcup\!\sqcup 21 = \{1243, 1324, 1342, 1423, 1432, 2134, 2314, 2341, 2413, 2431,$$
$$3124, 3142, 3214, 3241, 3421, 4123, 4132, 4213, 4231, 4312\}.$$

We are now ready to define the f-UNSHUFFLE-PERMUTATION problem we focus on in this paper.

Definition 6 (f-Unshuffle-Permutation). *Given a permutation $\pi \in S_{2n}$ and a bijective mapping $f : S_n \to S_n$, the f-UNSHUFFLE-PERMUTATION problem is to decide whether there exists a permutation $\sigma \in S_n$ such that $\pi \in \sigma \sqcup\!\sqcup f(\sigma)$.*

3   Generalized Square Permutations and Enumerative Properties

The bijections c, i, and r are involutions and can be seen as particular elements of the dihedral group D_4 encoding all the symmetries of the square. Before explaining why, let us recall a definition for D_4.

The group D_4 is generated by two elements **a** and **b** subjected exactly to the nontrivial relations

$$\mathbf{aa} = \epsilon, \quad \mathbf{bbbb} = \epsilon, \quad \mathbf{abab} = \epsilon, \tag{1}$$

where ϵ is the unit of D_4. One can think of element **a** (resp. **b**) acting on a square by performing a symmetry through the vertical axis (resp. a 90° clockwise rotation). Now, we can regard, for any $n \in \mathbb{N}$, each element of D_4 as a map $\phi : S_n \to S_n$ such that $\phi(\pi)$, $\pi \in S_n$, is the permutation obtained by performing on its permutation matrix the transformations specified by ϕ. The maps c, i, and r can thus be expressed as

$$\mathbf{c} = \mathbf{bba}, \quad \mathbf{i} = \mathbf{ba}, \quad \mathbf{r} = \mathbf{a}. \tag{2}$$

Figure 1 shows the Cayley graph of this group. The main interest of seeing the three trivial bijections on S_n as elements of D_4 lies in the fact that all the other bijections obtained by composing the trivial ones can be expressed by compositions of the two elements **a** and **b**. This will allow us to gain concision in some proofs presented in the sequel.

We now turn to defining *identity squares*, *complement squares*, *reverse squares* and *inverse squares* that are natural generalizations of square permutations to trivial bijections from S_n to itself.

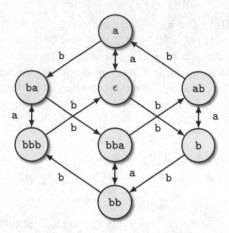

Fig. 1. The Cayley graph of the group D_4.

Definition 7 (Square permutations for trivial bijections). *For $n \in \mathbb{N}$, define*

$$S_{2n}^{\epsilon} = \{\pi \in S_{2n} : \exists \sigma \in S_n, \text{ such that } \pi \in \sigma \sqcup\!\sqcup \sigma\},$$
$$S_{2n}^{c} = \{\pi \in S_{2n} : \exists \sigma \in S_n \text{ such that } \pi \in \sigma \sqcup\!\sqcup \mathrm{c}(\sigma)\},$$
$$S_{2n}^{r} = \{\pi \in S_{2n} : \exists \sigma \in S_n \text{ such that } \pi \in \sigma \sqcup\!\sqcup \mathrm{r}(\sigma)\},$$
$$S_{2n}^{i} = \{\pi \in S_{2n} : \exists \sigma \in S_n \text{ such that } \pi \in \sigma \sqcup\!\sqcup \mathrm{i}(\sigma)\}.$$

We begin by proving that applying any trivial bijection is in some sense compatible with the shuffle operator.

Lemma 1 (Trivial bijections and shuffle). *Let π, σ_1 and σ_2 be permutations such that $\pi \in \sigma_1 \sqcup\!\sqcup \sigma_2$. Then*

- $\mathrm{c}(\pi) \in \mathrm{c}(\sigma_1) \sqcup\!\sqcup \mathrm{c}(\sigma_2)$;
- $\mathrm{r}(\pi) \in \mathrm{r}(\sigma_1) \sqcup\!\sqcup \mathrm{r}(\sigma_2)$;
- $\mathrm{i}(\pi) \in \mathrm{i}(\sigma_1) \sqcup\!\sqcup \mathrm{i}(\sigma_2)$.

Proof. Let n be the size of π. Since π is a shuffle of σ_1 and σ_2, there is a subset $I = \{i_1 < \cdots < i_k\}$ of $[n]$ such that $\pi_{|I} \simeq \sigma_1$ and $\pi_{|\bar{I}} \simeq \sigma_2$.

By setting $J = \{n - i_k + 1 < \cdots < n - i_1 + 1\}$, we have

$$\mathrm{r}(\pi)_{|J} = (\pi(n) \ldots \pi(1))_{|J} = \pi(i_k) \ldots \pi(i_1) \simeq \mathrm{r}(\sigma_1). \tag{3}$$

For the same reason, we have also $\mathrm{r}(\pi)_{|\bar{J}} \simeq \mathrm{r}(\sigma_2)$. Hence, $\mathrm{r}(\pi)$ is a shuffle of $\mathrm{r}(\sigma_1)$ and $\mathrm{r}(\sigma_2)$.

Now, set $J = \{\pi(i) : i \in I\} = \{\pi(\ell_1) < \cdots < \pi(\ell_k)\}$ where $\ell_1, \ldots, \ell_k$ are elements of I. We have

$$\mathrm{i}(\pi)_{|J} = \left(\pi^{-1}(1) \ldots \pi^{-1}(n)\right)_{|J} = \pi^{-1}(\pi(\ell_1)) \ldots \pi^{-1}(\pi(\ell_k)) = \ell_1 \ldots \ell_k \simeq \mathrm{i}(\sigma_1). \tag{4}$$

For the same reason, we have also $i(\pi)_{|J} \simeq i(\sigma_2)$. Hence, $i(\pi)$ is also shuffle of $i(\sigma_1)$ and $i(\sigma_2)$.

Finally, since $b = i \circ r$ in D_4, the map c can be expressed as compositions involving r and i. Hence, $c(\pi)$ is a shuffle of $c(\pi_1)$ and $c(\pi_2)$.   ∎

The following lemma is well-known and is a consequence of interpreting c, r, and i as compositions involving a and b according to (2) and the relations (1) between a and b.

Lemma 2 (Compositions of trivial bijections). *For any permutation π,*

- $(c \circ r)(\pi) = (r \circ c)(\pi);$
- $(i \circ c)(\pi) = (r \circ i)(\pi);$
- $(i \circ r)(\pi) = (c \circ i)(\pi).$

According to Lemma 2 (and the fact that the group D_4 has exactly eight elements), we thus only need to consider the following compositions in the rest of the paper.

Definition 8 (Square permutations for compositions of trivial bijections). *For $n \in \mathbb{N}$, define*

$$\mathcal{S}_{2n}^{\mathrm{rc}} = \{\pi \in S_{2n} : \exists \sigma \in S_n \text{ such that } \pi \in \sigma \amalg (r \circ c)(\sigma)\},$$
$$\mathcal{S}_{2n}^{\mathrm{ci}} = \{\pi \in S_{2n} : \exists \sigma \in S_n \text{ such that } \pi \in \sigma \amalg (c \circ i)(\sigma)\},$$
$$\mathcal{S}_{2n}^{\mathrm{ic}} = \{\pi \in S_{2n} : \exists \sigma \in S_n \text{ such that } \pi \in \sigma \amalg (i \circ c)(\sigma)\},$$
$$\mathcal{S}_{2n}^{\mathrm{irc}} = \{\pi \in S_{2n} : \exists \sigma \in S_n \text{ such that } \pi \in \sigma \amalg (i \circ r \circ c)(\sigma)\}.$$

We hence obtain the eight sets $\mathcal{S}_{2n}^{\epsilon}$, $\mathcal{S}_{2n}^{\mathrm{c}}$, $\mathcal{S}_{2n}^{\mathrm{r}}$, $\mathcal{S}_{2n}^{\mathrm{i}}$, $\mathcal{S}_{2n}^{\mathrm{rc}}$, $\mathcal{S}_{2n}^{\mathrm{ci}}$, $\mathcal{S}_{2n}^{\mathrm{ic}}$, and $\mathcal{S}_{2n}^{\mathrm{irc}}$ of square permutations provided by Definitions 7 and 8. Let us now investigate some easy bijective and enumerative properties satisfied by these sets.

Proposition 1 (Bijection between $\mathcal{S}_{2n}^{\mathrm{c}}$ and $\mathcal{S}_{2n}^{\mathrm{r}}$). *For any $n \in \mathbb{N}$, $i(\mathcal{S}_{2n}^{\mathrm{c}}) = \mathcal{S}_{2n}^{\mathrm{r}}$.*

Proof. Let $\pi \in S_{2n}$. We have

$$\pi \in \mathcal{S}_{2n}^{\mathrm{c}} \Leftrightarrow \exists \sigma \in S_n, \pi \in \sigma \amalg c(\sigma)$$
$$\Leftrightarrow \exists \sigma \in S_n, i(\pi) \in i(\sigma) \amalg (i \circ c)(\sigma) \qquad \text{(Lemma 1)}$$
$$\Leftrightarrow \exists \sigma \in S_n, i(\pi) \in i(\sigma) \amalg (r \circ i)(\sigma) \qquad \text{(Lemma 2)}$$
$$\Leftrightarrow \exists \sigma \in S_n, i(\pi) \in \sigma \amalg r(\sigma)$$
$$\Leftrightarrow i(\pi) \in \mathcal{S}_{2n}^{\mathrm{r}}.$$

∎

Proposition 2 (Bijection between $\mathcal{S}_{2n}^{\mathrm{ic}}$ and $\mathcal{S}_{2n}^{\mathrm{ci}}$). *For any $n \in \mathbb{N}$, $c\left(\mathcal{S}_{2n}^{\mathrm{ic}}\right) = \mathcal{S}_{2n}^{\mathrm{ci}}$, $i\left(\mathcal{S}_{2n}^{\mathrm{ic}}\right) = \mathcal{S}_{2n}^{\mathrm{ci}}$, and $r\left(\mathcal{S}_{2n}^{\mathrm{ic}}\right) = \mathcal{S}_{2n}^{\mathrm{ci}}$.*

Proof. Let $\pi \in S_{2n}$. We have

$$
\begin{aligned}
\pi \in \mathcal{S}_{2n}^{ic} &\Leftrightarrow \exists \sigma \in S_n, \ \pi \in \sigma \sqcup (i \circ c)(\sigma) \\
&\Leftrightarrow \exists \sigma \in S_n, \ c(\pi) \in c(\sigma) \sqcup (c \circ i \circ c)(\sigma) \quad &\text{(Lemma 1)} \\
&\Leftrightarrow \exists \sigma \in S_n, \ c(\pi) \in \sigma \sqcup (c \circ i)(\sigma) \\
&\Leftrightarrow c(\pi) \in \mathcal{S}_{2n}^{ci}.
\end{aligned}
$$

Moreover, we have

$$
\begin{aligned}
\pi \in \mathcal{S}_{2n}^{ic} &\Leftrightarrow \exists \sigma \in S_n, \ \pi \in \sigma \sqcup (i \circ c)(\sigma) \\
&\Leftrightarrow \exists \sigma \in S_n, \ i(\pi) \in i(\sigma) \sqcup (i \circ i \circ c)(\sigma) \quad &\text{(Lemma 1)} \\
&\Leftrightarrow \exists \sigma \in S_n, \ i(\pi) \in i(\sigma) \sqcup (i \circ r \circ i)(\sigma) \quad &\text{(Lemma 2)} \\
&\Leftrightarrow \exists \sigma \in S_n, \ i(\pi) \in \sigma \sqcup (i \circ r)(\sigma) \\
&\Leftrightarrow \exists \sigma \in S_n, \ i(\pi) \in \sigma \sqcup (c \circ i)(\sigma) \quad &\text{(Lemma 2)} \\
&\Leftrightarrow i(\pi) \in \mathcal{S}_{2n}^{ci}.
\end{aligned}
$$

Finally, the fact that $i \circ c \circ i = r$ implies that r is also a bijection between $\mathcal{S}_{2n}^{ic}$ and $\mathcal{S}_{2n}^{ci}$. ∎

Proposition 3 (Bijection between $\mathcal{S}_{2n}^{i}$ and $\mathcal{S}_{2n}^{irc}$). *For any $n \in \mathbb{N}$, $(i \circ c)\left(\mathcal{S}_{2n}^{irc}\right) = \mathcal{S}_{2n}^{i}$.*

Proof. Let $\pi \in S_{2n}$. We have

$$
\begin{aligned}
\pi \in \mathcal{S}_{2n}^{irc} &\Leftrightarrow \exists \sigma \in S_n, \ \pi \in \sigma \sqcup (i \circ r \circ c)(\sigma) \\
&\Leftrightarrow \exists \sigma \in S_n, \ i(\pi) \in i(\sigma) \sqcup (r \circ c)(\sigma) \quad &\text{(Lemma 1)} \\
&\Leftrightarrow \exists \sigma \in S_n, \ (r \circ i)(\pi) \in (r \circ i)(\sigma) \sqcup c(\sigma) \quad &\text{(Lemma 1)} \\
&\Leftrightarrow \exists \sigma \in S_n, \ (i \circ c)(\pi) \in (i \circ c)(\sigma) \sqcup c(\sigma) \quad &\text{(Lemma 2)} \\
&\Leftrightarrow \exists \sigma \in S_n, \ (i \circ c)(\pi) \in i(\sigma) \sqcup \sigma \\
&\Leftrightarrow (i \circ c)(\pi) \in \mathcal{S}_{2n}^{i}.
\end{aligned}
$$

∎

To simplify notations, we write s_{2n}^{ϵ}, s_{2n}^{c}, s_{2n}^{r}, s_{2n}^{i}, s_{2n}^{rc}, s_{2n}^{ci}, s_{2n}^{ic}, and s_{2n}^{irc} for the cardinalities of the sets $\mathcal{S}_{2n}^{\epsilon}$, $\mathcal{S}_{2n}^{c}$, $\mathcal{S}_{2n}^{r}$, $\mathcal{S}_{2n}^{i}$, $\mathcal{S}_{2n}^{rc}$, $\mathcal{S}_{2n}^{ci}$, $\mathcal{S}_{2n}^{ic}$, and $\mathcal{S}_{2n}^{irc}$, respectively. Table 1 shows the first cardinalities. We note that, at this time, none of the corresponding integer sequence does appear in OEIS [10].

Lemma 3 (Upper bound for the number of generalized squares). *For any $n \in \mathbb{N}$,*

$$
|Q_{2n}| \le \binom{2n-1}{n-1}\binom{2n}{n} n!, \tag{5}
$$

where Q_{2n} is any of the sets $\mathcal{S}_{2n}^{\epsilon}$, $\mathcal{S}_{2n}^{c}$, $\mathcal{S}_{2n}^{r}$, $\mathcal{S}_{2n}^{i}$, $\mathcal{S}_{2n}^{rc}$, $\mathcal{S}_{2n}^{ci}$, $\mathcal{S}_{2n}^{ic}$, or $\mathcal{S}_{2n}^{irc}$.

Table 1. Cardinalities of the sets of square permutations of sizes $0 \leq 2n \leq 10$.

n	0	1	2	3	4	5
s_{2n}^{rc}	1	2	20	472	18 988	1 112 688
$s_{2n}^{c} = s_{2n}^{r}$	1	2	20	480	19 744	1 185 264
$s_{2n}^{i} = s_{2n}^{irc}$	1	2	20	488	20 250	1 229 858
s_{2n}^{ϵ}	1	2	20	504	21 032	1 293 418
$s_{2n}^{ci} = s_{2n}^{ic}$	1	2	20	586	27 990	2 044 596
$(2n)!$	1	2	24	720	40 320	3 628 800

Proof. This is a consequence of the fact that to construct a permutation π of Q_{2n}, we can proceed by first choosing a set of n letters among the alphabet $[2n]$ ($\binom{2n}{n}$ choices), then by specifying an order from left to right for these letters ($n!$ choices), and finally by deploying the chosen letters onto a set I of n positions among the set of all possible positions $[2n]$ ($\binom{2n}{n}$ choices). The empty positions are filled by the unique authorized completion such that $\pi_{|I} \simeq \phi\left(\pi_{|\bar{I}}\right)$ where ϕ is the considered bijection on S_n. For this reason, there is no more than $\binom{2n}{n}^2 n!$ elements in Q_{2n}. The tighter bound (5) is the consequence of the fact that we can assume that the first position can always be fixed. ∎

Let us denote by $\mathcal{S}^{\bullet}{}_{2n}$ the union of all the sets $\mathcal{S}_{2n}^{\epsilon}$, $\mathcal{S}_{2n}^{c}$, $\mathcal{S}_{2n}^{r}$, $\mathcal{S}_{2n}^{i}$, $\mathcal{S}_{2n}^{rc}$, $\mathcal{S}_{2n}^{ci}$, $\mathcal{S}_{2n}^{ic}$, and $\mathcal{S}_{2n}^{irc}$. We show that S_{2n} and $\mathcal{S}^{\bullet}{}_{2n}$ do not coincide.

Proposition 4 (Existence of non-square permutations). *For any $n \in \mathbb{N}$, $n \geq 9, S_{2n} \neq \mathcal{S}^{\bullet}{}_{2n}$.*

Proof. By Lemma 3 (and using its notations), we have

$$|\mathcal{S}^{\bullet}{}_{2n}| \leq 8\,|Q_{2n}| \leq 8\binom{2n-1}{n-1}\binom{2n}{n}n!. \tag{6}$$

The proportion of the elements of $\mathcal{S}^{\bullet}{}_{2n}$ among all the permutations of S_{2n} admits as upper bound, when $n \geq 1$,

$$\frac{|\mathcal{S}^{\bullet}{}_{2n}|}{|S_{2n}|} \leq \frac{8\binom{2n-1}{n-1}\binom{2n}{n}n!}{(2n)!} = 4\binom{2n}{n}\frac{1}{n!}. \tag{7}$$

But $4\binom{2n}{n}\frac{1}{n!} < 1$ for $n \geq 9$, thereby proving the result. ∎

Proposition 5 (Square permutations in some intersections). *Let $n \in \mathbb{N}$ such that $2n \equiv 0 \pmod 4$. Then,*
the permutation $\pi = 12\ldots(n-1)(n)\ (2n)(2n-1)\ldots(n+2)(n+1)$
belongs to $\mathcal{S}_{2n}^{\epsilon} \cap \mathcal{S}_{2n}^{r} \cap \mathcal{S}_{2n}^{c} \cap \mathcal{S}_{2n}^{i} \cap \mathcal{S}_{2n}^{ci} \cap \mathcal{S}_{2n}^{ic}$.

Proof. Let $I = \{1, 3, 5, \ldots, 2n - 1\}$ and $\bar{I} = [2n] \setminus I$. Then, since

$$\pi_{|I} \simeq 12 \ldots \frac{n}{2} \; n(n-1) \ldots \left(\frac{n}{2} + 1\right) \simeq \pi_{|\bar{I}}, \tag{8}$$

we have $\pi_{|I} \simeq \pi_{|\bar{I}} \simeq \mathrm{i}(\pi_{|\bar{I}})$. This shows that π belongs to $\mathcal{S}_{2n}^{\epsilon} \cap \mathcal{S}_{2n}^{\mathrm{i}}$.

Now, let $J = \{1, 2, \ldots, n\}$ and $\bar{J} = [2n] \setminus J$. Then, since

$$\pi_{|J} \simeq 12 \ldots n \quad \text{and} \quad \pi_{|\bar{J}} \simeq n \ldots 21, \tag{9}$$

we have $\pi_{|J} \simeq \mathrm{r}(\pi_{|\bar{J}}) \simeq \mathrm{c}(\pi_{|\bar{J}}) \simeq (\mathrm{i} \circ \mathrm{c})(\pi_{|\bar{J}}) \simeq (\mathrm{c} \circ \mathrm{i})(\pi_{|\bar{J}})$. This shows that π belongs to $\mathcal{S}_{2n}^{\mathrm{r}} \cap \mathcal{S}_{2n}^{\mathrm{c}} \cap \mathcal{S}_{2n}^{\mathrm{ic}} \cap \mathcal{S}_{2n}^{\mathrm{ci}}$. ∎

4 Recognizing Generalized Square Permutations

This section is devoted to proving that deciding membership to $\mathcal{S}_{2n}^{\mathrm{r}}$, $\mathcal{S}_{2n}^{\mathrm{c}}$ and $\mathcal{S}_{2n}^{\mathrm{rc}}$ is NP-complete. The approach relies on constraint directed matchings on permutations.

Definition 9 (Directed matching on permutations). *Let $\pi \in S_n$. A directed matching on π is a set $\mathcal{M}$ of pairwise disjoint arcs (i, j), $1 \leq i, j \leq n$ and $i \neq j$, that connect pairs of elements of π. The directed matching $\mathcal{M}$ is perfect if every element of π is the source or the sink of an arc of $\mathcal{M}$.*

Definition 10 (Two-arcs pattern occurrences). *A* two-arc pattern *is a perfect directed matching on the set* [4]. *A perfect directed matching $\mathcal{M}$ on a permutation π of size $2n$ admits an occurrence of a two-arc pattern $\mathcal{A}$ if there is an increasing map $\phi : [4] \to [2n]$ (i.e., $i < i'$ implies $\phi(i) < \phi(i')$) such that, if (i, i') is an arc of $\mathcal{A}$, then $(\phi(i), \phi(i'))$ is an arc of $\mathcal{M}$. We say moreover that $\mathcal{M}$* avoids *$\mathcal{A}$ if $\mathcal{A}$ does not admits any occurrence of $\mathcal{A}$.*

We shall draw two-arcs patterns by unlabeled graphs wherein vertices are implicitly indexed from 1 to 4 from left to right. Figure 2 shows an example of a directed perfect matching on a permutation. The directed matching does contain the patterns ⟅⟆, ⟅⟆, ⟅⟆ ⟅⟆ and ⟅⟆ and does avoid the patterns ⟅⟆, ⟅⟆, ⟅⟆ ⟅⟆, ⟅⟆ ⟅⟆, ⟅⟆ and ⟅⟆.

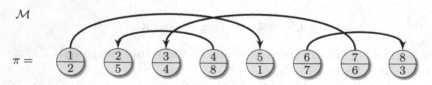

Fig. 2. A directed perfect matching $\mathcal{M} = \{(1, 5), (4, 2), (7, 3), (6, 8)\}$ on the permutation $\pi = 25481763$. Recall that arcs refer to positions in π.

In case π is written as a concatenation of contiguous patterns $\pi = \pi_1 \pi_2 \cdots \pi_k$, we write $\mathcal{M}_{\pi_i \to \pi_j}$ for the subset of arcs of $\mathcal{M}$ with source index in π_i and sink index in π_j. Hence,

$$\mathcal{M} = \bigsqcup_{\substack{i \in [k] \\ j \in [k]}} \mathcal{M}_{\pi_i \to \pi_j}. \tag{10}$$

The following properties will prove useful for defining equivalences between squares and restricted directed perfect matchings.

Definition 11 (Property $\mathbf{P_1}$—Order isomorphism). *Let π be a permutation. A directed perfect matching $\mathcal{M}$ on π is said to have property $\mathbf{P_1}$ if, for any two distinct arcs (i, i') and (j, j') of $\mathcal{M}$, we have $\pi(i) < \pi(j)$ if and only if $\pi(i') < \pi(j')$.*

Definition 12 (Property $\mathbf{P_2}$—Order anti-isomorphism). *Let π be a permutation. A directed perfect matching $\mathcal{M}$ on π is said to have property $\mathbf{P_2}$ if, for any two distinct arcs (i, i') and (j, j') of $\mathcal{M}$, we have $\pi(i) < \pi(j)$ if and only if $\pi(i') > \pi(j')$.*

Definition 13 (Property $\mathbf{Q_1}$—First start first terminal). *Let π be a permutation. A directed perfect matching $\mathcal{M}$ on π is said to have property $\mathbf{Q_1}$ if it avoids the following set of two-arcs patterns:*

$$\mathcal{Q}_1 = \left\{ \text{⌢⌢}, \text{⌢⌢}, \text{⌢⌢}, \text{⌢⌢}, \text{⌢⌢}, \text{⌢⌢} \right\}.$$

Definition 14 (Property $\mathbf{Q_2}$ —First start last terminal). *Let π be a permutation. A directed perfect matching $\mathcal{M}$ on π is said to have property $\mathbf{Q_2}$ if it avoids the following set of two-arcs patterns:*

$$\mathcal{Q}_2 = \left\{ \text{⌢⌢}, \text{⌢⌢}, \text{⌢⌢}, \text{⌢⌢}, \text{⌢⌢}, \text{⌢⌢} \right\}.$$

Observe that $\mathcal{Q}_1 \sqcup \mathcal{Q}_2$ is the set of all two-arcs patterns.

Theorem 1 ([4]). *Let $\pi \in S_{2n}$. The two following statements are equivalent.*

1. *$\pi \in \mathcal{S}_{2n}^{\epsilon}$.*
2. *There exists a directed perfect matching $\mathcal{M}$ on π that satisfies properties $\mathbf{P_1}$ and $\mathbf{Q_1}$.*

The following two definitions will intervene in the next constructions.

Definition 15 (Lifting). *Let $\pi = \pi(1)\,\pi(2)\,\ldots\,\pi(n)$ be a permutation of size n and k be a positive integer. We call k-lifting of π, denoted $\pi[k]$, the permutation $(k + \pi(1))\,(k + \pi(2))\,\ldots\,(k + \pi(n))$ on the alphabet $\{k+1, \ldots, k+n\}$.*

Definition 16 (Monotone). *For any positive integer k, we let $\nearrow_k$ stand for the increasing permutation $1\,2\,\ldots\,k$ and $\searrow_{k}$ stand for the decreasing permutation $k\,(k-1)\ldots 1$.*

Lemma 4 (One shot lemma). *Let $\pi \in S_{2n}$ and $\mathcal{M}$ be a perfect directed matching on π that satisfies properties $\mathbf{P}_i$ and $\mathbf{Q}_j$, $i \in \{1,2\}$ and $j \in \{1,2\}$. Then, the perfect directed matching $\mathcal{M}'$ obtained by reversing each arc of $\mathcal{M}$ satisfies properties $\mathbf{P}_i$ and $\mathbf{Q}_j$.*

Lemma 5. *Let $I^1 = \{i_j^1 : j \in [n]\}$, $i_1^1 < i_2^1 < \cdots < i_n^1$, and $I^2 = \{i_j^2 : j \in [n]\}$, $i_1^2 < i_2^2 < \cdots < i_n^2$, be such that $I^1 \cap I^2 = \emptyset$. The directed perfect matching $\mathcal{M} = ([2n], E)$ defined by $E = \big\{(i_j^1, i_j^2) : j \in [n]\big\}$ avoids the patterns* ⌢⌢, ⌢⌢, ⌢⌢, ⌢⌢, ⌢⌢, *and* ⌢⌢.

Proof. Suppose, aiming at a contradiction, that $\mathcal{M}$ does contain the pattern ⌢⌢, ⌢⌢, ⌢⌢, ⌢⌢, ⌢⌢, or ⌢⌢, say for arcs (i_j^1, i_j^2) and (i_k^1, i_k^2). Wlog, assume $i_j^1 < i_k^1$ (see Fig. 3). This is a contradiction since $i_j^1 < i_k^1$ implies $j < k$, and hence, $i_j^2 < i_k^2$. ∎

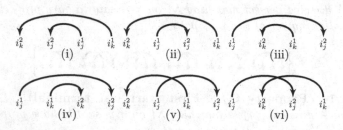

Fig. 3. Arcs (i_j^1, i_j^2) and (i_k^1, i_k^2) with $i_j^1 < i_k^2$.

Lemma 6. *Let $I^1 = \{i_j^1 : j \in [n]\}$, $i_1^1 < i_2^1 < \cdots < i_n^1$, and $I^2 = \{i_j^2 : j \in [n]\}$, $i_1^2 < i_2^2 < \cdots < i_n^2$, be such that $I^1 \cap I^2 = \emptyset$. The directed perfect matching $\mathcal{M} = ([2n], E)$ defined by $E = \big\{(i_j^1, i_{n-j+1}^2) : j \in [n]\big\}$ avoids the patterns* ⌢⌢, ⌢⌢, ⌢⌢, ⌢⌢, ⌢⌢, *and* ⌢⌢.

Proof. Suppose, aiming at a contradiction, that $\mathcal{M}$ does contain the pattern ⌢⌢, ⌢⌢, ⌢⌢, ⌢⌢, ⌢⌢, or ⌢⌢, say for arcs (i_j^1, i_{n-j+1}^2) and (i_k^1, i_{n-k+1}^2). Wlog, assume $i_j^1 < i_k^1$ (see Fig. 4). This is a contradiction since $i_j^1 < i_k^1$ implies $j < k$, and hence, $i_{n-j+1}^2 > i_{n-k+1}^2$. ∎

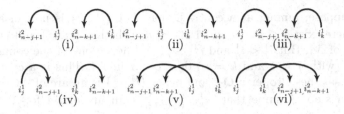

Fig. 4. Arcs (i_j^1, i_{n-j+1}^2) and (i_k^1, i_{n-k+1}^2) with $i_j^1 < i_k^2$.

Proposition 6. *Let $\pi \in S_{2n}$. The two following statements are equivalent.*

1. $\pi \in \mathcal{S}_{2n}^r$.
2. *There exists a directed perfect matching $\mathcal{M}$ on π that satisfies properties $\mathbf{P_1}$ and $\mathbf{Q_2}$.*

Proof. $(1 \Rightarrow 2)$ Let $\pi \in \mathcal{S}_{2n}^r$. Then, there exists $\sigma \in S_n$ such that $\pi \in \sigma \, \shuffle \, \mathrm{r}(\sigma)$, and hence there exist two disjoint set of indexes $I^1 = \{i_j^1 : j \in [n]\}$, $i_1^1 < i_2^1 < \cdots < i_n^1$, and $I^2 = \{i_j^2 : j \in [n]\}$, $i_1^2 < i_2^2 < \cdots < i_n^2$, such that $\pi(i_1^1)\pi(i_2^1)\cdots\pi(i_n^1)$ is order-isomorphic to σ and $\pi(i_1^2)\pi(i_2^2)\cdots\pi(i_n^2)$ is order-isomorphic to $\mathrm{r}(\sigma)$. Let $\mathcal{M} = (V, E)$ be the directed graph defined by $V = [2n]$ and $E = \{(i_j^1, i_{n-j+1}^2) : j \in [n]\}$. Clearly, $\mathcal{M}$ is a directed perfect matching since $I^1 \cap I^2 = \emptyset$ and $I^1 \cup I^2 = [2n]$. According to Lemma 6, $\mathcal{M}$ does avoid the patterns ⌢⌢, ⌢⌢, ⌢⌢, ⌢⌢, ⌢⌢⌢, and ⌢⌢⌢. Finally, for any two distinct arcs (i_j^1, i_{n-j+1}^2) and (i_k^1, i_{n-k+1}^2) of $\mathcal{M}$, we have $\pi\left(i_j^1\right) < \pi\left(i_k^1\right)$ if and only if $\pi\left(i_{n-j+1}^2\right) < \pi\left(i_{n-k+1}^2\right)$ since $\pi(i_1^1)\pi(i_2^1)\cdots\pi(i_n^1)$ is order-isomorphic to σ and $\pi(i_1^2)\pi(i_2^2)\cdots\pi(i_n^2)$ is order-isomorphic to $\mathrm{r}(\sigma)$. Therefore, $\mathcal{M}$ satisfies properties $\mathbf{P_1}$ and $\mathbf{Q_2}$.

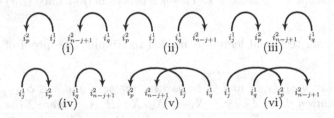

Fig. 5. All possible configurations considered in Proof of Proposition 6 $(2 \Rightarrow 1)$.

$(2 \Rightarrow 1)$ Let $\mathcal{M}$ be a directed perfect matching that satisfies properties $\mathbf{P_1}$ and $\mathbf{Q_2}$. Let $I^1 = \{i_j^1 : j \in [n]\}$, $i_1^1 < i_2^1 < \cdots < i_n^1$, be the set of the sources of the arcs of $\mathcal{M}$ and let $I^2 = \{i_j^2 : j \in [n]\}$, $i_1^2 < i_2^2 < \cdots < i_n^2$, be the set of the sinks of the arcs of $\mathcal{M}$. We first show that, for every $j \in [n]$, $\left(i_j^1, i_{n-j+1}^2\right)$ is an arc of $\mathcal{M}$. The proof is by induction on $j = 1, 2, \ldots, n$.

– Base. Suppose, aiming at a contradiction that (i_1^1, i_n^2) is not an arc of $\mathcal{M}$. Then, there exist $1 \le p < n$ and $1 < q \le n$ such that (i_1^1, i_p^2) and (i_q^1, i_n^2) are two arcs of $\mathcal{M}$. But $i_1^1 < i_q^1$ and $i_p^2 < i_n^2$, and hence one of the configurations of Fig. 5 (with $j = 1$ and $k = n$) does occur in $\mathcal{M}$. This is a contradiction since $\mathcal{M}$ satisfies Property $\mathbf{Q_2}$, and hence (i_1^1, i_n^2) is an arc of $\mathcal{M}$.

– Induction step. Assume that (i_k^1, i_{n-k+1}^2) is an arc of $\mathcal{M}$ for $1 \le k < j$, and suppose, aiming at a contradiction, that (i_j^1, i_{n-j+1}^2) is not an arc of $\mathcal{M}$. Then, there exist $1 \le p < n - j + 1$ and $j < q \le n$ such that (i_j^1, i_p^2) and (i_q^1, i_{n-j+1}^2) are two arcs of $\mathcal{M}$. But $i_j^1 < i_q^1$ and $i_p^2 < i_{n-j+1}^2$, and hence one of the configurations of Fig. 5 does occur in $\mathcal{M}$. This is a contradiction since $\mathcal{M}$ satisfies Property $\mathbf{Q_2}$, and hence (i_j^1, i_{n-j+1}^2) is an arc of $\mathcal{M}$.

Now, let σ_1 be the pattern of π induced by the sources of $\mathcal{M}$ and σ_2 be the pattern of π induced by the sinks of $\mathcal{M}$ (i.e., $\sigma_1 = \pi\left(i_1^1\right)\pi\left(i_2^1\right)\ldots\pi\left(i_n^1\right)$ and $\sigma_2 = \pi\left(j_1^2\right)\pi\left(j_2^2\right)\ldots\pi\left(i_n^2\right)$). Clearly σ_1 and σ_2 are disjoint in π (since $\mathcal{M}$ is a matching) and cover π (since $\mathcal{M}$ is perfect). Moreover, the fact that $\mathcal{M}$ satisfies $\mathbf{P_1}$ implies immediately that reduce(σ_1) is order-isomorphic to (reduce $\circ$ r)(σ_2), and hence this shows that $\pi \in \mathcal{S}_{2n}^r$. ∎

Lemma 7. *Let $\pi \in S_{2n}$ and $\mathcal{M}$ be a perfect directed matching on π that avoids the patterns* ⋂⋂, ⋂⋂, ⋂⋂, *and* ⋂⋂. *Then, for any arc $(i,j) \in \mathcal{M}$, either $1 \le i \le n < j \le 2n$ or $1 \le j \le n < i \le 2n$.*

Proof. We only prove the case $1 \le i \le n < j \le 2n$ since the proof for $1 \le j \le n < i \le 2n$ is similar.

Suppose, aiming at a contradiction, that there exists an arc $(i,j) \in \mathcal{M}$ with $1 \le i < j \le n$. Since $\mathcal{M}$ avoids the patterns ⋂⋂, ⋂⋂, ⋂⋂ and ⋂⋂, there is no arc $(k, \ell) \in \mathcal{M}$ with $j < k \le 2n$ and $j < \ell \le 2n$, $k \ne \ell$. Then it follows that there exist $2n - j \ge n$ distinct arcs $(k, \ell) \in \mathcal{M}$ with $1 \le \min\{k, \ell\} \le j - 1$ and $j + 1 \le \max\{k, \ell\} \le 2n$, $k \ne \ell$. This is a contradiction since $(i,j) \in \mathcal{M}$ with $1 \le i < j \le n$. ∎

The following technical lemma is needed to simplify the proof of upcoming Proposition 7.

Lemma 8. *Let $k, \ell_1, \ell_2, \ell_3$ and ℓ_4 be positive integers and $\pi \in \mathcal{S}_{2k+\ell_1+\ell_2+\ell_3+\ell_4}^r$ be the permutation defined by $\pi = X_1 \ L \ X_2 \ X_3 \ R \ X_4$, where $X_1 = \mu_1[k+\ell_2]$, $L = \nearrow_k [\ell_2]$, $X_2 = \mu_2$, $X_3 = \mu_3[k + \ell_1 + \ell_2]$, $R = \searrow_k[k + \ell_1 + \ell_2 + \ell_3]$ and $X_4 = \mu_4[2k + \ell_1 + \ell_2 + \ell_3]$ for some permutations $\mu_i \in S_{\ell_i}$, $1 \le i \le 4$ (see Fig. 6). Let $\mathcal{M}$ be a directed perfect matching on π that satisfies properties $\mathbf{P_1}$ and $\mathbf{Q_2}$. If $\mathcal{M}_{L \to R} \ne \emptyset$ or $\mathcal{M}_{R \to L} \ne \emptyset$, then either $|\mathcal{M}_{L \to R}| = k$ or $|\mathcal{M}_{R \to L}| = k$.*

Proof. Suppose first, aiming at a contradiction, that $\mathcal{M}_{L \to R} \ne \emptyset$ and $\mathcal{M}_{R \to L} \ne \emptyset$. Let $(i,j) \in \mathcal{M}_{L \to R}$ and $(i', j') \in \mathcal{M}_{R \to L}$. By construction, we have $\pi(i) < \pi(i')$ and $\pi(j) > \pi(j')$ thereby contradicting Property $\mathbf{P_1}$. Then it follows that either $\mathcal{M}_{L \to R} \ne \emptyset$ or $\mathcal{M}_{R \to L} \ne \emptyset$, but not both.

Suppose now that $\mathcal{M}_{L \to R} \neq \emptyset$ and $\mathcal{M}_{R \to L} = \emptyset$ (the case $\mathcal{M}_{L \to R} = \emptyset$ and $\mathcal{M}_{R \to L} \neq \emptyset$ can be proved with the same arguments). Suppose, aiming at a contradiction, that $|\mathcal{M}_{L \to R}| \neq k$ and let $(i, j) \in \mathcal{M}_{L \to R}$. According to Lemma 7, we are left with considering the four following cases.

- There exists $(i', j') \in \mathcal{M}_{L \to \mu_3}$. Since $\mathcal{M}$ avoids ⁀⁀, we have $i < i'$, and hence $\pi(i) < \pi(i')$ and $\pi(j) > \pi(j')$. This is a contradiction since $\mathcal{M}$ satisfies Property $\mathbf{P_1}$.
- There exists $(i', j') \in \mathcal{M}_{\mu_3 \to L}$ and hence $\pi(i) < \pi(i')$ and $\pi(j) > \pi(j')$. This is a contradiction since $\mathcal{M}$ satisfies Property $\mathbf{P_1}$.
- There exists $(i', j') \in \mathcal{M}_{L \to \mu_4}$. Since $\mathcal{M}$ avoids ⁀⁀, we have $i' < i$, and hence $\pi(i) > \pi(i')$ and $\pi(j) < \pi(j')$. This is a contradiction since $\mathcal{M}$ satisfies Property $\mathbf{P_1}$.
- There exists $(i', j') \in \mathcal{M}_{\mu_4 \to L}$ and hence $\pi(i) < \pi(i')$ and $\pi(j) > \pi(j')$. This is a contradiction since $\mathcal{M}$ satisfies Property $\mathbf{P_1}$.

Therefore, $|\mathcal{M}_{L \to R}| = k$. ∎

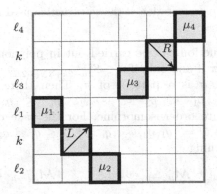

Fig. 6. Illustration of Lemma 8, where $\pi = X_1 \ L \ X_2 \ X_3 \ R \ X_4$.

Proposition 7. r-UNSHUFFLE-PERMUTATION *is* NP-*complete.*

Proof. We reduce from PERMUTATION PATTERN which, given two permutations $\pi \in S_n$ and $\sigma \in S_k$ with $k \leq n$, asks whether σ is a pattern of π. PERMUTATION PATTERN is known to be NP-complete [1]. Let $\pi \in S_n$ and $k \in S_k$ be an instance of PERMUTATION PATTERN. Set $N_2 = n + k + 1$ and $N_1 = 2N_2$. Define a target permutation $\mu \in S_{2n+2k+2N_1+2N_2}$ by $\mu = \mu_\pi \ \mu_2 \ \mu_1 \ \mu_\sigma \ \mu'_2 \ \mu'_\pi \ \mu'_1 \ \mu'_\sigma$, where

$$\mu_\sigma = \sigma, \qquad\qquad \mu_\pi = \pi[k + N_1 + N_2],$$
$$\mu'_\pi = \mathrm{r}(\pi)[2n + k + 2N_1 + 2N_2] \qquad \mu'_\sigma = \mathrm{r}(\sigma)[n + k + N_1 + 2N_2],$$
$$\mu_1 = \nearrow_{N_1}[k], \qquad\qquad \mu'_1 = \searrow_{N_1}[2n + k + N_1 + 2N_2],$$
$$\mu_2 = \nearrow_{N_2}[k + N_1], \qquad\qquad \mu'_2 = \searrow_{N_2}[n + k + N_1 + N_2].$$

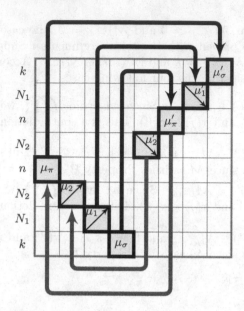

Fig. 7. Schematic representation of the construction used in Proposition 7.

Clearly, the construction can be carried out in polynomial time. We claim that σ is a pattern of π if and only if $\mu \in \mathcal{S}^{\mathrm{r}}_{2n+2k+2N_1+2N_2}$ (Fig. 7).

($\Rightarrow$) Suppose that σ is a pattern of π. Then there exist indices $P = \{p_1, p_2, \ldots, p_k\}$, $1 \le p_1 < p_2 < \cdots < p_k \le n$, such that σ and $\pi(p_1)\,\pi(p_2)\cdots\pi(p_k)$ are order-isomorphic. For the sake of simplification, let $Q = [n] \setminus P$ and write $Q = \{q_1, q_2, \cdots, q_{n-k}\}$, $1 \le q_1 < q_2 < \cdots < q_{n-k} \le n$. Define a directed matching

$$\mathcal{M} = \mathcal{M}_{\mu_\sigma \to \mu'_\pi} \uplus \mathcal{M}_{\mu'_\pi \to \mu_\pi} \uplus \mathcal{M}_{\mu_\pi \to \mu'_\sigma} \uplus \mathcal{M}_{\mu_1 \to \mu'_1} \uplus \mathcal{M}_{\mu_2 \to \mu'_2}$$

on μ as follows.

$$\mathcal{M}_{\mu_\sigma \to \mu'_\pi} = \{(n + N_1 + N_2 + i, n + k + N_1 + 2N_2 + p_i) : 1 \le i \le k\}$$
$$\mathcal{M}_{\mu'_\pi \to \mu_\pi} = \{(2n + k + N_1 + 2N_2 - q_i + 1, q_i) : 1 \le i \le n - k\}$$
$$\mathcal{M}_{\mu_\pi \to \mu'_\sigma} = \{(p_i, 2n + 2k + 2N_1 + 2N_2 - i) : 1 \le i \le k\}$$
$$\mathcal{M}_{\mu_1 \to \mu'_1} = \{(n + N_2 + i, 2n + k + 2N_1 + 2N_2 - i + 1) : 1 \le i \le N_1\}$$
$$\mathcal{M}_{\mu'_2 \to \mu_2} = \{(n + k + N_1 + 2N_2 - i + 1, n + i) : 1 \le i \le N_2\}.$$

Informally,

- $\mathcal{M}_{\mu_\sigma \to \mu'_\pi}$ is the directed (left-to-right) matching describing the (reversed) occurrence of σ in $\mathrm{r}(\pi)$,
- $\mathcal{M}_{\mu'_\pi \to \mu_\pi}$ is the directed (right-to-left) matching connecting the elements of $\mathrm{r}(\pi)$ and π that are not part of the occurrence of σ in π,
- $\mathcal{M}_{\mu_\pi \to \mu'_\sigma}$ is the directed (left-to-right) matching describing the (reversed) occurrence of $\mathrm{r}(\sigma)$ in π,

- $\mathcal{M}_{\mu_1 \to \mu_1'}$ is a directed (left-to-right) matching that fully connects μ_1 to μ_1' and
- $\mathcal{M}_{\mu_2' \to \mu_2}$ is a directed (right-to-left) matching that fully connects μ_2' to μ_2.

It is now a simple matter to check that $\mathcal{M}$ is a directed matching on μ that satisfies properties $\mathbf{P_1}$ and $\mathbf{Q_2}$. Therefore, according to Proposition 6, $\mu \in \mathcal{S}_{2n+2k+2N_1+2N_2}^{\mathrm{r}}$.

($\Leftarrow$) Suppose that $\mu \in \mathcal{S}_{2n}^{\mathrm{r}}$. Therefore, according to Proposition 6, there exists a directed perfect matching $\mathcal{M}$ on μ that satisfies properties $\mathbf{P_1}$ and $\mathbf{Q_2}$. We begin with a sequence of claims that help defining the general structure of $\mathcal{M}$.

Claim. We may assume that $|\mathcal{M}_{\mu_1 \to \mu_1'}| = N_1$.

Proof. Combining Lemma 7 and $N_1 = 2N_2 > n + k + N_2$, we conclude that $\mathcal{M}_{\mu_1 \to \mu_1'} \neq \emptyset$ or $\mathcal{M}_{\mu_1' \to \mu_1} \neq \emptyset$. Thus, applying Lemma 8, we obtain $|\mathcal{M}_{\mu_1 \to \mu_1'}| = N_1$ or $|\mathcal{M}_{\mu_1' \to \mu_1}| = N_1$. Therefore, by Lemma 4 (One shot lemma), we may assume that $\mathcal{M}_{\mu_1 \to \mu_1'} \neq \emptyset$. ∎

Claim. Assuming $|\mathcal{M}_{\mu_1 \to \mu_1'}| = N_1$, we have $|\mathcal{M}_{\mu_2' \to \mu_2}| = N_2$.

Proof. Combining Claim 4, Lemma 7 and $N_2 > n + k$, we conclude that $\mathcal{M}_{\mu_2 \to \mu_2'} \neq \emptyset$ or $\mathcal{M}_{\mu_2' \to \mu_2} \neq \emptyset$. Applying Lemma 8, we obtain $|\mathcal{M}_{\mu_2 \to \mu_2'}| = N_2$ or $|\mathcal{M}_{\mu_2' \to \mu_2}| = N_2$. The claim now follows from $\mathcal{M}_{\mu_1 \to \mu_1'} \neq \emptyset$ (Claim 4) and the fact that $\mathcal{M}$ avoids 🙶🙸 (Property $\mathbf{Q_2}$). ∎

From the above two claims and the fact that $\mathcal{M}$ avoids 🙶🙸 (Property $\mathbf{Q_2}$), we conclude that $|\mathcal{M}_{\mu_\sigma \to \mu_\pi'}| = k$. But $\mathcal{M}$ satisfies Property $\mathbf{P_1}$ and hence there exists an occurrence of $\mathrm{r}(\sigma)$ in $\mathrm{r}(\pi)$. It follows that σ is a pattern of π. ∎

Proposition 8. *Let $\pi \in S_{2n}$. The two following statements are equivalent.*

1. $\pi \in \mathcal{S}_{2n}^{\mathrm{c}}$.
2. *There exists a directed perfect matching $\mathcal{M}$ that satisfies properties $\mathbf{P_2}$ and $\mathbf{Q_1}$.*

Proof. ($\mathbf{1} \Rightarrow \mathbf{2}$) Let $\pi \in \mathcal{S}_{2n}^{\mathrm{r}}$. Then, there exists $\sigma \in S_n$ such that $\pi \in \sigma \sqcup \mathrm{c}(\sigma)$, and hence there exist two disjoint set of indices $I^1 = \{i_j^1 : j \in [n]\}$, $i_1^1 < i_2^1 < \cdots < i_n^1$, and $I^2 = \{i_j^2 : j \in [n]\}$, $i_1^2 < i_2^2 < \cdots < i_n^2$, such that $\pi(i_1^1)\pi(i_2^1)\cdots\pi(i_n^1)$ is order-isomorphic to σ and $\pi(i_1^2)\pi(i_2^2)\cdots\pi(i_n^2)$ is order-isomorphic to $\mathrm{c}(\sigma)$. Let $\mathcal{M} = (V, E)$ be the directed graph defined by $V = [2n]$ and $E = \{(i_j^1, i_j^2) : j \in [n]\}$. Clearly, $\mathcal{M}$ is a directed perfect matching since $I^1 \cap I^2 = \emptyset$ and $I^1 \cup I^2 = [2n]$. According to Lemma 5, the directed perfect matching $\mathcal{M}$ does avoid the patterns 🙶🙸, 🙶🙸, 🙶🙸, 🙶🙸, 🙶🙸, 🙶🙸. Finally, for any two distinct arcs (i_j^1, i_j^2) and (i_k^1, i_k^2) of $\mathcal{M}$, we have $\pi(i_j^1) < \pi(i_k^1)$ if and only if $\pi(i_j^2) > \pi(i_k^2)$ since $\pi(i_1^1)\pi(i_2^1)\cdots\pi(i_n^1)$ is order-isomorphic to σ and $\pi(i_1^2)\pi(i_2^2)\cdots\pi(i_n^2)$ is order-isomorphic to $\mathrm{c}(\sigma)$ (and hence $\pi(i_j^2) = n - \pi(i_j^1)$ and $\pi(i_k^2) = n - \pi(i_k^1)$). Therefore, $\mathcal{M}$ satisfies properties $\mathbf{P_2}$ and $\mathbf{Q_1}$.

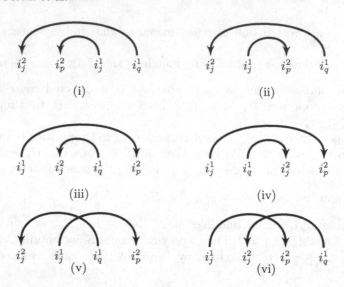

Fig. 8. All possible configurations considered in proof of proposition 8 ($\mathbf{2} \Rightarrow \mathbf{1}$).

($\mathbf{2} \Rightarrow \mathbf{1}$) Let $\mathcal{M}$ be a directed perfect matching that satisfies properties $\mathbf{P_2}$ and $\mathbf{Q_1}$. Let $I^1 = \{i_j^1 : j \in [n]\}$, $i_1^1 < i_2^1 < \cdots < i_n^1$, be the set the sources of the arcs of $\mathcal{M}$ and let $I^2 = \{i_j^2 : j \in [n]\}$, $i_1^2 < i_2^2 < \cdots < i_n^2$, be the set of the sinks of the arcs of $\mathcal{M}$. We first show that, for every $j \in [n]$, (i_j^1, i_j^2) is an arc of $\mathcal{M}$. The proof is by induction on $j = 1, 2, \cdots, n$.

- Base. Suppose, aiming at a contradiction that (i_1^1, i_1^2) is not an arc of $\mathcal{M}$. Then, there exist $1 < p \le n$ and $1 < q \le n$ such that (i_1^1, i_p^2) and (i_q^1, i_1^2) are two arcs of $\mathcal{M}$. But $i_1^1 < i_q^1$ and $i_1^2 < i_p^2$, and hence one of the configurations of Fig. 8 (with $j = 1$) does occur in $\mathcal{M}$. This is a contradiction since $\mathcal{M}$ satisfies Property $\mathbf{Q_1}$, and hence (i_1^1, i_1^2) is an arc of $\mathcal{M}$.
- Induction step. Assume that (i_k^1, i_{n-k+1}^2) is an arc of $\mathcal{M}$ for $1 \le k < j$, and suppose, aiming at a contradiction, that (i_j^1, i_j^2) is not an arc of $\mathcal{M}$. Then, there exist $j < p \le n$ and $j < q \le n$ such that (i_j^1, i_p^2) and (i_q^1, i_{n-j+1}^2) are two arcs of $\mathcal{M}$. But $i_j^1 < i_q^1$ and $i_j^2 < i_p^2$, and hence one of the configurations of Fig. 8 does occur in $\mathcal{M}$. This is a contradiction since $\mathcal{M}$ satisfies Property $\mathbf{Q_1}$, and hence (i_j^1, i_j^2) is an arc of $\mathcal{M}$.

Now, let σ_1 be the pattern of π induced by the sources of $\mathcal{M}$ and σ_2 be the pattern of π induced by the sinks of $\mathcal{M}$ (*i.e.*, $\sigma_1 = \pi\left(i_1^1\right)\pi\left(i_2^1\right)\ldots\pi\left(i_n^1\right)$ and $\sigma_2 = \pi\left(j_1^2\right)\pi\left(j_2^2\right)\ldots\pi\left(i_n^2\right)$). Clearly σ_1 and σ_2 are disjoint in π (since $\mathcal{M}$ is a matching) and cover π (since $\mathcal{M}$ is perfect). Moreover, the fact that $\mathcal{M}$ satisfies $\mathbf{P_2}$ implies immediately that reduce(σ_1) is order-isomorphic to (reduce $\circ$ c)(σ_2), and hence this shows that $\pi \in \mathcal{S}_{2n}^c$. ∎

Corollary 1. c-UNSHUFFLE-PERMUTATION *is* NP-*complete.*

Proof. Combine Proposition 7 with Proposition 1. ∎

Proposition 9. *Let $\pi \in S_{2n}$. The two following statements are equivalent.*

1. $\pi \in \mathcal{S}_{2n}^{\mathrm{rc}}$.
2. There exists a directed perfect matching $\mathcal{M}$ that satisfies properties $\mathbf{P_2}$ and $\mathbf{Q_2}$.

Proof. $(1 \Rightarrow 2)$ Let $\pi \in \mathcal{S}_{2n}^{\mathrm{rc}}$. Then, there exists $\sigma \in S_n$ such that $\pi \in \sigma \sqcup (\mathrm{r} \circ \mathrm{c})(\sigma)$, and hence there exist two disjoint set of indices $I^1 = \{i_j^1 : j \in [n]\}$, $i_1^1 < i_2^1 < \cdots < i_n^1$, and $I^2 = \{i_j^2 : j \in [n]\}$, $i_1^2 < i_2^2 < \cdots < i_n^2$, such that $\pi(i_1^1)\pi(i_2^1)\cdots\pi(i_n^1)$ is order-isomorphic to σ and $\pi(i_1^2)\pi(i_2^2)\cdots\pi(i_n^2)$ is order-isomorphic to $(\mathrm{r} \circ \mathrm{c})(\sigma)$. Let $\mathcal{M} = (V, E)$ be the directed graph defined by $V = [2n]$ and $E = \{(i_j^1, i_{n-j+1}^2) : j \in [n]\}$. Clearly, $\mathcal{M}$ is a directed perfect matching since $I^1 \cap I^2 = \emptyset$ and $I^1 \cup I^2 = [2n]$. According to Lemma 6, the directed perfect matching $\mathcal{M}$ does avoid the patterns $\curvearrowright\!\curvearrowleft$, $\curvearrowleft\!\curvearrowright$, $\curvearrowright\!\curvearrowright$, $\curvearrowleft\!\curvearrowleft$, and. Finally, for any two distinct arcs (i_j^1, i_{n-j+1}^2) and (i_k^1, i_{n-k+1}^2) of $\mathcal{M}$, we have $\pi(i_j^1) < \pi(i_k^1)$ if and only if $\pi(i_{n-j+1}^2) > \pi(i_{n-k+1}^2)$ since $\pi(i_1^1)\pi(i_2^1)\cdots\pi(i_n^1)$ is order-isomorphic to σ and $\pi(i_1^2)\pi(i_2^2)\cdots\pi(i_n^2)$ is order-isomorphic to $(\mathrm{r} \circ \mathrm{c})(\sigma)$. Therefore, $\mathcal{M}$ satisfies properties $\mathbf{P_1}$ and $\mathbf{Q_2}$.

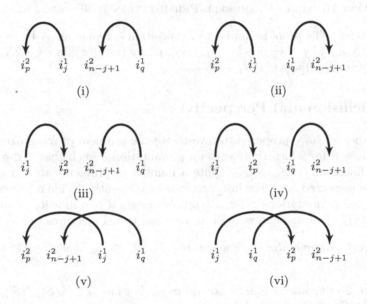

Fig. 9. All possible configurations considered in proof of proposition 9 $(2 \Rightarrow 1)$.

$(2 \Rightarrow 1)$ Let $\mathcal{M}$ be a directed perfect matching that satisfies properties $\mathbf{P_1}$ and $\mathbf{Q_2}$. Let $I^1 = \{i_j^1 : j \in [n]\}$, $i_1^1 < i_2^1 < \cdots < i_n^1$, be the set the sources of the arcs of $\mathcal{M}$ and let $I^2 = \{i_j^2 : j \in [n]\}$, $i_1^2 < i_2^2 < \cdots < i_n^2$, be the set of the sinks of the arcs of $\mathcal{M}$. We first show that, for every $j \in [n]$, (i_j^1, i_{n-j+1}^2) is an arc of $\mathcal{M}$. The proof is by induction on $j = 1, 2, \cdots, n$.

– Base. Suppose, aiming at a contradiction that $\left(i_1^1, i_n^2\right)$ is not an arc of $\mathcal{M}$. Then, there exist $1 \leq p < n$ and $1 < q \leq n$ such that $\left(i_1^1, i_p^2\right)$ and $\left(i_q^1, i_n^2\right)$ are two arcs of $\mathcal{M}$. But $i_1^1 < i_q^1$ and $i_p^2 < i_n^2$, and hence one of the configurations of Fig. 9 (with $j = 1$ and $k = n$) does occur in $\mathcal{M}$. This is a contradiction since $\mathcal{M}$ satisfies Property $\mathbf{Q_2}$, and hence $\left(i_1^1, i_n^2\right)$ is an arc of $\mathcal{M}$.

– Induction step. Assume that $\left(i_k^1, i_{n-k+1}^2\right)$ is an arc of $\mathcal{M}$ for $1 \leq k < j$, and suppose, aiming at a contradiction, that $\left(i_j^1, i_{n-j+1}^2\right)$ is not an arc of $\mathcal{M}$. Then, there exist $1 \leq p < n - j + 1$ and $j < q \leq n$ such that $\left(i_j^1, i_p^2\right)$ and $\left(i_q^1, i_{n-j+1}^2\right)$ are two arcs of $\mathcal{M}$. But $i_j^1 < i_q^1$ and $i_p^2 < i_{n-j+1}^2$, and hence one of the configurations of Fig. 9 does occur in $\mathcal{M}$. This is a contradiction since $\mathcal{M}$ satisfies Property $\mathbf{Q_2}$, and hence $\left(i_j^1, i_{n-j+1}^2\right)$ is an arc of $\mathcal{M}$.

Now, let σ_1 be the pattern of π induced by the sources of $\mathcal{M}$ and σ_2 be the pattern of π induced by the sinks of $\mathcal{M}$ (*i.e.*, $\sigma_1 = \pi\left(i_1^1\right)\pi\left(i_2^1\right)\ldots\pi\left(i_n^1\right)$ and $\sigma_2 = \pi\left(j_1^2\right)\pi\left(j_2^2\right)\ldots\pi\left(i_n^2\right)$). Clearly σ_1 and σ_2 are disjoint in π (since $\mathcal{M}$ is a matching) and cover π (since $\mathcal{M}$ is perfect). Moreover, the fact that $\mathcal{M}$ satisfies $\mathbf{P_2}$ implies immediately that reduce(σ_1) is order-isomorphic to (reduce $\circ$ r $\circ$ c)(σ_2), and hence this shows that $\pi \in \mathcal{S}_{2n}^{\mathrm{rc}}$. ∎

Proposition 10. (r $\circ$ c)-Unshuffle-Permutation *is* NP-*complete*.

Proof. (*sketch*) The proof is similar to Proposition 7, replacing μ_1' by $\nearrow_{N_1} [2n + k + N_1 + 2N_2]$, μ_2' by $\nearrow_{N_2} [n + k + N_1 + N_2]$, μ_π' by (r $\circ$ c)(π)$[2n + k + 2N_1 + 2N_2]$ and μ_σ' by (r $\circ$ c)(σ)$[n + k + N_1 + 2N_2]$. ∎

5   Conclusion and Perspectives

In this paper we have proposed to investigate the problem of recognizing those permutations π for which there exists a permutation σ such that $\pi \in \sigma \amalg f(\sigma)$ for some bijection $f : S_n \to S_n$. Quite a number of problems are left open by the results presented here. For instance, can we efficiently decide membership to $\mathcal{S}_{2n}^{\mathrm{i}}$ (*i.e.*, the permutations π for witch there exists a permutation σ such that $\pi \in \sigma \amalg \mathrm{i}(\sigma)$)? To conclude, we wish to mention two conjectures.

Conjecture 1 (Enumeration). For any $n \in \mathbb{N}$, $s_{2n}^{\mathrm{rc}} \leq s_{2n}^{\mathrm{c}} = s_{2n}^{\mathrm{r}} \leq s_{2n}^{\mathrm{irc}} = s_{2n}^{\mathrm{i}} \leq s_{2n}^{\epsilon} \leq s_{2n}^{\mathrm{ci}} = s_{2n}^{\mathrm{ic}}$.

Conjecture 2 (Ubiquitous generalized squares). For any $n \in \mathbb{N}$, $\mathcal{S}_{2n}^{\epsilon} \cap \mathcal{S}_{2n}^{\mathrm{r}} \cap \mathcal{S}_{2n}^{\mathrm{c}} \cap \mathcal{S}_{2n}^{\mathrm{i}} \cap \mathcal{S}_{2n}^{\mathrm{ci}} \cap \mathcal{S}_{2n}^{\mathrm{ic}} \cap \mathcal{S}_{2n}^{\mathrm{rc}} \cap \mathcal{S}_{2n}^{\mathrm{irc}} \neq \emptyset$.

References

1. Bose, P., Buss, J.F., Lubiw, A.: Pattern matching for permutations. Inf. Process. Lett. **65**(5), 277–283 (1998)
2. Buss, S., Soltys, M.: Unshuffling a square is NP-hard. J. Comput. Syst. Sci. **80**(4), 766–776 (2014)

3. Choffrut, C., Karhumäki, J.: Combinatorics of words. In: Rozenberg, G., Salomaa, A. (eds.) Handbook of Formal Languages, pp. 329–438. Springer, Heidelberg (1997). https://doi.org/10.1007/978-3-642-59136-5_6

4. Giraudo, S., Vialette, S.: Algorithmic and algebraic aspects of unshuffling permutations. Theor. Comput. Sci. **729**, 20–41 (2018)

5. Knuth, D.E.: The Art of Computer Programming: Volume III: Sorting and Searching. Addison-Wesley, Boston (1973)

6. van Leeuwen, J., Nivat, M.: Efficient recognition of rational relations. Inf. Process. Lett. **14**(1), 34–38 (1982)

7. Mansfield, A.: On the computational complexity of a merge recognition problem. Discrete Appl. Math. **5**(1), 119–122 (1983)

8. Neou, B.E., Rizzi, R., Vialette, S.: Pattern matching for separable permutations. In: Inenaga, S., Sadakane, K., Sakai, T. (eds.) SPIRE 2016. LNCS, vol. 9954, pp. 260–272. Springer, Cham (2016). https://doi.org/10.1007/978-3-319-46049-9_25

9. Rizzi, R., Vialette, S.: On recognizing words that are squares for the shuffle product. In: Proceedings of the 8th International Symposium in Computer Science - Theory and Applications, pp. 235–245 (2013)

10. Sloane, N.J.A.: The on-line encyclopedia of integer sequences. https://oeis.org/

11. Stankova, Z.: Forbidden subsequences. Discrete Math. **132**(1–3), 291–316 (1994)

12. Vargas, Y.: Hopf algebra of permutation pattern functions. In: 26th International Conference on Formal Power Series and Algebraic Combinatorics, pp. 839–850 (2014)

Continuous Team Semantics

Åsa Hirvonen[1], Juha Kontinen[1], and Arno Pauly[2]

[1] Department of Mathematics and Statistics, University of Helsinki, Helsinki, Finland
{asa.hirvonen,juha.kontinen}@helsinki.fi
[2] Department of Computer Science, Swansea University, Swansea, UK
Arno.M.Pauly@gmail.com

Abstract. We study logics with team semantics in computable metric spaces. We show how to define approximate versions of the usual independence/dependence atoms. For restricted classes of formulae, we show that we can assume w.l.o.g. that teams are closed sets. This then allows us to import techniques from computable analysis to study the complexity of formula satisfaction and model checking.

Keywords: Team semantics · Continuous logic · Computable analysis · Independence Logic · Dependence logic

1 Introduction

Team semantics is a semantical framework for logics of dependence and independence. Team semantics was originally invented by Hodges [11] and later systematically developed and made popular by Väänänen by the introduction of Dependence Logic [22]. In team semantics formulas are evaluated using sets of assignments (called teams) rather than single assignments as in first-order logic. Therefore, it is not surprising that the expressive power of many of the logics studied in team semantics exceeds the expressive power of first-order logic. The introduction of Independence Logic (FO($\perp$)) in [9] and inclusion ($x \subseteq y$) and exclusion atoms ($x \mid y$) (and the corresponding logics) [8] demonstrated the versatility of the framework and have led to several studies on the applications of team semantics in areas such as database theory, model theory, and quantum information theory (see, e.g. [1,6,10,12,13,16]).

In this article we explore and apply team semantics in a metric context. The expressive power that team semantics makes available to us in the form of dependence and independence atoms comes at the prize that in the definitions, we have to quantify over the powerset of our structure. On the logical level, this lets the expressivity exceed first-order logic, and in many cases, reach existential second-order logic. From an algorithmic perspective, this involves an exponential blow-up in the relevant search space, and causes a number of hardness results.

This research was partially supported by the Royal Society International Exchange Grant 170051 "Continuous Team Semantics: On dependence and independence in a continuous world" and grant 308712 of the Academy of Finland.

T. V. Gopal and J. Watada (Eds.): TAMC 2019, LNCS 11436, pp. 262–278, 2019.
https://doi.org/10.1007/978-3-030-14812-6_16

In a metric context, we are facing having to deal with sets such as the power-set of the unit interval. This would seem to destroy the hope of any algorithmic approach, and very much opens the door to the specter of independence of ZFC. Fortunately, our first main results show that for certain well-behaved classes of sentences, it is safe to assume that all teams are closed. The hyperspace of closed subsets of a (compact separable) metric space is much better understood, and been made accessible to algorithmic approaches through the development of computable analysis. We make use of the opportunity, and show that model-checking and satisfaction are semidecidable for the aforementioned well-behaved formulae. This is the best possible result in full generality.

In the metric context, it is very natural to consider approximate versions of the usual dependence and independence atoms (see, e.g., [6,15,23] for related previous work on so-called metric functional dependencies and approximate dependencies in (non-metric) team semantics). As an added bonus, the approximate versions are compatible with our notions of well-behaved formulae, and thus greatly increase what we can express directly without leaving the realm of tameness. We conclude our investigation (for now) by considering the translatability between the approximate versions, and contrast these to the established translability results regarding the exact versions.

2   Definitions

We are working with a fixed structure, which here is a (compact separable[1]) metric space $(\mathbf{X}, d)$ together with certain predicates, i.e. subsets of $\mathbf{X}^n$ for $n \in \omega$. To simplify notation, we will usually not explicitly mention the structure, but simply take it for granted. We then proceed to define when $T \models \phi$ holds, where $T \subseteq \mathbf{X}^n$ is a *team*, and ϕ is a positive formula involving both basic predicates and certain special primitives. These definitions are completely standard, see [8,22].

Variables are assumed to correspond to specific dimensions. We write $\pi_{\overline{x}}$ for the projection to the dimensions corresponding to the variables comprising the tuple $\overline{x}$; and π_{-x} for the projection to all dimensions except the one corresponding to the variable x. Note that the order n which the variables appear in $\overline{x}$ impacts the meaning of $\pi_{\overline{x}}$, e.g. $\pi_{xy}A$ and $\pi_{yx}A$ are related by $(a, b) \in \pi_{xy}A$ iff $(b, a) \in \pi_{yx}A$. We allow for the case of variables appearing multiple times in a tuple, this means the corresponding dimension will be duplicated in the result. By $\times$ we denote the usual cartesian product and, for R over $\overline{xz}$ and R' over $\overline{zy}$, the join $R \bowtie R'$ of R and R' is defined by

$$R \bowtie R' = \{\overline{xzy} \mid \overline{xz} \in R \text{ and } \overline{zy} \in R'\}.$$

$T \models P(x_1, \ldots, x_k)$ if $\forall (x_1, \ldots, x_k) \in T$ it holds that $P(x_1, \ldots, x_k)$.
$T \models \phi \wedge \psi$ if both $T \models \phi$ and $T \models \psi$.

---

[1] These requirements are used for the proofs, but are not strictly needed for our definitions to make sense.

$T \models \phi \vee \psi$ if there are T_1, T_2 such that $T_1 \models \phi$, $T_2 \models \psi$ and $T = T_1 \cup T_2$.

$T \models \forall x \ \phi$ if $(\mathbf{X} \times T) \models \phi$.

$T \models \exists x \ \phi$ if there is T' such that $\pi_{-x}T' = T$ and $T' \models \phi$.

$T \models = (\overline{x}, \overline{y})$ if for any $s, s' \in T$ it holds that if $s(\overline{x}) = s'(\overline{x})$ then $s(\overline{y}) = s'(\overline{y})$.

$T \models \overline{x} \perp \overline{y}$ if $(\pi_{\overline{x}}T) \times (\pi_{\overline{y}}T) = \pi_{\overline{xy}}T$.

$T \models \overline{x} \perp_{\overline{z}} \overline{y}$ if $(\pi_{\overline{xz}}T) \bowtie (\pi_{\overline{zy}}T) = \pi_{\overline{xzy}}T$.

$T \models \overline{x} \subseteq \overline{y}$ if $(\pi_{\overline{x}}T) \subseteq (\pi_{\overline{y}}T)$.

There is one case of a primitive where we will deviate from its usual definition. Usually, one would define

$T \models \overline{x}|\overline{y}$ if $\pi_{\overline{x}}T \cap \pi_{\overline{y}}T = \emptyset$ (**Classical definition**)

However, in a metric context *apartness* seems to be a far more natural notion than *disjointness* – and they obviously coincide in the traditional setting of finite models. We thus chose:

$T \models \overline{x}|\overline{y}$ if $d(\pi_{\overline{x}}T, \pi_{\overline{y}}T) > 0$ (**Our modified definition**)

We point out that in the semantics for $\exists$ and $\vee$, we are quantifying over teams. The precise scope of this quantification will vary in our investigation. We consider the case where the quantification ranges over the entire powerset of $\mathbf{X}$, the case where only closed teams are permitted, and briefly also the case where only open teams are permitted. These options are compared in Sect. 3.

2.1   Approximate Dependence/Independence Atoms

We can use the metric available as part of our structure to define approximate versions of the dependence/independence atoms. As mentioned in the introduction, this has precedence in database theory, related to data cleaning. In many cases, there are two independent parameters describing how exactly we approximate the atoms. Typically, one parameter corresponds to relaxing the atom, the other to strengthening it. Depending on whether we chose strict or non-strict inequalities, one gets both open and closed versions of the atoms[2]. We denote the closed versions with $\bar{\ }$, the open versions are the non-decorated ones.

$T \models =_{\delta}^{\varepsilon} (\overline{x}, \overline{y})$ if for any $s, s' \in T$ it holds that if $d(s(\overline{x}), s'(\overline{x})) \leq \delta$ then $d(s(\overline{y}), s'(\overline{y})) < \varepsilon$.

$T \models =_{\delta}^{\varepsilon}(\overline{x}, \overline{y})$ if for any $s, s' \in T$ it holds that if $d(s(\overline{x}), s'(\overline{x})) < \delta$ then $d(s(\overline{y}), s'(\overline{y})) \leq \varepsilon$.

$T \models \overline{x} \perp_{\overline{z}}^{\delta, \varepsilon} \overline{y}$ if for all $s, s' \in T$, if $d(s(\overline{z}), s'(\overline{z})) \leq \delta$ then there is $s'' \in T$ such that $d(s''(\overline{xz}), s(\overline{xz})) < \varepsilon$ and $d(s''(\overline{zy}), s'(\overline{zy})) < \varepsilon$.

$T \models \overline{x}\perp_{\overline{z}}^{\delta, \varepsilon}\overline{y}$ if for all $s, s' \in T$, if $d(s(\overline{z}), s'(\overline{z})) < \delta$ then there is $s'' \in T$ such that $d(s''(\overline{xz}), s(\overline{xz})) \leq \varepsilon$ and $d(s''(\overline{zy}), s'(\overline{zy})) \leq \varepsilon$.

$T \models \overline{x} \subseteq^{\varepsilon} \overline{y}$ if $(\pi_{\overline{x}}T) \subseteq B(\pi_{\overline{y}}T, \varepsilon)$.

$T \models \overline{x}\subseteq^{\varepsilon}\overline{y}$ if $(\pi_{\overline{x}}T) \subseteq \overline{B}(\pi_{\overline{y}}T, \varepsilon)$.

$T \models \overline{x}|^{\varepsilon}\overline{y}$ if $d(\pi_{\overline{x}}T, \pi_{\overline{y}}T) > \varepsilon$.

$T \models \overline{x}|^{\varepsilon}\overline{y}$ if $d(\pi_{\overline{x}}T, \pi_{\overline{y}}T) \geq \varepsilon$.

---

[2] Of course, we could also mix the cases. However, part of the overall theme of this article is to control topological complexity, so this seems undesirable.

3 Restricting Teams to Closed Sets

In this section, we show that for restricted formulae the semantics allowing arbitrary teams and the semantics allowing only closed teams coincide, and then give some examples how this breaks down for arbitrary formulae.

3.1 Closed Formulae

Theorem 1. *For positive sentences involving closed basic predicates,* $\perp$, $\subseteq$, $=_\delta^\varepsilon(\cdot,\cdot)$, $\perp^{\delta,\varepsilon}$, $\subseteq^\varepsilon$ *and* $|^\varepsilon$*, the usual team semantics and the teams-are-closed sets semantics agree.*

Proof. This is a special case of Corollary 1 below.

Lemma 1. *For an arbitrary team* T*, we find that*

1. $T \models P(\overline{x})$ *implies* $\overline{T} \models P(\overline{x})$*, where* P *is a basic closed predicate*
2. $T \models \overline{x} \perp \overline{y}$ *implies* $\overline{T} \models \overline{x} \perp \overline{y}$
3. $T \models \overline{x} \subseteq \overline{y}$ *implies* $\overline{T} \models \overline{x} \subseteq \overline{y}$
4. $T \models =_\delta^\varepsilon(\overline{x},\overline{y})$ *implies* $\overline{T} \models =_\delta^\varepsilon(\overline{x},\overline{y})$
5. $T \models \overline{x}\perp^{\delta,\varepsilon}_{\overline{z}}\overline{y}$ *implies* $\overline{T} \models \overline{x}\perp^{\delta,\varepsilon}_{\overline{z}}\overline{y}$
6. $T \models \overline{x}\subseteq^\varepsilon\overline{y}$ *implies* $\overline{T} \models \overline{x}\subseteq^\varepsilon\overline{y}$
7. $T \models \overline{x}|^\varepsilon\overline{y}$ *implies* $\overline{T} \models \overline{x}|^\varepsilon\overline{y}$

Proof. For 1–3, we only need that closure and projection commute. For 4, 5 we note that since the premise of the implication in the definition has a strict inequality and the conclusion a non-strict one, taking the closure of the team has no impact. For 6, 7 we are using the distance to the team, which is invariant under taking the closure.

Corollary 1. *Let* ϕ *be a positive formula involving basic closed predicates,* $\perp$, $\subseteq$, $=_\delta^\varepsilon(\cdot,\cdot)$, $\perp^{\delta,\varepsilon}$, $\subseteq^\varepsilon$ *and* $|^\varepsilon$*. For an arbitrary team* T*, we find that* $T \models \phi$ *implies that* $\overline{T} \models \phi$*.*

Proof. Induction over the structure of ϕ. Lemma 1 provides the base case. The only non-trivial steps are $\exists$ and $\vee$, where we use that projection commutes with closure for the former, and that $T = T_1 \cup T_2$ implies $\overline{T} = \overline{T}_1 \cup \overline{T}_2$ for the latter.

3.2 Open Formulae

Theorem 2. *For positive sentences involving basic open predicates and* $|$*, the following all agree:*

1. *the semantics allowing arbitrary sets as teams*
2. *the semantics demanding teams to be open sets*
3. *the semantics demanding teams to be closed sets*

Proof. By Lemma 2, truth in (1) implies truth in (2). By Lemma 3, truth in (2) implies truth in (3). That truth in (3) implies truth in (1) is trivial.

Lemma 2. *Let ϕ be a positive formula involving basic open predicates and $|$, let $T \models \phi$ and let $\overline{x} \in T$. Then there is some $\varepsilon > 0$ such that $T \cup B(\overline{x}, \varepsilon) \models \phi$, with only open teams being used as witnesses.*

Proof. We proceed by induction over the structure of ϕ. Universal and existential quantifier are trivial. For conjunctions, we use that the relevant formulae are downwards-closed, and take the minimum ε from both sides. For disjunctions, we note that from $\overline{x} \in T = T_1 \cup T_2$ we find $\overline{x} \in T_1$ or $\overline{x} \in T_2$, proceed to use the induction hypothesis on the relevant side, and then use that $T \cup B(\overline{x}, \varepsilon) = (T_i \cup B(\overline{x}, \varepsilon)) \cup T_{3-i}$.

The base cases for the induction are the basic open predicates and $|$, in both cases the claim follows from the definition.

Lemma 3. *Let ϕ be a positive formula involving basic open predicates and $|$, let $T \models \phi$ for open T, let $n \in \mathbb{N}$. Define $T_n = \{\overline{x} \in T \mid d(\overline{x}, T^C) \geq 2^{-n}\}$. Then $T_n \models \phi$, with only closed teams used as witnesses.*

Proof. Straight-forward from downwards-closure.

Downwards-closure of $|$ makes the proofs of Lemmas 2, 3 very simple, but this proof does not extent to the open approximate atoms that are not downwards-closed. We conjecture that downwards-closure is not actually needed for the statements to hold, but leave this for future work at this stage.

3.3   Counterexamples

We proceed to give some counterexamples showing that if we allow using both open and closed basic predicates in a formula, then the semantics for arbitrary teams and the semantics where we allow only closed teams differ. Using closed teams only lets us express some topological properties of the carrier metric space, and reveals some similarities to constructive mathematics.

Example 1. The formula

$$\forall x \forall y \ (x = y) \vee (x \neq y)$$

is a tautology for arbitrary teams, but expresses that the space is discrete for closed teams. Note that this is in line with how the formula works in constructive mathematics.

Example 2. The formula

$$\exists x \forall y \ (x = y) \vee (x \neq y)$$

is a tautology for arbitrary teams, but expresses that the space contains some isolated point for closed teams. Note that this is in line with how the formula works in constructive mathematics.

Example 3. The formula

$$\forall x \exists y \exists z \; ((x = y \lor x = z) \land y|z)$$

holds over $\mathbf{X} = [0,1]$ if arbitrary teams are allowed, but not if teams have to be closed sets. The reason is that the y and the z values have to be disjoint non-empty sets covering $\mathbf{X}$. For closed teams, this formula expresses that the space is disconnected.

Example 4. The formula

$$\forall x \exists y \; (x \neq y) \land = (x, y)$$

holds over every model with at least two elements, if arbitrary teams are allowed. If teams are restricted to closed sets, it asserts the negation of the fixed-point-property for $\mathbf{X}$.

4 Background on Computability on Metric Spaces and for Closed Sets

We wish to study the algorithmic properties of questions such as satisfiability and model checking for our continuous team semantics. The algorithmic aspects of logics with team semantics have been studied extensively over finite structures (see the survey [7]). This requires notions of effectivity and computability for separable metric spaces, for hyperspaces of closed subsets of metric spaces, and finally for the entire collection of compact separable metric spaces. The field of computable analysis provides all of these notions. The standard reference is [24], but we follow [18]. Another short introduction to the area is [3].

As we lack the space for a rigorous development of the area, we will restrict our undertaking to a cursory description of the needed notions and special cases. The foundational concept in computable analysis is a *represented space*, which is just a set X together with a partial surjection $\delta :\subseteq \mathbb{N}^{\mathbb{N}} \to X$. Here δ tells us how the elements of interest are coded. We can then lift the usual notion of computation on $\mathbb{N}^{\mathbb{N}}$ to any represented space by letting our machine model act on names for elements.

A class of represented spaces of particular interest for us are the *computable metric spaces*: We take a separable metric space with a designated dense sequence $(a_n)_{n \in \mathbb{N}}$ such that $s < d(a_n, a_m) < t$ is recursively enumerable in $n, m \in \mathbb{N}$, $s, t \in \mathbb{Q}$. Then a point x is coded by giving a sequence p of indices such that $d(a_{p(n)}, x) < 2^{-n}$. A computable metric space $\mathbf{X}$ is *computably compact*, if the set of finite sequences $(n_0, r_0), \ldots, (n_\ell, r_\ell)$ such that $\mathbf{X} \subseteq \bigcup_{i \leq \ell} B(a_{n_i}, 2^{-r_i})$ is recursively enumerable. Computable metric spaces always have two further properties we will use; they are computably Hausdorff and computably overt.

We use several hyperspace constructions, i.e. constructions of certain spaces of subsets of a given represented space. We have the space $\mathcal{O}(\mathbf{X})$ of *open subsets*, the space $\mathcal{A}(\mathbf{X})$ of closed subsets and the space $\mathcal{V}(\mathbf{X})$ of compact subsets.

The open subsets are characterized by $x \in U$ being semidecidable (recognizable) in $x \in \mathbf{X}$ and $U \in \mathcal{O}(\mathbf{X})$. For a computable metric space $\mathbf{X}$, this can concretely be achieved by coding $U \in \mathcal{O}(\mathbf{X})$ as $\langle p, q \rangle$ with $U = \bigcup_{\{n|p(n) \neq 0\}} B(a_{p(n)+1}, 2^{-q(n)})$. The closed subsets are the formal complements of the open sets, i.e. the codes for $A \in \mathcal{A}(\mathbf{X})$ are just the codes for $(X \setminus A) \in \mathcal{O}(\mathbf{X})$.

The overt subsets $\mathcal{V}(\mathbf{X})$ are assumed to be closed extensionally, but have different codes and subsequently very different associated computable operations from $\mathcal{A}(\mathbf{X})$. The overt subsets are characterized by $U \cap A \neq \emptyset$ being semidecidable (recognizable) in $U \in \mathcal{O}(\mathbf{X})$ and $A \in \mathcal{V}(\mathbf{X})$. In a computable metric space they can be coded as a list of all basic open balls intersecting them.

Since $\mathcal{A}(\mathbf{X})$ and $\mathcal{V}(\mathbf{X})$ pointwise have the same elements, we can also construct $\mathcal{A} \wedge \mathcal{V}(\mathbf{X})$, where a set is coded by the combination of its $\mathcal{A}(\mathbf{X})$-code and its $\mathcal{V}(\mathbf{X})$-code. Over a computably compact computable metric space $\mathbf{X}$, the space $\mathcal{A} \wedge \mathcal{V}(\mathbf{X})$ corresponds to the space of closed subsets equipped with the Hausdorff metric. In this case, the space is characterized by making $d : \mathbf{X} \times (\mathcal{A} \wedge \mathcal{V}(\mathbf{X})) \to \mathbb{R}$ computable.

4.1 Overtness and Compactness

The relevance of the notions of compactness and overtness for spaces in general, and for our purposes in particular, is exhibit by the following lemmas tying it in to the preservation of the complexity of formula under quantification.

Lemma 4. *The following are equivalent for a represented space* $\mathbf{X}$:

1. $\mathbf{X}$ *is computably overt.*
2. $\exists : \mathcal{O}(\mathbf{X} \times \mathcal{Y}) \to \mathcal{O}(\mathbf{Y})$ *is computable for all represented spaces* $\mathbf{Y}$ *(respectively some space containing a computable point)*
3. $\forall : \mathcal{A}(\mathbf{X} \times \mathcal{Y}) \to \mathcal{A}(\mathbf{Y})$ *is computable for all represented spaces* $\mathbf{Y}$ *(respectively some space containing a computable point)*

Proof. The equivalence of 1 and 2 is a special case of [18, Proposition 40]. The equivalence of 2 and 3 is by duality.

Lemma 5. *The following are equivalent for a represented space* $\mathbf{X}$:

1. $\mathbf{X}$ *is computably compact.*
2. $\forall : \mathcal{O}(\mathbf{X} \times \mathcal{Y}) \to \mathcal{O}(\mathbf{Y})$ *is computable for all represented spaces* $\mathbf{Y}$ *(respectively some space containing a computable point)*
3. $\exists : \mathcal{A}(\mathbf{X} \times \mathcal{Y}) \to \mathcal{A}(\mathbf{Y})$ *is computable for all represented spaces* $\mathbf{Y}$ *(respectively some space containing a computable point)*

Proof. The equivalence of 1 and 2 is a special case of [18, Proposition 42]. The equivalence of 2 and 3 is by duality.

4.2   Computable Operations on the Closed and Overt Sets

We proceed to recall or establish the basic properties of the space $\mathcal{A} \wedge \mathcal{V}(\mathbf{X})$ which we shall use in the following.

Theorem 3 (Park, Park, Park, Seon and Ziegler [17]). *For a computably compact computable metric space $\mathbf{X}$, the space $\mathcal{A} \wedge \mathcal{V}(\mathbf{X})$ is a computably compact computable metric space again.*

Corollary 2. *For a computably compact computable metric space $\mathbf{X}$, the space $\mathcal{A} \wedge \mathcal{V}(\mathbf{X})$ is computably overt.*

Corollary 3. *For a computably compact computable metric space $\mathbf{X}$, the space $\mathcal{A} \wedge \mathcal{V}(\mathbf{X})$ is computably compact.*

Lemma 6. *The following maps are computable for computably compact $\mathbf{Y}$, and countably-based $\mathbf{X}, \mathbf{Y}$:*

1. $\pi_x : \mathcal{A} \wedge \mathcal{V}(\mathbf{X} \times \mathbf{Y}) \to \mathcal{A} \wedge \mathcal{V}(\mathbf{X})$
2. $\times : \mathcal{A} \wedge \mathcal{V}(\mathbf{X}) \times \mathcal{A} \wedge \mathcal{V}(\mathbf{Y}) \to \mathcal{A} \wedge \mathcal{V}(\mathbf{X} \times \mathbf{Y})$

Proof. 1. We can show separately that $\pi_x : \mathcal{A}(\mathbf{X} \times \mathbf{Y}) \to \mathcal{A}(\mathbf{X})$ and $\pi_x : \mathcal{V}(\mathbf{X} \times \mathbf{Y}) \to \mathcal{V}(\mathbf{X})$ are computable. The former is [18, Proposition 8 (8)] (using computable compactness of $\mathbf{Y}$), the latter is [18, Proposition 21 (6)].

2. Again, this can be shown separately for $\mathcal{A}$ and $\mathcal{V}$. The former is [18, Proposition 6 (8)]. For the latter, we use the fact that $\mathbf{X}, \mathbf{Y}$ being countably-based implies that $\mathcal{O}(\mathbf{X} \times \mathbf{Y})$ effectively is the product topology, i.e. that there is a computable multi-valued operation $\text{Decompose} : \mathcal{O}(\mathbf{X} \times \mathbf{Y}) \rightrightarrows \mathcal{C}(\mathbb{N}, \mathcal{O}(\mathbf{X}) \times \mathcal{O}(\mathbf{Y}))$ such that $(U_i, V_i)_{i \in \mathbb{N}} \in \text{Decompose}(O)$ iff $O = \bigcup_{n \in \mathbb{N}} U_n \times V_n$. Now if $(U_i, V_i)_{i \in \mathbb{N}} \in \text{Decompose}(O)$ we find that O intersects $A \times B \subseteq \mathbf{X} \times \mathbf{Y}$ iff $\exists n \in \mathbb{N} \; U_n \cap A \neq \emptyset \wedge V_n \cap B \neq \emptyset$. By currying, this is all we need.

Lemma 7. *In a computably compact computable metric space $\mathbf{X}$, the map $(A, \varepsilon) \mapsto \overline{B}(A, \varepsilon) : \mathcal{A}(\mathbf{X}) \times \mathbb{R}_+ \to \mathcal{A}(\mathbf{X})$ is computable.*

Proof. We have $y \notin \overline{B}(A, \varepsilon)$ iff $\overline{B}(y, \varepsilon) \cap A = \emptyset$. The latter is an open property by computable compactness.

Lemma 8. *The following are closed predicates on $\mathcal{A} \wedge \mathcal{V}(\mathbf{X}) \times \mathcal{A} \wedge \mathcal{V}(\mathbf{X})$:*

1. $\subseteq$
2. $=$

Proof. Note that $A \subseteq B$ iff $A \cap B^C = \emptyset$, and that by definition of $\mathcal{A}$ and $\mathcal{V}$, $A \cap B^C \neq \emptyset$ is already an open predicate on $\mathcal{V}(\mathbf{X}) \times \mathcal{A}(\mathbf{X})$. For 2., just observe that $A = B$ iff $A \subseteq B \wedge B \subseteq A$.

Lemma 9. $\cup : \mathcal{A} \wedge \mathcal{V}(\mathbf{X}) \times \mathcal{A} \wedge \mathcal{V}(\mathbf{X}) \to \mathcal{A} \wedge \mathcal{V}(\mathbf{X})$ *is computable.*

Proof. This is the combination of [18, Proposition 6(3)] and [18, Proposition 21(3)].

4.3   Compact Metric Structures

We shall now discuss the connection of the theory of computable metric subspace and the induced hyperspaces to the notion of a structure as used to interpret logical formulas. First, we note that we have the represented space Pol of Polish spaces and the represented space KPol of compact separable metric spaces. A similar hyperspace of countably-based spaces was introduced and studied in [20]. In the space Pol, we code a separable space $\mathbf{X}$ by presupposing $\mathbb{N}$ as a dense set, and then providing all distances $d_{\mathbf{X}}(n, m)$. This uniquely determines a Polish space by considering the completion. The space KPol is not merely the subspace of Pol restricted to compact spaces, but here we additionally code all finite covers $B(n_0, 2^{-k_0}) \cup \ldots \cup B(n_\ell, 2^{-k_\ell})$ into the name of the space. We point out that all arguments given above[3] regarding the properties of computably compact computable metric spaces hold uniformly in a compact metric space given as an element of KPol.

To introduce the notion of a structure, we first need the notion of a signature. A *signature* consists of function and relation symbols, each with some finite arity. Once a signature is fixed, we define a structure to be an underlying set A, together with a function $f_i : A^{n_i} \to A$ for each function symbol of arity n_i, and a subset $R_i \subseteq A^{n_i}$ for each relation symbol of arity n_i. Note that, contrary to convention, we do not make equality available for free. Instead, we can only use equality in our formula if it is provided as a relation by the signature/structure. We lift this to compact metric spaces as follows:

Definition 1. *A compact metric closed (respectively open) structure (over a given signature) consists of a compact separable metric space $\mathbf{X}$ as carrier, a continuous function $f_i : \mathbf{X}^{n_i} \to \mathbf{X}$ for each function symbol of arity n_i in the signature, and a closed (respectively open) subset $R_i \subseteq \mathbf{X}^{n_i}$ for each relation symbol of arity n_i in the signature.*

We write ACMS (respectively OCMS) for the represented space of compact metric closed (respectively open) structures, where the carrier is given as $\mathbf{X} \in$ KPol, the functions as $f_i \in \mathcal{C}(\mathbf{X}^{n_i}, \mathbf{X})$ and the relations as $R_i \in \mathcal{A}(\mathbf{X}^{n_i})$ (respectively as $R_i \in \mathcal{O}(\mathbf{X}^{n_i})$).

5   Topological Complexity

We proceed to study the topological complexity of the atoms, of formula satisfaction and of model checking. Since we have a topology on the space of closed and overt sets we are using for teams, and on the spaces of structures, these are all well-defined notions.

---

[3] In fact, example of statements that are computable for each computable metric space, yet are not computable uniformly in the metric space are very rare in the literature. See [19, Proposition 14] for such a rare example.

5.1 Topological Complexity of Dependence Atoms

Proposition 1. *The following are closed predicates in the team:*

1. $T \models P(\overline{x})$, where P is a basic closed predicate
2. $T \models \overline{x} \perp \overline{y}$
3. $T \models \overline{x} \subseteq \overline{y}$
4. $T \models =^{\varepsilon}_{\delta}(\overline{x}, \overline{y})$
5. $T \models \overline{x} \perp^{\delta,\varepsilon}_{\overline{z}} \overline{y}$
6. $T \models \overline{x} \subseteq^{\varepsilon} \overline{y}$
7. $T \models \overline{x} |^{\varepsilon} \overline{y}$

Proof. By Lemmas 6, 8, 1–3 follow immediately from the definitions. For 4–5, we have a universal quantification over the team, and then a closed property, which makes for a closed property by Lemma 5. For 6, note this is obtained by combining Lemmas 6, 7 and 8. Finally, 7. just follows from the continuity of the Hausdorff distance on $\mathcal{A} \wedge \mathcal{V}(\mathbf{X})$.

Proposition 2. *The following are open predicates in the team:*

1. $T \models P(\overline{x})$, where P is a basic open predicate
2. $T \models \overline{x} | \overline{y}$
3. $T \models =^{\varepsilon}_{\delta}(\overline{x}, \overline{y})$
4. $T \models \overline{x} \perp^{\delta,\varepsilon}_{\overline{z}} \overline{y}$
5. $T \models \overline{x} \subseteq^{\varepsilon} \overline{y}$
6. $T \models \overline{x} |^{\varepsilon} \overline{y}$

Proof. For 1., note that this means $\pi_{\overline{x}} T \subseteq P$. By Lemma 6, we can compute $\pi_{\overline{x}} T$ as a closed set, which we also have as a compact set due to the fact that we are working in a compact space. By definition of compact sets, this makes the predicate open. Claim 2. follows immediately from the definition by Lemmas 6, 8. Items 3., 4., 5. and 6. are analogous to their closed counterparts in Proposition 1.

Proposition 3. *The following are Π^0_2-complete predicates in the team:*

1. $T \models =(\overline{x}, y)$
2. $T \models \overline{x} \perp_{\overline{z}} \overline{y}$

Proof. We obtain a lower bound for $T \models =(\overline{x}, y)$ from Lemma 10, and an upper bound for $T \models \overline{x} \perp_{\overline{z}} \overline{y}$ by noting that $\forall \varepsilon \exists \exists \delta T \models \overline{x} \delta \perp^{\delta,\varepsilon}_{\overline{z}} \overline{y}$ is equivalent to $T \models \overline{x} \perp_{\overline{z}} \overline{y}$. These are then linked by noting that $=(\overline{x}, y) \equiv y \perp_{\overline{x}} y$.

Lemma 10. $T \models =(x, y)$ *is Π^0_2-hard over $\{0, 1\}^{\mathbb{N}}$.*

Proof. Given some $p \in \{0, 1\}^{\mathbb{N}}$, we compute some $T_p \in \mathcal{A} \wedge \mathcal{V}(\{0, 1\}^{\mathbb{N}} \times \{0, 1\}^{\mathbb{N}})$ such that $T_p \models =(x, y)$ iff p contains infinitely many 1s. We point out that a set $A \in \mathcal{A} \wedge \mathcal{V}(\{0, 1\}^{\mathbb{N}})$ can be represented as a sequence $(W_k)_{k \in \mathbb{N}}$ where $W_k \subseteq \{0, 1\}^{2^k}$ satisfy that $\forall w \in W_k \exists u \in \{0, 1\}^{2^k} \, wu \in W_{k+1}$ and $q \in A \Leftrightarrow \forall k \, q_{\leq 2^k} \in W_k$.

We define our sequence inductively, and take into account the standard bijection $\{0,1\}^{\mathbb{N}} \times \{0,1\}^{\mathbb{N}} \cong \{0,1\}^{\mathbb{N}}$. We start with $W_1 = \{00, 01, 10, 11\}$. Whenever $p(k) = 1$, then we let $W_{k+1} = \{\langle wu, uu \rangle \mid \langle w, u \rangle \in W_k\}$. Whenever $p(k) = 0$, then $W_{k+1} = \{wu \mid w \in W_k \wedge u \in \{0,1\}^{2^k}\}$.

To argue that this construction works as intended, let us first consider the case where p has only finitely many 1s. Let K be sufficiently large that $p(j) = 0$ for all $j \geq K$. Pick some $\langle w, u \rangle \in W_K$. Now the construction ensures that $w\{0,1\}^{\mathbb{N}} \times u\{0,1\}^{\mathbb{N}} \subseteq T_p$, hence $T_p \not\models {=}(x, y)$.

Conversely, assume for the sake of a contradiction that p contains infinitely many 1s, yet $T_p \not\models {=}(x, y)$. Pick witnesses a, b_1, b_2 for the latter, i.e. satisfying that $(a, b_1) \in T_p$ and $(a, b_2) \in T_p$, yet $b_1 \neq b_2$. Pick K such that $p(K) = 1$ and $(b_1)_{\leq 2^K} \neq (b_2)_{\leq 2^K}$. We must have that $\langle a_{\leq 2^{K+1}}, (b_1)_{\leq 2^{K+1}} \rangle \in W_{K+1}$ and $\langle a_{\leq 2^{K+1}}, (b_2)_{\leq 2^{K+1}} \rangle \in W_{K+1}$, but these cannot both be of form $\langle wu, uu \rangle$, contradiction.

5.2   Complexity of Formula Satisfaction

Theorem 4. *Let ϕ be a positive formula involving closed basic predicates, $\bot$, $\subseteq$, $\overline{=}^{\varepsilon}_{\delta}(\cdot, \cdot)$, $\bot^{\delta, \varepsilon}.$, $\overline{\subseteq}^{\varepsilon}$ and $\overline{|}^{\varepsilon}$. Then $T \models \phi$ defines a closed predicate in the team (uniformly in ϕ).*

Proof. By induction on the structure of ϕ. The base cases are provided by Proposition 1.

Dealing with $\wedge$ is trivial.

Let $\phi = \phi_1 \vee \phi_2$. By induction hypothesis, $T_i \models \phi_i$ is a closed predicate in T_i. Then $T = T_1 \cup T_2 \wedge T_1 \models \phi_1 \wedge T_2 \models \phi_2$ is a closed predicate in (T, T_1, T_2), since $\cup$ is computable on $\mathcal{A} \wedge \mathcal{V}(\mathbf{X})$ by Lemma 9 and $\mathcal{A} \wedge \mathcal{V}(\mathbf{X})$ is Hausdorff by Lemma 8. By Corollary 3, Lemma 5 applies and lets us conclude that quantifying existentially over T_1 and T_2 still leaves us with a closed predicate.

For $\phi = \forall x\ \psi$, we note that $T \mapsto \mathbf{X} \times T$ is computable, and that $T \models \phi$ is preimage of the closed predicate $T' \models \psi$ under that map.

For $\phi = \exists x\ \psi$, we invoke Lemma 5 by means of Corollary 3.

Since we have not yet established whether the semantics for arbitrary teams and closed teams still agree when also the approximate open predicates are permitted, for now we study the complexity of satisfaction only in the case where satisfaction is unambiguous:

Theorem 5. *Let ϕ be a positive formula involving basic open predicates and $|$. Then $T \models \phi$ defines a open predicate in the team (uniformly in ϕ). As a consequence, if $T \models \phi$, then we can effectively find some $n \in \mathbb{N}$ such that any T' with $d(T, T')$ satisfies $T' \models \phi$.*

Proof. By induction on the structure of ϕ. The base cases are provided by Proposition 2.

Dealing with $\wedge$ is trivial.

Let $\phi = \phi_1 \vee \phi_2$. By induction hypothesis, $T_i \models \phi_i$ is an open predicate in T_i. Given T_0, T_1 with $T_i \models \phi_i$, we can find some $n \in \mathbb{N}$ such that if $d(T_i, T_i')$, then $T_i' \models \phi_i$. The map $T_i \mapsto OT_i := \{x \in \mathbf{X}^k \mid d(x, T_i) < 2^{-n}\} :\subseteq \mathcal{A} \wedge \mathcal{V}(\mathbf{X}^k) \to \mathcal{O}(\mathbf{X}^k)$ is computable. We can extend this to a computable total map (i.e. define it also for $T_i \not\models \phi$ by setting $OT_i = \emptyset$ in that case. Now $T_1 \models \phi_1 \wedge T_2 \models \phi_2 \wedge T \subseteq OT_1 \cup OT_2$ is an open predicate in (T, T_1, T_2). Clearly, whenever the T_i are suitable witnesses for $T \models \phi$, this predicate is satisfied. Conversely, if the predicate is satisfied, consider $T_i' = \{x \in T \mid d(x, T_i) \leq d(x, T_{2-i})\}$ and note that $d(T_i, T_i') < 2^{-n}$, hence $T_i' \models \phi$. Thus, our modified predicate is equivalent to the existence of witnesses for $T \models \phi$. Corollary 2 lets us invoke Lemma 4 to remove the existential quantifier over T_i.

For $\phi = \forall x\, \psi$, we note that $T \mapsto \mathbf{X} \times T$ is computable, and that $T \models \phi$ is preimage of the open predicate $T' \models \psi$ under that map.

Let $\phi = \exists x\, \psi$. Similar to the argument above, the map $T' \mapsto OT' : \mathcal{A} \wedge \mathcal{V}(\mathbf{X}^{k+1}) \to \mathcal{O}(\mathbf{X}^k)$ is computable; mapping T' with $T' \models \psi$ to $OT' = \{y \in \mathbf{X}^k \mid d(x, \pi_{-x}T') < 2^{-n}\}$, where n is chosen such that if $d(T', T'') < 2^{-n}$, then $T'' \models \psi$; and mapping $T' \not\models \psi$ to $OT' = \emptyset$. Now $T' \models \psi \wedge T \subseteq OT'$ is an open predicate in (T, T'), and as above, equivalent to the existence of a witness for $T \models \phi$. Corollary 2 lets us invoke Lemma 4 to remove the existential quantifier over T'.

5.3 The Complexity of Model Checking

As the proofs of Theorems 4 and 5 are fully uniform, we can obtain a classification of the model checking problem. We shall write $\mathcal{L}_+(\bot, \subseteq, \overline{=}^\varepsilon_\delta(\cdot, \cdot), \overline{\bot^{\delta, \varepsilon}}., \overline{\subseteq^\varepsilon}, \overline{|^\varepsilon})$ for the set of positive sentences involving basic predicates, $\bot$, $\subseteq$, $\overline{=}^\varepsilon_\delta(\cdot, \cdot)$, $\overline{\bot^{\delta, \varepsilon}}.$, $\overline{\subseteq^\varepsilon}$ and $\overline{|^\varepsilon}$. Likewise, we write $\mathcal{L}_+(|)$ for the set of positive sentences involving basic predicates and $|$. These sets are coded in the obvious way, including the real-valued parameters. Note that formulae from $\mathcal{L}_+(\bot, \subseteq, \overline{=}^\varepsilon_\delta(\cdot, \cdot), \overline{\bot^{\delta, \varepsilon}}., \overline{\subseteq^\varepsilon}, \overline{|^\varepsilon})$ can use all potential choices for ε and δ, but that no quantification over these parameters is available. We recall our convention that equality is not automatically available, but would need to be provided by interpreting some binary relation symbol accordingly. We then find:

Corollary 4. *It is semidecidable whether a formula* $\mathcal{L}_+(\bot, \subseteq, \overline{=}^\varepsilon_\delta(\cdot, \cdot), \overline{\bot^{\delta, \varepsilon}}., \overline{\subseteq^\varepsilon}, \overline{|^\varepsilon})$ *does* **not hold** *in a structure* $\mathfrak{S} \in \mathrm{ACMS}$.

Proof. From Theorem 4. Note that a sentence ϕ is satisfied in a structure with carrier $\mathbf{X}$ iff $\{1\} \models \phi$ for the trivial non-empty team $\{1\} \subseteq \mathbf{X}^0$.

Corollary 5. *It is semidecidable whether a formula* $\phi \in \mathcal{L}_+(|)$ **holds** *in a structure* $\mathfrak{S} \in \mathrm{OCMS}$.

Proof. From Theorem 5. Note that a sentence ϕ is satisfied in a structure with carrier $\mathbf{X}$ iff $\{1\} \models \phi$ for the trivial non-empty team $\{1\} \subseteq \mathbf{X}^0$.

A priori, having even decidability may seem desirable. This, however, is completely out of the question:

Proposition 4. *It is undecidable whether $\forall x\, R(x)$ holds in a structure $\mathfrak{S} \in$ AMCS or $\mathfrak{S} \in$ OMCS, even if we restrict to the case where the carrier space is the one-point space $\mathbf{1}$.*

Proof. In the restricted case, the question becomes whether R is interpreted as the universal predicate or as the empty predicate. This is undecidable for $R \in \mathcal{O}(\mathbf{1})$ or $R \in \mathcal{A}(\mathbf{1})$.

6   Translations Between Approximate Atoms

In classical dependence logic the expressive power of logics appended with various combinations of dependence/independence atoms has been studied. The comparisons rely on translations of the atoms. In an approximate setting we don't get exact translations, but can 'sandwich' atoms between parameterised variants of a formula using some other atoms.

In the translations we use as underlying logic first order logic without equality, and replace equality by either open or closed metric predicates.

We give three 'translations' between dependency atoms. We only show the open versions here, but the closed counterparts are proved similarly.

Proposition 5. *1. If $\varepsilon > \delta \geq 0$, then $=_\delta^\varepsilon (\overline{x}, \overline{y}) \quad \Rightarrow \quad \overline{y} \perp_{\overline{x}}^{\delta,\varepsilon} \overline{y}$.*
2. For any $\delta \geq 0, \varepsilon > 0$, $\overline{y} \perp_{\overline{x}}^{\delta,\varepsilon/2} \overline{y} \quad \Rightarrow \quad =_\delta^\varepsilon (\overline{x}, \overline{y})$.

Proof. For the first, assume $T \models =_\delta^\varepsilon (\overline{x}, \overline{y})$ and let $s, s' \in T$ be such that $d(s(\overline{x}), s'(\overline{x})) \leq \delta$. Then $d(s(\overline{y}), s'(\overline{y})) < \varepsilon$, so s satisfies the independence witness requirement $d(s(\overline{xy}), s(\overline{xy})) < \varepsilon$ and $d(s(\overline{xy}), s'(\overline{xy})) < \varepsilon$.

For the second claim, assume $T \models \overline{y} \perp_{\overline{x}}^{\delta,\varepsilon/2} \overline{y}$ and let $s, s' \in T$ be such that $d(s(\overline{x}), s'(\overline{x})) \leq \delta$. Then there is $s'' \in T$ such that $d(s''(\overline{xy}), s(\overline{xy})) < \varepsilon$ and $d(s''(\overline{xy}), s'(\overline{xy})) < \varepsilon$, and the claim follows by the triangle inequality.

The next proposition is a metric modification of Galliani's proof from [8]. The remarkable thing is, that the Boolean encoding he uses can be made to work in this metric setting.

Proposition 6. *Assuming all models considered have diameter at least D,*

$$\overline{x} \subseteq^\varepsilon \overline{y} \quad \Rightarrow \quad \forall v_1 \forall v_2 \forall \overline{z} ($$
$$(d(\overline{z}, \overline{x}) > \delta/2 \wedge d(\overline{z}, \overline{y}) > \varepsilon) \vee$$
$$(d(v_1, v_2) < d_2 + \delta \wedge d(v_1, v_2) > d_1 - \delta) \vee$$
$$(d(v_1, v_2) > d_2 \wedge d(\overline{z}, \overline{y}) > \varepsilon) \vee$$
$$((d(v_1, v_2) < d_1 \vee d(\overline{z}, \overline{y}) < \varepsilon + \delta/2) \wedge \overline{z} \perp^\delta v_1 v_2))$$

for any $d_1 < d_2 < D$ and $0 < \delta < d_1, D - d_2$.

Proof. Assume $T \models \overline{x} \subseteq^{\varepsilon} \overline{y}$. Let $T' = T[M/v_1][M/v_2][M/\overline{z}]$.[4] Let

$$T_1 = \{s \in T' : d(s(\overline{z}), s(\overline{x})) > \delta/2 \ \& \ d(s(\overline{z}), s(\overline{y})) > \varepsilon\},$$
$$T_2 = \{s \in T' : d(s(v_1), s(v_2)) < d_2 + \delta \ \& \ d(v_1, v_2) > d_1 - \delta\},$$
$$T_3 = \{s \in T' : d(s(v_1), s(v_2)) > d_2 \ \& \ d(s(\overline{z}), s(\overline{y})) > \varepsilon\},$$
$$T_4 = T' \backslash (T_1 \cup T_2 \cup T_3).$$

So we need to show that anything not in $T_1 \cup T_2 \cup T_3$ satisfies the fourth disjunct. Now, if $s \in T_4$ is such that $d(s(v_1), s(v_2)) \geq d_1 > d_1 - \delta$, then (as it is not in T_2) $d(s(v_1), s(v_2)) \geq d_2 + \delta > d_2$. Thus (since $s \notin T_3$) $d(s(\overline{z}), s(\overline{y})) \leq \varepsilon < \varepsilon + \delta/2$. So the first conjunct is satisfied.

Next consider $s, s' \in T_4$. If $d(s(\overline{z}), s(\overline{y})) \leq \varepsilon$, then $s'' = s[s'(v_1 v_2)/v_1 v_2] \in T_4$ and it witnesses the independence atom with respect to s and s'. If, on the other hand, $d(s(\overline{z}), s(\overline{y})) > \varepsilon$, then (by $s \notin T_1 \cup T_2 \cup T_3$) $d(s(\overline{z}), s(\overline{x})) \leq \delta/2$ and $d(s(v_1), s(v_2)) \leq d_2$, and thus $d(s(v_1), s(v_2)) < d_1$. Since T, and thus also T' satisfies $\overline{x} \subseteq^{\varepsilon} \overline{y}$, there is $s^+ \in T'$ such that $d(s^+(\overline{y}), s(\overline{x})) < \varepsilon$, so $d(s^+(\overline{y}), s(\overline{z})) < \varepsilon + \delta/2$. Let $s'' = s^+[s'(v_1 v_2)/v_1 v_2][s(\overline{x})/\overline{z}]$. Then $d(s''(\overline{z}), s(\overline{z})) = d(s(\overline{x}), s(\overline{z})) \leq \delta/2 < \delta$ and $s''(v_1 v_2) = s'(v_1 v_2)$ so we are done if we can show $s'' \in T_4$. But by the values for $v_1 v_2$, $s'' \notin T_2$, and $d(s''(\overline{z}), s''(\overline{y})) = d(s(\overline{x}), s^+(\overline{y})) < \varepsilon$ so $s'' \notin T_1 \cup T_3$.

Proposition 7.

$$\overline{x} \subseteq^{\varepsilon + \delta} \overline{y} \quad \Leftarrow \quad \forall v_1 \forall v_2 \forall \overline{z} ($$
$$(d(\overline{z}, \overline{x}) > \delta/2 \wedge d(\overline{z}, \overline{y}) > \varepsilon) \vee$$
$$(d(v_1, v_2) < d_2 + \delta \wedge d(v_1, v_2) > d_1 - \delta) \vee$$
$$(d(v_1, v_2) > d_2 \wedge d(\overline{z}, \overline{y}) > \varepsilon) \vee$$
$$((d(v_1, v_2) < d_1 \vee d(\overline{z}, \overline{y}) < \varepsilon + \delta/2) \wedge \overline{z} \perp^m v_1 v_2))$$

for any $d_1 < d_2 < D$ *and* $0 < \delta < d_1, D - d_2$ *and with* $m = \min\{\frac{d_2 - d_1}{2}, \frac{\delta}{2}\}$.

Proof. Assume T satisfies the formula on the right hand side and let $s \in T$ be arbitrary. Let $T' = T[M/v_1][M/v_2][M/\overline{z}]$ and choose $s' \in T'$ such that $s'(\overline{z}) = s'(\overline{x}) = s(\overline{x})$ and $d(s'(v_1), s(v_2)) < d_1$ (exists by universal quantification). Further choose $s^+ \in T'$ such that $d(s^+(\overline{z}), s^+(\overline{y})) \leq \varepsilon$ and $d(s^+(v_1), s^+(v_2)) \geq d_2 + \delta$. Then neither of s' and s^+ satisfies any of the first three disjuncts of the right hand side formula, so both must be in the part of T' satisfying the fourth. Thus there is s'' in this part satisfying $d(s''(\overline{z}), s'(\overline{z})) < m$, $d(s''(v_1 v_2), s^+(v_1 v_2)) < m$. So $d(s''(v_1), s''(v_2)) > d_2 - 2m > d_1$, and we must have $d(s''(\overline{z}), s''(\overline{y})) \leq \varepsilon + \delta/2$. So $d(s''(\overline{y}), s(\overline{x})) = d(s''(\overline{y}), s'(\overline{z})) < \varepsilon + \delta/2 + m \leq \varepsilon + \delta$.

Note that in the following translation we only have a translation for the dependence atom $=^{\varepsilon}_{\varepsilon}(\cdot)$ and not the general form $=^{\varepsilon}_{\delta}(\cdot)$. We show the proof for

---

[4] $T[M/x] = \{s(a/x) : s \in T, a \in M\}$ denotes the team one gets by adding every possible value for x to each assignment of T.

the open forms, but the closed ones go through practically verbatim. Here the closed version of the dependence atom may feel a bit more natural, as the open form talks about a contraction.

Proposition 8. *1.* $=_\varepsilon^\varepsilon (\bar{x}, y) \quad \Rightarrow \quad \forall z(d(z, y) < 2\varepsilon \vee \bar{x}z|^\varepsilon \bar{x}y)$.
2. $=_\varepsilon^\varepsilon (\bar{x}, y) \quad \Leftarrow \quad \forall z(d(z, y) < \varepsilon \vee \bar{x}z|^\varepsilon \bar{x}y)$.

Proof. For the first direction assume $T \models =_\varepsilon^\varepsilon (\bar{x}, y)$ and let $T' = T[M/z]$. Let $Y_1 = \{s \in T' : d(s(z), s(y)) < 2\varepsilon\}$ and $Y_2 = T'\backslash Y_1$. Now if $s, s' \in Y_2$ we cannot have $d(s(\bar{x}z), s'(\bar{x}y)) \leq \varepsilon$, as then we would have $d(s(y), s'(y)) < \varepsilon$ and thus $d(s(z), s(y)) \leq d(s(z), s'(y)) + d(s'(y), s(y)) < 2\varepsilon$, a contradiction. Thus Y_2 satisfies the second disjunct.

For the other direction, assume T satisfies the right hand side, and $s, s' \in T$ are such that $d(s(\bar{x}), s'(\bar{x})) \leq \varepsilon$. If $d(s(y), s'(y)) \geq \varepsilon$, then $s^+ := s[s'(y)/z]$ and $s'^+ := s'[s(y)/z]$ cannot satisfy the first disjunct on the right hand side, so we must have $d(s^+(\bar{x}z), s'^+(\bar{x}y)) > \varepsilon$, a contradiction.

7　Outlook

We have introduced team semantics for logics with both exact and approximate dependence/independence atoms in the setting of metric spaces. We have shown that for compact carrier spaces, requiring teams to be closed sets leads to very nice behaviour, provided that our formulae contain only atoms of a certain type. While not all atoms are permitted in a single formula for this, we have approximate version of all atoms available. For formulae using the *open* variants of the approximate atoms, some questions remain open for now, but we believe that this will be straight-forward to settle.

There are some potential connections to other areas of logic we wish to explore in the future. On the one hand, as observed in Subsect. 3.3 requiring teams to be closed adds a flavour of constructive math to the resulting statements. Concretely, there could be some relationship between satisfaction of formula interpreted via team semantics with closed teams, and provability in systems such as RCA_0 + WKL from reverse math [21] or BISH + WKL from intuitionistic reverse math (e.g. [5,14]).

On the other hand, the translations in Sect. 6 do not really give us a true comparison of logics, as they don't contain a measure of accuracy of the translations. A remedy seems to be considering many-valued logics, e.g., continuous first order logic from [2], that have a built-in grading of the strength of implications. Such a logic opens up a plethora of questions of the right choice of semantics, as it allows both for new connectives and enables new ways of aggregating truth values over a team.

References

1. Abramsky, S., Kontinen, J., Väänänen, J., Vollmer, H. (eds.): Dependence Logic: Theory and Applications. Springer, Cham (2016). https://doi.org/10.1007/978-3-319-31803-5
2. Yaacov, I.B., Berenstein, A., Henson, C.W., Usvyatsov, A.: Model theory for metric structures. In: Chatzidakis, Z., Macpherson, D., Pillay, A., Wilkie, A. (eds.) Model Theory with Applications to Algebra and Analysis, Vol. II. London Math Society Lecture Note Series, vol. 350, pp. 315–427. Cambridge University Press, Cambridge (2008)
3. Brattka, V., Hertling, P., Weihrauch, K.: A tutorial on computable analysis. In: Cooper, B., Löwe, B., Sorbi, A. (eds.) New Computational Paradigms: Changing Conceptions of What is Computable, pp. 425–491. Springer, New York (2008). https://doi.org/10.1007/978-0-387-68546-5_18
4. Crosilla, L., Schuster, P. (eds.): From Sets and Types to Topology and Analysis: Towards Practicable Foundations for Constructive Mathematics. Oxford Logic Guides, vol. 48. Clarendon, Oxford (2005)
5. Diener, H.: Constructive reverse mathematics. arXiv:1804.05495 (2018)
6. Durand, A., Hannula, M., Kontinen, J., Meier, A., Virtema, J.: Approximation and dependence via multiteam semantics. Ann. Math. Artif. Intell. **83**(3–4), 297–320 (2018). https://doi.org/10.1007/s10472-017-9568-4
7. Durand, A., Kontinen, J., Vollmer, H.: Expressivity and complexity of dependence logic. In: Abramsky, S., Kontinen, J., Väänänen, J., Vollmer, H. (eds.) Dependence Logic: Theory and Applications, pp. 5–32. Springer, Cham (2016). https://doi.org/10.1007/978-3-319-31803-5_2
8. Galliani, P.: Inclusion and exclusion in team semantics: on some logics of imperfect information. Ann. Pure Appl. Logic **163**(1), 68–84 (2012)
9. Grädel, E., Väänänen, J.: Dependence and independence. Stud. Log. **101**(2), 399–410 (2013)
10. Hannula, M., Kontinen, J.: A finite axiomatization of conditional independence and inclusion dependencies. Inf. Comput. **249**, 121–137 (2016). https://doi.org/10.1016/j.ic.2016.04.001
11. Hodges, W.: Compositional semantics for a language of imperfect information. Logic J. IGPL **5**, 539–563 (1997)
12. Hyttinen, T., Paolini, G., Väänänen, J.: Quantum team logic and Bell's inequalities. Rev. Symb. Logic **8**(4), 722–742 (2015). https://doi.org/10.1017/S1755020315000192
13. Hyttinen, T., Paolini, G., Väänänen, J.: A logic for arguing about probabilities in measure teams. Arch. Math. Log. **56**(5–6), 475–489 (2017). https://doi.org/10.1007/s00153-017-0535-x
14. Ishihara, H.: Constructive reverse mathematics: compactness properties. In: [4], pp. 245–267 (2005)
15. Koudas, N., Saha, A., Srivastava, D., Venkatasubramanian, S.: Metric functional dependencies. In: Proceedings of the 2009 IEEE International Conference on Data Engineering, pp. 1275–1278. ICDE 2009. IEEE Computer Society (2009). https://doi.org/10.1109/ICDE.2009.219
16. Paolini, G., Väänänen, J.: Dependence logic in pregeometries and ω-stable theories. J. Symb. Logic **81**(1), 32–55 (2016). https://doi.org/10.1017/jsl.2015.16
17. Park, C., Park, J., Park, S., Seon, D., Ziegler, M.: Computable operations on compact subsets of metric spaces with applications to Fréchet distance and shape optimization. arXiv 1701.08402 (2017). http://arxiv.org/abs/1701.08402

18. Pauly, A.: On the topological aspects of the theory of represented spaces. Computability **5**(2), 159–180 (2016). https://doi.org/10.3233/COM-150049, http://arxiv.org/abs/1204.3763

19. Pauly, A., Fouché, W.: How constructive is constructing measures? J. Logic Anal. **9** (2017). http://logicandanalysis.org/index.php/jla/issue/view/16

20. Rettinger, R., Weihrauch, K.: Products of effective topological spaces and a uniformly computable Tychonoff Theorem. Log. Methods Comput. Sci. **9**(4) (2013). https://doi.org/10.2168/LMCS-9(4:14)2013

21. Simpson, S.: Subsystems of Second Order Arithmetic. Perspectives in Logic. Cambridge University Press, Cambridge (2009)

22. Väänänen, J.: Dependence Logic: A New Approach to Independence Friendly Logic. Cambridge University Press, Cambridge (2007)

23. Väänänen, J.: The logic of approximate dependence. In: Başkent, C., Moss, L.S., Ramanujam, R. (eds.) Rohit Parikh on Logic, Language and Society. OCL, vol. 11, pp. 227–234. Springer, Cham (2017). https://doi.org/10.1007/978-3-319-47843-2_12

24. Weihrauch, K.: Computable Analysis. Springer, Heidelberg (2000). https://doi.org/10.1007/978-3-642-56999-9

Exact Satisfiabitity with Jokers

Gordon Hoi[1], Sanjay Jain[1(✉)], Sibylle Schwarz[2], and Frank Stephan[3]

[1] School of Computing, National University of Singapore,
13 Computing Drive, Block COM1, Singapore 117417, Republic of Singapore
e0013185@u.nus.edu, sanjay@comp.nus.edu.sg
[2] Fakultät Informatik, Mathematik und Naturwissenschaften,
Hochschule für Technik, Wirtschaft und Kultur Leipzig,
Gustav-Freytag-Str. 42a, 04277 Leipzig, Germany
sibylle.schwarz@htwk-leipzig.de
[3] Department of Mathematics and School of Computing,
National University of Singapore, 10 Lower Kent Ridge Road,
Block S17, Singapore 119076, Republic of Singapore
fstephan@comp.nus.edu.sg

Abstract. The XSAT problem asks for solutions of a set of clauses where for every clause exactly one literal is satisfied. The present work investigates a variant of this well-investigated topic where variables can take a joker-value (which is preserved by negation) and a clause is satisfied iff either exactly one literal is true and no literal has a joker value or exactly one literal has a joker value and no literal is true. While JX2SAT is in polynomial time, the problem becomes NP-hard when one searches for a solution with the minimum number of jokers used and the decision problem X3SAT can be reduced to the decision problem of the JX2SAT problem with a bound on the number of jokers used. JX3SAT is in both cases, with or without optimisation of the number of jokers, NP-hard and X3SAT can be reduced to it without increasing the number of variables. Furthermore, the general JXSAT problem can be solved in the same amount of time as variable-weighted XSAT and the obtained solution has the minimum amount of number of jokers possible.

1 Introduction

Consider a propositional formula φ with n variables, say $x_1, x_2, ..., x_n$. A well-studied question is which values for x_i from $\{0, 1\}$ one can take so that φ evaluates to 1. This problem is also known as the Boolean satisfiability problem, SAT, and was shown to be the first NP complete problem.

Since then, many variants of Boolean satisfiability problem have been introduced. One such variant is the 3SAT problem, where every clause must have

S. Jain was supported in part by NUS grant C252-000-087-001; furthermore, S. Jain and F. Stephan have been supported in part by the Singapore Ministry of Education Academic Research Fund Tier 2 grant MOE2016-T2-1-019/R146-000-234-112. Part of the work was done while S. Schwarz was on sabbatical leave to the National University of Singapore.

© Springer Nature Switzerland AG 2019
T. V. Gopal and J. Watada (Eds.): TAMC 2019, LNCS 11436, pp. 279–294, 2019.
https://doi.org/10.1007/978-3-030-14812-6_17

exactly 3 literals. Another variant is the Exact Satisfiability problem, XSAT, where we require that exactly 1 literal can take on the value 1 and the other literals must take on 0. The problems X2SAT and X3SAT are special instances of XSAT where we restrict the clauses to 2 and 3 literals respectively. It is known that the decision problem of 2SAT, X2SAT are in the class **P**. On the other hand, the decision problems 3SAT [5], XSAT [12] and X3SAT [14] are known to be **NP**-complete. Furthermore, one can also deviate from the decision problems as mentioned above to talk about optimisation problems. If we wish to maximise the number of satisfied clauses in an 2SAT problem, also known as Max2SAT, the complexity of this problem increases and is now **NP**-hard as compared to its decision counterpart [1,9]. Here Max2SAT asks how many clauses of (perhaps weighted) 2SAT-formulas can be satisfied in the best case and this problem has been widely studied [6,8,13]. Max2SAT can be generalised to Max-2-CSP where one has to satisfy as many constraints as possible over a domain, in this case a binary domain [3]. Chen [2] shows that various natural classes of CSP-problems have either a complement solvable in logarithmic time or satisfy one of the conditions **LOGSPACE**-complete, ⊕**LOGSPACE**-complete, **NLOGSPACE**-complete, **P**-complete and **NP**-complete.

The Boolean satisfiability problem with two truth values are indeed well-studied and instead of asking, how many clauses can be satisfied, one can also ask how many special third values are needed in order to satisfy all clauses. So in this paper, we introduce another truth value, "J" for joker, into our framework. The value "J" can be treated as an alternative to the truth value 1 with some slight changes. In this case, our satisfying condition can be either the values 1 or "J". The goal of this paper is to investigate if having more truth values can change the structural complexity of known existing problems. If one allows more truth values in the framework, will some existing **NP**-hard problem be brought down to **P** or vice versa.

We are in fact, not the first to introduce an additional truth value in our framework and study them. For example, Lardeux, Saubion and Hao, like us, introduce a new truth value into the system [7] and then study Boolean satisfiability from there. Their goal, unlike what we do, is to introduce a new truth value in order to define a local search procedure and to study the fundamental mechanisms behind it.

We show the following results. The decision problem JX2SAT is in the class **P** while the decision problem JX3SAT is **NP**-complete; in these problems we allow the usage of joker-values without making any statement on their number. The optimisation variant MinJX2SAT is shown to be **NP**-hard when asked to find the minimum amount of jokers or **NP**-complete when asked whether the number of jokers can be brought below a certain threshold. Finally, we give an exponential time algorithm to solve MinJXSAT in $O(1.185^n)$ time.

This algorithm builds on one of Porschen [10]; we refer the interested reader to the textbook of Fomin and Kratsch [4] for an overview on exponential time algorithms.

2 The Underlying Logic and Joker Assignments

Given a set $V = \{x_1, \ldots, x_n\}$ of variables, propositional formulas are defined inductively with the usual connectives like $\neg, \vee, \wedge, \rightarrow, \leftrightarrow$.

In this paper, we only use formulas in conjunctive normal form, that is, conjunctions of disjunctions (clauses) of literals. k-CNFs are formulas in CNF where every clause contains at most k literals. For XSAT-formulas and JXSAT-formulas, we make however the convention that a clause is satisfied if and only if exactly one literal in the clause is not 0; thus clauses like $x_1 \vee x_2 \vee \neg x_3$ have a different meaning in XSAT than in normal SAT, see below for more details.

The semantics of propositional formulas in XSAT problems with joker values is defined by assignments $\beta : V \rightarrow \{0, 1, J\}$. Negation maps 0 to 1, 1 to 0 and J to J.

In SAT problems, an assignment $\beta : V \rightarrow \{0, 1, J\}$ satisfies a k-clause $l_1 \vee \cdots \vee l_k$ iff there is *at least one* $i \in \{1, \ldots, k\}$ such that $\beta(l_i) \in \{1, J\}$.

In XSAT problems, an assignment $\beta : V \rightarrow \{0, 1, J\}$ satisfies a k-clause $l_1 \vee \cdots \vee l_k$ iff there is *exactly one* $i \in \{1, \ldots, k\}$ such that $\beta(l_i) \in \{1, J\}$ and $\beta(l_j) = 0$ for all other literals.

An assignment $\beta : V \rightarrow \{0, 1, J\}$ satisfies a CNF φ iff it satisfies every clause in φ. The problem JSAT is defined as follows:

Instance: CNF φ.
Question: Is there an assignment $\beta : \text{vars}(\varphi) \rightarrow \{0, 1, J\}$ such that β maps at least one literal in each clause of φ to 1 or J?

The problem JXSAT is defined as follows:

Instance: CNF φ.
Question: Is there an assignment $\beta : \text{vars}(\varphi) \rightarrow \{0, 1, J\}$ such that β maps exactly one literal in each clause of φ to 1 or J and the other literals in the clause to 0?

For CNFs containing only positive literals, the problems JSAT and JXSAT coincide with SAT and XSAT. In our work we mainly concentrate on JXSAT, JX2SAT and JX3SAT as well as MinJXSAT, MinJX2SAT and MinJX3SAT.

Proposition 1. *For every negation-free CNF φ, $\varphi \in$ JXSAT iff $\varphi \in$ XSAT.*

Proof. We translate each assignment $\beta : \text{vars}(\varphi) \rightarrow \{0, 1, J\}$ to the assignment $\beta' : \text{vars}(\varphi) \rightarrow \{0, 1\}$ where

$$\forall x_i \in \text{vars}(\varphi) : \quad \beta'(x_i) = \left\{ \begin{array}{l} 0 \text{ if } \beta(x_i) = 0 \\ 1 \text{ if } \beta(x_i) \in \{1, J\} \end{array} \right\}.$$

Then β' satisfies φ iff β satisfies φ. $\square$

CNFs containing a variable x_i only negated can be translated to another CNF by replacing each negative literal $\neg x_i$ by positive literal x_i'. Since the translated formula does not contain x_i, the number of variables in the formula does not increase.

Hence on the set of all CNFs containing each variable only positive or only negated, the problem JXSAT coincides with XSAT. Note that negation-free SAT is trivially in **P**, as one can make all literals true, since no literal comes in two opposed forms; thus also here the negation-free JSAT and SAT problems are equivalent. So the interesting case is negation-free XSAT and that is as hard as normal XSAT, as one can translate every XSAT instance in polynomial time into a negation-free XSAT instance where the number of variables and the number of clauses does not increase. For that reason, JXSAT is at least as hard as XSAT. A direct proof, not relying on this fact, for the hardness of JXSAT can also be obtained by the direct transfer of the reduction of Theorem 7 from X3SAT to JX3SAT into a reduction from XSAT to JXSAT.

3   Encodings of Truth Values and Clauses

3.1   Encoding of Truth Values as Pairs of Booleans

In the following, we will use the encoding $e_2 : \{0, 1, J\} \rightarrow \{0, 1\}^2$ of the truth values $x \in \{0, 1, J\}$ as a pair $(n, p) \in \{0, 1\}^2$ defined by

$$e_2(0) = (1, 0), \quad e_2(1) = (0, 1), \quad e_2(J) = (0, 0). \tag{1}$$

For each variable $x_i \in V$, n_i and p_i denote the first and second element of this pair, respectively. This encoding is a bijection between the sets $\{0, 1, J\}$ and $\{(1, 0), (0, 1), (0, 0)\}$. The inverse of e_2 is $e_2^{-1} : \{(1, 0), (0, 1), (0, 0)\} \rightarrow \{0, 1, J\}$.

$$e_2^{-1}(1, 0) = 0, \quad e_2^{-1}(0, 1) = 1, \quad e_2^{-1}(0, 0) = J. \tag{2}$$

3.2   Encoding of 2CNFs

Every 2-clause $l_i \vee l_j$ with variables in $\{x_1, \ldots, x_n\}$ is translated to a conjunction of two 2-clauses with variables in $\{n_1, \ldots, n_n, p_1, \ldots, p_n\}$ by

$$
\begin{aligned}
e_{2c}(x_i \vee x_j) &= (n_i \vee n_j) \wedge (\neg n_i \vee \neg n_j), \\
e_{2c}(x_i \vee \neg x_j) &= (n_i \vee p_j) \wedge (\neg n_i \vee \neg p_j), \\
e_{2c}(\neg x_i \vee x_j) &= (p_i \vee n_j) \wedge (\neg p_i \vee \neg n_j), \\
e_{2c}(\neg x_i \vee \neg x_j) &= (p_i \vee p_j) \wedge (\neg p_i \vee \neg p_j).
\end{aligned}
\tag{3}
$$

To prevent the pair $(1, 1)$ that is not an encoding of a truth value from $\{0, 1, J\}$, we add a clause $\neg n_i \vee \neg p_i$ for each variable x_i to the resulting 2CNF.

This encoding transforms every 2CNF $\varphi = \bigwedge_{i \in \{1, \ldots, m\}} c_i$ to a 2CNF

$$e_{2c}(\varphi) = \bigwedge_{i \in \{1, \ldots, m\}} e_{2c}(c_i) \wedge \bigwedge_{i \in \{1, \ldots, n\}} (\neg n_i \vee \neg p_i) \tag{4}$$

with $2n$ variables and $2m + n$ clauses that will be interpreted as 2SAT instance.

Proposition 2. *For every set C of 2-clauses with variables from $\{x_1, \ldots, x_n\}$, there is an assignment*

$$\beta : \{x_1, \ldots, x_n\} \to \{0, 1, J\}$$

such that for each clause $c = (l_i \vee l_j)$ in C exactly one of $\beta(l_i), \beta(l_j)$ is 0 iff there is an assignment

$$\beta' : \{n_1, \ldots, n_n, p_1, \ldots, p_n\} \to \{0, 1\}$$

such that for each clause $c = (l_i \vee l_j)$ in C, at least one of $\beta'(l_i), \beta'(l_j)$ is 1 in each clause of $e_{2c}(c)$ as well as $(\neg n_i \vee \neg p_i)$ for every variable x_i occurring in c.

Proof. ($\Rightarrow$) Given an assignment $\beta : V \to \{0, 1, J\}$, we define the assignment $\beta' : \{n_1, \ldots, n_n, p_1, \ldots, p_n\} \to \{0, 1\}$ for $e_{c2}(c)$ by the encoding of the truth values in $\{0, 1, J\}$ given in Eq. 1, i.e., for each $i \in \{1, \ldots, n\}$, $\beta'(n_i)$ and $\beta'(p_i)$ are defined by

$$(\beta'(n_i), \beta'(p_i)) = e_2(\beta(x_i)).$$

Let β be a satisfying assignment for C which satisfies exactly one literal in each clause in C. By definition of e_2 (Eq. 1), we have $\neg n_i \vee \neg p_i$ for every variable x_i occurring in c.

We show that β' satisfies at least one literal in each clause in $e_{2c}(c)$, for all $c \in C$.

For every type $c = l_i \vee l_j$ of clauses we show that if β satisfies exactly one literal in c then β' satisfies $e_{2c}(c)$ defined in Eq. 3.

- In clause $c = (x_i \vee x_j)$, exactly one of x_i, x_j is satisfied in each of the following cases:
  - In case $\{\beta(x_i), \beta(x_j)\} = \{0, 1\}$,
    We have $\{(\beta'(n_i), \beta'(p_i)), (\beta'(n_j), \beta'(p_j))\} = \{(1, 0), (0, 1)\}$.
  - In case $\{\beta(x_i), \beta(x_j)\} = \{0, J\}$,
    We have $\{(\beta'(n_i), \beta'(p_i)), (\beta'(n_j), \beta'(p_j))\} = \{(1, 0), (0, 0)\}$.
  In both cases, we have $\{\beta'(n_i), \beta'(n_j)\} = \{0, 1\}$ and therefore β' satisfies at least one literal in each clause of $e_{2c}(c) = (n_i \vee n_j) \wedge (\neg n_i \vee \neg n_j)$.
- In clause $c = x_i \vee \neg x_j$, exactly one of $x_i, \neg x_j$ is satisfied in each of the following cases:
  - In case $\beta(x_i) = \beta(x_j) \in \{0, 1\}$,
    We have $(\beta'(n_i), \beta'(p_i)) = (\beta'(n_j), \beta'(p_j)) \in \{(1, 0), (0, 1)\}$.
  - In case $\beta(x_i) = 0$ and $\beta(x_j) = J$,
    We have $(\beta'(n_i), \beta'(p_i)) = (1, 0)$ and $(\beta'(n_j), \beta'(p_j)) = (0, 0)$.
  - In case $\beta(x_i) = J$ and $\beta(x_j) = 1$,
    We have $(\beta'(n_i), \beta'(p_i)) = (0, 0)$ and $(\beta'(n_j), \beta'(p_j)) = (0, 1)$.
  In all cases, we have $\{\beta'(n_i), \beta'(p_j)\} = \{0, 1\}$ and therefore β' satisfies at least one literal in each clause of $e_{2c}(c) = (n_i \vee p_j) \wedge (\neg n_i \vee \neg p_j)$.
- The case $c = \neg x_i \vee x_j$ is similar to the previous case.
- In clause $c = \neg x_i \vee \neg x_j$, exactly one of $\neg x_i, \neg x_j$ is satisfied in each of the following cases:

- In case $\{\beta(x_i), \beta(x_j)\} = \{0,1\}$,
  We have $\{(\beta'(n_i), \beta'(p_i)), (\beta'(n_j), \beta'(p_j))\} = \{(1,0),(0,1)\}$.
- In case $\{\beta(x_i), \beta(x_j)\} = \{1, J\}$,
  We have $\{(\beta'(n_i), \beta'(p_i)), (\beta'(n_j), \beta'(p_j))\} = \{(0,1),(0,0)\}$.

In both cases, we have $\{\beta'(p_i), \beta'(p_j)\} = \{0,1\}$ and therefore β' satisfies at least one literal in each clause of $e_{2c}(c) = (p_i \vee p_j) \wedge (\neg p_i \vee \neg p_j)$.

($\Leftarrow$) Let $\beta' : \{n_1, \ldots, n_n, p_1, \ldots, p_n\} \to \{0,1\}$ be an assignment such that at least one of $\beta'(l_i), \beta'(l_j)$ is 1 in each clause of $e_{2c}(c)$ and in $\neg n_i \vee \neg p_i$ for every variable x_i occurring in c, for each $c \in \mathcal{C}$.

We define $\beta : \{x_1, \ldots, x_n\} \to \{0, 1, J\}$ by the inverse encoding defined in Eq. 2:

$$\beta(x_i) = e_{2c}^{-1}(\beta'(n_i), \beta'(p_i)).$$

For each clause $c = (l_i \vee l_j) \in \mathcal{C}$, we show that exactly one of $\beta(l_i), \beta(l_j)$ is 0.

- In case $c = x_i \vee x_j$,
  β' satisfies $e_{2c}(c) = (n_i \vee n_j) \wedge (\neg n_i \vee \neg n_j)$, that is, $\{\beta'(n_i), \beta'(n_j)\} = \{0,1\}$.
  - In case $\beta'(n_i) = 0, \beta'(n_j) = 1$,
    $(n_i, p_i) \in \{(0,1),(0,0)\}$ and $(n_j, p_j) = (1,0)$ and hence $\beta(x_i) \in \{1, J\}$ and $\beta(x_j) = 0$.
  - In case $\beta'(n_i) = 1, \beta'(n_j) = 0$ (similar to the above case),
    we have $(n_i, p_i) = (1,0)$ and $(n_j, p_j) \in \{(0,1),(0,0)\}$ and hence $\beta(x_i) = 0$ and $\beta(x_j) \in \{1, J\}$.

  In each case, β maps exactly one of the literals x_i, x_j to 0.
- In case $c = x_i \vee \neg x_j$,
  β' satisfies $e_{2c}(x_i \vee \neg x_j) = (n_i \vee p_j) \wedge (\neg n_i \vee \neg p_j)$, that is, $\{\beta'(n_i), \beta'(p_j)\} = \{0,1\}$.
  - In case $\beta'(n_i) = 0, \beta'(p_j) = 1$,
    We have $(n_i, p_i) \in \{(0,0),(0,1)\}$ and $(n_j, p_j) = (0,1)$ and hence $\beta(x_i) \in \{1, J\}$ and $\beta(x_j) = 1$, that is, $\beta(\neg x_j) = 0$.
  - In case $\beta'(n_i) = 1, \beta'(p_j) = 0$,
    We have $(n_i, p_i) = (1,0)$ and $(n_j, p_j) \in \{(1,0),(0,0)\}$ and hence $\beta(x_i) = 0$ and $\beta(x_j) \in \{0, J\}$, that is, $\beta(\neg x_j) \in \{1, J\}$.

  In each case, β maps exactly one of the literals $x_i, \neg x_j$ to 0.
- The case $c = \neg x_i \vee x_j$ is similar to the previous case.
- In case $c = \neg x_i \vee \neg x_j$,
  β' satisfies $e_{2c}(\neg x_i \vee \neg x_j) = (p_i \vee p_j) \wedge (\neg p_i \vee \neg p_j)$, that is, $\{\beta'(p_i), \beta'(p_j)\} = \{0,1\}$.
  - If $\beta'(p_i) = 0, \beta'(p_j) = 1$,
    Then $(\beta'(n_i), \beta'(p_i)) \in \{(0,0),(1,0)\}$, that is, $\beta(x_i) \in \{0, J\}$ and $\beta(\neg x_i) \in \{1, J\}$, and $(\beta'(n_j), \beta'(p_j)) = (0,1)$, that is, $\beta(x_j) = 1$ and $\beta(\neg x_j) = 0$.
  - If $\beta'(p_i) = 1, \beta'(p_j) = 0$ (similar to the above case),
    Then $(\beta'(n_i), \beta'(p_i)) = (0,1)$, that is, $\beta(x_i) = 1$, and $\beta(\neg x_i) = 0$, and $(\beta'(n_j), \beta'(p_j)) \in \{(1,0),(0,0)\}$, that is, $\beta(x_j) \in \{0, J\}$ and $\beta(\neg x_j) \in \{1, J\}$

  In each case, β maps exactly one of the literals $\neg x_i, \neg x_j$ to 0.

This completes the proof. $\square$

3.3   Encoding of Truth Values as Triples of Booleans

In the following, we will use the encoding $e_3 : \{0, 1, J\} \rightarrow \{0, 1\}^3$ of the truth values $x \in \{0, 1, J\}$ as a triple $(n, p, o) \in \{0, 1\}^3$ defined by

$$e_3(0) = (1, 0, 0), \quad e_3(1) = (0, 1, 0), \quad e_3(J) = (0, 0, 1). \tag{5}$$

For each variable $x_i \in V$, n_i, p_i and o_i denote the first, second and third element of this triple, respectively. This encoding is a bijection between the sets $\{0, 1, J\}$ and $\{(1, 0, 0), (0, 1, 0), (0, 0, 1)\}$. The inverse of e_3 is $e_3^{-1} : \{(1, 0, 0), (0, 1, 0), (0, 0, 1)\} \rightarrow \{0, 1, J\}$:

$$e_3^{-1}(1, 0, 0) = 0, \quad e_3^{-1}(0, 1, 0) = 1, \quad e_3^{-1}(0, 0, 1) = J. \tag{6}$$

3.4   Encoding of 3CNFs

Every 3-clause $l_i \vee l_j \vee l_k$ with variables in $\{x_1, \ldots, x_n\}$ is translated to a clause with variables in $\{n_1, \ldots, n_n, p_1, \ldots, p_n, o_1, \ldots, o_n\}$. Because the semantic interpretation of $\vee$ is symmetric, it suffices to define e_{3c} in the following cases:

$$\begin{aligned}
e_{3c}(x_i \vee x_j \vee x_k) &= (\neg n_i \vee \neg n_j \vee \neg n_k), \\
e_{3c}(\neg x_i \vee x_j \vee x_k) &= (\neg p_i \vee \neg n_j \vee \neg n_k), \\
e_{3c}(\neg x_i \vee \neg x_j \vee x_k) &= (\neg p_i \vee \neg p_j \vee \neg n_k), \\
e_{3c}(\neg x_i \vee \neg x_j \vee \neg x_k) &= (\neg p_i \vee \neg p_j \vee \neg p_k).
\end{aligned} \tag{7}$$

For every variable x_i, we add the 3-clause $p_i \vee n_i \vee o_i$ to guarantee that every satisfying assignment maps each variable x_i to a unique truth value in $\{0, 1, J\}$. This encoding transforms every 3CNF $\varphi = \bigwedge_{i \in \{1, \ldots, m\}} c_i$ to a 3CNF

$$e_{3c}(\varphi) = \bigwedge_{i \in \{1, \ldots, m\}} e_{3c}(c_i) \wedge \bigwedge_{i \in \{1, \ldots, n\}} (n_i \vee p_i \vee o_i) \tag{8}$$

with $3n$ variables and $m + n$ clauses that will be interpreted as X3SAT instance.

Proposition 3. *For every set of clauses C, with variables from $\{x_1, \ldots, x_n\}$, there is an assignment $\beta : \{x_1, \ldots, x_n\} \rightarrow \{0, 1, J\}$ such that exactly one of $\beta(l_i), \beta(l_j), \beta(l_k)$ is in $\{1, J\}$ for every clause $(l_i \vee l_j \vee l_k)$ in C iff there is an assignment $\beta' : \{n_1, \ldots, n_n, p_1, \ldots, p_n, o_1, \ldots, o_n\} \rightarrow \{0, 1\}$ such that exactly one of $\beta'(l_i), \beta'(l_j), \beta'(l_k)$ is 1 in each clause $(l_i \vee l_j \vee l_k)$ of $e_{3c}(c)$ for each $c \in C$ as well as for each $(n_i \vee p_i \vee o_i)$ for each variable x_i occurring in clauses of C.*

Proof. Note that $\beta(x_i) \in \{1, J\}$ iff $\beta'(\neg n_i) = 1$, and $\beta(\neg x_i) \in \{1, J\}$ iff $\beta'(\neg p_i) = 1$. The Proposition now follows from the definition of e_{3c}. $\square$

4   The Decision Problem JX2SAT

Instance: 2CNF φ.
Question: Is there an assignment $\beta : \text{vars}(\varphi) \rightarrow \{0, 1, J\}$ such that β maps exactly one literal in each clause of φ to 1 or J and the other literal to 0?

The encoding defined in Sect. 3.2, Eqs. 3 and 4, maps JX2SAT instances φ to 2SAT instances $e_{2c}(\varphi)$.

Lemma 4. *For every 2CNF φ we have*

$$\varphi \in \text{JX2SAT} \ iff \ e_{2c}(\varphi) \in \text{2SAT}.$$

Proof. Suppose φ is in JX2SAT as witnessed by assignment β to the variables. From the proof of Proposition 2, we know how to transform β to an assignment $\beta' : \{n_1, \ldots, n_n, p_1, \ldots, p_n\} \rightarrow \{0, 1\}$ that satisfies the corresponding 2SAT formula $e_{2c}(c)$, for each clause c in φ, as well as $\neg n_i \vee \neg p_i$ for every variable x_i occurring in c. Hence β' satisfies the 2SAT instance $e_{2c}(\varphi)$.

Suppose β' witnesses that $e_{2c}(\varphi) \in$ 2SAT. Then, from the proof of Proposition 2, we know how to transform β' to an assignment β such that β satisfies exactly one literal in each clause of φ. Thus, $\varphi \in$ JX2SAT. □

Theorem 5. *The decision problem JX2SAT is in **P**.*

Proof. By Lemma 4, the encoding in Eq. 3 is a reduction from JX2SAT to 2SAT. Since 2SAT is in **P** and the reduction is linear, JX2SAT is also in the class **P**. □

5   The Optimisation Problem MinJX2SAT

MinJX2SAT is the problem to decide if 2CNF has a satisfying assignment with a given number of jokers.

Instance: (φ, h), where φ is a 2CNF and $h \in \mathbb{N}$.
Question: Is there an assignment $\beta : \text{vars}(\varphi) \rightarrow \{0, 1, J\}$ such that $|\beta^{-1}(J)| \leq h$ and
β maps exactly one literal in each clause of φ to 1 or J and the other literals in the clause to 0?

The minimum number of jokers required to satisfy a 2CNF φ can be found by iteratively increasing h in (φ, h), beginning with $h = 0$. We note that the problem, as stated above, is even in **NP**; however, if one wants to determine the h by iterated solving of the problem, then the problem to get the optimal h from an JX2SAT instance is **NP**-hard, but there is no known method to reduce this numerical problem to a decision problem in **NP** with one query only.

Theorem 6. MinJX2SAT *is **NP**-hard.*

Proof. We will show this via a reduction of X3SAT to MinJX2SAT. Then we have **NP**-completeness of MinJX2SAT by **NP**-completeness of X3SAT [14].

First, we translate 3CNFs $\varphi = \bigwedge_{i \in \{1,...,m\}} c_i$ with $\text{vars}(\varphi) = \{x_1, \ldots, x_n\}$, to MinJX2SAT instances (φ', m) where m is the number of clauses in φ. Then we show that $\varphi \in$ X3SAT iff $(\varphi', m) \in$ MinJX2SAT.

$\varphi' = e_{32c}(\varphi)$ is defined by encoding each clause $c_i = \{l_{i1} \lor l_{i2} \lor l_{i3}\}$ in φ by a conjunction of six clauses

$$
\begin{aligned}
e_{32c}(c_i) = \; & (l_{i1} \lor y_{i1}) & (c_{i1}) \\
& \land (l_{i2} \lor \neg y_{i1}) & (c_{i2}) \\
& \land (l_{i2} \lor y_{i2}) & (c_{i3}) \\
& \land (l_{i3} \lor \neg y_{i2}) & (c_{i4}) \\
& \land (l_{i3} \lor y_{i3}) & (c_{i5}) \\
& \land (l_{i1} \lor \neg y_{i3}) & (c_{i6})
\end{aligned}
\tag{9}
$$

$e_{32c}(\varphi)$ contains $n + 3m$ variables $\text{vars}(e_{32c}(\varphi)) = \{x_1, \ldots, x_n\} \cup \bigcup_{i \in \{1,...,m\}} \{y_{i1}, y_{i2}, y_{i3}\}$. The auxiliary variables y_{ij} guarantee that in every assignment that satisfies $e_{32c}(\varphi)$, exactly one literal is true in each clause of φ.

For every variable x_i, we add the clause $x_i \lor \neg x_i$ to avoid assignments mapping x_i to J.

We get the transformation of X3SAT instances $\varphi = \bigwedge_{i \in \{1,...,m\}} c_i$ to a MinJX2SAT instance $(e_{32c}(\varphi), m)$ where

$$
e_{32c}(\varphi) = \bigwedge_{i \in \{1,...,m\}} e_{32c}(c_i) \land \bigwedge_{i \in \{1,...,n\}} (x_i \lor \neg x_i).
\tag{10}
$$

We show that an assignment β satisfies exactly one literal in each clause of φ iff there is an assignment β' with $|\beta'^{-1}(J)| \leq m$ mapping exactly one literal in each clause of $e_{32c}(\varphi)$ to 1 or J and the others to 0.

($\Rightarrow$) Let $\beta : \{x_1, \ldots, x_n\} \to \{0, 1\}$ be an assignment that satisfies exactly one literal in each clause $l_{i1} \lor l_{i2} \lor l_{i3}$ in φ.

We extend the assignment β to an assignment

$$
\beta' : \{x_1, \ldots, x_n\} \cup \{y_{ij} \mid i \in \{1, \ldots, m\}, j \in \{1, 2, 3\}\} \to \{0, 1, J\}
$$

with $|\beta'^{-1}(J)| = m$ and show that β' satisfies the encoded 2CNF defined in Eq. 9.

For every clause c_i in φ, we define the undirected graph $G(c_i) = (V, E)$ where

$$
\begin{aligned}
V = \; & \{l_{i1}, l_{i2}, l_{i3}, y_{i1}, y_{i2}, y_{i3}, \neg y_{i1}, \neg y_{i2}, \neg y_{i3}\} \text{ and} \\
E = \; & \{\{y_{ij}, \neg y_{ij}\} \mid j \in \{1, 2, 3\}\} \\
& \cup \{\{l_{ij}, l\} \mid (l_{ij} \lor l) \text{ is a clause in } e_{32c}(c_i)\};
\end{aligned}
$$

In this graph we have that two neighbouring nodes are either in a JX2SAT-clause or a pair of negated literals. Now $G(c_i) = l_{i1} - y_{i1} - \neg y_{i1} - l_{i2} - y_{i2} - \neg y_{i2} - l_{i3} - y_{i3} - \neg y_{i3} - l_{i1}$ is a circle of length nine and thus has no 2-colouring using

0 and 1. Hence every satisfying assignment β' for φ' maps at least one literal in $e_{32c}(c)$ to J. Because β' satisfies $\bigwedge_{i \in \{1,\ldots,n\}}(x_i \vee \neg x_i)$, it maps each x_i to 0 or 1. Therefore, $\beta'(y_{ij}) = J$ for one or more variables among y_{i1}, y_{i2}, y_{i3}.

The following construction shows that it suffices to map exactly one of the y_{ij} to J for each clause $c_i = l_{i1} \vee l_{i3} \vee l_{i3}$ in φ. Let $\beta'(x_i) = \beta(x_i)$.

Without loss of generality, we can assume $\beta(l_{i1}) = 1$ and $\beta(l_{i2}) = \beta(l_{i3}) = 0$, because the semantic interpretation of $\vee$ is symmetric.

Since $\beta'(l_{i1}) = \beta(l_{i1}) = 1$, we have $\beta'(y_{i1}) = 0$ and $\beta'(y_{i3}) = 1$ to satisfy clauses c_{i1} and c_{i6} in Eq. 9. Since $\beta'(l_{i2}) = \beta'(l_{i3}) = 0$, clauses c_{i2} and c_{i5} are satisfied by β' as well.

The remaining clauses c_{i3} and c_{i4} contain the variable y_{i2} as positive and negative literal, respectively. Because $\beta(l_{i2}) = \beta(l_{i3}) = 0$, neither y_{i2} nor $\neg y_{i2}$ can be mapped to 0 by a satisfying assignment. To satisfy both clauses, β' maps y_{i2} to J.

Applied to each clause in the original formula φ, this proves that β' satisfies $e_{32c}(\varphi)$ using exactly one joker for each clause of the original formula φ.

We have defined a construction of an exact satisfying assignment β' for $e_{32c}(\varphi)$, given an exact satisfying assignment β for φ. Hence we have shown that exact satisfiability of φ implies exact satisfiability of $e_{32c}(\varphi)$.

($\Leftarrow$) Let $\beta' : \{x_1, \ldots, x_n\} \cup \bigcup_{i \in \{1,\ldots,m\}} \{y_{i1}, y_{i2}, y_{i3}\} \rightarrow \{0, 1, J\}$ be an assignment that satisfies exactly one literal of each clause in $e_{32c}(\varphi)$. We show that the restriction of β' to the domain $\{x_1, \ldots, x_n\}$ is an assignment $\beta : \{x_1, \ldots, x_n\} \rightarrow \{0, 1\}$ that maps exactly one literal of each clause $l_{i1} \vee l_{i2} \vee l_{i3}$ in φ to 1.

Because $e_{32c}(\varphi)$ contains a clause $x_i \vee \neg x_i$ for each original variable $x_i \in \{x_1, \ldots, x_n\}$, there is no $x_i \in \{x_1, \ldots, x_n\}$ such that $\beta(x_i) = \beta'(x_i) = J$.

For each clause $c_i = l_{i1} \vee l_{i2} \vee l_{i3}$ in φ we know that β' satisfies the encoding $e_{32c}(c_i)$ defined in Eq. 9.

We show that β maps exactly one of the literals l_{ij} to 1 and the others to 0.

1. We show that β maps at least one of the literals l_{ij} to 1.
   Assume $\beta(l_{i1}) = \beta(l_{i2}) = \beta(l_{i3}) = 0$.
   Then all clauses $c_{i1}, \ldots, c_{i6}$ can only be satisfied if $\beta'(y_{i1}) = \beta'(y_{i2}) = \beta'(y_{i3}) = J$. But in this case, β' does not satisfy the condition $|\beta'^{-1}(J)| \leq m$ since the encoding of each clause requires at least one J.
   A contradiction to the assumption; thus β maps at least one of the literals l_{ij} to 1.
2. We show that β maps at most one of the literals l_{ij} to 1.
   Assume there are distinct k and k' in $\{1, 2, 3\}$ such that $\beta(l_{ik}) = \beta(l_{ik'}) = 1$. Without loss of generality, we can assume $k = 1$ and $k' = 2$, e.g., $\beta(l_{i1}) = \beta(l_{i2}) = 1$.
   Then $e_{32c}(c_i)$ contains the pair of clauses $(l_{i1} \vee y_{i1})$ and $(l_{i2} \vee \neg y_{i1})$. There is no value $\beta'(y_{i1}) \in \{0, 1, J\}$ such that exactly one literal in each clause of c_i is satisfied.

A contradiction to the assumption; thus β maps at most one of the literals l_{ij} to 1.

We have shown that the restriction of β' to $\{x_1, \ldots, x_n\}$ satisfies exactly one literal in each clause of φ and therefore φ. Therefore exact satisfiability of $e_{32c}(\varphi)$ implies exact satisfiability of φ, and we have shown both directions of the equivalence $\varphi \in$ X3SAT $\Leftrightarrow e_{32c}(\varphi) \in$ MinJX2SAT. $\qquad\square$

6 The Decision Problem JX3SAT

Instance: 3CNF φ.

Question: Is there an assignment $\beta : \text{vars}(\varphi) \rightarrow \{0, 1, J\}$ such that β maps exactly one literal in each clause of φ to 1 or J and all other literals to 0?

Theorem 7. *The decision problem JX3SAT is **NP**-hard.*

Proof. We show this by reduction of X3SAT to JX3SAT. Then JX3SAT is **NP**-hard because X3SAT is **NP**-hard [14].

X3SAT instances φ are translated to JX3SAT instances φ' by adding clauses $x_i \vee \neg x_i \vee F$ for each variable x_i occurring in φ. This translates each 3CNF φ with n variables and m clauses to a 3CNF

$$\varphi' = \varphi \wedge \bigwedge_{x_i \in \text{vars}(\varphi)} (x_i \vee \neg x_i \vee F)$$

with $n+1$ variables and $m+n$ clauses. In above, F is a new variable which would always take value 0 in any satisfying assignment. We show that $\varphi \in$ X3SAT iff $\varphi' \in$ JX3SAT.

($\Rightarrow$) Assume $\beta : \{x_1, \ldots, x_n\} \rightarrow \{0, 1\}$ satisfies exactly one literal in each clause of φ. Take F to be 0. Because β maps every variable x_i to 0 or 1, it satisfies exactly one of the literals in $x_i \vee \neg x_i \vee F$, for each variable x_i. Therefore it satisfies exactly one literal in each clause of φ'.

($\Leftarrow$) Assume $\beta' : \{x_1, \ldots, x_n\} \rightarrow \{0, 1, J\}$ maps exactly one literal in each clause of φ' to 1 or J. φ' contains a clause $x_i \vee \neg x_i \vee F$ for each variable x_i and β' satisfies φ', e.g. β' maps exactly one literal in each clause of φ to 1 or J.

Assume $\beta'(x_i) = J$ for a variable $x_i \in \text{vars}(\varphi)$. Then $\beta'(\neg x_i) = J$ and β' satisfies more than one literal in $x_i \vee \neg x_i \vee F$. Therefore β' maps every variable x_i to a value in $\{0, 1\}$ and satisfies exactly one literal in each clause of φ. $\qquad\square$

7 An Exponential Algorithm to Solve MinJXSAT

Earlier, we defined the optimisation problem MinJX2SAT and showed that it is **NP**-hard. Similarly, we define the **NP**-hard optimisation problem MinJXSAT.

Instance: (φ, h), where φ is in CNF and $h \in \mathbb{N}$.

Question: Is there an assignment $\beta : \text{vars}(\varphi) \rightarrow \{0, 1, J\}$ such that $|\beta^{-1}(J)| \leq h$ and β maps exactly one literal in each clause of φ to 1 or J and the other literals in the clause to 0?

Note that since the MinJXSAT problem is more general than MinJX2SAT, having an algorithm to solve it would also mean having an algorithm to solve MinJX2SAT. Here, we first reduce the MinJXSAT problem to the MinXSAT problem, which is known to be **NP**-hard. Before presenting the reduction, we define the MinXSAT problem.

Instance: (φ, ω, h), where φ is in CNF and ω are weights assigned to variables when they take on certain values and $h \in \mathbb{N}$.

Question Is there an assignment satisfying φ such that exactly one literal in each clause is true—the rest are false and, in addition, the sum of weights of variables assigned true in this assignment is at most h?

Theorem 8. MinJXSAT *instances can be reduced to* MinXSAT *instances, using at most the same number of variables and clauses.*

Proof. Consider a MinJXSAT instance P having variables $x_1, \ldots, x_n$ with at most h jokers. h remains the same for all instances constructed. Start with MinXSAT instance G_0 as follows:

(a) Variables of G_0 are u_i, v_i, w_i for $i \in \{1, 2, \ldots, n\}$, where the intended meaning is the following: $u_i = 1$ iff $x_i = J$; $v_i = 1$ iff $x_i = 0$; $w_i = 1$ iff $x_i = 1$.

(b) Clauses of G_0 are given by (b.1) and (b.2).

(b.1) Clauses $(u_i \vee v_i \vee w_i)$, for $i = 1, \ldots, n$ Note that this ensures exactly one of u_i, v_i, w_i is one. Thus, we can have our intended interpretation of u_i, v_i, w_i with respect to the values of x_i.

(b.2) For each clause $(x_{i_1} \vee x_{i_2} \vee \ldots \vee x_{i_k} \vee \neg x_{j_1} \vee \neg x_{j_2} + \ldots \vee \neg x_{j_r})$ in P, the following clause is in G_0:
$(\neg v_{i_1} \vee \neg v_{i_2} \vee \ldots \vee \neg v_{i_k} \vee \neg w_{j_1} \vee \neg w_{j_2} \vee \ldots \vee \neg w_{j_r})$.

(c) Weights of u_i are 1 and weights of v_i, w_i are 0.

Note that if exactly one of $x_{i_1}, x_{i_2}, \ldots, x_{i_k}, \neg x_{j_1}, \neg x_{j_2}, \ldots, \neg x_{j_r}$ is non-zero then using the values for u_i, v_i, w_i as in (a), exactly one of $\neg v_{i_1}, \neg v_{i_2}, \ldots, \neg v_{i_k}, \neg w_{j_1}, \neg w_{j_2}, \ldots, \neg w_{j_r}$ is non-zero. Similarly, if exactly one of $\neg v_{i_1}, \neg v_{i_2}, \ldots, \neg v_{i_k}, \neg w_{j_1}, \neg w_{j_2}, \ldots, \neg w_{j_r}$ is non-zero then using the substitution of values as in (a), exactly one of $x_{i_1}, x_{i_2}, \ldots, x_{i_k}, \neg x_{j_1}, \neg x_{j_2}, \ldots, \neg x_{j_r}$ is non-zero.

Thus, any solution for G_0 as a XSAT instance with weight w gives a solution for P as a JXSAT instance using w jokers (using the interpretation in (a)). Similarly, any solution for P as an JXSAT instance using w jokers gives a solution for G_0 using weight w.

Note that G_0 has $3n$ variables. Below we will progressively construct $G_1, G_2, \ldots, G_n$ to reduce the number of variables. The following invariants are maintained for $k = 0, 1, \ldots, n$:

(I): In G_k, for $i = 1, \ldots, k$, at most one of u_i, v_i, w_i appears in any clause. If u_k does not appear: then any solution for G_k is a solution for G_{k-1} by fixing $u_k = 0$ and $v_k = \neg w_k$. Furthermore, the minimal weight solution for G_{k-1} has $u_k = 0$, and any solution for G_{k-1} with $u_k = 0$ is also a solution for G_k.

(II): For all i with $k < i \leq n$: Except for the clauses $(u_i \vee v_i \vee w_i)$, the variables v_i, w_i only appear as $\neg v_i, \neg w_i$ in the clauses in G_k. In G_k, $u_i, \neg u_i$, only appear in the clause $(u_i \vee v_i \vee w_i)$.

(III): If $k > 0$, the number of clauses in G_k is at most the number of clauses in G_{k-1}. The variables used in G_k are a subset of the variables used in G_{k-1}.

We now describe how to obtain G_k from G_{k-1} for $k = 1, \ldots, n$:

(1) Suppose all the clauses which have $\neg v_k$ or $\neg w_k$ in them are $\neg v_k \vee \alpha_i$ for $i \in \{1, \ldots, s\}$ and $\neg w_k \vee \beta_j$ for $j \in \{1, \ldots, r\}$, where α_i, β_j are disjunctive formula over variables.

Without loss of generality assume that no clause has both $\neg v_k$ and $\neg w_k$ (as otherwise, for the clause $(\neg v_k \vee \neg w_k \vee \alpha)$ along with $(u_k \vee v_k \vee w_k)$ we trivially have that exactly one v_k and w_k is 1 and the other 0; thus u_k must be 0 and all literals in α must be 0; we can thus simplify the clauses in G_{k-1}, and drop the clause $(\neg v_k \vee \neg w_k \vee \alpha)$).

(2) Case 1: $s > 0$ and $r > 0$, then form G_k from G_{k-1} as follows:
   (a) Drop $\neg v_k \vee \alpha_i$ for $i \in \{1, \ldots, s\}$,
   (b) Drop $\neg w_k \vee \beta_j$ for $j \in \{1, \ldots, r\}$,
   (c) Drop $(u_k \vee v_k \vee w_k)$,
   (d) Add $(\alpha_i \vee u_k \vee \beta_1)$ for $i \in \{1, \ldots, s\}$,
   (e) Add $(\alpha_1 \vee u_k \vee \beta_j)$ for $j \in \{1, \ldots, r\}$.
   Note that the old clauses of type (a), (b) and (c) imply $v_k \equiv \alpha_1 \equiv \ldots \equiv \alpha_s$ and $w_k \equiv \beta_1 \equiv \ldots \equiv \beta_r$ and that the new clauses of types (d) and (e) imply $\alpha_1 \equiv \alpha_2 \equiv \ldots \equiv \alpha_s$ and $\beta_1 \equiv \beta_2 \equiv \ldots \equiv \beta_r$. The above modification removes the variables v_k and w_k. Each solution for G_{k-1} is also a solution for G_k as we must have $v_k \equiv \alpha_i, w_k \equiv \beta_j$ in G_{k-1}. Furthermore, any solution for G_k gives a solution for G_{k-1}, by taking $v_k \equiv \alpha_1, w_k \equiv \beta_1$ in G_{k-1}.

(3) Case 2: At most one of s, r is 0:
   Then, fix $u_k = 0$ and let $v_k \equiv \neg w_k$, and drop the clause $(u_k \vee v_k \vee w_k)$ from G_{k-1} to form G_k.
   Note that any solution of G_k is also a solution for G_{k-1} by taking $u_k = 0$ and $v_k \equiv \neg w_k$. Also, any solution for G_{k-1} with $u_k = 0$ is also a solution for G_k. Note here that in the minimal weight solution of G_{k-1}, u_k must be 0.

From the above analysis, it follows that MinXSAT solution for G_n gives a MinJXSAT solution for the P and vice versa. □

Porschen [10] provided an algorithm which solves variable-weighted XSAT in time $O(1.185^n)$. Although Porschen and Pagge [11] provided a faster algorithm for the special case of variable-weighted X3SAT, this algorithm cannot be used above, as the elimination of variables increases the clause-length and therefore also JX2SAT and JX3SAT are only translated to weighted XSAT. For Corollary 9 (b), note that one can use the elimination procedure of the previous theorem to eliminate from any XSAT-instance variables occurring both positive and negative without increasing the number of clauses and then use Proposition 1.

Corollary 9

(a) MinJXSAT *can be solved in time* $O(1.185^n)$.
(b) JXSAT *has the same time complexity (except for additional polynomial time) as* XSAT *when measured in dependence of the number of variables and the number of clauses.*

8 Conclusion

In this paper, we studied the structural complexity of satisfiability with a specific joker value besides the usual two Boolean values; the joker value is preserved by negation and can be used to satisfy an XSAT formula if all other literals in the clause are 0; however, two joker values or a joker value plus a 1 do not satisfy the clause. The aim of introducing the joker value was to allow to solve also instances which would normally be unsatisfiable in the XSAT model: for example $(x \lor y) \land (\neg x \lor y) \land (\neg y)$ can be solved by choosing the values $x = J$ and $y = 0$ but cannot be solved using binary values. However, one would still want that the exception-value J is used as rarely as possible.

For this specific model of satisfiability with a joker value, our results show that their main structural properties are similar: The decision problem JX2SAT is shown to be in the class **P** by reducing it to 2SAT while the decision problem JX3SAT is **NP**-complete. However, when trying to minimise the number of joker values, we can show hardness of the problem. In order to keep it a decision problem, we formalised MinJXSAT with a bound, that is, the input is an instance plus a bound and solvable when there is a solution for which the number of jokers does not exceed the bound. Now this problem MinJX2SAT is **NP**-complete, similarly also MinJX3SAT and MinJXSAT are **NP**-complete.

This contrasts to the binary case where minimum variable weight X2SAT is in **P**: Each clause links two variables as $x = y$ or $x = \neg y$ and after all clauses are processed, the variables are split into linked groups of variables. If there is now a solution, that is, if the process of linking did not lead to a contradiction like $(x = y) \land (x = \neg y)$, then one checks such linked group, which of the two possible values would give the lower weight and assigns the variables accordingly. So variable-weighted JX2SAT behaves more like variable-weighted 2SAT than variable-weighted X2SAT.

Finally, we also gave an exponential time algorithm to solve MinJXSAT in time $O(1.185^n)$ by reducing the problem, with some preprocessing, to an algorithm to solve variable-weighted MinXSAT. Questions left open in our research are the following ones:

(a) Can the preprocessing of the MinJX2SAT algorithm be improved so that it produces an MinX3SAT problem rather than a MinXSAT problem? We did not find any way to do so, but perhaps there is some way on improving the current algorithm.
(b) Does randomness help in the algorithm? In other words, if one allows the usage of randomness, does this help to reduce the exponential time complexity at least a little bit? Though we expect that also the randomised algorithm will use exponential time, perhaps it can do it in time $O(\alpha^n)$ for some $\alpha < 1.185$.
(c) Chen [2] showed that for natural classes in CSP-solving, they are either in the complement of **NLOGTIME** or **LOGSPACE**-complete or **NLOGSPACE**-complete or $\oplus$**LOGSPACE**-complete or **P**-complete or **NP**-complete. So it would be worth looking into this for classes defined using the present or other notions of joker values for X2SAT, X3SAT and XSAT.

References

1. Accorsi, R.: MAX2SAT is NP-complete. WINS-ILLC, University of Amsterdam (1999)
2. Chen, H.: A rendezvous of logic, complexity, and algebra. SIGACT News (Logic Column). ACM (2006)
3. Gaspers, S., Sorkin, G.B.: Separate, measure and conquer: faster polynomial-space algorithms for Max 2-CSP and counting dominating sets. ACM Trans. Algorithms (TALG) **13**(4), 44:1–44:36 (2017)
4. Fomin, F.V., Kratsch, D.: Exact Exponential Algorithms. Springer, Heidelberg (2010)
5. Karp, R.M.: Reducibility among combinatorial problems. In: Miller, R.E., Thatcher, J.W., Bohlinger, J.D. (eds.) Complexity of Computer Computations, pp. 85–103. Springer, Boston (1972). https://doi.org/10.1007/978-1-4684-2001-2_9
6. Kojevnikov, A., Kulikov, A.S.: A new approach to proving upper bounds for MAX-2-SAT. In: Proceedings of the Seventeenth Annual ACM-SIAM Symposium on Discrete Algorithms, SODA 2006, pp. 11–17. ACM (2006)
7. Lardeux, F., Saubion, F., Hao, J.-K.: Three truth values for the SAT and MAX-SAT problems. In: International Joint Conference on Artificial Intelligence, IJCAI 2005, vol. 19, p. 187. Lawrence Erlbaum Associates Ltd. (2005)
8. Niedermeier, R., Rossmanith, P.: New upper bounds for maximum satisfiability. J. Algorithms **36**, 63–88 (2000)
9. Papadimitriou, C.H.: Computational Complexity. Addison-Wesley, Reading (1994)
10. Porschen, S.: On variable-weighted exact satisfiability problems. Ann. Math. Artif. Intell. **51**(1), 27–54 (2007)
11. Porschen, S., Plagge, G.: Minimizing variable-weighted X3SAT. In: Proceedings of the International Multiconference of Engineers and Computer Scientists, IMECS 2010, Hongkong, 17–19 March 2010, vol. 1, pp. 449–454 (2010)
12. Porschen, S., Schmidt, T.: On some SAT-variants over linear formulas. In: Nielsen, M., Kučera, A., Miltersen, P.B., Palamidessi, C., Tůma, P., Valencia, F. (eds.) SOFSEM 2009. LNCS, vol. 5404, pp. 449–460. Springer, Heidelberg (2009). https://doi.org/10.1007/978-3-540-95891-8_41

13. Raible, D., Fernau, H.: A new upper bound for max-2-SAT: a graph-theoretic approach. In: Ochmański, E., Tyszkiewicz, J. (eds.) MFCS 2008. LNCS, vol. 5162, pp. 551–562. Springer, Heidelberg (2008). https://doi.org/10.1007/978-3-540-85238-4_45
14. Schaefer, T.J.: The complexity of satisfiability problems. In: Proceedings of the Tenth Annual ACM Symposium on Theory of Computing, STOC 1978, pp. 216–226 (1978)

Theoretical Model of Computation and Algorithms for FPGA-Based Hardware Accelerators

Martin Hora[1], Václav Končický[2], and Jakub Tětek[2(✉)]

[1] Computer Science Institute, Charles University,
Malostranské nám. 25, 11800 Prague, Czech Republic
mhora@iuuk.mff.cuni.cz
[2] Department of Applied Mathematics, Charles University,
Malostranské nám. 25, 11800 Prague, Czech Republic
{koncicky,jtetek}@kam.mff.cuni.cz

Abstract. While FPGAs have been used extensively as hardware accelerators in industrial computation [20], no theoretical model of computation has been devised for the study of FPGA-based accelerators. In this paper, we present a theoretical model of computation on a system with conventional CPU and an FPGA, based on word-RAM. We show several algorithms in this model which are asymptotically faster than their word-RAM counterparts. Specifically, we show an algorithm for sorting, evaluation of associative operation and general techniques for speeding up some recursive algorithms and some dynamic programs. We also derive lower bounds on the running times needed to solve some problems.

1 Introduction

While Moore's law has dictated the increase in processing power of processors for several decades, the current technology is hitting its limits and the single core performance is largely staying the same. A new paradigm is needed to speed up computation. Many alternative architectures have been made, such as multicore, manycore (specifically GPU) and FPGA, which is the one we consider in this paper. FPGA (Field-programmable gate array) is a special hardware for efficient execution of boolean circuits. While there exists a theoretical model which allows theoretical treatment of algorithms in the multicore and manycore model [17], no such model has been devised for computation with FPGA-based accelerators, despite their rising popularity. In this paper, we consider heterogeneous systems which have a conventional CPU and an FPGA. This is especially relevant since Intel has recently introduced combined CPU with an FPGA [13], which should lower the latency between CPU and FPGA. We define a theoretical model for

This work was carried out while the authors were participants in the DIMATIA-DIMACS REU exchange program at Rutgers University.
The work was supported by the grant SVV–2017–260452.

T. V. Gopal and J. Watada (Eds.): TAMC 2019, LNCS 11436, pp. 295–312, 2019.
https://doi.org/10.1007/978-3-030-14812-6_18

such systems and show several asymptotically fast algorithms on this model solving some fundamental problems.

1.1 Previous Work

While there is no theory of FPGA algorithms, there are experimental results which show that FPGA can be used to speed up the computation by one to two orders of magnitude and decrease the power consumption by up to three orders of magnitude. Chrysos et al. [6] and Mahram [18] show how FPGA can be used to speed up computations in bioinformatics. FPGAs have also been used to speed up genetic algorithms, as shown by Alam [2]. In the following paper, the authors discuss what causes the performance gains of using an FPGA [10]. Che et al. [4] and Grozea et al. [9] showed a comparison with GPU and multicore CPU. Several algorithms have been implemented on FPGA and benchmarked, including matrix inversion [16], AES [5] and k-means [14].

1.2 Our Contributions

The main contribution of this paper is the definition of a model of computation which captures computation with FPGA-based accelerators (Sect. 3). In this model, we speed up classical algorithms for some fundamental problems, specifically sorting (Sect. 4.2), evaluation of associative operation (Sect. 4.3), speed up some dynamic programming problems including longest common subsequence and edit distance (Sect. 4.4) and speed up some recursive algorithms including Strassen algorithm, Karatsuba algorithm as well as many state-of-the-art algorithms for solving NP-hard problems (Sect. 4.5). In Sect. 5 we show some lower bounds which follow from a relation to cache-aware algorithms.

2 Preliminaries

2.1 Boolean Circuits

Definition 1 (Boolean circuit). *A boolean circuit with n inputs and m outputs is a labeled directed acyclic graph. It has n vertices $x_1, \cdots, x_n$ with no incoming edges, which we call the input nodes, and m vertices $y_1, \cdots, y_m$ with no outgoing edges, which we call output nodes. We call the vertices $x_1, \cdots x_n, y_1, \cdots, y_m$ the I/O nodes. Each non-input vertex in this graph corresponds to a boolean function from the set $\{\vee, \wedge, \neg, id\}$ and is called a gate. All gate vertices have indegree equal to the arity of their boolean function.*

The circuit computes a function $\{0,1\}^n \to \{0,1\}^m$ by evaluating this graph. At the beginning of computation, an input node x_i is set to the i-th bit of input. In each step of computation, gates which received their input calculate the function and send their output to the connected vertices. When all output nodes receive their value, the circuit outputs the result with the value of y_i being the i-th bit of output.

The order of a circuit is the number of its input nodes. Note that in Sect. 4.5 we introduce a different notion of order of a circuit, which does not equal the number of circuit's input nodes.

We consider two complexity measures of circuits – *circuit size*, defined as the number of gates, and *depth*, defined as the length of the longest path from an input node to an output node.

We denote the size and depth of a circuit C by $size(C)$ and $del(C)$, respectively. Size complexity and depth of a family of circuits C is denoted by $size_C(n)$ and $del_C(n)$, respectively, where n is the order of the circuit.

Definition 2 (Synchronous circuit). *A circuit is synchronous if for every gate g, all paths from input nodes to g have the same length and all output nodes are in the same depth.*

While any circuit can be made synchronous by adding intermediate identity gates, this can asymptotically increase the number of gates used. Note that our notion of synchronous circuits is stronger than the one introduced by Harper [12]. The original definition does not require the output gates to be in the same depth.

For more detailed treatment of this topic, refer to [25].

2.2 Field Programmable Gate Arrays

Field Programmable Gate Array, abbreviated as FPGA, is hardware which can efficiently execute boolean circuits. It has programmable gates, which can be programmed to perform a specific function. The gates can then be connected via interconnects into the desired circuit. Usual FPGA consists of a grid of gates with interconnects between them. The architecture of FPGAs only allows for the execution of synchronous circuits, but multiple synchronous circuits of different depths can be realised on an FPGA. There are techniques for delaying signals which can be used for execution of asynchronous circuits. However, only limited number of signals can be delayed. For this reason, we limit the model to synchronous circuits.

Let G denote the number of gates of the FPGA. Then we assume that any set of circuits with a total of at most G gates can be realised by the interconnects. This assumption tends to be mostly correct in practice, at least asymptotically. The FPGA communicates with other hardware using I/O pins, which are usually placed at the edges of the FPGA. Let I denote the number of I/O pins. It usually holds that $I \approx \sqrt{G}$ since the I/O pins are placed at the edges of the chip.

In the theoretical model, we count execution of a circuit on FPGA as a single operation, not accounting for the memory transfers. This is to capture the increased throughput of FPGA compared to a conventional CPU. Moreover, the speed of RAM is on the same order of magnitude as the speed with which the data can be read by the FPGA. For example, Intel claims that their new FPGA embedded on CPU has I/O throughput of 160 Gbs [13], while modern

RAM memories have both reading and writing speed at least 160 Gbs (for dual-channel) and up to 480 Gbs (for quad-channel) [22].

The computation of the FPGA takes place in discrete time-steps and can be pipelined. Each time-step, output signals of layer i of the circuit go to the layer $i + 1$, while layer i receives and processes the output signals of layer $i - 1$. This means that in each time-step, we can give the FPGA an input and we get the output delayed by a number of steps proportional to the depth of the circuit. If we run the FPGA at least as many times as is the depth of the circuit, the amortized time spent per input is, therefore, $\mathcal{O}(1)$.

Reprogramming an FPGA can take up a significant amount of time. For this reason, we require in the model that the circuits used in an algorithm are constructed beforehand and stay fixed during the computation.

3   Model of Computation

Throughout this paper, we use the following model of computation. We call the model Pipelining Circuit RAM, abbreviated as PCRAM. The computer consists of word-RAM of word-size w together with a circuit execution module with G circuit gates and I circuit inputs/outputs. A program in the PCRAM model consists of two parts. First is a program in the word-RAM running in time polynomial in G and I, taking no input and outputting a sequence of circuits $\mathcal{C} = C_1, \cdots, C_m$ for some m, in a standard adjacency list representation (i.e. for each gate, the incoming edges are specified) with a total of at most G gates and I I/O nodes. Since the algorithm generating the circuits does not take any input except G and I, the sequence $\mathcal{C}$ only depends on this parameters. The time needed for generating $\mathcal{C}$ does not count towards the time complexity of the algorithm. After the end of the first phase, the sequence of circuits $\mathcal{C}$ cannot be changed. Second part is the main program, working in a modified word-RAM:

At the beginning of execution of the main program, the values G and I are stored in memory. We always assume that $G \geq I$. Additionally, one might assume that $G \in \Theta(I^2)$ because it is usually the case in practice as explained in Sect. 2.2. This assumption can greatly simplify sharing of resources, as is described below.

The word-RAM has an additional instruction RUNCIRCUIT(i, s, t) that starts computation of circuit C_i on specified input. The instruction has three parameters. Parameter $i \in [m]$ identifies which circuit should be run. Parameter s is the source address of the input data. The address s has to be aligned to whole words. If the circuit has l inputs then the contiguous block of l bits starting from address s is used as input for the circuit. Similarly, t is the address where the output of the circuit is stored after the circuit computes its output. The output is also stored as a contiguous block of bits and, as with the input, the address t has to be aligned to whole words. However, note that variable shift operation by at most w bits can be easily implemented in depth $\mathcal{O}(\log \log w)$, making it possible to use unaligned inputs and outputs. Instruction RUNCIRCUIT lasts one time-step and it only starts execution of the circuit and does not wait for the results. The computation time of a circuit is proportional to its depth. If we are evaluating instruction RUNCIRCUIT(i, s, t) then we get the output starting at address t exactly del(C_i) steps after we executed the instruction. This can result in concurrent writes. While concurrent writes are allowed, the resulting value is undefined and the memory cells which are in the intersection of the writes can have any value. The model allows for use of the RUNCIRCUIT instruction without waiting for the results of the previous RUNCIRCUIT instruction. This is called pipelining and it is a critical feature of the PCRAM model which makes it possible to speed up many algorithms.

We also consider a randomized version of PCRAM with an extra operation which assigns to a memory cell a value chosen uniformly at random independently from other calls of the operation.

As in the case of (word) RAM, we measure the time complexity by the number of operations executed in a worst case. We express the complexity of an algorithm in terms of the input size n, the word size w, the number of available circuit inputs and outputs I and the number of available gates G. The complexity of a query on a data structure is measured in time, defined as the number of steps executed on the RAM, and delay, defined as the number of steps when the RAM is waiting for an output of a circuit. The reason for devising two complexity measures is that when performing multiple queries, the processor time is sequential and cannot be pipelined, whereas the execution of the circuit can potentially be pipelined, not necessarily resulting in the delay adding up.

Note that in general, it is not possible to use multiple circuits and assume that each can use all G gates and I I/O nodes. However, if all circuits use at most polynomial number of gates in their number of input/output nodes, the speedup caused by the circuits is at most polynomial in the number of their I/O nodes. Shrinking the circuits by a constant factor (that is, taking circuits for values of I and G smaller by a constant factor) will, therefore, incur at most constant slowdown. In this paper, whenever we use multiple circuits in an algorithm, they all use a polynomial number of gates and we do not further discuss the necessity to share resources between circuits.

We assume that we can copy $\mathcal{O}(I)$ bits in memory in time $\mathcal{O}(1)$. We call this the COPY instruction. For the case when G is polynomial in I, this follows from the following theorem (however, in practice it would not make sense to use FPGA for copying data).

Theorem 1. *In the PCRAM model, if $G \in \mathcal{O}(poly(I))$, a contiguous block of $\mathcal{O}(I)$ bits can be copied in time $\mathcal{O}(1)$ without use of the COPY instruction.*

Proof. Let CP_k be a copy circuit of size k that is formed by k identity gates. CP_k copies a contiguous block of k bits in time $\mathcal{O}(1)$ and it uses $2k$ I/O pins. Next, let K denote the greatest power of two such that $8K \leq I$. We put $\mathcal{C} = \{CP_1, CP_2, CP_4, \ldots, CP_K\}$. In total the circuits in $\mathcal{C}$ use $2K - 1 < G$ gates and $4K - 2 < I$ I/O pins.

For every contiguous block of $\mathcal{O}(I)$ bits there exists $s \in \{2^0, 2^1, 2^2, \ldots, K\}$ such that the block can be covered by $\mathcal{O}(1)$ contiguous intervals of size s, not necessarily pairwise disjoint. The block can then be copied by $\mathcal{O}(1)$ calls of RUNCIRCUIT(CP_s) instruction in time $\mathcal{O}(1)$.

At most one half of available gates and I/O gates are used for the COPY circuit. We set $G' = \frac{1}{2}G$ and $I' = \frac{1}{2}I$ as the parameters of the circuit module for the other circuits. Since $G \in \mathcal{O}(\mathrm{poly}(I))$, we incur at most constant slowdown by a constant decrease in the model's parameters.   □

The running time of the algorithms will often depend on "number of elements that can be processed on the circuit at once". We usually call this number k and define it when needed.

4 Algorithmic Results

In this section, we show several algorithms that are asymptotically faster than the best known algorithms in the word-RAM.

It would also be possible to use in PCRAM a circuit to speed up data structures. However, since efficient data structures often depend on doing random access in memory, it will be difficult to improve the running time of the data structure by more than a factor of $\Theta(\log(I))$, as speedups gained from circuits depend on sequential access to data. It is likely that this improvement will not be enough to justify practical use of an FPGA. In Sect. 5, we give lower bounds in the PCRAM model, including lower bound on search trees.

4.1 Aggregation

We show an efficient algorithm performing aggregation operation on an array. While this operation cannot be done in constant time on word-RAM, pipelining can be used to get constant time per operation when amortized over great enough number of instances in the PCRAM model. In Sect. 4.2, we use this operation for sorting numbers.

Problem 1 (Aggregation)

Input: array A of n numbers, n-bit mask M
Output: t – the number of ones in M and array B – a permutation of A such that if $M[i] = 1$ and $M[j] = 0$, then $A[i]$ precedes $A[j]$ in array B for all i, j.

In this section, we use $k \in \Theta\big(\min\big(\frac{I}{w}, \frac{G}{w}/\log^2 \frac{G}{w}\big)\big)$.

Theorem 2. *Aggregation of n numbers can be computed on PCRAM in time* $\mathcal{O}\big(\frac{n}{k} + \log^2 k\big)$.

Proof. Without loss of generality, we suppose that n is divisible by k. Otherwise, we use a circuit to extend the input to the smallest greater size divisible by k.

We construct a circuit that can aggregate k numbers using a sorting network of order k [3]. We call this the aggregator circuit. We use a sorting network to sort M in descending order. The sorting network uses modified comparators that do not only compare and potentially swap values in M, but also swap the corresponding items in array A when swapping values in M. We use a synchronous sorting network called the bitonic sorter [3]. Bitonic sorter of order k uses $\mathcal{O}(k \log^2 k)$ comparators and has depth $\mathcal{O}(\log^2 k)$. The modified comparator that also moves items in array A can be implemented using $\mathcal{O}(w)$ gates in depth $\mathcal{O}(1)$, since we are only comparing one-bit values. Therefore, our modification of the sorting network uses $\mathcal{O}(kw \log^2 k)$ gates, its depth is $\mathcal{O}(\log^2 k)$ and uses $\mathcal{O}(kw)$ I/O nodes.

Number t has $\lceil \log k \rceil$ bits and each of them can be derived independently on others from the position of the last occurrence of bit 1 in sorted array M. First, the circuit determines in parallel for every position of sorted M whether it is the last occurrence of 1. This can be done in constant depth with $\mathcal{O}(k)$ gates. Then, for every bit of t, the circuit checks if the last 1 is at one of the positions that imply that the bit of t is equal to 1. The value of one bit of t can be computed in depth $\mathcal{O}(\log k)$ with $\mathcal{O}(k)$ gates using a binary tree of OR gates, therefore we need $\mathcal{O}(k \log k)$ gates to compute the value of t from sorted M in depth $\mathcal{O}(\log k)$.

In total the circuit uses $\mathcal{O}(kw \log^2 k)$ gates, it takes $k(w+1)$ input bits, produces $kw + \lceil \log k \rceil$ output bits and its depth is $\mathcal{O}(\log^2 k)$.

The algorithm runs in two phases:

In the *first phase*, split the input into chunks $(A_1, M_1), (A_2, M_2), \ldots (A_{n/k}, M_{n/k})$ of size k and use the aggregator circuit to separately aggregate elements in each block. The circuit produces outputs $t_1, t_2, \ldots, t_{n/k}$ and $B_1, B_2, \ldots, B_{n/k}$. It may be necessary to wait for the delay after this phase.

In the *second phase*, for every $i \in 1, 2, \ldots, n/k$ split array B_i into two arrays P_i and S_i, where P_i is the prefix of B_i of length t_i and S_i is the rest of B_i. Concatenate the arrays $P_1, P_2, \ldots, P_{n/k}, S_1, S_2, \ldots, S_{n/k}$ using the COPY instruction. In total, the algorithm runs in time $\mathcal{O}\left(\frac{n}{k} + \log^2 k\right)$. □

The construction can likely be asymptotically improved by using a sorting network with $\mathcal{O}(n \log n)$ comparators and depth $\mathcal{O}(\log n)$ [1]. However, such sorting networks are complicated and impractical due to large multiplicative constants.

Theorem 3. *The aggregation operation on m arrays with a total length n can be computed in time $\mathcal{O}\left(\frac{n}{k} + m + \log^2 k\right)$.*

Proof. We use the algorithm from the proof of Theorem 2 and interleave its executions to decrease delay.

First, we run the first phase of the algorithm from proof of Theorem 2 for all the arrays and then run the second phases. This way, we only wait for the delay before the second phase at most once. □

4.2 Sorting

We show an asymptotically faster modification of randomized Quicksort with expected running time $\mathcal{O}\left(\frac{n}{k} \log n + \mathrm{polylog}(n, k, w)\right)$. We achieve this by improving the time complexity of the partition subroutine. Throughout this section, we use $k \in \Theta\left(\min\left(\frac{I}{w}, \frac{G}{w \log w} \big/ \log^2 \frac{G}{w \log w}\right)\right)$.

Problem 2 (Pivot Partition)

Input: array A of n numbers, pivot p
Output: arrays A_1, A_2 and the integer $|A_1|$, such that $A_1 \uplus A_2 = A$ and A_1 consists exactly of elements $a \in A$, s.t. $a \leq p$, where $\uplus$ is the disjoint union.

Theorem 4. *Pivot partition of n numbers can be computed in the PCRAM model in time $\mathcal{O}(n/k + \log^2 k + \log w)$.*

Proof. We reduce the pivot partition problem to aggregation. For each $0 \leq i < n$ we set $M[i] = [A[i] \leq p]$ and perform aggregation with input (A, M), getting output (t, B). A_1 consists of the first t elements of B and A_2 of the rest of B. The array length $|A_1|$ is then equal to t.

The reduction can be done on PCRAM in time $\mathcal{O}(n/k + \log w)$ by using a circuit to compute k values of array M at once in depth $\mathcal{O}(\log w)$ with $\mathcal{O}(kw)$ gates. Together with aggregation, this gives the desired running time.   □

Similarly to aggregation, pivot partition problems can be solved in bulk. The next theorem follows from Theorem 3 by the same reduction as in Theorem 4.

Theorem 5. *We can solve m independent pivot partition problems for arrays of total length n in the PCRAM model in time $\mathcal{O}(n/k + m + \log^2 k + \log w)$.*

Theorem 6. *There is a randomized algorithm in the PCRAM model which sorts n numbers in expected time $\mathcal{O}\left(\frac{n}{k}\log n + \text{polylog}(n, k, w)\right)$.*

Proof. We show a modification of randomized Quicksort. We use bulk pivot partitioning from Theorem 5 and a bitonic sorter [3] of order k sorting w-bit numbers. Each comparator can be synchronously implemented by $\mathcal{O}(w \log w)$ gates in depth $\mathcal{O}(\log w)$ by taking a standard comparator of depth $\mathcal{O}(\log w)$ and adding indentity gates to make it synchronous. Therefore, the bitonic sorter has $\mathcal{O}(kw \log^2 k \log w)$ gates and depth $\mathcal{O}(\log^2 k \log w)$. Note that we can also use the bitonic sorter of order k to sort less than k numbers if we fill the unused inputs by the maximum w-bit value and it is therefore enough to only have sorting circuit of one size. The algorithm traverses the recursive tree of Quicksort algorithm in BFS manner. Whenever a subproblem has size at most k, it is sorted by the bitonic sorter. Each layer of the recursion tree corresponds to a sequence of blocks $B_1, \ldots, B_m$ for some integer m to be sorted satisfying that if $b_1 \in B_j, b_2 \in B_\ell$, such that $1 \leq j < \ell \leq m$, then $b_1 \leq b_2$. In each layer, blocks of length at most k are sorted using the bitonic sorter and the remaining blocks are partitioned into two by the bulk pivot partition algorithm from Theorem 5 with pivots being chosen uniformly at random for each of the blocks.

In each layer, there can be at most n/k blocks with more than k elements. Thus, from Theorem 5 follows that the partition requires $\mathcal{O}(n/k + \log^2 k + \log w)$ time. Furthermore, there can be at most $2n/k$ blocks of size at most k. We do not have to wait for the delay of the sorting circuit before processing of the next layer. Therefore, we spend $\mathcal{O}(n/k + \log^2 k + \log w)$ time per layer.

The expected depth of the Quicksort recursion tree is $\mathcal{O}(\log n)$ [19] (note that Quicksort recursion depth is the same as the depth of a random binary search tree, which is proven to be $\mathcal{O}(\log n)$ in the cited paper). It follows that the total expected time complexity of our algorithm is $\mathcal{O}((n/k + \log^2 k + \log w) \log n + \log^2 k \log w)$. $\square$

4.3 Associative Operation

Let $\otimes : \{0,1\}^w \times \{0,1\}^w \to \{0,1\}^w$ be an associative operation (such as maximum or addition over a finite group). Let C_1 be a synchronous boolean circuit with g' gates, $2w$ inputs and depth d' which computes the function $\otimes$. We want to efficiently compute $\bigotimes_{i=1}^{n} a_i$ for input numbers $a_1, \ldots, a_n \in \{0,1\}^w$.

We can assume without loss of generality that there is a neutral element e satisfying $(\forall x \in \{0,1\}^w)\, (x \otimes e = e \otimes x = x)$. If there is no such e, we extend $\otimes$ to have one. This only changes the size and depth of C_1 by a constant factor.

Problem 3 ($\otimes$-sum)

Input: array A of n numbers $a_1, \ldots, a_n \in \{0,1\}^w$
Output: value $\bigotimes_{i=0}^{n-1} A[i]$

The idea of the algorithm is to split the input into blocks that can be processed by the circuit simultaneously by pipelining the computation. In order to do this, we arrange copies of circuit C_1 into a full binary tree. For $j \geq 1$ we inductively define circuit C_{j+1} that consists of two copies of C_j whose outputs go into a copy of C_1.

In this section, we use $k \in \Theta\big(\min\big(\frac{I}{w}, \frac{G}{g'}\big)\big)$. We assume that k is a power of two.

Theorem 7. *$\otimes$-sum of n w-bit numbers can be evaluated on PCRAM in time* $\mathcal{O}(\frac{n}{k} + d' \log(kd'))$.

Proof. The algorithm repeatedly calls subroutine SUMBLOCKS which splits the input into blocks of size k and computes the $\otimes$-sum of the block using the circuit $C_{\log k}$. The results from $C_{\log k}$ are collected in a buffer that is used as the input for the next call of the subroutine. Each call of SUMBLOCKS reduces the input size k times, therefore we have to call the subroutine $\lceil \log_k n \rceil$ times.

If the length of the input of subroutine SUMBLOCKS is not divisible by k we fill the last block to the full size with neutral element e. By an argument similar to that in Theorem 1, this can be done in constant time by having $\Theta(\log n)$ circuits, the i-th of which has size $w2^i$, has no inputs and produces 2^i copies of e.

For an input of length n', SUMBLOCKS invokes $\lceil \frac{n'}{k} \rceil$ calls of $C_{\log k}$ and runs in time $\Theta\big(\frac{n'}{k}\big)$ with delay $\Theta(d' \log k)$.

The size of the problem decreases exponentially, so the first call of SUMBLOCKS dominates in terms of time complexity. The delay is the same for each call of SUMBLOCKS. However, if the input has length of $\Omega(kd' \log k)$ words, the individual values are computed by the circuit before they are needed in the current call of SUMBLOCKS and it is, therefore, not necessary to wait for the delay. We only wait for the delay the last $\mathcal{O}(\log_k(kd' \log k)) = \mathcal{O}(\log_k(kd'))$ executions of SUMBLOCKS, resulting in total waiting time of $\mathcal{O}(d' \log k \log_k(kd')) = \mathcal{O}(d' \log(kd'))$.

4.4   Dynamic Programs

A *dynamic program* is a recursive algorithm where the computed values are being memoized, meaning that the function is not computed if it has already been computed with the same parameters. Many dynamic programs used in real-world applications have the property that the subproblems form a multidimensional grid. We then call the individual subproblems cells. For simplicity, we will focus only on dynamic programs with a rectangular grid. However, it is possible to generalise the described technique to higher dimensions.

Many dynamic programs such as those solving the longest common subsequence, shortest common supersequence and edit distance problems or the Needleman-Wunsch algorithm satisfy the following three properties:

1. Each cell depends only on a constant number of other cells.
2. Each cell C depends only on cells which are in the upper-left quadrant with C in the corner (as shown in Fig. 1).
3. For any cell C, the cells that C depends on have the same relative position with respect to C (as shown in Fig. 2).

We call such dynamic programs CULD (constant upper-left dependency).

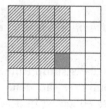

Fig. 1. The grey cell can only depend on the hatched cells in a CULD dynamic program

A CULD dynamic program has linear number of cells that are near the left or top edge of the grid and, therefore, should have a dependency on a cell that does not exist (because it would have at least one coordinate negative). These cells have to have their values computed in some other way and are called base cases. As it is usually trivial to compute the values of these cells, we will not deal with computing them in this paper.

Dynamic programs can be sped up on the PCRAM model using the circuit for (1) speeding up the computation of the individual values (it has to take more than constant time to compute each value for this to be possible) or (2)

parallelizing the computation of the values (usually when the individual values can be computed in constant time). Computation of individual values of the function can be sped up in the special case of associative operation using the algorithm described in Sect. 4.3. In this section, we show how to use the second approach to speed up the computation of the CULD dynamic programs.

Theorem 8. *Let P be a CULD dynamic program with dependency grid of size $M \times N$. Let g' and d' be the number of gates and delay, respectively, of a circuit which computes value of a cell from the values that the cell depends on. After all base cases of the dynamic program have been computed, the remaining values in the dependency grid can be computed in time $\mathcal{O}\left(\frac{MN}{k} + M + N + kd'^2\right)$ where $k \overset{\text{def}}{=} \min\left(\frac{I}{w}, \frac{G}{g'}\right)$.*

Proof. Instead of storing the grid in a row-major or column-major layout, we store it in antidiagonal-major layout. That is, the antidiagonals are stored contiguously in memory.

Let d be the number of cells that a cell depends on. We calculate the values of k cells in parallel using a circuit. Notice that subsequent cells in the layout depend on at most d contiguous sequences of cells (see Fig. 2). First, we move these cells to one contiguous chunk of memory. This can be used as input to a circuit which computes the values of the k cells.

This process can be done repeatedly for all antidiagonals, computing values of all cells in the grid. Note that it is necessary to wait for the delay only if the antidiagonal that is being computed has $\mathcal{O}(kd')$ cells. The total waiting time is, therefore, $\mathcal{O}(kd'^2)$ and the total time complexity $\mathcal{O}\left(\frac{MN}{k} + M + N + kd'^2\right)$. $\qquad \square$

The circuit computing the cell value can be obtained either by using the general technique of simulating a RAM, which we show in Sect. 4.5, or an ad-hoc circuit can be used.

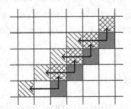

Fig. 2. Contiguous sequence of cells (grey) depend on a constant number (in this case two) of contiguous sequences of cells (hatched)

4.5   Recursive Algorithms

In this section, we describe how the circuit can be used to speed up recursive algorithms in which the amount of work is increasing exponentially in the recursion level. This includes some algorithms whose time complexity can be analyzed by the Master theorem [7] as well as many exponential time algorithms. Our approach does not result in a substantial increase of memory usage.

We use the circuit to solve the problem for small input sizes. We do this by using the concept of totally computing circuits, which we introduce. Totally computing circuit is a circuit that computes a given family of functions for all inputs of length bounded by some constant. This is a natural concept in PCRAM since we often want to use the circuit to compute a function on arbitrary input that is short enough and not only for inputs of a specific length.

Definition 3 (Totally computing circuit). *Let $\{f_k\}_{k\in\mathbb{N}}$ be a family of functions such that $f_k : \{0,1\}^k \to \{0,1\}^k$ for every k. Boolean circuit C_n is a totally computing circuit of order n of sequence $\{f_k\}_{k\in\mathbb{N}}$ if it has $n + \lceil \log_2 n \rceil$ inputs, interpreting the first $\lceil \log_2 n \rceil$ bits of input as a binary representation of number n' and it outputs on the first k outputs the value of function $f_{n'}$ evaluated on the next n' bits of the input.*

A similar definition could be made, where the circuit would work in ternary logic with one of the values meaning "no input", requiring that the input is contiguous and starts at the first I/O node. We stick with the definition mentioned above because this notion of totally computing circuits can be easily simulated by totally computing circuits according to Definition 3.

Definition 4 ((f, G)-recursive algorithm with independent branches). *We call an algorithm A in word-RAM model to be (f, G)-recursive with independent branches if it is of the form $A(0) = 0$ (the recursion base case) and $A(x) = f(A(g_1(x)), \ldots, A(g_k(x)))$ otherwise, where the functions f and g_i have no side effects, can use randomness and $G = \{g_i\}_{i=1}^k$ is the family of functions g_i.*

Theorem 9. *Let R be an (f, G)-recursive algorithm with independent branches for some f, G in which the time spent on recursion level i increases exponentially with i. Let $\mathcal{C}_R$ be a family of circuits computing the same function as R such that the circuits in $\mathcal{C}_R$ can be generated by a Turing machine in time polynomial in their size. Then there exists an algorithm in the PCRAM that runs in time equal to the total time spent by R on problems of size smaller than $k \overset{\text{def}}{=} \min(size_{\mathcal{C}_R}^{-1}(G), K)/w$, where K is the greatest number such that $2K + \lceil \log_2 K \rceil \leq I$, plus $del_{\mathcal{C}_R}(kw)$.*

Proof. Notice that k is the maximum subproblem size which can be solved using the circuit. We run a recursion similar to R while using the circuit for solving subproblems of size at most k. The algorithm has two phases and it uses a buffer to store results from the first phase.

In the *first phase*, we compute results of subproblems of size at most k. We run R with the modification that results of subproblems of size at most k are computed using the circuit and stored in the buffer.

In the *second phase*, we compute results of subproblems of size more than k. We run R with the modification that when subproblem size is at most k, we obtain the result computed in phase 1 from the buffer and return it.

Both phases take asymptotically same amount of time which is spent in algorithm R on subproblems of size more than k. The second phase might have to wait for the delay of the circuit before its execution.   □

The memory consumption of the algorithm can be high, as we store all the results for subproblems of size at most k in the buffer. This can be avoided by running both phases in parallel with the second phase delayed by the number of time-steps equal to the delay caused by the circuit. Then, at any moment, the number of items in the buffer does not exceed the delay of the circuit and it is, therefore, enough to use a buffer of size equal to the delay of the circuit.

Corollary 1. *Let R be a recursive algorithm with independent branches whose time complexity can be analysed by the Master theorem, running in time $t(n)$. If the time spent on recursion level i of R is increasing exponentially in i, then there is an equivalent algorithm R' in the PCRAM model that runs in time $\mathcal{O}(t(n/k) + del(C_k))$ where C_k is the circuit used do solve subproblems of size at most k.*

Such problems include the Karatsuba algorithm [15] and the Strassen algorithm [23]. Other algorithms which are not analyzed by master theorem but can be sped up using this technique include some exponential recursive algorithms for solving 3-SAT, maximum independent set, as well as other exponential algorithms based on the branch-and-reduce method [21, 26].

Note that it may be possible to avoid using totally computing circuits by ensuring that the subproblems which are solved using the circuit have all the same size or only have constant number different sizes.

Simulation of RAM or a Turing Machine can be used to get totally computing circuits. We say that a circuit of order n simulates given Turing Machine (word-RAM program), if it computes the same function as the Turing machine (word-RAM program) for all inputs of size n.

Observation 1. *Let A be an algorithm that runs on RAM or Turing machine and computes function $f(x)$, where x is the input. Then for any n, there is in an algorithm A' which takes the first $\lceil \log_2 n \rceil$ inputs as the binary representation of n' and computes the function f on next n' inputs. Moreover, there is such A' that has the same asymptotic complexity as A.*

This observation can be used as a basis for creation of totally computing circuits as it is sufficient to simulate the algorithm A' to get a totally computing circuit.

The following theorems say that it is possible to simulate a Turing machine or RAM on a circuit. For proof of Theorem 10, see the Appendix A. Theorem 11 is well known and its proof can be found in lecture notes by Zwick and Gupta [27].

Theorem 10. *A Word RAM with word size w running in time $t(n)$ using $m(n)$ memory can be simulated by a circuit of size $\mathcal{O}(t(n)m(n)w \log(m(n)w))$ and depth $\mathcal{O}(t(n) \log(m(n)w))$.*

Theorem 11. *A Turing Machine running in time $t(n)$ using $m(n)$ memory can be simulated by a synchronous circuit of size $\mathcal{O}(t(n)m(n))$ and depth $\mathcal{O}(t(n))$.*

5 Lower Bounds

In this section, we use the I/O model of computation to obtain several lower bounds for the PCRAM model.

5.1 I/O Model

The computer in the I/O model is a word-RAM [11] with two levels of memory – cache and main memory. Both the main memory and the cache are partitioned into memory blocks. A block consists of B aligned consecutive words. The processor can only work with data stored in the cache. If a word that is accessed is not in the cache, its whole block is loaded into the cache from the main memory. When a word which is already in the cache is accessed, the computer does not access the main memory. Only M words fit into the cache. When the cache is full and a block is to be loaded into it, a block which is stored in the cache has to be evicted from it. This is done according to a specified eviction strategy, usually to evict the least recently used item (abbreviated as LRU).

The *I/O complexity* is defined to be the function f such that $f(n)$ is equal to the maximum number of memory blocks transferred from the main memory to the cache over all inputs of length n.

A *cache-aware algorithm* is an algorithm in the word-RAM model with the knowledge of B and M. The measure of I/O complexity is usual for cache-aware algorithms.

5.2 Lower Bounds

There are several lower bounds known on the I/O complexity of cache-aware algorithms. We show a general theorem which gives lower bounds in PCRAM from lower bounds in the cache-aware model. Recall that the set of circuits $\mathcal{C}$ is generated by a polynomial algorithm in word-RAM. Throughout this section, we use $t_{\text{gen}}(G, I)$ to denote the time complexity of this program.

Theorem 12. *Let A be an algorithm in the PCRAM model running in time $t(n, G, I, w)$ where n is the input size. Then there is a cache-aware algorithm A' in the I/O model, simulating algorithm A with I/O complexity $\mathcal{O}(t(n, M, Bw, w) + t_{gen}(M, Bw))$.*

Note that since B and M is the number of words of cache-line and cache, respectively, they have Bw and Mw bits, respectively.

Proof. We simulate algorithm A using $\mathcal{O}(1)$ amortized I/O operations per instruction. We can achieve this by simulating the circuits of A in the cache. Instructions of PCRAM model that do not work with circuits can be trivially performed with $\mathcal{O}(1)$ I/O operations. There exists $c < 1$ such that synchronous circuits with at most cM gates and Bw I/O nodes can be simulated in the cache. At the beginning of the simulation, we generate the set of circuits $\mathcal{C}$ and load them into the cache. This takes time $t_{gen}(M, Bw)$. Any time we simulate circuit execution, we charge the I/O operations necessary for loading the circuit into the cache to the operation which caused eviction of the blocks which have to be loaded. This proves a bound of $\mathcal{O}(1)$ amortized I/O operation per simulated operation.

We can, therefore, simulate A with I/O complexity $\mathcal{O}(t(n, cM, Bw, w) + t_{gen}(cM, Bw))$. Since the speedup in the PCRAM model and t_{gen} depend on the number of gates at most polynomially then $\mathcal{O}(t(n, cM, Bw, w) + t_{gen}(cM, Bw)) = \mathcal{O}(t(n, M, Bw, w) + t_{gen}(M, Bw))$. $\qquad\square$

Corollary 2. *If there is a lower bound on a problem of $\Omega(t(n, M, B, w))$ in the cache-aware model, then the lower bound of $\Omega(t(n, G, I/w, w) - t_{gen}(G, I))$ holds in the PCRAM.*

Some lower bounds in the cache-aware model hold even if the cache can have arbitrary content independent of the input at the start of the computation. Such lower bounds in the cache-aware model imply a lower bound of $\Omega(t(n, G, I/w, w))$ in the PCRAM. This is the case for the lower bound used in Corollary 3.

The following lower bound follows from Corollary 2 and a lower bound shown in a survey by Demaine [8].

Corollary 3. *In the PCRAM model, given a set of numbers, assuming that we can only compare them, the search query takes $\Omega(\log_{I/w} n)$ time.*

More lower bounds on cache-aware algorithms are known, but, like in the Corollary 3, they often make further assumptions about operations that we can perform, often requiring that we treat the input items as indivisible objects. Problems for which a non-trivial lower bound is known include sorting, performing permutation and FFT [24].

6 Open Problems

There are many problems waiting to be solved. An important direction of possible future work is to implement the algorithms described in this paper and compare experimental results with predictions based on time complexities in our model. The soundness of the model of computation needs to be empirically verified.

There are many theoretical problems that are not yet solved, including the following:

Open problem 1 (Fast Fourier Transform). *Is there an asymptotically fast algorithm performing the Fast Fourier Transform in the PCRAM model?*

Open problem 2 (Data structures). *Are there any data structures in the PCRAM model with asymptotically faster queries than their word-RAM counterparts?*

A   Simulation of Word-RAM

Theorem 10. *A Word RAM with word size w running in time $t(n)$ using $m(n)$ memory can be simulated by a circuit of size $\mathcal{O}(t(n)m(n)w\log(m(n)w))$ and depth $\mathcal{O}(t(n)\log(m(n)w))$.*

Proof. We first construct an asynchronous circuit. In the proof, we will be using the following two subcircuits for writing to and reading from the RAM's memory.

Memory read subcircuit gets as input nw bits consisting of $m(n)$ words of length w together with a number k which fits into one word when represented in binary. It returns the k'th group. There is such circuit with $\mathcal{O}(m(n)w)$ gates and depth $\mathcal{O}(\log(m(n)w))$.

Memory write subcircuit gets as input $m(n)w$ bits consisting of $m(n)$ words of length w and additional numbers k and v, both represented in binary, each fitting into a word. The circuit outputs the $m(n)w$ bits from input with the exception of the k'th word, which is replaced by value v. There is such circuit with $\mathcal{O}(m(n)w)$ gates in depth $\mathcal{O}(\log w)$.

The circuit consists of $t(n)$ layers, each of depth $\mathcal{O}(\log(m(n)w))$. Each layer executes one step of the word-RAM. Each layer gets as input the memory of the RAM after execution of the previous instruction and the instruction pointer to the instruction which is to be executed and outputs the memory after execution of the instruction and a pointer to the next instruction. Each layer works in two phases.

In the *first phase*, we retrieve from memory the values necessary for execution of the instruction (including the address where the result is to be saved, in case of indirect access). We do this using the memory read subcircuit (or possibly two of them coupled together in case of indirect addressing). This can be done since the addresses from which the program reads can be inferred from the instruction pointer.

In the *second phase*, we execute all possible instruction on the values retrieved in phase 1. Note that all instructions of the word-RAM can be implemented by a circuit of depth $\mathcal{O}(\log w)$. Each instruction has an output and optional wires for outputting the next instruction (which is used only by jump and conditional jump – all other instructions will output zeros). The correct instruction can be inferred from the instruction pointer, so we can use a binary tree to get the output from the correct instruction to specified wires. This output value is then stored in memory using the memory store subcircuit.

The first layer takes as input the input of the RAM. The last layer outputs the output of the RAM.

Every signal has to be delayed for at most $\mathcal{O}(\log(m(n)w))$ steps. The number of gates is, therefore, increased by a factor of at most $\mathcal{O}(\log(m(n)w))$.   □

References

1. Ajtai, M., Komlós, J., Szemerédi, E.: An 0(n log n) sorting network. In: Proceedings of the Fifteenth Annual ACM Symposium on Theory of Computing, STOC 1983, pp. 1–9. ACM, New York (1983). http://doi.acm.org/10.1145/800061.808726

2. Alam, N.: Implementation of genetic algorithms in FPGA-based reconfigurable computing systems. Master's thesis, Clemson University (2009). https://tigerprints.clemson.edu/all_theses/618/?utm_source=tigerprints.clemson.edu%2Fall_theses%2F618&utm_medium=PDF&utm_campaign=PDFCoverPages

3. Batcher, K.E.: Sorting networks and their applications. In: Proceedings of the Spring Joint Computer Conference, 30 April–2 May 1968, AFIPS 1968 (Spring), pp. 307–314. ACM, New York (1968). http://doi.acm.org/10.1145/1468075.1468121

4. Che, S., Li, J., Sheaffer, J.W., Skadron, K., Lach, J.: Accelerating compute-intensive applications with GPUs and FPGAs. In: 2008 Symposium on Application Specific Processors, pp. 101–107, June 2008

5. Chodowiec, P., Gaj, K.: Very compact FPGA implementation of the AES algorithm. In: Walter, C.D., Koç, Ç.K., Paar, C. (eds.) CHES 2003. LNCS, vol. 2779, pp. 319–333. Springer, Heidelberg (2003). https://doi.org/10.1007/978-3-540-45238-6_26

6. Chrysos, G., et al.: Opportunities from the use of FPGAs as platforms for bioinformatics algorithms. In: 2012 IEEE 12th International Conference on Bioinformatics Bioengineering (BIBE), pp. 559–565, November 2012

7. Cormen, T.H., Leiserson, C.E.: Introduction to Algorithms, 3rd edn. MIT Press, Cambridge (2009)

8. Demaine, E.: Cache-oblivious algorithms and data structures. EEF Summer Sch. Massive Data Sets 8(4), 1–249 (2002)

9. Grozea, C., Bankovic, Z., Laskov, P.: FPGA vs. multi-core CPUs vs. GPUs: hands-on experience with a sorting application. In: Keller, R., Kramer, D., Weiss, J.-P. (eds.) Facing the Multicore-Challenge LNCS, vol. 6310, pp. 105–117. Springer, Heidelberg (2010). https://doi.org/10.1007/978-3-642-16233-6_12

10. Guo, Z., Najjar, W., Vahid, F., Vissers, K.: A quantitative analysis of the speedup factors of FPGAs over processors. In: Proceedings of the 2004 ACM/SIGDA 12th International Symposium on Field Programmable Gate Arrays, FPGA 2004, pp. 162–170. ACM, New York (2004). http://doi.acm.org/10.1145/968280.968304

11. Hagerup, T.: Sorting and searching on the word RAM. In: Morvan, M., Meinel, C., Krob, D. (eds.) STACS 1998. LNCS, vol. 1373, pp. 366–398. Springer, Heidelberg (1998). https://doi.org/10.1007/BFb0028575

12. Harper, L.H.: An $n \log n$ lower bound on synchronous combinational complexity. Proc. Am. Math. Soc. 64(2), 300–306 (1977). http://www.jstor.org/stable/2041447

13. Huffstetler, J.: Intel processors and FPGAs-better together, May 2018. https://itpeernetwork.intel.com/intel-processors-fpga-better-together/

14. Hussain, H.M., Benkrid, K., Seker, H., Erdogan, A.T.: FPGA implementation of k-means algorithm for bioinformatics application: an accelerated approach to clustering microarray data. In: 2011 NASA/ESA Conference on Adaptive Hardware and Systems (AHS), pp. 248–255, June 2011

15. Karatsuba, A., Ofman, Y.: Multiplication of many-digital numbers by automatic computers. In: Dokl. Akad. Nauk SSSR, vol. 145, pp. 293–294 (1962). http://mi.mathnet.ru/dan26729

16. Karkooti, M., Cavallaro, J.R., Dick, C.: FPGA implementation of matrix inversion using QRD-RLS algorithm. In: Conference Record of the Thirty-Ninth Asilomar Conference on Signals, Systems and Computers 2005, pp. 1625–1629 (2005)

17. Ma, L., Agrawal, K., Chamberlain, R.D.: A memory access model for highly-threaded many-core architectures. Future Gener. Comput. Syst. **30**, 202–215 (2014). http://www.sciencedirect.com/science/article/pii/S0167739X13001349, special Issue on Extreme Scale Parallel Architectures and Systems, Cryptography in Cloud Computing and Recent Advances in Parallel and Distributed Systems, ICPADS 2012 Selected Papers

18. Mahram, A.: FPGA acceleration of sequence analysis tools in bioinformatics (2013). https://open.bu.edu/handle/2144/11126

19. Reed, B.: The height of a random binary search tree. J. ACM **50**(3), 306–332 (2003). https://doi.org/10.1145/765568.765571

20. Romoth, J., Porrmann, M., Rückert, U.: Survey of FPGA applications in the period 2000–2015 (Technical report) (2017)

21. van Rooij, J.M., Bodlaender, H.L.: Exact algorithms for dominating set. Discrete Appl. Math. **159**(17), 2147–2164 (2011). http://www.sciencedirect.com/science/article/pii/S0166218X11002393

22. Sklavos, D.: DDR3 vs. DDR4: raw bandwidth by the numbers, September 2015. https://www.techspot.com/news/62129-ddr3-vs-ddr4-raw-bandwidth-numbers.html

23. Strassen, V.: Gaussian elimination is not optimal. Numer. Math. **13**(4), 354–356 (1969). https://doi.org/10.1007/BF02165411

24. Vitter, J.S.: Algorithms and data structures for external memory. Found. Trends Theor. Comput. Sci. **2**(4), 54–63 (2008). https://doi.org/10.1561/0400000014

25. Vollmer, H.: Introduction to Circuit Complexity: A Uniform Approach. Springer, Heidelberg (1999). https://doi.org/10.1007/978-3-662-03927-4

26. Woeginger, G.J.: Exact algorithms for NP-hard problems: a survey. In: Jünger, M., Reinelt, G., Rinaldi, G. (eds.) Combinatorial Optimization - Eureka, You Shrink!. LNCS, vol. 2570, pp. 185–207. Springer, Heidelberg (2003). https://doi.org/10.1007/3-540-36478-1_17. http://dl.acm.org/citation.cfm?id=885909

27. Zwick, U., Gupta, A.: Concrete complexity lecture notes, lecture 3 (1996). www.cs.tau.ac.il/~zwick/circ-comp-new/two.ps

On the Complexity of and Algorithms for Min-Max Target Coverage On a Line Boundary

Peihuang Huang[1], Wenxing Zhu[2], and Longkun Guo[2(✉)]

[1] College of Physics and Information Engineering, Fuzhou University, Fuzhou, China
peihuang.huang@foxmail.com
[2] College of Mathematics and Computer Science, Fuzhou University, Fuzhou, China
{wxzhu,lkguo}@fzu.edu.cn

Abstract. Given a set of sensors distributed on the plane and a set of Point of Interests (POIs) on a line segment, a primary task of the mobile wireless sensor network is to schedule a coverage of the POIs by the sensors, such that each POI is monitored by at least one sensor. For balancing the energy consumption, we study the min-max line barrier target coverage (LBTC) problem which aims to minimize the maximum movement of the sensors from their original positions to final positions for the coverage. We first proved that when the radius of the sensors are non-uniform integers, even 1-dimensional LBTC (1D-LBTC), a special case of LBTC in which the sensors are distributed on the line segment instead of the plane, is $\mathcal{NP}$-hard. The hardness result is interesting, since the continuous version of LBTC of covering a given line segment instead of the POIs is known polynomial solvable [2]. Then we presented an exact algorithm for LBTC with sensors of uniform radius distributed on the plane, via solving the decision version of LBTC. We showed that our algorithm always finds an optimal solution in time $O(mn(\log m + \log n))$ to LBTC when there exists any, where m and n are the numbers of POIs and sensors.

1 Introduction

In the past decades, wireless sensor networks have brought tremendous changes to human society and proposed many technique challenges. Among them, the coverage topic including area coverage [10] and barrier coverage [8] is one of the hot spots that attract lots of research interest. In area coverage, the task is to schedule the new positions of the sensors, such that each point in the given target region is covered by at least one sensor. Differently, in barrier cover the task is to monitor only the boundary of a given region, and the aim is to guarantee that intruders can be found when they are crossing the barrier. Comparing to area

The research is supported by Natural Science Foundation of China (Nos. 61772005, 61672005, 61300025) and Natural Science Foundation of Fujian Province (No. 2017J01753).

coverage, barrier coverage has an advantage of using significantly less sensors and hence is scalable for large scale wireless sensor networks (WSN). Furthermore, some applications only require a set of Points Of Interest (POIs) along the boundary to be monitored. In the context, a problem arises how to guarantee every POI on the barrier to be covered. The current-state-of-art method is to firstly cover POIs using the stationary sensors, and secondly use mobile sensors to cover every not-yet covered POI along the boundary. For the second phase, we traditionally have the following assumptions for the modeling: (1) Sensors are acquired with mobile ability; (2) The initial positions of the sensors are distributed on the plane, and the POIs are distributed along a line segment (Although the shape of the boundary can be various, most researches nonetheless focus on line boundary since curves of other shapes can be considered as a variable); (3) The aim of the sensor network is to prolong the lifetime. This arises the min-max 2D Line Boundary Target Coverage problem in (min-max 2D-LBTC) as follows:

Definition 1. *Let Ψ and Γ be respectively a set of POIs distributed in a line segment $[0, M]$ and a set of mobile sensors distributed on the plane, where $j \in \Psi$ has a position $(p_j, 0)$ and $i \in \Gamma$ has a position (x_i, y_i) and a positive sensing radius r_i. The min-max 2D-LBTC problem aims to compute a new position $(x_i', 0)$ for each sensor $i \in \Gamma$, such that each POI $j \in \Psi$ is covered by at least one sensor, i.e. for each POI $j \in \Psi$ there exists a sensor $i \in \Gamma$ with position $(x_i', 0)$ that $x_i' - r_i \leq p_i \leq x_i' + r_i$, and the maximum movement of the sensors from their original positions to the new positions is minimized, i.e. $\max_{i \in \Gamma} \left\{ \sqrt{(x_i - x_i')^2 + y_i^2} \,\middle|\, i \in \Gamma \right\}$ is minimized.*

When no confusion arises, we shall use LBTC short for the min-max 2D-LBTC problem for the sake of briefness. In particular, we use one dimensional min-max Line Boundary Target Coverage problem (1D-LBTC) to denote the special case of LBTC when the initial positions of all the sensors are also distributed on the line boundary $[0, M]$. Moreover, the decision version of LBTC (decision LBTC for short) is, for a given movement bound D, to determine whether there exists a feasible coverage with each sensor's movement bounded by D. Besides, when the aim is to cover the line boundary itself instead of the POIs thereon, we respectively have the min-max Line Boundary Coverage (LBC) problem and one-dimensional-LBC (1D-LBC) problem, which have already been well studied and a number of algorithms have been developed.

1.1 Related Works

To the best of our knowledge, Kumar et al. [8] were the first to consider the barrier coverage problem using sensors against a closed curve (i.e., a moat), via transforming the coverage problem to the path problem of determining whether there exists a path between two specified nodes, although the research of barrier coverage started from early 90s in the last century due to Gage [7]. The algorithm from Kumar et al. is scalable and can also be extended to solve the k-coverage problem by transforming to the k-disjoint path problem, but can only be used to determine whether a coverage exists using the deployed stationary sensors. A problem for stationary sensors is that, after deployment there might exist no coverage over all targets. For the case, a state-of-art solution is to employ mobile sensors to fill the gaps between the stationary sensors. In the scenario, the WSN applications would require to maximize the minimum lifetime of the mobile sensors or to minimize the total energy consumption. For the former, the aim is to schedule new positions for the mobile sensors such that the barrier is completely covered, and that the maximum movement of the sensors is minimized as to prolong the lifetime of the WSN. When the sensors are on the line of the barrier, the 1D-LBC problem is shown optimally solvable in $O(n^2)$ time for uniform radius in Paper [4]. The same paper has also proposed an algorithm with $O(n)$ time for uniform radii and $\sum_i r_i \leq L$, and with $x_1 \leq \cdots \leq x_n$ for the sensor $\Gamma = \{s_1, \cdots, s_n\}$, where L is the length of the barrier, n is the number of the sensors. Later, Chen et al. have improved the time complexity to $O(n \log n)$ for uniform sensor radii and proposed an $O(n^2 \log n)$ time algorithm for non-uniform radii in paper [2]. Besides straight line barrier, circle/simple polygon barriers has been studied and two algorithms have been given developed by Bhattacharya et al. in [1], which have an $O(n^{3.5} \log n)$ time relative to cycle barriers and an $O(mn^{3.5} \log n)$ time relative to polygon barriers, in which m is the number of the edges on the polygon. The later time complexity was then decreased to $O(n^{2.5} \log n)$ in [12]. For the more generalized case in which the sensors are distributed on the plane, the LBC problem is known to be strongly $\mathcal{NP}$-hard for sensors with general integral sensing radius [6], while LBC using uniform radius sensors is shown solvable in $O(n^3 \log n)$ time [9].

Other than the Min-Max case, there are also applications require min-sum coverage that is to minimize the total energy consumption, which is to minimize the total movement of the mobile sensors. For this objective, both Min-Sum LBC and LBTC, which aim to minimize the sum of the movements of all the sensors, were studied in literature. Min-Sum LBC was shown $\mathcal{NP}$-complete for arbitrary radius while solvable in time $O(n^2)$ for uniform radii by Czyzowicz et al. [5]. The Min-Num relocation problem of minimizing the number of sensors moved, is also proven $\mathcal{NP}$-complete for arbitrary radii and polynomial solvable for uniform radii by Mehrandish et al. [11]. A PTAS has been developed for the Min-Sum relocation problem against circle/simple polygon barriers by Bhattacharya et al. [1], which was later improved to an $O(n^4)$ time exact algorithm by Tan and Wu [12]. For covering targets with Min-Sum movement, the most recent result is a factor-$\sqrt{2}$ approximation algorithm for covering targets along a barrier using uniform-radius sensors, aiming to minimize the sum of the movement [3]. However, it remains open whether the min-sum LBC problem is $\mathcal{NP}$-hard.

1.2 Our Results

In this paper, we first show that 1D-LBTC is $\mathcal{NP}$-hard when the sensors are with non-uniform integral radii by proposing a reduction from the 3-partition problem that is known strongly $\mathcal{NP}$-complete. This hardness result is surprising, because

1D-LBC, the continuous version of 1D-LBTC, is shown solvable in polynomial time $O(n^2 \log n)$.

Then, we propose a sufficient and necessary condition to determine whether there exists a feasible cover for the barrier under the relocation distance bound D. Based on the condition, we propose a simple greedy approach that outputs "infeasible" if $D < D^*$, and otherwise computes a feasible solution under the movement bound D, such that new positions for the sensors wrt which each target is monitored by at least a sensor. We show that the decision algorithm is with a runtime $O(n \log n)$. By employing the binary search technique, we propose an algorithm using the decision algorithm as a routine which takes $O(n \log n \log(d_{max} + L))$ time to actually find a minimum integral movement bound $D = D^*$, where d_{max} is the maximum distance between the sensors and the POIs, and L is the length of the line segment.

For instances with large d_{max} and L, we propose another algorithm that employs the binary search method against $O(mn)$ possible values of D^* instead of the continuous value range. This improves the runtime of the algorithm to $O(mn(\log m + \log n))$, which is the time needed to sort the $O(mn)$ values. The later algorithm remains correct even when D is any real number. In contrast, the former algorithm only works for integral D^*.

1.3   Organization

The following paragraphs will be organized as below: Sect. 2 gives the $\mathcal{NP}$-completeness proof; Sect. 3 presents the algorithm for Decision LBTC with uniform sensor radii, and shows that it always produces an optimal solution; Sect. 4 actually solves the LBTC problem by employing the binary search method, and then improve the runtime to $O(mn(\log m + \log n))$; Sect. 5 concludes the paper.

2 $\mathcal{NP}$-Completeness of Decision 1D-LBTC

In this section, we shall show the Decision LBTC problem is $\mathcal{NP}$-complete when the sensors are with non-uniform integral radii by giving a reduction from the 3-partition problem. In the 3-partition problem that is known strongly $\mathcal{NP}$-complete, we are given a set of integers $\mathcal{U} = \{a_1, \ldots, a_{3n}\}$ with $\sum_{i=1}^{3n} a_i = Bn$ for an integer $B > 0$. The aim is to determine whether $\mathcal{U}$ can be divided into n subsets, such that each subset is with an equal sum B.

Theorem 2. *Decision 1D-LBTC is $\mathcal{NP}$-complete when the sensors are with non-uniform integral radii.*

The key idea of the proof is to construct a reduction from 3-Partition to the decision LBTC problem. For a given instance of 3-Partition, the construction of the corresponding instance of decision LBTC is simply as below:

1. Construct a line barrier with length $(2n - 1)B$;
2. Place $2nB$ targets on the line barrier composed by n sections, where in the ith *section*, $i = 0, \ldots, n - 1$, the targets are with positions $2iB + j + j\epsilon$ and $2iB + j + 1 - (B - j)\epsilon$, for $j = 0, \ldots, B - 1$;
3. Place $3n$ sensors on position $(0, 0)$, where sensor i is with radii $\frac{a_i}{2}$;
4. The maximum movement is $D = (2n - 1)B$.

Note that, the instance of decision 1D-LBTC constructed above contains $2nB$ POIs and $3n$ sensors. Anyhow, 3-Partition is known strongly $\mathcal{NP}$-complete, which means, 3-Partition remains $\mathcal{NP}$-complete even when B is polynomial to n. Therefore, the construction can be done in polynomial time for B being polynomial to n.

The main idea behind the construction is to construct a relationship between the number of covered targets and the diameters of the sensors that are actually the integers in $\mathcal{U}$. More precisely, the property on the relationship is as in the following:

Proposition 3. *Against a 1D-LBTC instance produced by the above construction, a sensor with diameter $2r$ can cover at most $4r$ targets.*

Proof. When a sensor is with a diameter 2, apparently it can cover at most 4 targets. Suppose the proposition is true for sensors with diameter smaller than $2r$. Then, let $r_1 + r_2 = r$ be two positive integers smaller than r. By induction, we have that sensors with diameters $2r_1$ and $2r_2$ can cover upto $4r_1$ and $4r_2$ targets, respectively. In addition, the two sensors with radii r_1 and r_2 can cover as many POIs as a sensor with a radii $r = r_1 + r_2$ does. Therefore, the sensor with diameter $2r$ can cover no more than $4r_1 + 4r_2 = 4r$ targets. This completes the proof.   □

Lemma 4. *An instance of 3-Partition is feasible if and only if the corresponding 1D-LBTC instance is feasible.*

Proof. Suppose the instance of 3-Partition is feasible. Without loss of generality, we assume that $\{U_i | i = 0, \ldots, n - 1\}$ is a solution to the 3-Partition instance which divides $\mathcal{U}$ to a collection of n sets, among which $U_i = \{a_{l_i+1}, \ldots, a_{l_{i+1}}\}$ and $l_0 = 0$. Since $D = (2n - 1)B$ equals the length of the barrier and the original position of each sensor is $(0, 0)$, each sensor can be moved any point of the barrier. Then we need only to use the sensors in U_i, which are with radius $a_{i_j}, \ldots, a_{i_{j+1}}$ and with a sum exactly B, to cover the segment from $2iB$ to $(2i + 1)B$. That apparently results in a coverage for all the targets in the ith section.

Conversely, suppose the corresponding LBTC instance is feasible. Then since sensor j with radii $\frac{a_j}{2}$ can at most cover $2a_j$ continuous targets, and each section contains exactly $2B$ targets, so the diameter sum of the sensors for each section is at least B. Then because the diameter sum of all the sensors is Bn, and there are n sections, the diameter sum of the sensors for each section is exactly B. Therefore, the diameters for the sensors for the sections is a solution to the corresponding instance of 3-Partition.   □

From the fact that 3-Partition is strongly $\mathcal{NP}$-complete, and following a similar idea of the above proof for Theorem 2, we immediately have the following hardness for LBTC:

Corollary 5. *Decision 1D-LBTC is strongly $\mathcal{NP}$-complete.*

3   A Greedy Algorithm for 2D-LBTC with Uniform Sensors

The basic idea of the algorithm is to cover the target from left to right, preferably using sensors that are likely less useful for later coverage. More precisely, let $[l_i, g_i]$ be the possible coverage range of sensor i, where l_i and g_i are respectively the positions of the leftmost and the rightmost targets, with respect to the given distance D. That is, l_i and g_i are the leftmost and the rightmost targets sensor i can cover within movement D. Then the key idea of our algorithm is to cover the targets from left to right, using the sensor that can cover the leftmost uncovered target within movement D and is with minimum g_i.

The algorithm is first to compute its possible coverage range $[l_i, g_i]$ for each sensor i with respect to the movement constraint D. Apparently, $(x_i, 0)$ is the projective point of sensor i on the line, so we have $l_i = x_i - \sqrt{D^2 - y_i^2} - D$ and $g_i = x_i + \sqrt{D^2 - y_i^2}$ for each sensor i. Then, the algorithm starts from point $s = (0, 0)$, to cover the line from left to right. The algorithm prefers using the sensor with a small g_i, since a sensor with a large g_i would has a better potential to cover the targets on the right part of the line.

Let s be the position the uncovered leftmost target on the line barrier. Then among the set of sensors $\{i | l_i \leq s \leq g_i\}$, the algorithm repeats selecting the sensor with minimum g_i to cover the uncovered targets of the line barrier starting at s. Note that $\{i | l_i \leq s \leq g_i\}$ is exactly the set of sensors that can monitor a set of uncovered targets starting at s by relocating at most D distance. The algorithm terminates either the set of targets are completely covered, or the instance is found infeasible (i.e. there exists no unused sensor i with $l_i \leq s \leq g_i$ while the coverage is not yet done). The detailed algorithm is formally as in Algorithm 1.

Note that Algorithm 1 takes $O(n)$ time to compute l_i and g_i for all the sensors in Steps 2–3, and takes $O(n \log n)$ time to assign the sensors to cover the targets on the line barrier in Steps 4–15. Therefore, we have the time complexity of the algorithm:

Lemma 6. *Algorithm 1 runs in time $O(n \log n)$.*

Before proving the correctness of Algorithm 1, we need the following lemma stating the existence of a special coverage for a feasible LBTC instance.

Proposition 7. *Let (x_j, y_j) be the position of sensor j in the plane. Assume $p_1(s, 0)$, $p_2(x'_j, 0)$ and $p_3(x''_j, 0)$ are three points on a line segment. If $s \leq x''_j \leq x'_j$, then $d(j, p_3) \leq \max\{d(j, p_1), d(j, p_2)\}$ holds. That is, the distance between the sensor and the middle point is not larger than the larger distance between the sensor and the other two points.*

Algorithm 1. A simple greedy algorithm for decision LBTC.

Input: A movement distance upper bound $D \in \mathbb{Z}^+$, a set of sensors $\Gamma = \{1, \ldots, n\}$ with original position $\{(x_i, y_i) | i \in [n]^+\}$ and r being the sensing radii, a set of POIs $P = \{1, \ldots, m\}$ with positions $p_1 \preceq p_2 \preceq \cdots \preceq p_m$, where p_j is the position for $j \in P$;
Output: New positions $\{x_i' | i \in [n]^+\}$ for the sensors or return "infeasible".

1: Set $\mathcal{I} := \Gamma$, $s := p_1$; /*s is the leftmost point of the uncovered part of the barrier.*/
2: **For** each sensor i **do**
3:     Compute the leftmost position l_i and the rightmost position g_i, both of which
        sensor i can monitor;
4: **While** $\mathcal{I} \neq \emptyset$ **do**
5:     **If** there exists $i' \in \mathcal{I}$, such that $l_{i'} \leq s \leq g_{i'}$ **then**
6:         Select sensor $i \in \mathcal{I}$ for which $g_i = \min_{i' : l_{i'} \leq s \leq g_{i'}} \{g_{i'}\}$;
        /* Select the sensor with minimum g_i among all the sensors $\{i' | l_{i'} \leq s \leq g_{i'}\}$.*/
7:         Set $t := \min\{s + 2r, g_i\}$, $\mathcal{I} := \mathcal{I} \setminus \{i\}$, $x_i' := t - r$;
8:         **If** $\{p | p > t, p \in P\} = \emptyset$ **then** /*All targets are covered. */
9:             Return "feasible" together with the new positions $\{x_i' | i \in \Gamma\}$;
10:       **Endif**
11:       Set $s := \min\{p_j | p_j > t\}$;
12:     **Else**
13:       Return "infeasible";
14:     **Endif**
15: **Endwhile**

Lemma 8. *If an instance of LBTC is feasible, then there must exist a coverage in which the sensors are s-ordered.*

Proof. The key idea of the proof is that, any coverage of LBTC that is not s-ordered, can be converted to an s-ordered coverage by re-scheduling the sensors of covering the POIs.

Suppose there exist two sensors i and j, such that $g_i > g_j$ but $x_i' < x_j'$. Then we need only to swap the final positions of i and j, i.e. to simply set the new final positions x_i'' and x_j'' of sensor i and j as below: If $x_i' - r \geq s$, then set $x_i'' := x_j'$ and later $x_j'' := x_i'$; Otherwise set $x_i'' := x_j'$ and $x_j'' := s + r$.

Apparently, the POIs exclusively covered by i are now covered by sensor j, and *vice versa*. So after the swap the sensors will remains a coverage for the POIs on the line. It remains to show the swap will not increase the maximum movement. Recall that the leftmost and the rightmost points sensor j can cover are respectively l_j and g_j. Because sensor j can move to x_j' under the movement bound D, we have

$$l_j \leq x_j' - r \leq x_j' + r \leq y_j \leq g_i. \tag{1}$$

On the other hand, in either case of the swap, we have $x_i'' = x_j' \geq x_i'$. So combining Inequality (1), we have $l_i \leq x_i'' - r \leq x_i'' + r \leq g_i$. That means

$$l_i + r \leq x_i'' \leq g_i - r.$$

Then following Proposition 7, the distance between sensor i and its new position x_i'' is bounded by $D = \max\{d(i, (l_i + r, 0)), d(i, (g_i - r, 0))\}$. The case for the new position of sensor j is similar except that the distance between sensor j and its new position x_i'' is bounded by $D = \max\{d(j, (\max\{s, l_j + r\}, 0)), d(i, (g_i - r, 0))\}$. This completes the proof.   □

Based on Lemma 8, given a feasible instance of LBTC, we can assume there exists an s-ordered coverage, say $\Gamma' = \{s_1, \ldots, s_k\}$ which is the set of sensors used to compose the coverage with j_i being the rightmost target covered by s_i. Then we have the following lemma, which leads to the correctness of the algorithm:

Lemma 9. *When running against a feasible LBTC instance, Algorithm 1 covers the targets $\{1, \ldots, j_i\}$ without using any sensor in $\{s_{i+1}, \ldots, s_k\}$.*

Proof. We shall prove this claim by induction. When $i = 1$, the lemma is obviously true, as we need only s_1 to cover the targets $\{1, \ldots, j_1\}$. Suppose the lemma holds for $i = h$, then it remains only to show the case for $i = h + 1$. By induction, Algorithm 1 covers the targets $\{1, \ldots, j_h\}$ without using any sensor in $\{s_{h+1}, \ldots, s_k\}$. Then Algorithm 1 can simply cover targets $\{j_h + 1, \ldots, j_{h+1}\}$ by using sensor s_{h+1}. Combining with the induction, we covers $\{1, \ldots, j_{h+1}\}$ without using any sensor in $\{s_{h+2}, \ldots, s_k\}$. This completes the proof.   □

We can now prove the following theorem to get the correctness of Algorithm 1:

Theorem 10. *Algorithm 1 returns "feasible" iff the targets can be completely covered by the sensors within relocation distance D.*

Proof. Suppose Algorithm 1 returns "feasible", then obviously the produced solution $\{x_i' | i \in \Gamma\}$ is truly a coverage, because in the solution the movement of each sensor is bounded by D and all the targets are covered by at least one sensor.

Conversely, suppose there is a coverage for the instance. Then by Lemma 8, there must exist an s-ordered coverage, say $\Gamma' = \{s_1, \ldots, s_k\}$ which is the set of sensors used to compose the coverage. Following Lemma 9, Algorithm 1 covers targets $\{1, \ldots, j_i\}$ without using any sensor in $\{s_{i+1}, \ldots, s_k\}$ for every $i \in [1, k]$. So the algorithm can always find sensors for further coverage, and in the worst case use s_{i+1} to cover the targets $\{j_i + 1, \ldots, j_{i+1}\}$. Therefore, the algorithm will eventually find a feasible coverage. This completes the proof.   □

4   The Complete Algorithms

In this section, we will show how to employ Algorithm 1 to really compute D^* the minimum movement bound for LBTC. Firstly, when only considering integral D^*, we can find it simply by employing the binary search method against a large range that contains D^*; Secondly, for real number D^*, we construct a set of size $O(mn)$ which arguably contains D^*, and then eventually finds D^* in the set again by the binary search method.

Algorithm 2. The whole algorithm for optimal LBTC.

Input: A movement distance upper bound $D \in \mathbb{Z}^+$, a set of sensors $\Gamma = \{1, \ldots, n\}$ with original position $\{(x_i, y_i) | i \in [n]^+\}$ and r being the sensing radii, a set of POIs $P = \{1, \ldots, m\}$ with positions $p_1 \preceq p_2 \preceq \cdots \preceq p_m$, where p_j is the position for $j \in P$;

Output: The minimized maximum movement of the sensors together with their new positions $\{x'_i | i \in [n]^+\}$.

1: Set $upper := d_{max}$ and $lower := 1$, where d_{max} is the maximum distance between the sensors and the POIs;

2: **If** there exists no coverage by calling Algorithm 1 wrt $D = d_{max}$ **then**

3:      Return "infeasible";

4: **EndIf**

5: Set $temp := \left\lceil \frac{lower+upper}{2} \right\rceil$;

6: **While** $upper - lower > 1$ **do**

7:      **If** there exists no coverage by calling Algorithm 1 wrt $D = temp$ **then**

8:          Set $lower := temp$ and then $temp := \left\lceil \frac{lower+upper}{2} \right\rceil$;

9:      **Else**

10:          Set $upper := temp$ and then $temp := \left\lfloor \frac{lower+upper}{2} \right\rfloor$

11:      **EndIf**

12: **EndWhile**

13: Return the result of calling Algorithm 1 wrt $D = temp$ and terminate.

4.1   A Simple Binary Search Based Algorithm

The algorithm is simply applying the binary search method to find D^* within the range of $[1, d_{max}]$, where d_{max} is the maximum distance between the targets and the sensors. The main observation is as the following proposition whose correctness is easy to prove:

Proposition 11. *If LBTC is feasible, then we have $D^* \leq d_{max}$.*

The detailed algorithm is as in Algorithm 2.

For the correctness and time complexity of Algorithm 2, we immediately have the following lemma:

Lemma 12. *Using binary search and employing Algorithm 1 for $O(\log D_{max})$ times, Algorithm 2 will compute the optimum movement D^* within time complexity $O(n \log n \log D_{max})$.*

4.2   An Improved Algorithm via Discretized Binary Search

In this subsection, we shall show the time complexity of our algorithm can be further improved via a more sophisticated implementation over the binary search. The key observation is that, we need only to apply a binary search over a set of discrete values which arguably contain the optimum min-max movement D^*. Let $\{c_1, \ldots, c_t\}$ be the set of possible combinations. Let d_{ij} be the minimum movement using sensor i to cover combination c_j, where c_j is a set of POIs which can be exactly covered by a sensor. Then we have the following lemma:

Algorithm 3. A fast algorithm for LBTC.

Input: A set of sensors $\Gamma = \{1, \ldots, n\}$ with original position $\{(x_i, y_i) | i \in [n]^+\}$ and an identical sensing radii r, a set of POIs $P = \{1, \ldots, m\}$ on the line segment with positions $p_1 \preceq p_2 \preceq \cdots \preceq p_m$, where p_j is the position for $j \in P$;
Output: Minimum movement bound D under which the sensors can be relocated to covered all the POIs of P.

0: Set $\Psi := \emptyset$ and compute the collection of combinations $\Phi := \{c_1, \ldots, c_t\}$;
1: **For** each sensor i **do**
2:    **For** each combination $c_j \in \Phi$ **do**
3:        Compute d_{ij}, the minimum movement needed to using sensor i to cover c_j;
4:        Set $\Psi := \Psi \cup \{d_{ij}\}$;
5:    **EndFor**
6: **EndFor**
7: Sort Ψ in a non-decreasing order and set $lb := 1$ and $ub := |\Psi|$;
8: Use $\Psi[1]$ as the movement bound (i.e. D) to call Algorithm 1;
   /*$\Psi[1]$ is the smallest element in Ψ. */
9: **If** there exists a feasible coverage under movement bound $\Psi[1]$ **then**
10:    Return $\Psi[1]$ as the optimum movement bound;
11: **Endif**
12: **While** $ub - lb > 1$ **do**
13:    Set $idx := \lceil \frac{lb+ub}{2} \rceil$;
14:    Use $\Psi[idx]$ as the movement bound (i.e. D) and call Algorithm 1;
       /*$\Psi[idx]$ is the idx smallest element in Ψ. */
15:    **If** there exists a feasible coverage under movement bound $\Psi[idx]$ **then**
16:        Set $ub := idx$;
17:    **Else**
18:        Set $lb := idx$;
19:    **Endif**
20: **Endwhile**
21: Return $\Psi[idx]$ as the optimum movement bound.

Lemma 13. *Let d_{opt} be an optimal solution to the uniform 2D-LBTC problem. Then $d_{opt} \in \Psi = \{d_{ij} | i \in \Gamma, c_j \in \{c_1, \ldots, c_t\}\}$.*

Proof. Suppose the lemma is not true. Then let $d_{max} = \max_d\{d \mid d \in \Psi, d < d_{opt}\}$. First we show that under maximum distance d_{max} and d_{opt}, every sensor i covers an identical collection of combinations. That is because every POI, which sensor i can cover under movement bound d_{opt}, can also be covered by sensor i under movement bound d_{max} (as $d_{ij} \leq d_{max}$ iff $d_{ij} < d_{opt}$), and conversely every POI, which cannot be covered by sensor i under d_{max}, can not be covered by the same sensor within the movement bound d_{opt} ($d_{ij} > d_{max}$ iff $d_{ij} > d_{opt}$). Therefore, a feasible coverage solution under maximum movement d_{opt} would also remain feasible under d_{max}. This together with $d_{max} < d_{opt}$ contradicts with the fact that d_{opt} is an optimal solution to the problem.   □

Our algorithm will first compute the collection of distances between the combinations and the sensors, say $\Psi = \{d_{ij} | i \in \Gamma, c_j \in \{c_1, \ldots, c_t\}\}$, and then sort the distance in Ψ in non-decreasing order. Then by applying the binary search method to Ψ and using Algorithm 1 as a subroutine, we find a minimum d_{ij} under which there exists a relocation of the sensors such that all the targets can be covered. The detailed algorithm is as in Algorithm 3.

Lemma 14. *The time complexity of Algorithm 3 is* $O(mn(\log m + \log n))$.

Proof. Apparently, $|\Phi| = O(m)$, so we have $|\Psi| = O(mn)$. Then sorting the elements in Ψ takes $O(|\Psi|\log|\Psi|) = O(mn\log mn) = O(mn(\log m + \log n))$ time. Besides, the while-loop from Step 12 to Step 20 will be repeated for at most $O(\log m + \log n)$ times, each of which takes $O(n\log n)$ time to run Algorithm 1. Therefore, the total time complexity of the algorithm is $O(mn(\log m + \log n))$.
$\square$

Theorem 15. *Algorithm 3 produces an optimum solution to the LBTC problem.*

5   Conclusion

In this paper, we first proved that 1D-LBTC is $\mathcal{NP}$-hard when the radius of the sensors are not identical, in contrast with the known result that 1D-LBC problem can be efficiently solved in a polynomial time. Then, we designed an algorithm for decision 2D-LBTC with uniform radius, and consequently proposed an algorithm for really solving 2D-LBTC based on the binary search method. Moreover, we improved the binary search method to a runtime $O(mn(\log m + \log n))$ by observing that the optimum movement bound is within the set of distances between combinations of POIs and the sensors. We are currently investigating how to further improve the runtime of the algorithm.

References

1. Bhattacharya, B., Burmester, M., Hu, Y., Kranakis, E., Shi, Q., Wiese, A.: Optimal movement of mobile sensors for barrier coverage of a planar region. Theor. Comput. Sci. **410**(52), 5515–5528 (2009)
2. Chen, D., Yan, G., Li, J., Wang, H.: Algorithms on minimizing the maximum sensor movement for barrier coverage of a linear domain. Discrete Comput. Geom. **50**(2), 374–408 (2013)
3. Cherry, A., Gudmundsson, J., Mestre, J.: Barrier coverage with uniform radii in 2D. In: Fernández Anta, A., Jurdzinski, T., Mosteiro, M.A., Zhang, Y. (eds.) ALGO-SENSORS 2017. LNCS, vol. 10718, pp. 57–69. Springer, Cham (2017). https://doi.org/10.1007/978-3-319-72751-6_5
4. Czyzowicz, J., et al.: On minimizing the maximum sensor movement for barrier coverage of a line segment. In: Ruiz, P.M., Garcia-Luna-Aceves, J.J. (eds.) ADHOC-NOW 2009. LNCS, vol. 5793, pp. 194–212. Springer, Heidelberg (2009). https://doi.org/10.1007/978-3-642-04383-3_15
5. Czyzowicz, J., et al.: On minimizing the sum of sensor movements for barrier coverage of a line segment. In: Nikolaidis, I., Wu, K. (eds.) ADHOC-NOW 2010. LNCS, vol. 6288, pp. 29–42. Springer, Heidelberg (2010). https://doi.org/10.1007/978-3-642-14785-2_3

6. Dobrev, S., et al.: Complexity of barrier coverage with relocatable sensors in the plane. Theor. Comput. Sci. **579**, 64–73 (2015)
7. Gage, D.W.: Command control for many-robot systems. Technical report, Naval Command Control and Ocean Surveillance Center RDT and E Div, San Diego, CA (1992)
8. Kumar, S., Lai, T.H., Arora, A.: Barrier coverage with wireless sensors. In: Proceedings of the 11th Annual International Conference on Mobile Computing and Networking, pp. 284–298. ACM (2005)
9. Li, S., Shen, H.: Minimizing the maximum sensor movement for barrier coverage in the plane. In: IEEE Conference on Computer Communications (INFOCOM), pp. 244–252. IEEE (2015)
10. Li, X., Frey, H., Santoro, N., Stojmenovic, I.: Localized sensor self-deployment with coverage guarantee. ACM SIGMOBILE Mob. Comput. Commun. Rev. **12**(2), 50–52 (2008)
11. Mehrandish, M., Narayanan, L., Opatrny, J.: Minimizing the number of sensors moved on line barriers. In: IEEE Wireless Communications and Networking Conference (WCNC), pp. 653–658. IEEE (2011)
12. Tan, X., Wu, G.: New algorithms for barrier coverage with mobile sensors. In: Lee, D.-T., Chen, D.Z., Ying, S. (eds.) FAW 2010. LNCS, vol. 6213, pp. 327–338. Springer, Heidelberg (2010). https://doi.org/10.1007/978-3-642-14553-7_31

Online Travelling Salesman Problem
on a Circle

Vinay A. Jawgal[✉], V. N. Muralidhara, and P. S. Srinivasan

IIIT Bangalore, Bangalore, India
{vinay.jawgal,ps.srinivasan}@iiitb.org, murali@iiitb.ac.in

Abstract. In the online version of Travelling Salesman Problem, requests to the server (salesman) may be presented in an online manner i.e. while the server is moving. In this paper, we consider a special case in which requests are located only on the circumference of a *circle* and the server moves only along the circumference of that circle. We name this problem as online Travelling Salesman Problem on a circle (OLTSP-C). Depending on the minimization objective, we study two variants of this problem. One is the *homing* variant called H-OLTSP-C in which the objective is to minimize the time to return to the origin after serving all the requests. The other is the *nomadic* variant called N-OLTSP-C in which after serving all the requests, it is not required to end the tour at the origin. The objective is to minimize the time to serve the last request. For both the problem variants, we present online algorithms and lower bounds on the competitive ratios. An online algorithm is said to be zealous if the server that is used by the online algorithm does not *wait* when there are unserved requests. For N-OLTSP-C, we prove a lower bound of $\frac{28}{13}$ on the competitive ratio of any zealous online algorithm and present a 2.5-competitive zealous online algorithm. For H-OLTSP-C, we show how the proofs of some of the known results of OLTSP on general metric space and on a line metric, can be adapted to get lower bounds of $\frac{7}{4}$ and 2 on the competitive ratios of any zealous and non-zealous online algorithms, respectively.

Keywords: Online algorithms · Competitive ratio ·
Travelling Salesman Problem

1 Introduction

In the classical Travelling Salesman Problem (TSP), a salesman (server) is given a designated origin O and a set of n requests (points in some metric space). The goal is to find a tour that starts and ends at the origin and minimizes the total distance travelled or the completion time, by serving each point at least once. So, the nature of classical TSP is "offline", in the sense that at time $t = 0$ when the server is located at the origin, it has all the information about the input. TSP and many of its variants have been studied extensively in this offline setting, where all the information about the requests is known to the server before starting

© Springer Nature Switzerland AG 2019
T. V. Gopal and J. Watada (Eds.): TAMC 2019, LNCS 11436, pp. 325–336, 2019.
https://doi.org/10.1007/978-3-030-14812-6_20

the tour. But in many practical situations, we do not have all the information a-priori. This is the motivation to consider the online version of TSP called OLTSP, where requests are revealed online while the server is moving. Each request is associated with a release time, only after which information about the request becomes known and can it be served.

In this paper, we study a special case of the online Travelling Salesman Problem in which the requests are located on the circumference of a *circle* and the server moves only along the circumference of that circle. We call this problem as OLTSP-C.

A standard technique which has been used for analysing many problems in an online setting is *competitive analysis* in which the performance of an online algorithm is compared against an optimal offline algorithm which knows all the input data in advance. More formally, over all possible input sequences, consider the ratio between the objective function value produced by an online algorithm and the objective function value produced by an optimal offline algorithm. Such a ratio is called *competitive ratio*. Therefore, an online algorithm is said to be ρ-competitive if for every input its objective value is at most ρ times the optimal offline objective value for the same input.

1.1 Preliminaries

Depending on the minimization objective, we consider two variants of OLTSP-C. When the objective is to minimize the time to serve the last request, it is the nomadic variant called N-OLTSP-C. When there is an additional requirement to end the tour at the origin, then the objective is to minimize the time to reach the origin after serving all the requests. This is the homing variant called H-OLTSP-C. Thus,

N-OLTSP-C Minimize the time it takes for the online server to all the requests.
H-OLTSP-C Minimize the time it takes for the online server to serve all the requests and return to the origin.

In this paper, we restrict ourselves to deterministic online algorithms. The server used by a deterministic online algorithm can move at any speed varying between 0 and 1 (*maximum* speed). Blom *et al.* [5] proposed a particular class of algorithms called *zealous algorithms*.

Definition 1. *[Zealous Algorithm] An algorithm ALG for the OLTSP is called zealous, if it satisfies the following conditions:*

- *If there are still unserved requests, then the direction of the server operated by ALG changes only if a new request becomes known, or the server is either in the origin or at a request that has just been served.*
- *At any time when there are unserved requests, the server operated by ALG either moves towards an unserved request or to the origin at maximum (i.e. unit) speed. (The latter case is only allowed if the server operated by ALG is not yet in the origin.)*

Notation: Consider the circle C of radius r. Let the origin O be the *top-most* point on the circle where the online server is initially located at time $t = 0$. Each request $\sigma = (t, x)$ is identified by its release time t and a location x on the circumference of the circle. A request can be served only at or after its release time. The offline version of OLTSP-C is the problem known as *TSP With Release Dates* in which the release dates of the requests are known to the offline server at time $t = 0$. For OLTSP-C, let Z^{ol} and Z^* denote the objective function values of an online algorithm and the corresponding optimal offline algorithm, respectively. Such an online algorithm is said to be ρ-competitive if there exists a constant ρ such that for any request sequence, $Z^{ol} \leq \rho Z^*$.

1.2 Related Work

Ausiello *et al.* [3] introduced OLTSP and proved lower bound results and competitive ratios of online algorithms for both the nomadic and homing versions of OLTSP on general graphs and on the real line. Blom *et al.* [5] introduced the notion of *zealous algorithms* and *fair adversaries* and studied the homing version of OLTSP on a half-line. They proved that non-zealous algorithms are strictly better than the zealous ones. Lipmann [8] presents many results on OLTSP on the line and half-line with sophisticated mathematical analysis of the non-zealous algorithms. More recently, tight bounds and competitive analysis of OLTSP on real line have been presented [4]. Feuerstein [6] and Ascheuer [2] have studied the generalized version of OLTSP called online Dial-a-Ride Problem where online requests should be picked up from a source and delivered to a destination. Jaillet and Wegner [7] and Albers [1] presents survey about online routing and online algorithms in general, respectively.

1.3 Our Contribution

In the seminal paper of Ausiello *et al.* [3], the authors introduced and analysed OLTSP from the perspective of competitive analysis for the first time. In the concluding section of their paper, the authors pointed out that it would be interesting to studying OLTSP on special metrics like the circle. This motivated us to consider OLTSP on a circle. To the best of our knowledge, we believe that the results in this paper, for the first time, give a formal treatment to OLTSP-C.

For N-OLTSP-C, we prove a lower bound of $\frac{28}{13}$ on the competitive ratio of any *zealous* online algorithm and present a 2.5-competitive *zealous* online algorithm. We consider this as our main contribution.

For H-OLTSP-C, we show how the proofs of some of the known results of OLTSP on general metric space and on a line metric, can be adapted to get lower bounds of $\frac{7}{4}$ and 2 on the competitive ratios of any zealous and non-zealous online algorithm, respectively.

Though various bounds are known for OLTSP on the line and half-line, the problem becomes interesting to analyse on a circle due to the following reasons.

– In a circle, from any point, the online server can reach the origin O in two ways (clockwise or anti-clockwise). While in the case of a line or halfline, there is only one direction by which online server can reach the origin.
– A circle is finite but a line/half-line is infinite. So, in case of a circle, there is a boundary within which the adversary can release requests.

It is therefore important to understand that not all the results known for line directly apply for circle due to the reasons stated above. Some of the lower bound results for OLTSP on a line rely on the fact that the line is of infinite length on either sides of the origin. Thus, there arises a need to rework some of these proofs for OLTSP on the circle.

Outline: Structure of this paper is as follows. We discuss lower bound results on the competitive ratios for H-OLTSP-C and N-OLTSP-C in Sect. 2. Algorithms for H-OLTSC-C and N-OLTSC-C are presented in Sect. 3. Section 4 concludes with the summary.

2   Lower Bound Proofs

In this section, we present the lower bounds on the competitive ratios of zealous and non-zealous online algorithms, respectively.

Some lower bound proofs for competitive ratios of algorithms for TSP on the line can be extended to the circular case by assuming that the size of the circle is relatively large and all the requests occur close to the origin thereby imitating a line on the circle. But certain lower bound proofs on a line which relies on the property that the line is of infinite length on either sides of the origin do not extend trivially to a circle, as in the case of a circle we cannot assume an infinite length path on either sides of the origin.

Lemma 1. *Any ρ-competitive deterministic zealous online algorithm for N-OLTSP-C has $\rho \geq \frac{28}{13}$.*

Proof. For the ease of notations, let the point $\frac{\pi}{2}$ radians from the origin O in the clockwise direction be denoted as A, the point diametrically opposite to the origin O be denoted as B and the point diametrically opposite to A be denoted as C. Let us also assume, without loss of generality, that the circumference of the circle is of length 8 units. Denote by D as the point $\frac{\pi}{4}$ radians away from the origin in the clockwise direction and denote by E as the point $\frac{\pi}{4}$ radians away from the origin in the anticlockwise direction. Denote by X and Y as the mid-points of the minor arcs OD and OE, respectively (Fig. 1).

Suppose two requests $\sigma_1 = (0, X)$ and $\sigma_2 = (\frac{1}{2}, O)$ are released. After serving these requests, the online server should be at the origin at time 1. At time 1, two requests $\sigma_3 = (1, D)$ and $\sigma_4 = (1, E)$ are released. At time 3, the online server is again at the origin after serving either σ_3 or σ_4. Assume WLOG, that it has served σ_3.

At time 3, another request $\sigma_5 = (3, D)$ is presented. Therefore, at time $t = 3$, the online server is at the origin and has to serve σ_4 and σ_5 at locations D and

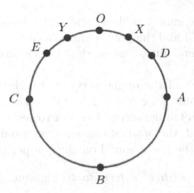

Fig. 1. LB instance for (zealous) N-OLTSP-C.

E, respectively. Assume, WLOG, that it decides to serve σ_5 as both the requests are symmetric. At time $t = 4$, the online server reaches the location of σ_5 i.e D. On the other hand, the offline server can serve this request sequence in the order (E, Y, O, X, D, A) and can be in location A at time 4.

Therefore, at time 4, the offline server is at location A and the online server is at location D.

The online server, in order to serve σ_4 at location E, starts moving towards the origin from location D at time 4 (if instead the online server continues to move in the clockwise direction at time 4 in order to serve σ_4 then the adversary can stop releasing any further requests and the algorithm performs bad). On the other hand, suppose that at time 4, the offline server, located at A, starts moving away from the origin and continues to move in the same direction till it reaches a point P such that current position of the online server is exactly the mid-point of the path with end points P and the location of σ_4 i.e E. The path we consider here is the one that passes through the origin. Let x be the length of the minor arc PA. Note that this is the same distance that the online server has covered, starting from the location of σ_5 i.e. D. We can calculate x as follows.

$$1 + 2 + x = 2(1 + (1 - x)) \implies x = \frac{1}{3}$$

At time $t = (4 + \frac{1}{3} = \frac{13}{3})$, a new request σ_5 is presented at location $2 + \frac{1}{3} = \frac{7}{3}$ i.e. $\sigma_6 = (\frac{13}{3}, \frac{7}{3})$. Completion time of the offline server is $\frac{13}{3}$. On the other hand, completion time of the online server which has to serve both requests σ_4 and σ_6 is $\frac{13}{3} + 3(1 + (1 - \frac{1}{3})) = \frac{13}{3} + 5$. Therefore, the competitive ratio is given by $\frac{(\frac{13}{3} + 5)}{\frac{13}{3}} \approx 2.153$. ∎

Lemma 2. *Any ρ-competitive deterministic online algorithm for N-OLTSP-C has $\rho \geq 2$.*

Proof. We can trivially extend Theorem 3.1 of Ausiello et al. [3] for the line to that of a circle, since the proof does not require a line of infinite length. For the ease of notations, let the point $\frac{\pi}{2}$ radians away from the origin O in the clockwise direction be denoted as A and the point diametrically opposite to A be demoted as C. Without loss of generality, assume that the circumference of the circle is of length 4 units.

At time 1, if position of the online server is on the minor arc OC, then a request is released in point A i.e $\sigma = (1, A)$. Otherwise, the request is released in point C i.e $\sigma = (1, C)$. Online server, in order to serve σ, takes $1 + 1 = 2$ time units. On the other hand, the optimal offline server would be at the location of σ at time 1. Therefore, the lower bound on the competitive ratio is $\frac{2}{1} = 2$. ∎

Lemma 3. *Any ρ-competitive deterministic zealous online algorithm for H-OLTSP-C has $\rho \geq \frac{7}{4}$.*

Proof. For the ease of notations, let the point $\frac{\pi}{2}$ radians from the origin O in the clockwise direction be denoted as A, the point diametrically opposite to the origin O be denoted as B and the point diametrically opposite to A be denoted as C. Denote by D as the point $\frac{\pi}{4}$ radians away from the origin in the clockwise direction. Let us also assume, without loss of generality, that the circumference of the circle is of length 4 units.

At time 0, the request $\sigma_1 = (0, D)$ is released. Online server, after serving σ_1, reaches the origin at time 1. At this time, two new requests $\sigma_2 = (1, A)$ and $\sigma_3 = (1, C)$ are released. Due to zealousness, at time 3, the online server would be located in either O or B after having served either σ_2 or σ_3. Without loss of generality, assume that the online server has served σ_2 before time 3. On the other hand, the optimal offline server, after serving σ_3 at time 1, would be located in A at time 3. That means that at time 3, the offline server has served both the requests σ_2 and σ_3.

At time 3, a new request $\sigma_4 = (3, A)$ is released. Irrespective of whether the online server is located in O or B at time 3, it shall take an additional 4 units of time to return to the origin after serving the unserved requests i.e. $Z^{ol} = 3 + 4 = 7$. On the other hand, the optimal offline server is already present at A at time 3. Thus, $Z^* = 3 + 1 = 4$. Therefore, the lower bound on the competitive ratio is $\frac{7}{4}$. ∎

Lemma 4. *Any ρ-competitive deterministic online algorithm for H-OLTSP-C has $\rho \geq 2$.*

Proof. Consider a circle with the top-most point as the origin O. Suppose, WLOG, that the circumference of the circle is 4 units. Denote the point diametrically opposite to O as O'. The position of the online server at time t is denoted by p_t^{ol}. Define *distance* between any two points on the circle as the length of the minor arc between them. Call the two semi-circles with the diameter OO' as *left half* and *right half*, respectively.

At time 0, multiple requests are released all along the circumference of the circle with a distance of ϵ between any consecutive requests. The number of requests can be arbitrarily large with small ϵ.

We first show that for some $\delta \in [0, 2]$, at time $2 + \delta$ any on-line server must be in one of the two points at distance $2 - \delta$ from the origin (not necessarily requested points). To formalize this notion, consider a function $f : [0, 2] \to [0, 2]$ such that $f(\delta)$ is the distance of the position of the online server from the point O at time $2+\delta$. Define another function $g(\delta) = f(\delta) - \delta$. Clearly, $g(0) = f(0) - 0 \geq 0$ and $g(2) = f(2) - 2 \leq 0$. Since g is continuous, there exists $\delta_0 \in [0, 2]$ such that $g(\delta_0) = 0$ i.e. $f(\delta_0) = \delta_0$. Consider smallest such δ_0. Therefore, at time $2 + \delta_0$, the online server is at a distance of $2 - \delta_0$ from the origin O (Fig. 2).

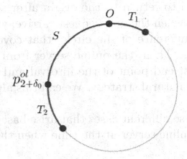

Fig. 2. LB instance for (non-zealous) H-OLTSP-C.

Next, we prove that there exists a segment of length atleast 2 such that none of the requests in that segment have been touched at time $2 + \delta_0$. Let S be the segment corresponding to the minor arc with end points as the origin O and p_t^{ol}, respectively. Let T_1 be the segment starting from the origin O in the clockwise direction to some point such that the requests in T_1 are served before time $2 + \delta_0$. Similarly, let T_2 be the segment starting from the position of the online server at time $2 + \delta_0$, i.e. $p_{2+\delta_0}^{ol}$, in the anti-clockwise direction to some point such that requests in T_2 are served before time $2 + \delta_0$. Clearly, we have

$$|S| + 2|T_1| + 2|T_2| \leq 2 + \delta_0$$
$$2 - \delta_0 + 2|T_1| + 2|T_2| \leq 2 + \delta_0$$
$$|T_1| + |T_2| \leq \delta_0$$

Therefore, $|S| + |T_1| + |T_2| \leq 2$. The arc excluding segments S, T_1 and T_2 has length atleast 2. Therefore, such an arc is a segment of length atleast 2 such that none of the requests in that segment have been touched at time $2 + \delta_0$.

At time $2 + \delta_0$, a new set of requests (with a distance of ϵ between them) are released in the segment S. The optimal completion time, $Z^* = 4$ since the optimal offline server can tour the cycle in the anti-clockwise direction and serve all the requests at or after their release times. The completion time of the online server $Z^{ol} \geq 2 + \delta_0 + 4 + 2 - \delta_0 = 8$. Therefore, the competitive ratio is $\frac{Z^{ol}}{Z^*} \geq 2$. ∎

3   Online Algorithms

In this section, we present the online algorithms for H-OLTSP-C and N-OLTSP-C, respectively.

3.1   Algorithm for N-OLTSP-C

First, we present some intuition behind the algorithm and its analysis. Here, the objective is to minimize the time at which the last request is served and the server is not required to return to the origin after serving all the requests. Therefore, at any time t, when the new requests arrive, we first find the smallest interval, along the circumference of the circle, that covers all the released but yet to be served requests. Then, the online server from its current position at time t goes to the nearest end point of the interval and serves all the requests. This is a very simple and natural strategy. We call this algorithm ERF (Extreme Request First).

In the proof, we analyse different cases that arise based on the interval length and the position of the online server at the time when the interval is calculated.

Algorithm 1. Extreme Request First (ERF)

The online server is located at the origin at time $t = 0$. At the moment when the first request(s) are released, find the smallest interval I along the circumference, that covers all the released requests. Serve I through the shortest route. New requests may arrive while the server is moving.

1. If the new request(s) can be served on the current route then we can safely "ignore" them, as they will be anyway served.

2. Otherwise, recalculate the shortest interval I which covers released but not yet served requests. Go to the extreme (end point) of I that is nearest to the server's current position and then serve all of I.

Theorem 1. *ERF is 2.5-competitive.*

Proof. Let t be the release time of last request and p_t^{ol} be the position of the online server at time t. Denote by I as the shortest interval consisting of all the unserved requests at time t. Suppose X_l and X_r denote the end points of I such that $I = [X_l, X_r]$. Denote by $d^s(a, b)$ and $d^l(a, b)$ as the lengths of minor arc and major arc with end points a and b, respectively. We proceed with the proof by considering the different cases that arise depending on the length of I and the current position of the online server, p_t^{ol}.

1. $|I| > \pi r$

   – $p_t^{ol} \in I$

     In this case, for the offline server we have $Z^* \geq t$ and $Z^* \geq |I|$. Assume WLOG that the nearest extreme from p_t^{ol} is X_r. The upper bound on Z^{ol} is given by the following

$$Z^{ol} \leq t + d^s(p_t^{ol}, X_r) + |I|$$
$$\leq t + \frac{|I|}{2} + |I| \leq 2.5Z^*$$

$- p_t^{ol} \notin I$

In this case, for the offline server we have $Z^* \geq t$, $Z^* \geq |I|$ and $Z^* \geq \pi r$. Assume WLOG that the nearest extreme from p_t^{ol} is X_r. The upper bound on Z^{ol} is given by the following

$$Z^{ol} \leq t + d^s(p_t^{ol}, X_r) + |I|$$
$$\leq t + \frac{(2\pi r - |I|)}{2} + |I|$$
$$= t + \pi r + \frac{|I|}{2}$$
$$\leq Z^* + Z^* + \frac{Z^*}{2} = 2.5Z^*$$

2. $|I| \leq \pi r$

$- p_t^{ol} \in I$

The offline server has to wait till the release time of the last request, so $Z^* \geq t$. Also, the offline server should cover requests in I. Therefore, $Z^* \geq |I|$. Assume, WLOG, that the extreme of I that is nearest to p_t^{ol} is X_r. The upper bound on Z^{ol} is given by the following

$$Z^{ol} \leq t + d^s(p_t^{ol}, X_r) + |I|$$
$$\leq t + \frac{|I|}{2} + |I| < 2.5Z^*$$

$- p_t^{ol} \notin I$

Let $X_l X_l^{'}$ and $X_r X_r^{'}$ be the diameters passing through X_l and X_r respectively. Given that $p_t^{ol} \notin I$, there are three regions where p_t^{ol} could be in. Define $R1$ as the region covering the minor arc with end points X_l and $X_r^{'}$, $R2$ as the region covering the minor arc with end points $X_r^{'}$ and $X_l^{'}$ and $R3$ as the region covering the minor arc with end points $X_l^{'}$ and X_r. Depending on the region where p_t^{ol} lies, we analyse the upper bound on Z^{ol}.

- $p_t^{ol} \in R2$

  * $R2$ does not contain the origin

  If online server has entered $R2$ from a different region, then claim that there must exist some request(s) in $R2$. We prove this by contradiction. Assume WLOG, that the online server entered $R2$ from $R1$. We know that at time t, there were no requests in $R3$. If there were no request(s) in $R2$ also, then the semi-circle with end points X_r and $X_r^{'}$ containing $R2$ and $R3$ had no requests. In such a situation, the online server, always seeking the shortest paths, would have never moved from $X_r^{'}$ along $R2$. This means, it should not have been located anywhere in this semi-circle at time t. This is a contradiction since the current position is in $R2$. Therefore, if p_t^{ol} is in $R2$, then there must exist some request(s) in $R2$.

So, in order to serve requests in I and request(s) in $R2$, the offline server would have to cover at least the length of one half circle i.e. $Z^* \geq \pi r$. Also, we have $Z^* \geq t$ and $Z^* \geq |I|$.

Let us assume that for the online server WLOG, that the nearest extreme from p_t^{ol} is X_r. Then, the upper bound on Z^{ol} is given by the following.

$$Z^{ol} \leq t + d^s(p_t^{ol}, X_r) + |I|$$

$$\leq t + \frac{(2\pi r - |I|)}{2} + |I| = t + \pi r + \frac{|I|}{2}$$

$$\leq Z^* + Z^* + \frac{Z^*}{2} = 2.5 Z^*$$

* $R2$ contains the origin

In this case, the online server is already in $R2$. In order to serve the interval I, the optimal offline server has to travel at least πr distance. So we have $Z^* \geq \pi r$, $Z^* \geq t$ and $Z^* \geq |I|$. For the online server, we have $Z^{ol} \leq t + d^s(p_t^{ol}, X_r) + |I|$. These are the same inequalities that we have in the previous sub-case. so we get the bound of 2.5-competitiveness.

- $p_t^{ol} \in R1$

  * $R1$ does not contain the origin

  If $R1$ does not contain the origin then there must be some request(s) in the semi-circle $\mathbb{C}$ with X_l and X_l' as the end-points i.e. there must be some request in either $R1$ or $R2$, since otherwise the online server is not on the shortest path. Let F be the farthest request from the origin in $\mathbb{C}$ that is ever presented.

  If F is in $R2$ then by the same arguments as discussed in the previous sub-case hold here as well and we get 2.5 competitiveness.

  If F is in $R1$, then F should be in the minor arc with end points p_t^{ol} and X_r'. Therefore, $Z^* \geq d^s(F, X_l) + |I|$. Upper bound on Z^{ol} is the following

$$Z^{ol} \leq t + d^s(p_t^{ol}, X_l) + |I|$$

$$\leq t + d^s(F, X_l) + |I|$$

$$\leq Z^* + Z^* = 2 Z^*$$

  * $R1$ contains the origin

  Here, $R1$ contains the origin and the online server is already present in $R1$ at time t (meaning the online server did not enter $R1$ from outside). Within $R1$, if p_t^{ol} is between the origin O and X_l then we have $Z^* \geq d^s(O, X_l) + I$ and $Z^* \geq t$. Clearly, we have $d^s(p_t^{ol}, X_l) \leq d^s(O, X_l)$. Upper bound on Z^{ol} is the following.

$$Z^{ol} \leq t + d^s(p_t^{ol}, X_l) + |I|$$
$$\leq t + d^s(O, X_l) + |I|$$
$$\leq Z^* + Z^* = 2Z^*$$

But if p_t^{ol} is in the part between the origin O and X_r' then there must exist some request(s) in that part (since otherwise there is no reason for the online server to move towards X_r', away from the interval I). Of all such request(s), let F be the position of the request that is farthest from the origin. For the optimal offline server, we have $Z^* \geq t$ and $Z^* \geq d^s(F, X_l) + |I|$. Also, $d^s(p_t^{ol}, X_l) \leq d^s(F, X_l)$. Upper bound on Z^{ol} is the following

$$Z^{ol} \leq t + d^s(p_t^{ol}, X_l) + |I|$$
$$\leq t + d^s(F, X_l) + |I|$$
$$\leq Z^* + Z^* = 2Z^*$$

- $p_t^{ol} \in R3$
  Proof for this case is analogous to the previous sub-case where p_t^{ol} is in $R1$. ■

3.2   Algorithm for H-OLTSP-C

Ausiello *et al.* [3] presented a 2-competitive online algorithm called PAH (Plan At Home) for H-OLTSP on any general metric space. PAH is the best possible online algorithm for H-OLTSP-C (since the lower bound is 2, see Lemma 4).

The intuition behind PAH is that since the online server returns to the origin after serving all the requests, it can ignore any new online requests that are "closer" to the origin. Such ignored requests can be served in a separate tour. A new online request is considered to be "closer" to the origin if at the time of it's release, it's distance from the origin (measured as shortest path) is less than the distance of the online server from the origin. For completeness, we present the algorithm here but omit the analysis (see [3] for the complete proof).

Algorithm 2. Plan At Home (PAH)

---

1. Whenever the server is at the origin, it starts to follow an optimal route that serves all the requests yet to be served and goes back to the origin.
2. If at time t a new request is presented at point x , then it takes one of two actions depending on its current position p :
   (a) If $d(x, O) > d(p, O)$, then the server goes back to the origin (following the shortest path from p) where it appears in the Case 1 situation.
   (b) If $d(x, O) \leq (p, O)$ then the server ignores it until it arrives at the origin, where again it reenters Case 1.

Theorem 4.2 Ausiello et al. [3]. PAH is 2-competitive.

4   Concluding Remarks

We have studied the homing and nomadic versions of online TSP for the special case in which the points/requests lie on a circle. We have presented lower bound proofs and an online algorithm for both homing and nomadic variants of this problem.

For N-OLTSP-C, there is work to be done in closing the gap between the lower bound of $\frac{28}{13}$ and the 2.5-competitive ratio of the zealous online algorithm that we have presented. It would also be interesting to study generalizations of TSP such as Dial-A-Ride-Problem in this setting.

Acknowledgements. We are thankful to Shyam K.B for being part of technical discussions that led to this work and for reviewing this paper.

References

1. Albers, S.: Online algorithms: a survey. Math. Program. **97**(1), 3–26 (2003)
2. Ascheuer, N., Krumke, S.O., Rambau, J.: Online dial-a-ride problems: minimizing the completion time. In: Reichel, H., Tison, S. (eds.) STACS 2000. LNCS, vol. 1770, pp. 639–650. Springer, Heidelberg (2000). https://doi.org/10.1007/3-540-46541-3_53
3. Ausiello, G., Feuerstein, E., Leonardi, S., Stougie, L., Talamo, M.: Algorithms for the on-line travelling salesman 1. Algorithmica **29**(4), 560–581 (2001)
4. Bjelde, A., et al.: Tight bounds for online tsp on the line. In: Proceedings of the Twenty-Eighth Annual ACM-SIAM Symposium on Discrete Algorithms, SODA 2017, Philadelphia, PA, USA, pp. 994–1005 (2017)
5. Blom, M., Krumke, S.O., de Paepe, W.E., Stougie, L.: The online tsp against fair adversaries. INFORMS J. Comput. **13**(2), 138–148 (2001)
6. Feuerstein, E., Stougie, L.: On-line single-server dial-a-ride problems. Theoret. Comput. Sci. **268**(1), 91–105 (2001). On-line Algorithms 1998
7. Jaillet, P., Wagner, M.R.: Online vehicle routing problems: a survey. In: Golden, B., Raghavan, S., Wasil, E. (eds.) The Vehicle Routing Problem: Latest Advances and New Challenges. Operations Research/Computer Science Interfaces, pp. 221–237. Springer, Boston (2008). https://doi.org/10.1007/978-0-387-77778-8_10
8. Lipmann, M.: On-line routing (2003)

Second-Order Linear-Time Computability with Applications to Computable Analysis

Akitoshi Kawamura[1], Florian Steinberg[2], and Holger Thies[1(✉)]

[1] Kyushu University, Fukuoka, Japan
info@holgerthies.com
[2] Inria Saclay, Palaiseau, France

Abstract. In this work we put forward a complexity class of type-two linear-time. For such a definition to be meaningful, a detailed protocol for the cost of interactions with functional inputs has to be fixed. This includes some design decisions the defined class is sensible to and we carefully discuss our choices and their implications. We further discuss some properties and examples of operators that are and are not computable in linear-time and nearly linear-time and some applications to computable analysis.

Keywords: Computational complexity · Linear time · Computable analysis

1 Introduction

Classical computability and complexity theory is concerned with programs whose inputs and outputs are finite strings over a finite alphabet. Mathematical structures are operated on by use of encodings. Since the set of finite strings is countable, the mathematical structures that can be treated have to be countable. However, many of the computational branches of mathematics like numerical analysis use objects of continuum cardinality in an essential way. The set of real numbers, e.g., is uncountable and thus it is impossible to uniquely identify each real number by a finite string. On the other hand, any real x can be encoded by a function that gives arbitrarily exact rational approximations to x. Since rational numbers can be encoded by finite strings, real numbers can be encoded by elements of Baire-space. To compute on real numbers one may thus fix a model of computation for operators on Baire space, i.e., do higher order computability theory.

Since it is essential that Baire-space is uncountable, restricting to computable elements and operating on indices is not an option. The accepted model of computation on the full Baire space is that of oracle Turing machines. Besides their name, oracle Turing machines provide a realistic model of computation where any computation is finitary and the access to functional inputs is provided via

© Springer Nature Switzerland AG 2019
T. V. Gopal and J. Watada (Eds.): TAMC 2019, LNCS 11436, pp. 337–358, 2019.
https://doi.org/10.1007/978-3-030-14812-6_21

calls for the function values. The field of research dealing with computation on the real numbers in this setting is known as **computable analysis** and already dates back to Turing [22]. The theory of computing real valued functions has further been developed by Grzegorczyk [7], Lacombe [15] and others, and a more general theory of computation on continuous structures via encodings called **representations** was developed by Weihrauch and Kreitz [23]. Computable analysis is an active field of research [19, 20, and many more] and the model of computation used in computable analysis has applications in many other areas such as for instance machine learning [3, 4].

While computability theory asks whether tasks are algorithmically solvable, it is clear that in practice not only correctness but also performance is important. Computational complexity theory is concerned with judging the quality of algorithms. The most common measure for the quality of an algorithm is the number of steps a Turing machine carrying out this algorithm takes. Of course, this number depends on the input and is usually bounded by a function in the size of the input. The computationally feasible functions are identified with those functions whose runtime can be bounded by a polynomial. Note that this makes essential use of the classical setting, where all inputs are finite strings and no functional inputs are present.

For computations at type level two, that is when some inputs may reach over Baire space, the class of basic feasible functionals is widely accepted as the natural class of feasible operations [9]. This class was originally introduced by Mehlhorn [16] by means of a bounded recursion scheme, and accepted to capture feasibility after Kapron and Cook [10] gave an characterization in terms of oracle machines whose runtimes are bounded polynomially in the sizes of the inputs. The characterization lead to many applications in computable analysis [11, 14, 17, 21] as well as in other fields [4–6].

In practice, a proof that an algorithm runs in polynomial time does not necessarily imply that it performs well, but should be understood as evidence that it does not perform unreasonably bad for large inputs. Traditionally, subpolynomial complexity classes are used for more refined quality judgements. Examples of such classes use logarithmic bounds on the memory, restrict the degree of a polynomial runtime, or use different classes of sub-polynomial runtimes such as quasi-linear ones which are linear up to a logarithmic factor.

Unfortunately, this part of complexity theory does not translate to a higher order setting very well: In Kapron and Cook's characterization, the size of a functional input is a function on the natural numbers. Thus, the runtimes, and also the class of polynomial runtimes, consists of higher-order functions for which the structure theory is considerably more complicated. For instance, different notions of the 'degree' of second-order polynomials have been suggested by different authors [13, 24] and none of them seem fully satisfactory. As a result, the existent work on sub-polynomial classes in a higher order setting is mostly concerned with restrictions of a different nature. Somewhat surprisingly, since restrictions to memory consumption are known to be problematic with respect to relativization [2], space-based complexity classes such as L and NC have recently been extended to a type-two setting by Kawamura and Ota [12].

The present paper is concerned with linear-time computability. In classical complexity theory, **linear-time** or **real-time** computation is a two sided sword. On one hand, proofs of linear time computability are considered highly desirable in applications and there is a well-developed theory for the model of multi-tape Turing machines [18]. On the other hand, linear time computability is subtle due to its lack of robustness under minor changes in the computational model [8], and the fact that the best known universal Turing machines introduce a logarithmic overhead in execution times.

The reason we still decided to look into linear-time computability in a higher-order setting in more detail is that a candidate class of linear running times is readily available by omitting the multiplication from the generation scheme of second-order polynomials. The linear second-order polynomials introduced in this way share many of the closure properties making second-order polynomials useful. It should be noted that this approach is somewhat orthogonal to notions of "hyper-linearity" that can be obtained from restricting the different notions of degree of the second-order polynomials [1, 24]. This is because the notions of degree are mostly concerned with restricting the number of iterations of the functional argument. Our approach does not restrict this quantity as this necessarily leads to a loss of desirable closure properties.

To give a meaningful definition of linear-time computable operators on Baire-space as those computable by an oracle machine whose runtime is bounded by a linear second-order polynomial, a protocol for oracle interactions and some other details about the machine model have to be fixed. Thus, Sect. 2 recalls some definitions in some detail and discusses some of the choices we are faced with and some of their implications. Its last part discusses the classical proof that polynomial-time computable operations on Baire-space are closed under composition.

Section 3 introduces linear second-order polynomials and type-two linear-time computability and subsequently investigates closure under composition: We prove that the linear-time computable operators take linear-time computable inputs to linear-time computable outputs but that a composition of linear-time computable operators need not be linear time computable. We also investigate the closure of the linear-time computable operators under composition and call operators from this class nearly linear-time computable. These operators still have the important property that they preserve linear-time computability.

We give an explicit example of linear-time computable operators such that there does not exist any machine that computes their composition in linear time. This counter-example works mostly independently of the details of the oracle interaction protocols. We believe that the failure of closure under composition is due to the complexity of the operation of composing oracle machines and not an artifact of our definitions. It should be noted that in many use-cases one is not interested in composing general oracle machines but only those relevant for a fixed application and that often optimized composition schemes can be specified. It should also be noted that the failure of closure under composition is foreshadowed by a formula for bounding the runtime of the composition derived in Sect. 2: It includes a product that, while clearly being a coarse overestimation, is impossible to remove completely.

Finally in Sect. 4 we apply our theory to problems from computable analysis. One important property we prove is a very strong equivalence of the signed digit and the Cauchy representation: The notion of linear time computable real numbers does not change when switching between these representations. We also prove that the standard representation of the continuous functions on the unit interval as introduced by Kawamura and Cook allows for a linear-time evaluation algorithm. We establish that a function has a linear-time computable name in this representation if and only if it has a realizer from our class of linear-time computable operators. Furthermore, we prove that the linear-time computable functions on the reals are closed under composition, i.e., that there is an efficient way to compose those linear-time computable operators that appear as realizers of real-valued functions.

2 Machines

Let $\Sigma := \{0, 1\}$ be the binary alphabet, Σ^* the set of finite binary strings and denote the elements of the later by $\mathbf{a}, \mathbf{b}, \ldots$. Let $\mathcal{B} := \Sigma^* \to \Sigma^*$ denote the space of all functions from finite binary strings to finite binary strings and denote its elements by $\varphi, \psi, \ldots$. In this paper we choose the basic devices that carry out computations to be Turing machines with an arbitrary but fixed finite number of memory tapes. Let K be a finite list of instructions, i.e., a program for such a machine. For a finite string $\mathbf{a}$ written to the input tape of the machine, one may run the machine on the set of instructions K. The machine has a designated return state to indicate that the computation has terminated. If after finite time the machine enters the return state, we interpret the collections of digits on the output tape (up to the first empty cell) to be the return value and denote it by $K(\mathbf{a})$. K computes a partial function $f_K \colon \subseteq \Sigma^* \to \Sigma^*$, where $\mathrm{dom}(f_K)$ are the elements such that $K(\mathbf{a})$ is defined and $f_K(\mathbf{a}) = K(\mathbf{a})$ in this case. Note that, due to being partial, this function need not be an element of Baire space.

Interaction with a functional input can be modeled by oracle Turing machines. In the literature, such machines are most commonly given access to a subset of the finite binary strings, i.e., they may write a finite string to a tape and call an external procedure to test a condition on this string. We consider a slightly different model and allow the return value not only to be a Boolean but a finite binary string again: An oracle machine is a Turing machine that has an additional instruction called the oracle query instruction and two designated working tapes for oracle input and output. For a given string function φ and string $\mathbf{a}$ an oracle machine with a set of instructions M behaves just like a Turing machine with one exception. Whenever entering the oracle state, the state of the machine is altered as follows: Let $\mathbf{b}$ be the content of the oracle input tape up to the first empty cell. Change the oracle output tape to contain $\varphi(\mathbf{b})$ followed by empty cells. Erase the oracle input tape and set the heads on the oracle input and output tapes to the first position. If the oracle machine on functional input φ and string input $\mathbf{a}$ eventually enters the return state, let $M^\varphi(\mathbf{a})$ denote the content of the output tape up to the first empty cell. In this case, we denote the set of oracle queries asked during the computation by $Q_M(\varphi, \mathbf{a})$.

As we are mostly interested in applications to computable analysis, we adapt the interpretation of what an oracle Turing machine computes with a specific instruction set to the conventions common in computable analysis. That is, we understand each M to correspond to an operation that takes inputs from Baire space and returns elements of Baire space. Fix some function $\varphi \in \mathcal{B}$. Note that the function $\mathbf{a} \mapsto M^\varphi(\mathbf{a})$ need not be total. In the case where it is total, denote the corresponding element of Baire space by M^φ. The partial operator $F_M : \subseteq \mathcal{B} \to \mathcal{B}$ corresponding to M is defined as follows: The set $\mathrm{dom}(F_M)$ consists of all φ such that M^φ is defined and the value is given by $F_M(\varphi) := M^\varphi$. We consider an operator $F : \subseteq \mathcal{B} \to \mathcal{B}$ to be computable if there exists an M such that F_M extends F, i.e., such that $F_M|_{\mathrm{dom}(F)} = F$. In accordance with this we say that M computes F if F_M extends F.

2.1   Runtimes and Input Sizes

For a set of instructions K for a regular Turing machine and an input string $\mathbf{a}$ such that $K(\mathbf{a})$ is defined, the runtime $\mathrm{time}_K(\mathbf{a}) \in \mathbb{N}$ is defined to be the number of applications of the transition function needed until termination. If some higher-level programming language is used to specify the transition function, one may instead count the number of commands executed, provided that the commands are chosen from a finite list of possible commands each of which takes finite time to carry out. The runtime is specific to the exact model of computation used. For more stability under switching between models of computation we use $\mathcal{O}$-notation: For $t, t' : A \to \mathbb{N}$ by $t \in \mathcal{O}(t')$ we mean that there exists a $C \in \mathbb{N}$ such that $t(a) \leq Ct'(a) + C$ for all $a \in A$. We leave the type of the input open for reasons that become apparent shortly, by abuse of notation we sometimes use the $\mathcal{O}$-notation with the arguments filled in.

For oracle machines we use very similar conventions: We denote by $\mathrm{time}_M(\varphi, \mathbf{a})$ the number of instructions executed in the run of M on oracle φ and input $\mathbf{a}$. In particular, executing the oracle instruction is considered to take one time step. As we require the oracle tape to be emptied on execution of an oracle query, and only one bit may be written per time step it follows that

$$\sum_{\mathbf{b} \in Q_M(\varphi, \mathbf{a})} |\mathbf{b}| \in \mathcal{O}(\mathrm{time}_M(\varphi, \mathbf{a})). \tag{Q}$$

(Where φ and $\mathbf{a}$ are the inputs the bigo-notation reaches over.) We use this inequality several times throughout the paper. While the dependency on the details of the time-counting conventions of non-oracle instructions of oracle machines is taken care of by the $\mathcal{O}$-notation, the protocol of interaction with the functional input is essential for its validity. For instance, if the query tape was not erased, a machine could ask a linear number of queries of linear size with coinciding initial segments in linear time. We believe that for any reasonable convention this inequality should hold. Some additional discussion as to why we believe this can be found in Sect. 3.2.

For a set of instructions K for a regular Turing machine, the time function has type $\text{time}_K \colon \Sigma^* \to \mathbb{N}$. In complexity theory one is usually interested in bounding the time function by means of the size of the input, where the size of a binary string $\mathbf{a}$ is considered to be the number $|\mathbf{a}| \in \mathbb{N}$ of digits of the string. That is, one looks for slowly growing functions $t \colon \mathbb{N} \to \mathbb{N}$ such that for all valid inputs $\mathbf{a}$ it holds that $\text{time}_K(\mathbf{a}) \le t(|\mathbf{a}|)$. The most common notion of being slowly growing is to consider polynomials and the corresponding set of functions are called the polynomial-time computable functions.

It might seem like we are diverging from the most common definitions by allowing the time constraint to only hold on a specified set of 'valid inputs'. However, under some weak additional assumptions one can easily modify an instruction set to fulfill the time-constraint globally. Namely the time bounds considered should be time constructible and one should only be interested in time consumption up to multiplicative and additive constants. Thus, for regular Turing machines our notions are essentially equivalent to the standard notions.

For a set M of instructions for an oracle machine, the time function has type $\text{time}_M \colon \mathcal{B} \times \Sigma^* \to \mathbb{N}$. To be able to bound the time depending on the size of inputs, it is necessary to say what the size of an element of Baire-space is. The usual choice for this size function is the worst-case rate of increase in size from input to output: For any function $\varphi \in \mathcal{B}$ let $|\varphi| \colon \mathbb{N} \to \mathbb{N}$ denote the function

$$|\varphi|(n) := \max\{|\varphi(\mathbf{a})| \colon |\mathbf{a}| \le n\}.$$

For instance, if K is a set of instructions for a Turing machine that computes φ, i.e., $f_M = \varphi$, then for any input $\mathbf{a}$ writing the output requires time and thus $|\varphi(\mathbf{a})| \in \mathcal{O}(\text{time}_K(\mathbf{a}))$. We assume that writing each bit of the output takes one time step. Thus the following lemma holds.

Lemma 1 (size and time). *Let $\varphi \in \mathcal{B}$ be computable in time $t \colon \mathbb{N} \to \mathbb{N}$. Then*

$$\forall n, |\varphi|(n) \le t(n).$$

A function $T \colon \mathbb{N}^{\mathbb{N}} \times \mathbb{N} \to \mathbb{N}$ is called a time-bound for M if for all valid inputs φ and $\mathbf{a}$ it holds that $\text{time}_M(\varphi, \mathbf{a}) \le T(|\varphi|, |\mathbf{a}|)$.

One of the main sources of additional difficulties over the situation for regular Turing machines is the infeasibility of the size function: We have to deal with the fact that the size is not polynomial-time computable.

2.2   Second-Order Polynomials

The class of second order polynomials is the smallest class of functions of type $\mathbb{N}^{\mathbb{N}} \times \mathbb{N} \to \mathbb{N}$ that contains the constant zero and one function, the function $(l, n) \mapsto n$ and is closed under point wise addition and multiplication as well as the operation $P \mapsto P^+$, where $P^+(l, n) := l(P(l, n))$. An example of a second-order polynomial is $(l, n) \mapsto l(n^2 + 4) \cdot l(l(n)^2 + n) + n^4$. We denote the function $n \mapsto P(l, n)$ by $P(l, \cdot)$.

A machine is said to run in polynomial time on a set $A \subseteq \mathcal{B}$ of valid inputs if there exists a second-order polynomial P such that $\text{time}_M(\varphi, \mathbf{a}) \leq P(|\varphi|, |\mathbf{a}|)$ for all $\varphi \in A$ and $\mathbf{a} \in \Sigma^*$. An operator $F : \subseteq \mathcal{B} \to \mathcal{B}$ is computable in polynomial time if there is a machine that computes F and runs in polynomial time on $\text{dom}(F)$. Restricted to the total operators, this class of polynomial-time computable operations is well investigated, has been characterized in several different ways and is established as the correct one for classifying feasibility in a type-two setting.

We list some well known properties of second-order polynomials:

Lemma 2. *The following properties hold for second-order polynomials.*

1. *If P is a second-order polynomial and p is a polynomial, then $P(p, \cdot)$ is a polynomial.*
2. *If P is a second-order polynomial, $l, k \colon \mathbb{N} \to \mathbb{N}$ are non-decreasing functions and k is point wise bigger than l, then the same is true for the functions $P(l, \cdot)$ and $P(k, \cdot)$.*

Note that the natural inputs to second-order polynomials are sizes of elements of Baire-space and that exactly the non-decreasing functions turn up as such. As we are mostly interested in second-order polynomials as runtimes, the important class of functions are those that are point wise majorized by second-order polynomials.

There are two canonical ways to compose second-order polynomials: Given P and Q one may hand the value of the later polynomial as input to the first polynomial, i.e., set $(P \star Q)(l, n) := P(l, Q(l, n))$ or use the function $Q(l, \cdot)$ as input for P, i.e., $(P \circ Q)(l, n) := P(Q(l, \cdot), n)$.

Both of these operations result in second-order polynomials again. For example, for $P(l, n) = Q(l, n) = l(n) + n$ the compositions are given by $P \circ Q(l, n) = Q(l, n) + n = l(n) + 2n$ and $P \star Q(l, n) = P(l, l(n) + n) = l(l(n) + n) + n$. In particular, $\star$ may introduce iterations of the functional argument.

2.3   Composition of Machines

Given operators $F, G \colon \subseteq \mathcal{B} \to \mathcal{B}$ let $F \circ G$ denote their composition, i.e., the operator whose domain $\text{dom}(F \circ G)$ is the set of φ from $\text{dom}(G)$ such that $G(\varphi) \in \text{dom}(F)$ and whose value on such a φ is given by $(F \circ G)(\varphi) := F(G(\varphi))$. One of the first questions one may ask about the class of polynomial-time computable operators is whether this class is closed under composition. That is, whether $F \circ G$ is a polynomial-time computable operator whenever F and G are polynomial-time computable operators. Indeed it is possible to answer this question positively in a fully uniform sense: We can specify a way to combine two sets of instructions N and M fully uniformly to a new set of instructions $_N M$

such that the corresponding operator F_{NM} extends the composition $F_M \circ F_N$ of the operators corresponding to N and M respectively.

Intuitively it is fairly obvious what the program $_NM$ should look like. This program runs on a machine whose number of tapes is the sum of the numbers of tapes needed by N and M separately and replaces each oracle query instruction that appears in N by inlining M. Of course references to memory tapes have to be corrected and some management tasks have to be done: Following the inlining of a copy of M for N to correctly continue its run, it is necessary to erase the memory tape that it writes to instead of the oracle query tape and return its head to the beginning of the tape. The cost for doing so is bounded by a constant times the length of the oracle query that was asked plus the time executing M took. Furthermore, some cleanup of the work tapes the machine M uses is needed to make the next execution of M run smoothly. The cost for cleaning up is again bounded by a constant times the time the execution of M took. In total we end up with an estimation for the runtime of $_NM$ as follows:

$$\text{time}_{NM}(\varphi, \mathbf{a}) \in \mathcal{O}\Big(\text{time}_N(M^\varphi, \mathbf{a}) + \sum_{\mathbf{b} \in Q_N(M^\varphi, \mathbf{a})} (\text{time}_M(\varphi, \mathbf{b}) + |\mathbf{b}|) \Big).$$

Here, the constant hidden in the $\mathcal{O}$-notation may depend on M, i.e. M is not considered an input but fixed. Using the inequality from equation (Q), the formula can be further simplified to

$$\text{time}_{NM}(\varphi, \mathbf{a}) \in \mathcal{O}\Big(\text{time}_N(M^\varphi, \mathbf{a}) + \sum_{\mathbf{b} \in Q_N(M^\varphi, \mathbf{a})} \text{time}_M(\varphi, \mathbf{b}) \Big). \tag{T}$$

Let P_M and P_N be second-order polynomials that bound the runtimes of M and N on sets $A_M, A_N \subseteq \mathcal{B}$ respectively. Note that the existence of the time bound guarantees that M^φ is defined whenever $\varphi \in A_M$. For ease of writing we say that φ is valid if $\varphi \in A_M$ and $M^\varphi \in A_N$. To obtain a polynomial runtime bound for $_NM$ from the above inequality first note that for any $\varphi \in A_M$

$$|M^\varphi|(n) = \max\{|M^\varphi(\mathbf{a})| : |\mathbf{a}| \leq n\} \leq \max\{\text{time}_M(\varphi, \mathbf{a}) : |\mathbf{a}| \leq n\} \leq P_M(|\varphi|, n),$$

where we used the monotonicity of second-order polynomials from Lemma 2 to remove the maximum. The above leads to an estimate for the first summand on the right hand side of the inequality from equation (T): For any valid φ

$$\text{time}_N(M^\varphi, \mathbf{a}) \leq P_N(|M^\varphi|, |\mathbf{a}|) \leq P_N(P_M(|\varphi|, \cdot), |\mathbf{a}|).$$

The second inequality again uses the monotonicity property to substitute the estimate for $|M^\varphi|$ in the argument. Note that the outcome can be expressed as $P_N \circ P_M(|\varphi|, |\mathbf{a}|)$ by using one of the compositions for second-order polynomials.

To estimate the sum in the second summand on the right of the inequality in equation (T), two additional arguments about oracle queries are necessary. Firstly note that writing each oracle query consumes time and thus, whenever φ is valid, each $\mathbf{b} \in Q_N(M^\varphi, \mathbf{a})$ is in size bounded by the time taken by N. I.e. $|\mathbf{b}| \leq P_N \circ P_M(|\varphi|, |\mathbf{a}|)$. For valid φ and $\mathbf{b} \in Q_N(M^\varphi, \mathbf{a})$ it follows that

$$\text{time}_M(\varphi, \mathbf{b}) \leq P_M(|\varphi|, |\mathbf{b}|) \leq P_M(|\varphi|, (P_N \circ P_M)(|\varphi|, |\mathbf{a}|)),$$

where the outcome $P_M \star (P_N \circ P_M)(|\varphi|, |\mathbf{a}|)$ involves the other composition for second-order polynomials from the last section. Secondly, note that each execution of the oracle instruction takes time and thus the number $\#Q_N(M^\varphi, \mathbf{a})$ of oracle queries done by N can be bounded whenever φ is valid:

$$\#Q_N(M^\varphi, \mathbf{a}) \leq \mathrm{time}_N(M^\varphi, \mathbf{a}) \leq (P_N \circ P_M)(|\varphi|, |\mathbf{a}|).$$

The uniform bound on each summand together with a bound on their number proves that

$$\mathrm{time}_{N}M \in \mathcal{O}((P_N \circ P_M) \cdot (1 + (P_M \star (P_N \circ P_M)))), \tag{P}$$

where the domain of the time-function reaches over all valid φ, i.e., such that $\varphi \in A_M$ and $M^\varphi \in A_N$. This bound is a polynomial by the closure of second-order polynomials under the two kinds of composition, addition and multiplication. To complete the argument for polynomial-time computability: Let F and G be polynomial time computable operators. Then there exist M and N that compute F and G and run in time polynomial on $\mathrm{dom}(F)$ and $\mathrm{dom}(G)$. Note that $F_M \circ F_N$ extends $F \circ G$ and thus $_N M$ computes $F \circ G$. Furthermore the above specifies a polynomial runtime bound of $_N M$ on the set of $\varphi \in \mathrm{dom}(G)$ such that $M^\varphi = G(\varphi) \in \mathrm{dom}(F)$, i.e., on the natural domain of the composition.

As a final remark of this section note that the formula for the runtime of the composition of two machines includes the two compositions of second-order polynomials and additionally a product that originates from the coarse overestimation: The assumed worst case would be that a machine queries in each of its steps a string as long as the total number of steps it takes and will clearly not be assumed in realistic situations.

3   A Notion of Type-Two Linear-Time

Linear-time computability in a type-two setting is in particular appealing as the class of runtime bounds can be chosen as a subset of the polynomials that is closed under the two kinds of composition. The class of second-order polynomials allows for a straight forward modification in its production rules to exclude any non-linear operations we use to obtain a candidate for the linear higher-order runtimes:

Definition 1. *Let the class of linear second-order polynomials be the smallest class of functions of type $\mathbb{N}^\mathbb{N} \times \mathbb{N} \to \mathbb{N}$ that contains the constant zero and one function, the function $(l, n) \mapsto n$ and is closed under point wise addition as well as the operation $P \mapsto P^+$, where $P^+(k, n) := l(P(l, n))$.*

By definition, the linear second-order polynomials form a subset of the second-order polynomials. Many of the properties of second-order polynomials have straight forward counterparts with identical proofs for the linear second-order polynomials:

Lemma 3 (linearity). *The following properties hold for linear second-order polynomials.*

1. *If P is a linear second-order polynomial and p is a linear function, then $P(p, \cdot)$ is a linear function.*
2. *Let the runtime of M be bounded by a linear second-order polynomial P on a set A. Then for each $\varphi \in A$ of linear size, M^φ is of linear size.*
3. *Let P and Q be linear second-order polynomials, then so are the functions $P \circ Q$ and $P \star Q$ defined by $(P \circ Q)(l, n) := P(Q(l, \cdot), n)$ and $(P \star Q)(l, n) := P(l, Q(l, n))$.*

The following lemma is useful for proving that certain functions cannot be majorized by a linear second-order polynomials:

Lemma 4 (constant values). *For each linear second-order polynomial P there exist constants K_1, K_2, K_3 such that for all $c, n \in \mathbb{N}$*

$$P(m \mapsto c, n) = K_1 \cdot c + K_2 \cdot n + K_3.$$

Proof. The constants are given by $K_3 = P(m \mapsto 0, 0)$, $K_2 = P(m \mapsto 0, 1) - K_3$ and $K_1 := P(m \mapsto 1, 0) - K_3$, as can be verified by an easy induction.

Lemma 5 (counter-examples). *Let $f \colon \mathbb{N} \to \mathbb{N}$ be an unbounded function and let $g \colon \mathbb{N} \to \mathbb{N}$ be arbitrary. The function $T(l, n) := f(n) \cdot l(g(n))$ can not be majorized by a linear second-order polynomial, not even on the constant functions.*

Proof. Let P be a linear second-order polynomial and let K_1, K_2, K_3 be the constants from the previous lemma. Since f is unbounded we may pick a large enough n and a constant c such that $K_1 \cdot c + K_2 \cdot n + K_3 < f(n) \cdot c$.

Instances of the previous lemma are that the functions $(l, n) \mapsto n \cdot l(0)$ and $(l, n) \mapsto \lceil \frac{n}{\log n} \rceil l(\log n)$ can not be majorized by linear second-order polynomials.

3.1   Counter Examples and Composition

First let us formally state our definition of a linear-time computable operator: We say that an oracle machine M runs in linear time, if there exists a linear second-order polynomial bounding its runtime.

Definition 2. *We call an operator $F \colon \subseteq B \to B$ **linear-time computable** if there exists a machine that computes it and runs in linear time.*

Examples for linear-time computable operators are the identity on Baire-space, that can be computed in time $\mathcal{O}(n + l(n))$ and the operator $F(\varphi) := \varphi \circ \varphi$, which can be computed in time $\mathcal{O}(n + l(n) + l(l(n)))$. As an example of an important operator that is not linear-time computable, consider the operator $F \colon B \to B$ defined by $F(\varphi)(\mathbf{a}) := 1^{\max\{|\varphi(\mathbf{b})| \colon \mathbf{b} \subseteq \mathbf{a}\}}$, where $\mathbf{b} \subseteq \mathbf{a}$ means that $\mathbf{b}$ is an initial segment of $\mathbf{a}$. We call this operator the *maximization operator*. The straight forward algorithm runs in polynomial time. To see that F is not linear-time computable, let φ_ϵ be the constant function always returning the empty string and consider the run of M on φ_ϵ. Since M computes F it must also hold that $M^{\varphi_\epsilon}(\mathbf{a}) = \epsilon$. Argue that the query set $Q_M(\varphi_\epsilon, \mathbf{a})$ contains all initial segments of $\mathbf{a}$: If $\mathbf{b}$ was an initial segment that is not queried we could modify φ_ϵ in $\mathbf{b}$ without the machine noticing. That is, we could replace φ_ϵ by the oracle φ' that is identical to φ_ϵ except that $\varphi'(\mathbf{b}) = 1$. Since $\mathbf{b}$ is not queried, the run of M on oracle φ' would be identical and we arrive at a contradiction. Note that the sum of the sizes of the initial segments is quadratic in the size of $\mathbf{a}$ and that our conventions for oracle access force each bit of an oracle query to consume a time step leading to a quadratic lower bound on the runtime. It is possible to prove in a rather straight forward way that linear time computable operators preserve linear time computability.

Proposition 1. *Let $\varphi \in \mathcal{B}$ be linear-time computable. If M runs in linear time then M^φ is linear-time computable.*

Note that attempts to generalize this statement to the general case of composition of two oracle machines fail. Indeed, it is fairly easy to see that no such proof can be obtained when the composition procedure we specified is used. One may for instance use a machine that asks the input as query, and returns two copies of the answer written after one another as top machine and let the lower procedure pose the length of its string input many queries of the empty word. The composed machine evaluates each of the oracle queries separately and this leads to a time investment of about $|\mathbf{a}| \cdot |\varphi|(0)$ steps, which is not bounded by any linear second-order polynomial by Lemma 5.

One is tempted to say that this behaviour is due to a lack of care when implementing the composition and that the behaviour described above can easily be circumvented. However, the next example shows that in general there is no strategy to solve the problem: The product that appeared in the estimate for the runtime of the composition of two operators from equation (P) can indeed not fully be removed.

Example 1. For a string $\mathbf{a}$ let $\mathbf{a}_\vee$ be the bit-wise or over all the digits, i.e. $\mathbf{a}_\vee = 0$ if all digits of $\mathbf{a}$ are zero and $\mathbf{a}_\vee = 1$ otherwise. Let $\mathbf{b}_i$ denote the smallest binary encoding of the number i. The operator defined by

$$F(\varphi)(\mathbf{a}) := \varphi(\mathbf{b}_{|\mathbf{a}| - \lceil \frac{|\mathbf{a}|}{\lceil \log(|\mathbf{a}|) \rceil} \rceil})_\vee \cdots \varphi(\mathbf{b}_{|\mathbf{a}|})_\vee$$

is not computable in linear time but can be written as a composition of two linear-time computable operators.

Proof. First argue, that the operator can be written as a composition of two linear-time computable operators. For this let $\mathbf{a}_0$ denote the leftmost digit of the string $\mathbf{a}$. Let G and G' be the operators defined by

$$G(\varphi)(\mathbf{b}) := \varphi(\mathbf{b})_\vee \quad \text{and} \quad G'(\psi)(\mathbf{a}) := \varphi(\mathbf{b}_{|\mathbf{a}| - \lceil \frac{|\mathbf{a}|}{\lceil \log(|\mathbf{a}|) \rceil} \rceil})0 \cdots \varphi(\mathbf{b}_{|\mathbf{a}|})0.$$

It is easy to verify that $G' \circ G = F$ and that the operator G is computable in time $\mathcal{O}(n + l(n))$. To argue that also G' is computable in time bounded by a linear second-order polynomial, note that $\lceil \log(|\mathbf{a}|) \rceil$ can be computed in linear time by converting $|\mathbf{a}|$ to binary and that the standard algorithm for integer division can be used to compute $\lceil |\mathbf{a}| / \lceil \log(|\mathbf{a}|) \rceil \rceil$ in time sub quadratic in the size of $|\mathbf{a}|$, i.e. logarithmic in the size of $\mathbf{a}$.

To see that the operator itself can not be computed in linear time let M be a machine to compute F. By a similar argument as used to show that the maximization operator is not linear-time computable, we can show that for any φ_0 that only returns strings of 0s, the set $Q_M(\varphi_0, \mathbf{a})$ includes all of the strings $\mathbf{b}_i$ for $i \in \{|\mathbf{a}| - \lceil |\mathbf{a}| / \lceil \log(|\mathbf{a}|) \rceil \rceil, \ldots, |\mathbf{a}|\} =: I_{|\mathbf{a}|}$. As the machine has to read each digit of each answer at least once (otherwise we could change that very digit) we get the lower bound $\sum_{i \in I_{|\mathbf{a}|}} |\varphi_0(\mathbf{b}_i)| \leq \mathrm{time}_M(\varphi, \mathbf{a})$.

By restricting the possible φ_0 to those functions that satisfy $|\varphi_0(\mathbf{a})| \leq |\varphi_0(\mathbf{b})|$ whenever $|\mathbf{a}| \leq |\mathbf{b}|$ and noting that for $|\mathbf{a}| \geq 2$ at least half of the queries have size at least $\lfloor \log(|\mathbf{a}|) \rfloor - 1$, obtain $\lceil |\mathbf{a}| / 2\lceil \log(|\mathbf{a}|) \rceil \rceil \cdot |\varphi_0|(\lfloor \log(|\mathbf{a}|) \rfloor - 1) \leq \mathrm{time}_M(\varphi, \mathbf{a})$.

Finally note, that any non-decreasing function can be obtained as the length of a length-increasing function only returning zeros. Thus, using the counter-example formula from Lemma 5 the function on the left hand side can not be majorized by a linear second-order polynomial already on constant inputs, let alone on non-decreasing inputs. It follows, that the time consumption of M can not be bounded by a linear second-order polynomial.

The runtimes of the component operators are both $\mathcal{O}(n + l(n))$, in particular they do not contain iterations of the length function. The lower bound on the runtime of their composition involves a product of something that is almost linear in the input size and the length function evaluated in a non-constant term. The argument that the operator G' should not be linear-time computable because the time for writing the oracle return value should be accounted for in the runtime can easily be countered by making the operator G' abort the computation once it encounters the first oracle answer that is longer than one digit.

The example can be modified to show that the notion of linear-time computability of operators is not stable under seemingly minor changes in the model: Replacing the or operation with just returning the last bit allows for the same argumentation as the head on the answer tape is always reset to the first position but the same operator will be computable efficiently if the head is always put in the last position of the string instead. A candidate for a more stable notion is the closure under composition of linear-time computable operators.

Definition 3. *We call an operator **nearly linear-time computable** if it can be written as a composition of linear-time computable operators.*

It is reasonable to ask how much taking the closure under composition increases the size of the class. The following lemma immediately implies that the class is still strictly contained in the class of polynomial-time computable operators.

Lemma 6. *Any nearly linear-time computable operator maps linear-time computable functions to linear-time computable functions.*

Proof. The proof is an easy induction, where the induction step is taken care of by linear-time computability being preserved pointwise by Lemma 3.

In the same way the preservation of linear sizes from Lemma 3 carries over to all nearly linear-time computable operators.

Lemma 7. *Any nearly linear-time computable operator maps functions of linear size to functions of linear size.*

3.2 Quantitative Continuity

It is well known, that computability and complexity considerations about operators on Baire space are tightly connected to quantitative notions of continuity. This chapter makes this connection more explicit for the notion of nearly linear-time computability introduced in the last section. As an application we prove that the maximization operator is indeed also not nearly linear-time computable.

An operator $F \colon \subseteq \mathcal{B} \to \mathcal{B}$ is called continuous if for all $\varphi \in \mathrm{dom}(F)$ and all $\mathbf{a} \in \Sigma^*$ there exists a finite list $L(\varphi, \mathbf{a})$ of strings such that $F(\varphi)(\mathbf{a})$ and $F(\psi)(\mathbf{a})$ coincide whenever ψ coincides with φ on $L(\varphi, \mathbf{a})$, i.e., whenever $\mathbf{b} \in L(\varphi, \mathbf{a})$ implies that $\psi(\mathbf{b}) = \varphi(\mathbf{b})$. We call such a list $L(\varphi, \mathbf{a})$ a certificate for φ and $\mathbf{a}$ and a function L that returns certificates for any $\varphi \in \mathrm{dom}(F)$ and $\mathbf{a} \in \Sigma^*$ a modulus of continuity. Any computable operator is continuous: The function Q_M that assigns to φ and $\mathbf{a}$ the finite list of oracle queries that are posed when M is run with oracle φ and input $\mathbf{a}$ is a modulus of continuity. We have implicitly used this fact a couple of times in proofs. Indeed, one may make the case that any computable operator has a computable modulus of continuity, as one may follow the run of the computation to obtain the list of queries that were done.

For a list L of finite binary strings, let $|L|$ be defined as the sum of the sizes of its elements. One may ponder the dependency of the size of certificates on the sizes of the inputs of an operator. Note that the requirement for a reasonable model of oracle interaction that we imposed in equation (Q) can directly be interpreted as a statement about the sizes of certificates. It requires the model to be chosen such that the size of certificates of an operator can be bounded from the runtime of a machine computing that operator.

Recall that the maximization operator defined in Sect. 3.1 is not computable in linear time. The core of this argument was a quadratic lower bound on sizes of queries a machine computing the operator has to do. Indeed, the main part of the argument is involved with proving the query set of a machine computing the operator to be a modulus of continuity and then using that whenever L is a modulus of continuity of F, then

$$\frac{|\mathbf{a}|(|\mathbf{a}| - 1)}{2} \leq |L(\varphi, \mathbf{a})|. \tag{L}$$

It is not a priori clear how to obtain bounds on the modulus of continuity of a nearly linear-time computable operator. This is because the nearly linear-time computable operators are the closure under composition of the linear-time computable operators which contains operators that are not linear-time computable. However, an argument very similar to the point-wise preservation of nearly linear-time computability can be used to prove that a linear bound can be obtained for all possible inputs of size bounded by a fixed linear function. To see this, we first prove a technical lemma about compositions of machines. Recall that an oracle φ is called valid if in a computation of a composition of machines on function input φ all intermediate results are contained in the sets in which the runtime bounds are valid.

Lemma 8. *For $r \in \mathbb{N}$ let M be the composition of machines $M_1, \ldots, M_r$ that each run in time bounded by a linear second-order polynomial on sets $A_1, \ldots, A_r$. For any linear function p there exists a linear function q_p such that for valid φ*

$$|\varphi| \leq p \quad \Rightarrow \quad \forall \mathbf{a} \in \Sigma^*, \mathrm{time}_M(\varphi, \mathbf{a}) \leq q_p(|\mathbf{a}|).$$

Proof. The proof proceeds by induction on the number r of machines involved. For $r = 1$ the statement is immediate since linear second-order polynomials map linear functions to linear functions and are monotone. For the induction step let N be the composition of the machines $M_2, \ldots, M_r$. Then M is the composition of M_1 with N. Fix some linear function p and let $q_p(n) = Cn + D$ be the linear function that is guaranteed to exist by the induction hypothesis. The formula for the time-function of the composition of machines from equation (T) on page 8 and the linear time-bound from the induction hypothesis provides that any valid φ with $|\varphi| \leq p(n)$ satisfies

$$\mathrm{time}_M(\varphi, \mathbf{a}) \leq C_N \cdot \Big(\mathrm{time}_{M_1}(N^\varphi, \mathbf{a}) + \sum_{\mathbf{b} \in Q_{M_1}(N^\varphi, \mathbf{a})} (C \cdot |\mathbf{b}| + D) \Big).$$

Using the linearity of summation, and estimating both the sum over the length of queries in the second term and the number of summands in the third term by the time M_1 takes, we end up with

$$\mathrm{time}_M(\varphi, \mathbf{a}) \leq C_N \cdot (C + D)\, \mathrm{time}_{M_1}(N^\varphi, \mathbf{a}) \leq (C + D) \cdot P(|N^\varphi|, |\mathbf{a}|),$$

where P is a linear second-order polynomial that bounds the runtime of M_1. The statement now follows from the fact that linear time computable operators take inputs of linear length to outputs of linear length and that linear second-order polynomials preserve linearity.

This lemma fixes the input φ to have size bounded by a linear function and only talks about linearity of the size in the size of the string input. This is for a reason: Both in its proof as well as in the proof that linear second-order polynomials preserve linearity, the constants appearing in the linear functions are multiplied.

Corollary 1. *For any nearly linear-time computable $F : \subseteq \mathcal{B} \to \mathcal{B}$ there exists a modulus of continuity L such that for any $\varphi \in \mathrm{dom}(F)$ of linear size there is a linear function p_φ bounding the size of the certificates, i.e., $|L(\varphi, \mathbf{a})| \leq p_\varphi(|\mathbf{a}|)$.*

We now again consider the maximization operator defined in Sect. 3.1.

Theorem 1. *The maximization operator is not nearly linear-time computable.*

Proof. Towards a contradiction assume the operator was nearly linear-time computable. This means that by Corollary 1 there exists a modulus of continuity L such that for any function φ of linear size there is a linear function p_φ with $|L(\varphi, \mathbf{a})| \leq p_\varphi(|\mathbf{a}|)$. Pick any specific function φ of linear size, for instance the constant function returning the empty string. Since in equation (L) on page 13 a quadratic lower bound on the size of any certificate was provided, we have just bounded a quadratic function by a linear function, which is a contradiction.

4 Operating on Real Numbers and Functions

In this chapter we discuss some applications in computable analysis. Computable analysis considers representations to carry out computations on infinite structures. There are two different models that are popular and known to be computationally equivalent. From the point of view of polynomial-time complexity, the models are known to be distinct. However, as long as the space computed on is not too large for one of the frameworks to lead to a good complexity theory it can be shown that the models coincide. The first question we ask is whether the models can be proven equivalent in our refined setting in a meaningful way.

We first describe the model that fits in with the way we have talked about complexity so far most conveniently: A representation of a set X is a partial surjective mapping $\delta \colon \subseteq \mathcal{B} \to X$. An element $\varphi \in \mathcal{B}$ is called a δ-name of an element $x \in X$ if $\delta(\varphi) = x$. Note that surjectivity of the representation means that each element of X is given at least one name. A pair $\mathbf{X} := (X, \delta_\mathbf{X})$ of a set together with a representation is called a represented space. An element of a represented space is called computable resp. polynomial-time or linear time computable, if it has a name with this property. A representation should be understood to induce notions of computability and complexity and whether or not an element is computable, polynomial-time computable or linear-time computable depends on the details of the representation.

Example 2 (Cauchy representation). The dyadic numbers are the rational numbers of the form $z \cdot 2^{-n}$ for some $z \in \mathbb{Z}$ and $n \in \mathbb{N}$. Denote the set of all dyadic numbers by $\mathbb{D}$. Let $\nu_\mathbb{D} \colon \Sigma^* \to \mathbb{D}$ be a surjective encoding of the dyadic numbers by binary strings and let the corresponding Cauchy representation δ_C of the real numbers be defined by

$$\delta_C(\varphi) = x \quad \Leftrightarrow \quad \forall \mathbf{a} \in \Sigma^*, |\nu_\mathbb{D}(\varphi(\mathbf{a})) - x| \leq 2^{-|\mathbf{a}|}.$$

We are interested in comparing different representations. Given two representations δ and δ' of a set X, we say that an operator $F \colon \subseteq \mathcal{B} \to \mathcal{B}$ is a translation of δ into δ' if whenever its argument is a δ-name, its return value is a δ'-name, i.e., if for all $\varphi \in \mathrm{dom}(\delta)$ it holds that $\delta(\varphi) = \delta'(F(\varphi))$. Two representations are called computably, polynomial-time or linear-time equivalent if a computable, polynomial-time computable, or linear-time computable translations in both directions exist. Note that there are point-wise preservation theorems for each of these classes. Thus, the notion of computable elements is stable under switching between computably equivalent representations, the notion of polynomial-time computable elements is stable under polynomial-time equivalence and the notion of linear-time computable elements under linear-time equivalence.

Next let us describe the second model, which is also popular for doing computable analysis. Let $\mathcal{C} := \Sigma^\omega$ denote the set of infinite strings over a finite alphabet. Here $\mathcal{C}$ is for Cantor-space. A Cantor-space representation of a set X is a partial surjective mapping $\rho: \subseteq \mathcal{C} \to X$. Again, an infinite string χ is called a ρ-name, or just name of $x \in X$ if $\rho(\chi) = x$. To define computability and complexity of an element of a Cantor-space represented set it is necessary to specify how to compute on infinite strings: An infinite string $\chi \in \mathcal{C}$ is computable if there is a Turing machine with a write only output tape that does not take any inputs and successively writes the digits of an infinite sequence to the output tape. Computably, this notion is equivalent to decidability of the set that has χ as a characteristic function. However, the above description suggests a specific notion of resource consumption: χ is said to be computable in time $t: \mathbb{N} \to \mathbb{N}$ if there is a machine producing the n-th digit within the first $t(n)$ steps of its run. In particular, this leads to notions of polynomial- and linear-time computability.

Example 3 (signed digits). It is a well know fact that encoding real numbers by their binary expansions does not lead to a well-behaved Cantor-space representation. Instead consider the three element alphabet $\Sigma := \{-1, 0, 1\}$, and define the signed digit representation ρ_{sd} of the unit interval by

$$\rho_{sd}(\chi) = x \quad \Leftrightarrow \quad x = \sum_{i=0}^{\infty} \chi_i \cdot 2^{-i},$$

where the elements of the alphabet are interpreted as the obvious real numbers and the infinite sum is defined as the limit of its partial sums. In this case the represented space is not all real numbers but the unit interval.

The way to make a connection between the two models that is suitable for our purposes is as follows:

Definition 4. *For a Cantor-space representation $\rho: \subseteq \mathcal{C} \to X$ of a set X let the representation $\delta_\rho: \subseteq \mathcal{B} \to X$ be defined by*

$$\delta_\rho(\varphi) = x \quad \Leftrightarrow \quad \exists \chi \in \mathcal{C}, \rho(\chi) = x \wedge \forall \mathbf{a} \in \Sigma^*, \varphi(\mathbf{a}) = \chi^{\leq |\mathbf{a}|}.$$

By being a little clever about the instances to run algorithms on, one can now establish the desired property:

Proposition 2. *For any Cantor-space representation ρ the representation δ_ρ has the same computable, polynomial-time computable and linear-time computable elements.*

Proof. To see the equality of computable and polynomial-time computable elements, the straight-forward translations of the algorithms work. For the linear-time case first assume we have an algorithm to compute a ρ-name of x in linear-time. An algorithm to compute a δ_ρ-name in linear time can be obtained by producing the digits using the linear-time algorithm for the ρ-name and counting down the length of the input to check when enough digits have been produced.

For the other direction, given an algorithm to compute a δ_ρ-name produce a diverging program as follows: Successively execute the given linear-time program on inputs 1^{2^i} and each time after it finishes copy the second half of the resulting string to the output tape to extend what was written there before. The exponential increase in the size of queries leads to a geometric series in and thus to a linear bound of the time consumption.

4.1 Signed-Digit and Cauchy Representation Compared

The goal of this section is to prove that the Cauchy representation from Example 2 and the signed digit representation from Example 3 are equivalent in a very strong sense. Since the Cauchy representation is a representation of the real numbers and the signed digit representation only of the unit interval, we take the subspace representation, i.e., a range restriction for the Cauchy representation. As the signed digit and the Cauchy representation use different models of computability and complexity we use the translation from Definition 4.

Theorem 2. *The range restriction of the Cauchy-representation to the unit interval is linear-time equivalent to the representation $\delta_{\rho_{sd}}$ obtained form the signed-digit representation as described in Definition 4.*

Proof. First specify an algorithm that translates from the Cauchy representation to the signed digit representation. Let φ be a δ_C-name of some x and **a** an input. Recall that we need to produce the string of the $|\mathbf{a}|$ first digits of a fixed ρ_{sd} name of x. The important part in this direction is to make sure that the digits of later approximations do not contradict digits of earlier approximations. To ensure this consistency, query the oracle for $\varphi(1^{2^i})$ for each i such that $2^{i-1} \leq |\mathbf{a}|$. Read approximation of length linear in 2^i off the oracle answer tape. Note that the returned strings may be long, but the reading head is reset to the inital position during an oracle query and thus the first bits are readily available. The obtained values are approximations in binary, but it need not be the case that the previous ones are initial segments of the later ones. This can be accounted for by comparing the last two digit of the previous approximation with the corresponding digits of current approximation. Due to the use of signed digits, it is possible to correct the difference by changing the first digit of the second half of the current approximation if neccessary and then copying it to the output tape to extend the previous approximation. This algorithm produces consistent intial segments of a fixed ρ_{sd} name. The exponentially growing queries lead to a geometric series when counting the steps and to a time consumption that depends linearly on the size of **a**.

That the translation in the other direction is linear time computable is a direct consequence of the existence of a linear time translation from integers represented as sequences of signed digits to integers represented in binary.

As an immediate consequence of Theorem 2 we note that:

Corollary 2. *The signed-digit representation and the Cauchy representation have the same linear-time computable elements.*

As a non-uniform result, this remains true if the signed-digit representation is extended to all of the real numbers.

4.2 Operating on Continuous Functions

Once the representations are fixed, one can also define what it means for functions between spaces to be computable: An operator $F: \subseteq \mathcal{B} \to \mathcal{B}$ is called a realizer of a function $f: \mathbf{X} \to \mathbf{Y}$ between represented spaces if it maps each name of some $x \in \mathbf{X}$ to a name of $f(x) \in \mathbf{Y}$, i.e. if for all $\varphi \in \mathrm{dom}(\delta_{\mathbf{X}})$ it holds that $\delta_{\mathbf{Y}}(F(\varphi)) = f(\delta_{\mathbf{X}}(\varphi))$. In particular the domain of $\delta_{\mathbf{X}}$ should be included in the domain of F. A function between represented spaces is called computable if it has a computable realizer. We may go ahead and call a function between represented spaces polynomial-time computable or linear-time computable if there is a realizer that has this property. Computable analysis usually makes the additional assumptions that the names come from a fixed subset of Baire-space where the length function is computable in polynomial-time. While the representations we consider do not only use names from this fixed set, it is easy to check that they make the size information easily accessible for a machine.

For the space $C([0,1])$ of continuous functions $f : [0,1] \to \mathbb{R}$, we use the standard representation introduced in [11]: A $\varphi \in \mathcal{B}$ is a name for $f \in C([0,1])$ if it has the following two properties:

1. For any $n \in \mathbb{N}$, $\varphi(1^n) = 1^{\mu(n)}$ where $\mu : \mathbb{N} \to \mathbb{N}$ is a modulus of continuity of the function f, i.e., $|x - y| \leq 2^{-\mu(n)} \Rightarrow |f(x) - f(y)| \leq 2^{-n}$.
2. Whenever $\mathbf{a}$ is an encoding of a dyadic rational number from $[0,1]$ then $|\nu_{\mathbb{D}}(\varphi(01^n\mathbf{a})) - f(\nu_{\mathbb{D}}(\mathbf{a}))| \leq 2^{-n}$ for all $n \in \mathbb{N}$.

Lemma 9. *A function $f : [0,1] \to \mathbb{R}$ is linear-time computable if and only if it has a linear-time computable name.*

Proof. First assume there is a linear-time machine K that computes a name for $f : [0,1] \to \mathbb{R}$ and its running time is bounded by a linear function $l_K : \mathbb{N} \to \mathbb{N}$. We describe an oracle machine M that computes f, i.e., on input 1^n and a name φ of $x \in [0,1]$ returns an approximation of $f(x)$ with error at most 2^{-n}. To achieve this first follow what M does on input 1^{n+1} to get a string 1^m such that for all $x_1, x_2 \in [0,1]$, $|x_1 - x_2| \leq 2^{-m} \Rightarrow |f(x_1) - f(x_2)| \leq 2^{-(n+1)}$. Note that $m \leq l_M(n+2)$ as writing symbols to the output tape consumes time. Thus, the string 1^m can be produced on the oracle tape in linear time. An oracle query results in an encoding $\mathbf{a}$ of a dyadic approximation on the oracle answer tape.

We follow the instructions of K with input $\mathtt{01}n^{n+1}\mathtt{0a}$ to get an approximation of $f(\nu_\mathbb{D}(\mathbf{a}))$ with error at most $2^{-(n+1)}$ in time linear in $n + |\varphi| (n + 1) + 3$. Using the assumptions that K computes a name of f and φ is a name of x, a simple triangle inequaltiy argument proves that this is a valid return value.

Next, assume that there is an oracle machine M computing f with a linear time bound P. By definition of the Cauchy representation, there exists a linear function l such that each element of the unit interval has a name of length l and one can verify that $P(l, \cdot + 1)$ is a modulus of continuity of f. Let C and D be the constants of the linear function $P(l, \cdot + 1)$. A machine K that computes a name of f in linear may act as follows: On input $\mathtt{1}^n$ evaluate $C \cdot n + D$ in linear time and return its value in unary. On input $\mathtt{01}^n\mathtt{0a}$ follow the run of the machine M on input $\mathtt{1}^n$, where $\mathbf{a}$ is put on the oracle answer tape and the oracle answers are ignored. I.e. where M receives $\mathbf{a}$ as answer to each oracle call. Since the constant function returning $\mathbf{a}$ is a valid name of the dyadic number $\mathbf{a}$ encodes interpreted as a real number, this leads to a valid return value. For the time consumption let K_1, K_2 and K_3 be the constants from Lemma 4. Then, the time consumption of M is bounded by $K_1 \cdot |\mathbf{a}| + K_2 \cdot n + K_3$, which is a linear bound in the size $n + |\mathbf{a}| + 3$ of the input.

The proof of the first implication is uniform and can be modified to prove the following:

Theorem 3. *Evaluation* $\mathcal{C}([0,1]) \times [0,1], (f, x) \mapsto f(x)$ *is linear time computable.*

Theorem 4. *Let* $f : \mathbb{R} \to \mathbb{R}$ *and* $g : \mathbb{R} \to \mathbb{R}$ *be linear-time computable. Then* $f \circ g$ *is linear-time computable.*

Proof. Let M_1 and M_2 be machines that compute f and g in linear time. It follows from the Definition of the Cauchy representation that there exists constants C, D such that each real number x has a name of size $n + D \cdot |\varphi|(0) + C$ and that such a name can be obtained from an arbitary name in linear time by truncating. Let P be the linear time-bound of M_1. An easy induction shows that there exist constants E', D' and C' such that for φ such that $|\varphi|(n) \leq n + D \cdot |\varphi|(0) + C$, it holds that $P(|\varphi|, n) \leq E' \cdot n + D' \cdot |\varphi|(0) + C'$. To evaluate the composition, assume that x is some input and φ is a name of x that is as specified above. Furthermore let Q be a linear time bound of the machine M_2. To get an approximation to $f(g(x))$ in linear time, first evaluate $k := Q(m \mapsto E' \cdot n + D' \cdot |\varphi|(0) + C', n)$. Since the constants E', D', C' and the linear second order polynomial Q are fixed, this amounts to evaluating a linear function and can be done in linear time. Then run the machine M_1 on input $\mathtt{1}^k$ to obtain an encoding $\mathbf{a}$ of an 2^{-k}-approximation to $g(x)$. Finally run M_2 on input $\mathtt{1}^n$ while simulating the constant oracle with value $\mathbf{a}$. Since the number of steps M_2 takes is bounded by k the machine can only ask oracle queries with less than k bits. Furthermore, since $\mathbf{a}$ encodes an approximation to $g(x)$, $g(x)$ has a name starting in the constant function returning $\mathbf{a}$ on all strings smaller than k. As each machine is only evaluated once on inputs of linear size, the overall time-consumption is linear.

5   Conclusion

This paper puts forward a class for linear-time computability in the setting of type-2 computations by naturally restricting the second-order polynomials used as time-bounds for the running time of oracle machines. This class is not stable under composition and we investigated some examples that indicate that this property can not be expected in a setting as general as the one we work in. We also investigated the closure of this class under composition and provide some tools that allow to prove that an operator is not included in this wider class. As an application we proved that the maximization operator can not be written as a composition of linear-time computable operators.

The motivation for looking at linear-time computability in a higher-order setting came from the availability of a class of linear functions that is suggested by the definition of the second-order polynomials used in type-two complexity theory and the availability of a canonical benchmark problem: Can the product that arises in the standard proof of closure under composition, and is clearly a coarse overestimation, be removed? We gave a negative answer to this question. Since the type-two class that was considered is even less stable than the type-one linear-time class that is already considered fairly fragile, it seems desirable to have a type-two analogue of almost linear-time computability. Unfortunately, in a type-two setting it is not clear how to find an appropriate class of 'almost linear' running times and we did not succeed to prove closure under composition for any of the candidates we considered. Thus, finding such a class has to be left open for future research.

We also took a look at some applications in computable analysis. We detailed a translation between the two most common models of computation used in computable analysis that preserves linear time computability of names. Using this translation we proved that the most commonly used representations in the corresponding models have the same linear-time computable names and we made this statement uniform when only the unit interval is considered. Finally we proved that the standard representation of continuous functions is well-behaved with respect to linear-time computability in the sense that it gives linear-time computable names exactly to linear time computable functions and that linear time computable functions on the real numbers are closed under composition.

While the results are satisfactory, there is a lack of examples. The real numbers that are known to be linear-time computable are constructed particularly for this purpose and for even the most basic real numbers of practical relevance it is an open problem whether there are linear time algorithms. For instance, there is no known algorithm for computing π in linear time. The same holds true for functions on the real numbers computable in linear time. Finally note, that many of the results from the last section depend on removing the higher-order asspects when computing on real numbers. A notable exception to this is the linear-time computability of the evaluation operator.

Acknowledgements. This work was supported by JSPS KAKENHI Grant Numbers JP18H03203 and JP18J10407, by the Japan Society for the Promotion of Science (JSPS), Core-to-Core Program (A. Advanced Research Networks), by the ANR project FastRelax (ANR-14-CE25-0018-01) of the French National Agency for Research and by EU-MSCA-RISE project 731143 "Computing with Infinite Data" (CID).

References

1. Brausse, F., Steinberg, F.: A minimal representation for continuous functions. arXiv preprint arXiv:1703.10044 (2017)
2. Buss, J.F.: Relativized alternation and space-bounded computation. J. Comput. Syst. Sci. **36**(3), 351–378 (1988)
3. Case, J.: Resource restricted computability theoretic learning: illustrative topics and problems. Theory Comput. Syst. **45**(4), 773–786 (2009)
4. Case, J., Kötzing, T., Paddock, T.: Feasible iteration of feasible learning functionals. In: Hutter, M., Servedio, R.A., Takimoto, E. (eds.) ALT 2007. LNCS (LNAI), vol. 4754, pp. 34–48. Springer, Heidelberg (2007). https://doi.org/10.1007/978-3-540-75225-7_7
5. Férée, H., Hainry, E., Hoyrup, M., Péchoux, R.: Interpretation of stream programs: characterizing type 2 polynomial time complexity. In: Cheong, O., Chwa, K.-Y., Park, K. (eds.) ISAAC 2010. LNCS, vol. 6506, pp. 291–303. Springer, Heidelberg (2010). https://doi.org/10.1007/978-3-642-17517-6_27
6. Fournet, C., Kohlweiss, M., Strub, P.-Y.: Modular code-based cryptographic verification. In: Proceedings of the 18th ACM Conference on Computer and Communications Security, pp. 341–350. ACM (2011)
7. Grzegorczyk, A.: On the definitions of computable real continuous functions. Fund. Math. **44**, 61–71 (1957)
8. Gurevich, Y., Shelah, S.: Nearly linear time. In: Meyer, A.R., Taitslin, M.A. (eds.) Logic at Botik 1989. LNCS, vol. 363, pp. 108–118. Springer, Heidelberg (1989). https://doi.org/10.1007/3-540-51237-3_10
9. Irwin, R.J., Royer, J.S., Kapron, B.M.: On characterizations of the basic feasible functionals, part i. J. Funct. Program. **11**(1), 117–153 (2001)
10. Kapron, B.M., Cook, S.A.: A new characterization of type-2 feasibility. SIAM J. Comput. **25**(1), 117–132 (1996)
11. Kawamura, A., Cook, S.: Complexity theory for operators in analysis. ACM Trans. Comput. Theory **4**(2), Article 5 (2012)
12. Kawamura, A., Ota, H.: Small complexity classes for computable analysis. In: Csuhaj-Varjú, E., Dietzfelbinger, M., Ésik, Z. (eds.) MFCS 2014. LNCS, vol. 8635, pp. 432–444. Springer, Heidelberg (2014). https://doi.org/10.1007/978-3-662-44465-8_37
13. Kawamura, A., Pauly, A.: Function spaces for second-order polynomial time. In: Beckmann, A., Csuhaj-Varjú, E., Meer, K. (eds.) CiE 2014. LNCS, vol. 8493, pp. 245–254. Springer, Cham (2014). https://doi.org/10.1007/978-3-319-08019-2_25
14. Kawamura, A., Steinberg, F., Thies, H.: Parameterized complexity for uniform operators on multidimensional analytic functions and ODE solving. In: Moss, L.S., de Queiroz, R., Martinez, M. (eds.) WoLLIC 2018. LNCS, vol. 10944, pp. 223–236. Springer, Heidelberg (2018). https://doi.org/10.1007/978-3-662-57669-4_13

15. Lacombe, D.: Sur les possibilités d'extension de la notion de fonction récursive aux fonctions d'une ou plusieurs variables réelles. In: Le raisonnement en mathématiques et en sciences expérimentales, Colloques Internationaux du Centre National de la Recherche Scientifique, LXX. Editions du Centre National de la Recherche Scientifique, Paris, pp. 67–75 (1958)
16. Mehlhorn, K.: Polynomial and abstract subrecursive classes. In: Proceedings of the Sixth Annual ACM Symposium on Theory of Computing, pp. 96–109. ACM (1974)
17. Pauly, A., Steinberg, F.: Comparing representations for function spaces in computable analysis. Theory Comput. Syst. **62**(3), 557–582 (2018)
18. Regan, K.W.: Machine models and linear time complexity. ACM SIGACT News **24**(3), 5–15 (1993)
19. Schröder, M.: Extended admissibility. Theor. Comput. Sci. **284**(2), 519–538 (2002)
20. Schröder, M.: Admissible representations for continuous computations. Fernuniv., Fachbereich Informatik (2003)
21. Steinberg, F.: Computational complexity theory for advanced function spaces in analysis. Ph.D. thesis, Technische Universität (2017)
22. Turing, A.M.: On computable numbers, with an application to the Entscheidungsproblem. Proc. London Math. Soc. **2**(1), 230–265 (1936)
23. Weihrauch, K.: Computable Analysis. Springer, Berlin/Heidelberg (2000). https://doi.org/10.1007/978-3-642-56999-9
24. Ziegler, M.: Hyper-degrees of 2nd-order polynomial-time reductions. Measuring the Complexity of Computational Content: Weihrauch Reducibility and Reverse Analysis (Dagstuhl Seminar 15392)

Consistency as a Branching Time Notion

Astrid Kiehn[✉] and Mohnish Pattathurajan

Indian Institute of Technology Mandi,
Kamand, Mandi 175001, Himachal Pradesh, India
astrid@iitmandi.ac.in

Abstract. Superposed on a distributed computation, a snapshot algorithm computes a global state, the so-called snapshot, consistent with the underlying computation. On intuive grounds it can be argued that the snapshot is a global state which would have been obtained when processes would have been frozen at the time their local states were recorded (their checkpoints). This interpretation is stronger than consistency as it also considers the nondeterministic behaviour of processes up to their checkpoints.

In this paper, we provide a formal setup to study the latter interpretation. In particular, we introduce the notion of a *freeze bisimulation* which allows us to express the above intuition formally as *progressive bi-consistency*. This definition does not only capture branching time behaviour but also applies to snapshot algorithms that compute a partial snapshot and of which the intermediate behaviour is of equal interest as the final snapshot (eg in case of approximations or eventual consistency). We establish fundamental properties of progressive bi-consistency which includes a theorem similar to the CAP theorem for distributed databases: it is impossible to have a partial snapshot algorithm which computes a minimal number of checkpoints, is progressively bi-consistent and non-inhibiting at the same time.

We illustrate our results by evaluating snapshot algorithms modelled in Milner's process algebra CCS. This includes the seminal algorithms by Chand/Lamport [9] and Lai/Yang [14] and a recent snapshot algorithm for data stream processing systems [8].

Keywords: Snapshot · Consistency · Distributed systems

1 Introduction

Snapshot algorithms (see [2,13,17]) are used to compute a consistent global state of an ongoing distributed application computation. Consistency, in its traditional sense, means the global state computed, the snapshot, is a state the distributed application computation could be reset to. Formally, this is expressed by looking at the application computation retrospectively and by arguing that the snapshot computed posesses this property. This is the classical approach. It is linear time as the application computation is looked at out of context, not considering events which could have been enabled at some time but had not been executed.

© Springer Nature Switzerland AG 2019
T. V. Gopal and J. Watada (Eds.): TAMC 2019, LNCS 11436, pp. 359–377, 2019.
https://doi.org/10.1007/978-3-030-14812-6_22

In this paper we take a different approach and argue that a snapshot is a global state which would have been obtained when application processes would have been frozen at the time their local states were recorded to be part of the snapshot. This interpretation is stronger as it takes into account the nondeterministic behavior of processes up to their state recording. It is forward looking and branching time.

The difference between these two approaches becomes apparent when the quality of a snapshot algorithm is to be established. A snapshot algorithm consists of a set of routines superposed on local processes. The routines act as an interface between application processes and the channels of the system. They may block [8,18], delay [11] or modify [14] application messages when these are to be received or sent. The routines may also send their own messages like requests to record the local state [9]. Such features – though not all of them together – are required to ensure that there are no two processes of which the recorded state of one of them shows a message as having been received while at the recorded state of the other process that message had not yet been sent. With these features, it is easy to influence the behavior of the application computation in some way. A snapshot algorithm that computes correct snapshots according to the linear time notion, still may influence the course of the application computation by enforcing it to proceed in a certain direction (by blocking some of the communication channels eg). Correctness with respect to the branching time notion, however, would guarantee that all this would not happen.

We express the branching time notion of consistency by what we call progressive bi-consistency. It uses concepts of concurrency theory, in particular, of bisimilarity [15]. It ensures that the application computation does not display new behavior while its local states are recorded, and that messages which in principle can be delivered to the application computation can indeed be received by it. In a similar way, we define what it means for a snapshot algorithm to be non-inhibiting. A snapshot algorithm is non-inhibiting if, superposed an application computation, the resulting system is observation equivalent with the application computation without superposed snapshotting. This refined definition allows to detect blocked channels.

With these definitions at hand, we are able to express an interesting result for partial snapshot algorithms which only take checkpoints of some of the processes of an application computation: no partial snapshot algorithm exists which is non-inhibiting, progressively bi-consistent while taking a minimal number of checkpoints, only. This refines an earlier result in [5].

In the second part of the paper we investigate existing snapshot algorithms. Depending on the nature of the snapshot routines, one distinguishes between coordinated snapshot algorithms [9], communication induced checkpointing [14] and combinations of these. We also explore partial snapshot algorithms which only checkpoint some and not all of the processes [4,10–12]. With [8] a recent checkpoint algorithm for data stream processing systems is considered. All these algorithms except mutable checkpointing [4] are progressively bi-consistent. This shows that progressive bi-consistency is a natural concept.

All snapshot algorithms are formulated in Milner's process algebra CCS, [15]. They are contained in Sect. 4. Section 2 provides the formal model and necessary background definitions, and Sect. 3 the definition and results of progressive (bi-)consistency.

2   Basic Definitions

Throughout the paper, by a distributed system we understand a set of sequential application processes A_i, $i \in I$, without shared data communicating by message passing over reliable FIFO channels C_{ij}, $i, j \in I$, $i \neq j$. Processes and channels are modelled as terms of the process algebra CCS, cf. [15]. Every process A_i has its own action set Act_i, disjoint from all Act_j, $j \neq i$. Process A_i sends and receives messages to and from A_j via the actions in $Comm_i = \{\overline{send_{ij}(msg)}, receive_{ji}(msg) \mid msg \in MSG, j \in I \setminus \{i\}\}$ a subset of Act_i where MSG is the set of application messages. All such events are given by $L_1 = \bigcup_{i \in I} Comm_i$. We use App to denote the collection of application processes of an application computation, that is $App = \prod_{i \in I} A_i$.

We explicitly model the interface between application processes and channels by Int which simply passes on messages sent by the application to channels and, respectively, receives messages from channels and delivers them to the application. So $Int = \prod_{i \in I} U_i$ where

$$U_i = \sum_{j \in I, i \neq j, msg \in MSG} send_{ij}(msg).\overline{i\_send_{ij}(msg)}.U_i + \sum_{j \in I, i \neq j} i\_receive_{ji}(msg).\overline{receive_{ji}(msg)}.U_i.$$

$msg, msg' \in MSG$. With $\overline{i\_send_{ij}(msg)}$ and $i\_receive_{ji}(msg)$ the message is put into or removed from the channel leading from P_i to P_j, respectively. Channels, however, may transfer other kind of items as well. With MSG_{int} we denote the set of all items that may be put into a channel. This includes application messages, manipulated application messages and coordination messages. Channels are modelled by

$$C_{ij}(\varepsilon) := \sum_{m \in MSG_{int}} i\_send_{ij}(m).C_{ij}(m)$$

$$C_{ij}(s.m) := \overline{i\_receive_{ij}(m)}.C_{ij}(s) + \sum_{m \in MSG_{int}} i\_send_{ij}(m').C_{ij}(m'.s.m)$$

where $m \in MSG_{int}$. The complete system consisting of application processes, interface and channels, thus, is

$$\left(\prod_{i \in I} A_i \mid \left(\prod_{i \in I} U_i \mid \prod_{i, j \in I, i \neq j} C_{ij}(\varepsilon) \right) \setminus L_2 \right) \setminus L_1.$$

where $L_2 = \{i\_send_{ij}(m), \overline{i\_receive_{ji}(m)} \mid m \in MSG_{int}\}$. The restriction by $L_1 \cup L_2$ ensures that processes cannot send and receive messages without synchronizing with the interface and respective channels. To ease readability, we render it by $\prod A_i \parallel (\prod U_i \parallel \prod C_{ij}(\varepsilon))$ assuming that indices i, j are understood. Further $R \parallel S$ stands for $(R \mid S) \setminus X$ where X is the set of actions over which R and S can communicate. The infrastructure or network of the system is given by $IntSYS = \prod U_i \parallel \prod C_{ij}(\epsilon)$ which in variations will be studied in the sequel. Further, we sometimes abbreviate $\prod C_{ij}(\epsilon)$ by *Channels*.

A distributed computation is a sequence of transitions

$$\pi = S^0 \xrightarrow{e_1} S^1 \xrightarrow{e_2} S^2 \ldots \xrightarrow{e_n} S^n$$

where each transition $S^i \xrightarrow{e_{i+1}} S^{i+1}$ is obtained from the semantics of CCS. With π, $trace(\pi) = (t_1, \ldots, t_{|I|})$ is obtained where each t_i is the restriction of $e_1 \ldots e_n$ to the (visible) events of process A_i. We say, $trace(\pi)$ leads to S^n. Note, that the trace does not record events of the network. Differing from the classical CCS semantics (see next section) we observe send and receive events of application processes by *SEND* and *RECEIVE*. Hence, the states of application processes at state S^j can be retrieved from the trace of the computation up to S^j and this trace is a prefix of $trace(\pi)$. A tuple $(h_1, \ldots, h_{|I|})$ is a prefix of $(t_1, \ldots, t_{|I|})$ if $h_i \sqsubseteq t_i$ for all $i \in I$ ($\sqsubseteq$ being the prefix relation on strings). However, not every prefix of $trace(\pi)$ corresponds to a reachable state. For this, additionally consistency is required.

Definition 1. *A trace* $(h_1, \ldots, h_{|I|})$ *is consistent with trace* $(t_1, \ldots, t_{|I|})$ *if*

1. $h_j \sqsubseteq t_j$ *for all* $j \in I$,
2. *whenever* $RECEIVE_{ij}(msg)$ *is contained in* h_j, $\overline{SEND_{ij}(msg)}$ *is contained in* h_i.

Traditional correctness proofs of snapshot algorithms are based on this consistency notion. It is shown that for a given snapshot computation π, the trace identified with the snapshot is consistent with $trace(\pi)$, see eg. [9].

2.1 Operational Semantics

Processes are specified in CCS, [15]. We recall the rules of its operational semantics as they are slightly modified but without changing the overall behaviour. For a sequential process Q two kind of transitions are possible. If $Q = \mu.R$ where μ can be any action, unconditionally, $Q \xrightarrow{\mu} R$. If $Q = \sum_{i \in M} Q_i$, $M \subseteq I$, the nondeterministic choice expressed by the sum operator, is resolved with any transition $Q_i \xrightarrow{\mu} R_i$, $i \in M$:

$$\frac{Q_i \xrightarrow{\mu} R_i}{\sum_{i \in M} Q_i \xrightarrow{\mu} R_i}$$

Transitions of a entire distributed system are stated in a compact way. In a product like $\prod A_i \parallel (\prod U_i \parallel \prod C_{ij}(\alpha_{ij}))$ we only show the processes that change with a transition. Transitions of channels depend on the contents, only.

That is, $C_{ij}(\alpha) \xrightarrow{\overline{i\_receive_{ij}(msg)}} C_{ij}(\beta)$ if and only if $\alpha = \beta.msg$. Similarly, $C_{ij}(\alpha) \xrightarrow{i\_send_{ij}(msg)} C_{ij}(msg.\alpha)$ is always possible. We use the shorter expressions in the rules where $\alpha, \beta \in MSG^*$. With these conventions we obtain the following compact rules.

Transitions Involving Application Processes

The main difference in our semantics compared to the classical CCS semantics is that we keep send and receive events visible though they are communicating with the channel.

– internal events μ, $\mu \notin L$:

$$\frac{A_i \xrightarrow{\mu} A_i'}{\ldots \mid A_i \mid \ldots \parallel \ldots \xrightarrow{\mu} \ldots \mid A_i' \mid \ldots \parallel \ldots}$$

– send events of application process A_i:

$$\frac{A_i \xrightarrow{send_{ij}(msg)} A_i' \qquad P_i \xrightarrow{send_{ij}(msg)} P_i'}{\ldots \mid A_i \mid \ldots \parallel \ldots \mid P_i \mid \ldots \xrightarrow{\overline{SEND_{ij}(msg)}} \ldots \mid A_i' \mid \ldots \parallel \ldots \mid P_i' \mid \ldots}$$

– receive events of application process A_i:

$$\frac{A_i \xrightarrow{receive_{ji}(msg)} A_i' \qquad P_i \xrightarrow{receive_{ji}(msg)} P_i'}{\ldots \mid A_i \mid \ldots \parallel \ldots \mid P_i \mid \ldots \xrightarrow{RECEIVE_{ji}(msg)} \ldots \mid A_i' \mid \ldots \parallel \ldots \mid P_i' \mid \ldots}$$

Snapshot algorithms consist of a set or routines superposed on application processes. We model a snapshot algorithm Snp as a special interface which can trigger checkpoints (that is, the recording of local states), pass on, buffer, modify, withhold application messages or send its own messages (so-called coordination messages) over the channels. The checkpointing of a process by P_i is indicated by the event cp_taken_i.

Internal Transitions of the Interface

Messages sent may be application or coordination messages, or a combination of both of them.

– Adding a message to a channel:

$$\frac{P_i \xrightarrow{i\_send_{ij}(m)} P_i'}{\ldots \mid P_i \mid \ldots \parallel \ldots \mid C_{ij}(\alpha) \mid \ldots \xrightarrow{\tau} \ldots \mid P_i' \mid \ldots \parallel \ldots \mid C_{ij}(m.\alpha) \mid \ldots}$$

In Sect. 4 we use a broadcasting of messages as well. With $i\_send_{iM}(m)$, $M \subseteq I$, message m is added to all channels C_{ij}, $j \in M$. The concerned channels are updated as in the rule given here.

- Removing a message from a channel:

$$\frac{P_i \xrightarrow{i\_receive_{ji}(m)} P_i' \qquad \alpha = \beta.m \qquad \text{does not trigger checkpoint}}{\ldots \mid P_i \mid \ldots \parallel \ldots \mid C_{ji}(\alpha) \mid \ldots \xrightarrow{\tau} \ldots \mid P_i' \mid \ldots \parallel \ldots \mid C_{ji}(\beta) \mid \ldots}$$

$$\frac{P_i \xrightarrow{i\_receive_{ji}(m)} P_i' \qquad \alpha = \beta.m \qquad \text{triggers checkpoint}}{\ldots \mid P_i \mid \ldots \parallel \ldots \mid C_{ji}(\alpha) \mid \ldots \xrightarrow{cp\_taken_i} \ldots \mid P_i' \mid \ldots \parallel \ldots \mid C_{ji}(\beta) \mid \ldots}$$

Note that whether an event triggers a checkpoint or not can be derived from the state of the executing process. In other words, it is a predicate over process states.

- initiating the checkpointing (WLOG we assume this to be done by P_1):

$$\frac{P_1 \xrightarrow{cp\_taken_1} P_1'}{\ldots \parallel \ldots \mid P_1 \mid \ldots \xrightarrow{cp\_taken_1} \ldots \parallel \ldots \mid P_1' \mid \ldots}$$

Each of these rules can be derived from the original rules of CCS if we ignore the observability of communications of application processes.

A snapshot algorithm Snp takes a checkpoint of each application process. So a completed snapshot computation

$$\pi = S^0 \xrightarrow{e_1} S^1 \xrightarrow{e_2} S^2 \ldots \xrightarrow{e_n} S^n$$

of $App \parallel (Snp \parallel Channels)$ contains a cp_taken_i event for each process A_i. Let h_i be the restriction of the prefix $e_1 \ldots cp\_taken_i$ of $e_1 \ldots e_n$ to the visible events of A_i. The state of App identified with $(h_1, \ldots, h_{|I|})$ is the snapshot computed.

Definition 2. *A snapshot algorithm Snp is correct (in the classical sense) if for every application App and every snapshot computation π of App $\parallel$ (Snp $\parallel$ Channels) with completed checkpointing, the trace identified with the computed snapshot is consistent with trace(π).*

3   Progressive (Bi-)Consistency

The intention behind the notion of progressive bi-consistency is to capture the branching time behavior of a system before a snapshot. Up to the snapshot the system with overlaid snapshot routines should be weakly bisimilar to the original system. That is, the application processes should be able to communicate (send and receive messages) in the same way as they would do without the superlaid checkpoint routines. To capture this independently of any particular application, we model the snapshot algorithm together with the network (FIFO channels) and analyze it independently. We require system SYS_{Snp} consisting of snapshot algorithm Snp and channels to be weakly bisimilar (up to checkpointing) to the system in which communications are disabled (frozen) at the time a checkpoint is taken. The intuition is that in the context of an application process, if a process takes a checkpoint then from this local state it will not further proceed.

The interface system in which communications for individual processes may be frozen autonomously defines the freeze system. Its individual processes are defined by:

$$F_i = \sum_{j \in I, i \neq j, msg \in MSG} send_{ij}(msg).\overline{i\_send_{ij}(msg)}.F_i +$$
$$\sum_{j \in I, i \neq j} i\_receive_{ji}(msg).\overline{receive_{ji}(msg)}.F_i + freeze_i.Frozen_i$$

The initial state is $FSYS = \prod F_i \parallel Channels$.

The comparison of $SYS_{Snp} = Snp \parallel Channels$ and $FSYS$ is done via a freeze bisimulation. It is a variation of a bisimulation but sensitive to processes having taken a checkpoint and to frozen processes. If a checkpoint is taken, this has to be matched with freezing the respective process in the freeze system. A frozen process cannot perform any action nor can it be un-frozen. However, it is not deemed to match any action.

We use predicates $cp\_taken_i^S$ to indicate whether a checkpoint has been taken by P_i at state S and, respectively, $frozen_i^S$ to indicate whether Q_i has been frozen at S. As usual, $\overset{\nu}{\Longrightarrow} = (\overset{\nu}{\rightarrow})^*$.

Definition 3. $R \subseteq STATES(SYS_{Snp}) \times STATES(FSYS)$ *is a* freeze bisimulation *if for all* $(P, Q) \in R$ *the following holds*

1. *(a) if* $P \overset{a}{\rightarrow} P'$ *where* $a \in \{send_{ij}(msg), \overline{receive_{ij}(msg)} \mid msg \in MSG, j \in I\}$ *then*
   $$\begin{cases} \neg cp\_taken_i^P : \exists Q' : Q \overset{a}{\Longrightarrow} Q' \text{ and } (P', Q') \in R \\ cp\_taken_i^P : frozen_i(Q) \text{ and } (P', Q) \in R \end{cases}$$
   (b) if $P \overset{\tau}{\rightarrow} P'$ *then* $\exists Q' : Q \overset{\epsilon}{\Longrightarrow} Q'$ *and* $(P', Q') \in R$ *or* $(P', Q) \in R$,
   (c) if $P \xrightarrow{cp\_taken_i} P'$ *then* $\exists Q' : Q \overset{freeze_i}{\Longrightarrow} Q'$ *and* $(P', Q') \in R$,

2. *(a) if* $Q \overset{a}{\rightarrow} Q'$ *where* $a \in \{send_{ij}(msg), \overline{receive_{ij}(msg)} \mid msg \in MSG, j \in I\}$, *then* $\neg cp\_taken_i^P$ *and* $\exists P' : P \overset{a}{\Longrightarrow} P'$, *and* $(P', Q') \in R$,
   (b) if $Q \overset{\tau}{\rightarrow} Q'$ *then* $\exists P' : P \overset{\epsilon}{\Longrightarrow} P'$ *and* $(P', Q') \in R$.

If R satisfies item 1. only, it is called a freeze simulation.

Definition 4 (Progressive (Bi-)Consistency). *A Snapshot Algorithm Snp is called* progressively bi-consistent *if there is freeze bisimulation R containing* $(SYS_{Snp}, FSYS)$.

A Snapshot Algorithm Snp is progressively consistent *if there is freeze simulation R containing* $(SYS_{Snp}, FSYS)$.

The name progressive (bi-)consistency is justified with the following lemma.

Lemma 1. *Let Snp be a progressively bi-consistent or progressively consistent snapshot algorithm. Let $App = \prod A_i$ be any application. If T is the trace of a snapshot computation of $App \parallel SYS_{Snp}$ and H is the trace of π up to the checkpoints taken, then H is consistent with T.*

Proof. It is sufficient to consider progressively consistent snapshot algorithms, only. Let

$$\pi = A^0 \parallel S^0 \xrightarrow{e_1} A^1 \parallel S^1 \cdots \xrightarrow{e_n} A^n \parallel S^n$$

be a snapshot computation of $App \parallel SYS_{Snp}$, not necessarily completed (where A^0 and S^0 denotes the initial state of App and SYS_{Snp}, respectively). The prefix of $trace(\pi)$ up to the checkpoints taken is defined by $(h_1, \ldots, h_{|I|})$ where h_i is the restriction of $e_0 \ldots e_n$ to events of A_i up to cp_taken_i. If $e_0 \ldots e_n$ does not contain cp_taken_i then it is simply the restriction to events of A_i. Note, that cp_taken_i will not appear in h_i or t_i as it is not an event of A_i but of P_i.

As Snp is progressively consistent there is a freeze simulation R containing $(SYS_{Snp}, FSYS)$. By means of R we inductively built up a computation of $App \parallel FSYS$

$$\pi_{frozen} = A^0 \parallel S^0 \xRightarrow{\hat{e}_1} \hat{A}^1 \parallel \hat{S}^1 \cdots \xRightarrow{\hat{e}_n} \hat{A}^n \parallel \hat{S}^n$$

in which each transition of SYS_{Snp} is matched according to clause 1.(a), 1.(b) or 1.(c) of Definition 4. Let π^j and π_{frozen}^j be the prefix of π up to the jth computation step, and let $T^j = (t_1^j, \ldots, t_{|I|}^j)$ and $H^j = (h_1^j, \ldots, h_{|I|}^j)$ be the traces of π^j and π_{frozen}^j, respectively. In the induction step, the following properties are assumed to hold for π^j and π_{frozen}^j and need to be shown for π^{j+1} and π_{frozen}^{j+1}:

1. $h_i^j = t_i^j$ and $A_i^j = \hat{A}_i^j$ if A_i has not been checkpointed yet,
2. $h_i^j \sqsubseteq t_i^j$ if the checkpoint of A_i has been taken,
3. if $RECEIVE_{ki}(msg)$ is contained in h_i^j then $\overline{SEND_{ki}(msg)}$ is contained in h_k^j,
4. h_i^j is the restriction of $e_0 \ldots e_j$ to events of A_i up to cp_taken_i (if $e_0 \ldots e_j$ does not contain cp_taken_i then it is simply the restriction to events of A_i),
5. $(S^j, S'^j) \in R$.

For S^0 the statement is obviously true.

For the induction step we inspect how π_{frozen}^j was extended to π_{frozen}^{j+1}.

- Case: e_{j+1} is an internal event of some A_i^j.

  If A_i^j has not been checkpointed yet, $\hat{A}_i^j = A_i^j$ by IH and $\hat{A}_i^j$ can perform the same transition. So, $A'^{j+1}_i = A_i^{j+1}$, $\hat{e}_{j+1} = e_{j+1}$ and $h_i^{j+1} = h_i^j.e_i = t_i^j.e_i = t_{i+1}^j$. Property 3. holds as e_{j+1} is an event.

  If A_i^j has been checkpointed then $\hat{e}_{j+1} = \varepsilon$ and $\hat{A}_i^{j+1} = \hat{A}_i^j$ and all the properties are satisfied.

- Case: e_{j+1} is a $\overline{SEND_{ij}(msg)}$ event of some A_i^j.
  This case is similar to the previous case, but this time SYS_{Snp} has to participate with a $send_{ij}(msg)$ transition (as A_i^j performed a $\overline{send_{ij}(msg)}$ event). If A_i has not been checkpointed yet, then F^j can match the move by 1.(a) of Definition 4 and $(S^{j+1}, F^{j+1}) \in R$. So, $\hat{e}_{j+1} = e_{j+1}$ and conditions 1.-3. hold. If A_i has been checkpointed then $\hat{e}_{j+1} = \varepsilon$ and $\hat{A}_i^{j+1} = \hat{A}_i^j$ and all the properties are satisfied.
- Case: e_{j+1} is a $RECEIVE_{ji}(msg)$ event of some A_i.
  This case is similar to the previous case. However, property 3. needs to be verified. It follows from the fact, that π_{frozen}^j is a computation of $App \parallel FSYS$. Messages can only be received if they have earlier been sent.
- Case: e_{j+1} is a τ event of S^j. In this case neither A^j nor the traces get modified, so the properties follow from induction hypothesis and 1.(b) of Definition 4.
- Case: e_{j+1} is cp_taken_i of some P_i^j. In this case traces do not change, as $\hat{e}_{j+1} = frozen_i$. All the properties are satisfied. In particular, note that $h_i^{j+1} = h_i^j = t_i^j = t_i^{j+1}$, so the history of A_i^{j+1} equals that of $\hat{A}_i^{j+i}$ up to the checkpoint.

From conditions 3. and 4. we obtain consistency of H^j with T^j according to Definition 1. □

As a corollary of Lemma 1 we obtain correctness of progressively consistent snapshot algorithms in the classical sense.

Theorem 1. *A snapshot computed by a progressively consistent or progressively bi-consistent snapshot algorithm is consistent.*

Proof. Let $App = \prod A_i$ be any application and let π be a snapshot computation of $App \parallel SYS_{Snp}$ in which every process has taken a checkpoint. The trace of π up to checkpointing is the trace of the computed snapshot. By Lemma 1 it is consistent with the trace of π. □

A snapshot algorithm correct in the classical sense, is not necessarily progressively consistent. A counterexample is Mutable Checkpointing algorithm, see Sect. 4.

For snapshot algorithms it is desirable that they are not inhibiting the application computation. Intuitively, that means that if all actions related to the ongoing checkpointing are made transparent, the snapshot computation should not be distinguishable from an application computation without overlaid snapshot algorithm. This property can naturally be expressed as weak bisimilarity ($\approx$) with $IntSYS$.

Definition 5 (Non-Inhibitance). *A snapshot algorithm Snp is called non-inhibiting if $(SYS_{Snp} \parallel CK) \approx IntSYS$ where $CK = \sum \overline{cp\_taken_i}.CK$.*

Note, that non-inhibitance would simply reduce to bisimilarity with $IntSYS$ if cp_taken_i transitions would be invisible (that is treated as τ transitions).

It is easy to see that a progressively bi-consistent snapshot algorithm does not need to be non-inhibiting. For example, an algorithm may suspend a process after it has taken a checkpoint. Conversely, non-inhibitance does not imply progressive bi-consistency as shown next.

Theorem 2. *Progressive consistency and non-inhibitance do not imply progressive bi-consistency.*

Proof. A counterexample is given by a modification of algorithm CL/LY, the combined algorithms of Chandy-Lamport and Lai-Yang given in Sect. 4.1. The modification concerns the initiating process only of which the propagation of checkpoint requests is done before the checkpoint is taken. With this modification, P_1 can broadcast the checkpoint requests without having taken the initial checkpoint. This move can be matched by $FSYS$, with an ϵ move (doing nothing). But now, $FSYS$ could proceed with a $send_{12}(msg)$ event, which cannot be matched by P_1 without taking the checkpoint prior to that. Hence, the modified algorithm is not progressively bi-consistent. That the modified algorithm is progressively consistent and non-inhibiting can be shown similarly as for the unmodified algorithm. □

Snapshot algorithms, traditionally, take a snapshot of the entire system. More recently, snapshot algorithms have been considered which only take checkpoints of processes relevant to keep consistently of the process initiating the snapshotting, like mutable checkpointing [4,10], blocking queue algorithms [12,18] and partial snapshotting [11]. The motivation behind them is that in a large distributed system, taking an overall snapshot is expensive and not always required. Indeed, the intention is to keep the number of checkpoints as small as possible. A checkpoint needs to be taken if there is a Z-dependency from the process checkpointed to the initiating process. A z-dependency covers causal dependencies and related scenarios which would result in an orphan (a message received but not sent) if the checkpoint is not taken.

A snapshot algorithms is called *minimal* if for every checkpoint, either it is the initiating checkpoint or there is a Z-dependency from the respective process to the initiator. Z-dependencies are induced by the flow of application messages. If a process P_i before taking a checkpoint had received an application message from P_j sent before a checkpoint was taken, then there is a Z-dependency from P_j to P_i. The Z-dependency relation is transitive, see [13] for more details.

Theorem 3. *There is no non-inhibiting, progressively bi-consistent or progressively consistent, minimal snapshot algorithm for systems with at least two processes.*

Proof. It suffices to consider progressive consistency.

Suppose Snp is a non-inhibiting, progressively consistent, minimal snapshot algorithm. We consider a computation up to a state in which P_i has taken a checkpoint and afterwards sent a message to P_j. The computation can progress as described by π:

$$\pi = SYS_{Snp} \overset{\varepsilon}{\Longrightarrow} \xrightarrow{cp\_taken_i} S_1 \overset{\varepsilon}{\Longrightarrow} \xrightarrow{send_{ij}(msg)} S_2$$

The weak ε moves may arise from internal computation steps of Snp. The $send_{ij}(msg)$ event is justified by non-inhibitance of Snp (as $IntSYS$ can perform $send_{ij}(msg)$ as a first event). $FSYS$ has as matching sequence of events only

$$FSYS \xrightarrow{freeze_i} F^1$$

with $(S^1, F^1), (S^2, F^1) \in R$ where R is the relation justifying progressive consistency of Snp. By non-inhibitance of Snp, P_j needs to be able to receive msg. According to algorithm Snp

1. a checkpoint of P_j is taken before receiving msg, or
2. a checkpoint of P_j is not taken before receiving msg.

In the first case, $S^2 \xoverset{\varepsilon}{\Longrightarrow} \xrightarrow{cp\_taken_2} S^3 \xoverset{\varepsilon}{\Longrightarrow} \xrightarrow{\overline{receive_{ij}(msg)}} S^4$. So, $FSYS$ will match the checkpointing with $F^1 \xoverset{\varepsilon}{\Longrightarrow} \xrightarrow{freeze_2} F^2$, $(S^3, F^2) \in R$. However, this means that a checkpoint of P_j was taken without there being a Z-dependency from P_j to P_i, contradicting that Snp is minimal.

In the second case, $S^2 \xoverset{\varepsilon}{\Longrightarrow} S^3 \xrightarrow{receive_{ij}(msg)} S^4$. This can be matched by $FSYS$ only with $F^1 \xoverset{\varepsilon}{\Longrightarrow} \xrightarrow{receive_{ij}(msg)} F^2$. But then the trace of F^2 in the context of a suitable application is $(\varepsilon, RECEIVE_{ij}(msg))$ which is not consistent, contradicting Lemma 1. As in both cases we obtained a contradiction, $FSYS$ cannot match the last move, so Snp cannot be progressively consistent if it is minimal and non-inhibiting. □

Theorem 3 relates to an earlier result in [5] which states that there is no minimal coordinated snapshot algorithm which is not blocking the computation of processes. The possibility of blocking only channels is not considered and not expressible in their linear time framework. Our result applies to a larger class of snapshot algorithms (like communication induced checkpointing) and refines non-blocking to non-inhibiting. Non-inhibitance cannot be expressed in a linear time framework.

4   Snapshot Algorithms Analyzed

For the analysis we translated snapshot algorithms into CCS assuming channels as in Sect. 2. In this way we modeled Chandy/Lamport's algorithm [9], its modification discussed in the proof of Theorem 2, Lai/Yang's algorithm [14], the combined Chandy/Lamport-Lai/Yang algorithm, and Mutable checkpointing [4]. To provide an example for data stream processing, we also analyzed the ABS algorithm of [8].

The results are summarized in Fig. 1. Mutable Checkpointing (column M) is not progressively bi-consistent as this algorithm takes checkpoints on a tentative basis and later possibly discards them (see [10] for details, a concept of unfreezing would be required). We included results on blocking queue algorithms (column B) [12,18] and partial snapshotting (column PS), [11]. These algorithms

property	CL	LY	CL/LY	CL/LYmod.	M	BQ	PS	ABS
prog. consistent	√	√	√	√	-	√	√	√
prog. b-consistent	√	√	√	-	-	√	√	√
non-inhibiting	√	√	√	√	√	–	–	–
minimal	–	–	–	–	√	√	√	–

Fig. 1. Properties of snaphshot algorithms

– additionally to coordination messages and flags– use the concept of message buffering. The algorithms are minimal and progressively consistent, proved in a setting that does not separate application and snapshot computation. By Theorem 3 they are not non-inhibiting. Intuitively, they are not non-inhibiting as both algorithms temporarily buffer messages (that is prevent them from being delivered) after a checkpoint. However, they are progressively bi-consistent stated here without a proof. The ABS algorithm shows similar features.

We show in detail the proofs for the combined Chandy/Lamport-Lai/Yang algorithm and the ABS algorithm. The former exhibits coordination induced and communication induced checkpointing. The latter gives an example for a snapshot algorithm buffering messages temporarily.

4.1   The Combined Algorithm of Chandy/Lamport and Lai/Yang

Algorithm CL/LY is a combination of Chandy/Lamport's and of Lai/Yang's checkpoint algorithms. It combines coordinated checkpointing with communication induced checkpointing. The process initiating the snapshotting (assumed to be P_1) sends out checkpoint requests (that is coordination messages) to all other processes. Having received such a request when a checkpoint had not been taken, a process takes its checkpoint. From then onward it attaches a flag to its messages indicating these have been sent after a checkpoint. If a flagged message is received before a checkpoint had been taken then a checkpoint is taken instantly before the reception of the message (a communication induced checkpoint). The CCS model of this algorithm is given in Fig. 2. The CCS model of the CL/LY algorithm is thus $\prod_{i \in I} P_i$ and superposed an application $\prod_{i \in I} A_i$ within the channels connecting processes it is

$$\left(\prod_{i \in I} (A_i \mid P_i) \setminus L_1 \mid \prod_{i,j \in I, i \neq j} C_{ij}(\varepsilon) \right) \setminus L_2$$

which is bisimilar to

$$\left(\prod_{i \in I} A_i \mid \left(\prod_{i \in I} P_i \mid \prod_{i,j \in I, i \neq j} C_{ij}(\varepsilon) \right) \setminus L_2 \right) \setminus L_1$$

where

$$L_1 = \{ send_{ij}(msg), receive_{ij}(msg) \mid i \neq j, msg \in MSG \}$$
$$L_2 = \{ i\_send_{ij}(m), i\_receive_{ij}(m) \mid i \neq j, m \in MSG_{int} \}.$$

snaphot initiating process:

$$P_1 = \sum_{j\neq 1} send_{1j}(msg).\overline{i\_send_{1j}(msg,0)}.P_1$$
$$+ \sum_{j\neq 1} i\_receive_{j1}(msg,0).\overline{receive_{j1}(msg)}.P_1$$
$$+ \overline{cp\_taken_1}.P_1^{cpr}$$
$$P_1^{cpr} = \overline{i\_send_{1I}(cpr)}.P_1^{cpt}$$
$$P_1^{cpt} = \sum_{j\neq 1} send_{1j}(msg).\overline{i\_send_{1j}(msg,1)}.P_1^{cpt}$$
$$+ \sum_{j\neq 1} i\_receive_{j1}(msg,*).\overline{receive_{j1}(msg)}.P_1^{cpt}$$

non-initiator processes:

$$P_i = \sum_{j\neq i} send_{ij}(msg).\overline{i\_send_{ij}(msg,0)}.P_i$$
$$+ \sum_{j\neq i} i\_receive_{ji}(msg,0).\overline{receive_{ji}(msg)}.P_i$$
$$+ i\_receive_{1i}(cpr).P_i^{cpt} + \sum_{j\neq i} i\_receive_{ji}(msg,1).P_i^{cpt}$$
$$P_i^{cpt} = \sum_{j\neq i} send_{ij}(msg).\overline{i\_send_{ij}(msg,1)}.P_i^{cpt}$$
$$+ \sum_{j\neq i} i\_receive_{ji}(msg,*).\overline{receive_{ji}(msg)}.P_i^{cpt}$$
$$+ i\_receive_{1i}(cpr).P_i^{cpt}$$

where $* \in \{0,1\}$.

Fig. 2. The CCS model of the CL/LY snapshot algorithm

In short, this is $App \parallel SYS_{\mathrm{CL/LY}}$. A communication induced checkpoint is triggered with $i\_receive_{ji}(msg,1)$ executed by P_i, $i \in I$.

Theorem 4. *The combined algorithm of Chandy/Lamport and Lai/Yang is progressively bi-consistent.*

Proof. To set up a freeze bisimulation establishing progressive bi-consistency, we integrate the history into process terms. Transitions of a sequential process $P \xrightarrow{\alpha} P'$, $\alpha \neq \tau$, are replaced by their history extended version $h :: P \xrightarrow{\alpha} h\alpha :: P'$ where h is the history of visible events of P. For channels we keep the original semantics. So for example,
$(U_1|U_2) \parallel (C_{12}(\varepsilon)|C_{21}(\varepsilon))$
$\xrightarrow{send_{12}(msg_1)} (send_{12}(msg_1) :: \overline{i\_send_{12}(msg_1)}.U_1|U_2) \parallel (C_{12}(\varepsilon)|C_{21}(\varepsilon))$
$\xrightarrow{\tau} (send_{12}(msg_1).\overline{i\_send_{12}(msg_1)} :: U_1|U_2) \parallel (C_{12}(msg_1)|C_{21}(\varepsilon))$
$\xrightarrow{\tau} (send_{12}(msg_1).i\_send_{12}(msg_1) :: U_1|i\_receive_{12}(msg_1) :: receive_{12}(msg_1).U_2)$
$(\parallel C_{12}(\varepsilon)|C_{21}(\varepsilon))$
The set-up is similar to the location sensitive semantics underlying location equivalence and related work [3,6]. Let $\mathcal{R}_h(SYS_{\mathrm{CL/LY}})$ denote the set of reachable states with histories. For the freeze bisimulation we construct the pair $\langle [A], \widehat{A} \rangle$ out of $A \in \mathcal{R}_h(SYS_{\mathrm{CL/LY}})$ where $[]$ removes the histories from a term. With $\widehat{}$ processes and channel contents is modified as follows:

Q	$[Q]$	$\widehat{Q}$
$h :: P_1$	P_1	F_1
$h :: \overline{i\_send_{1j}(msg, 0)}.P_1$	$\overline{i\_send_{1j}(msg, 0)}.P_1$	$\overline{i\_send_{1j}(msg)}.F_1$
$h :: receive_{j1}(msg).P_1$	$receive_{j1}(msg).P_1$	$receive_{j1}(msg).F_1$
$h :: P_1^{cpr}$	P_1^{cpr}	$Frozen_1$
$h :: P_1^{cpt}$	P_1^{cpt}	$Frozen_1$
$h :: \overline{i\_send_{1j}(msg, 1)}.P_1^{cpt}$	$\overline{i\_send_{1j}(msg, 1)}.P_1^{cpt}$	$Frozen_1$
$h :: \overline{receive_{j1}(msg)}.P_1^{cpt}$	$\overline{receive_{j1}(msg)}.P_1^{cpt}$	$Frozen_1$
$h :: receive_{j1}(msg, 1).P_1^{cpt}$	$receive_{j1}(msg).P_1^{cpt}$	$Frozen_1$
$h :: P_i$	P_i	U_i
$h :: \overline{i\_send_{ij}(msg, 0)}.P_i$	$\overline{i\_send_{ij}(msg, 0)}.P_i$	$\overline{i\_send_{ij}(msg)}.U_i$
$h :: receive_{ji}(msg).P_i$	$receive_{ji}(msg).P_i$	$receive_{ji}(msg).F_i$
$h :: P_i^{cpt}$	P_i^{cpt}	$Frozen_i$
$h :: \overline{i\_send_{ij}(msg, 1)}.P_i^{cpt}$	$\overline{i\_send_{ij}(msg, 1)}.P_i^{cpt}$	$Frozen_i$
$h :: \overline{receive_{ji}(msg)}.P_i^{cpt}$	$\overline{receive_{ji}(msg)}.P_i^{cpt}$	$Frozen_i$
$h :: \overline{receive_{ji}(msg)}.P_i^{cpt}$	$\overline{receive_{ji}(msg)}.P_i^{cpt}$	$Frozen_i$

In case of channel $C_{ij}(\alpha)$, the modification of α to β is based on the checkpoint status of the two processes which the channel is connecting. In all cases flags and cpr messages must be removed (indicated by delineation in $\|\ \|$). The status is one of the following:

1. If both the processes have not taken the checkpoint, $\beta = \|\alpha\|$.
2. If P_i has not and P_j has taken the checkpoint, then let γ denote the sequence of messages which are received from P_i by P_j after having taken the checkpoint in sending order. Then $\beta = \|\alpha\gamma\|$.
3. If P_i has taken and P_j has not taken the checkpoint, let γ be the largest suffix of α not including cpr or a flagged message. Then $\beta = \|\gamma\|$.
4. If both P_i and P_j have taken the checkpoint then let γ_1 be the largest suffix of α not including cpr or a flagged message, and γ_2 be the messages received by P_j after having taken the checkpoint, sent by P_i before having taken the checkpoint, in sending order. Then $\beta = \|\gamma_1\gamma_2\|$.

Strings γ, γ_1 and γ_2 can be determined from the histories of the concerned processes.

The relation $R = \{\langle [A], \widehat{A} \rangle \mid A \in \mathcal{R}_h(SYS_{\mathrm{CL/LY}})\}$ can then shown to be a freeze bisimulation. For details, confer [16].   □

Theorem 5. *The combined algorithm of Chandy/Lamport and Lai/Yang is non-inhibiting.*

Proof. The relation $R = \{\langle A, A' \rangle \mid A \in Reachable(SYS_{\mathrm{CL/LY}} \parallel CK)\}$ can be shown a weak bisimulation, where $A' = \prod U_i \parallel \prod C_{ij}(\|\alpha_{ij}\|)$ and α_{ij} being the contents of channel C_{ij} in A.   □

4.2   A Snapshot Algorithms for Data Stream Processing

A comparably new application of snapshot algorithms is distributed data stream processing, [7,8]. A data stream is pipelined through a set of tasks which perform simple operations on it (like counting keywords etc.). Snapshots of the tasks are taken not only to recover from failures but also to reallocate tasks or to adjust to reconfigurations.

In this section we model the ABS (Asynchronous Barrier Snapshotting) algorithm of [8]. It assumes an acyclic infrastructure, that is the tasks together with their connecting channels form an acyclic graph. In our model tasks, correspond to application processes and data items to application messages. So, MSG corresponds to $DATA$. Due to the acyclic streaming, a task obtains input from one set of tasks and outputs to another set of tasks, denoted by $Input_i$ and $Output_i$, respectively, for task i. Earlier we assumed full connectivity, that is, $Input_i = Output_i = I \setminus \{i\}$.

The basic idea behind the ABS algorithms is as follows. The snapshotting is started by injecting so-called barriers into the input data streams. Barriers correspond to checkpoint request messages. If a task receives a barrier from one of its input channels, it will block that channel by buffering all data items coming from that channel until it has received a barrier from all its other input channels. If barriers have been received from all input channels, the task is checkpointed and barriers are sent out on all output channels. As the infrastructure is acyclic, the checkpointing moves like a wave through the system.

ABS processes are of the form $\mathrm{ABS}_{i,buffered}^{blocked}$ where $blocked \subseteq I$ denotes the input channels of P_i currently blocked (that is C_{ji} for $j \in blocked$), and $buffered = (\alpha_j)_{j \in Input_i}$, $\alpha_j \in DATA^*$, which gives the contents of the channel buffers. Initially, $blocked = \emptyset$ and $\alpha_j = \varepsilon$ for all $j \in Input_i$. The CCS model of $\mathrm{ABS}_{i,buffered}^{blocked}$ is given in Fig. 3. The updating of a buffer j with contents α is denoted by $buffered[\alpha/\alpha_j]$ (α replaces α_j).

None of the ABS processes is a designated snapshot initiator. However, additionally to the ABS processes, there is a source and a sink process. The source process So creates the data stream which we model by sending data to output channels. The snapshotting will be induced by injecting barriers to the data streams of all channels. The task processes of the data stream processing system are thus reacting to the reception of barriers, only. They will not initiate a snapshotting by themselves.

$$So = \sum_{j \in Output_{So}} send_{So,j}(d).\overline{i\_send}_{So,j}(d).So$$
$$+ \; i\_send_{So,Output_{So}}(b).So^{cpt}$$
$$So^{cpt} = \sum_{j \in Output_{So}} send_{So,j}(d).\overline{i\_send}_{So,j}(d).So^{cpt}$$

The sink process Si absorbs all data items sent through the data processing system. The initial state of the sink process is $Si^{\emptyset}$. With the superscript the task identifiers are recorded from which a barrier has already been received.

$$Si_{buffered}^{blocked} = \sum_{j \notin blocked} i\_receive_{j,Si}(d).\overline{receive_{j,Si}(d)}.Si_{buffered}^{blocked}$$
$$+ \sum_{j \in blocked} i\_receive_{j,Si}(d).Si_{buffered[d.\alpha_j/\alpha_j]}^{blocked}$$
$$+ \sum_{blocked \cup \{j\} \neq OUT} i\_receive_{j,Si}(b).Si_{buffered}^{blocked \cup \{j\}}$$
$$+ \sum_{blocked \cup \{j\} = Input_{Si}} i\_receive_{j,Si}(b).Si_{buffered}^{cpt}$$

$$Si_{buffered}^{cpt} = + \sum_{\alpha_j = \varepsilon} i\_receive_{ji}(d).\overline{receive_{ji}(d)}.Si_{buffered}^{cpt}$$
$$+ \sum_{\alpha_j = \alpha.d} \overline{receive_{ji}(d)}.Si_{buffered[\alpha/\alpha_j]}^{cpt}$$

$$ABS_{i,buffered}^{blocked} = \sum_{j \in Output_i} send_{ij}(d).\overline{i\_send_{ij}(d)}.ABS_{i,buffered}^{blocked}$$
$$+ \sum_{j \notin blocked} i\_receive_{ji}(d).\overline{receive_{ji}(d)}.ABS_{i,buffered}^{blocked}$$
$$+ \sum_{j \in blocked} i\_receive_{ji}(d).ABS_{i,buffered[d.\alpha_j/\alpha_j]}^{blocked}$$
$$+ \sum_{blocked \cup \{j\} \neq Input_i} i\_receive_{ji}(b).ABS_{i,buffered}^{blocked \cup \{j\}}$$
$$+ \sum_{blocked \cup \{j\} = Input_i} i\_receive_{ji}(b).ABS_{i,buffered}^{cp}$$

$$ABS_{i,buffered}^{cp} = \overline{i\_send_{iOutput_i}(b)}.ABS_{i,buffered}^{cpt}$$

$$ABS_{i,buffered}^{cpt} = \sum_{j \in Output_i} send_{ij}(d).\overline{i\_send_{ij}(d)}.ABS_{i,buffered}^{cpt}$$
$$+ \sum_{\alpha_j = \varepsilon} i\_receive_{ji}(d).\overline{receive_{ji}(d)}.ABS_{i,buffered}^{cpt}$$
$$+ \sum_{\alpha_j = \alpha.d} \overline{receive_{ji}(d)}.ABS_{i,buffered[\alpha/\alpha_j]}^{cpt}$$

Fig. 3. The CCS model of the ABS algorithm for snapshotting data stream processing

The entire system is thus

$$SYS_{ABS} = (So \mid (\prod_{i \in I} ABS_{i,init}^{\emptyset}) \mid Si) \parallel Channels$$

where *init* is the initial buffered channel contents which is empty. Superposed a task system it is

$$\left((So \mid (\prod_{i \in I} T_i \mid ABS_{i,init}^{\emptyset}) \mid Si) \setminus L_1 \mid \prod_{i,j \in I, i \neq j} C_{ij}(\varepsilon) \right) \setminus L_2$$

SYS_{ABS} is bisimilar to $Task \parallel SYS_{ABS}$ where $Task = \prod_{i \in I} T_i$. A checkpoint is triggered with $i\_receive_{ji}(b)$ by $ABS_{i,buffered}^{blocked}$ if C_{ji} is the last input channel to deliver a barrier ($blocked \cup \{j\} = Input_i$).

To establish progressive bi-consistency of ABS, we need to adapt the freeze system for data stream processing. What is required is the addition of a source and a sink process:

$$F_{So} = \sum_{j \in Output_{So}} send_{So,j}(d).\overline{i\_send}_{So,j}(d).F_{So}$$
$$+ freeze_{So}.Frozen_{So}$$

$$F_{Si} = \sum_{j \in Input_{Si}} i\_receive_{j,Si}(d).\overline{receive}_{j,Si}(d).F_{Si}$$
$$+ freeze_{Si}.F_{Si}$$

So the freeze system for data processing is $FSYS_{DS} = (F_{So} \mid \prod_{i \in I} F_i \mid F_{Si}) \parallel$ Channels.

Theorem 6. *The ABS snapshot algorithm for data stream processing is progressively bi-consistent.*

Proof. We define a freeze simulation containing $(SYS_{ABS}, FSYS_{DS})$. This proves progressive consistency of ABS.

For the freeze simulation we construct the pair $\langle A, \widehat{A} \rangle$ out of $A \in \mathcal{R}(SYS_{ABS})$ where $\mathcal{R}(SYS_{ABS})$ is the set of reachable states of SYS_{ABS}. With $\frown$ processes and channel contents are modified as follows:

Q	$\widehat{Q}$
So	F_{So}
$\overline{i\_send}_{So,j}(d).So$	$\overline{i\_send}_{So,j}(d).F_{So}$
So^{cpt}	$Frozen_{So}$
$\overline{i\_send}_{So,j}(d).So^{cpt}$	$Frozen_{So}$
$Si^{blocked}_{buffered}$	F_{Si}
$\overline{receive}_{j,Si}(d).Si^{blocked}_{buffered}$	$\overline{receive}_{j,Si}(d).F_{Si}$
$Si^{cpt}_{buffered}$	$Frozen_{Si}$
$\overline{receive}_{j,Si}(d).Si^{cpt}_{buffered}$	$Frozen_{Si}$
$ABS^{blocked}_{ibuffered}$	F_i
$\overline{receive}_{ji}(d).ABS^{blocked}_{i,buffered}$	$\overline{receive}_{ji}(d).F_i$
$\overline{i\_send}_{ij}(d).ABS^{blocked}_{i,buffered}$	$\overline{i\_send}_{ij}(d).F_i$
$ABS^{cp}_{ibuffered}$	$Frozen_i$
$\overline{receive}_{ji}(d).ABS^{cpt}_{i,buffered}$	$Frozen_i$
$\overline{i\_send}_{ij}(d).ABS^{cpt}_{i,buffered}$	$Frozen_i$

In case of channel $C_{ij}(\alpha)$, the modification of α to β is based on the checkpoint status of the tasks which the channel is connecting. Recall, that for ABS we assume that the communication graph is acyclic. So if we consider the channel from process i to process j, and process i has not been checkpointed, then task j cannot have been checkpointed either nor can channel C_{ij} be blocked. ABS_i has taken a checkpoint if it is in a state with superscript cp or cpt.

For the remaining cases, the following states need to be considered.

1. If both processes have not taken a checkpoint, $\beta = \alpha$.
2. If ABS_i has taken a checkpoint but ABS_j has not, ABS_j is in a state of the form $\text{ABS}_{i,buffered}^{blocked}$. If $i \notin blocked$, β is the largest suffix of α without a barrier. If $i \in blocked$, $\beta = \varepsilon$.
3. If both, ABS_i and ABS_j have taken a checkpoint then $\beta = \varepsilon$.

The relation $R = \{\langle A, \widehat{A} \rangle \mid A \in \mathcal{R}(SYS_{\text{ABS}})\}$ is a freeze simulation.

5   Conclusions

With progressive bi-consistency we have proposed a refined correctness notion for snapshot algorithms. Apart from ensuring consistency of the computed snapshot, it also takes into account the partially computed snapshot and guarantees consistency in the context of the ongoing computation at any point of the computation. In preliminary work and a different setting, two particular algorithms –partial snapshotting [11] and a blocking queue algorithm [12]– have been shown progressively consistent. However, the notions given in this paper is detached from any particular snapshot algorithm and uses the well-established concepts of bisimilarity and observation equivalence [15]. As a novel feature, it allows to check whether a superposed snapshot algorithm preserves and does not inhibit the behavior of the underlying computation. So it can detect whether the course of the underlying computation is forced into a particular direction once the checkpointing has started (eg by disabling certain communications). This is a branching time property which cannot be expressed in the traditional, linear time correctness notion.

We further established the fundamental result that it is impossible to have a progressively (bi-)consistent and non-inhibiting partial snapshot algorithm which takes a minimal number of checkpoints. This result makes more precise an earlier statement in [5] which says that a minimal partial snapshot cannot be computed without *blocking* processes. As shown here, *blocking* must be understood as blocking channels or –but not necessarily– suspending processes. Existing minimal partial snapshot algorithms are either not progressively consistent (mutable checkpointing [4]) or they are inhibiting (blocking queue algorithms [12,18], partial snapshotting [11]).

We tested our definitions by establishing for well-known existing algorithms whether they are progressively (bi-)consistent and non-inhibiting or not. The results add to the understanding of these algorithms and show the applicability of our notion for snapshot algorithms using features like coordination message, communication induced checkpointing or message buffering.

From a broader point of view, we have shown how algorithms superposed on other algorithms can formally be verified using well-established notions of concurrency theory. We hope that this will benefit formal verification techniques for such algorithms. An example for a related verification can be found in [1].

References

1. Andriamiarina, M.B., Mery, D., Singh, N.K.: Revisiting snapshot algorithms by refinement-based techniques. Comput. Sci. Inf. Syst. **11**, 251–270 (2014)
2. Babaoglu, Ö., Marzullo, K.: Consistent global states of distributed systems: fundamental concepts and mechanisms. Distrib. Syst. **53** (1993)
3. Boudol, G., Castellani, I., Hennessy, M., Kiehn, A.: Observing localities. Theor. Comput. Sci. **114**(1), 31–61 (1993)
4. Cao, G., Singhal, M.: Checkpointing with mutable checkpoints. Theor. Comput. Sci. **290**(2), 1127–1148 (2003)
5. Cao, G., Singhal, M.: On coordinated checkpointing in distributed systems. IEEE Trans. Parallel Distrib. Syst. **9**(12), 1213–1225 (1998)
6. Castellani, I.: Process algebras with localities. In: Handbook of Process Algebra. North-Holland (2001)
7. Carbone, P., Ewen, S., Fora, G., Heif, S., Richter, S., Tzoumas, K.: State management in Apache Flink - consistent stateful distributed stream processing. VLDB Endowment **10**(12), 1718–1729 (2017)
8. Carbone, P., Fora, G., Ewen, S., Haridi, S., Tzoumas, K.: Lightweight asynchronous snapshots for distributed dataflows. arXiv preprint arXiv:1506.08603 (2015)
9. Chandy, K.M., Lamport, L.: Distributed snapshots: determining global states of distributed systems. ACM Trans. Comput. Syst. (TOCS) **3**(1), 63–75 (1985)
10. Kiehn, A., Aggarwal, D.: A study of mutable checkpointing and related algorithms. Sci. Comput. Program. **160**, 78–92 (2018)
11. Kiehn, A., Mittal, A.: Partial snapshotting: checkpoint dissemination and termination. Technical report TR-IITMANDI-2018-1, IIT Mandi (2018)
12. Kiehn, A., Raj, P., Singh, P.: A causal checkpointing algorithm for mobile computing environments. In: Chatterjee, M., Cao, J., Kothapalli, K., Rajsbaum, S. (eds.) ICDCN 2014. LNCS, vol. 8314, pp. 134–148. Springer, Heidelberg (2014). https://doi.org/10.1007/978-3-642-45249-9_9
13. Kshemkalyani, A.D., Singhal, M.: Distributed Computing: Principles, Algorithms, and Systems. Cambridge University Press, Cambridge (2008)
14. Lai, T.H., Yang, T.H.: On distributed snapshots. Inf. Process. Lett. **25**, 153–158 (1987)
15. Milner, R.: Communication and Control. Prentice Hall, Upper Saddle River (1989)
16. Pattathurajan, M.: Expresssing linear-time properties of distributed algorithms in a branching-time setting. MSc thesis, IIT Mandi (2018)
17. Raynal, M.: Distributed Algorithms for Message-passing Systems. Springer, Heidelberg (2013). https://doi.org/10.1007/978-3-642-38123-2
18. Singh, P., Cabillic, G.: A checkpointing algorithm for mobile computing environment. In: Conti, M., Giordano, S., Gregori, E., Olariu, S. (eds.) PWC 2003. LNCS, vol. 2775, pp. 65–74. Springer, Heidelberg (2003). https://doi.org/10.1007/078-3-540-39867-7_6

Finite Choice, Convex Choice and Sorting

Takayuki Kihara[1] and Arno Pauly[2,3](✉) (iD)

[1] Department of Mathematical Informatics, Nagoya University, Nagoya, Japan
kihara@i.nagoya-u.ac.jp
[2] Department of Computer Science, Swansea University, Swansea, UK
Arno.M.Pauly@gmail.com
[3] Department of Computer Science, University of Birmingham, Birmingham, UK

Abstract. We study the Weihrauch degrees of closed choice for finite sets, closed choice for convex sets and sorting infinite sequences over finite alphabets. Our main result is that choice for finite sets of cardinality $i + 1$ is reducible to choice for convex sets in dimension j, which in turn is reducible to sorting infinite sequences over an alphabet of size $k + 1$, iff $i \leq j \leq k$. Our proofs invoke Kleene's recursion theorem, and we describe in some detail how Kleene's recursion theorem gives rise to a technique for proving separations of Weihrauch degrees.

Keywords: Computable analysis · Weihrauch reducibility ·
Closed choice

1 Introduction

The Weihrauch degrees are the degrees of non-computability for problems in computable analysis. In the wake of work by Brattka, Gherardi, Marcone and Pauly [3,4,16,24] they have become a very active research area in the past decade. A recent survey is found as [7].

We study the Weihrauch degrees of closed choice for finite sets, closed choice for convex sets and sorting infinite sequences over finite alphabets. The closed choice operators have turned out to be a useful scaffolding in that structure: We often classify interesting operations (for example linked to existence theorems) as being equivalent to a choice operator, and then prove separations for the choice operators, as they are particularly amenable for many proof techniques. Examples of this are found in [2,3,5,6,9–11,17,21]. Convex choice in particular captures the degree of non-computability of finding fixed points of non-expansive mappings via the Goehde-Browder-Kirk fixed point theorem [21].

Kihara's research was partially supported by JSPS KAKENHI Grant 17H06738, 15H03634, and the JSPS Core-to-Core Program (A. Advanced Research Networks). This project has received funding from the European Union's Horizon 2020 research and innovation programme under the Marie Skłodowska-Curie grant agreement No. 731143, *Computing with Infinite Data*.

T. V. Gopal and J. Watada (Eds.): TAMC 2019, LNCS 11436, pp. 378–393, 2019.
https://doi.org/10.1007/978-3-030-14812-6_23

The present article is a continuation of [20] by Le Roux and P., which already obtained some results on the connections between closed choice for convex sets and closed choice for finite sets. We introduce new proof techniques and explore the connection to the degree of sorting infinite sequences. Besides laying the foundations for future investigations of specific theorems, we are also addressing a question on the complexity caused by dimension: Researchers have often wondered whether there is a connection between the dimension of the ambient space and the complexity of certain choice principles. An initial candidate was to explore closed choice for connected subsets, but it turned out that the degree is independent of the dimension, provided this is at least 2 [10]. As already shown in [20], this works for convex choice. One reason for this was already revealed in [20]: We need n dimensions in order to encode a set of cardinality $n+1$. We add another reason here: Each dimension requires a separate instance of sorting an infinite binary sequence in order to find a point in a convex set.

Structure of the Paper. Most of our results are summarized in Fig. 1 on Page 5. Section 2 provides a brief introduction to Weihrauch reducibility. In Sect. 3 we provide formal definitions of the principles under investigation, and give a bit more context. We proceed to introduce our new technique to prove separations between Weihrauch degrees in Sect. 4; it is based on Kleene's recursion theorem. The degree of sorting an infinite binary sequence is studied in Sect. 5, including a separation technique adapted specifically for this in Subsect. 5.1, its connection to convex choice in Subsect. 5.2 and a digression on the task of finding connected components of countable graphs in Subsect. 5.3. Section 6 is constituted by Theorem 5 and its proof, establishing the precise relationship between finite choice and sorting. Finally, in Sect. 7 we introduce a game characterizing reducibility between finite choice for varying cardinalities.

2   Background on Weihrauch Reducibility

Weihrauch reducibility is a quasiorder defined on multi-valued functions between represented spaces. We only give the core definitions here, and refer to [25] for a more in-depth treatment. Other sources for computable analysis are [8,29].

Definition 1. *A represented space* $\mathbf{X}$ *is a set* X *together with a partial surjection* $\delta_{\mathbf{X}} :\subseteq \mathbb{N}^{\mathbb{N}} \to X$.

A partial function $F :\subseteq \mathbb{N}^{\mathbb{N}} \to \mathbb{N}^{\mathbb{N}}$ is called a *realizer* of a function $f :\subseteq \mathbf{X} \to \mathbf{Y}$ between represented spaces, if $f(\delta_{\mathbf{X}}(p)) = \delta_{\mathbf{Y}}(F(p))$ holds for all $p \in \mathrm{dom}(f \circ \delta_{\mathbf{X}})$. We denote F being an realizer of f by $F \vdash f$. We then call $f :\subseteq \mathbf{X} \to \mathbf{Y}$ *computable* (respectively *continuous*), iff it has a computable (respectively continuous) realizer.

Represented spaces can adequately model most spaces of interest in *everyday mathematics*. For our purposes, we are primarily interested in the construction of the hyperspace of closed subsets of a given space.

The category of represented spaces and continuous functions is cartesian-closed, by virtue of the UTM-theorem. Thus, for any two represented spaces

X, **Y** we have a represented spaces $\mathcal{C}(\mathbf{X}, \mathbf{Y})$ of continuous functions from **X** to **Y**. The expected operations involving $\mathcal{C}(\mathbf{X}, \mathbf{Y})$ (evaluation, composition, (un)currying) are all computable.

Using the Sierpiński space $\mathbb{S}$ with underlying set $\{\top, \bot\}$ and representation $\delta_{\mathbb{S}} : \mathbb{N}^{\mathbb{N}} \rightarrow \{\top, \bot\}$ defined via $\delta_{\mathbb{S}}(\bot)^{-1} = \{0^{\omega}\}$, we can then define the represented space $\mathcal{O}(\mathbf{X})$ of *open* subsets of **X** by identifying a subset of **X** with its (continuous) characteristic function into $\mathbb{S}$. Since countable *or* and binary *and* on $\mathbb{S}$ are computable, so are countable union and binary intersection of open sets.

The space $\mathcal{A}(\mathbf{X})$ of closed subsets is obtained by taking formal complements, i.e. the names for $A \in \mathcal{A}(\mathbf{X})$ are the same as the names of $X \setminus A \in \mathcal{O}(\mathbf{X})$ (i.e. we are using the negative information representation). Intuitively, this means that when reading a name for a closed set, this can always shrink later on, but never grow. It is often very convenient that we can alternatively view $A \in \mathcal{A}(\{0,1\}^{\mathbb{N}})$ as being represented by some tree T via $[T] = A$ (here $[T]$ denotes the set of infinite paths through T).

We can now define Weihrauch reducibility. Again, we give a very brief treatment here, and refer to [7] for more details and references.

Definition 2 (Weihrauch reducibility). *Let f, g be multivalued functions on represented spaces. Then f is said to be* Weihrauch reducible *to g, in symbols $f \leq_W g$, if there are computable functions $K, H :\subseteq \mathbb{N}^{\mathbb{N}} \rightarrow \mathbb{N}^{\mathbb{N}}$ such that $(p \mapsto K\langle p, GH(p)\rangle) \vdash f$ for all $G \vdash g$.*

The Weihrauch degrees (i.e. equivalence classes of $\leq_W$) form a distributive lattice, but we will not need the lattice operations in this paper. Instead, we use two kinds of products. The usual cartesian product induces an operation $\times$ on Weihrauch degrees. We write f^k for the k-fold cartesian product with itself. The compositional product $f \star g$ satisfies that

$$f \star g \equiv_W \max_{\leq_W} \{f_1 \circ g_1 \mid f_1 \leq_W f \wedge g_1 \leq_W g\}$$

and thus is the hardest problem that can be realized using first g, then something computable, and finally f. The existence of the maximum is shown in [12] via an explicit construction, which is relevant in some proofs. Both products as well as the lattice-join can be interpreted as logical *and*, albeit with very different properties.

We'll briefly mention a further unary operation on Weihrauch degrees, the finite parallelization f^*. This has as input a finite tuple of instances to f and needs to solve all of them.

As mentioned in the introduction, the closed choice principles are valuable benchmark degrees in the Weihrauch lattice:

Definition 3. *For a represented space **X**, the closed choice principle $C_{\mathbf{X}} :\subseteq \mathcal{A}(\mathbf{X}) \rightrightarrows \mathbf{X}$ takes as input a non-empty closed subset A of **X** and outputs some point $x \in A$.*

3   The Principles Under Investigation

We proceed to give formal definitions of the three problems our investigation is focused on. These are *finite choice*, the task of selecting a point from a closed subset (of $\{0,1\}^{\mathbb{N}}$ or $[0,1]^n$) which is guaranteed to have either exactly or no more than k elements; *convex choice*, the task of selecting a point from a convex closed subset of $[0,1]^k$; and *sorting* an infinite sequence over the alphabet $\{0,1,\ldots,k\}$ in increasing order. Our main result is each task forms a strictly increasing chain in the parameter k, and these chains are perfectly aligned as depicted in Fig. 1. For finite choice and convex choice, this was already established in [20]. Our Theorem 5 implies the main theorem from [20] with a very different proof technique.

Definition 4 ([20, Definition 7]). *For a represented space* $\mathbf{X}$ *and* $1 \leq n \in \mathbb{N}$, *let* $C_{\mathbf{X}, \sharp=n} := C_{\mathbf{X}}|_{\{A \in \mathcal{A}(\mathbf{X}) \mid |A|=n\}}$ *and* $C_{\mathbf{X}, \sharp\leq n} := C_{\mathbf{X}}|_{\{A \in \mathcal{A}(\mathbf{X}) \mid 1 \leq |A| \leq n\}}$.

It was shown as [20, Corollary 10] that for every computably compact computably rich computable metric space $\mathbf{X}$ we find $C_{\mathbf{X}, \sharp=n} \equiv_{\mathrm{W}} C_{\{0,1\}^{\mathbb{N}}, \sharp=n}$ and $C_{\mathbf{X}, \sharp\leq n} \equiv_{\mathrm{W}} C_{\{0,1\}^{\mathbb{N}}, \sharp\leq n}$. This in particular applies to $\mathbf{X} = [0,1]^d$. We denote this Weihrauch degree by $C_{\sharp=n}$ respectively $C_{\sharp\leq n}$.

Definition 5 ([20, Definition 8]). *By* XC_n *we denote the restriction of* $C_{[0,1]^n}$ *to convex sets.*

Since for subsets of $[0,1]$ being an interval, being convex and being connected all coincide, we find that XC_1 is the same thing as one-dimensional connected choice CC_1 as studied in [10] and as interval choice C_{I} as studied in [3].

Definition 6. *Let* $\mathrm{Sort}_d : d^{\omega} \to d^{\omega}$ *be defined by* $\mathrm{Sort}_d(p) = 0^{c_0} 1^{c_1} \ldots k^{\infty}$, *where* $|\{n \mid p(n) = 0\}| = c_0$, $|\{n \mid p(n) = 1\}| = c_1$, *etc, and* k *is the least such that* $|\{n \mid p(n) = k\}| = \infty$. *We write just* Sort *for* Sort_2.

Sort was introduced and studied in [22], and then generalized to Sort_k in [9]. Note that the principle just is about sorting a sequence in order without removing duplicates. In [26] it is shown that $\mathrm{Sort}_{n+1} \equiv_{\mathrm{W}} \mathrm{Sort}^n$; it follows that $\mathrm{Sort}^* \equiv_{\mathrm{W}} \mathrm{Sort}_d^* \equiv_{\mathrm{W}} \coprod_{d \in \mathbb{N}} \mathrm{Sort}_d$. The degree Sort^* was shown in [22] to capture the strength of the strongly analytic machines [13,15], which in turn are an extension of the BSS-machines [1]. Sort is equivalent to Thomae's function; and to the translation of the standard representation of the reals into the continued fraction representation [28]. In [17], Sort is shown to be equivalent to certain projection operators.

There are some additional Weihrauch problems we make passing reference to. *All-or-unique choice* captures the idea of a problem either having a unique solution, or being completely undetermined:

Definition 7. *Let* $\mathrm{AoUC}_{\mathbf{X}}$ *be the restriction of* $C_{\mathbf{X}}$ *to* $\{\{x\} \mid x \in \mathbf{X}\} \cup \{\mathbf{X}\}$.

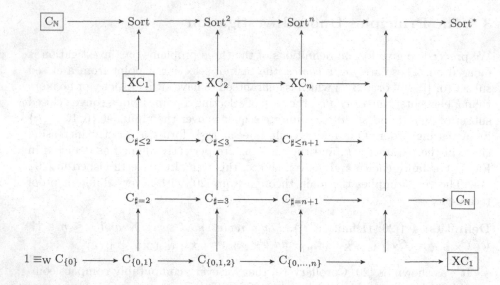

Fig. 1. Overview of our results; extending [20, Fig. 1] by the top row. The diagram depicts all Weihrauch reductions between the stated principles up to transitivity. Boxes mark degrees appearing in two places in the diagram. Our additional results are provided as Theorems 3 and 5.

A prototypical example (which is equivalent to the full problem) is solving $ax = b$ over $[0,1]$ with $0 \leq b \leq a$: Either there is the unique solution $\frac{b}{a}$, or $b = a = 0$, and any $x \in [0,1]$ will do. The degree of $\mathrm{AoUC}_{\mathbf{X}}$ is the same for any computably compact computably rich computable metric space, in particular for $\mathbf{X} = \{0,1\}^{\mathbb{N}}$ or $\mathbf{X} = [0,1]^d$. We just write AoUC for that degree. This problem was studied in [19,23] where it is shown that AoUC^* is the degree of finding Nash equilibria in bimatrix games and of executing Gaussian elimination.

4   Proving Separations via the Recursion Theorem

A core technique we use to prove our separation results invokes Kleene's recursion theorem in order to let us prove a separation result by proving computability of a certain map (rather than having to show that no computable maps can witness a reduction). We had already used this technique in [19], but without describing it explicitly. Since the technique has proven very useful, we formally state the argument here as Theorem 1 after introducing the necessary concepts to formulate it.

Definition 8. *A representation δ of $\mathbf{X}$ is* precomplete, *if every computable partial $f :\subseteq 2^{\omega} \to \mathbf{X}$ extends to a computable total $F : 2^{\omega} \to \mathbf{X}$.*

Proposition 1. *For effectively countably-based $\mathbf{X}$, the space $\mathcal{O}(\mathbf{X})$ (and hence $\mathcal{A}(\mathbf{X})$) is precomplete.*

Proof. It suffices to show this for $\mathcal{O}(\mathbb{N})$, where it just follows from the fact that we can delay providing additional information about a set as long as we want; and will obtain a valid name even if no additional information is forthcoming.

The preceding proposition is a special case of [27, Theorem 6.5], which shows that many pointclasses have precomplete representations.

Proposition 2. *The subspaces of $\mathcal{A}([0,1]^n)$ consisting of the connected respectively the convex subsets are computable multi-valued retracts, and hence precomplete.*

Proof. For the connected sets, this follows from [10, Proposition 3.4]; for convex subsets this follows from computability of the convex hull operation on $[0,1]^n$, see e.g. [20, Proposition 1.5] or [30].

By $\mathcal{M}(\mathbf{X}, \mathbf{Y})$ we denote the represented space of strongly continuous multi-valued functions from $\mathbf{X}$ to $\mathbf{Y}$ studied in [12]. The precise definition of strong continuity is irrelevant for us, we only need every partial continuous function on $\{0,1\}^{\mathbb{N}}$ induces a minimal strongly continuous multivalued function that it is a realizer of; and conversely, every strongly continuous multivalued function is given by a continuous partial realizer.

Theorem 1. *Let $\mathbf{X}$ have a total precomplete representation. Let $f : \mathbf{X} \rightrightarrows \mathbf{Y}$ and $g : \mathbf{U} \rightrightarrows \mathbf{V}$ be such that there exists a computable $e : \mathbf{U} \times \mathcal{M}(\mathbf{V}, \mathbf{Y}) \rightrightarrows \mathbf{X}$ such that if $x \in e(u, k)$ and $v \in g(u)$, then $k(v) \not\subseteq f(x)$. Then $f \not\leq_W g$.*

Proof. Assume that $f \leq_W g$ via computable H, K. Let computable E be a realizer of e. Let $(\phi_n :\subseteq \mathbb{N} \to \mathbb{N})_{n\in\mathbb{N}}$ be a standard enumeration of the partial computable functions. By assumption, we can consider each ϕ_n to denote some element in $\mathbf{X}$. Let λ be a computable function such that $\phi_{\lambda(n)} = E(H(\phi_n), (v \mapsto K(\phi_n, v)))$. By Kleene's recursion theorem, there is some n_0 with $\phi_{n_0} = \phi_{\lambda(n_0)}$. Inputting ϕ_{n_0} to f fails the assumed reduction witnesses.

As simple sample application for how to prove separations of Weihrauch degrees via the recursion theorem, we shall point out that XC_1 already cannot solve some simple products. For contrast, however, note that $\mathrm{C}_2^* \leq_W \mathrm{XC}_1$ was shown as [10, Proposition 9.2].

Theorem 2. $\mathrm{C}_2 \times \mathrm{AoUC} \not\leq_W \mathrm{XC}_1$.

Proof. Given a convex tree $T \subseteq 2^{<\omega}$ and a partial continuous function $\psi :\subseteq \{0,1\}^{\mathbb{N}} \to 2 \times \{0,1\}^{\mathbb{N}}$, we compute set $S \in \mathcal{A}(\{0,1\})$ and $V \in \mathcal{A}(\{0,1\}^{\mathbb{N}})$ such that $S \neq \emptyset$, and $V = \{0,1\}^{\mathbb{N}}$ or $V = \{p\}$ for some $p \in \{0,1\}^{\mathbb{N}}$. Our construction ensures that $\exists p \in [T] \; \phi(p) \notin S \times V$.

Initially, $S = \{0,1\}$ and $V = \{0,1\}^{\mathbb{N}}$.

We first search for s such that for any $\sigma \in T$ of length s, the first value of $\phi(\sigma)$ is determined. If we never find one, then $S = \{0,1\}$ and $V = \{0,1\}^{\mathbb{N}}$ work as desired.

Next, we search for some $\tau \in \{0,1\}^s$ such that $P_\tau := [T] \cap \bigcup_{j<2} \phi^{-1}(j, [\tau])$ is such that any interval contained in P_τ is contained in some $[\sigma]$ for $\sigma \in \{0,1\}^s$. Note that if $(J_i)_{i \in I}$ is a collection of pairwise disjoint intervals in $\{0,1\}^{\mathbb{N}}$ such that every J_i intersects with at least two cylinders $[\sigma]$ and $[\sigma']$ for some strings $\sigma \neq \sigma'$ of length s, then the size of I is at most $2^s - 1$. Hence, if ϕ is defined on $[T]$, such a τ has to exist. Once we have found it, we set $V = \{\tau 0^\omega\}$.

Either we are already done (since we would have that $\exists p \in [T]\ \phi(p) \notin S \times V$), or it holds that $[T] \subseteq [\sigma]$ for some $\sigma \in \{0,1\}^s$. In that case, by choice of s we find that $\exists j \in \{0,1\}\ \pi_0 \phi(p) = j$ for all $p \in [T]$. We can set $S = \{1 - j\}$, and have obtained the desired property that $\exists p \in [T]\ \phi(p) \notin S \times V$. By Theorem 1, the claim follows.

5   Some Observations on Sort

5.1   Displacement Principle for $Sort_k$

The basic phenomenon that the number of parallel copies of Sort being used corresponds to a dimensional feature can already by a result similar in feature to the displacement principle from [10]:

Proposition 3. $C_2 \times f \leq_W Sort_{k+1}$ *implies* $f \leq_W Sort_k \times C_{\mathbb{N}}$.

Proof. Let the reduction $C_2 \times f \leq_W Sort_{k+1}$ be witnessed by computable H, K_1, K_2. Assume, for the sake of a contradiction, that for some input x to f and a name p for $\{0,1\}$ it holds that $H(p,x)$ contains infinitely many 0s. In that case, $Sort_k(H(p,x)) = 0^\omega$, and hence K_1 is defined as either 0 or 1 on $p, x, 0^\omega$. But then there is some $k \in \mathbb{N}$ such that K_1 already outputs the answer on reading some prefix $p_{\leq k}, x_{\leq k}, 0^k$. Additionally, we can chose some $k' \geq k$ such that H writes at least k' 0s upon reading the prefixes $p_{\leq k'}, x_{\leq k'}$. By changing p after k' to be a name of $\{1 - K_1(p, x, 0^\omega)\}$ shows the contradiction.

Now we note that $x \mapsto H(p,x)$ and K_2 witness a reduction from f to the restriction of $Sort_{k+1}$ to inputs containing only finitely many 0s. But this restriction is reducible to $Sort_k \times C_{\mathbb{N}}$: In parallel, call $Sort_k$ on the sequence obtained by skipping 0s and decrementing every other digit by 1, and using $C_{\mathbb{N}}$ to determine the original number of 0s.

Corollary 1. *Let f be a closed fractal. Then $C_2 \times f \leq_W Sort_{k+1}$ implies $f \leq_W Sort_k$.*

Corollary 2. $C_2 \times C_{\sharp \leq 2}^n \not\leq_W Sort_{n+1}$.

Corollary 3. $C_2 \times XC_1^n \not\leq_W Sort_{n+1}$.

We also get an alternative proof of the following, which was previously shown in [22] using the squashing principle from [14]:

Corollary 4. $Sort_{k+1} \not\leq_W Sort_k$.

5.2   Sort and Convex Choice

The one-dimensional case of the following theorem was already proven as [9, Proposition 16]:

Theorem 3. $XC_n \leq_W \mathrm{Sort}_{n+1}$.

Proof. Let $(H_i^d)_{i \in \mathbb{N}}$ be an effective enumeration of the d-dimensional rational hyperplanes for each $d \leq n - 1$. Given $A \in \mathcal{A}([0, 1]^n)$, we can recognize that $A \cap H_i^d = \emptyset$ by compactness of $[0, 1]^n$. We proceed to compute an input p to Sort_{n+1} as follows:

We work in stages $(\ell_0, \ldots, \ell_{n-1})$. We simultaneously test whether $A \cap H_{\ell_0}^{n-1} = \emptyset$, whether $A \cap H_{\ell_0}^{n-1} \cap H_{\ell_1}^{n-2} = \emptyset$, ..., and whether $A \cap H_{\ell_0}^{n-1} \cap \ldots \cap H_{\ell_{n-1}}^{1} = \emptyset$.

If we find a confirmation for a query involving ℓ_k as the largest index, we write a k to p, increment ℓ_k by 1, and reset any ℓ_i for $i > k$. All tests of smaller indices are continued (and hence will eventually fire if true before a largest index test interferes). In addition, we write ns to p all the time to ensure an infinite result.

Now consider the output $\mathrm{Sort}_{n+1}(p)$. If this is 0^ω, then A does not intersect any $n - 1$-dimensional rational hyperplane at all. As a convex set, A has to be a singleton. Thus, as long as we read 0s from $\mathrm{Sort}_{n+1}(p)$, we can just wait until A shrinks sufficiently to produce the next output approximation. If we ever read a 1 in $\mathrm{Sort}_{n+1}(p)$ at position t, we have thus found a $n - 1$-dimensional hyperplane H_t^{n-1} intersecting A. We can compute $A \cap H_t^{n-1} \in \mathcal{A}([0, 1]^n)$, and proceed to work with that set. By retracing the computation leading up to the observation that $A \cap H_{t-1}^{n-1} = \emptyset$, we can find out how many larger-index tests were successful before that. We disregard their impact on $\mathrm{Sort}_{n+1}(p)$. Now as long as we keep reading 1s, we know that $A \cap H_{t-1}^{n-1}$ is not intersecting $n - 2$-dimensional rational hyperplanes (and hence could be singleton). Finding a 2 means we have identified a $n - 2$-dimensional hyperplane $H_{t'}^{n-2}$ intersecting $A \cap H_{t-1}^{n-1}$, and we proceed to work with $A \cap H_{t-1}^{n-1} \cap H_{t'}^{n-2}$. Continuing this process, we always find that either our set has been collapsed to singleton (from which we can extract the point), or we will be able to reduce its dimension further (which can happen only finitely many times). □

5.3   A Digression: Sort and Finding Connected Components of a Graph

On a side note, we explore how Sort relate to the problem FCC of finding a connected component of a countable graph with only finitely many connected components. Here the graph (V, E) is given via the characteristic functions of $V \subseteq \mathbb{N}$ and $E \subseteq \mathbb{N} \times \mathbb{N}$, and the connected component is to be produced likewise as its characteristic function. In addition, we have available to us an upper bound for the number of its connected components. In the reverse math context, this problem was studied in [18] and shown to be equivalent to Σ_2^0-induction.

Theorem 4. *The following are equivalent:*

1. *FCC*
2. *Sort**

Proof. **FCC $\leq_{\mathbf{W}} \bigsqcup_{k \in \mathbb{N}} \mathrm{Sort}_k$**

We are given $n \in \mathbb{N}$ and a graph with at most n connected components. For each $2 \leq i \leq n$, we pick some standard enumeration $(V_j^i)_{j \in \mathbb{N}}$ of the i-element subsets of $\mathbb{N}$. As soon as we learn that none of the V_j^i with $j \leq l$ is an independent set, we write the l-th symbol $i - 2$ on the input to Sort_{n-1}. We write an $n - 1$ occasionally to ensure that the output is actually infinite. Now assume we have access to the corresponding output q of Sort_{n-1}. This will be 0^ω iff the graph had a single connectedness component, and of the form $0^l 1 p$ else where V_l^2 is an independent pair. We can thus start computing the connectedness component of 0 by searching in parallel whether $q \neq 0^\omega$ and searching for a path from 0 to the current number. Either search will terminate. In the latter case, we can answer yes. In the former, we now search for paths to the two vertices in the pair (and thus might be answer to correctly no). Simultaneously we investigate the remnant p whether $p = 1^\omega$ (and thus the graph has 2 connectedness components, and any vertex is linked to either member of V_l^2), or find an independent set of size 3, etc.

$\mathrm{Sort}_k \equiv_{\mathbf{W}} \mathrm{Sort}^{k-1}$

This was shown in [26].

Sort $\leq_{\mathbf{W}}$ FCC

We compute a graph with at most 2 connectedness components. The graph will be bipartite, with the odd and even numbers being separate components. All odd numbers are connected to 0, and at any stage there will be some even number $2n$ not yet connected to 0, which represents some number i such that we have not yet read i times 0 in the input p to Sort. If we read the i-th 0 in p at time t, we connect $2t + 1$ to both 0 and $2n$. If we read a 1 at time t, then $2t + 1$ gets connected to 0 and $2t$.

If p contains infinitely many 0s, then we end up with a single connectedness component. Otherwise we obtain either the connectedness component of 0, or equivalently, its complement. Once we see that e.g. 2 is in this connectedness component, then we can output 0. Moreover, then 2 must be linked to 0 via some $2t + 1$ (which we can exhaustively search for), and whether $2t$ is in the connectedness component tells us whether the next bit of the output is 1 (and then continuous as 1^ω), or 0 again, in which case we need to search for the next significant digit.

FCC $\times$ FCC $\leq_{\mathbf{W}}$ FCC

Just use the product graph.

6 Finite Choice and Sorting

Theorem 5. $C_{\#\leq k+1} \not\leq_{\mathbf{W}} \mathrm{Sort}_k$.

Proof. By the recursion theorem, it suffices to describe an effective procedure which, given $\alpha \in k^\omega$ and Φ, constructs an instance C of $C_{\#\leq k+1}$ such that there is a solution q to $\mathrm{Sort}_k(\alpha)$ such that $\Phi(q)$ is not a solution to C.

For a finite tree T of height s, we say that $\sigma \in T$ is *extendible* if there is a leaf $\rho \in T$ of height s which extends σ. Note that an instance of $\mathsf{C}_{\#\leq k+1}$ is generated by an increasing sequence $(T_s)_{s\in\omega}$ of finite binary trees satisfying the following conditions for every s.

(I) T_s is of height s, and T_s has at least one, and at most $k+1$ extendible leaves.
(II) Every node $\sigma \in T_{s+1} \setminus T_s$ is of length $s+1$, and extends an extendible leaf of T_s.

More precisely, for such a sequence (T_s), the union $T = \bigcup_s T_s$ forms a (T_s)-computable tree which has at most $k+1$ many infinite paths. Therefore, the set of all infinite paths through T is an instance of $\mathsf{C}_{\#\leq k+1}$.

For $\eta \in k^{<\omega}$ and $u < k$, let $N[\eta, u]$ be the number of the occurrences of u's in η, i.e., $N[\eta, u] = \#\{i : \eta(i) = u\}$. We define the *$u$-partial sort of η* as the following string:

$$(\eta)^{\mathrm{sort}}_u = 0^{N[\eta,0]} 1^{N[\eta,1]} 2^{N[\eta,2]} \ldots (u-1)^{N[\eta,u-1]}.$$

Our description of an effective procedure which, given an instance α of Sort_k, returns a sequence $(T_s)_{s\in\omega}$ of finite trees generating an instance of $\mathsf{C}_{\#\leq k+1}$ is subdivided into k many strategies $(\mathcal{S}_u)_{u<k}$. At stage s, the u-th strategy $\mathcal{S}_u$ for $u < k$ believes that u is the least number occurring infinitely often in a given instance α of Sort_k, and there is no $i \geq s$ such that $\alpha(i) < u$. In other words, the strategy $\mathcal{S}_u$ believes that $(\alpha \restriction s)^{\mathrm{sort}}_u {}^\frown u^\omega$, the u-partial sort of the current approximation of α followed by the infinite constant sequence u^ω, is the right answer to the instance α of Sort_k. Then, the strategy $\mathcal{S}_u$ waits for $\Phi((\alpha \restriction s)^{\mathrm{sort}}_u {}^\frown u^\omega)$ being a sufficiently long extendible node ρ of T_s, and then make a branch immediately after an extendible leaf $\rho_u \in T_s$ extending ρ, where this branch will be used for diagonalizing $\Phi((\alpha \restriction s)^{\mathrm{sort}}_u {}^\frown u^\omega)$. This action injures all lower priority strategies $(\mathcal{S}_v)_{u<v<k}$ by initializing their states and letting ρ_v be undefined.

More precisely, each strategy $\mathcal{S}_u$ has a state, $\mathtt{state}_s(u) \in \{0,1,2\}$, at each stage s, which is initialized as $\mathtt{state}_0(u) = 0$. We also define a partial function $u \mapsto \rho^s_u$ for each s, where ρ^s_u is extendible in T_s if it is defined. Roughly speaking, ρ^s_u is the stage s approximation of the diagonalize location for the u-th strategy as described above. We assume that ρ^0_u is undefined for $u > 0$, for any $s \in \omega$, ρ^s_0 is defined as an empty string, and ρ^s_u is a finite string whenever it is defined.

At the beginning of stage $s+1$, inductively assume that a finite tree T_s of height s and a partial function $u \mapsto \rho^s_u$ has already been defined. Moreover, we inductively assume that if $\mathtt{state}_s(u) = 1$ then ρ^s_u is defined, and $\rho^s_u{}^\frown i$ is extendible in T_s for each $i < 2$. At substage u of stage $s+1$, the strategy $\mathcal{S}_u$ acts as follows:

1. If $(\alpha \restriction s+1)^{\mathrm{sort}}_u \neq (\alpha \restriction s)^{\mathrm{sort}}_u$, then initialize the strategy, that is, put $\mathtt{state}_{s+1}(u) = 0$, and let ρ^{s+1}_u be undefined. Then go to the next substage $u+1$ if $u < k$; otherwise go to the next stage $s+2$.
2. If $(\alpha \restriction s+1)^{\mathrm{sort}}_u = (\alpha \restriction s)^{\mathrm{sort}}_u$ and $\mathtt{state}_s(u) = 0$, then ask if $\Phi((\alpha \restriction s)^{\mathrm{sort}}_u {}^\frown u^\omega)[s]$ is an extendible node $\rho \in T_s$ such that for any $v < u$, if ρ^s_v is defined, then $\rho \not\preceq \rho^s_v$ holds.

(a) If yes, define ρ_u^{s+1} as the leftmost extendible leaf of T_s extending such a ρ, and put $\mathtt{state}_{s+1}(u) = 1$. Injure all lower priority strategies, that is, put $\mathtt{state}_{s+1}(v) = 0$ and let ρ_v^{s+1} be undefined for any $u < v < k$. Then go to the next stage $s + 2$.

(b) If no, go to the next substage $u + 1$ if $u < k$; otherwise go to the next stage $s + 2$.

3. If $(\alpha \upharpoonright s + 1)_u^{\mathrm{sort}} = (\alpha \upharpoonright s)_u^{\mathrm{sort}}$ and $\mathtt{state}_s(u) = 1$, then ask if $\Phi((\alpha \upharpoonright s)_u^{\mathrm{sort}} {}^\frown u^\omega)[s]$ is an extendible node $\rho \in T_s$ which extends $\rho_u^s {}^\frown i$ for some $i < 2$.

(a) If yes, define $\rho_u^{s+1} = \rho_u^s {}^\frown (1 - i)$ for such i, and put $\mathtt{state}_{s+1}(u) = 2$. Injure all lower priority strategies, that is, put $\mathtt{state}_{s+1}(v) = 0$ and let ρ_v^{s+1} be undefined for any $u < v < k$. Then go to the next stage $s + 2$.

(b) If no, go to the next substage $u + 1$ if $u < k$; otherwise go to the next stage $s + 2$.

4. If not mentioned, set $\mathtt{state}_{s+1}(u) = \mathtt{state}_s(u)$ and $\rho_u^{s+1} = \rho_u^s$.

At the end of stage $s + 1$, we will define T_{s+1}. Consider the downward closure T_{s+1}^* of the following set:

$$\{\rho_u^{s+1} {}^\frown i : \mathtt{state}(u) = 1 \text{ and } i < 2\} \cup \{\rho_u^{s+1} : \mathtt{state}(u) = 2\}.$$

Let $T_{s+1}^{*,\mathrm{leaf}}$ be the set of all leaves of T_{s+1}^*. Note that every element of $T_{s+1}^{*,\mathrm{leaf}}$ is extendible in T_s since ρ_u^{s+1} is extendible in T_s. For each leaf $\rho \in T_{s+1}^{*,\mathrm{leaf}}$, if $|\rho| = s + 1$ then put $\eta_\rho = \eta$; otherwise choose an extendible leaf $\eta \in T_s$ extending ρ, and define $\eta_\rho = \eta {}^\frown 0$.

Let T_0 be an empty tree. We define T_{s+1} as follows:

$$T_{s+1} = T_s \cup \{\eta_\rho : \rho \in T_{s+1}^{*,\mathrm{leaf}}\}.$$

Note that the extendible nodes in T_{s+1} are exactly the downward closure of $\{\eta_\rho : \rho \in T_{s+1}^{*,\mathrm{leaf}}\}$, and every element of T_{s+1}^* is extendible in T_{s+1}, that is,

- If $\mathtt{state}_{s+1}(u) = 1$, then $\rho_u^{s+1} {}^\frown i$ is extendible in T_{s+1} for each $i < 2$.
- If $\mathtt{state}_{s+1}(u) = 2$, then ρ_u^{s+1} is extendible in T_{s+1}.

Our definition of $(T_s)_{s \in \omega}$ clearly satisfies the property (II) mentioned above. Concerning the property (I), one can see the following:

Lemma 1. T_{s+1} has at least one, and at most $k + 1$ extendible leaves.

Proof. The former assertion trivially holds since ρ_0^s is always defined as an empty string for any $s \in \omega^\omega$. For the latter assertion, it suffices to show that any branching extendible node of T_{s+1} is of the form ρ_u^{s+1} for some $u < k$. This is because T_s is binary, and then the above property automatically ensures that T_s has at most $k + 1$ extendible leaves.

Let σ be a branching extendible node of T_{s+1}. If $|\sigma| = s$, since T_s is of height s, σ is of the form ρ_u^{s+1} by our definition of T_{s+1}. If $|\sigma| < s$, then it is also a branching extendible node of T_s by the property (II) of our construction, and thus it is of the form ρ_u^s by induction. If $\rho_u^s = \rho_u^{s+1}$ for any u, then our Lemma clearly holds. If $\rho_u^s \neq \rho_u^{s+1}$, then it can happen at (2a) or (3a), and thus, there is $v \leq u$ such that the v-th strategy has acted at stage $s+1$. We claim that for any $\rho \in T_{s+1}^*$ we have $\rho_u^s \not\prec \rho$. This claim implies that ρ_u^s is not a branching extendible node in T_{s+1}, which is a contradiction, and therefore we must have $\rho_u^s = \rho_u^{s+1}$.

To show the claim, note that ρ_w^{s+1} is undefined for $w > v$. If $w < u$ and ρ_w^s is defined then $\rho_u^s \not\preceq \rho_w^s$ by S_u's action at (2a). If $w < v$ then $\rho_w^{s+1} = \rho_w^s$. For $w = v$, if $\mathsf{state}_{s+1}(v) = 1$ then S_v reaches at (2a) at stage $s+1$ and $\rho_u^s \not\preceq \rho_u^s$ by S_v's action. If $\mathsf{state}_{s+1}(v) = 2$ then S_v reaches at (3a) at stage $s+1$, and thus ρ_v^{s+1} is a successor of ρ_u^{s+1} and thus $\rho_u^s \not\prec \rho_v^{s+1}$. Hence, there is no $\rho \in T_{s+1}^*$ such that $\rho_u^s \prec \rho$ as desired.

Lemma 2. *If* $\mathsf{state}_{s+1}(u) = 2$, *then* $\Phi((\alpha \restriction s+1)_u^{\mathrm{sort}} \frown u^\omega)$ *is not extendible in* T_{s+1}.

Proof. If $\mathsf{state}_{s+1}(u) = 2$, then there is stage $t \leq s+1$ such that $(\alpha \restriction t)_u^{\mathrm{sort}} = (\alpha \restriction s+1)_u^{\mathrm{sort}}$ and the u-th strategy S_u arrives at (2a) at stage s and (3a) at $s+1$, and the u-th strategy is not injured by any higher priority strategy during stages between t and $s+1$, and in particular, $\rho_u^t = \rho_u^s$. By our action (3a), $\Phi((\alpha \restriction s+1)_u^{\mathrm{sort}} \frown u^\omega)$ extends the sister of ρ_u^{s+1}. If $v > u$ then ρ_u^{s+1} is undefined. If $v < u$ and ρ_v^t is undefined, then since no injury happens below u during stages between t and $s+1$, we have $\rho_u^s = \rho_u^t \not\preceq \rho_v^t = \rho_v^{s+1}$, which implies that ρ_u^{s+1} does not extend the sister of ρ_u^{s+1}. Hence the sister of ρ_u^{s+1} does not extend to a leaf of T_{s+1}^*. Therefore, $\Phi((\alpha \restriction s+1)_u^{\mathrm{sort}} \frown u^\omega)$ is not extendible in T_{s+1}. $\square$

We now verify our construction. Put $T = \bigcup_k T_k$. By Lemma 1, since our construction of $(T_\partial)_{\partial \subset \omega}$ satisfies the conditions (I) and (II), the set $[T]$ of all infinite paths through T is an instance of $\mathsf{C}_{\#\leq k+1}$. Let α be an instance of Sort_k.

Lemma 3. $\Phi(\mathsf{Sort}_k(\alpha)) \notin [T]$.

Proof. By pigeonhole principle, there exists u such that $\alpha(i) = u$ for infinitely many i. Let u be the least such number. Then there exists s such that $(\alpha)^{\mathrm{sort}} := (\alpha \restriction s)_u^{\mathrm{sort}} \frown u^\omega$ is the right answer to the instance α of Sort_k, that is, it is the result by sorting α. Then, for any $v \leq u$, the v-partial sort of α stabilizes after s, that is, $(\alpha \restriction t+1)_v^{\mathrm{sort}} = (\alpha \restriction t)_v^{\mathrm{sort}}$ for all $t \geq s$. After the v-partial sort of α stabilizes, the v-th strategy S_v can injure lower priority strategies at most two times, i.e., at (2a) and (3a). Therefore, there is stage $s_0 \geq s$ such that the u-th strategy S_u is never injured by higher priority strategies after s_0. Then, $\mathsf{state}_t(u)$ converges to some value.

Case 1. $\lim_t \mathsf{state}_t(u) = 0$. By our choice of s_0, S_u always goes to (2b), and never goes to (2a) after s_0. However, if $\Phi((\alpha)^{\mathrm{sort}})$ is an infinite string, then the strategy must go to (2a) since $\{\rho_v^s : v < u\}$ is finite. Hence, $\Phi((\alpha)^{\mathrm{sort}})$ cannot be an infinite path through T.

Case 2. $\lim_t \mathsf{state}_t(u) = 1$. Let $s_1 \geq s_0$ be the least stage such that S_u reaches (2a) with some ρ. We claim that if an extendible node in T_t extends ρ, then it also extends ρ_u^t for any $t > s_1$. According to the condition of $\mathcal{S}_u$'s strategy (2), for any $v < u$, we have $\rho \not\leq \rho_v^{s_1} = \rho_v^{s_0}$. By injury in (2a), $\rho_v^{s_1}$ is undefined for any $v > u$. Therefore, any extendible node of T_{s_1+1} extends ρ_v^t or $\rho_v^t{}^\frown i$ for some $v \leq u$ and $i < 2$. Hence, if an extendible node in T_{s_1+1} extends ρ, then it also extends $\rho_u^{s_1+1} = \rho_u^t$. By the property (II) of our construction, the claim follows. Now, by our assumption, $\mathcal{S}_u$ always goes to (3b), and never goes to (3a). This means that $\Phi((\alpha \restriction t)_u^{\text{sort}}{}^\frown u^\omega)$ extends ρ, but does not extend ρ_u^t for any $t > s_1$. Therefore, $\Phi((\alpha \restriction t)_u^{\text{sort}}{}^\frown u^\omega)$ is not extendible in T_t for any $t > s_1$. Consequently, $\Phi((\alpha)^{\text{sort}}) \notin [T]$.

Case 3. $\lim_t \text{state}_t(u) = 2$. Let $s_2 \geq s_0$ be the least stage such that $\mathcal{S}_u$ reaches (3a). Then by Lemma 2, $\Phi((\alpha \restriction s_2)_u^{\text{sort}}{}^\frown u^\omega)$ is not extendible in T_{s_2}. Since $\mathcal{S}_u$ is not injured after s_0, we conclude $\Phi((\alpha)^{\text{sort}}) \notin [T]$.

By the recursion theorem, this obviously implies the desired assertion.

7   The Comparison Game for Products of Finite Choice

In this section we consider the question when finite choice for some cardinality is reducible to some finite product of finite choice operators. We do not obtain an explicit characterization, but rather an indirect one. We introduce a special reachability game (played on a finite graph), and show that the winner of this game tells us whether the reduction holds. This in particular gives us a decision procedure (which so far has not been implemented yet, though).

Our game is parameterized by numbers k, and $n_0, n_1, \ldots, n_\ell$. We call the elements of $\bigcup_{i \leq \ell} \{i\} \times n_i$ *colours*, and the elements of $\Pi_{i \leq \ell} n_i$ *tokens*. A token w has colour (i, c), if $w_i = c$.

The current board consists of up to k boxes each of which contains some set of tokens, with no token appearing in distinct boxes. If there ever is an empty box, then Player 1 wins. If the game continues indefinitely without a box becoming empty, Player 2 wins. The initial configuration is chosen by Player 1 selecting the number of boxes, and by Player 2 distributing all tokens into these boxes.

The available actions are as follows:

Remove Player 1 taps a box b. Player 2 selects some colours C such that every token in b has a colour from C. Then the box b and all tokens with a colour from C are removed.

Reintroduce colour Player 2 picks two 'adjacent' colours (i, c) and (i, d), such that no token on the board has colour (i, d). For every box b, and every token $w \in b$ having colour c, he then adds a token w' to b that is identical to w except for having colour (i, d) rather than (i, c).

Split box If there are less than k boxes on the board, Player 1 can select a box b to be split into two boxes b_0 and b_1. Player 2 can chose how to distribute the tokens from b between b_0 and b_1. Moreover, Player 2 can do any number of *Reintroduce colour* moves before the *Split box*-move takes effect.

Theorem 6. $C_{\sharp\leq k} \leq_W C_{\sharp\leq n_0} \times \ldots C_{\sharp\leq n_\ell}$ *iff Player 2 wins the comparison game for parameters* $k, n_0, \ldots, n_\ell$.

The proof proceeds via Lemmas 4 and 5 below. We observe that the game is a reachability game played on a finite graph. In particular, it is decidable who wins the game for a given choice of parameters. We have only considered the case $n_i = 2$ so far, and know:

Proposition 4

1. *Player 2 wins for* $k + 1 \leq \ell$.
2. *Player 1 wins for* $k + 1 \geq 2^{\ell-1}$.

Proof. The first claim follows from Theorem 6 in conjunction with [20, Proposition 3.9] stating that $C_{\sharp\leq n+1} \leq_W C^n_{\sharp\leq 2}$. The second is immediate when analyzing the game.

Lemma 4. *From a winning strategy of Player 2 in the comparison game we can extract witnesses for the reduction* $C_{\sharp\leq k} \leq_W C_{\sharp\leq n_0} \times \ldots C_{\sharp\leq n_\ell}$.

Proof. We recall that the input to $C_{\sharp\leq k}$ can be seen as an infinite binary tree having at most k vertices on each level. We view this tree as specifying a strategy for Player 1 in the comparison game: The boxes correspond to the paths existing up to the current level of the tree. If a path dies out, Player 1 taps the corresponding box. If a path splits into two, Player 1 splits the corresponding box.

Which tokens exist at a certain time tells us how the instances to $C_{\sharp\leq n_0}, \ldots, C_{\sharp\leq n_\ell}$ are built. The colour (i, j) refers to the j-path through the i-th tree at the current approximation. If a colour gets removed, this means that the corresponding path dies out. If a colour gets reintroduced, we split the path corresponding to the duplicated colour into two.

It remains to see how the outer reduction witness maps infinite paths through these trees back to an infinite path through the input tree. If we are currently looking at some finite approximation of the input tree and the query trees, together with an infinite path through each query tree, then the infinite paths indicates some token which never will be removed. That means that any box containing that token never gets tapped, i.e. that certain prefixes indeed can be continued to an infinite path.

Lemma 5. *From a winning strategy of Player 1 in the comparison game we can extract a witness for the non-reduction* $C_{\sharp\leq k} \not\leq_W C_{\sharp\leq n_0} \times \ldots C_{\sharp\leq n_\ell}$ *according to Theorem 1.*

Proof. We need to describe a procedure that constructs an input for $C_{\sharp\leq k}$ given inputs to $C_{\sharp\leq n_0}, \ldots, C_{\sharp\leq n_\ell}$ and an outer reduction witness. Inverting the procedure from Lemma 4, we can view the given objects as describing a strategy of Player 2 in the game. We obtain the input tree to $C_{\sharp\leq k}$ by observing how the winning strategy of Player 1 acts against this. When Player 1 taps the i-th box, we let the i-th path through the tree die out. When Player 1 splits the i-th box, we let both children of the i-th vertex present at the current layer be present at the subsequent layer. Otherwise, we keep the left-most child of any vertex on the previous layer.

Since Player 1 is winning, we will eventually reach an empty box. At that point, we let all other paths die out, and only keep the one corresponding to the empty box. This means that any path selected by the outer reduction witness we obtained Player 2's strategy from will fall outside the tree, and thus satisfy the criterion of Theorem 1.

Acknowledgement. We are grateful to Stéphane Le Roux for a fruitful discussion leading up to Theorems 2 and 3.

References

1. Blum, L., Cucker, F., Shub, M., Smale, S.: Complexity and Real Computation. Springer, New York (1998). https://doi.org/10.1007/978-1-4612-0701-6
2. Brattka, V., de Brecht, M., Pauly, A.: Closed choice and a uniform low basis theorem. Ann. Pure Appl. Log. **163**(8), 968–1008 (2012). https://doi.org/10.1016/j.apal.2011.12.020
3. Brattka, V., Gherardi, G.: Effective choice and boundedness principles in computable analysis. Bull. Symb. Log. **17**, 73–117 (2011). https://doi.org/10.2178/bsl/1294186663. arXiv:0905.4685
4. Brattka, V., Gherardi, G.: Weihrauch degrees, omniscience principles and weak computability. J. Symb. Log. **76**, 143–176 (2011). arXiv:0905.4679
5. Brattka, V., Gherardi, G., Hölzl, R.: Probabilistic computability and choice. Inf. Comput. **242**, 249–286 (2015). https://doi.org/10.1016/j.ic.2015.03.005. http://arxiv.org/abs/1312.7305
6. Brattka, V., Gherardi, G., Hölzl, R., Pauly, A.: The vitali covering theorem in the Weihrauch lattice. In: Day, A., Fellows, M., Greenberg, N., Khoussainov, B., Melnikov, A., Rosamond, F. (eds.) Computability and Complexity. LNCS, vol. 10010, pp. 188–200. Springer, Cham (2017). https://doi.org/10.1007/978-3-319-50062-1_14
7. Brattka, V., Gherardi, G., Pauly, A.: Weihrauch complexity in computable analysis. arXiv:1707.03202 (2017)
8. Brattka, V., Hertling, P., Weihrauch, K.: A tutorial on computable analysis. In: Cooper, S.B., Löwe, B., Sorbi, A. (eds.) New Computational Paradigms: Changing Conceptions of What is Computable, pp. 425–491. Springer, New York (2008). https://doi.org/10.1007/978-0-387-68546-5_18
9. Brattka, V., Hölzl, R., Kuyper, R.: Monte Carlo computability. In: Vollmer, H., Vallée, B. (eds.) 34th Symposium on Theoretical Aspects of Computer Science (STACS 2017). Leibniz International Proceedings in Informatics (LIPIcs), vol. 66, pp. 17:1–17:14. Schloss Dagstuhl-Leibniz-Zentrum fuer Informatik, Dagstuhl, Germany (2017). https://doi.org/10.4230/LIPIcs.STACS.2017.17. http://drops.dagstuhl.de/opus/volltexte/2017/7016
10. Brattka, V., Miller, J., Le Roux, S., Pauly, A.: Connected choice and Brouwer's fixed point theorem. J. Math. Log. (20XX, accepted for publication). arXiv:1206.4809
11. Brattka, V., Pauly, A.: Computation with advice. Electron. Proc. Theor. Comput. Sci. **24**, 41–55 (2010). http://arxiv.org/html/1006.0551, cCA 2010

12. Brattka, V., Pauly, A.: On the algebraic structure of Weihrauch degrees. Log. Methods Comput. Sci. **14**(4) (2018). https://doi.org/10.23638/LMCS-14(4:4)2018. http://arxiv.org/abs/1604.08348
13. Chadzelek, T., Hotz, G.: Analytic machines. Theor. Comput. Sci. **219**, 151–167 (1999)
14. Dorais, F.G., Dzhafarov, D.D., Hirst, J.L., Mileti, J.R., Shafer, P.: On uniform relationships between combinatorial problems. Trans. AMS **368**, 1321–1359 (2016). https://doi.org/10.1090/tran/6465. arXiv:1212.0157
15. Gärtner, T., Hotz, G.: Computability of analytic functions with analytic machines. In: Ambos-Spies, K., Löwe, B., Merkle, W. (eds.) CiE 2009. LNCS, vol. 5635, pp. 250–259. Springer, Heidelberg (2009). https://doi.org/10.1007/978-3-642-03073-4_26
16. Gherardi, G., Marcone, A.: How incomputable is the separable Hahn-Banach theorem? Notre Dame J. Form. Log. **50**(4), 393–425 (2009). https://doi.org/10.1215/00294527-2009-018
17. Gherardi, G., Marcone, A., Pauly, A.: Projection operators in the Weihrauch lattice. Computability (20XX, accepted for publication). arXiv:1805.12026
18. Gura, K., Hirst, J.L., Mummert, C.: On the existence of a connected component of a graph. Computability **4**(2), 103–117 (2015). https://doi.org/10.3233/COM-150039
19. Kihara, T., Pauly, A.: Dividing by zero - how bad is it, really? In: Faliszewski, P., Muscholl, A., Niedermeier, R. (eds.) 41st International Symposium on Mathematical Foundations of Computer Science (MFCS 2016). Leibniz International Proceedings in Informatics (LIPIcs), vol. 58, pp. 58:1–58:14. Schloss Dagstuhl (2016). https://doi.org/10.4230/LIPIcs.MFCS.2016.58
20. Le Roux, S., Pauly, A.: Finite choice, convex choice and finding roots. Log. Methods Comput. Sci. (2015). https://doi.org/10.2168/LMCS-11(4:6)2015. http://arxiv.org/abs/1302.0380
21. Neumann, E.: Computational problems in metric fixed point theory and their Weihrauch degrees. Log. Methods Comput. Sci. **11**(4) (2015). https://doi.org/10.2168/LMCS-11(4:20)2015
22. Neumann, E., Pauly, A.: A topological view on algebraic computations models. J. Complex. **44** (2018). https://doi.org/10.1016/j.jco.2017.08.003. http://arxiv.org/abs/1602.08004
23. Pauly, A.: How incomputable is finding Nash equilibria? J. Univers. Comput. Sci. **16**(18), 2686–2710 (2010). https://doi.org/10.3217/jucs-016-18-2686
24. Pauly, A.: On the (semi)lattices induced by continuous reducibilities. Math. Log. Q. **56**(5), 488–502 (2010). https://doi.org/10.1002/malq.200910104
25. Pauly, A.: On the topological aspects of the theory of represented spaces. Computability **5**(2), 159–180 (2016). https://doi.org/10.3233/COM-150049. http://arxiv.org/abs/1204.3763
26. Pauly, A., Tsuiki, H.: T^ω-representations of compact sets. arXiv:1604.00258 (2016)
27. Selivanov, V.L.: Total representations. Log. Methods Comput. Sci. **9**(2) (2013)
28. Weihrauch, K.: The degrees of discontinuity of some translators between representations of the real numbers. Informatik Berichte 129, FernUniversität Hagen, Hagen, July 1992
29. Weihrauch, K.: Computable Analysis. Springer, Heidelberg (2000). https://doi.org/10.1007/978-3-642-56999-9
30. Ziegler, M.: Computable operators on regular sets. Math. Log. Q. **50**, 392–404 (2004)

The Number of Languages
with Maximum State Complexity

Bjørn Kjos-Hanssen[(⊠)] and Lei Liu

University of Hawaii at Manoa, Honolulu, USA
bjoernkh@hawaii.edu

Abstract. Champarnaud and Pin (1989) found that the minimal deterministic automaton of a language $L \subset \Sigma^n$, where $\Sigma = \{0, 1\}$, has at most

$$\sum_{i=0}^{n} \min(2^i, 2^{2^{n-i}} - 1)$$

states, and for each n there exists L attaining this bound. Câmpeanu and Ho (2004) have shown more generally that the tight upper bound for Σ of cardinality k and for complete automata is

$$\frac{k^r - 1}{k - 1} + \sum_{j=0}^{n-r} (2^{k^j} - 1) + 1$$

where $r = \min\{m : k^m \geq 2^{k^{n-m}} - 1\}$. (In these results, requiring totality of the transition function adds 1 to the state count.) Câmpeanu and Ho's result can be viewed as concerning functions $f : [k]^n \to [2]$ where $[k] = \{0, \ldots, k-1\}$ is a set of cardinality k. We generalize their result to arbitrary function $f : [k]^n \to [c]$ where c is a positive integer.

Let O_i be the number of functions from $[b^i]$ to $[c^{b^{n-i}}]$ that are onto $[c^{b^{n-i}} - 1]$. Câmpeanu and Ho stated that it is very difficult to determine the number of maximum-complexity languages. Here we show that it is equal to O_i, for the least i such that $O_i > 0$.

For monotone languages a tightness result seems harder to obtain. However, we show that the following upper bound is attained for all $n \leq 10$.

$$\sum_{i=0}^{n} \min(2^i, M(n-i) - 1),$$

where $M(k)$ is the kth Dedekind number.

1 Introduction

The function $+$ on $\mathbb{Z}/5\mathbb{Z}$ may seem rather complicated as functions on that set go. On the other hand, $f(x, y, z) = x + y + z \mod 5$ is less so, in that we can decompose it as $(x + y) + z$, so that after seeing x and y, we need not remember the pair (x, y) but only their sum. Out of the 5^{5^3} ternary functions

© Springer Nature Switzerland AG 2019
T. V. Gopal and J. Watada (Eds.): TAMC 2019, LNCS 11436, pp. 394–409, 2019.
https://doi.org/10.1007/978-3-030-14812-6_24

on a 5-element set, at most $5^{2 \cdot 5^2}$ can be decomposed as $(x *_1 y) *_2 z$ for some binary functions $*_1$, $*_2$. In Sect. 2 we make precise a sense in which such are not the most complicated ternary functions. We do this by extending a result of Câmpeanu and Ho [3] to functions taking values in a set of size larger than two.

Rising to an implicit challenge posed by Câmpeanu and Ho, we give a formula for the number of maximally complex languages in Sect. 2.2.

A motivation from finance will be felt in Sects. 3 and 4. The complexity of financial securities came into focus with the 2008 financial crisis. While Arora et al. [1] obtained NP-hardness results for the pricing of a security, here we look at the automatic complexity associated with executing a given trading strategy. The possibility of exercising early leads to a less complex option in our sense, as is easy to see. Thus we shall restrict attention to options which are European insofar as they can only be exercised at the final time n.

2 Complexity of Languages and Operations

Definition 2.1. A deterministic finite automaton (DFA) [9] M is a 5-tuple, $(Q, \Sigma, \delta, q_0, F)$, where

- Q is a finite set of states,
- $q_0 \in Q$ is the start state,
- $F \subseteq Q$ is the set of accept states,
- Σ is a finite set of input symbols and
- $\delta : Q \times \Sigma \longrightarrow Q$ is the transition function.

If δ is not required to be total then we speak of a *partial deterministic finite automaton (PDFA)*.

Definition 2.2. Let $\Sigma = \{0, 1\}$, let $n \in \mathbb{Z}^+$ and $X \subseteq \Sigma^{\leq n}$. Define $A_-(X)$ to be the minimum $|Q|$ over all PDFAs $M = (Q, \Sigma, \delta, q_0, F)$ for which $L(M)$, the language recognized by M, equals X. We call a PDFA $M = (Q, \Sigma, \delta, q_0, F)$ *minimal for X* if

$$|Q| = A_-(X).$$

2.1 Operations

Champarnaud and Pin [4] obtained the following result.

Theorem 2.3 ([4, Theorem 4]). *The minimal PDFA of a language $L \subset \{0, 1\}^n$ has at most*

$$\sum_{i=0}^{n} \min(2^i, 2^{2^{n-i}} - 1)$$

states, and for each n there exists L attaining this bound.

Theorem 2.3 was generalized by Câmpeanu and Ho [3]:

Theorem 2.4 ([3, Corollary 10]). *For $k \geq 1$, let $[k] = \{0, \ldots, k-1\}$. Let $l \in \mathbb{N}$ and let M be a minimal DFA for a language $L \subseteq [k]^l$. Let Q be the set of states of M. Then we have:*

(i) $\#Q \leq \frac{k^r - 1}{k-1} + \sum_{j=0}^{l-r}(2^{k^j} - 1) + 1$, where $r = \min\{m \mid k^m \geq 2^{k^{l-m}} - 1\}$;
(ii) there is an M such that the upper bound given by (i) is attained.

Both of these results involve an upper bound which can be viewed as a special case of Theorem 2.7 below.

Definition 2.5. *Let b, n, and c be positive integers. We say that a PDFA M accepts a function $f : [b]^n \to [c]$ if there are $c-1$ many special states $q_1, \ldots, q_{c-1}$ of M such that for all $\vec{x} \in [b]^n$,*

- *for $i > 0$, $f(\vec{x}) = i$ iff M on input $\vec{x}$ ends in state q_i; and*
- *$f(\vec{x}) = 0$ iff M does not end in any of the special states on input $\vec{x}$.*

Definition 2.5 generalizes the case $b = 2$ studied by Champarnaud and Pin. We write A^B for the set of all functions from B to A.

Definition 2.6. *Let $[c]^{[b]^n}$ be the set of n-ary functions $f : [b]^n \to [c]$. Let b and c be positive integers and let $\mathfrak{C} \subseteq [c]^{[b]^n}$. The Champarnaud–Pin family of $\mathfrak{C}$ is the family of sets $\{\mathfrak{C}_k\}_{0 \leq k \leq n}$, where $\mathfrak{C}_k \subseteq [c]^{[b]^{n-k}}$, $0 \leq k \leq n$, given by*

$$\mathfrak{C}_k = \{g \in [c]^{[b]^{n-k}} : \exists f \in \mathfrak{C}, \vec{d} \in [b]^k \quad \forall \vec{x} \quad g(\vec{x}) = f(\vec{d}, \vec{x})\}.$$

So $\mathfrak{C}_0 = \mathfrak{C}$, $\mathfrak{C}_1$ is obtained from $\mathfrak{C}_0$ by plugging in constants for the first input, and so forth. We write $\mathfrak{C}_n^- = \{f \in \mathfrak{C}_n : f \neq 0\}$ in order to throw out the constant zero function. Note that $|\mathfrak{C}_n^-| \geq |\mathfrak{C}_n| - 1$.

Theorem 2.7. *Let b and c be positive integers. Let $\mathfrak{C} \subseteq [c]^{[b]^n}$. An upper bound on the minimal number of states of PDFAs accepting members of $\mathfrak{C}$ is given by*

$$\sum_{i=0}^{n} \min(b^i, |\mathfrak{C}_i^-|).$$

The proof will be apparent from the proof of the next result, which is a generalization of Câmpeanu and Ho's theorem.

Theorem 2.8. *Let b and c be positive integers. For the minimal number of states of PDFAs M accepting functions $f : [b]^n \to [c]$, the upper bound*

$$\sum_{i=0}^{n} \min(b^i, c^{b^{n-i}} - 1)$$

is attained.

Proof. Let $\log = \log_b$. The critical point for this result is the pair of values (i, k) with $i + k = n$ such that $b^i \leq c^{b^k} - 1$ (i.e., $i < b^k \log c$) and $b^{i+1} > c^{b^{k-1}} - 1$ (i.e., $b^{i+1} \geq c^{b^{k-1}}$, i.e., $(i+1) \geq b^{k-1} \log c$), which can be summarized as

$$b^{k-1} \log c - 1 \leq i < b^k \log c. \tag{1}$$

$$b^{k-1} \log c \leq i + 1 \leq b^k \log c.$$

We shall define a set A of k-ary functions of size $(c^{b^{k-1}})/b$ which when using the b many transitions (substitutions for say p_1) maps onto each of the $c^{b^{k-1}}$ many $k - 1$-ary functions α. This will suffice if

$$c^{b^{k-1}}/b \leq b^i$$

which does hold for all b by (1). The construction is similar to that of [3, Figure 1 and Theorem 8]; we shall be slightly more explicit than they were. Let $s = c^{b^{k-1}} - 1$. Let $f_0, \ldots, f_{s-1}$ the set of all nonzero $k - 1$-ary functions. As s may not be divisible by b, let us write $s = qb + r$ with quotient $q \geq 0$ and remainder $0 \leq r < b$. For j with $0 \leq j \leq q - 1$, let g_j be given by

$$g_j(i, \vec{x}) = f_{jb+i}(\vec{x})$$

for each $i \in [b]$ and $\vec{x} \in [b]^{k-1}$. Let g_q be given by $g_q(i, \vec{x}) = f_{qb+i}(\vec{x})$ for each $0 \leq i \leq r - 1$, and let $g_q(i, \vec{x})$ be arbitrary for $r \leq i < b$. Finally, extend the set of functions $g_0, \ldots, g_q$ to b^i many k-ary functions in an arbitrary way, obtaining functions h_σ for $\sigma \in [b]^i$. Then our function attaining the bound is given by

$$H(\sigma, \tau) = h_\sigma(\tau).$$

□

When $b = 2$ and c is larger, Theorem 2.8 corresponds to automatic complexity of equivalence relations on binary strings as studied in [6]. When $b = c$, we have the case of n-ary operations on a given finite set, which is of great interest in universal algebra.

2.2   The Number of Maximally Complex Languages

Definition 2.9. Let b and c be positive integers and let $0 \leq i \leq n$. Let $O_i = O_i^{(b,c,n)}$ be the number of functions from $[b^i]$ to $[c^{b^{n-i}}]$ that are onto $[c^{b^{n-i}} - 1]$. That is, functions $f : [b^i] \to [c^{b^{n-i}}]$ such that for each $y \in [c^{b^{n-i}} - 1]$ there is an $x \in [b^i]$ with $f(x) = y$.

Câmpeanu and Ho lamented that it seemed very difficult to count the number of maximum-complexity languages. Here we show

Theorem 2.10. *Let b and c be positive integers and let $n \geq 0$. The number of maximum complexity functions $f : [b]^n \to [c]$ is O_i, where $0 \leq i \leq n$ is minimal such that $O_i > 0$.*

Proof. Champarnaud and Pin, and Câmpeanu and Ho, and the present authors in Theorem 2.8, all found a maximal complexity by explicitly exhibiting the general automaton structure of a maximal-complexity language: we start with states corresponding to binary strings and end with strings corresponding to Boolean functions, and there is a crossover point in the middle where, in order that all states be used, we need an onto function exactly as specified in the definition of O_i. The crossover point occurs for the least i such that $O_i > 0$, which is when the value of the minimum of $(b^i, c^{b^{n-i}} - 1)$ switches from the first to the second coordinate. The number of such functions is then the number of such onto functions. Since we do not require totality and do not use a state for output 0 ("reject") we omit the constant 0 Boolean function in the range of our onto maps. □

Note that the number of onto functions is well known in terms of Stirling numbers of the second kind. Let $O_{m,n}$ be the number of onto functions from $[m]$ to $[n]$. Then

$$O_{m,n} = n! \left\{ \begin{matrix} m \\ n \end{matrix} \right\},$$

where $\left\{ \begin{matrix} m \\ n \end{matrix} \right\}$, the number of equivalence relations on $[m]$ with n equivalence classes, is a Stirling number of the second kind.

Note also that the number of functions from $[a]$ to $[b]$ that are onto the first $b - 1$ elements of $[b]$ is, in terms of the number m of elements going to the not-required element,

$$\sum_{m=0}^{a-(b-1)} \binom{a}{m} O_{a-m,b-1}.$$

Example 2.11. When $n = 3$ and $b = c = 2$, we have that O_i is the number of functions from 2^i to $2^{2^{3-i}}$ that are onto $2^{2^{3-i}} - 1$. In this case, $O_1 = 0$. However, O_2 is the number of functions from 4 to 4 that are onto 3. This is

$$\sum_{m=0}^{4-(4-1)} \binom{4}{m} O_{4-m,4-1} = O_{4,3} + 4 O_{3,3} = 36 + 24 = 60.$$

These 60 languages are shown in Table 1.

Table 1. All possible sets Z with $A_-(Z) = 7$.

Size	Z
$|Z| = 4$	$\{000, 001, 010, 101\}$, $\{000, 001, 010, 111\}$,
	$\{000, 001, 011, 100\}$, $\{000, 001, 100, 111\}$,
	$\{000, 001, 011, 110\}$, $\{000, 001, 101, 110\}$,
	$\{000, 010, 011, 101\}$, $\{000, 010, 011, 111\}$,
	$\{001, 010, 011, 100\}$, $\{010, 011, 100, 111\}$,
	$\{001, 010, 011, 110\}$, $\{010, 011, 101, 110\}$,
	$\{000, 011, 100, 101\}$, $\{000, 100, 101, 111\}$,
	$\{001, 010, 100, 101\}$, $\{010, 100, 101, 111\}$,
	$\{001, 100, 101, 110\}$, $\{011, 100, 101, 110\}$;
	$\{000, 011, 110, 111\}$, $\{000, 101, 110, 111\}$,
	$\{001, 010, 110, 111\}$, $\{010, 101, 110, 111\}$,
	$\{001, 100, 110, 111\}$, $\{011, 100, 110, 111\}$
$|Z| = 5$	$\{000, 001, 010, 100, 111\}$, $\{000, 001, 010, 101, 110\}$,
	$\{000, 001, 011, 100, 110\}$, $\{000, 001, 011, 101, 110\}$,
	$\{000, 001, 011, 100, 111\}$, $\{000, 001, 010, 101, 111\}$,
	$\{000, 010, 011, 100, 111\}$, $\{000, 010, 011, 101, 110\}$,
	$\{001, 010, 011, 100, 110\}$, $\{001, 010, 011, 101, 110\}$,
	$\{001, 010, 011, 100, 111\}$, $\{000, 010, 011, 101, 111\}$,
	$\{000, 010, 100, 101, 111\}$, $\{000, 011, 100, 101, 110\}$,
	$\{001, 010, 100, 101, 110\}$, $\{001, 011, 100, 101, 110\}$,
	$\{001, 010, 100, 101, 111\}$, $\{000, 011, 100, 101, 111\}$,
	$\{000, 010, 101, 110, 111\}$, $\{000, 011, 100, 110, 111\}$,
	$\{001, 010, 100, 110, 111\}$, $\{001, 011, 100, 110, 111\}$,
	$\{001, 010, 101, 110, 111\}$, $\{000, 011, 101, 110, 111\}$
$|Z| = 6$	$\{000, 001, 010, 011, 100, 111\}$,
	$\{000, 001, 010, 011, 101, 110\}$,
	$\{000, 001, 010, 100, 101, 111\}$,
	$\{000, 001, 011, 100, 101, 110\}$,
	$\{000, 001, 010, 101, 110, 111\}$,
	$\{000, 001, 011, 100, 110, 111\}$,
	$\{000, 010, 011, 100, 101, 111\}$,
	$\{001, 010, 011, 100, 101, 110\}$,
	$\{000, 010, 011, 101, 110, 111\}$,
	$\{001, 010, 011, 100, 110, 111\}$,
	$\{000, 011, 100, 101, 110, 111\}$,
	$\{001, 010, 100, 101, 110, 111\}$

Listing 1. Pseudocode for our variant of the Myhill–Nerode algorithm.

```
Input: Strings s and t, a set of strings L, and a max length n.
Output: The boolean of whether s and t are equivalent for L.
For u a binary string of length between 0 and n−1,
  if len(s+u), len(t+u) both at most n
  and exactly one of s+u, t+u is in the up−closure of L,
    return False
Return True.
```

2.3   Polynomial-Time Algorithm

It is perhaps worth pointing out that there is a polynomial-time algorithm for finding the minimal automaton of Boolean functions, based on essentially the Myhill–Nerode theorem [5,10]. In this subsection we detail that somewhat.

Definition 2.12. Given a language L, and a pair of strings x and y, define a *distinguishing extension* to be a string z such that exactly one of the two strings xz and yz belongs to L. Define a relation R_L on strings by the rule that xR_Ly if there is no distinguishing extension for x and y.

As is well known, R_L is an equivalence relation on strings, and thus it divides the set of all strings into equivalence classes.

Theorem 2.13 (Myhill–Nerode). *A language L is regular if and only if R_L has a finite number of equivalence classes. Moreover, the number of states in the smallest deterministic finite automaton (DFA) recognizing L is equal to the number of equivalence classes in R_L. In particular, there is a unique DFA with minimum number of states.*

The difference is that for us we require $|xz| \leq n$ and $|yz| \leq n$, see Listing 1.

3   Monotone Boolean Functions

The main theoretical results of the paper are in Sect. 2. The present, longer section deals with a more computational and exploratory investigation: what happens if we try to prove that the natural upper bound on complexity is attained in restricted settings such as monotone functions?

Definition 3.1. An isotone map is a function φ with $a \leq b \implies \varphi(a) \leq \varphi(b)$.

The Online Encyclopedia of Integer Sequences (OEIS) has a tabulation of Dedekind numbers, i.e., the number $M(n)$ of monotone functions [12], which is also the number of elements of the free distributive lattice on n generators and the number of antichains of subsets of $[n]$.

Definition 3.2. For an integer $n \geq 0$, F_n is the set of monotone Boolean functions of n variables (equivalently, the free distributive lattice on n generators, allowing 0 and 1 to be included), and $F_n^- = F_n \setminus \{0\}$ where 0 is the constant 0 function.

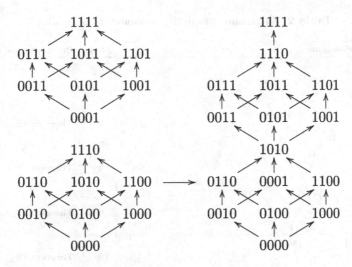

Fig. 1. Isotone 1:1 map from 2^4 to F_3^-.

The illustrative case $n = 3$ is shown in Fig. 4.

Theorem 3.3. *The minimal automaton of a monotone language $L \subset \{0,1\}^n$ has at most*

$$\sum_{i=0}^{n} \min(2^i, |F_{n-i}| - 1)$$

states. This bound is attained for $n \leq 10$.

Proof. The upper bound follows from Theorem 2.7. The sharpness results are obtained in a series of theorems tabulated in Table 2. □

Thinking financially, an option is monotone if whenever s is pointwise dominated by t and $s \in L$ then $t \in L$, where L is the set of exercise situations for the option. This is the case for common options like call options or Asian average-based options and makes financial sense if a rise in the underlying is always desirable and always leads to a higher option value.

Example 3.4 (Asian option; Shreve [11, Exercise 1.8]). This is the example that in part motivates our looking at monotone options. Let $n = 3$ and consider a starting capital $S_0 = 4$, up-factor $u = 2$, down-factor $d = \frac{1}{2}$. Let $Y_i = \sum_{k=0}^{i} S_k$. The payoff at time $n = 3$ is $(\frac{1}{4}Y_3 - 4)^+$. To fit this example into our framework in the present paper, let us look at which possibilities lead to exercising, i.e., $\frac{1}{4}Y_3 - 4 > 0$ or $Y_3 > 16$. Computation shows that the set of exercise outcomes is $\{011, 100, 101, 110, 111\}$. The complexity is 6 (Fig. 3), so it is maximally complex for a monotone option.

For $n = 3$ we are looking at isotone functions from $\{0,1\}$ to the family of monotone functions on two variables p and q. For the Asian option in Example 3.4 $\{0,1\}$ are mapped to $\{p \wedge q, 1\}$. For the majority function, $\{0,1\}$ are mapped to $\{p \wedge q, p \vee q\}$ (Fig. 3).

Table 2. Maximum complexity of monotone securities.

n	Adequacy diagram	#States	Proof/witness
0	1 ↓ (1)	1	
1	1 (2) ↓ (2) ⇉ 1	2	Theorem 3.9
2	1 2 (4) ↓ (2) ⇉ 1	4	Theorem 3.9
3	1 2 (4) ↓ (5) ⇉ 2 1	6	Theorem 3.9; Example 3.4
4	1 2 4 (8) ↓ (5) ⇉ 2 1	10	Theorem 3.9
5	1 2 4 (8) ↓ (19) ⇉ 5 2 1	15	Theorem 3.9
6	1 2 4 8 (16) ↓ (19) ⇉ 5 2 1	23	Theorem 3.10
7	1 2 4 8 16 (32) ↓ (19) ⇉ 5 2 1	39	Figure 1; Theorem 3.10
8	1 2 4 8 16 (32) ↓ (167) ⇉ 19 5 2 1	58	Theorem 3.8
9	1 2 4 8 16 (32) (64) ↓ (167) ⇉ 19 5 2 1	90	Theorem 3.13
10	1 2 4 8 16 32 64 (128) ↓ (167) ⇉ 19 5 2 1	154	Theorem 3.12

The sets $\{p \wedge q, p \vee q\}$ and $\{p \wedge q, 1\}$ both have the desirable property (from the point of view of increasing the complexity) that by substitution we obtain a full set of nonzero monotone functions in one fewer variables, in this case $\{p, 1\}$.

Definition 3.5. Let us say that a set of monotone functions on variables $p_1, \ldots, p_n$ is *adequate* if by substitutions of values for $p_1 \in \{0, 1\}$ they contain all monotone nonzero functions on $p_2, \ldots, p_n$. If one value for p_1 suffices then we say *strongly adequate*.

Let us write $\mathbf{2}^i$ for the set $\{0, 1\}^i$ with the product ordering.

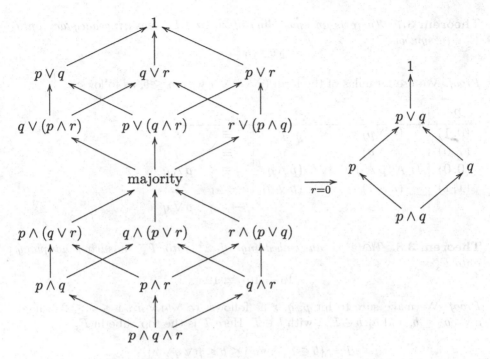

Fig. 2. Adequacy in the proof that $2^4 \to 19 \rightrightarrows 5$.

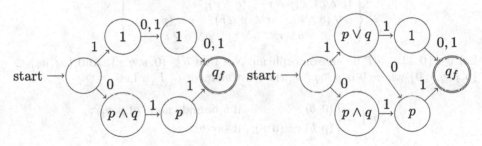

Fig. 3. Asian option, and European call option (corresponding to the majority function).

Definition 3.6. If there is an embedding of 2^i into F_j^- ensuring adequacy onto F_{j-1}^-, in the sense that we map *into* F_{j-1} (so self-loops may be used in the automaton), and we map *onto* F_{j-1}', then we write

$$2^i \to |F_j^-| \rightrightarrows |F_{j-1}^-|.$$

It is crucial to note that in Sect. 2, adequacy was automatic: the concept of function is much more robust than that of a monotone function, meaning that functions can be combined in all sorts of ways and remain functions. As an example of the unusual but convenient notation of Definition 3.6, we have:

Theorem 3.7. *There is an embedding of 2^2 into F_3^- ensuring adequacy onto F_2^-. In symbols,*

$$4 \to 19 \rightrightarrows 5.$$

Proof. We use formulas of the form $(r \wedge b) \vee a$ with $a \leq b$, as follows:

2^2	F_3^-				F_2^-
$(0,1)$	$(r \wedge p)$	$\vee$	p	$\equiv$	p
$(1,0)$	$(r \wedge q)$	$\vee$	q	$\equiv$	q
$(0,0)$	$(r \wedge (p \wedge q))$	$\vee$	$(p \wedge q)$	$\equiv$	$p \wedge q$
$(1,1)$	$(r \wedge 1)$	$\vee$	$(p \vee q)$	$\mapsto_{r=1}$	1
				$\mapsto_{r=0}$	$p \vee q$

$\square$

Theorem 3.8. *There is an embedding of 2^4 into F_4^- ensuring adequacy onto F_3^-:*

$$16 \to 167 \rightrightarrows 19$$

Proof. We make sure to hit p, q, r as follows: $(r \wedge b) \vee a_i$, $1 \leq i \leq 2$, where $a_1 < a_2 \leq b$, and $a_i, b \in F_3^-$, with $b \in T$. Here T is the top cube in F_3^-,

$$T = \{b \in F_3^- : \mathrm{maj} \leq b \leq p \vee q \vee r\}$$

$$= \left\{ \begin{array}{l} (p \wedge q) \vee (p \wedge r) \vee (q \wedge r), \\ p \vee (q \wedge r), \quad q \vee (p \wedge r), \quad r \vee (p \wedge q), \\ p \vee q, \quad p \vee r, \quad q \vee r, \quad p \vee q \vee r \end{array} \right\}.$$

Let $\psi : \{0,1\}^3 \to T$ be an isomorphism. Not that $b \notin \{\hat{0}, p, q, r\}$. And $\{a_1, a_2\} \subset \{b, p, q, r, \hat{0}\}$ where $\hat{0}$ is $p \wedge q \wedge r$, the least element of F_3^-. Let

$$(a_1, a_2) = \begin{cases} (\hat{0}, b) & \text{if } b \text{ bounds none of } p, q, r; \\ (p, b) \text{ or } (\hat{0}, p) & \text{if say } b > p; \end{cases}$$

By Lemma above, $(r \wedge \psi(x)) \vee a_i \leq (r \wedge \psi(y)) \vee c_i$ iff $x \leq y$ and $a_i \leq c_i$. $\square$

We can consider whether $u \to v \rightrightarrows w$ whenever the numbers are of the form 2^m, $|F_n^-| \in \{1, 2, 5, 19, 167, \ldots\}$, $|F_{n-1}^-|$, and $u \leq v$ and $w \leq 2u$ (as u increases, being 1:1 becomes harder but being adequate becomes easier). In the case of strong adequacy witnessed by $p = p_0$ we write simply $u \to v \to_{p_0} w$; this can only happen when $w \leq u$.

Theorem 3.9. *We have the following adequacy calculations:*

1. $2^0 \to 2 \to 1$
2. $2^1 \to 2 \to 1$
3. $2^0 \to 5 \rightrightarrows 2$
4. $2^1 \to 5 \to 2$
5. $2^2 \to 5 \to 2$

We omit the trivial proof of Theorem 3.9.

Theorem 3.10. $2^3 \to 19 \rightrightarrows 5$ *and* $2^4 \to 19 \rightrightarrows 5$.

Proof. The map in Fig. 1 is onto $F_3^- \setminus \{p, q, r\}$ so it works. As shown in Fig. 2, if we restrict that map to the top cube, mapping onto $T \cup \{1\} \setminus \{p \vee q \vee r\}$, and set $r = 0$ then we map onto F_2^-. $\qquad\square$

Lemma 3.11. *Let* a_1, a_2, b_1, b_2 *be Boolean functions of* p, q, r *and let*

$$f_{a_i b_i}(p, q, r, s) = [s \wedge b_i] \vee [\neg s \wedge a_i].$$

Then $f_{a_1 b_1} \leq f_{a_2 b_2} \iff a_1 \leq a_2$ *and* $b_1 \leq b_2$.

Proof. By definition,

$$f_{a_1 b_1} \leq f_{a_2 b_2} \iff [p_4 \wedge b_1] \vee [\neg p_4 \wedge a_1] \leq [p_4 \wedge b_2] \vee [\neg p_4 \wedge a_2].$$

Clearly, $a_1 \leq a_2$ and $b_1 \leq b_2$ implies this, so we just need the converse. If $a_1 \nleq a_2$ then any assignment that makes p_4 false, a_1 true, and a_2 false will do. Similarly if $b_1 \nleq b_2$ then any assignment that makes p_4 true, b_1 true, and b_2 false will do. $\qquad\square$

Theorem 3.12. *There is an injective isotone map from* 2^6 *into* F_4, *and in fact*

$$2^6 \to 167 \rightrightarrows 19.$$

Proof. We start with a monotone version of the simple equation $2^{2^n} = (2^{2^{n-1}})^2$. Namely, a pair of monotone functions g, h of $n - 1$ variables, with $g \leq h$, gives another monotone function via

$$\begin{aligned} f(p_1, \ldots, p_n) &= [p_n \wedge f(p_1, \ldots, p_{n-1}, 1)] \vee [\neg p_n \wedge f(p_1, \ldots, p_{n-1}, 0)] \\ &= [p_n \wedge h(p_1, \ldots, p_{n-1})] \vee [\neg p_n \wedge g(p_1, \ldots, p_{n-1})] \\ &= [p_n \wedge h(p_1, \ldots, p_{n-1})] \vee g(p_1, \ldots, p_{n-1}). \end{aligned}$$

Now consider elements a of the bottom hypercube in F_3 and b of the top hypercube in F_3 in Fig. 2. So we must have $a \leq b$ since the bottom is below the top (and $a = b$ can happen since the two hypercubes overlap in the majority function). Let $f_{ab} = [p_4 \wedge b] \vee [\neg p_4 \wedge a]$. Since $a \leq b$, f_{ab} is monotonic. By Lemma 3.11, these functions f_{ab} are ordered as $2^6 = 2^3 \times 2^3$.

Finally, in order to ensure adequacy we modify this construction to reach higher in F_4^-, replacing the top cube in the lower half by a cube formed from the upper half. In more detail, consider $(r \wedge b) \vee a$ with $a \leq b$ from F_3^-, where the a's are chosen from the bottom cube of F_3, and the b's from the top cube, except that when a is the top of the bottom cube we let b be the top cube with the top replaced by 1, and when a is the bottom of the bottom cube we let b be the cube

$$\{p, q, r, p \wedge q, p \wedge r, q \wedge r, p \vee q, p \vee r, q \vee r, p \wedge q \wedge r, p \vee q \vee r\}.$$

$\qquad\square$

406   B. Kjos-Hanssen and L. Liu

Theorem 3.13. $2^5 \to 167 \rightrightarrows 19$.

Proof. A small modification of Theorem 3.12; only use bottom, top and two intermediate "cubes" within the cube.   □

Open Problem. For $n = 11$ we need to determine whether the following holds, which has so far proved too computationally expensive:

$$2^7 \to 167 \rightrightarrows 19?$$

That is, is there an isotone map from the 128-element lattice $\mathbf{2}^7$ into F_4^-, the set of nonzero monotone functions in variables p, q, r, s, such that upon plugging in constants for p, we cover all of F_3^-, the set of nonzero monotone functions in q, r, s?

Probability

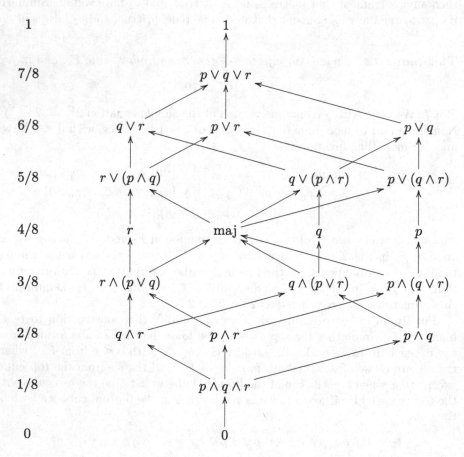

Fig. 4. The lattice F_3 of all monotone Boolean functions in three variables p, q, r.

4 Early-Monotone Functions and Complete Simple Games

In this section we take the financial ideas from Sect. 3 one step further, by noting that the Asian option (Example 3.4) has the added property that earlier bits matter more. In economics terms, we have what is called a *complete simple game*: there is a set of goods linearly ordered by intrinsic value. You get some of the goods and there are thresholds for how much value you need to win.

Definition 4.1. Let $e_i \in \{0,1\}^n$ be defined by $e_i(j) = 1$ if and only if $j = i$. An n-ary Boolean function f is *early* if for all $0 \leq i < j < n$ and all $y \in \{0,1\}^n$ with $y(i) = y(j) = 0$, if $f(y + e_j) = 1$ then $f(y + e_i) = 1$.

The number of early (not necessarily monotone) functions starts

$$2, 4, 12, 64, 700, 36864, \ldots$$

If a function is early and monotone we shall call it early-monotone. Early-monotonicity encapsulates an idea of time-value-of-money; getting paid now is better than next week, getting promoted now is better than next decade, etc.

In the early context one needs the map from 2^m into the early functions to be "early", i.e., the function mapped to by 100 should dominate the one mapped to by 010 etc. That is, the map must be order-preserving from 2^m with the majorization lattice order into the complete simple games.

The number of early-monotone functions on n variables, including zero, is

$$2, 3, 5, 10, 27, 119, 1173, \ldots$$

which appears in OEIS A132183 as the number of "regular" Boolean functions in the terminology of Donald Knuth. He describes them also as the number of order ideals (or antichains) of the binary majorization lattice with 2^n points.

Definition 4.2. The binary majorization lattice E_n is the set $\{0,1\}^n$ ordered by $(a_1, \ldots, a_n) \leq (b_1, \ldots, b_n)$ iff $a_1 + \cdots + a_k \leq b_1 + \cdots + b_k$ for each k.

The lattice E_5 for $n = 5$ is illustrated in [7, Fig. 8, Volume 4A, Part 1]. The basic properties of this lattice are discussed in [7, Exercise 109 of Section 7.1.1]. The majorization order is obtained by representing e.g. 1101 as $(1, 2, 4, \infty)$, showing where the kth 1 appears (the ∞ signifying that there is no fourth 1 in 1101), and ordering these tuples by majorization. OEIS cites work of Stefan Bolus [2] who calls the "regular" functions *complete simple games* [8], a term from the economics and game theory literature. There, arbitrary monotone functions are called *simple games*, and "complete" refers to the fact that the positions have a complete linear ordering (in the finance application, earlier positions are most valuable). Figure 5 shows that in the complete-simple-games setting we have

$$1 \rightarrow 2 \rightarrow 4 \rightarrow 8 \rightarrow 16$$
$$\downarrow$$
$$26 \rightrightarrows 9 \rightarrow 4 \rightarrow 2 \rightarrow 1$$

for a total maximal complexity of 47 for complete simple games at $n = 8$. This contrasts with Theorem 3.8 which shows that for arbitrary simple games the complexity can reach 58 at $n = 8$.

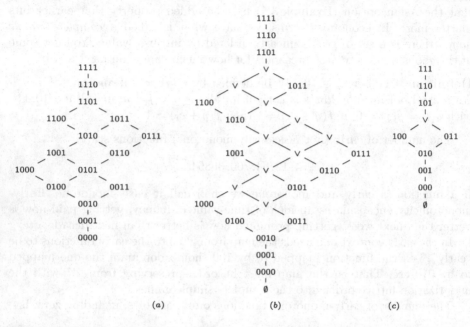

Fig. 5. (a) The majorization lattice E_4 on 4 variables. (b) The 27 complete simple games C_4 on 4 variables. The symbol ∨ denotes an element that is join-reducible. Red and blue denote the image under the first map $E_4 \rightarrow C_4$ and blue in particular denotes some elements sufficient for the second map $C_4 \rightarrow C_3$ to be onto. (c) The 10 complete simple games C_3 on 3 variables. (Color figure online)

References

1. Arora, S., Barak, B., Brunnermeier, M., Ge, R.: Computational complexity and information asymmetry in financial products. Commun. ACM **54**(5), 101–107 (2011)
2. Bolus, S.: Power indices of simple games and vector-weighted majority games by means of binary decision diagrams. Eur. J. Oper. Res. **210**(2), 258–272 (2011)
3. Câmpeanu, C., Ho, W.H.: The maximum state complexity for finite languages. J. Autom. Lang. Comb. **9**(2–3), 189–202 (2004)
4. Champarnaud, J.-M., Pin, J.-E.: A maxmin problem on finite automata. Discrete Appl. Math. **23**(1), 91–96 (1989)
5. Hopcroft, J.E., Ullman, J.D.: Introduction to Automata Theory, Languages, and Computation. Addison-Wesley Series in Computer Science. Addison-Wesley Publishing Co., Reading (1979)

6. Kjos-Hanssen, B.: On the complexity of automatic complexity. Theory Comput. Syst. **61**(4), 1427–1439 (2017)
7. Knuth, D.E.: The Art of Computer Programming. Combinatorial Algorithms, Part 1, vol. 4A. Addison-Wesley, Upper Saddle River (2011)
8. Kurz, S., Tautenhahn, N.: On Dedekind's problem for complete simple games. Int. J. Game Theory **42**(2), 411–437 (2013)
9. Linz, P.: An Introduction to Formal Language and Automata. Jones and Bartlett Publishers Inc., Burlington (2006)
10. Nerode, A.: Linear automaton transformations. Proc. Am. Math. Soc. **9**, 541–544 (1958)
11. Shreve, S.E.: Stochastic Calculus for Finance I: The Binomial Asset Pricing Model. Springer Finance Textbooks. Springer, New York (2004)
12. Sloane, N.J.A.: The online encyclopedia of integer sequences (2018). Sequence A000372

Deterministic Coresets for Stochastic Matrices with Applications to Scalable Sparse PageRank

Harry Lang[1], Cenk Baykal[1(✉)], Najib Abu Samra[2], Tony Tannous[2],
Dan Feldman[2], and Daniela Rus[1]

[1] MIT CSAIL, Cambridge, USA
{harry1,baykal,rus}@mit.edu
[2] Computer Science Department, University of Haifa, Haifa, Israel
najib.as1990@gmail.com, tonytanios1994@gmail.com,
dannyf.post@gmail.com

Abstract. The PageRank algorithm is used by search engines to rank websites in their search results. The algorithm outputs a probability distribution that a person randomly clicking on links will arrive at any particular page. Intuitively, a node in the center of the network should be visited with high probability even if it has few edges, and an isolated node that has many (local) neighbours will be visited with low probability. The idea of PageRank is to rank nodes according to a stable state and not according to the previous local measurement of inner/outer edges from a node that may be manipulated more easily than the corresponding entry in the stable state.

In this paper we present a deterministic and completely parallelizable algorithm for computing an ε-approximation to the PageRank of a graph of n nodes. Typical inputs consist of millions of pages, but the average number of links per page is less than ten. Our algorithm takes advantage of this sparsity, assuming the out-degree of each node at most s, and terminates in $O(ns/\varepsilon^2)$ time. Beyond the input graph, which may be stored in read-only storage, our algorithm uses only $O(n)$ memory. This is the first algorithm whose complexity takes advantage of sparsity. Real data exhibits an average out-degree of 7 while n is in the millions, so the advantage is immense. Moreover, our algorithm is simple and robust to floating point precision issues. Our sparse solution (core-set) is based on reducing the PageRank problem to an ℓ_2 approximation of the Carathéodory problem, which independently has many applications

Lang, Baykal, and Rus thank NSF 1723943, NSF 1526815, and The Boeing Company. This research was supported by Grant No. 2014627 from the United States-Israel Binational Science Foundation (BSF) and by Grant No. 1526815 from the United States National Science Foundation (NSF). Dan Feldman is grateful for the support of the Simons Foundation for part of this work that was done while he was visiting the Simons Institute for the Theory of Computing.
H. Lang and C. Baykal—contributed equally to this work.

© Springer Nature Switzerland AG 2019
T. V. Gopal and J. Watada (Eds.): TAMC 2019, LNCS 11436, pp. 410–423, 2019.
https://doi.org/10.1007/978-3-030-14812-6_25

such as in machine learning and game theory. We hope that our approach will be useful for many other applications for learning sparse data and graphs.

Algorithm, analysis, and open code with experimental results are provided.

1   Introduction

Matrix Notation. For an integer $n \geq 1$, let $[n] = \{1, \cdots, n\}$. We denote by $\mathbb{R}^{n \times n}$ the set of $n \times n$ real matrices. The jth column of G is denoted by $G_{\bullet j}$ and the entry on its ith row is G_{ij}. The ith entry of a column vector $v \in \mathbb{R}^n$ is denoted by v_j. We write 0_n for the n-dimensional zero vector, and $\mathcal{I}_n$ for the $n \times n$ identity matrix. Let e_j denote the jth column of $\mathcal{I}_n$.

Stochastic Matrix. A *distribution vector* $z \in [0,1]^n$ is a non-negative vector whose sum is 1. A *column-stochastic matrix* is a matrix such that every one of its columns is a distribution vector. The input matrix G is called the *transition matrix* of a graph if it is equal to the adjacency matrix but with each non-zero column scaled to have unit sum. Every positive column-stochastic matrix $A \in \mathbb{R}^{n \times n}$ (i.e. whose entries are positive) has a distribution vector $z^* \in \mathbb{R}^n$ such that $Az^* = z^*$ by the Perron-Frobenius theorem. This vector is called the *stable state* of A.

Problem Setup. The input to the PageRank problem is a non-negative square matrix $G \in [0,1]^{n \times n}$ that represents the scaled adjacency matrix of a graph, i.e. the entry G_{ij} represents the probability of moving from node i to node j, and each column of G is scaled so that its sum is 1. For simplicity, we assume (only for the moment) that there is no node with out-degree 0.

Hence, G is a column-stochastic matrix where each of its columns defines a distribution, i.e., a vector whose entries are non-negative and sum to 1. For example, its second column $G_{\bullet 2}$ defines the visited node after taking a single step (random walk) from the second node to its random neighbour. For a given vector $z \in [0,1]^n$ that defines a distribution over the currently visited nodes, the multiplication $y = Gz$ yields the distribution y of the visited nodes after taking a single random step from the current state. In this sense, G describes a Markov Chain. If $Gz^* = z^*$ for some distribution z^*, then z^* is called a *stable state* of G. Hence, given an initial distribution z^*, the distributed on the next visited node will not change.

Stable State and Page's Rank. The stable state z^* is important because it can be also be proved roughly that the probability of stopping on the ith node after a sufficiently long random walk (when the initial visited node has almost no influence), approaches z_i^*. Intuitively, a node in the center of the network should be visited with high probability even if it has few edges, and an isolated node that has many (local) neighbours will be visited with low probability.

The idea of PageRank is to rank nodes according to this stable state z^* and not according to the previous local measurement of inner/outer edges from a node, since these may be manipulated more easily than the corresponding entry in the stable state. Intuitively, this is because the stable state measures a global property that depends on all the nodes in the graph (connecting a single edge in one side of the graph changes the rank of a node even of the other side of the graph). Unfortunately, this is also why the problem of computing the stable state of G is hard. In fact, this vector is not unique.

Damping Factor. To solve this issue, the PageRank algorithm uses a given a parameter $d \in (0,1)$ which is called the *damping factor*. A common value is $d = 0.15$. Let $\mathbf{1}$ denote the $n \times n$ matrix whose all entries are 1, and let $A = (1-d)G + \frac{d}{n} \cdot \mathbf{1}$. For example, suppose that in $G_{ji} = G_{ki} = 1/2$ for the transitions $i \to j$ and $i \to k$. In the matrix A, the probability of moving to any other node increases from 0 to d/n, and the probability $1/2$ of moving from i to j or k is changed to $\frac{1-d}{2} + \frac{d}{n}$. The new sum of probabilities is still 1, so A is now a positive column-stochastic matrix. In the context of surfing the internet, if each column of G represents links in a web-page, in A, with probability d we will visit a new random page that is not linked from our current web-page.

More generally, in the v-*Personalized PageRank* we get an additional input distribution vector $v \in (0,1)^n$ whose entries are strictly positive and sum to one, and define $A = (1-d)G + d \cdot v \cdot (1, \cdots, 1)$. In this case, v defines a distribution over the webpages to visit in case that no web-link is chosen in the existing page. For the uniform distribution $v = (1/n, \cdots, 1/n)^\top$ the v-Personalized PageRank is the same as PageRank.

Unlike G, which is a non-negative matrix, A is a *positive* column-stochastic matrix. By this fact and the Perron-Frobenius theorem, A has a unique stable state distribution z^* such that $Az^* = z^*$. Hence, the PageRank problem reduced to the problem of computing this vector. A common way to compute the largest eigenvector of such a matrix A is to use the power method. This iterative technique computes $z^{(0)} = A \cdot (1/n, \cdots, 1/n)^T$, and $z^{(i)} = Az^{(i-1)}$ for every $i \geq 1$. It can be proven that the approximation error $\left\| Az^{(i)} - z^{(i)} \right\|_2 \leq \varepsilon$ after $O(\log(n)/\varepsilon)$ iterations. Recall that $\|Az^* - z^*\|_2 = 0$ since $Az^* = z^*$.

Our Result. In this paper we suggest a *deterministic* algorithm for computing a provable ε-approximation z to the PageRank z^* of a graph with n nodes. The exact running time and memory depends on the sparsity s of each column of G, i.e. the maximal number of neighbours or, more precisely, the out-degree, of each node in G. This number is small in social or web networks, where most of the nodes have few related links to neighbor nodes, compared to the total number n of nodes in the network. In fact, real datasets show $s \approx 7$ and $n > 10^6$ [8], making the improvement from $O(n^2)$ to $O(ns)$ quite drastic.

Formally, we present an algorithm that computes a distribution vector z such that $\|Az - z\|_2 \leq \varepsilon$, compared to the optimal solution where $\|Az^* - z^*\| = 0$. The algorithm runs in $O(ns/\varepsilon^2)$ time and requires only $O(n)$ read-write memory in addition to reading the input graph (which is sparsely represented using $O(ns)$ memory).

Our output z is a sparse vector with $O(1/\varepsilon^2)$ non-zero entries. Intuitively, the non-zero entries of z are the "heavy hitters" or important nodes, while small ranks are rounded to 0. Our algorithm is iterative, and the user can stop it after finding the top desired ranks, or after observing that the last returned rank is already small.

Our Techniques. We show that the PageRank problem can be reduced to the Approximated Carathéodory, which was recently used in applications such as machine learning, and game theory [5]. In this problem we wish to approximate a point in the convex hull of n points by a convex combination of a small subset of these points.

The problem of computing z^* such that $Az^* = z^*$ is the same as computing z^* such that $(A - \mathcal{I}_n)z^* = 0$. We define $B = A - \mathcal{I}_n$ and seek a distribution vector z such that $\|Bz\|_2 \leq \varepsilon$. Viewing the columns of B as points in $\mathbb{R}^d$, the problem is reduced to finding a convex combination of points close to the origin. We instantiate the Frank-Wolfe algorithm to iteratively add columns of B until we arrive sufficiently close to the origin. At most $O(1/\varepsilon^2)$ columns are needed before we arrive at a sufficient solution.

While the input G is has sparse columns, the matrix B is not sparse because of the damping factor d and distribution vector v. In fact, B has no zeros. We show a technique to cache previous computations and extract the relevant information from G, d, and v without every actually computing B.

For simplicity, we prove the main claim for the PageRank application, but the proof is written in more general notation that we hope will be used by other researchers for other functions, such as estimators that are robust to outliers as the m-estimators, that usually have the required smoothness condition.

Related Work. A frequently explored technique for solving PageRank involves some variant of random walks, appearing in results such as [2,6,12,16,17] A sampling based approach is suggest in [1]. Multipass streaming algorithms for approximating the rank (i.e. value in the stable state) of a single node have been explored [15]. Under some assumptions, it is also possible to return a list of the top-k ranked nodes [3].

In the distributed setting, [17] present a $O(\frac{\log n}{d})$-round algorithm that approximates each entry of the stable state up to a multiplicative factor of $(1 + \varepsilon)$. However, they treat computation time as an unlimited resource and the runtime is not explicitly bounded.

Extensions of PageRank where the stable state may fluctuate in time have been explored [4,13,14,18]. When considering location-based networks, an algorithm was presented by [9].

Classical techniques include the power method (iterating until the stable state is approached asymptotically) and explicit computation (involving finding a matrix inverse). These classical techniques are deterministic but require $\omega(n^2)$ time. All prior $O(n^2)$ time algorithms require randomization, making ours the first fast algorithm to be entirely deterministic. Moreover, we take advantage of sparsity and therefore run in optimal $O(n)$ time on real datasets whose average out-degree is around 7 [8].

2 Preliminaries

The network of n pages is represented by an $n \times n$ matrix G. The ith column represents the links on the ith page. If a page contains no links, its column is filled with zeros. Otherwise, the column's entries sum to 1 where the jth entry represents the probability that a user who clicks on a link from page i will follow a link to page j.

Definition 1 (Unprocessed PageRank Matrix). *The input to the PageRank problem is a non-negative $n \times n$ matrix G where each column represents the links from a page. If the page is a sink node with no links to other pages, the column is all zeros. Otherwise, the column represents the transition probabilities to other pages and has unit ℓ_1 norm. We call such a matrix the* unprocessed PageRank matrix.

With probability $d \in (0,1)$, a user may jump to page without clicking a link from the current page. The distribution vector v represents the probabilities that a user will randomly jump to a particular page. The matrix Ψ, introduced in the next definition, accounts for this possibility of jumping without following any link. The entry Ψ_{ij} is the probability that a user on page j will move to page i.

Definition 2 (Approximate PageRank). *Let $G \in \mathbb{R}^{n \times n}$ be an unprocessed PageRank matrix. Given a damping factor $d \in (0,1)$ and a positive distribution vector v, the processed PageRank matrix $\Psi(G, d, v)$ is defined column-by-column as:*

$$\Psi(G, d, v)_{\bullet i} = \begin{cases} v & \text{if } G_{\bullet i} = 0 \\ dG_{\bullet i} + (1-d)v & \text{if } \|G_{\bullet i}\|_1 = 1 \end{cases}$$

Note that $\Psi(G, d, v)$ is a positive column-stochastic matrix, so there exists a unique distribution vector z^* such that $\Psi(G, d, v)z^* = z^*$. A distribution vector x is called an ε-approximation if $\|(\Psi(G, d, v) - \mathcal{I}_n)x\|_2 \leq \varepsilon$.

We relate PageRank to a well-studied geometric problem called the Carathéodory problem. In what follows, 0_d denotes the origin in $\mathbb{R}^d$, and $\mathcal{C}(P)$ denotes the convex hull of a point-set P.

Definition 3. *Let P be a set of n points in $\mathbb{R}^d$ such that $0_d \in \mathcal{C}(P)$. An ε-approximation to the ℓ_2-Carathéodory Problem is a multiset Q of T points from P such that:*

$$\left\| \frac{1}{T} \sum_{q \in Q} q \right\|_2 \leq \varepsilon$$

In the previous definition, the unweighted average is taken over Q. However, Q is a multiset meaning that it may contain points with multiplicity. One can consider a point with multiplicity m to have weight $\frac{m}{T}$. As a simple example, let $P \subset \mathbb{R}^2$ be three points whose convex hull contains the origin. Depending on the location of the origin, an arbitrarily large Q is required for sufficiently small ε despite the fact that P contains only three distinct points.

Definition 4 (Smoothness). *A continuously differentiable function* $f : \mathbb{R}^d \to \mathbb{R}$ *is said to be* β-*smooth with respect to norm* $\|\cdot\|$ *on domain* $D \subset \mathbb{R}^d$ *if:*

$$\|\nabla f(x) - \nabla f(y)\| \le \beta \|x - y\|$$

for every $x, y \in D$.

For $p \ge 1$ we write $\|x\|_p$ to denote the ℓ_p-norm of x which is $(x_1^p + \ldots + x_d^p)^{\frac{1}{p}}$. Observe that the function $\|x\|_p^p$ is convex and continuously differentiable.

Lemma 1. *The function* $f(x) = \|x\|_2^2$ *is 2-smooth with respect to any norm on* $\mathbb{R}^d$.

Proof.

$$\nabla(\|x\|_2^2) = \nabla(x_1^2 + \ldots + x_d^2)$$
$$= (2x_1 \quad ,\ldots \quad 2x_d)^\top$$
$$= 2x$$

Therefore $\|\nabla f(x) - \nabla f(y)\| = 2 \|x - y\|$ where we have pulled out the 2 by the scalability property of a norm.

Outline of Presentation. In the next section, we present the well-known Frank-Wolfe algorithm (Algorithm 1 and Theorem 1). We show that Frank-Wolfe can be used to solve the Carathéodory problem (Corollary 1). We will transform the PageRank problem so it can be solved by Frank-Wolfe, presenting a simple deterministic algorithm (Algorithm 2 and Lemma 3). Finally, we make modifications to the algorithm so that it takes advantage of the sparsity of the input (Algorithm 3 and Theorem 2).

3   Algorithm

The following algorithm was first presented by Frank and Wolfe in 1956 [7]. As written, it loops infinitely. In practice, we terminate the loop after a sufficient number of iterations given by gaurantees such as in Theorem 1.

We begin by setting x to be an arbitrary point of P. In each iteration, we select a point y which is a vertex of the convex hull of P. We then redefine x as a weighted average of itself and y, where each iteration gives less weight to the new y.

The first observation is that the $\{x^{(t)}\}$ are simply averages of the $\{y^{(t)}\}$.

Algorithm 1. Input: continuously differentiable function $f : \mathbb{R}^d \to \mathbb{R}$ and point set $P = \{p_1, \dots, p_n\} \subset \mathbb{R}^d$

1: $x^{(0)} \leftarrow p_1$
2: $t \leftarrow 1$
3: **while** true **do**
4: $\quad \eta^{(t)} \leftarrow 1/t$
5: $\quad y^{(t)} \leftarrow \arg\min_{y \in P} y^\top \nabla f(x^{(t-1)})$
6: $\quad x^{(t)} \leftarrow (1 - \eta^{(t)}) x^{(t-1)} + \eta^{(t)} y^{(t)}$
7: $\quad t \leftarrow t + 1$

Lemma 2. $x^{(t)} = \frac{1}{t}(y^{(1)} + \dots + y^{(t)})$ *for every* $t \geq 1$.

Proof. Observe directly from Line 6 that $x^{(1)} = y^{(1)}$. Now assume inductively that $x^{(t-1)} = \frac{1}{t-1}(y^{(1)} + \dots + y^{(t-1)})$. $\eta^{(t)} = \frac{1}{t}$ and so:

$$
x^{(t)} = \left(1 - \frac{1}{t}\right) x^{(t-1)} + \frac{1}{t} y^{(t)}
$$

$$
= \left(\frac{t-1}{t}\right)\left(\frac{1}{t-1}\right)(y^{(1)} + \dots + y^{(t-1)}) + \frac{1}{t} y^{(t)}
$$

$$
= \frac{1}{t}(y^{(1)} + \dots + y^{(t)})
$$

The following theorem governs the behavior of the Frank-Wolfe algorithm, and was presented in the original paper.

Theorem 1 ([7]). *Fix a norm* $\|\cdot\|$. *If Algorithm 1 is run for* T *iterations on input* (f, P) *where* $\max_{p \in P} \|p\| \leq R$ *and* f *is a* β-*smooth convex function with respect to* $\|\cdot\|$ *on* $\mathcal{C}(P)$, *then:*

$$
f(x_T) - \min_{x \in \mathcal{C}(P)} f(x) \leq \frac{2\beta R^2}{T+1}
$$

We arrive at the following corollary by applying the previous theorem with basic manipulations. In this form, it is apparent how the Carathéodory problem may be solved using Frank-Wolfe.

Corollary 1. *Let* $f(x) = \|x\|_2^2$ *and* $T = \lceil \frac{8}{\varepsilon^2} - 1 \rceil$. *If Algorithm 1 is run for* T *iterations on input* (f, P) *where* $\max_{p \in P} \|p\|_2 \leq \sqrt{2}$ *and* $0 \in \mathcal{C}(P)$, *then:*

$$
\left\| x^{(T)} \right\|_2 \leq \varepsilon
$$

We now show that Algorithm 1 provides an algorithm for PageRank, forging a connection between the two problems. Observe that when we form a $d \times n$ matrix B whose columns are the points of P, then $\mathcal{C}(P)$ is the union of Bz over all distribution vectors $z \in \mathbb{R}^n$.

Algorithm 2 outputs an ε-approximation for Personalized PageRank. When adapting Algorithm 1, we have set $f(x) = \|x\|_2^2$ and so $\nabla f(x^{(t-1)})$ is simply $2x^{(t-1)}$. We also introduce a new vector z which will be the approximate stable state.

Lemma 3. *Algorithm 2 outputs a distribution vector z that satisfies $\|Bz\|_2 \leq \varepsilon$.*

Proof. It is clear that z is a distribution vector since Line 10 is executed T times so z is non-negative and $\|z\|_1 = 1$. It remains to show the bound $\|Bz\|_2 \leq \varepsilon$.

Given the positive column-stochastic matrix $B = \Psi(G, d, v) - \mathcal{I}_n$, we wish to find a vector z such that $\|z\|_1 = 1$ and $\|Bz\|_2 \leq \varepsilon$. Let $P \subset \mathbb{R}^n$ be the columns of B. By the Perron-Frobenius theorem, there exists a unique stable state z^* such that $Bz^* = 0$. This implies that the convex hull of P contains the origin, since the values of z^* define weightings of a convex combination of the columns of B. The reader can verify that $\|A_{\bullet i}\|_1 = 1$ implies $\|B_{\bullet i}\|_2 \leq \sqrt{2}$. Therefore by Corollary 1, Algorithm 1 will satisfy that $\left\|x^{(T)}\right\|_2 \leq \varepsilon$ since we have set $T = \lceil \frac{8}{\varepsilon^2} - 1 \rceil$ on Line 4.

It now suffices to show that $Bz = x^{(T)}$. Observe that $Be_i = B_{\bullet i}$. Let $j^{(t)}$ be the value set on Line 8 during the tth iteration of the while-loop. $y^{(t)} = B_{\bullet j^{(t)}} = Be_{j^{(t)}}$. By Line 10, $z = \frac{1}{T} \sum_{t \leq T} e_{j^{(t)}}$ when the algorithm terminates. Therefore:

$$Bz = \frac{1}{T} \sum_{t \leq T} Be_{j^{(t)}}$$

$$= \frac{1}{T} \sum_{t \leq T} y^{(t)}$$

$$= x^{(T)}$$

where the last line follows from Lemma 2.

Algorithm 2. Input: an unprocessed $n \times n$ PageRank matrix G, damping factor $d \in (0, 1)$, positive distribution vector $v \in (0, 1)^n$, approximation factor $\varepsilon > 0$

1: $B \leftarrow \Psi(G, d, v) - \mathcal{I}_n$
2: $x^{(0)} \leftarrow B_{\bullet 1}$
3: $t \leftarrow 1$
4: $T \leftarrow \lceil \frac{8}{\varepsilon^2} - 1 \rceil$
5: $z \leftarrow 0_n$
6: **while** $t \leq T$ **do**
7:     $\eta^{(t)} \leftarrow 1/t$
8:     $j \leftarrow \arg\min_{i \in [n]} B_{\bullet i}^{\top} x^{(t-1)}$
9:     $y^{(t)} \leftarrow B_{\bullet j}$
10:     $z_j \leftarrow z_j + \frac{1}{T}$
11:     $x^{(t)} \leftarrow (1 - \eta^{(t)}) x^{(t-1)} + \eta^{(t)} y^{(t)}$
12:     $t \leftarrow t + 1$
13: Output z

Theorem 3 gaurantees an ε-approximation, but the matrix $B = \Psi(G, d, v) - \mathcal{I}_n$ is not sparse. In fact, B does not have any zero entries. As written, Algorithm 2 runs in $O(n^2/\varepsilon^2)$ time. In the next section, we will improve the runtime by taking advantage of the fact that although B is not sparse, it can be represented as a function of the sparse matrix G.

4   Sparse Runtime

Each column of G has at most s non-zero entries. We can leverage this in the computation to improve the $O(n^2/\varepsilon^2)$ runtime of Algorithm 2 to the vastly improved $O(ns/\varepsilon^2)$ time. Recall that the average out-degree of a page is around 7 while the total number of pages is at least in the millions, so this causes a drastic improvement.

To take advantage of the sparsity, we assume that the columns of G are stored in memory with an *s-sparse representation* which means a list of at most s index-value pairs. In contrast, by a *standard representation* we mean a list of n values including the zeros such that the value for any index can be retrieved in $O(1)$ time.

Fact 1. *Let u and w be two vectors in $\mathbb{R}^n$. If u has an s-sparse representation, then $u^\top w$ can be computed in $O(s)$ time and $O(1)$ space if w has either an s-sparse representation or a standard representation.*

The proof of Fact 1 is clear. When u is represented as a list of s index-value pairs, we need not even bother to read the values of any entries of v that are not in the list of u. Since these entries are zero in u, their value in v is irrelevant.

We begin with a lemma on how to compute the dot product efficiently between columns of B. This is important because we must compute (see Line 8 of Algorithm 2) the value of i that minimizes the dot product $B_{\bullet i}^\top x^{(t-1)}$. Recall from Lemma 2 that $x^{(t-1)} = \frac{1}{t-1}(y^{(1)} + \ldots + y^{(t-1)})$. From Line 9, each $y^{(i)}$ is a column of B. Therefore we may distribute and write this dot product as a sum of dot products between columns of B.

Lemma 4. *Assume that the value $v^\top v$ is known. For any $1 \leq i, j \leq n$, the value $B_{\bullet i}^\top B_{\bullet j}$ can be computed in $O(s)$ time and $O(1)$ memory.*

Proof. We break into three cases about the two columns $G_{\bullet i}$ and $G_{\bullet j}$: (1) both are zero-columns, (2) only $G_{\bullet j}$ is a zero-column, (3) neither are zero-columns. Note that Case (2) has no loss of generality since if only $G_{\bullet i}$ is a zero-column then we can simply swap i and j before proceeding.

In what follows, we assume that $i \neq j$. If $i = j$ we simply add 1 due to the term $e_i^\top e_j$. Otherwise $e_i^\top e_j = 0$.

Case 1:

$$B_{\bullet i}^\top B_{\bullet j} = (v - e_i)^\top (v - e_j)$$
$$= v^\top v - v_i - v_j$$

Case 2:

$$B_{\bullet i}^{\mathsf{T}} B_{\bullet j} = (dG_{\bullet i} + (1-d)v - e_i)^{\mathsf{T}} (v - e_j)$$
$$= dG_{\bullet i}^{\mathsf{T}} v + (1-d)v^{\mathsf{T}} v - v_i - dG_{ji} + (1-d)v_j$$

Case 3:

$$B_{\bullet i}^{\mathsf{T}} B_{\bullet j} = (dG_{\bullet i} + (1-d)v - e_i)^{\mathsf{T}} (dG_{\bullet j} + (1-d)v - e_j)$$
$$= d^2 G_{\bullet i}^{\mathsf{T}} G_{\bullet j} + d(1-d)(G_{\bullet i}^{\mathsf{T}} v + G_{\bullet j}^{\mathsf{T}} v)$$
$$- d(G_{ij} + G_{ji}) - (1-d)(v_i + v_j)$$
$$+ (1-d)^2 v^{\mathsf{T}} v$$

Given the input tuplet (G, d, v, X, i, j) where $X = v^{\mathsf{T}} v$, the time and space complexities are proven as follows. Case 1 is simple since we return X along with entries for v (retrievable in $O(1)$ time from the standard representation). For Case 2, we must also use Fact 1 to address the $G_{\bullet i}^{\mathsf{T}} v$ term. For the G_{ji} term, this can be found in $O(s)$ time (not necessarily $O(1)$ time since $G_{\bullet i}$ has an s-sparse representation). Case 3 follows using the same principles.

We write $\mathtt{DOT}(G, d, v, X, i, j)$ to denote the $O(s)$ time procedure from Lemma 4 which computes $B_{\bullet i}^{\mathsf{T}} B_{\bullet j}$ given that $X = v^{\mathsf{T}} v$. We use this procedure in Algorithm 3 which is an efficient implementation of Algorithm 2.

Algorithm 3. Input: an unprocessed $n \times n$ PageRank matrix G, damping factor $d \in (0, 1)$, positive distribution vector $v \in (0, 1)^n$, approximation factor $\varepsilon > 0$

1: $T \leftarrow \lceil \frac{8}{\varepsilon^2} 1 \rceil$
2: $X \leftarrow v^{\mathsf{T}} v$
3: $j \leftarrow 1$
4: $\mathtt{CACHE} \leftarrow 0_n$
5: $z \leftarrow 0_n$
6: **for** $t \in [T]$ **do**
7: $\quad j \leftarrow \arg\min_{i \in [n]} \mathtt{CACHE}_i + \mathtt{DOT}(G, d, v, X, i, j)$
8: $\quad z_j \leftarrow z_j + \frac{1}{T}$
9: $\quad$ **for** $i \in [n]$ **do**
10: $\quad\quad \mathtt{CACHE}_i \leftarrow \mathtt{CACHE}_i + \mathtt{DOT}(G, d, v, X, i, j)$
11: Output z

The vectors $x^{(t)}$ and $y^{(t)}$, which were central to Algorithms 1 and 2, are only implicitly present in Algorithm 3. At the end of the tth iteration of the while-loop, $y^{(t)} = B_{\bullet j}$ and $x^t = \frac{T}{t} Bz$.

The vector $\mathtt{CACHE}$ stores previous computations of $B_{\bullet i}^{\mathsf{T}} x^{(t)}$ so we can simply compute $B_{\bullet i}^{\mathsf{T}} y^{(t+1)}$ in $O(s)$ time instead of re-computing this entire quantity in $O(ts)$ time. Without caching the computation, we would have an overall runtime of $O(ns/\varepsilon^4)$.

Lemma 5. *In Algorithm 3, $\text{CACHE}_i = tB_{\bullet i}^\top x^{(t)}$ after the tth iteration of the while-loop. Here $x^{(t)}$ is the value from Algorithm 2.*

Proof. Let $\text{CACHE}^{(t)}$ denote the value of CACHE after the tth iteration of the while-loop. Then $\text{CACHE}^{(0)} = 0_n$. By Lemma 2, $tx^{(t)} = y^{(1)} + \ldots + y^{(t)}$. By induction, it suffices to show that $\text{CACHE}_i^{(t)} = \text{CACHE}_i^{(t-1)} + B_{\bullet i}^\top y^{(t)}$.

After Line 7 of the tth iteration, j is such that $y^{(t)} = B_{\bullet j}$. Lemma 4 establishes that $\text{DOT}(G, d, v, X, i, j) = B_{\bullet i}^\top B_{\bullet j}$. Combining these statements shows that Line 10 sets $\text{CACHE}_i^{(t)} = \text{CACHE}_i^{(t-1)} + B_{\bullet i}^\top y^{(t)}$ as desired.

We now present our main theoretical result. Algorithm 3 is simple, deterministic, parallelizable, and takes advantage of the sparsity of the input.

Theorem 2. *Algorithm 3 is deterministic, terminates in $O(ns/\varepsilon^2)$ time, requires $O(n)$ memory in addition to the read-only input, and returns an ε-approximation for Personalized PageRank of an input (G, d, v). Moreover, the algorithm is parallelizable with $P \le n$ processors with runtime $O((\frac{n}{P} + \log P)s/\varepsilon^2)$ and the same memory requirement.*

Proof. It is clear that the algorithm is deterministic. Correctness follows from equivalence to Algorithm 2 which is straightforward to verify.

For runtime, Line 2 takes $O(n)$ time. On Line 7, there are n invocations of DOT, each of which takes $O(s)$ time by Lemma 4. Therefore this line takes $O(ns)$ time. The loop has $O(\frac{1}{\varepsilon^2})$ iterations, resulting in the runtime of $O(ns/\varepsilon^2)$.

For space, besides constant-size variables, only two vectors CACHE and z are stored in memory. It has been established in Lemma 4 that the DOT procedure requires $O(1)$ memory, so $P \le n$ processors could compute Lines 2 and 7 in parallel in $O(n/P + \log P)$ time. The only other line of code requiring $\omega(1)$ time is Line 10, which can clearly be parallelized to $O(n/P)$ time.

5   Results

In this section, we evaluate the practical effectiveness of our scalable and sparse PageRank algorithm on real world, benchmark data sets [11]. In particular, we evaluated our algorithm on the following data sets:

1. Youtube Social Network and Ground Truth Communities *(youtube)*—network representation of the Youtube social network. Contains 1,134,890 nodes and 2,987,624 edges.
2. California Road Network *(roadNet-CA)*—A road network of California consisting of 1,965,206 nodes and 2,766,607 edges.
3. Wikipedia Talk Network *(wiki-Talk)*—representation of the activity and discussions (edges) between the users (nodes) on the Wikipedia page. A directed edge (u, v) between two users u and v exists if user u edited a page of user v. The graph contains 2,394,385 nodes and 5,021,410 edges.

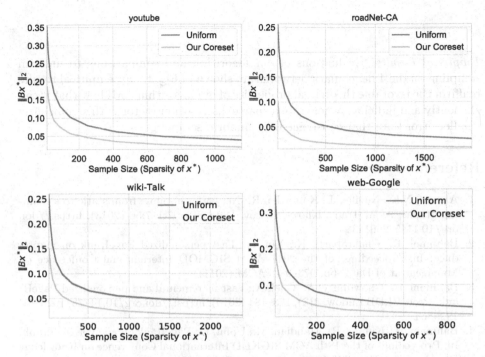

Fig. 1. Evaluation of the relative error $\|Bx^*\|_2 = \|(\Psi(G, d, v) - \mathcal{I}_n)x^*\|_2$ of the sparse, approximate distribution vector x^*.

4. Google Web Graph *(web-Google)*—Network composed of 875,713 nodes and 5,105,039 edges denoting web pages and the hyperlinks between them, respectively.

We compared the performance of our Frank-Wolfe-based approach in generating a sparse solution x that minimizes $\|Bx\|_2$ to that of constructing an x via uniform sampling for various values of sparsity (i.e., sample size). All algorithms were implemented in Python using the Sparse package of the SciPy library [10] and simulations were conducted on a computer with a 2.60 GHz Intel i9-7980XE processor (18 cores total) and 128 GB RAM.

Experimental Setup. For each data set represented by the scaled adjacency matrix $G \in \mathbb{R}^{n \times n}$, we constructed a set of $k = 10$ geometrically-spaced subsample sizes $S = \{S_1, \ldots, S_m\} \subseteq [\log n, \sqrt{n}]^k$ and for each sample size $m \in S$, we invoked each algorithm to construct an approximate distribution vector x with sparsity m. The results were averaged across 100 trials for each subsample size. More specifically, for each $m \in S$, we ran our algorithm for m iterations to construct an m-sparse vector x. The uniform sampling procedure was implemented in a similar iterative manner, where at each iteration a random index $j \in [n]$ was selected and the vector x (initially all zeros) was modified to be

$$x_j \leftarrow x_j + \frac{1}{T}.$$

Empirical Results. Evaluations of our algorithm and comparisons to uniform sampling on the 4 benchmark networks are shown in Fig. 1. Our empirical results reaffirm the favorable theoretical properties of our algorithm and show that it can efficiently and judiciously generate a sparse distribution vector with significantly smaller error than that constructed by uniform sampling.

References

1. Ahmed, N.K., Neville, J., Kompella, R.: Network sampling: from static to streaming graphs. ACM Trans. Knowl. Discov. Data **8**(2), 7:1–7:56 (2013). https://doi.org/10.1145/2601438
2. Bahmani, B., Chakrabarti, K., Xin, D.: Fast personalized PageRank on mapreduce. In: Proceedings of the 2011 ACM SIGMOD International Conference on Management of Data, pp. 973–984. ACM (2011)
3. Bahmani, B., Chowdhury, A., Goel, A.: Fast incremental and personalized PageRank. Proc. VLDB Endow. **4**(3), 173–184 (2010). https://doi.org/10.14778/1929861.1929864
4. Bahmani, B., Kumar, R., Mahdian, M., Upfal, E.: PageRank on an evolving graph. In: Proceedings of the 18th ACM SIGKDD International Conference on Knowledge Discovery and Data Mining, pp. 24–32. ACM (2012)
5. Barman, S.: Approximating nash equilibria and dense bipartite subgraphs via an approximate version of caratheodory's theorem. In: Proceedings of the Forty-Seventh Annual ACM on Symposium on Theory of Computing, pp. 361–369. ACM (2015)
6. Das Sarma, A., Nanongkai, D., Pandurangan, G.: Fast distributed random walks. In: Proceedings of the 28th ACM Symposium on Principles of Distributed Computing, PODC 2009, pp. 161–170. ACM, New York (2009). https://doi.org/10.1145/1582716.1582745
7. Frank, M., Wolfe, P.: An algorithm for quadratic programming. Naval Res. Logist. Q. **3**(1–2), 95–110 (1956)
8. Haveliwala, T., Kamvar, A., Klein, D., Manning, C., Golub, G.: Computing PageRank using power extrapolation, August 2003
9. Jin, Z., Shi, D., Wu, Q., Yan, H., Fan, H.: LBSNRank: personalized PageRank on location-based social networks. In: Proceedings of the 2012 ACM Conference on Ubiquitous Computing, pp. 980–987. ACM (2012)
10. Jones, E., Oliphant, T., Peterson, P., et al.: SciPy: Open source scientific tools for Python (2001). http://www.scipy.org/. Accessed
11. Leskovec, J., Sosič, R.: Snap: a general-purpose network analysis and graph-mining library. ACM Trans. Intell. Syst. Technol. (TIST) **8**(1), 1 (2016)
12. Mitliagkas, I., Borokhovich, M., Dimakis, A.G., Caramanis, C.: FrogWild!: fast • PageRank approximations on graph engines. Proc. VLDB Endow. **8**(8), 874–885 (2015)
13. Rossi, R.A., Gleich, D.F.: Dynamic PageRank using evolving teleportation. In: Bonato, A., Janssen, J. (eds.) WAW 2012. LNCS, vol. 7323, pp. 126–137. Springer, Heidelberg (2012). https://doi.org/10.1007/978-3-642-30541-2_10

14. Rozenshtein, P., Gionis, A.: Temporal PageRank. In: Frasconi, P., Landwehr, N., Manco, G., Vreeken, J. (eds.) ECML PKDD 2016. LNCS (LNAI), vol. 9852, pp. 674–689. Springer, Cham (2016). https://doi.org/10.1007/978-3-319-46227-1_42
15. Sarma, A.D., Gollapudi, S., Panigrahy, R.: Estimating pageRank on graph streams. J. ACM **58**(3), 13:1–13:19 (2011). https://doi.org/10.1145/1970392.1970397
16. Sarma, A.D., Molla, A.R., Pandurangan, G.: Near-optimal random walk sampling in distributed networks. arXiv preprint arXiv:1201.1363 (2012)
17. Das Sarma, A., Molla, A.R., Pandurangan, G., Upfal, E.: Fast distributed PageRank computation. In: Frey, D., Raynal, M., Sarkar, S., Shyamasundar, R.K., Sinha, P. (eds.) ICDCN 2013. LNCS, vol. 7730, pp. 11–26. Springer, Heidelberg (2013). https://doi.org/10.1007/978-3-642-35668-1_2
18. Yu, W., Lin, X., Zhang, W.: Fast incremental simrank on link-evolving graphs. In: 2014 IEEE 30th International Conference on Data Engineering (ICDE), pp. 304–315. IEEE (2014)

GPU Based Horn-Schunck Method to Estimate Optical Flow and Occlusion

Vanel Lazcano[1]([✉])[iD] and Francisco Rivera[2]

[1] Núcleo de Matemática, Física y Estadística, Universidad Mayor,
Manuel Montt 318, 7500628 Providencia, Santiago, Chile
vanel.lazcano@umayor.cl
[2] Universidad Mayor, Manuel Montt 318, Providencia, Santiago, Chile
francisco.riveras@umayor.cl

Abstract. Optical flow is the apparent motion pattern of pixels in two consecutive images. Optical flow has many applications: navigation control of autonomous vehicles, video compression, noise suppression, and others. There are different methods to estimate the optical flow, where variational models are the most frequently used. These models state an energy model to compute the optical flow. These models may fail in presence of occlusions and illumination changes. In this work is presented a method that estimates the flow from the classical Horn-Schunk method and the incorporation of an occlusion layer that gives to the model the ability to handle occlusions. The proposed model was implemented in an Intel i7, 3.5 GHz, GPU GeForce NVIDIA-GTX-980-Ti, using a standard webcam. Using images of 320×240 pixels we reached 4 images per second, i.e. this implementation can be used in an application like an autonomous vehicle.

Keywords: Optical flow · Occlusion estimation · Variational model

1 Introduction

Optical flow is defined as the apparent motion pattern of the objects in an image sequence. There are different methods to estimate the optical flow. Frequently, these methods propose an energy model to estimate the motion field. Generally, this model or variational model is composed of two terms: a data term and a regularization term. The data term considers the similarity between the intensity of the pixels of consecutive images. The regularity term imposes spatial smoothness to the estimated motion field. The flow $\mathbf{v}$ that minimizes both terms: the data term and the regularization term is the estimated optical flow of the video sequence.

In general, variational models are nonlinear and non-convex, therefore the minimization of the proposed energy cannot be performed using a descent gradient.

T. V. Gopal and J. Watada (Eds.): TAMC 2019, LNCS 11436, pp. 424–437, 2019.
https://doi.org/10.1007/978-3-030-14812-6_26

Some optical flow estimation methods use a linear approximation of the data term transforming the model in a convex one [5]. The seminal work of Horn-Schunck [5] has been successfully proved in video sequences of images that present small displacements.

This work presents a method for estimating the optical flow that integrates this two ideas: (i) linearization of the data term and (ii) an occlusion estimation.

In order to handle occlusion, we will follow the method proposed in [2]. The authors proposed a model that considers three consecutive frames (previous frame, the current frame, and the next frame). Their approach incorporates information that allows detecting occluded pixels ($\chi(\mathbf{x})$ function). The authors assumed that occluded pixels in the current frame are visible ($\chi(\mathbf{x}) = 1$) in the previous frame. The optical flow on non-occluded pixels ($\chi(\mathbf{x}) = 0$) is estimated forward and the flow in occluded pixels is estimated backward [2].

In [6] the authors extended the model given in [2] to color images and to handle illumination changes. The authors proposed a model which is able to deal with the usual drawback of variational methods related to large displacements of objects in the scene which are larger than the object itself. The addition of a term that balances gradient and intensities increases the robustness of the model to illumination changes. The inclusion of supplementary information coming from matching obtained by exhaustive search helps to follow large displacements.

The proposed model presented in this article was implemented in a GPU NVIDIA improving its processing time and obtaining a near real-time performance.

In Sect. 2 is presented the methodology to estimate the optical flow and the occlusion. In Sect. 3 we present a proposition that let us compute the optical flow and the occlusion layer. In Sect. 4 is presented the implementation of the methodology. In Sect. 5 is presented the database used to perform experiments. Results are presented in Sect. 6 and finally, conclusions are presented in Sect. 7.

2   Methodology

Our proposed model estimates jointly the occlusion and the optical flow of three consecutive images following the main ideas are given in [2] but adapted to the Horn-Schunck method.

The proposed model uses a binary occlusion layer $\xi(\mathbf{x})$ that indicates the location where pixels, visible in the current frame, are not visible in the next frame. Assuming that occluded pixels are visible in the previous frame, the optical flow on visible pixels is forward estimated while optical flow in the occluded pixels is backward estimated.

Our proposed model is based on the L-2 total variation regularization of the flow and in the L-2 norm of the data fidelity term. These terms consider non-linearities, which are decoupled by introducing an auxiliary variable. These decoupling variables w_1, w_2, also represent the forward and backward flow. This new energy is minimized by means of a numerical scheme based on [5].

2.1 Horn-Schunck Method

This method estimates the optical flow $\mathbf{v}(\mathbf{x}) = (v_1(\mathbf{x}), v_2(\mathbf{x}))$ assuming that two consecutive images are given $I_0, I_1 : \Omega \subset \mathbb{Z} \times \mathbb{Z} \to \mathbb{R}$. The optical flow $v(\mathbf{x})$ is estimated minimizing the following functional:

$$J(\mathbf{v}(\mathbf{x})) = J_{data}(\mathbf{v}(\mathbf{x})) + J_{reg}(\mathbf{v}(\mathbf{x})), \tag{1}$$

where

$$J_{data}(\mathbf{v}(\mathbf{x})) = \frac{1}{2} \int_{\Omega} |I_x(\mathbf{x})v_1(\mathbf{x}) + I_y(\mathbf{x})v_2(\mathbf{x}) + I_t(\mathbf{x})|^2 \, d\mathbf{x} \tag{2}$$

and

$$J_{reg}(\mathbf{v}(\mathbf{x})) = \int_{\Omega} \alpha |\nabla v_1(\mathbf{x})|^2 + \alpha |\nabla v_2(\mathbf{x})|^2 \, d\mathbf{x}, \tag{3}$$

where $\nabla v_1(\mathbf{x})$ is the gradient of the horizontal component of the estimated optical flow $\mathbf{v}(\mathbf{x})$ i.e. $\nabla v_1(\mathbf{x}) = \left(\dfrac{\partial v_1(\mathbf{x})}{\partial x}, \dfrac{\partial v_2(\mathbf{x})}{\partial y} \right)$. Analogously, for vertical component $\nabla v_2(\mathbf{x})$ and α is a real constant value.

Values $I_x(\mathbf{x})$, $I_y(\mathbf{x})$ and $I_t(\mathbf{x})$ are the partial derivatives and temporal derivative of the considered images, respectively.

The *Euler-Lagrange* equation are computed in order to estimate $\mathbf{v}(\mathbf{x})$ obtaining the following set of equations:

$$\begin{aligned} (I_x(\mathbf{x})v_1(\mathbf{x}) + I_y(\mathbf{x})v_2(\mathbf{x}) + I_t(\mathbf{x}))I_x(\mathbf{x}) - \alpha div(\nabla v_1(\mathbf{x})) = 0 \\ (I_x(\mathbf{x})v_1(\mathbf{x}) + I_y(\mathbf{x})v_2(\mathbf{x}) + I_t(\mathbf{x}))I_y(\mathbf{x}) - \alpha div(\nabla v_2(\mathbf{x})) = 0, \end{aligned} \tag{4}$$

where $div(\nabla v_1(\mathbf{x}))$ is the *Laplacian* operator $\Delta v_1(\mathbf{x}) = \dfrac{\partial^2 v_1(\mathbf{x})}{\partial x^2} + \dfrac{\partial^2 v_1(\mathbf{x})}{\partial y^2}$. This operator is approximated as $\Delta v_1(\mathbf{x}) = (\bar{v}_1(\mathbf{x}) - v_1(\mathbf{x}))$, where $\bar{v}_1(\mathbf{x})$ is a filtered version of $v_1(\mathbf{x})$, $\bar{v}_1(\mathbf{x}) = G * v_1(\mathbf{x})$. The G matrix is defined as:

$$G = \begin{pmatrix} 1/12 & 1/6 & 1/12 \\ 1/6 & 0 & 1/6 \\ 1/12 & 1/6 & 1/12 \end{pmatrix}$$

Taking into account the *Laplacian* approximation given in [5], the following set of equation is obtained:

$$\begin{aligned} I_x(\mathbf{x})^2 v_1 + I_x(\mathbf{x})I_y(\mathbf{x})v_2(\mathbf{x}) + I_x(\mathbf{x})I_t(\mathbf{x}) - \alpha(\bar{v}_1(\mathbf{x}) - v_1(\mathbf{x})) = 0 \\ I_x(\mathbf{x})I_y(\mathbf{x})v_1(\mathbf{x}) + I_y(\mathbf{x})^2 v_2(\mathbf{x}) + I_y(\mathbf{x})I_t(\mathbf{x}) - \alpha(\bar{v}_2(\mathbf{x}) - v_2(\mathbf{x})) = 0. \end{aligned} \tag{5}$$

Solving this equation system in Eq. 5 for $v_1(\mathbf{x})$ and $v_2(\mathbf{x})$:

$$\begin{aligned} v_1(\mathbf{x}) = \bar{v}_1(\mathbf{x}) - I_x(\mathbf{x}) \frac{I_x(\mathbf{x})v_1(\mathbf{x}) + I_y(\mathbf{x})v_2(\mathbf{x}) + I_t(\mathbf{x})}{\alpha + I_x(\mathbf{x})^2 + I_y(\mathbf{x})^2} \\ v_2(\mathbf{x}) = \bar{v}_2(\mathbf{x}) - I_y(\mathbf{x}) \frac{I_x(\mathbf{x})v_1(\mathbf{x}) + I_y(\mathbf{x})v_2(\mathbf{x}) + I_t(\mathbf{x})}{\alpha + I_x(\mathbf{x})^2 + I_y(\mathbf{x})^2}. \end{aligned} \tag{6}$$

2.2   A Model for Estimating Occlusions

In the proposed model $\xi : \Omega \to [0,1]$ is the occlusion layer. If $\xi(\mathbf{x}) = 1$ means that visible pixel in $I_0(\mathbf{x})$ is occluded pixel in $I_1(\mathbf{x})$.

In [2] is assumed that pixels not visible in $I_1(\mathbf{x})$ are visible in the previous frame I_{-1}. If $\xi(\mathbf{x}) = 0$, $\nabla I_1(\mathbf{x})$ with $I_t(\mathbf{x})$ are compared, in the other hand, if $\xi(\mathbf{x}) = 1$, $\nabla I_{-1}(\mathbf{x})$ with $-I_{t-1}(\mathbf{x})$ are compared. Where $I_{x-1}(\mathbf{x}) = \dfrac{(I_{-1}(\mathbf{x}) + I_0(\mathbf{x})) * G_x}{4}$, $I_{y-1}(\mathbf{x}) = \dfrac{(I_{-1}(\mathbf{x}) + I_0(\mathbf{x})) * G_y}{4}$ and $I_{t-1}(\mathbf{x}) = \dfrac{(I_{-1}(\mathbf{x}) - I_0(\mathbf{x})) * G_t}{4}$. The occluded regions, where $\xi(\mathbf{x}) = 1$ as in [2] are considered as regions where the $div(\mathbf{v}(\mathbf{x})) < 0$. The following model is proposed:

$$J(\mathbf{v}, \xi) = J_{data}(\mathbf{v}, \xi) + J_{reg}(\mathbf{v}, \xi) + 2\beta \int \xi \, div(\mathbf{v}) dx, \qquad (7)$$

where,

$$J_{data}(\mathbf{v}, \xi) = \int_\Omega (1 - \xi)|I_x v_1 + I_y v_2 - I_t|^2 + \xi|I_{x-1} v_1 + I_{y-1} v_2 + I_{t-1}|^2 dx, \quad (8)$$

and,

$$J_{reg}(\mathbf{v}, \xi) = \int_\Omega (\alpha|\nabla v_1|^2 + \alpha|\nabla v_2|^2 + \delta|\nabla \xi|^2) dx, \qquad (9)$$

where α and δ are real constants, and for simplicity we omitted the explicit dependency of $\mathbf{x}$ in the images and in the flow. The arguments $\mathbf{v}$ that minimize Eq. 7 is the estimated optical flow of the image sequence.

The drawback of the above model is that it cannot handle large displacements. An approach to tackle this problem is supposing that the flow $\mathbf{v}$ is already computed and a small variation $\mathbf{dv}$ can be estimated using Horn-Schunck model. Let us define warped image $\tilde{I}_1(\mathbf{x})$ and warped image $\tilde{I}_{-1}(\mathbf{x})$ as $\tilde{I}_1(\mathbf{x}) = I_1(\mathbf{x}+\mathbf{v})$ and $\tilde{I}_{-1}(\mathbf{x}) = I_{-1}(\mathbf{x}-\mathbf{v})$ respectively. Using these definitions is possible to state:

$$\tilde{I}_1(\mathbf{x}) = \nabla \tilde{I}_1(\mathbf{x})^T d\mathbf{v} + I_1(\mathbf{x}), \qquad (10)$$

and

$$\tilde{I}_{-1}(\mathbf{x}) = \nabla \tilde{I}_{-1}(\mathbf{x})^T d\mathbf{v} - I_{-1}(\mathbf{x}). \qquad (11)$$

Taking into account this consideration the proposed model become:

$$J(\mathbf{dv}, \xi) = J_{data}(\mathbf{dv}, \xi) + J_{reg}(\mathbf{dv}, \xi) + 2\beta \int \xi \, div(\mathbf{v} + \mathbf{dv}) dx, \qquad (12)$$

where,

$$J_{data}(\mathbf{dv}, \xi) = \int_\Omega (1 - \xi)|\nabla \tilde{I}^T \mathbf{dv} + I_t|^2 + \xi|\nabla \tilde{I}_{-1}^T \mathbf{dv} - I_{t-1}|^2 dx, \qquad (13)$$

and,

$$J_{reg}(\mathbf{dv}, \xi) = \int_\Omega (\alpha|\nabla dv_1|^2 + \alpha|\nabla dv_2|^2 + \delta|\nabla \xi|^2) dx. \qquad (14)$$

2.3   Justification of the Proposed Model

The use of the divergence of the optical flow $\mathbf{v}(\mathbf{x})$ in order to estimates the occlusion layer was proposed originally by [7]. The authors argued that the divergence of the flow is positive for not-occluded boundaries, negative for occluding boundaries, and near to zero for shear boundaries. The authors defined a one-sided divergence function $d(\mathbf{x})$:

$$d(\mathbf{x}) = \begin{cases} div(\mathbf{v}(\mathbf{x})) & div(\mathbf{v}(\mathbf{x})) < 0, \\ 0 & \text{otherwise.} \end{cases}$$

this d function is included in the data term of their optical flow model.

3   Solving the Model

Two decoupling variables $\mathbf{w_1}$ and $\mathbf{w_2}$ are introduced. These variables represent the forward and backward optical flow respectively. A deviation w.r.t. $\mathbf{dv}$ is penalized, therefore the new functional to be minimized is given by:

$$J(\mathbf{dv}, \mathbf{w_1}, \mathbf{w_2}, \xi) = J_{data}(\mathbf{w_1}, \mathbf{w_2}, \xi) + J_{reg}(\mathbf{dv}, \xi) + 2\beta \int \xi \, div(\mathbf{v} + \mathbf{dv}) d\mathbf{x}$$

$$+ \frac{1}{\theta} \int_\Omega (1 - \xi)|\mathbf{dv} - \mathbf{w_1}|^2 d\mathbf{x} + \frac{1}{\theta} \int_\Omega \xi |\mathbf{dv} + \mathbf{w_2}|^2 d\mathbf{x},$$

$$(15)$$

This energy depends on the variables $\mathbf{dv}$, $\mathbf{w_1}$, $\mathbf{w_2}$ and ξ, $\theta > 0$. In order to solve this model one proposition is stated:

Proposition 1. *The minimum of model (15) w.r.t. $\mathbf{dv}(\mathbf{x})$ is:*

$$\mathbf{dv}(\mathbf{x}) = \frac{\alpha\theta\overline{\mathbf{dv}}(\mathbf{x}) + (1 - \xi(\mathbf{x}))\mathbf{w_1}(\mathbf{x}) - \xi(\mathbf{x})\mathbf{w_2}(\mathbf{x}) + \beta\theta\nabla\xi(\mathbf{x})}{1 + \alpha\theta}. \quad (16)$$

Minimum of (15) w.r.t, $\mathbf{w_1}(\mathbf{x})$, $\mathbf{w_2}(\mathbf{x})$ is given by:

$$\mathbf{w_1}(\mathbf{x}) = \left(I + \theta\nabla I_1 I_1^T\right)^{-1} \left(\mathbf{dv}(\mathbf{x}) - \theta\nabla I_1(\mathbf{x})^T I_t(\mathbf{x})\right) \quad (17)$$

$$\mathbf{w_2}(\mathbf{x}) = \left(I + \theta\nabla I_{-1}(\mathbf{x})I_{-1}(\mathbf{x})^T\right)^{-1} \left(\mathbf{dv}(\mathbf{x}) - \theta\nabla I_{-1}(\mathbf{x})^T I_{t-1}(\mathbf{x})\right) \quad (18)$$

Given $\mathbf{w_1}(\mathbf{x})$, $\mathbf{w_2}(\mathbf{x})$ and $\mathbf{v}$, the minimum of the model in (15) with respect to $\xi(\mathbf{x})$ can be obtained by:

$$A = |\rho|^2 - |\rho_{-1}|^2 + \frac{(\mathbf{dv}(\mathbf{x}) - \mathbf{w_1}(\mathbf{x}))^2}{\theta} - \frac{(\mathbf{dv}(\mathbf{x}) + \mathbf{w_2}(\mathbf{x}))^2}{\theta} - 2\beta \, div(\mathbf{v} + \mathbf{dv}), \quad (19)$$

$$\xi(\mathbf{x}) = \begin{cases} 1 & \xi(\mathbf{x}) + \frac{1}{2\delta}A \geq 1 \\ \xi(\mathbf{x}) + \frac{1}{2\delta}A & \text{other case} \\ 0 & \xi(\mathbf{x}) + \frac{1}{2\delta}A \leq 0 \end{cases} \quad (20)$$

where $\rho = \nabla\tilde{I}(\mathbf{x})^T\mathbf{w_1}(\mathbf{x}) + I_t(\mathbf{x})$ and $\rho = \nabla\tilde{I}_{-1}(\mathbf{x})^T\mathbf{w_2}(\mathbf{x}) + I_{t-1}(\mathbf{x})$.

4   Implementation

4.1   Derivative Computation

Derivatives are computed using the following centered difference kernels G_x, G_y and G_t:

$$G_x = \begin{pmatrix} -1 & 8 & 0 & -8 & 1 \end{pmatrix}$$

$$G_y = \begin{pmatrix} -1 \\ 8 \\ 0 \\ -8 \\ 1 \end{pmatrix}$$

$$G_t = \begin{pmatrix} 1 & 1 \\ 1 & 1 \end{pmatrix}$$

$I_x = \dfrac{(I_1 + I_0) * G_x}{12}$, $I_y = \dfrac{(I_1 + I_0) * G_y}{12}$ and $I_t = \dfrac{(I_1 - I_0) * G_t}{4}$, where " $*$ " is the convolution operator.

4.2   Divergence Computation

The estimation of the divergence is computed as the dual operator of derivatives. Those points at borders of the image are mirrored. The divergence operator is given by:

$$div_x \xi_{i,j} = \begin{cases} \xi_{2,j} - 7\xi_{1,j} & i = 1, \\ 8\xi_{2,j} - \xi_{3,j} & i = 2, \\ \xi_{4,j} - 8\xi_{3,j} - \xi_{1,j} & i = 3, \\ \xi_{i+2,j} - 8\xi_{i+1,j} + 8\xi_{i-1,j} - \xi_{i-2,j} & otherwise, \\ -8\xi_{N-1,j} + 8\xi_{N-3,j} - \xi_{N-4,j} & i = N - 2, \\ 8\xi_{N-2,j} - \xi_{N-3,j} & i = N - 1, \\ 7\xi_{N-1,j} - \xi_{N-2,j} & i = N. \end{cases}$$

and,

$$div_y \xi_{i,j} = \begin{cases} \xi_{i,2} - 7\xi_{i,1} & j = 1, \\ 8\xi_{i,2} - \xi_{i,3} & j = 2, \\ \xi_{i,4} - 8\xi_{i,3} - \xi_{i,1} & j = 3, \\ \xi_{i,j+2} - 8\xi_{i,j+1} + 8\xi_{i,j-1} - \xi_{i,j-2} & otherwise, \\ -8\xi_{i,M-1} + 8\xi_{i,M-3} - \xi_{i,M-4} & j = M - 2, \\ 8\xi_{i,M-2} - \xi_{i,M-3} & j = M - 1, \\ 7\xi_{i,M-1} - \xi_{i,M-2} & j = M. \end{cases}$$

where N and M are the column and row numbers of the image, respectively.

4.3   Pseudo-code

In order to minimize the proposed model in Eq. (15), a numerical procedure is employed. In this case, the proposed energy is convex but this model is valid only if the displacements are small. Hence, the energy minimization is embedded in a coarse-to-fine approach. The pseudo code is presented in $Algorithm 1$. Inputs to the algorithm are three consecutive images I_{-1} I_0, and I_1 the balance parameter between data term and regularization, α, scales number N_{scales}, iterations number per each scale $MaxIter$ and, finally, warping number $MaxWarp$.

> $Input$: three consecutive gray level images I_{-1}, I_0, I_1.
> $Parameters$ α, θ , β, δ, $MaxIter$, β, $Number_{scales}$, $Number_{warpings}$.
> $Output$: optical flow $v = (u_1, u_2)$ and occlusion mask ξ
> Down-scale I_{-1}, I_0, I_1,
> Initialization $u_1 = u_2 = 0; n_j = n_i = 0$
> for $scales \leftarrow Number_{scales}$ to 1
>   compute $I_{1w}^{scales}(x + v^{scales})$, $I_{-1w}^{scales}(x - v^{scales})$
>   compute $\nabla I_{1w}^{scales}(x + v^{scales})$, $\nabla I_{-1w}^{scales}(x - v^{scales})$
> for $nj \leftarrow Number_{warpings}$ to 1
>   compute $\mathbf{dv^{scales}}$, equation 16.
>   compute $\mathbf{w_i}$, equation 17 and 18.
>   compute A, equation 19
>   compute ξ equation 20.
> $endfor$
>   update $\mathbf{v^{scales}} = \mathbf{v^{scales}} + \mathbf{dv^{scales}}$
>   up-sample $\mathbf{v^{scales}}$
> $endfor$
> Out $\mathbf{v}$.

4.4   GPU Implementation

We have implemented the proposed model in a GeForce NVIDIA GTX GPU 980 Ti, based on the code presented in [8]. With this implementation (in images of 320×240 pixels) we processed 965 images and we processed on average 4.074 images per second with a variance of 0.008 s.

5   Database and Experiments

The performance of the proposed model is evaluated in Middlebury [1] database. This database contains video sequences that present moving objects and also camera movements. For some of these sequences, the Middlebury database has ground-truth. This ground-truth is used to compute the performance of the proposed optical flow method. The performance of the method is evaluated considering EPE (endpoint error) and AAE [3] (average angular error). Figure 1 shows examples of image sequences of the Middlebury. In these sequences the

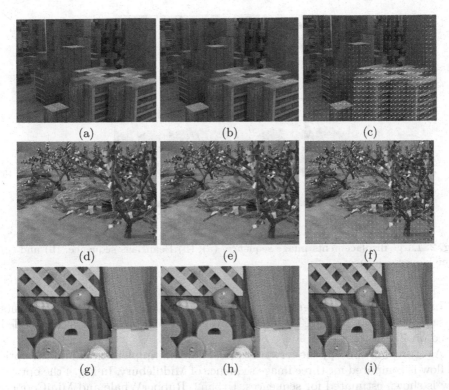

Fig. 1. Image sequences of the Middlebury database. (a), (b): Urban2 sequence, (c) the displacements is shown with white arrows. (d) and (e): Grove2 sequence. (f) displacements of the pixels represented with white arrows. (g) and (h): RubberWhale sequence. (i) displacement represented with white arrows.

ground-truth is available and we show it using white arrows in Fig. 1(c), (f) and (i) for Urban2, Grove3, and RubberWhale respectively.

Additionally, Middlebury includes large displacements image sequences as is shown in Fig. 2. In these sequences, the ground-truth is not available. And only qualitative evaluations can be performed.

Let $\mathbf{g}(\mathbf{x}) = (g_1(\mathbf{x}), g_2(\mathbf{x}))$ and $\mathbf{v}(\mathbf{x}) = (v_1(\mathbf{x}), v_2(\mathbf{x}))$ be the ground truth and the estimated optical flow respectively. Let N be the number of points of the image. EPE and AAE is defined:

$$
\begin{aligned}
EPE &= \frac{1}{N} \sum_{i=1}^{N} \sqrt{(g_{1i} - v_{1i})^2 + (g_{2i} - v_{2i})^2} \\
AAE &= \frac{1}{N} \sum_{i=1}^{N} cos^{-1} \left(\frac{1 + g_{1i} v_{1i} + g_{2i} v_{2i}}{\sqrt{1 + g_{1i}^2 + g_{2i}^2} \sqrt{1 + v_{1i}^2 + v_{2i}^2}} \right)
\end{aligned}
\tag{21}
$$

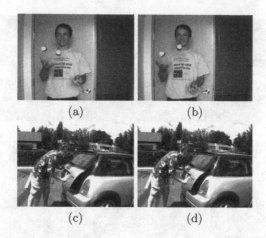

Fig. 2. Large displacements image sequence. (a), (b) BeanBags sequence. (b) and (c) MiniCooper sequence.

In Fig. 3 is shown the ground-truth of the Middlebury small displacements for sequence Urban2, Grove3, and RubberWhale. The groud-truth is color-coded based on Fig. 3(a).

As a proof of concept the forward optical flow was implemented and the optical flow is computed for three image sequence of Middlebury. In Fig. 4 the optical flow is shown estimated for sequences: Urban2, RubberWhale and MiniCooper.

In [9] The occlusion layer is determined using the largest values of the data term. The data term is compared with a threshold and pixels that present errors larger than the threshold that are considered occluded. In [9] is proposed the function $\rho_{occ}(\mathbf{v})$ for detecting occlusions as:

$$\rho_{occ}(\mathbf{v}) = \begin{cases} 1 & (I_0(\mathbf{x}) - I_1(\mathbf{x} + \mathbf{v}))^2 > \epsilon_I, \\ 0 & \text{otherwise} \end{cases}$$

where ϵ_I is the threshold, $\rho_{occ}(\mathbf{v}) = 1$ means that the pixel is occluded and $\rho_{occ}(\mathbf{v}) = 0$ the pixel is visible. The ϵ_I value is set in $\epsilon_I = 0.005$.

In [2] the negative values divergence of the optical flow is used as an estimator of the occlusion. The value of the divergence of the flow is compared with a threshold θ_{occ}. In this case, was set in $\theta_{occ} = -0.5$. In Fig. 5 is shown the occlusion estimations using the proposal given in [9]. It is observed that the occlusion in (a) is very sparse, the occlusion in (b) is not well estimated, and in (c) presents many false estimations. In Fig. 5(d), (e) and (f) are shown the occlusion estimated using negative values of the divergence. It is observed that in (d) the occlusion is overestimated but it presents a good correlation with the correct one. In (e) the occlusion is well estimated and also overestimated. In (f) the occlusion is estimated well and overestimated presenting good correlation with the correct one. In (g), (h) and (i) the estimated occlusion is overlapped in red on the current image.

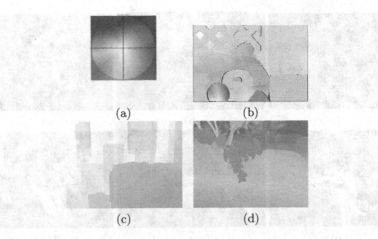

Fig. 3. Color coded ground-truth. (a) color code, (b) RubberWhale, (c) Urban2 and (d) Grove2 sequences. (Color figure online)

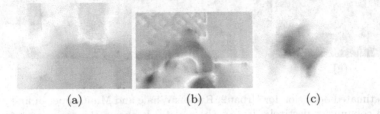

Fig. 4. Estimated optical flow color, coded for sequences: (a) Urban2, (b) RubberWhale and (c) MiniCooper. (Color figure online)

Our model uses directly the divergence of the flow as we presented in Eqs. (16), (19) and (20).

6 Results

6.1 Results in Real-Time Estimation

We present some results in image sequences acquired with a webcam of 320×480 pixels. In Fig. 6 we show images of captures with the webcam where an individual moves his arms upward and downward. Our proposal was compared with the results obtained by [6] and [2], which implementation is available in [4].

In Fig. 6(a) and (b) the individual moves the arm on the right downward. In (c) we show the estimated occlusion. In (d) we show the estimated occlusion obtained by [6]. In (e) is shown the occlusion estimated by [2]. In Fig. 6(f) and (g) the individual moves the arm on the right upward. In Fig. 6(h) the occlusion is overestimated due to illumination changes. In (i) estimated occlusion given by [6]. (j) The occlusion layer given by [2]. In Fig. 6(k) and (l) the individual moves

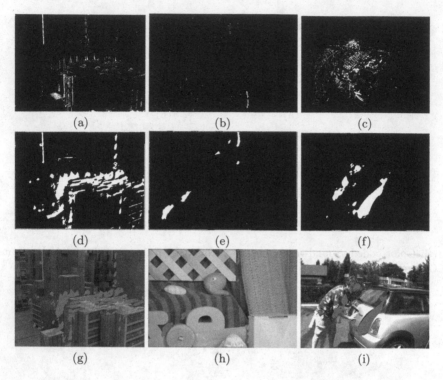

Fig. 5. Estimated occlusion for Urban2, RubberWhale and MiniCooper in first, second and third column, respectively. In (a), (b) and (c) is shown the occlusion estimation using larger error of the data term according to [9]. Estimated occlusions using negative values of the divergence of the flow are shown in (d), (e), (f). Overlapped occlusion (in red) (d), (e) and (f) on the original reference image. (Color figure online)

the arms on the left upward and the arm on the right downward. In Fig. 6(m) the occlusion is estimated correctly. In Fig. 6(n) and (o) is shown results obtained by [6] and [2], respectively. In (p) and (q) the individual moves the arm on the left downward and the arm on the right upward. In Fig. 6(r) we show our occlusion estimation which is estimated correctly. Figure 6(s) and (t) present the results given by [6] and [2], respectively.

Comparing Fig. 6(c), (d) and (e) we observe that the estimation in (d) and (e) is very sparse and present many false detections in comparison to (c) which is very regular. Estimation in (h) presents false detection on the right arm. If we compare figure in (h) with figures (i) and (j), the estimation in (h) is very regular and the estimation in (i) and (j) presents many false detection on the wall due to illumination changes.

Obtained results show that our proposal outperforms [6] and [2] estimating the occlusion layer for these image sequences. Our proposal performs better in small images with low resolution obtained by a webcam. In order to be fairer we

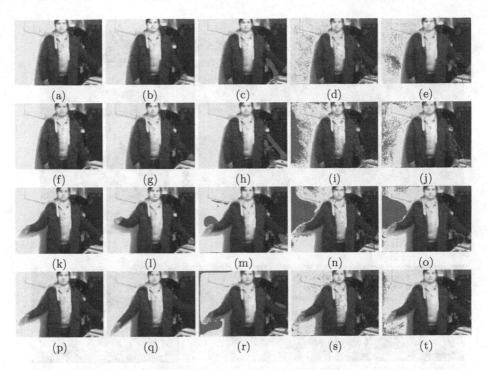

Fig. 6. Image Sequences processed in real-time. (a) image66 and (b) image67. (c) Estimated occlusion superimposed to the images66 (in red). (d) estimated occlusion using [6] and (e) occlusion estimation given by [2]. (f) image68 and (g) image69. (h) Estimated occlusion superimposed to image68.(i) Estimated occlusion layer obtained by [6]. (j) Estimated occlusion given by [2]. (k) image90 and (l) image91. (m) Estimated occlusion superimposed to image90. (n) Estimated occlusion given by [6]. (o) results given by [2]. (p) image92 and (q) image93 respectively. (r) Estimated occlusion superimposed to image92. (s) and (t) Estimated occlusion layer obtained by [6] and [2] respectively. (Color figure online)

present results obtained in a sequence of much better quality an larger size. We show in Fig. 7 results obtained by our model and results obtained by [6] and [2].

In Fig. 7 we show obtained results in three sequences: Urban2, RubberWhale and Mini. These sequences are part of the Middlebury database. Figure 7(a), (b) and (c) shows our results for the three image sequences. In (d), (e) and (f) We show the results obtained by [2] and finally (g), (h) and (j) show the results obtained by [6]. If we compare the results obtained by these three methods for the sequence Urban2 we observe that our proposal overestimated the occlusion in (a). The estimation in (d) given by [2] underestimates the occlusion and in (g) a better estimation is obtained. Comparing images in (b), (e) and (h) we observe that the occlusion in (b) is near the wheel and above the RubberWhale. This occlusion is overestimated but well located, in (e) the occlusion is underestimated but in (h) the occlusion is very sparse. In MiniCooper sequence in (c) is very

Fig. 7. Results obtained by our method and results obtained by [6] and [2] in sequences Urban2, RubberWhale, and Mini. (a), (b) and (c) results obtained by our method in these three sequences. (d), (e), and (f) results obtained by [2]. (g), (h), and (j) results obtained by [6].

well located but the width of the occlusion is overestimated, in (f) the occlusion is underestimated and (j) the method in [6] is better estimates but presents false detection in the sky.

6.2 Long-Term Experiment

As an empirical proof of the stability of our proposal, we performed an experiment in order to detect instability of the proposed model. In our experiment, we processed 10000 images obtained by a webcam. In this experiment, our prototype performed in a similar way to short-term experiments. In this long term experiment, we estimated the optical flow and the occlusion layer for three consecutive images.

7   Conclusions

We have extended the Horn-Schunck model to handle and to estimate occlusions. We evaluated the model qualitatively and the results show that the model estimates the occlusion correctly. The proposed model considers three gray level images estimating the correspondences in occluded pixels backward as other works in the literature. The proposal considers a TVL-2 regularization term and the obtained results show a very regular and well-located occlusion layer in comparison with two similar methods that use the TVL-1 regularization term. Our implementation runs in a GPU reaching 4 images per second which is useful in real-time applications as in a control of autonomous vehicles. As a future work is necessary to develop a methodology to compensate illumination changes in order to improve occlusion estimation.

References

1. Baker, S., Scharstein, D., Lewis, J., Roth, S., Black, M., Szelensky, R.: A database and evaluation methodology for optical flow. Int. J. Comput. Vis. **92**, 1–31 (2011)
2. Ballester, C., Garrido, L., Lazcano, V., Caselles, V.: A TV-L1 optical flow method with occlusion detection. In: Pinz, A., Pock, T., Bischof, H., Leberl, F. (eds.) DAGM/OAGM 2012. LNCS, vol. 7476, pp. 31–40. Springer, Heidelberg (2012). https://doi.org/10.1007/978-3-642-32717-9_4
3. Barron, J., Fleet, D., Beauchemin, S.: Performance of optical flow technics. Int. J. Comput. Vis. **12**(1), 43–47 (2011)
4. Garamendi, J.F., Ballester, C., Garrido, L., Lazcano, V.: Joint TV-L1 optical flow and occlusion estimation. IPOL J. - Image Process. On Line, preprint (2014)
5. Horn, B., Schunck, B.G.: Determining optical flow. Artif. Intell. **17**, 185–204 (1981)
6. Lazcano, V., Garrido, L., Ballester, C.: Jointly optical flow and occlusion estimation for images with large displacements. In: SciTePress (ed.) Proceedings of the 13th International Joint Conference on Computer Vision Imaging and Computer Graphics Theory and Applications, pp. 588–595. INSTICC (2018)
7. Sand, P., Teller, S.: Particle video: long-range motion estimation using point trajectory. Int. J. Comput. Vis. **80**(1), 72 (2008)
8. Smirnov, M.: Optical flow estimation with CUDA. NVIDIA white papers (2012)
9. Xiao, J., Cheng, H., Sawhney, H., Rao, C., Isnardi, M.: Bilateral filtering-based optical flow estimation with occlusion detection. In: Leonardis, A., Bischof, H., Pinz, A. (eds.) ECCV 2006. LNCS, vol. 3951, pp. 211–224. Springer, Heidelberg (2006). https://doi.org/10.1007/11744023_17

Robot Computing for Music Visualization

Pei-Chun Lin[1]([⊠]), David Mettrick[2]([⊠]), Patrick C. K. Hung[2]([⊠]),
and Farkhund Iqbal[3]([⊠])

[1] Department of Information Engineering and Computer Science,
Feng Chia University, Taichung, Taiwan
peichunpclin@gmail.com
[2] Faculty of Business and IT, University of Ontario Institute of Technology,
Oshawa, ON, Canada
{david.mettrick,patrick.hung}@uoit.ca
[3] College of Technological Innovation, Zayed University, Dubai, UAE
farkhund.iqbal@zu.ac.ae

Abstract. This paper presents an algorithm design of Music Visualization on Robot (MVR) which could automatically link the flashlight, color, and emotion through music. We call this algorithm as MVR algorithm that composed by two analyses. First, we focus on Music Signal Analysis. Second, we focus on Music Sentiment Analysis. We integrate two analysis results and implement the MVR algorithm on a robot called Zenbo which is released from ASUS Company. We perform the Zenbo Robot in luminous environments. The MVR system not only could be used in Zenbo robot but also could extend to other fields of Artificial Intelligent (AI) equipment in the future.

Keywords: Music Visualization on Robot (MVR) · Beat tracking ·
Music Sentiment Analysis · Music Information Retrieval ·
Music Signal Analysis · Robot computing

1 Introduction and Literature Review

Music visualization is a feature in electronic music visualizers' and media players' software that generates animated images based on a piece of music. Images are typically generated during live playback, rendered and synchronized with music. Some music visualization systems such as Geiss' MilkDrop [1–3] are created different visualizations for each song or audio. Music visualization can be implemented in a 2D or 3D coordinate system where adding color, intensity and transparency can increase the dimension to 4th, 5th and even up to 6 dimensions.

Visualization techniques range from simple (i.e. only oscilloscope display) to the complex technology which typically comprises a plurality of synthetic effects. Music loudness and spectral change are also the key properties to the visualizations. Effective music visualization is intended to achieve a high degree of visual correlation between the spectral characteristics of the soundtrack (e.g. frequency and amplitude) and the objects or components of the visual image being rendered and displayed.

To realize the music visualization on Robot, it not only needs to know the music signal analysis but the emotions concerned with color. There are many types of

© Springer Nature Switzerland AG 2019
T. V. Gopal and J. Watada (Eds.): TAMC 2019, LNCS 11436, pp. 438–447, 2019.
https://doi.org/10.1007/978-3-030-14812-6_27

research works on **Music Signal Analysis (MSA)**. For example, Muller *et al.* [4], provided an overview of some signal analysis techniques specific to the musical dimension, such as harmony, melody, rhythm, and tone. They studied how specific characteristics of music signals affect and determine these techniques, and highlight their techniques to handle novel music analysis and retrieval tasks. Moreover, McFee *et al.* [5] provides a document describing Librosa version 0.4.0 which is a Python package for audio and music signal processing. At a higher level, Librosa provides an implementation of various common functions used in the field of Music Information Retrieval (MIR). In this package, they provided a brief overview of the capabilities of Librosa, as well as design goals, software development practices, and explanations of symbolic conventions. The documents are consistent with our design process of MVR system. Hence we decided to use Librosa to import our Musical Instrument Digital Interface (MIDI) music.

Music Sentiment Analysis (MSEA) is a kind of sentiment analysis method that analyzes the music features. Martín-Gómez and Cáceres [6] said that listening to music can affect people emotions. They propose a sentiment analysis method based on data mining to analyze the musical features. Also, Shukla *et al.* [7] present a review on how lyrics can prove its usefulness in mood classification utilizing the features like linguistic lyric feature set and text stylistic feature set. Jamdar et al. [8] propose to detect the emotion of a song based on its lyrical and audio features. They use feature weighting and stepwise threshold reduction based on the k-Nearest Neighbors (KNN) algorithm and fuzziness in the classification. Combing those technical methods, Abboud and Tekli [9] propose a MUsical Sentiment Expression (MUSE) system which combined the MSA methods and fuzzy statistical methods to import a MIDI music file and producing output a sentiment vector which describing 6 primary emotions (i.e., anger, fear, joy, love, sadness, and surprise). The process is more consistent with our concept. We will adopt this concept to build our MVR system in this paper.

In the field of design, color is always an indispensable element to meet the needs of human beings. Color designers must understand the target customers' feelings in many industrial fields such as architecture, beauty, advertising, and automobiles [10]. The color can also be correlated to other sensations or emotions. Colors can evoke emotions, but without a face, it is always hard to express what the robot wants to say. Different colors can influence personal feelings. For example, despite that blue can represent sadness, sometimes, blue might represent happiness if blue is their favorite color. There are many studies presented on the research of colors and emotions. For example, the color red may correlate to anger, excitation, and arousal, whereas blue might represent sadness, and green to be refreshing or natural [11].

Kim *et al.* showed that the intensity of emotions can be expressed in terms of color and flicker. These are all applicable to the robot through LED lights [12]. Their results found that color and blinking can increase four of the six emotions, two of which are fear and sadness. This highly suggests that with the use of lighting effects and other features from the robots such as gestures and sounds as mentioned, that a colorful robot can impart more meaningful emotions. In another research paper, Terada *et al.* provided a similar approach to robots by dynamically changing the color brightness of its body [13].

In view of those investigates, we have proposed an MVR prototype [14]. To consistently the topic of MVR, we provide the MVR algorithm in this paper. The structure of this paper is organized as follows. Section 2 describes the ASUS Zenbo robot characteristics. We give our MVR algorithm in Sect. 3. Section 4 describes a performance based on ASUS Zenbo robot. Section 5 gives conclusions and future work.

2 ASUS Zenbo

ASUS release a robot called Zenbo which is positioned as a small cute home robot to the consumers. Zenbo has 24 different facial expressions except for the emotion anger. The 24 facial expressions make Zenbo be one of the many upcoming social robots that can remove misconceptions about robots. Zenbo also has a couple of wheel LED indicators that are generally used to indicate the system's status. It can be programmable to the user's preference.

Therefore, Zenbo can be programmed to express emotions using colors along with its 24 different facial expressions. In addition to Zenbo's wheel LED indicators, Zenbo could also use gestures, movements, and sound to intensify the expression of emotion. A blinking red light could mean a warning, blinking various light might represent a party, soft breathing/fading effects could mean a romantic, serious, and/or calm situation depending on the color. We give a short explanation of Zenbo commands which concert with our system in Table 1. The commands can be executed by Zenbo's App Builder [15].

As Table 1 shown, there are 8 LED lights on each of Zenbo's wheels. This determines how many lights could be displayed on Zenbo's wheels. The developers of Zenbo have decided the colors in different group types: 255 (8 lights), 15 (4 lights), 7 (2 lights), 1 (1 light). The color is calculated by Red (R), Green (G) and Blue (B) Color System (RGB system) which forms all the colors of the RGB color combination. The R, G and B colors use 8 bits in each, and their integer values range from 0 to 255. This makes 256 * 256 * 256 = 16777216 possible colors. Each pixel in the LED monitor displays colors in this manner through a combination of R, G and B LEDs. When the red pixel is set to 0, the LED is turned off. When the red pixel is set to 255, the LED will be fully turned on. Any value between them sets the LED to partial light emission [23].

We integrate Zenbo Commands into our MVR algorithm and import two other systems which are Librosa packages [5] and Tone Analyzer [16] that is from IBM developer cloud. We give the detail of this algorithm in the next section.

Table 1. Zenbo commands from APP Builder [15]

Commands	Actions	Parameters
setExpression()	Change Zenbo's facial expression	• **Face_id** - The name or id that corresponds to the name of the facial expression - Can be the following: ▪ -2 or "hideface" ▪ 13 or "helpless" ▪ 1 or "interested" ▪ 14 or "serious" ▪ 2 or "doubting" ▪ 15 or "worried" ▪ 3 or "proud" ▪ 16 or "pretend-ing" ▪ 4 or "default" ▪ 5 or "happy" ▪ 17 or "lazy" ▪ 6 or "expecting" ▪ 18 or "aware_r" ▪ 7 or "shocked" ▪ 19 or "tired" ▪ 8 or "question-ing" ▪ 20 or "shy" ▪ 21 or "innocent" ▪ 9 or "impatient" ▪ 22 or "singing" ▪ 10 or "confident" ▪ 23 or "aware_l" ▪ 11 or "active" ▪ 24 or "de-fault_still" ▪ 12 or "pleased"
playMusic()	Play a music file	• **source_name** - String of the audio filename • **duration** - The duration for how many seconds Zenbo should play of the audio file - Can be an integer or float value • **volume** - The volume to play the audio file - Can be from 0 to 100 • **startfrom** - The time in the audio file that Zenbo should start playing from. I don't see any reason to not always keep this at 0 - Can be an integer or float value
ledWheel()	Light up Zenbo's wheel LED lights	• **side** - Determines which side of Zenbo's wheels will light up - Can be the following: o "SYNC_BOTH" o "ASYNC_LEFT" o "ASYNC_RIGHT" • **type** - Indicates a different pattern of lighting for Zenbo's wheel lights - Can be the following strings: o "blink" o "breathing" o "stable" o "marques" o "charging" o "rainbow"

(continued)

Table 1. *(continued)*

		• **led_color**
		- The color Zenbo's lights will use
		- Can be an HTML hex color code string such as "#00FF00", the color for lime green
		• **display**
		- There are 8 LEDs on each of Zenbo's wheels This determines how many are lit
		- Can be the following:
		   o 255 (8 lights)
		   o 15 (4 lights)
		   o 7 (2 lights)
		   o 1 (1 light)
		• **brightness**
		- The brightness of the led wheel lights
		- Can be from 0 to 100
		• **duration**
		- The amount of the time in seconds the light should be on
		- Can be an integer or float value

3 MVR Algorithm

In this section, we simply describe how to perform music visualization on Zenbo by programming.

3.1 Music to Beat Time

Music is always around us. Whenever we hear any music related to our hearts, we will lose it. Subconscious, we will hear the beats we hear. You may not notice that your feet have automatically followed the beat of the music. However, turning this cognitive process into an automated system that reliably applies to a variety of musical styles is a challenging task [4].

To overcome the above problem, we first simplify the problem as considering only the beat time to perform music on the robot. We display the emotion and color on beat time which could be found by considering the method of musical signal analysis. There are many types of research works on how to analyze the audio. Most of those research works are concerned with the field of MIR, which is the interdisciplinary science of retrieving information from music. MIR is a rapidly developing field of research in recent years, which contains many practical applications. Researchers involved in MIR may have many combinations of knowledge, such as machine learning, informatics, signal processing, intelligent computing, musicology, psychoacoustics, psychology, and academic music research, etc.

Considering the MVR algorithm should be combined with methods in different fields, and the Librosa software package [5] has included all the analysis methods which we need in this paper. We use this package in detecting the beat time of MIDI music. Librosa software packages have many topics of packages which are "core functionality," "spectral features," "display," "onsets, tempo, and beats," "structural analysis," "decompositions," "effects," "output," etc.

The tempo and beat positions for an input signal can be easily calculated by onsets', tempo', and beats' packages. The onset and beat sub-modules provides information on all aspects of time in the estimated music [5]. The onset module provides two functions: onset_strength and onset_detect. The onset_strength function calculates the threshold spectral flux operation on the spectrogram and returns a one-dimensional array. On the other hand, the onset_detect function selects the peak position from the initial intensity curve after the heuristic described by Böck et al. [17]. The beat module uses the method of Ellis [18] to provide a function that estimates the global tempo and positions of the beat events from the onset_strength function.

3.2 Music to Emotions

Emotions are defined as "a strong feeling deriving from one's circumstances, mood, or relationships with others" [19]. When talking about emotions pertaining to social robots, it can be about how people feel about the appearance, the tone, the movement, the interaction, and the communication of the robot. For social robots, emotion is crucial as it is one of the biggest design features. In order to be social, it must understand, and express emotions making it more companionable and home friendly.

While it is easy to understand emotions, such as feeling angered when frustrated, happy when excited, and depressed when sad. It is harder for the machine to understand that through code. Therefore, it is important for social robots to not only express emotions carefully but also to understand emotions. For example, Zenbo designers make a friendly robot on its cute faces design to convey the feeling from users. We divide the facial expressions into two parts: Optimistic/Positive (P) and Pessimistic/Negative (N). We hope that when we import music on the robot, the Zenbo face could express its feeling of this music automatically.

To realize this system, we use the Tone Analyzer [16] which is from IBM developer cloud. We first transfer the audio to text (lyrics) and then the Tone Analyzer will provide the degree of emotion results from lyrics for us. In Feature Parsing Components, we recognize the feature parsing components by considering its degree of emotions. Generally speaking, high-level symbolic features belong to Optimistic/Positive (P) emotions and Low-level frequency-domain features belong to Pessimistic/Negative (N) emotions.

We link the degrees of emotions with music and it's P/N emotions as the results shown in two-dimensional space's emotions (See Fig. 1).

3.3 Emotions with Colors

As mentioned above, colors are an excellent way to express but the latter a bit more difficult. Some ways for social robots to understand when people interact with it could be the pitch of the voice, tone of voice, facial expression, gestures, etc.

Facial expressions of emotions are one of the ways social robots could use to better understand emotion. With the camera and facial recognition, social robots can learn to react to the feelings of their "masters". Though, not completely effective, because the robot must recognize the expression each time. But over time if it can learn, perhaps through the use of colors, learning the pitch and tone of the voice as well as memorizing the faces it can effectively become a companionable and understanding robot.

There were many types of research works which have mentioned that colors are relative to emotions. Most of the researchers were making experiments designs to discover the relationships between "Colors and emotions" [20], "Music and emotions" [21], but they do not link these three relationships/factors (music, emotion, and color). Recently, Palmer et al. [22] discovers the music-color associations with emotion by a serious of experiments and reveal the relationship with three factors by using Analysis of variance (ANOVA) which is a statistical technique that allows us to test the null hypothesis of three or more populations, using their sample information, equal to the alternative hypothesis of non-equivalent means. We continue this experimental design and considering the complex thoughts of human psychology, we add fuzzy ANOVA test [23] in analyzing the relationships between music, emotions, and colors.

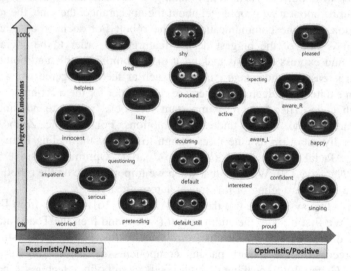

Fig. 1. Category of Zenbo facial expressions and its degrees of Emotions [14] (Zenbo facial expressions adopted from [15]) (Color figure online)

In this paper, we generally divide the emotions into two groups Optimistic/Positive (P) emotions and Pessimistic/Negative (N) emotions. To subdivide the P/N emotions, we divide them into 8 basic emotions (say, Angry, Anticipation, Joy, Trust, Fear, Surprise, Sadness, and Disgust. Here, Anticipation, Joy, Trust, and Surprise belong to P emotions, and Anger, Fear, Sadness, and Disgust belong to N emotions. After clear group the emotions, we link the 8 emotions with 8 basic colors (say, Red, Orange, Yellow, Chartreuse, Green, Cyan, Blue, Purple) based on RGB color system. The colors of LED lights will decide by the degree of two-dimensional emotions (as shown in Fig. 1). The 256 color codes we used in Zenbo developer's system can be found in reference [24].

Hence, we link the emotions, colors with music. Our MVR Algorithm is given in Fig. 2 and the system is put on GitHub [25]. We give a real case study for using MVR algorithm on Zenbo robot in the following section.

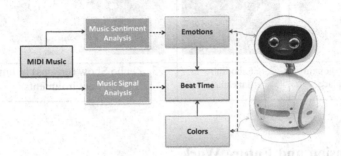

Fig. 2. MVR algorithm [14]

4 Performance

To design a humanity interface for Zenbo as an automatically dancer, we design a user-friendly system which could let the user have just a finger to input the music and the system could automatically output a performance. The performance will include the emotions on Zenbo's faces and LED lights on Zenbo's wheels. We give the process of using MVR algorithm for performance as follows:

1. Calculate the beat time through music by Librosa packages [5].

   We first calculate the beat time by using Librosa software packages. Every following action will be based on the beat time. To calculate the duration time of LED lights, we first calculate the waiting time which is the time between two beats. In our experiments, the duration time of the LED lights will be decided as half waiting time that will have better action on LED lights and do not overlap the lights on next beat.

2. Design the Emotion from Tone Analyzer [16].

   The results of Tone Analyzer have 7 emotions which are anger, fear, joy, sadness, analytical, confident, and tentative. We divide them into positive and negative emotions. The positive emotions are joy, analytical, and confident. The negative emotions are anger, fear, sadness, and tentative. Ever emotions will link the Zenbo 24 faces by different degree of their emotions.

3. Generate the file genzba.py [25] with total actions (faces, LED colors etc.).

   We integrate the code as genzba.py and provide it on Github [25]. We give three different situations to perform music with lights and emotions. The three kinds of different performances are shown in Table 2.

Table 2. Three kinds of different performances for Zenbo Show

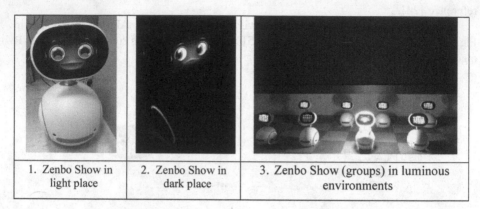

1. Zenbo Show in light place	2. Zenbo Show in dark place	3. Zenbo Show (groups) in luminous environments

5   Conclusion and Future Work

In this paper, we would like to display a performance of music visualization on the robot. We give an algorithm design to show that how to program the Zenbo to perform music automatically with LED lights on its wheels and express emotion on its face. The MVR prototype system contains two components of systems: MSA system (Librosa software packages), Tone Analyzer system. Based on ASUS Zenbo developer's APP, we link these two systems to control ASUS Zenbo developer's APP and hence, we could easily insert music and output a performance automatically. We hope that the MVR system not only could be used in Zenbo robot but also could extend to other fields of Artificial Intelligent (AI) equipment in the future.

Acknowledgment. This research work is supported by the Ministry of Education, R.O.C., under the grants of TEEP@AsiaPlus.

References

1. MilkDrop 1.04 for Windows 2000/NT/ME/98/95. Shareware Music Machine. Hitsquad Pty Ltd. Accessed 11 Oct 2010
2. MilkDrop Version History Archived 23 May 2007 at the Wayback Machine
3. MilkDrop preset authoring guide Archived 7 June 2007 at the Wayback Machine
4. Muller, M., Ellis, D.P.W., Klapuri, A., Richard, G.: Signal processing for music analysis. IEEE J. Sel. Top. Signal Process. 5(6), 1088–1110 (2011)
5. McFee, B., et al.: Librosa: audio and music signal analysis in python. In: Proceedings of the 14th Python in Science Conference, SCIPY 2015, pp. 18–25 (2015)
6. Gómez, L.M., Cáceres, M.N.: Applying data mining for sentiment analysis in music. In: De la Prieta, F., et al. (eds.) PAAMS 2017. AISC, vol. 619, pp. 198–205. Springer, Cham (2018). https://doi.org/10.1007/978-3-319-61578-3_20

7. Shukla, S., Khanna, P., Agrawal, K.K.: 2017 International Conference on Infocom Technologies and Unmanned Systems (Trends and Future Directions) (ICTUS), December 2017
8. Jamdar, A., Abraham, J., Khanna, K., Dubey, R.: Emotion analysis of songs based on lyrical and audio features. Int. J. Artif. Intell. Appl. (IJAIA) 6(3), 35–50 (2015)
9. Abboud, R., Tekli, J.: MUSE prototype for music sentiment expression. In: 2018 IEEE International Conference on Cognitive Computing, San Francisco, USA, 2–7 July 2018, pp. 106–109 (2018)
10. Wang, Q., Hai, Y., Shao, X., Gong, R.: Investigation on factors to influence color emotions and color preference responses. Optik 136, 71–78 (2017)
11. Kawabata, Y., Takahashi, F.: The association between colors and emotions for emotional words and facial expression. Wiley (2017)
12. Kim, M., Lee, H.S., Park, J.W., Jo, S.H., Chung, M.J.: Determining color and blinking to support facial expression of robot for conveying emotional intensity. In: IEEE International Symposium on Robot and Human Interactive Communication, Technische Universität München, Munich (2008)
13. Terada, K., Yamauchi, A., Ito, A.: Artificial emotion expression for a robot by dynamic color change. In: The 21st IEEE International Symposium on IEEE RO-MAN, Paris (2012)
14. Lin, P.-C., Mettrick, D., Hung, P.C.K., Iqbal, F.: IEEE International Conference on Artificial Intelligence and Virtual Reality (AIVR) 2018, Taichung, Taiwan, 10–12 December 2018 (2018)
15. https://zenbo.asus.com/developer/
16. https://tone-analyzer-demo.ng.bluemix.net/
17. Böck, S., Krebs, F., Schedl, M.: Evaluating the online capabilities of onset detection methods. In: 11th International Society for Music Information Retrieval Conference (ISMIR 2012), pp. 49–54 (2012)
18. Ellis, D.P.W.: Beat tracking by dynamic programming. J. New Music Res. 36(1), 51–60 (2007)
19. Oxford Dictionaries, "Emotion". Oxford University Press, Oxford. https://www.oxforddic tionaries.com/
20. Song, S., Yamada, S.: Exploring mediation effect of mental alertness for expressive lights. In: HAI 2017, Bielefeld, Germany, 17–20 October 2017 (2017)
21. Rubin, S., Agrawala, M.: Generating emotionally relevant musical scores for audio stories. In: UIST 2014, Honolulu, HI, USA, 5–8 October 2014 (2014)
22. Palmer, S.E., Schloss, K.B., Xu, Z., Prado-León, L.R.: Music-color associations are mediated by emotion. PNAS 110(22), 8836–8841 (2013)
23. Lin, P.C., Arbaiy, N., Hamid, I.R.A.: One-way ANOVA model with fuzzy data for consumer demand. In: Herawan, T., Ghazali, R., Nawi, N.M., Deris, M.M. (eds.) SCDM 2016. AISC, vol. 549, pp. 111–121. Springer, Cham (2017). https://doi.org/10.1007/978-3-319-51281-5_12
24. https://www.rapidtables.com/web/color/RGB_Color.html
25. https://github.com/RebeccaLab/Zenbo-Show

Combinatorial Properties of Fibonacci Arrays

Manasi S. Kulkarni[iD], Kalpana Mahalingam[✉][iD],
and Sivasankar Mohankumar[iD]

Department of Mathematics, Indian Institute of Technology Madras,
Chennai 600036, India
{ma16ipf01,ma16d028}@smail.iitm.ac.in, kmahalingam@iitm.ac.in

Abstract. The non-trivial extension of Fibonacci words to Fibonacci arrays was proposed by Apostolico and Brimkov in order to study repetitions in arrays. In this paper we investigate several combinatorial as well as formal language theoretic properties of Fibonacci arrays. In particular, we show that the set of all Fibonacci arrays is a 2D primitive language (under certain conditions), count the number of borders in Fibonacci arrays, and show that the set of all Fibonacci arrays is a non-recognizable language. We also show that the set of all square Fibonacci arrays is a two dimensional code.

Keywords: Fibonacci words · Fibonacci arrays ·
Recognizable picture language · Two dimensional code · Primitivity

1 Introduction

Pattern recognition and image processing have largely motivated the extension of concepts in formal language and combinatorics on words from one dimension to two dimensions, [9,14,25]. Since then, the field has witnessed a surge of activity, especially in generalizing formal language theoretic notions to two dimensions, to define formal models to recognize two dimensional (2D) languages, see [17] for a review on 2D languages. From combinatorial and coding theory perspective, concepts like codes [4,6,10], palindromes [9,16,19], periodicity [3,15], Fine and Wilf's theorem [24], Sturmian sequences [8], Fibonacci words [7,13], etc., have been generalized to their respective two dimensional counterparts.

The detection of repetitions in a string forms the very base of the field of combinatorics on words. These repetitions play a major role in the field of biology and computer science. Squares are the most fundamental of all repetitions. When extended to two dimensions, it is known that the number of square blocks in a 2D array of size $m \times n$ has the upper bound, $\mathcal{O}(m^2 n \log n)$, [7]. Moreover, Fibonacci arrays are shown to attain this general upper bound, [7].

The study of Fibonacci arrays, however, lacks the natural combinatorial perspective. In this paper, we aim to fill this gap by studying combinatorial and

T. V. Gopal and J. Watada (Eds.): TAMC 2019, LNCS 11436, pp. 448–466, 2019.
https://doi.org/10.1007/978-3-030-14812-6_28

formal language theoretic properties of Fibonacci arrays. The paper is organized as follows: In Sect. 2, we state the basic notations and definitions that will be used throughout the paper. In Sect. 3, we recall the definition of Fibonacci arrays from [7], and discuss some basic properties of such arrays including various decomposition properties, whereas in Sect. 4, we discuss the primitivity, non-recognizability property of the set of Fibonacci arrays, and also count the number of borders of any Fibonacci array $f_{m,n}$. In Sect. 5, we give an algorithm to check whether a given array over an alphabet Σ is a Fibonacci array or not. We end the paper with few concluding remarks in Sect. 6.

2   Preliminaries

An alphabet Σ is a finite non-empty set of symbols. By Σ^*, we denote the set of all words over Σ including the empty word λ, whereas Σ^+ denotes the set of all non-empty words over Σ. The length of a word $u \in \Sigma^*$ (i.e., the number of symbols in a word) is denoted by $|u|$. A word $u \in \Sigma^*$ is a prefix (suffix, respectively) of a word $w \in \Sigma^*$ if $w = uv$ ($w = vu$, respectively) for some $v \in \Sigma^*$. A word $w \in \Sigma^+$ is said to be bordered if there exists $x \in \Sigma^+$ such that $w = xy = zx$ where $y, z \in \Sigma^+$. A word is called unbordered if it is not bordered. Moreover, the set of all words with exactly i borders (including λ), where $i \geq 1$, is denoted by $D(i)$. The reversal of $u = a_1 a_2 \cdots a_n$ is defined to be the string $u^R = a_n \cdots a_2 a_1$, where $a_i \in \Sigma$ for $1 \leq i \leq n$. A word u is said to be a palindrome or a 1D palindrome if $u = u^R$. A word w is said to be primitive if $w = u^n$ implies $n = 1$ and $w = u$. For all other concepts in formal language theory and combinatorics on words, the reader is referred to [18, 20].

2.1   Arrays

Definition 1. *An array (also called a picture or two-dimensional word) $u = [u_{i,j}]_{1 \leq i \leq m, 1 \leq j \leq n}$ of size (m, n) over Σ is a two-dimensional rectangular finite arrangement of letters:*

$$u = \begin{matrix} u_{1,1} & u_{1,2} & \cdots & u_{1,n-1} & u_{1,n} \\ u_{2,1} & u_{2,2} & \cdots & u_{2,n-1} & u_{2,n} \\ \vdots & \vdots & \ddots & \vdots & \vdots \\ u_{m-1,1} & u_{m-1,2} & \cdots & u_{m-1,n-1} & u_{m-1,n} \\ u_{m,1} & u_{m,2} & \cdots & u_{m,n-1} & u_{m,n} \end{matrix}$$

Given an array u, by $|u|_{\text{row}}$ and $|u|_{\text{col}}$, we denote the number of rows and columns of u, respectively. An empty array is an array of size $(0, 0)$, and we denote it by λ, by the abuse of notation. The arrays of size $(m, 0)$ and $(0, m)$ for $m > 0$ are not defined. The set of all arrays over Σ including the empty array, λ, is denoted by Σ^{**}, whereas Σ^{++} is the set of all non-empty arrays over Σ.

We recall the following definition of the concatenation operation between two arrays, and two array languages.

Definition 2. [17] *Let* u, v *be arrays of sizes* (m_1, n_1) *and* (m_2, n_2), *respectively with* $m_1, n_1, m_2, n_2 > 0$, *over* Σ. *Then,*

1. *The column concatenation of* u *and* v, *denoted by* $\textcircled{1}$, *is a partial operation, defined if* $m_1 = m_2 = m$, *and is given by*

$$u \textcircled{1} v = \begin{matrix} u_{1,1} & \cdots & u_{1,n_1} & v_{1,1} & \cdots & v_{1,n_2} \\ \vdots & & \vdots & \vdots & & \vdots \\ u_{m,1} & \cdots & u_{m,n_1} & v_{m,1} & \cdots & v_{m,n_2} \end{matrix}$$

2. *The row concatenation of* u *and* v, *denoted by* $\ominus$, *is a partial operation, defined if* $n_1 = n_2 = n$, *and is given by*

$$u \ominus v = \begin{matrix} u_{1,1} & \cdots & u_{1,n} \\ \vdots & & \vdots \\ u_{m_1,1} & \cdots & u_{m_1,n} \\ v_{1,1} & \cdots & v_{1,n} \\ \vdots & & \vdots \\ v_{m_2,1} & \cdots & v_{m_2,n} \end{matrix}$$

It is clear that, similar to the concatenation operation in one dimension, the operations of column and row concatenations are associative but not commutative. Moreover, the column and row concatenation of u *and the empty array* λ *is always defined and* λ *is a neutral element for both the operations.*

Note that, the operations of column and row concatenation can be extended to languages in a similar fashion.

Definition 3. [19] *Let* $u \in \Sigma^{**}$. *An array* $v \in \Sigma^{**}$ *is said to be a prefix of* u *(suffix of* u, *respectively), denoted by* $v \leq_p^{2d} u$ *(*$v \leq_s^{2d} u$, *respectively) if* $u = (v \ominus x) \textcircled{1} y$ *(*$u = y \textcircled{1} (x \ominus v)$, *respectively) for some* $x, y \in \Sigma^{**}$. *Furthermore,* v *is said to be a proper prefix of* u *(proper suffix of* u, *respectively) denoted by* $v <_p^{2d} u$ *(*$v <_s^{2d} u$, *respectively) if either* $x \neq \lambda$ *or* $y \neq \lambda$, *or both* $x, y \in \Sigma^{++}$.

An array v is said to be a border of an array u if v is a prefix and as well a suffix of u. Note that, the empty array, λ, is always a border of any array. By $D_{2d}(i)$, let us denote the set of all arrays with exactly $i \geq 1$ borders.

If $x \in \Sigma^{++}$, then by $(x^{k_1 \textcircled{1}})^{k_2 \ominus}$ we mean that the array is constructed by repeating x, k_1 times column-wise and $x^{k_1 \textcircled{1}}$, k_2 times row-wise. An array $w \in \Sigma^{++}$ is said to be 2D *primitive* if $w = (x^{k_1 \textcircled{1}})^{k_2 \ominus}$ implies that $k_1 k_2 = 1$ and $w = x$, [15]. By Q_{2d}, let us denote the set of all 2D primitive arrays. Also, if $w = (x^{k_1 \textcircled{1}})^{k_2 \ominus}$ and x is 2D primitive, then x is said to be a 2D-*primitive root* of w denoted by $\rho_{2d}(w)$ which is always unique for a given array, [15].

Definition 4. *Let* $u = [u_{i,j}]$ *be an array of size* (m, n).

1. [9] *The reverse image of u, i.e.,*

$$u^R = \begin{matrix} u_{m,n} & u_{m,n-1} & \cdots & u_{m,2} & u_{m,1} \\ u_{m-1,n} & u_{m-1,n-1} & \cdots & u_{m-1,2} & u_{m-1,1} \\ \vdots & \vdots & \ddots & \vdots & \vdots \\ u_{2,n} & u_{2,n-1} & \cdots & u_{2,2} & u_{2,1} \\ u_{1,n} & u_{1,n-1} & \cdots & u_{1,2} & u_{1,1} \end{matrix}$$

Furthermore, if u is equal to its reverse image u^R, then u is said to be a two-dimensional palindrome, [9,16]. By P_{2d}, we denote the set of all 2D palindromes.

2. *The transpose of u, i.e., u^T is defined as:*

$$u^T = \begin{matrix} u_{1,1} & u_{2,1} & \cdots & u_{m-1,1} & u_{m,1} \\ \vdots & \vdots & \ddots & \vdots & \vdots \\ u_{1,n} & u_{2,n} & \cdots & u_{m-1,n} & u_{m,n} \end{matrix}$$

3   Fibonacci Arrays

Recall that, the Fibonacci numerical sequence $F(n)$ is defined recursively as $F(0) = 1$, $F(1) = 1$, $F(n) = F(n-1) + F(n-2)$ for $n \geq 2$. Similarly, for $\Sigma = \{a, b\}$, the sequence $\{f_n\}_{n\geq 0}$ ($\{f'_n\}_{n\geq 0}$, respectively) of Fibonacci words, is defined recursively by $f_0 = a$, $f_1 = b$, $f_n = f_{n-1}f_{n-2}$ for $n \geq 2$ ($f'_0 = a$, $f'_1 = b$, $f'_n = f'_{n-2}f'_{n-1}$, respectively). Moreover, $|f_n| = |f'_n| = F(n)$ for $n \geq 0$. Since $f_2 = f_1f_0$, $f_3 = f_2f_1$, $\cdots$, we note that all the Fibonacci words f_n, $n \geq 1$ over $\{f_0, f_1\}$ has f_1 as its prefix. Note that, throughout the paper, the Fibonacci words, f_n and f'_n, are always defined over the binary alphabet, $\{a, b\}$, unless otherwise specified.

In [7], authors have extended the concept of Fibonacci words to Fibonacci arrays.

Definition 5. [7] *Let $\Sigma = \{a, b, c, d\}$. The sequence of Fibonacci arrays, $\{f_{m,n}\}$ where $m, n \geq 0$, is defined as:*

1. $f_{0,0} = \beta, f_{0,1} = \gamma, f_{1,0} = \delta, f_{1,1} = \alpha$ *where α, β, γ and δ are symbols from Σ with some but not all, among α, β, γ and δ might be identical.*
2. *For $k \geq 0$ and $m, n \geq 1$,*

$$f_{k,n+1} = f_{k,n} \oplus f_{k,n-1}, \quad f_{m+1,k} = f_{m,k} \ominus f_{m-1,k}.$$

For the sake of convenience, we fix $f_{0,0} = a$, $f_{0,1} = b$, $f_{1,0} = c$, $f_{1,1} = d$, where some but not all of a, b, c and d might be identical. We call $f_{k,n+1} = f_{k,n} \oplus f_{k,n-1}$, the column-wise expansion and $f_{m+1,k} = f_{m,k} \ominus f_{m-1,k}$, the row-wise expansion. We demonstrate the definition of Fibonacci arrays with the following example.

Example 6. Let $\Sigma = \{a, b, c, d\}$. Then,
$$f_{2,3} = f_{1,3} \ominus f_{0,3} = (f_{1,2} \oplus f_{1,1}) \ominus (f_{0,2} \oplus f_{0,1})$$

$$= (f_{1,1} \oplus f_{1,0} \oplus f_{1,1}) \ominus (f_{0,1} \oplus f_{0,0} \oplus f_{0,1}).$$

It can also be obtained by column-wise expansion,
$$f_{2,3} = f_{2,2} \oplus f_{2,1} = f_{2,1} \oplus f_{2,0} \oplus f_{2,1}$$

$$= (f_{1,1} \ominus f_{0,1}) \oplus (f_{1,0} \ominus f_{0,0}) \oplus (f_{1,1} \ominus f_{0,1}).$$

Using the fact that $f_{0,0} = a, f_{0,1} = b, f_{1,0} = c, f_{1,1} = d$, $f_{2,3}$ is given by

$$f_{2,3} = \begin{matrix} d \ c \ d \\ b \ a \ b \end{matrix}$$

Observation: It can be observed that, when we perform the column-wise expansion of $f_{m,n}$ till the index j of each entry, $f_{i,j}$, becomes either 0 or 1, we get $F(n)$ number of columns. Furthermore, when the row-wise expansion is performed till the index i of each entry, $f_{i,j}$, becomes either 0 or 1, we get $F(m)$ number of rows. Thus, the size of Fibonacci array $f_{m,n}$ is $(F(m), F(n))$.

Similar to the definition of Fibonacci word f_n', we can define the sequence of Fibonacci arrays $\{f_{m,n}'\}_{m,n \geq 0}$ recursively as,

$$f_{m,n}' = f_{m,n-2}' \oplus f_{m,n-1}', \quad f_{m,n}' = f_{m-2,n}' \ominus f_{m-1,n}',$$

where $f_{0,0}' = a, f_{0,1}' = b, f_{1,0}' = c$ and $f_{1,1}' = d$ with some but not all of a, b, c and d might be identical. However, we study the properties of only $f_{m,n}$ in this paper, as equivalent properties of $f_{m,n}'$ can be obtained similarly. We state the relation between $f_{m,n}$ and $f_{m,n}'$ in Corollary 12.

Lemma 7. [7] *If $f_{0,0} \neq f_{0,1}$ and $f_{1,0} \neq f_{1,1}$, then for $m, n \geq 1$, each row of the Fibonacci array $f_{m,n}$, written as a 1D word is the Fibonacci word f_n over either $\{f_{0,0}, f_{0,1}\}$ (starting with $f_{0,1}$) or $\{f_{1,0}, f_{1,1}\}$ (starting with $f_{1,1}$). Also, each column of $f_{m,n}$, written as a 1D word is the Fibonacci word f_m over either $\{f_{0,0}, f_{1,0}\}$ (starting with $f_{1,0}$) or $\{f_{0,1}, f_{1,1}\}$ (starting with $f_{1,1}$).*

We recall the following result from [11].

Theorem 8. *Every integer $n \geq 0$ admits a representation as sum of distinct Fibonacci numbers, i.e.,*

$$n = F(k_r) + F(k_{r-1}) + \ldots + F(k_1), \quad (k_r > k_{r-1} > \ldots > k_1).$$

Furthermore, for the above representation, we associate the word $a_{k_r} \cdots a_1 a_0$ with $a_{k_r} = \ldots = a_{k_1} = 1$ and $a_i = 0$, otherwise. Such a binary word is called a Fibonacci representation of a given integer n. For example, the Fibonacci representations of 5 and 6(=5+1) are 1000 and 1001, respectively.

However, in the following, we define a new notion called the reduced representation of a given Fibonacci word, f_n, for $n \geq 2$, and the corresponding binary word as a Fibonacci reduced representation of n, namely FibRed(n). This definition will aid us in a much simpler way of constructing the Fibonacci array, $f_{m,n}$ for any $m, n \geq 0$, irrespective of the knowledge of the intermediate Fibonacci arrays.

Definition 9. *Given a Fibonacci word, f_n, for $n \geq 2$, if f_n is simplified using the recurrence relation $f_n = f_{n-1}f_{n-2}$ until f_n is expressed using only f_0 and f_1, then such an expression is called the reduced representation of f_n. Formally, $f_n = f_{i_1}f_{i_2}\cdots f_{i_{F(n)}}$ with $i_1, i_2, \ldots, i_{F(n)} \in \{0,1\}$, is called the reduced representation of f_n. Also, the binary string $i_1 i_2 \cdots i_{F(n)} = FibRed(n)$ is called the Fibonacci reduced representation of the integer n.*

Note that, FibRed(1) = 1 and FibRed(0) = 0 and also one can observe that, FibRed(5) = 10110101 and FibRed(6) = 1011010110110.

Lemma 10. *FibRed(n) ends with 10 if n is even and ends with 01 if n is odd. Also $\{FibRed(n)\}_{n \geq 1}$ are 1D Fibonacci words over $\{0,1\}$(starting with 1).*

Proof. If n is even, $n-2$, $n-4$, $\cdots$ are all even. Therefore, the expansion, $f_n = f_{n-1}f_{n-2} = f_{n-2}f_{n-3}f_{n-3}f_{n-4}$ ultimately becomes, $f_1 \cdots f_1 f_0$ and hence FibRed(n) ends with 10. Similarly, we can prove that, FibRed(n) ends with 01 if n is odd. Now, for $n \geq 1$, since $f_n = f_{i_1}f_{i_2}\cdots f_{i_{F(n)}}(i_1, i_2, \ldots, i_{F(n)} \in \{0,1\})$, is a 1D Fibonacci word over $\{f_0, f_1\}$ (starting with f_1), by replacing $f_{i_1}, f_{i_2}, \cdots f_{i_{F(n)}}$ in $f_{i_1}f_{i_2}\cdots f_{i_{F(n)}}$ by the respective suffixes $i_1, i_2, \ldots, i_{F(n)}$ we observe that FibRed(n) is a 1D Fibonacci word over $\{0,1\}$ (starting with 1).

Based on the Definition 9, we state the following result which yields $f_{m,n}$ for any given $m, n \geq 0$, using the Fibonacci reduced representation of m and n.

Lemma 11. *Let FibRed(m) = $i_1 i_2 \cdots i_{F(m)}$ and FibRed(n) = $j_1 j_2 \cdots j_{F(n)}$. Then the indices of the elements in $f_{m,n}$ ($f'_{m,n}$, respectively) are ordered pairs of the cartesian product $\{i_1, i_2, \ldots, i_{F(m)}\} \times \{j_1, j_2, \ldots, j_{F(n)}\}$ ($\{i_{F(m)}, \ldots, i_1\} \times \{j_{F(n)}, \ldots, j_1\}$, respectively). Further, $f_{n,m} = [\hat{f}_{m,n}]^T$ where $\hat{f}_{m,n}$ is the array obtained by replacing each entry $f_{i,j}$, $i,j \in \{0,1\}$ of $f_{m,n}$ by $f_{j,i}$.*

Proof. We prove the result only for $f_{m,n}$. Given $f_{m,n}$, we know that the size of $f_{m,n}$ is $(F(m), F(n))$. Thus, when we expand $f_{m,n}$ column wise till the column index of each element becomes either 0 or 1, the concatenation of the column indices (from left to right) of any row, results into the Fibonacci reduced representation of n, FibRed(n) = $j_1 j_2 \cdots j_{F(n)}$. Furthermore, when the row wise expansion is performed till the row index of each element becomes either 0 or 1, the concatenation of the row indices (from top to bottom) of any column, results into the Fibonacci reduced representation of m, FibRed(m) = $i_1 i_2 \cdots i_{F(m)}$. That is to say that $f_{m,n} = (f_{i_1,j_1} \oplus f_{i_1,j_2} \oplus \cdots \oplus f_{i_1,j_{F(n)}}) \ominus (f_{i_2,j_1} \oplus f_{i_2,j_2} \oplus \cdots \oplus f_{i_2,j_{F(n)}}) \ominus \cdots \ominus (f_{i_{F(m)},j_1} \oplus f_{i_{F(m)},j_2} \oplus \cdots \oplus f_{i_{F(m)},j_{F(n)}})$. Hence, the indices of elements in $f_{m,n}$ are the ordered pairs of the cartesian product $\{i_1, i_2, \ldots, i_{F(m)}\} \times \{j_1, j_2, \ldots, j_{F(n)}\}$.

Also, since $f_{n,m} = (f_{j_1,i_1} \oplus f_{j_1,i_2} \oplus \cdots \oplus f_{j_1,i_{F(m)}}) \ominus (f_{j_2,i_1} \oplus f_{j_2,i_2} \oplus \cdots \oplus f_{j_2,i_{F(m)}}) \ominus \cdots \ominus (f_{j_{F(n)},i_1} \oplus f_{j_{F(n)},i_2} \oplus \cdots \oplus f_{j_{F(n)},i_{F(m)}})$, we see that $f_{n,m} = [\hat{f}_{m,n}]^T$ where $\hat{f}_{m,n}$ is the array obtained by replacing each entry $f_{i,j}$, $i,j \in \{0,1\}$ of $f_{m,n}$ by $f_{j,i}$. $\qquad\qquad\square$

Using Lemma 11, we state the relation between Fibonacci arrays, $f_{m,n}$ and $f'_{m,n}$ in the following result. Let $F_{2d} = \{f_{m,n} : m,n \geq 0\}$ and $F'_{2d} = \{f'_{m,n} : m,n \geq 0\}$.

Corollary 12. *Let $f \in F_{2d}$ and $f' \in F'_{2d}$ be two Fibonacci arrays of the same size. Then $f' = f^R$.*

3.1   Decomposition of Fibonacci Arrays

Partitioning a word in to smaller segments (or sub words), which have relatively simpler properties, is one way of understanding the structure and properties of the given word. Analysis of these decompositions will help in two-dimensional string matching which is infact the core of any image processing algorithm [12]. In this section, we establish two possible decompositions of the Fibonacci array $f_{m,n}$ which will enhance our understanding of its structural properties.

Lemma 13. *[21] Any Fibonacci word f_n over $\{f_0, f_1\}$ (starting with f_1), can be written as a concatenation of a palindrome of length $F(n) - 2$, say α_n, and a suffix of length 2, say d_n, which is $f_1 f_0$ if n is even, and $f_0 f_1$ if n is odd.*

Lemma 14. *If $w \in \Sigma^{++}$ is such that every row and every column of w is a 2D palindrome, then $w \in P_{2d}$.*

Proposition 15. *Let $\{f_{m,n}\}_{m,n\geq 0}$ be the sequence of Fibonacci arrays over the alphabet $\Sigma = \{a,b,c,d\}$. For $m,n \geq 3$,*

$$f_{m,n} = (A \oplus B) \ominus (C \oplus D) \text{ with } |A|_{col} = |C|_{col}, \text{ such that}$$

1. *A is a 2D palindrome of size $(F(m) - 2, F(n) - 2)$;*
2. *B and C are arrays of size $(F(m) - 2, 2)$ and $(2, F(n) - 2)$, respectively. Moreover, if $m = n$, $[B]^T = \hat{C}$ where $\hat{C}$ is the array obtained from C by replacing each entry $f_{i,j}$ of C with $f_{j,i}$ for $i,j \in \{0,1\}$;*
3. *D is an array of size $(2,2)$ such that*
    (i) $(f_{1,1} \oplus f_{1,0}) \ominus (f_{0,1} \oplus f_{0,0})$ if both m and n are even.
    (ii) $(f_{0,0} \oplus f_{0,1}) \ominus (f_{1,0} \oplus f_{1,1})$ if both m and n are odd.
    (iii) $(f_{1,0} \oplus f_{1,1}) \ominus (f_{0,0} \oplus f_{0,1})$ if m is even and n is odd.
    (iv) $(f_{0,1} \oplus f_{0,0}) \ominus (f_{1,1} \oplus f_{1,0})$ if m is odd and n is even.

Proof. 1. By Lemma 7, we know that if $f_{0,0} \neq f_{0,1}$ and $f_{1,0} \neq f_{1,1}$ ($f_{0,0} \neq f_{1,0}$ and $f_{0,1} \neq f_{1,1}$, respectively), then each row (column, respectively) of the Fibonacci array $f_{m,n}$, written as a 1D word is the Fibonacci word f_n (f_m, respectively), and furthermore by Lemma 13, each of these rows (columns, respectively) has a palindromic prefix of length $F(n) - 2$ ($F(m) - 2$, respectively). Therefore, all rows and columns of the prefix array of size $(F(m) - 2, F(n) - 2)$ of $f_{m,n}$ will be palindromes and hence by Lemma 14, the prefix array itself is a palindrome. Note that, even if some of $f_{0,0}, f_{0,1}, f_{1,0}, f_{1,1}$ are identical, each row and column of the prefix array of size $(F(m) - 2, F(n) - 2)$ of $f_{m,n}$ is still a palindrome. Thus, $f_{m,n}$ has a palindromic prefix A of size $(F(m) - 2, F(n) - 2)$.

2. Since $|A|_{\text{row}} = |B|_{\text{row}}$ and $|A|_{\text{col}} = |C|_{\text{col}}$, the sizes of B and C are $(F(n) - 2, 2)$ and $(2, F(m) - 2)$, respectively. Further, if $m = n$, $F(m) = F(n)$ and the sizes of $[B]^T$ and C become the same. Hence, by Lemma 11, $[B]^T = \hat{C}$ where $\hat{C}$ is the array obtained from C by replacing each entry $f_{i,j}$ of C with $f_{j,i}$ for $i, j \in \{0, 1\}$.

3. We prove the result only when m and n are even. Since m and n are even, the Fibonacci reduced representations of both m and n end with 10. Thus, the elements of the suffix of size $(2, 2)$ of $f_{m,n}$ have the indices which are ordered pairs of the cartesian product $\{1, 0\} \times \{1, 0\}$. That is to say that such a suffix is nothing but $(f_{1,1} \oplus f_{1,0}) \ominus (f_{0,1} \oplus f_{0,0})$. $\qquad\square$

It is known from [21] that any Fibonacci word, f_n, can be uniquely written as a concatenation of two distinct palindromes of length $F(n - 1) - 2$ and $F(n - 2) + 2$. In the following proposition, we provide the unique decomposition of any Fibonacci array into four palindromic subarrays.

Lemma 16. [21] *For all $n > 4$, f_n is the product $u_n v_n$ of two uniquely determined palindrome words of Σ^+, $\Sigma = \{f_0, f_1\}$, whose lengths are $|u_n| = F(n - 1) - 2$ and $|v_n| = F(n - 2) + 2$.*

Proposition 17. *For $m, n > 4$ we have that*

$$f_{m,n} = (F_{tl} \oplus F_{tr}) \ominus (F_{bl} \oplus F_{br})$$

where $F_{tl}, F_{tr}, F_{bl}, F_{br}$ are all palindromic subarrays such that $size(F_{tl}) = (F(m - 1) - 2, F(n - 1) - 2)$, $size(F_{tr}) = (F(m - 1) - 2, F(n - 2) + 2)$, $size(F_{bl}) = (F(m - 2) + 2, F(n - 1) - 2)$ and $size(F_{br}) = (F(m - 2) + 2, F(n - 2) + 2)$.

Proof. Existence: If $f_{0,0} \neq f_{0,1}$ and $f_{1,0} \neq f_{1,1}$ ($f_{0,0} \neq f_{1,0}$ and $f_{0,1} \neq f_{1,1}$, respectively), then by Lemma 7, each row (column, respectively) of a Fibonacci array, written as a 1D word is a Fibonacci word. Furthermore, by Lemma 16, these Fibonacci words (over whatever symbols they may be), can be represented as a product of two palindromes of lengths $F(n - 1) - 2$ and $F(n - 2) + 2$ ($F(m - 1) - 2$ and $F(m - 2) + 2$, respectively). Thus, in the Fibonacci array, $f_{m,n}$, each row (column, respectively) can be represented as a column catenation (row catenation, respectively) of two palindromes of sizes $(1, F(n - 1) - 2)$ and $(1, F(n - 2) + 2)$ (($F(m - 1) - 2, 1)$ and $(F(m - 2) + 2, 1)$, respectively). The result then follows by Lemma 14. Note that, even if some of these a, b, c, d are identical, the result still holds.

Uniqueness: When none of a, b, c, d are identical, the uniqueness follows from the uniqueness result in one dimension. In all other cases, at least one row or one column will be a 1D Fibonacci word, and hence the representation is unique. $\qquad\square$

We have seen various unique decompositions of $f_{m,n}$ in this subsection. However, from the definition it is clear that the decomposition of $f_{m,n}$ in terms of other Fibonacci arrays is certainly not unique. The natural question that arises in this context is whether any array $u \in \Sigma^{**}$ can be decomposed uniquely into Fibonacci arrays. The general answer is "no" as demonstrated in the following example:

$$f_{m,n+3} \oplus f_{m,n} = f_{m,n+2} \oplus f_{m,n+1} \oplus f_{m,n} = f_{m,n+2} \oplus f_{m,n+2}.$$

We recall the following from [5].

Definition 18. *The domain of a picture p is the set of coordinates $dom(p) = \{1, 2, \cdots, |p|_{row}\} \times \{1, 2, \cdots, |p|_{col}\}$. A subdomain of $dom(p)$ is a set d of the form $\{i, i+1, \cdots, i'\} \times \{j, j+1, \cdots, j'\}$, where $1 \le i \le i' \le |p|_{row}, 1 \le j \le j' \le |p|_{col}$, also specified by the pair $[(i, j), (i', j')]$.*

Definition 19. *Let $X \subseteq \Sigma^{**}$. The tiling star of X, denoted by X^{**}, is the set of pictures p whose domain can be partitioned in to disjoint subdomains $\{d_1, d_2, \ldots, d_k\}$ such that, for any $h = 1, \cdots, k$, the subpicture p_h of p associated with the subdomain d_h belongs to X. Also if $p \in X^{**}$, the partition $t = \{d_1, d_2, \ldots, d_k\}$ of the domain of p, together with the corresponding pictures $\{p_1, p_2, \cdots, p_k\}$, is called a tiling decomposition of p in X.*

Definition 20. *A set $X \subseteq \Sigma^{**}$ is a code if any $p \in \Sigma^{**}$ has at most one tiling decomposition in X.*

Picture codes are an important class of codes in two dimensions, with polyomino codes being the more general class. Though many results pertaining to codes are available in one dimensional language theory, major results in a two dimensional context are still in demand [10]. Studies are carried out with an aim to construct efficient two dimensional codes which are useful in communication theory.

Our earlier demonstration, $f_{m,n+3} \oplus f_{m,n} = f_{m,n+2} \oplus f_{m,n+2}$, proves that, the language, $F_{2d} = \{f_{m,n} : m, n \ge 0\}$ is not a code. But it is interesting to observe the following theorem, in which, we prove that, when $m = n$ and none of a, b, c, d are identical, the language of Fibonacci arrays, $L = \{f_{n,n} : n \ge 0\}$ is a two dimensional code.

Theorem 21. *Let $\{f_{m,n}\}_{m,n \ge 0}$ be the sequence of Fibonacci arrays over $\Sigma = \{a, b, c, d\}$. Then when none of a, b, c, d are identical, the language $L = \{f_{n,n} : n \ge 0\}$ is a two-dimensional code.*

Proof. For the sake of contradiction, assume that the language L is not a code. Then, there exists a $w \in \Sigma^{++}$ such that w has two distinct decompositions in L, say t_1 and t_2.

Now, consider the bottom right blocks of t_1 and t_2, say f_{k_1,k_1} and f_{k_2,k_2}, respectively, such that $k_1 \ne k_2$. Let us assume that $k_2 < k_1$. Since both f_{k_1,k_1} and f_{k_2,k_2} are square arrays and $k_2 < k_1$, f_{k_2,k_2} must be a suffix of f_{k_1,k_1}, i.e.,

$f_{k_2,k_2} <_s^{2d} f_{k_1,k_1}$. Without loss of generality, let us assume that k_1 is even. Then, by Proposition 15, the suffix of f_{k_1,k_1} in t_1 is $(f_{1,1} \oplus f_{1,0}) \ominus (f_{0,1} \oplus f_{0,0})$, which is also a suffix of f_{k_2,k_2} in t_2, and thus k_2 has to be even as well. We now divide the proof into following two cases:

Case (1): Let $k_1 = k_2 + 2$. Then, for $k_3 = k_2 + 1$, we have, $F(k_1) = F(k_2)+F(k_3)$. Furthermore, since k_1 and k_2 are both even, the Fibonacci reduced representations of k_1 and k_2 end with 10, and hence the last column of f_{k_1,k_1} (and f_{k_2,k_2}) consists of $f_{1,0} = c$ and $f_{0,0} = a$ alone. Since k_3 is odd, the Fibonacci reduced representation of k_3 ends with 01. This further implies that, the entries in the last column of f_{k_1,k_1} above the last column of f_{k_2,k_2} must consist of $f_{0,1} = b$ and $f_{1,1} = d$, a contradiction.

Case (2): Let $k_1 > k_2 + 2$. Since k_1 and $k_2 + 2$ are both even, $f_{k_2,k_2} <_s^{2d} f_{k_1-2,k_1-2}$. Now, we can use the argument as that of Case 1 to arrive at a similar contradiction.

Since, both the cases lead to contradictions, we conclude that $k_1 = k_2$. We can extend this argument to all other blocks in t_1 and t_2, and arrive at a similar contradiction. The case when k_1 is odd can be proved similarly. Thus w has an unique decomposition in L. Hence L is a code. $\square$

4   Other Properties of Fibonacci Arrays

A study of primitive words in a language is required for a better understanding of that language [26]. Also, primitive words play a major role in designing algorithms for natural language processing and in computational biology. In addition, by locating and counting special patterns like palindromes, squares and borders in the words of a language, we can design efficient algorithms for text editing, information retrieval and data compression. In this section, first we prove that the set of Fibonacci arrays is 2D primitive (under certain conditions). Then, we prove the non-recognizability of Fibonacci arrays. And finally, an explicit formula for counting the number of borders in a given Fibonacci array $f_{m,n}$ is derived.

Theorem 22. *The set of Fibonacci arrays $f_{m,n}$ for $m, n > 1$ is 2D primitive except in the following cases:*

1. $f_{0,0} = f_{1,0}$ $(a = c)$ and $f_{0,1} = f_{1,1}$ $(b = d)$
2. $f_{0,0} = f_{0,1}$ $(a = b)$ and $f_{1,0} = f_{1,1}$ $(c = d)$

Furthermore, Fibonacci arrays $\{f_{0,0}, f_{0,1}, f_{1,0}, f_{1,1}\} \cup \{f_{1,n}, f_{m,1}\} \subseteq Q_{2d}$ for $m, n > 1$ if $b \neq d$ and $c \neq d$.

Proof. The Fibonacci arrays $f_{0,0}, f_{0,1}, f_{1,0}$ and $f_{1,1}$ are trivially 2D primitive.

First, let us assume that none of a, b, c, d are identical. Now, if $m = 1$ and $n > 1$ ($n = 1$ and $m > 1$, respectively), then $f_{1,n}$ ($f_{m,1}$, respectively), written as a 1D word is a Fibonacci word and hence primitive, [21].

Now, assume that, $f_{m,n}$ is not 2D primitive for some $m, n \geq 2$. Then, $f_{m,n} = (u^{k_1 \oplus})^{k_2 \ominus}$ where at least one of k_1 or k_2 is strictly greater than 1. Without loss of generality, let $k_1 = 1$ and $k_2 > 1$, then $f_{m,n} = u^{k_2 \ominus}$. This implies that, the first column of $f_{m,n}$ is $u_1^{k_2 \ominus}$ where $u_1 <_p^{2d} u$. However, this contradicts the fact that Fibonacci words are primitive. The cases when $k_1 > 1, k_2 = 1$ and $k_1, k_2 > 1$ can be proved similarly.

Now, from the definition of Fibonacci arrays, we know that, some of a, b, c, d might be identical. Thus, we have the following possible cases:

Case (1): Let $a = b, c \neq d$. Then, we have the following three subcases:

Case (1a): Let $a = d$. Then, we have $a = b = d \neq c$. This implies that, the first row (second column, respectively) of $f_{m,n}$ written as a 1D word is a Fibonacci word over $\{a, c\}$. Moreover, since the Fibonacci words are primitive, there cannot exist $u \in Q_{2d}$ such that $f_{m,n} = (u^{k_1 \oplus})^{k_2 \ominus}$ where either $k_1 > 1$ or $k_2 > 1$.

The subcases (1b): $a = b = c \neq d$ and (1c): $a = b \neq d \neq c$ can be proved similarly. Also, the cases when (2): $a = c, b \neq d$, (3): $a = d, b \neq c$, (4): $b = c, a \neq d$, (5): $b = d, a \neq c$ and (6): $c = d, a \neq b$, and its subcases can be proved similarly.

Case (7): Let $b = c, a = d$. Then, the first row and the first column of $f_{m,n}$ written as a 1D words are Fibonacci words over $\{a, b\}$ and the set of Fibonacci words is known to be primitive.

Now, let us consider the following cases:

Case (I): Let $a = b, c = d$. If $a = b \neq c$ then all the columns are identical. Moreover, all these columns are Fibonacci words over $\{a, c\}$, say u, and hence for $F(n) \geq 2$, $f_{m,n} = u^{F(n) \oplus}$, which is not 2D primitive.

The case (II): $a = c, b = d$ can be proved similarly.

Hence, the set of Fibonacci arrays $f_{m,n}$ for $m, n \geq 0$ are 2D primitive except in Case (I) and Case (II). $\qquad \square$

Recall that an array $x \in \Sigma^{**}$ is said to be a border of an array $u \in \Sigma^{++}$ if x is a prefix as well as a suffix of u. Moreover, the set of all arrays with exactly i borders (including λ), where $i \geq 1$, is denoted by $D_{2d}(i)$. We state the following result from [27] regarding the borders of 1D Fibonacci words.

Theorem 23. [27] *Let $\{f_n\}_{n \geq 1}$ be the sequence of 1D Fibonacci words over $\{a, b\}$. Then,*

1. $\{f_0, f_1, f_2\} \subseteq D(1)$
2. *For $i \geq 3$, $f_i \in D(j)$ where $j = \lceil \frac{i}{2} \rceil$.*

We state an analogous result for Fibonacci arrays in Theorem 24. Note that, since the indices of the elements of $f_{m,n}$ for any given $m, n \geq 0$, are cartesian product of $FibRed(m)$ and $FibRed(n)$, the borders of $f_{m,n}$ are created by the borders of $FibRed(m)$ and $FibRed(n)$.

Theorem 24. *Let $\{f_{m,n}\}_{m,n \geq 0}$ be the sequence of 2D Fibonacci arrays over $\{a, b, c, d\}$ such that none of a, b, c, d are identical. Then,*

1. $\{f_{0,0}, f_{0,1}, f_{1,0}, f_{1,1}, f_{1,2}, f_{2,1}\} \subseteq D_{2d}(1)$
2. $\{f_{1,n}\}_{n\geq 3} \subseteq D_{2d}(j)$, where $j = \lceil \frac{n}{2} \rceil$ and $\{f_{m,1}\}_{m\geq 3} \subseteq D_{2d}(j)$, where $j = \lceil \frac{m}{2} \rceil$
3. For $m, n \geq 2$, $f_{m,n} \in D_{2d}(j)$ where $j = \lceil \frac{m}{2} \rceil \lceil \frac{n}{2} \rceil$.

Proof. The statement (1) can be easily verified. Since, $\{f_{1,n}\}_{n\geq 3}$ and $\{f_{m,1}\}_{m\geq 3}$, written as 1D words are nothing but 1D Fibonacci words, statement (2) follows from Theorem 23.

Now, let $m, n \geq 2$ and let $FibRed(m) = i_1 i_2 \cdots i_{F(m)}$ and $FibRed(n) = j_1 j_2 \cdots j_{F(n)}$.

By Lemma 10, we note that $FibRed(m)$ and $FibRed(n)$ are 1D Fibonacci words over $\{0, 1\}$, starting with 1. Furthermore, since $m, n \geq 2$, by Theorem 23, both $FibRed(m)$ and $FibRed(n)$ will have non-empty borders. For any k, l such that $1 \leq k < F(m)$ and $1 \leq l < F(n)$, let $B = i_1 \cdots i_k$ and $B' = j_1 \cdots j_l$, be the borders of $FibRed(m)$ and $FibRed(n)$, respectively. Due to the fact that the indices of elements in $f_{m,n}$ are ordered pairs of the cartesian product $\{i_1, i_2, \ldots, i_{F(m)}\} \times \{j_1, j_2, \ldots, j_{F(n)}\}$, the prefix array of size (k, l) of $f_{m,n}$ is

$$x = (f_{i_1,j_1} \oplus \cdots \oplus f_{i_1,j_l}) \ominus \cdots \ominus (f_{i_k,j_1} \oplus \cdots \oplus f_{i_k,j_l}).$$

Since $i_1 \cdots i_k$ and $j_1 \cdots j_l$ are borders of $FibRed(m)$ and $FibRed(n)$ respectively, the same subarray x occurs as a suffix of $f_{m,n}$. Note that, these prefix and suffix subarrays can overlap with each other. Hence, the borders of $f_{m,n}$ are generated due to the borders of $FibRed(m)$ and $FibRed(n)$, and therefore, by Theorem 23 $f_{m,n}$ has $\lceil \frac{m}{2} \rceil \times \lceil \frac{n}{2} \rceil$ number of borders.   □

Table 1. Number of borders in $f_{m,n}$ under various cases.

Case →	1	2	3	4
	m odd	m odd	m even	m even
Subcase ↓	n odd	n even	n even	n odd
(a) $(a = b), (c = d)$	V_6	V_6	V_6	V_6
(b) $(a = c), (b = d)$	V_5	V_5	V_5	V_5
(c) $(a = d), (b = c)$	V_7	V_1	V_7	V_1
(d) $(a = b) \neq (c \neq d)$	V_1	V_1	V_1	V_1
(e) $(a = c) \neq (b \neq d)$	V_1	V_1	V_1	V_1
(f) $(a = d) \neq (b \neq c)$	V_1	V_1	V_7	V_1
(g) $(b = c) \neq (a \neq d)$	V_1	V_1	V_1	V_1
(h) $(b = d) \neq (a \neq c)$	V_2	V_1	V_1	V_2
(i) $(c = d) \neq (a \neq b)$	V_3	V_3	V_1	V_1
(j) $a \neq (b = c = d)$	V_4	V_3	V_1	V_2
(k) $b \neq (a = c = d)$	V_3	V_3	V_7	V_1
(l) $c \neq (a = b = d)$	V_2	V_1	V_7	V_2
(m) $d \neq (a = b = c)$	V_1	V_1	V_1	V_1

By definition of 2D Fibonacci words, we know that some of a, b, c, d can be identical. Thus, together with the fact that m, n can be even or odd, in total, 52 different cases arise. The number of borders in each of the 52 cases are listed in Table 1, where the values of V_i's for $1 \le i \le 7$ are as per Theorem 25.

Note that, in the proof of Theorem 25, for $s, t \in \Sigma$, we use $f_{k,\{s<t\}}$ to denote, by the abuse of notation, the kth Fibonacci word over $\{s, t\}$ starting with s.

Theorem 25. *Let $\{f_{m,n}\}_{m,n\ge 0}$ be the sequence of 2D Fibonacci arrays over $\{a, b, c, d\}$ such that some but not of all a, b, c, d are identical. Then, the number of borders of $f_{m,n}$ takes any of the values from the set $\{V_1, V_2, V_3, V_4, V_5, V_6, V_7\}$ where,*

1. $V_1 = \lceil \frac{m}{2} \rceil \lceil \frac{n}{2} \rceil$
2. $V_2 = \lceil \frac{m}{2} \rceil \lceil \frac{n}{2} \rceil + (F(m) - \lceil \frac{m}{2} \rceil)$
3. $V_3 = \lceil \frac{m}{2} \rceil \lceil \frac{n}{2} \rceil + (F(n) - \lceil \frac{n}{2} \rceil)$
4. $V_4 = \lceil \frac{m}{2} \rceil \lceil \frac{n}{2} \rceil + (F(m) - \lceil \frac{m}{2} \rceil) + (F(n) - \lceil \frac{n}{2} \rceil) - 1$
5. $V_5 = \lceil \frac{m}{2} \rceil \lceil \frac{n}{2} \rceil + \lceil \frac{n}{2} \rceil (F(m) - \lceil \frac{m}{2} \rceil)$
6. $V_6 = \lceil \frac{m}{2} \rceil \lceil \frac{n}{2} \rceil + \lceil \frac{m}{2} \rceil (F(n) - \lceil \frac{n}{2} \rceil)$
7. $V_7 = \lceil \frac{m}{2} \rceil \lceil \frac{n}{2} \rceil + 1$

Proof. We divide the proof in to four main cases depending on the values of m, n, and further in to subcases depending on the equality among a, b, c, d.

Case 1: m is odd, n is odd.

Case 1(a): Let $a = b, c = d$. Note that, all the columns written as 1D words, are Fibonacci words $f_{m,\{d<b\}}$. Thus $f_{m,n} = f_{m,1}^{F(n)\oplus}$.
In the case when none of the a, b, c, d are identical, we already have $\lceil \frac{m}{2} \rceil \lceil \frac{n}{2} \rceil = k$ borders. Thus $f_{m,n}$, in this case will surely have k borders. Since m is odd, along with the 1st row even the last row will be of the form $d^{F(n)\oplus}$. Thus every prefix of $d^{F(n)\oplus}$ becomes a border of $f_{m,n}$. Thus, we get $F(n) - \lceil \frac{n}{2} \rceil$ extra borders. Since every column is identical and we have $\lceil \frac{m}{2} \rceil$ borders for $f_{m,\{d<b\}}$, we get in total $\lceil \frac{m}{2} \rceil (F(n) - \lceil \frac{n}{2} \rceil)$ extra borders.

Case 1(b): Let $a = c, b = d$. Here $f_{m,n} = f_{1,n}^{F(m)\ominus}$ where $f_{1,n}$ written as a 1D word is $f_{n,\{d<c\}}$. This case can be proved similar to that of Case 1(a).

Case 1(c): Let $a = d, b = c$. Here, all the rows and the columns, written individually as 1D words are either $f_{n,\{d<c\}}$ or $f_{n,\{c<d\}}$, and either $f_{m,\{d<c\}}$ or $f_{m,\{c<d\}}$, respectively. Since m, n are odd, the prefix and the suffix of $f_{m,n}$, of size $(2,2)$ are $(d \oplus c) \ominus (b \oplus a)$ and $(a \oplus b) \ominus (c \oplus d)$, respectively. Now since $a = d, b = c$, both the prefix and suffix will become $(d \oplus c) \ominus (c \oplus d)$. Thus, this will contribute to an extra border of $f_{m,n}$. Hence the count.

Case 1(d): Let $(a = b) \ne (c \ne d)$. Recall that, the first two rows and the first two columns play a key role in contributing the borders of $f_{m,n}$. Since in this case 1st row and 1st column written as 1D words are $f_{n,\{d<c\}}$ and $f_{m,\{d<b\}}$ respectively, we will not get any extra borders other than the available $\lceil \frac{m}{2} \rceil \lceil \frac{n}{2} \rceil$ borders.

The sub cases 1(e-g) and 1(m) can be proved similarly as that of case 1(d).

Case 1(h): Let $(b = d) \neq (a \neq c)$. Here, since n is odd, the first and the last columns of $f_{m,n}$ are of the form $d^{F(m)\ominus}$. Thus, prefixes of of $d^{F(m)\ominus}$ become borders of $f_{m,n}$. Hence we get $F(m) - \lceil \frac{m}{2} \rceil$ extra borders.

Case 1(l): Proof is similar to that of Case 1(h).

Case 1(i): Let $(c = d) \neq (a \neq b)$. Here, since m is odd, the first and the last rows of $f_{m,n}$ are of the form $d^{F(n)\oplus}$ and hence the proof is similar to that of Case 1(h).

Case 1(k): Proof is similar to that of Case 1(i).

Case 1(j): Let $a \neq (b = c = d)$. Since m and n are odd, the first and last rows, and the first and last columns of $f_{m,n}$ are $d^{F(n)\oplus}$ and $d^{F(m)\ominus}$ respectively. Thus, we get $(F(m) - \lceil \frac{m}{2} \rceil) + (F(n) - \lceil \frac{n}{2} \rceil)$ extra borders. However, since the $(1,1)$ sized border 'd' is counted twice in the above count, we get V_4 as the number of borders.

Case 2: m is odd, n is even.

Case 2(a): Let $a = b, c = d$. Here $f_{m,n} = f_{m,1}^{F(n)\oplus}$ where $f_{m,1}$ written as a 1D word is $f_{m,\{d<b\}}$ and hence the proof is similar to that of Case 1(a).

Case 2(b): Let $a = c, b = d$. Here $f_{m,n} = f_{1,n}^{F(m)\ominus}$ where $f_{m,1}$ written as a 1D word is $f_{n,\{d<c\}}$. Also since n is even, $f_{m,n}$ has a prefix and a suffix of the form $(d \oplus c)^{F(m)\ominus}$. The argument similar to that of Case 1(b) can now be applied.

The proofs of cases 2(c-h), 2(l-m) are similar to that of 1(d), since in all these cases, the first row and the first column written as 1D words are $f_{n,\{d<c\}}$, and $f_{m,\{d<c\}}$ or $f_{m,\{d<b\}}$ respectively.

Cases 2(i-k): Since m is odd, the argument similar to that of Case 1(i) can be applied.

Case 3: m is even, n is even.

Case 3(a): Let $a = b, c = d$. The proof is similar to that of Case 1(a).

Case 3(b): $a = c, b = d$. Since n is even, the proof is similar to that of Case 2(b).

Case 3(c): $a = d, b = c$. Here, all the rows and the columns, written as 1D words, are either $f_{n,\{d<c\}}$ or $f_{n,\{c<d\}}$, and either $f_{m,\{d<c\}}$ or $f_{m,\{c<d\}}$, respectively. Since m, n are even, the prefix and the suffix of $f_{m,n}$, of size $(1,1)$ are 'd' and 'a', respectively. Now since $a = d$, we get an extra border of size $(1,1)$. Hence the count.

As in the cases 3(f), 3(k) and 3(l), $d = a$ with m and n being even, we get an an extra border of size $(1,1)$.

The proofs of cases 3(d), 3(e), 3(g-j) and 3(m) are similar to that of Case 1(d).

Case 4: m is even, n is odd.

Cases 4(a), 4(b), 4(h) and 4(l) are similar to that of cases 1(a), 1(b), 1(h) and 1(l), respectively.

Case 4(j): Since n is odd and $b = a$ the first and last columns are of the form $b^{F(m)\ominus}$ and hence the count.

Cases 4(c-g), 4(i), 4(k), 4(m): In these cases any one of the following situations may happen and hence there is no possibility of creation of extra borders. (I) The first row and the first column are 1D Fibonacci words (4(c,g,m): $f_{n,\{d<c\}}, f_{m,\{d<c\}}$, 4(d-f): $f_{n,\{d<c\}}, f_{m,\{d<b\}}$), (II) The first row is of the form $d^{F(n)}\oplus$, but the last row and the first column are over two-letter alphabet.   □

4.1 Non-recognizability 2D Fibonacci Language

In this subsection, we prove that the set of Fibonacci arrays is not tiling recognizable.

Let Γ and Σ be finite alphabets and let $w = [w_{i,j}]$ be an array of size (m,n). Let us call an array of size $(2,2)$ a tile. By $B_{2,2}(w)$, let us denote the set of all tiles/subarrays of w (of size $(2,2)$). Let $\widehat{w}$ be the word of size $(m+2, n+2)$ obtained from w by surrounding w with a special boundary symbol $\#$ such that $\# \notin \Sigma$.

For an array $w \in \Gamma^{**}$ of size (m,n), the projection by mapping $\pi : \Gamma \to \Sigma$ of w is an array $w' = [w'_{i,j}] \in \Sigma^{**}$ such that $w'_{i,j} = \pi(w_{i,j})$, for all $1 \leq i \leq m, 1 \leq j \leq n$. Similarly, the projection by mapping π of a 2D language L is the language $L' = \{w' : w' = \pi(w), \forall w \in L\} \subseteq \Sigma^{**}$.

Definition 26. [17] *Let Γ be a finite alphabet. A 2D language $L \subseteq \Gamma^{**}$ is local if there exists a finite set Θ of tiles over the alphabet $\Gamma \cup \{\#\}$ such that $L = \{w \in \Gamma^{**} : B_{2,2}(\widehat{w}) \subseteq \Theta\}$, whereas Γ is called a local alphabet.*

Definition 27. [17] *A tiling system (TS) is a 4-tuple $\tau = (\Sigma, \Gamma, \Theta, \pi)$, where Σ and Γ are two finite alphabets, Θ is a finite set of tiles over the alphabet $\Gamma \cup \{\#\}$ and $\pi : \Gamma \to \Sigma$ is a projection.*

A tiling system τ is said to recognize a language L', denoted by $L' = L'(\tau)$, if $L' = \pi(L)$ where L is the underlying local language. A language $L' \subseteq \Sigma^{**}$ is said to be *recognizable* if there exists a tiling system $\tau = (\Sigma, \Gamma, \Theta, \pi)$ such that $L' = L'(\tau)$.

The following lemma is analogous to the pumping lemma for regular languages.

Lemma 28 [17] *(Horizontal Iteration Lemma). Let L' be a recognizable two-dimensional language. Then, there is a function $\varphi : \mathbb{N} \to \mathbb{N}$ such that if p is an array in L' such that $|p|_{row} = n$ and $|p|_{col} > \varphi(n)$, we may write $p = x \oplus q \oplus y$ with $|x \oplus q|_{col} \leq \varphi(n)$ and $|y|_{col} \geq 1$; and for all $i \geq 0$, word $x \oplus q^{i\oplus} \oplus y$ is in L'. Furthermore, $\varphi(n) \leq \gamma^n$, $n \in \mathbb{N}$, where γ is the size of any local alphabet used to represent L'.*

Theorem 29. *The set $F_{2d} = \{f_{m,n} : m,n \geq 2\}$, is not a tiling recognizable 2D language.*

Proof. We prove the result by contradiction. Let us assume that F_{2d} is tiling recognizable. Then, there exists a tiling system $\tau = (\Sigma, \Gamma, \Theta, \pi)$ that recognizes F_{2d}. Let γ be the size of a local alphabet Γ.

For $p \in F_{2d}$, let $|p|_{row} = k$, and thus, k must be a Fibonacci number. Then, by horizontal iteration lemma, there exists a function $\varphi : \mathbb{N} \to \mathbb{N}$ such that $\varphi(k) \leq \gamma^k$ and $|p|_{col} > \varphi(k)$, and we can write $p = x \oplus q \oplus y$ with $|x \oplus q|_{col} \leq \varphi(k)$ and $|y|_{col} \geq 1$; and for all $i \geq 0$, the array $x \oplus q^{i \oplus} \oplus y$ is in F_{2d}. There certainly exists some $i \geq 0$ such that the number of columns of $x \oplus q^{i \oplus} \oplus y$ will not be a Fibonacci number, a contradiction, since the number of rows and columns of any array in the set F_{2d} must be a Fibonacci number. Hence, F_{2d} is not a tiling recognizable language. □

5   An Algorithm to Check an Array for Fibonacci

In this section, we provide an algorithm that decides whether any given array $w \in \Sigma = \{a, b, c, d\}$ is Fibonacci or not.

Let $\Sigma_1 = \{a, b\}$, $\Sigma_2 = \{c, d\}$, $\Sigma'_1 = \{a, c\}$ and $\Sigma'_2 = \{b, d\}$. We use the property that, any row (column, respectively) of $f_{m,n}$ written as a one dimensional word is a Fibonacci word over either Σ_1 or Σ_2 (Σ'_1 or Σ'_2, respectively), to decide whether a given w is a member of F_{2d}. In fact, we breakdown the given w in to its rows and columns and carry out the algorithm.

We use the sufficient condition given in [21] for a 1D word to be a Fibonacci word.

Proposition 30. [21] *Let $\{w_n\}, n \geq 1$, be a sequence of words of Σ^+ each of which contains atleast two different letters of the alphabet Σ(i.e. $alph(w_n) \geq 2$). Let us moreover suppose that for all $n \geq 5$,*

$$w_n = \alpha_n \beta_n = \gamma_n c_n$$

with $c_n \in \Sigma^*, \alpha_n = \alpha_n^R, \beta_n = \beta_n^R, \gamma_n = \gamma_n^R$ and $|\alpha_n| = F(n-1) - 2, |\beta_n| = F(n-2) + 2, |\gamma_n| = F(n) - 2$. If the word w_n begin always with the same letter then $w_n = f_n(n \geq 5)$.

Theorem 31. *We can check if a given word $w \in \{a, b\}$ of length N starting with b is a Fibonacci word in $O(N)$ time.*

Proof. We use Proposition 30 as a base to check whether a given word is Fibonacci or not. The algorithm takes input a word over a fixed binary alphabet $\Sigma = \{a, b\}$ starting with b. It is a well known fact that a given a number k is Fibonacci if and only if either $5k^2 + 4$ or $5k^2 - 4$ is a perfect square [1]. This property, and the algorithms given in [2] can be used to check if the given number is Fibonacci.

Algorithm 1. To check whether a given $w \in \Sigma^+$ belongs to $F_{a,b}$ or not

---

1 If $|w|$ is a Fibonacci number go to 2 else output NO
2 Find the sub words $\alpha_n, \beta_n, \gamma_n$ c_n with the appropriate lengths mentioned in the Proposition 30
3 Check whether $\alpha_n, \beta_n, \gamma_n$ are palindromes. If so output YES else output NO

Regarding the computational complexity of Algorithm 1, Step 1 is of order $O(ln(N))$ where $|w_n| = N$ [2]; Step 2 is of order $O(N)$), as checking a word of length k for palindrome is of order $O(k)$; so Algorithm 1 is of complexity $O(N)$.

□

Theorem 32. *For $\Sigma = \{a, b, c, d\}$, we can check if a given array $w \in \Sigma^{++}$ of size (M, N) having $(d \oplus c) \ominus (b \oplus a)$ as a prefix is a Fibonacci array in $O(MN)$ time.*

Proof. We mainly use Theorem 31 to check whether an given array is Fibonacci.

Algorithm 2. To decide whether a given w is in F_{2d} or not

---

1 Let $(M, N) = size(w)$. If both M and N are Fibonacci numbers go to Step 2, else output NO
2 Check whether the first column written as a 1D word is a Fibonacci word over Σ_2'. If so go to Step 3, else output NO
3 **for** $i=1$ to M **do**
4 | M_i = the word present in row i
5 | If M_i starts with a d go to step 6 and if M_i starts with b go to step 7
6 | If M_i is a 1D Fibonacci word over Σ_2 then break else output NO
7 | If M_i is a 1D Fibonacci word over Σ_1 then break else output NO
8 **end**
9 Output YES

Regarding the computational complexity of Algorithm 2, step 1 is of order $O(ln(max(M, N)))$; step 2 is of order $O(M)$, by Algorithm 1; steps 6 and 7 are executed M times. Note that step 2 and steps 6,7 are independent. And steps 6,7 call the standalone polynomial time algorithm, Algorithm 1, a fixed number of times, M. Hence, Algorithm 2 is of complexity $O(N)$. The complexity will be $O(M)$, if we first check the first row for a Fibonacci word and then check all the columns for Fibonacci words.

□

6   Conclusions and Future Works

In this paper some structural and combinatorial properties of Fibonacci arrays, a natural extension of 1D Fibonacci words have been studied. Various ways of decomposing the Fibonacci array $f_{m,n}$ are discussed. Primitive and non-recognizable characteristics of the set of Fibonacci arrays are proved. A formula to count the number of borders in a given Fibonacci array is also derived. An algorithm to check whether a randomly given array over an alphabet Σ is Fibonacci or not is also provided.

In our future works, further properties like balancedness and the number of squares (in general tandems) in Fibonacci arrays, will be studied. Two dimensional iterated morphisms, generating the Fibonacci arrays will also be analysed. In [23], the author has constructed a few one dimensional infinite and bi-infinite words using the geometry of tilings of the hyperbolic plane. In particular, the one dimensional infinite Fibonacci word is generated and how the construction is related to the grossone numeral system [22] is thoroughly explained. In line with these concepts, we will study and explore the possibilities of geometrically constructing an infinite Fibonacci array.

References

1. Alfred, S.P., Ingmar, L.: The Fabulous Fibonacci Numbers. Prometheus Books (2007)
2. Ali, D.: Twelve simple algorithms to compute Fibonacci numbers. CoRR. http://arxiv.org/abs/1803.07199 (2018)
3. Amir, A., Benson, G.: Two-dimensional periodicity in rectangular arrays. SIAM J. Comput. **27**(1), 90–106 (1998)
4. Anselmo, M., Giammarresi, D., Madonia, M.: Two dimensional prefix codes of pictures. In: Béal, M.-P., Carton, O. (eds.) DLT 2013. LNCS, vol. 7907, pp. 46–57. Springer, Heidelberg (2013). https://doi.org/10.1007/978-3-642-38771-5_6
5. Anselmo, M., Giammarresi, D., Madonia, M.: Prefix picture codes: a decidable class of two-dimensional codes. Int. J. Found. Comput. Sci. **25**(8), 1017–1031 (2014)
6. Anselmo, M., Madonia, M.: Two-dimensional comma-free and cylindric codes. Theor. Comput. Sci. **658**, 4–17 (2017)
7. Apostolico, A., Brimkov, V.E.: Fibonacci arrays and their two-dimensional repetitions. Theor. Comput. Sci. **237**(1–2), 263–273 (2000)
8. Berthé, V., Vuillon, L.: Tilings and rotations on the torus: a two-dimensional generalization of Sturmian sequences. Discret. Math. **223**(1–3), 27–53 (2000)
9. Berthé, V., Vuillon, L.: Palindromes and two-dimensional Sturmian sequences. J. Autom. Lang. Comb. **6**(2), 121–138 (2001)
10. Bozapalidis, S., Grammatikopoulou, A.: Picture codes. RAIRO-Theor. Inform. Appl. **40**(4), 537–550 (2006)
11. Carlitz, L.: Fibonacci representations II. Fibonacci Q. **8**, 113–134 (1970)
12. Chang, C., Wang, H.: Comparison of two-dimensional string matching algorithms. In: 2012 International Conference on Computer Science and Electronics Engineering, vol. 3, pp. 608–611. IEEE (2012)
13. Dallapiccola, R., Gopinath, A., Stellacci, F., Dal, N.L.: Quasi-periodic distribution of plasmon modes in two-dimensional Fibonacci arrays of metal nanoparticles. Opt. Express **16**(8), 5544–5555 (2008)
14. Fu, K.S.: Syntactic Methods in Pattern Recognition, vol. 112. Elsevier, Amsterdam (1974)
15. Gamard, G., Richomme, G., Shallit, J., Smith, T.J.: Periodicity in rectangular arrays. Inf. Process. Lett. **118**, 58–63 (2017)
16. Geizhals, S., Sokol, D.: Finding maximal 2-dimensional palindromes. In: The Proceedings of the 27th Annual Symposium on Combinatorial Pattern Matching, CPM 2016, vol. 54, no. 19, pp. 1–12. Dagstuhl (2016)
17. Giammarresi, D., Restivo, A.: Two-dimensional languages. In: Rozenberg, G., Salomaa, A. (eds.) Handbook of Formal Languages, pp. 215–267. Springer, Heidelberg (1997). https://doi.org/10.1007/978-3-642-59126-6_4
18. Hopcroft, J.E., Motwani, R., Ullman, J.D.: Introduction to Automata Theory, Languages and Computation, vol. 24. Pearson (2006)
19. Kulkarni, M.S., Mahalingam, K.: Two-dimensional palindromes and their properties. In: Drewes, F., Martín-Vide, C., Truthe, B. (eds.) LATA 2017. LNCS, vol. 10168, pp. 155–167. Springer, Cham (2017). https://doi.org/10.1007/978-3-319-53733-7_11

20. Lothaire, M.: Algebraic Combinatorics on Words, vol. 90. Cambridge University Press, Cambridge (2002)
21. Luca, A.: A combinatorial property of the Fibonacci words. Inf. Process. Lett. **12**(4), 193–195 (1981)
22. Margenstern, M.: An application of grossone to the study of a family of tilings of the hyperbolic plane. Appl. Math. Comput. **218**(16), 8005–8018 (2012)
23. Margenstern, M.: Fibonacci words, hyperbolic tilings and grossone. Commun. Nonlinear Sci. Numer. Simul. **21**(1–3), 3–11 (2015)
24. Mignosi, F., Restivo, A., Silva, P.V.: On Fine and Wilf's theorem for bidimensional words. Theor. Comput. Sci. **292**(1), 245–262 (2003)
25. Minsky, M., Papert, S.: Perceptrons. The MIT Press, Cambridge (1969)
26. Pal, D., Masami, I.: Primitive words and palindromes. In: Context-Free Languages and Primitive Words, pp. 423–435. World Scientific (2014)
27. Yu, S.S., Zhao, Y.K.: Properties of Fibonacci languages. Discret. Math. **224**(1–3), 215–223 (2000)

Watson-Crick Jumping Finite Automata

Kalpana Mahalingam[ID], Rama Raghavan[✉][ID], and Ujjwal Kumar Mishra[ID]

Department of Mathematics, Indian Institute of Technology Madras,
Chennai 600036, India
{kmahalingam,ramar}@iitm.ac.in, ma16d030@smail.iitm.ac.in

Abstract. In this paper, we introduce a new automata called Watson-Crick jumping finite automata, working on tapes which are double stranded sequences of symbols, similar to that of a Watson-Crick automata. This automata scans the double stranded sequence in a discontinuous manner (i.e.) after reading a double stranded string, the automata can jump over some subsequence and continue scanning, depending on the rule. We define some variants of such automata and compare the languages accepted by these variants with the language classes in Chomsky hierarchy. We also investigate some closure properties.

Keywords: Jumping Finite Automata · Watson-Crick automata · Watson-Crick jumping finite automata

1 Introduction

DNA Computing is a branch of natural computing. The research in this area has produced several interesting theoretical aspects of computing [9]. One such aspect is Watson-Crick automata (or *WKA* in short).

WKA is introduced as a counter part of the Sticker system [5]. This automata works on double strand. The symbols on the double strand are related by a complementary relation. The input strand is called as Watson-Crick tape used in a FIFO manner. The basic model of *WKA* scans the two strands separately but in a correlated manner. Variants of *WKA* can be seen in [3,4,7]. The variants of *WKA* introduced in the literature concentrate on Watson-Crick data structure. Several interesting studies that have emerged on the concept of *WKA* can be seen in [9].

General Jumping Finite Automata (*GJFA*) is introduced in [8] as a model that formalizes discontinuous information processing. Another variant of *GJFA* is defined in [2], where the read head is set to move only in one direction and in [10] the author further investigate the properties of *GJFA*. In order to preserve the probity of the genome in successive generation, sharp accurate method of DNA replication is required. In DNA replication, the double stranded molecule is replicated by replicating continuously the upper or lower strand, while the other strand in a discontinuous manner. This process is called a semi-continuous process. Also rearrangement of genes during molecular evolution often are regarded

© Springer Nature Switzerland AG 2019
T. V. Gopal and J. Watada (Eds.): TAMC 2019, LNCS 11436, pp. 467–480, 2019.
https://doi.org/10.1007/978-3-030-14812-6_29

as difficult operation [1]. However, there are simple ways to find such rearrangements. In order to understand the discontinuous reading or rearrangements of
DNA molecules, one requires the computational model that reads the strands
in a discontinuous manner. With this motivation, we define a theoretical model
called Watson-Crick jumping finite automata (or *WKJFA* in short). We introduce four variants. The variants are compared for their expressibility power.
We also compare the variants with Chomsky hierarchy. For basic definitions of
automata theory the reader is referred to [6].

This paper is divided into five sections. In, Sect. 2, we give the definitions
of *GJFA* and *WKA* with examples. Section 3 describes the new definition of
WKJFA and its variants. The model is illustrated with examples. In Sect. 4 we
compare the power of variants with Chomsky hierarchy. Section 5 compares the
power of variants of *WKJFA* among themselves. We investigate various closure
properties of the languages accepted by these variants in Sect. 6. We end with a
few concluding remarks.

2   Prelimanaries

We now recall the formal definition of *GJFA* introduced by Meduna et.al. in [8].

Definition 1. *A general jumping finite automaton (GJFA) is a quintuple*

$$M = (V, Q, q_0, F, R)$$

*where V is the input alphabet (finite set), Q is a finite set of states, $V \cap Q = \phi$,
$q_0 \in Q$ is the start state, $F \subseteq Q$ is a set of final states and $R \subseteq Q \times \Sigma^* \times Q$ is
a finite set of rules. A configuration of GJFA M is any string of $\Sigma^* \times Q \times \Sigma^*$.
After application of each rule the automaton changes its configuration. Sup-
pose the automaton is in a configuration (x, p, yz), then after application of the
rule $(p, y, q) \in R$ to the configuration the automaton goes to a new configuration (x', q, z'), where $xz = x'z'$, and deletes the string y from the configuration
(x, p, yz). Let $x, z, x', z' \in V^*$ such that $xz = x'z'$ and $(p, y, r) \in R$, then M
makes a jump from $xpyz$ to $x'qz'$, written as:*

$$xpyz \curvearrowright x'qz'.$$

*Let $\curvearrowright^+$ and $\curvearrowright^*$ denote the transitive and the reflexive-transitive closure of $\curvearrowright$,
respectively. The language accepted by M is defined as:*

$$L(M) = \{uv \ : \ u, v \in \Sigma^*, \ uq_0v \curvearrowright^* f, \ f \in F\}.$$

*We say a GJFA is of degree n if $|y| \leq n$ for all $(p, y, q) \in R$. A degree 1 GJFA
is called jumping finite automaton (JFA).*

Example 1. Consider the JFA

$$M = (\{a, b, c\}, \{q_0, q_1, q_2\}, q_0, \{q_0\}, R)$$

where $R = \{(q_0, a, q_1), (q_1, b, q_2), (q_2, c, q_0)\}$.

The language accepted by the automaton is

$$L(M) = \{w \in \{a, b, c\}^* \ : \ \mid w \mid_a = \mid w \mid_b = \mid w \mid_c\}$$

where $\mid w \mid_a$ denotes the number of occurrences of a in w.

We now recall the concept of Watson-Crick finite automata [9]. This automata works on tapes which are double stranded sequences of symbols related by a complementarity relation, similar to that of a DNA molecule. Let V be an alphabet set and ρ be a symmetric relation, which is also called complementarity relation, on V. A double strand string is represented as $\begin{pmatrix} w_1 \\ w_2 \end{pmatrix}$ over the set $\begin{pmatrix} V^* \\ V^* \end{pmatrix}$, where $\begin{pmatrix} w_1 \\ w_2 \end{pmatrix}$ and $\begin{pmatrix} V^* \\ V^* \end{pmatrix}$ are alternate notations of (w_1, w_2) and $V^* \times V^*$. We concatenate the elements of $\begin{pmatrix} V^* \\ V^* \end{pmatrix}$ component wise that is

$$\begin{pmatrix} x_1 \\ x_2 \end{pmatrix} \begin{pmatrix} z_1 \\ z_2 \end{pmatrix} = \begin{pmatrix} x_1 z_1 \\ x_2 z_2 \end{pmatrix}.$$

We denote $\begin{bmatrix} V \\ V \end{bmatrix}_\rho = \left\{ \begin{bmatrix} a \\ b \end{bmatrix} \mid a, b \in V, (a, b) \in \rho \right\}$ and $WK_\rho(V) = \begin{bmatrix} V \\ V \end{bmatrix}_\rho^*$. The set $WK_\rho(V)$ is called the Watson-Crick domain associated to V and ρ.

The essential difference between $\begin{pmatrix} w_1 \\ w_2 \end{pmatrix}$ and $\begin{bmatrix} w_1 \\ w_2 \end{bmatrix}$ is just that $\begin{pmatrix} w_1 \\ w_2 \end{pmatrix}$ is an alternative notation for the pair (w_1, w_2), whereas $\begin{bmatrix} w_1 \\ w_2 \end{bmatrix}$ implies that the strings w_1 and w_2 have the same length and the corresponding letters are connected by the complementarity relation.

Definition 2. *A Watson-Crick automaton is a tuple*

$$M = (V, \rho, K, q_0, F, R)$$

where V is a finite alphabet set, ρ is a symmetric relation on V, K is a finite set of states, $V \cap K = \phi$, $q_0 \in K$ is the initial state, $F \subseteq K$ is a set of final states and $R \subseteq K \times \begin{pmatrix} V^* \\ V^* \end{pmatrix} \times K$ is a finite set of rules.

We define transition in a Watson -Crick automaton as follows:

For $\begin{pmatrix} u_1 \\ u_2 \end{pmatrix}, \begin{pmatrix} v_1 \\ v_2 \end{pmatrix}, \begin{pmatrix} w_1 \\ w_2 \end{pmatrix} \in \begin{pmatrix} V^* \\ V^* \end{pmatrix}$ such that $\begin{bmatrix} u_1 v_1 w_1 \\ u_2 v_2 w_2 \end{bmatrix} \in WK_\rho(V)$ and $p, q \in Q$, we write

$$\begin{pmatrix} u_1 \\ u_2 \end{pmatrix} p \begin{pmatrix} v_1 \\ v_2 \end{pmatrix} \begin{pmatrix} w_1 \\ w_2 \end{pmatrix} \Rightarrow \begin{pmatrix} u_1 \\ u_2 \end{pmatrix} \begin{pmatrix} v_1 \\ v_2 \end{pmatrix} q \begin{pmatrix} w_1 \\ w_2 \end{pmatrix}$$

if and only if $(p, \begin{pmatrix} v_1 \\ v_2 \end{pmatrix}, q) \in R$. Let $\Rightarrow^+$ and $\Rightarrow^*$ denote the transitive and the reflexive -transitive closure of $\Rightarrow$, respectively. The language accepted by a

Watson-Crick automaton is

$$L(M) = \{w_1 \in V^* \mid q_0 \begin{bmatrix} w_1 \\ w_2 \end{bmatrix} \Rightarrow^* \begin{bmatrix} w_1 \\ w_2 \end{bmatrix} f, \ f \in F, w_2 \in V^*, \begin{bmatrix} w_1 \\ w_2 \end{bmatrix} \in WK_\rho(V)\}.$$

Example 2. Consider the Watson-Crick automaton

$$M = (\{a, b, c\}, \rho, \{q_0, q_1, q_2, q_f\}, q_0, \{q_f\}, R)$$

where ρ is the identity relation and R is the set of rules

$$\left\{ (q_0, \begin{pmatrix} a \\ \lambda \end{pmatrix}, q_0), \ (q_0, \begin{pmatrix} b \\ a \end{pmatrix}, q_1), \ (q_1, \begin{pmatrix} b \\ a \end{pmatrix}, q_1), \ (q_1, \begin{pmatrix} c \\ b \end{pmatrix}, q_2), \ (q_2, \begin{pmatrix} c \\ b \end{pmatrix}, q_2), \right.$$

$$\left. (q_2, \begin{pmatrix} \lambda \\ c \end{pmatrix}, q_f), \ (q_f, \begin{pmatrix} \lambda \\ c \end{pmatrix}, q_f) \right\}.$$

The language accepted by the automaton is

$$L(M) = \{a^n b^n c^n \ : \ n \geq 1\}.$$

3   Watson-Crick Jumping Finite Automata

In this section we define a new type of automata called as Watson-Crick jumping finite automata (*WKJFA* in short).

Definition 3. *A Watson-Crick jumping finite automaton is a tuple*

$$M = (V, \rho, K, q_0, F, R)$$

where V is a finite alphabet set, ρ is a symmetric relation on V, K is a finite set of states, $V \cap K = \phi$, $q_0 \in K$ is the initial state, $F \subseteq K$ is a set of final states and $R \subseteq K \times \begin{pmatrix} V^* \\ V^* \end{pmatrix} \times K$ is a finite set of rules.

Let $\begin{pmatrix} x_1 \\ x_2 \end{pmatrix}, \begin{pmatrix} z_1 \\ z_2 \end{pmatrix}, \begin{pmatrix} x_1' \\ x_2' \end{pmatrix}, \begin{pmatrix} z_1' \\ z_2' \end{pmatrix} \in \begin{pmatrix} V^* \\ V^* \end{pmatrix}$ and $(p, \begin{pmatrix} y_1 \\ y_2 \end{pmatrix}, q) \in R$ be a rule. Then M makes a jump from $\begin{pmatrix} x_1 \\ x_2 \end{pmatrix} p \begin{pmatrix} y_1 \\ y_2 \end{pmatrix} \begin{pmatrix} z_1 \\ z_2 \end{pmatrix}$ to $\begin{pmatrix} x_1' \\ x_2' \end{pmatrix} q \begin{pmatrix} z_1' \\ z_2' \end{pmatrix}$, written as

$$\begin{pmatrix} x_1 \\ x_2 \end{pmatrix} p \begin{pmatrix} y_1 \\ y_2 \end{pmatrix} \begin{pmatrix} z_1 \\ z_2 \end{pmatrix} \curvearrowright \begin{pmatrix} x_1' \\ x_2' \end{pmatrix} q \begin{pmatrix} z_1' \\ z_2' \end{pmatrix}, where \begin{pmatrix} x_1 z_1 \\ x_2 z_2 \end{pmatrix} = \begin{pmatrix} x_1' z_1' \\ x_2' z_2' \end{pmatrix}.$$

The language accepted by M is defined as:

$$L(M) = \left\{ uv \ : \ \begin{pmatrix} u \\ u' \end{pmatrix} q_0 \begin{pmatrix} v \\ v' \end{pmatrix} \curvearrowright^* f, f \in F, u, v, u', v' \in V^*, \begin{bmatrix} uv \\ u'v' \end{bmatrix} \in \right.$$

$$\left. WK_\rho(V) \text{ and } \begin{bmatrix} uv \\ u'v' \end{bmatrix} = \begin{pmatrix} u \\ u' \end{pmatrix} \begin{pmatrix} v \\ v' \end{pmatrix} \right\}.$$

The difference between WKA and $WKJFA$ lies in the parsing of the double stranded sequence when it is scanned by the automata. In WKA, given any input double stranded sequence, the sequence is parsed only once (in the beginning) and the string is scanned using the given set of rules. If the transition reaches one of the final states, the string is accepted. However, in $WKJFA$, the sequence is recombined and parsed after every transition step depending on the available set of rules. We illustrate with the following example.

Example 3. Consider the Watson-Crick jumping finite automaton

$$M = (\{a, b, c\}, \rho, \{q_0, q_1, q_f\}, q_0, \{q_f\}, R)$$

where ρ is the identity relation and

$$R = \left\{ \left(q_0, \begin{pmatrix} ab \\ \lambda \end{pmatrix}, q_0\right), \left(q_0, \begin{pmatrix} \lambda \\ bc \end{pmatrix}, q_1\right), \left(q_1, \begin{pmatrix} \lambda \\ bc \end{pmatrix}, q_1\right), \left(q_1, \begin{pmatrix} c \\ a \end{pmatrix}, q_f\right), \right.$$
$$\left. \left(q_f, \begin{pmatrix} c \\ a \end{pmatrix}, q_f\right) \right\}.$$

The acceptance of the word $aabbcc$ is as:

$$\begin{pmatrix} a \\ aabbcc \end{pmatrix} q_0 \begin{pmatrix} ab \\ \lambda \end{pmatrix} \begin{pmatrix} bcc \\ \lambda \end{pmatrix} \curvearrowright \begin{pmatrix} \lambda \\ aabbcc \end{pmatrix} q_0 \begin{pmatrix} ab \\ \lambda \end{pmatrix} \begin{pmatrix} cc \\ \lambda \end{pmatrix}$$
$$\curvearrowright \begin{pmatrix} cc \\ aab \end{pmatrix} q_0 \begin{pmatrix} \lambda \\ bc \end{pmatrix} \begin{pmatrix} \lambda \\ c \end{pmatrix}$$
$$\curvearrowright \begin{pmatrix} cc \\ aa \end{pmatrix} q_1 \begin{pmatrix} \lambda \\ bc \end{pmatrix}$$
$$\curvearrowright q_1 \begin{pmatrix} c \\ a \end{pmatrix} \begin{pmatrix} c \\ a \end{pmatrix} \curvearrowright q_f \begin{pmatrix} c \\ a \end{pmatrix} \curvearrowright q_f.$$

However, there are many ways to accept the same string, the above given transition steps is one of them. The language accepted by the automaton is $L(M) = \{a^n b^n c^n \mid n \geq 1\}^+$.

Example 4. Consider the Watson-Crick jumping finite automaton

$$M = (\{a, b, c\}, \rho, \{q_0, q_1, q_2, q_3, q_4\}, q_0, \{q_0, q_1, q_2, q_3, q_4\}, R)$$

where ρ is the identity relation and R is the set of rules

$$\left\{ \left(q_0, \begin{pmatrix} a \\ \lambda \end{pmatrix}, q_1\right), \left(q_1, \begin{pmatrix} b \\ \lambda \end{pmatrix}, q_2\right), \left(q_2, \begin{pmatrix} c \\ \lambda \end{pmatrix}, q_0\right), \left(q_0, \begin{pmatrix} \lambda \\ a \end{pmatrix}, q_3\right), \left(q_3, \begin{pmatrix} \lambda \\ b \end{pmatrix}, q_4\right), \right.$$
$$\left. \left(q_4, \begin{pmatrix} \lambda \\ c \end{pmatrix}, q_0\right) \right\}.$$

The language accepted by the automaton is

$$L(M) = \{w \in \{a, b, c\}^* \ : \ |w|_a = |w|_b = |w|_c\}.$$

Similar to that of the classical case, we also consider several variants of Watson-Crick jumping finite automata. We say that $M = (V, \rho, K, q_0, F, R)$ is

- *stateless*, if $K = F = \{q_0\}$;
- *all − final*, if $K = F$;
- *simple*, if all $\left(p, \begin{pmatrix} x_1 \\ x_2 \end{pmatrix}, q\right) \in R$ we have either $x_1 = \lambda$ or $x_2 = \lambda$;
- *1 − limited*, if for all $\left(p, \begin{pmatrix} x_1 \\ x_2 \end{pmatrix}, q\right) \in R$ we have $| x_1 x_2 | = 1$.

We denote by **AWKJFA, NWKJFA, FWKJFA, SWKJFA, 1WKJFA**, the families of languages recognized by Watson-Crick jumping finite automata which are arbitrary (A), stateless $(N,$ from "no state"), all-final (F), simple (S), 1-limited (1) respectively. We also consider combinations of such variants $\#_1\#_2 WKJFA$ where $\#_1, \#_2 \in \{N, 1, F, S, A\}$. Thus, $NSWKJFA$ is a 'No' state simple Watson-Crick jumping finite automaton. We now construct a $SWKJFA$ that accepts the language $L = \{ab\}^*$.

Example 5. Consider the $SWKJFA$

$$M = (\{a, b\}, \rho, \{q_0, q_1, q_f\}, q_0, \{q_0, q_f\}, R)$$

where ρ is the identity relation and
$$R = \left\{ (q_0, \begin{pmatrix} ab \\ \lambda \end{pmatrix}, q_0), (q_0, \begin{pmatrix} \lambda \\ ba \end{pmatrix}, q_0), (q_0, \begin{pmatrix} \lambda \\ a \end{pmatrix}, q_1), (q_1, \begin{pmatrix} \lambda \\ b \end{pmatrix}, q_f) \right\}.$$
The language accepted by the automaton is $L(M) = \{ab\}^*$.

We now compare the families **GJFA** and **JFA** with the family **WKJFA** and show that **GJFA** ⊂ **SWKJFA** and hence **JFA** ⊆ **1WKJFA**.

Proposition 1. GJFA ⊂ **SWKJFA**.

Proof. Let $L \in$ **GJFA**, then there exists a $GJFA$ $M = (\Sigma, Q, q_0, F, R)$ such that $L = L(M)$. Without loss of generality, we can assume that in the $GJFA$ there is no λ transition [8]. Now we construct an equivalent $SWKJFA$ M' such that $L = L(M')$. The construction of the $SWKJFA$ is as follows:

$$M' = (V = \Sigma, \rho, K, q_0, F' = F, R').$$

where ρ is the identity relation on V. Since R is finite, we can enumerate the elements of R. Corresponding to each rule, numbered, $i : (p, w, q)$, $i \geq 1$, we associate with it two rules in R' which are $(p, \begin{pmatrix} w \\ \lambda \end{pmatrix}, q_i), (q_i, \begin{pmatrix} \lambda \\ w \end{pmatrix}, q)$, $q_i \notin Q$. Let K be the union of the set of states Q and the newly introduced states q_i's. It is easy to show $L(M) = L(M')$. Hence **GJFA** ⊆ **SWKJFA**. It was shown in [10] that there does not exist a $GJFA$ that accepts the language $L = \{ab\}^*$. But by Example 5, $L = \{ab\}^* \in$ **SWKJFA**. Hence **GJFA** ⊂ **SWKJFA**.

Corollary 1. JFA ⊆ **1WKJFA**.

Let $perm(w)$ denote the set of all words obtained from taking all permutations of w and $perm(L) = \bigcup_{w \in L} perm(w)$. We have the following observations.

Proposition 2. If $M = (V, \rho, K, q_0, F, R)$ be a WKJFA with rules $\left(p, \begin{pmatrix} x \\ y \end{pmatrix}, q\right)$ where $\mid x \mid \leq 1$ and $\mid y \mid \leq 1$, then $L(M) = perm(L(M))$.

Proof. Clearly $L(M) \subseteq perm(L(M))$. Now let $w = a_1 a_2 \cdots a_n \in L(M)$. It is sufficient to show that $perm(w) \subseteq L(M)$. Since $w = a_1 a_2 \cdots a_n \in L(M)$, there exist $w' = b_1 b_2 \cdots b_n \in V^*$ such that

$$\begin{bmatrix} w \\ w' \end{bmatrix} = \begin{bmatrix} a_1 a_2 \cdots a_n \\ b_1 b_2 \cdots b_n \end{bmatrix} \in WK_\rho(V) \text{ and } \begin{bmatrix} a_1 \\ b_1 \end{bmatrix}; \begin{bmatrix} a_2 \\ b_2 \end{bmatrix}, \cdots, \begin{bmatrix} a_n \\ b_n \end{bmatrix} \in \begin{bmatrix} V \\ V \end{bmatrix}_\rho.$$

Therefore

$$\begin{bmatrix} a_{i_1} \\ b_{i_1} \end{bmatrix} \begin{bmatrix} a_{i_2} \\ b_{i_2} \end{bmatrix} \cdots \begin{bmatrix} a_{i_n} \\ b_{i_n} \end{bmatrix} = \begin{bmatrix} a_{i_1} a_{i_2} \cdots a_{i_n} \\ b_{i_1} b_{i_2} \cdots b_{i_n} \end{bmatrix} \in WK_\rho(V)$$

where $a_{i_1} a_{i_2} \cdots a_{i_n}$ and $b_{i_1} b_{i_2} \cdots b_{i_n}$ are permutations of w and w', respectively. Since for any rule $\left(p, \begin{pmatrix} x \\ y \end{pmatrix}, q\right)$ we have $\mid x \mid \leq 1$ and $\mid y \mid \leq 1$, the automaton will consume either $\begin{pmatrix} \lambda \\ \lambda \end{pmatrix}$ or $\begin{pmatrix} a_i \\ \lambda \end{pmatrix}$ or $\begin{pmatrix} \lambda \\ b_j \end{pmatrix}$ or $\begin{pmatrix} a_k \\ b_l \end{pmatrix}$ at each step of computation in order to accept $w = a_1 a_2 \cdots a_n$ equivalently in order to consume $\begin{bmatrix} a_1 a_2 \cdots a_n \\ b_1 b_2 \cdots b_n \end{bmatrix}$. Since, after each transition step the remaining strings of upper and lower strands get regrouped and in the next step we again parse the remaining double strand and because of jumping nature the automaton will consume

$$\begin{bmatrix} a_{i_1} a_{i_2} \cdots a_{i_n} \\ b_{i_1} b_{i_2} \cdots b_{i_n} \end{bmatrix}$$

using the rules which were used to accept w. Hence $a_{i_1} a_{i_2} \cdots a_{i_n} \in L(M)$ and $perm(w) \subseteq L(M)$.

Corollary 2. If $L \in$ **1WKJFA**, then $L = perm(L)$.

Corollary 3. There does not exist any 1WKJFA which accept the language $\{ab\}$.

It is not difficult to prove the following useful Lemma.

Lemma 1. There is no WKJFA that accepts $a^* b^*$.

It is known that ([9]) **REG** $\subset$ **AWK**, where **AWK** denotes the family of languages accepted by an arbitray Watson-Crick automaton and hence the language $a^* b^* \in$ **AWK**. But by Lemma 1, $a^* b^* \notin$ **AWKJFA**, hence **AWK** $\not\subseteq$ **AWKJFA**. Also $\{ab\}^* \in$ **AWK** $\cap$ **AWKJFA**, by Example 5 and **REG** $\subset$ **AWK**. Thus, we conclude that **AWK** $\cap$ **AWKJFA** $\neq \phi$.

4   Relations with Well-Known Language Families

In this section we investigate the relationship between the families of languages accepted by several variants of *WKJFA* described in the previous section and the families in the Chomsky hierarchy. The following result can be easily deduced from the definition of *WKJFA*, Corollary 3 and Example 4. One can easily verify that the language accepted in Example 4 is context sensitive and the respective automaton described is a *1WKJFA* as well as *SWKJFA*.

Lemma 2. *The following are true:*

1. **1WKJFA** *and* **FIN** *are incomparable.*
2. **1WKJFA** *and* **REG** *are incomparable.*
3. **1WKJFA** *and* **CF** *are incomparable.*

Proof. From Corollary 3, it is clear that **FIN, REG, CF** $\not\subseteq$ **1WKJFA**. Also, it follows from Example 4, that **1WKJFA** $\not\subseteq$ **FIN**, or **REG** or **CF**. Thus the family of languages accepted by a 1-limited Watson-Crick jumping finite automaton is not comparable with the families of Finite, Regular or Context-free languages in the Chomsky heirarchy.

As opposed to the previous result, we show that the class of all finite languages are subsets of the family of languages accepted by *SWKJFA*.

Lemma 3. FIN $\subset$ SWKJFA.

Proof. Let $K \in$ **FIN**. Since K is a finite language there exist an $n \geq 0$ such that $K = \{w_1, w_2, \ldots, w_n\}$. We construct a *SWKJFA* as follows:

$$M = (V, \rho, \{q_0, q_1, \cdots, q_n, q_f\}, q_0, \{q_f\}, R)$$

where $V = alph(K)$, ρ is the identity relation on V and

$$R = \left\{ (q_0, \binom{w_1}{\lambda}, q_1), (q_1, \binom{\lambda}{w_1}, q_f), \cdots, (q_0, \binom{w_n}{\lambda}, q_n), (q_n, \binom{\lambda}{w_n}, q_f) \right\}.$$

Clearly $L(M) = K$. Therefore **FIN $\subseteq$ SWKJFA**. Also, one can observe from Example 4 that, **SWKJFA $\setminus$ FIN $\neq \phi$**. Hence, **FIN $\subset$ SWKJFA**.

Similarly we can show that **FIN $\subset$ FWKJFA**, by making $K = F$ and ignoring λ string in the above construction.

In the following we construct a 'No' state Watson Crick jumping finite automata, (i.e.) the automata has exactly one state and all transition rules are from and to this unique state. Hence, when representing such an automaton, the presence of the unique state is understood and is not specified.

Example 6. Consider the *NWKJFA* $M = (\{a, b, c\}, \rho, R)$, where ρ is the identity relation and $R = \left\{ \binom{a}{b}, \binom{b}{c}, \binom{\lambda}{a}, \binom{c}{\lambda} \right\}$.

The language accepted is $L(M) = \{w \in \{a, b, c\}^* : |w|_a = |w|_b = |w|_c\}$.

We discuss the comparability of various language classes in Table 1. The results represented in the Table 1 can be shown using Lemmas 1, 2, and 3, Examples 4 and 6 and definition of $NWKJFA$, as proved in Lemma 2. In the Table 1, $\times$ represents incomparable, $\lrcorner$ represents that **FIN** $\subset$ **FWKJFA, SWKJFA**.

Table 1. Comparability of various families of languages.

	FIN	REG	CF
NWKJFA	$\times$	$\times$	$\times$
1WKJFA	$\times$	$\times$	$\times$
FWKJFA	$\lrcorner$	$\times$	$\times$
SWKJFA	$\lrcorner$	$\times$	$\times$

Now we recall some results ([9]) related to Watson-Crick automata:

Lemma 4. *([9])* **AWK(u)** $\subseteq$ **CS**.

Lemma 5. *([9]) Every one-letter language in* **AWK(u)** *is regular* .

We now show that the family of languages accepted by an arbitrary Watson-Crick jumping finite automaton is a proper subset of the family of context-sensitive languages.

Lemma 6. AWKJFA $\subset$ **CS**.

Proof. By Lemma 4, we have **AWK** $\subseteq$ **CS**. Moreover, jumps of an $AWKJFA$ can be simulated by context-sensitive grammars, hence **AWKJFA** $\subseteq$ **CS**. Also from Lemma 5, we have every one-letter language in **AWK(u)** is regular. However, in languages over one letter alphabet jumping transitions have no effect and hence **AWK** = **AWKJFA** = **REG** over one letter alphabet. Hence **AWKJFA** $\subset$ **CS**.

5   Relations Between Watson-Crick Jumping Finite Automata Language Families

In this section we investigate the relationship among the families of languages accepted by several variants of Watson-Crick jumping finite automata described in Sect. 3. We begin the section by constructing a 'No' state Watson-Crick jumping finite automaton that accepts the Dyck language.

Example 7. Consider the $NWKJFA$

$$M = (\{a, b\}, \{(a, a), (b, b)\}, \left\{ \begin{pmatrix} ab \\ \lambda \end{pmatrix}, \begin{pmatrix} \lambda \\ ab \end{pmatrix} \right\}).$$

The language accepted by the automaton is the Dyck language over $\{a, b\}$.

We have the following results.

Proposition 3. N1WKJFA ⊂ NSWKJFA.

Proof. It is clear from the definition that, **N1WKJFA** ⊆ **NSWKJFA**. The Dyck language L in Example 7 is accepted by a *NSWKJFA*. The language contains ab but it does not contain ba. Hence by Corollary 2, the Dyck language L cannot be accepted by any *N1WKJFA*. Thus **N1WKJFA** ⊂ **NSWKJFA**.

Using the same argument, we can show that **F1WKJFA** ⊂ **FSWKJFA**. Also, since *NWKJFA* always accepts infinite language, **NSWKJFA** ⊂ **FSWKJFA**. And hence **N1WKJFA** ⊂ **F1WKJFA** and **N1WKJFA** ⊂ **FSWKJFA**. By Lemma 3 and Corollary 3, **1WKJFA** ⊂ **SWKJFA**. We note that, **NSWKJFA** and **F1WKJFA** are incomparable, which can be shown using Example 7 and Corollary 2.

Proposition 4. SWKJFA = AWKJFA.

Proof. We have, by definition, **SWKJFA** ⊆ **AWKJFA**.
Now, let $M = (V, \rho, K, q_0, F, R)$ be a *AWKJFA*. We construct an equivalent *SWKJFA*, $M' = (V, \rho, K', q_0, F, R')$ as follows:

V, ρ, q_0, F are same in M and M'. Each rule of the form $\left(p, \binom{w}{\lambda}, q\right)$ or $\left(r, \binom{\lambda}{w'}, s\right)$ or $\left(t, \binom{\lambda}{\lambda}, u\right)$ of R are members of R'. All other types of rules of R are indexed from 1 to k, where $k \leq n$ and n is the number of rules in R. Let $K' = K \cup \{p_i \mid 1 \leq i \leq k\}$.

Consider a rule of R indexed as $i : \left(p, \binom{a_1 a_2 \ldots a_n}{b_1 b_2 \ldots b_m}, q\right)$, where $m, n \geq 1$. We construct two rules equivalent to i as follows:

$$i' : \left(p, \binom{a_1 a_2 \ldots a_n}{\lambda}, p_i\right), \quad i'' : \left(p_i, \binom{\lambda}{b_1 b_2 \ldots b_m}, q\right).$$

Then, R' consists of rules of R which are of the form $\left(p, \binom{w}{\lambda}, q\right)$ or $\left(r, \binom{\lambda}{w'}, s\right)$ or $\left(t, \binom{\lambda}{\lambda}, u\right)$ and newly introduced rules which are indexed by i' and i'' for $1 \leq i \leq k$. Cardinality of R' will be $(n + k)$. Consider a t-step derivation of M. This derivation may contain applications of t_1 rules of the form $\left(p, \binom{w}{\lambda}, q\right)$ or $\left(r, \binom{\lambda}{w'}, s\right)$ or $\left(t, \binom{\lambda}{\lambda}, u\right)$, where $t_1 \leq t$. This t–step derivation can be simulated by rules in R' in $(2t - t_1)$-steps, by the above constructed rules of R'. Note that the strands are regrouped every time after an application of a rule. Similarly, a $(2t - t_1)$-step derivation of M' can be simulated in t–steps by M. Hence $L(M) = L(M')$.

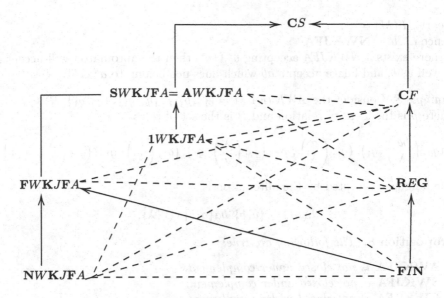

Fig. 1. Comparability of various classes of languages

The results, related to comparability of various families of languages, presented in this paper are consolidated in Fig. 1.

Dotted line between two families represents that the families are incomparable, whereas ↗, ↘ and ↑ represent that one family is strict subset of other and ⌈ represents below family is subset of above.

6 Closure Properties of WKJFA Family

Proposition 5. AWKJFA *is closed under union.*

Proof. Let $M_1 = (V_1, \rho_1, K_1, q_1, F_1, R_1)$ and $M_2 = (V_2, \rho_2, K_2, q_2, F_2, R_2)$ be two *AWKJFAs*. Without loss of generality, let $K_1 \cap K_2 = \phi$ and $q_0 \notin (K_1 \cup K_2)$. Define a *AWKJFA* $M = (V_1 \cup V_2, \rho_1 \cup \rho_2, K_1 \cup K_2 \cup \{q_0\}, q_0, F_1 \cup F_2, R_1 \cup R_2 \cup$
$\{(q_0, \begin{pmatrix} \lambda \\ \lambda \end{pmatrix}, q_1), (q_0, \begin{pmatrix} \lambda \\ \lambda \end{pmatrix}, q_2)\})$. Clearly, $L(M) = L(M_1) \cup L(M_2)$.

Using similar construction to the one defined in the Proposition 5 we can show the following:

Proposition 6. SWKJFA *and* **FWKJFA** *are closed under union.*

In the following we show that **NWKJFA** is not closed under union.

Proposition 7. NWKJFA *is not closed under union.*

Proof. Consider the *NWKJFA*

$$M = (\{a\}, \{(a, a)\}, \left\{ \begin{pmatrix} a \\ \lambda \end{pmatrix}, \begin{pmatrix} \lambda \\ a \end{pmatrix} \right\}).$$

Clearly, $L(M) = a^*$.
Hence $a^*, b^* \in$ **NWKJFA**.
If there exists a *NWKJFA* accepting $a^* \cup b^*$, then the automaton will accept a
as well as b, and hence accepts ab which does not belong to $a^* \cup b^*$.

Example 8. Consider the *AWKJFA* $M = (\{a,b\}, \rho, \{q_0, q_1\}, \{q_0, q_1\}, R)$
where ρ is the identity relation and R is the set of rules

$$\left\{ \left(q_0, \begin{pmatrix} ba \\ \lambda \end{pmatrix}, q_1\right), \left(q_1, \begin{pmatrix} a \\ \lambda \end{pmatrix}, q_1\right), \left(q_1, \begin{pmatrix} \lambda \\ a \end{pmatrix}, q_1\right), \left(q_1, \begin{pmatrix} b \\ \lambda \end{pmatrix}, q_1\right), \left(q_1, \begin{pmatrix} \lambda \\ b \end{pmatrix}, q_1\right) \right\}.$$

The language accepted by the automaton is

$$L(M) = \{a,b\}^* ba \{a,b\}^* \cup \{\lambda\}.$$

Proposition 8. *The following are true:*

1. **AWKJFA** *is not closed under complement.*
2. **SWKJFA** *is not closed under complement.*
3. **FWKJFA** *is not closed under complement.*

Proof. Complement of the language of Example 8 is $a^* b^* - \{\lambda\}$.
But by Lemma 1 we know that the complement language cannot be accepted by
any *WKJFA*. Moreover, the automaton in Example 8 is simple as well as all final.
Hence **AWKJFA, SWKJFA** and **FWKJFA** are not closed under complement.

Proposition 9. NWKJFA *is not closed under complement.*

Proof. Consider the *NWKJFA*

$$M = \left(\{a,b\}, \{(a,a), (b,b)\}, \left\{ \begin{pmatrix} a \\ b \end{pmatrix}, \begin{pmatrix} \lambda \\ a \end{pmatrix}, \begin{pmatrix} b \\ \lambda \end{pmatrix} \right\} \right).$$

The language accepted by the automaton is

$$L(M) = \{w \in \{a,b\}^* \mid \ \mid w \mid_a = \mid w \mid_b\}.$$

The complement of the language is $L(M)^c = \{w \in \{a,b\}^* \mid \ \mid w \mid_a \neq \mid w \mid_b\}$.
If there exist a *NWKJFA* M' accepting $L(M)^c$, then $a, b \in L(M')$ implying
that $ab \in L(M')$. But $ab \notin L(M)^c$. Hence there does not exist any *NWKJFA*
accepting $L(M)^c$. Hence **NWKJFA** is not closed under complement.

Proposition 10. *The following are true:*

1. **AWKJFA** *is not closed under concatenation.*
2. **SWKJFA** *is not closed under concatenation.*
3. **FWKJFA** *is not closed under concatenation.*

Proof. The languages a^* and b^* are accepted by all the above three variants but
by Lemma 1 we know that none of the above variant can accept the language
$a^* b^*$. Hence none of the above variants are closed under concatenation.

Proposition 11. NWKJFA *is not closed under concatenation.*

Proof. The languages a^* and b^* are accepted by *NWKJFA* but the language a^*b^* cannot be accepted by any *NWKJFA* because if there exist a *NWKJFA* accepting a^*b^* then it will accept ab and hence accept $abab$, using same rules again. But $abab \notin L(a^*b^*)$, where $L(a^*b^*)$ is language of a^*b^*. Hence **NWKJFA** is not closed under concatenation.

Proposition 12. 1WKJFA *is not closed under concatenation.*

Proof. Finite languages a and b are accepted by *1WKJFA* but ab cannot be accepted by any *1WKJFA* by Corollary 3. Hence **1WKJFA** is not closed under concatenation.

Proposition 13. NWKJFA *is closed under Kleene star.*

Proof. We will proof this proposition by proving that if $L \in$ **NWKJFA**, then $L = L^*$.
Clearly $L \subseteq L^*$.
Now take $w \in L^*$, then $w = w_1 w_2 \cdots w_n$, where $w_i \in L$ for $i = 1, 2 \cdots n$. Since $w_i \in L$ therefore there is at least one way to accept $w_1 w_2 \cdots w_n$ using the rules that accept each w_i in sequence. Hence $L^* \subseteq L$. Thus $L = L^*$

Proposition 14. 1WKJFA *is not closed under Kleene star.*

Proof. Finite language $L = \{ab, ba\}$ is accepted by *1WKJFA* but $L^* = \{ab, ba\}^*$ cannot be accepted by any *1WKJFA*. If there exist a *1WKJFA* accepting L^*, then it will also accept $aabb$ as $abba \in L^*$ by Corollary 2. But $aabb \notin L^*$. Hence **1WKJFA** is not closed under Kleene Star.

Proposition 15. *The following are true:*

1. **NWKJFA** *is not closed under homomorphism.*
2. **1WKJFA** *is not closed under homomorphism.*

Proof. Consider a homomorphism $h : \{a\} \to \{a, b\}^*$ defined as: $h(a) = ab$. Now consider the *NWKJFA* as well as *1WKJFA* $M = (\{a\}, \{(a, a), \left\{ \begin{pmatrix} a \\ \lambda \end{pmatrix}, \begin{pmatrix} \lambda \\ a \end{pmatrix} \right\}\})$.
The language accepted by the automaton is $L(M) = a^*$.
But the language $h(L(M)) = \{ab\}^*$ cannot be accepted by any *NWKJFA*. Since $ab \in \{ab\}^*$ therefore there will be rules to consume ab. If there exist a *NWKJFA* accepting $\{ab\}^*$ then it will accept $aabb$ because using the rules, consuming ab, first it will consume inner ab after that it will consume remaining ab. But $aabb \notin \{ab\}^*$.
The same language $h(L(M)) = \{ab\}^*$ also cannot be accepted by any *1WKJFA* because of the Corollary 2.
Hence **NWKJFA** and **1WKJFA** are not closed under homomorphism.

The results related to the closure properties of *WKJFA* and its variants are consolidated in Table 2. $\checkmark$ represents that the family is closed, $\times$ represents that the family is not closed whereas ? represents that the results are not known.

Table 2. Closure properties of WKJFA and its variants.

Union	Complement	Concatenation	Kleene star	Homomorphism	Intersection
✓ Arbitrary	× Arbitrary	× Arbitrary	? Arbitrary	? Arbitrary	? Arbitrary
✓ Simple	× Simple	× Simple	? Simple	? Simple	? Simple
✓ All final	× All final	× All final	? All final	? All final	? All final
✓ No state	× No state	× No state	✓ No state	× No state	? No state
✓ 1-limited	? 1-limited	× 1-limited	× 1-limited	× 1-limited	? 1-limited

7   Conclusion

In this paper a new model called *WKJFA* is introduced. The existing basic *WKA* is now set to process the double strand in a discontinuous manner and hence we call it *WKJFA*. Four variants are introduced and compared for their power. We also compare the variants with Chomsky hierarchy. It will be interesting to look for simulation of *WKJFA* with semi-discontinuous process in DNA replication. It will also be interesting to look in to the descriptional complexity aspects of *WKJFA*.

References

1. Caetano-Anollés, G.: Evolutionary Genomics and Systems Biology. John Wiley & Sons Inc., Hoboken (2010)
2. Chigahara, H., Fazekas, S.Z., Yamamura, A.: One-way jumping finite automata. Int. J. Found. Comput. Sci. **27**(3), 391–405 (2016)
3. Freund, R., Păun, Gh., Rozenberg, G., Salomaa, A.: Watson-Crick finite automata. In: The Proceedings of the 3rd Annual DIMACS Symposium on DNA Based Computers, Philadelphia, pp. 305–317 (1997)
4. Freund, R., Păun, G., Rozenberg, G., Salomaa, A.: Watson-Crick automata, Technical report 97–13. Leiden University, Department of Computer Science (1997)
5. Kari, L., Păun, G., Rozenberg, G., Salomaa, A., Yu, S.: DNA computing, sticker systems, and universality. Acta Informatica **35**(5), 401–420 (1998)
6. Krithivasan, K., Rama, R.: Introduction to Formal Languages Automata Theory and Computation. Pearson, India (2009)
7. Martin-Vide, C., Păun, Gh, Rozenberg, G., Salomaa, A.: Universality results for finite H systems and for Watson-Crick finite automata. In: Păun, Gh (ed.) Computing with Bio-Molecules, Theory and Experiments, pp. 200–220. Springer, Berlin (1998)
8. Meduna, A., Zemek, P.: Jumping finite automata. Int. J. Found. Comput. Sci. **23**(7), 1555–1578 (2012)
9. Păun, G., Rozenberg, G., Salomaa, A.: DNA Computing : New Computing Paradigms, 1st edn. Springer, Heidelberg (1998)
10. Vorel, V.: On basic properties of jumping finite automata. Int. J. Found. Comput. Sci. **29**(1), 1–15 (2018)

Dispersion of Mobile Robots: The Power of Randomness

Anisur Rahaman Molla[1]([✉]) [iD] and William K. Moses Jr.[2] [iD]

[1] Cryptology and Security Research Unit, Indian Statistical Institute,
Kolkata, 700108, India
molla@isical.ac.in
[2] Faculty of Industrial Engineering and Management,
Technion - Israel Institute of Technology, Haifa, Israel
wkmjr3@gmail.com

Abstract. We consider cooperation among insects, modeled as cooperation between mobile robots on a graph. Within this setting, we consider the problem of mobile robot dispersion on graphs. The study of mobile robots on a graph is an interesting paradigm with many interesting problems and applications. The problem of dispersion in this context, introduced by Augustine and Moses Jr. [4], asks that n robots, initially placed arbitrarily on an n node graph, work together to quickly reach a configuration with exactly one robot at each node. Previous work on this problem has looked at the trade-off between the time to achieve dispersion and the amount of memory required by each robot. However, the trade-off was analyzed for *deterministic algorithms* and the minimum memory required to achieve dispersion was found to be $\Omega(\log n)$ bits at each robot. In this paper, we show that by harnessing the power of *randomness*, one can achieve dispersion with $O(\log \Delta)$ bits of memory at each robot, where Δ is the maximum degree of the graph. Further, we show a matching lower bound of $\Omega(\log \Delta)$ bits for any *randomized algorithm* to solve dispersion. We further extend the problem to a general k-dispersion problem where $k > n$ robots need to disperse over n nodes such that at most $\lceil k/n \rceil$ robots are at each node in the final configuration.

Keywords: Nature-inspired computing · Mobile robots · Dispersion ·
Collective robot exploration · Scattering · Uniform deployment ·
Load balancing · Distributed algorithms · Randomized algorithms

---

A. R. Molla—Research supported in part by DST Inspire Faculty research grant
DST/INSPIRE/04/2015/002801.
W. K. Moses Jr.—Research supported in part by a grant of his postdoctoral fellowship
hosts from the Israeli Ministry of Science.

T. V. Gopal and J. Watada (Eds.): TAMC 2019, LNCS 11436, pp. 481–500, 2019.
https://doi.org/10.1007/978-3-030-14812-6_30

1   Introduction

1.1   Background and Motivation

The mobile robots paradigm has been used to study many types of systems, including those where simple insects cooperate with each other to accomplish some goal. These robots typically need to work together to solve some common problem such as shape formation or exploration of the environment or gathering at some common point. One of the primary motivations of this type of research is to understand how to use resource-limited robots to achieve some large task in a distributed manner.

Typically the environment that serves as a backdrop to these problems is a either a finite plane or a connected graph. However, a graph can just be thought of as a discretization of the space of a finite plane or in fact three dimensional space. Thus, using graphs as an environment allows, in some sense, for a more general study of a given problem.

The problem of dispersion on graphs was recently introduced by Augustine and Moses Jr. [4]. The initial version of the problem asks that n robots that are initially arbitrarily placed on a graph should work together to reach a final configuration such that there is exactly one robot on each node. We study this problem and the more general version of it where k robots (for any k) are initially arbitrarily placed and must reach a configurations such that at most $\lceil k/n \rceil$ robots are present on any given node. The study of dispersion is interesting and has practical applications to any problem where the cost of several robots sharing the same resource (node) far outweighs the cost of a robot finding a new resources (moving on the graph). One such example is when multiple electric cars must find a recharge station in an area where recharge stations are located close by. The time to charge the vehicle may be in the order of hours while the time to find another station would be in the order of minutes. Further, if the vehicles are "smart" and communicate with each other to exchange information about what stations are free or not, this problem is exactly modeled as dispersion. The study of dispersion is also interesting as it relates to the related problems of scattering on graphs, multi-robot exploration, and load balancing on graphs. Scattering on graphs asks that $k \leq n$ spread themselves out in an equidistance manner on symmetric graphs like rings or grids. This is just dispersion with an extra constraint of equi-spacing. Multi-robot exploration asks that k robots starting at the same node work together to visit each node of the graph as quickly as possible. It is clear that any solution to dispersion solves this problem. Finally, load balancing on graphs asks that nodes send and receive loads and evenly distribute these loads among themselves. Dispersion can be seen as flipping this model by having the loads (i.e., robots) move around and distribute themselves evenly among the nodes. The techniques used to solve load balancing in graphs are quite different from those used to solve problems in mobile robots and by studying dispersion, our hope is to build a bridge between the two areas for cross-pollination of ideas and techniques.

As mentioned earlier, one of the key aspects of mobile robots is that we are solving large tasks in a distributed manner with *resource limited mobile robots*. In previous work, the study of memory of robots and time to achieve dispersion was of great interest. This paper furthers that study and shows that the introduction of randomness in a novel way allows robots to achieve dispersion using much less memory than previously shown.

1.2 Our Results

Throughout this paper, we study the trade-off between memory required by robots and the time it takes to achieve dispersion of n robots on different types of graphs with n nodes and m edges. We denote the diameter of these graphs by D and maximum degree of any node of the graph by Δ. We present algorithms for increasingly general types of graphs that utilize randomness to allow robots, with typically $O(\log \Delta)$ bits of memory[1], to achieve dispersion. This is a substantial improvement over past algorithms which, while deterministic, required robots to have $\Omega(\log n)$ bits of memory each[2]. We also show a lower bound on the memory requirement that any randomized algorithm requires $\Omega(\log \Delta)$ bits to achieve dispersion, assuming all robots have the same amount of memory.

When we consider a *rooted graph* of a certain type, it implies that the topology of the graph is of that type and all robots start at one node, called the root, of the graph. We initially describe our algorithms for dispersion of n robots on n node graphs and subsequently generalize them to dispersion of any k robots on n node graphs. We assume that robots do not know the values of n, m, k, Δ, or D. However, in several instances, our algorithms require robots to have memory proportionate to either parameter Δ or D. This means that whatever memory supplied to the robot should be enough to satisfy the requirement, but the explicit knowledge of the parameter itself is not needed. Our upper bound results for dispersion of n robots on n node graphs are summarized in Table 1.

We first describe a primitive, *Local-Leader-Election*, that can be used by robots with access to randomness to choose one robot to settle down at a given node. This allows us to side-step the requirement of $\Omega(\log(k/n))$ bits of memory required by each robot for a unique label if we want robots to deterministically choose a robot to settle down at each node.

We then proceed to show how a simple algorithm for rooted rings, *Rooted-Ring*, that requires robots to have $O(\log \Delta)$ bits of memory and achieves dispersion in $O(n)$ rounds. This serves as a warm-up and allows readers to internalize the way we use *Local-Leader-Election* and how these sorts of algorithms (with reduced memory) need to operate.

We then develop the algorithm *Rooted-Tree* for rooted trees that requires robots to have $O(\log \Delta)$ bits of memory and achieves dispersion in $O(n)$. This algorithm contains the key ideas for our algorithm on general rooted graphs.

---

[1] Note that all log's that appear in this paper are to the base 2.

[2] Notice that our algorithms require only $O(1)$ bits memory at each robot on paths, rings, grids and any constant degree graphs, whereas the previous deterministic algorithms require $O(\log n)$ bits.

We then present two algorithms to achieve dispersion on rooted graphs, *Rooted-Graph-LogDelta-LogD* and *Rooted-Graph-Delta*, which require robots to have $O(\max\{\log \Delta, \log D\})$ and $O(\Delta)$ bits of memory respectively and both algorithms achieve dispersion in $O(m)$ rounds. Both algorithms are extensions of the *Rooted-Tree* algorithm and use different amounts of memory to handle the issue of dealing with cycles that may arise in the graph. We provide these two algorithms with contrasting memory requirements so that if an algorithm designer a priori knows which of the two memory requirements is less, they can program robots to use that algorithm.

Finally, we present an algorithm *Arbitrary-Graph*, which works on arbitrary graphs, and allows robots with $O(\log \Delta)$ bits of memory to achieve dispersion in the cover time of the graph with high probability. The algorithm is a Las Vegas type randomized algorithm in that robots will eventually achieve dispersion, but the exact running time is not fixed and but bounded with high probability. The "cost" of having robots use less memory is that we require robots to stay active after they settle down. Namely, each robot runs the algorithm until it settles down and then must stay active to inform other robots that come to the node that the node is settled. In this sense, the algorithm is non-terminating. Contrast this against other algorithms which allow robots to settle down and need not be active after certain conditions are met.

Table 1. Upper bound results of dispersion of n robots on an n node graph (with Δ maximum degree and diameter D) for different types of graphs along with the memory requirement of robot.

Serial no.	Type of graph	Memory requirement of each robot	Algorithm name	Time until dispersion achieved
1	Rooted ring	$O(1)$ bits	Rooted-Ring	$O(n)$ rounds
2	Rooted tree	$O(\log \Delta)$ bits	Rooted-Tree	$O(n)$ rounds
3	Rooted graph	$O(\max\{\log \Delta, \log D\})$ bits	Rooted-Graph-LogDelta-LogD	$O(m)$ rounds
4	Rooted graph	$O(\Delta)$ bits	Rooted-Graph-Delta	$O(m)$ rounds
5	Arbitrary graph	$O(\log \Delta)$ bits	*Arbitrary-Graph	Cover time of graph

*Arbitrary-Graph: In this algorithm, robots do not terminate execution of the algorithm, unlike the other algorithms.

The cover time of a graph lies in the range between $\Omega(n \log n)$ and $O(mn)$.

After presenting the above algorithms and analyzing them for dispersion of n robots on n nodes, we then show how to generalize them to achieve dispersion of k robots on n nodes, where k can be any positive integer value.

We then present our lower bound of $\Omega(\log \Delta)$ bits of memory needed by each robot to achieve dispersion when randomness is allowed.

1.3   Related Work

The problem of dispersion of mobile robots on a graph was introduced recently by Augustine and Moses Jr. [4] and studied in different graph classes. In the full version [5], the authors rectified and improved some of their dispersion algorithms. Improvements and rectifications were also independently performed by Kshemkalyani and Ali [21]. Both papers focused on the trade-off between time complexity and memory requirement of robots to solve dispersion deterministically. Our results improve over the previous works [5,21] in terms of memory requirement with the help of randomness. In particular, our randomized algorithms reduce the memory requirement from $O(\log n)$ bits to $O(\log \Delta)$ bits and the time complexity remains same or is faster (in some cases). While the algorithms in [4,5] chiefly rely on a timer to signal termination of the algorithm and as such require $\Omega(\log n)$ bits of memory, our algorithms are more event oriented and robots terminate when the termination condition is triggered.

Dispersion is closely connected to graph exploration by robots; a well-studied problem in the context of mobile robots. In the graph exploration problem, k robots are initially located at a node and the goal is to have the robots collectively visit all nodes in the graph. A number of papers have worked on this problem, however, most of the works are in specific graph classes, such as rings [14,22], trees [17,18,25], and grids [6,16]. Several papers consider exploration on general graphs [9,10,12,15,20]. However, the model assumptions or the goal of these papers are different from ours and they may produce inefficient solutions to dispersion. For example, the papers close to our model [12,20,27] only focus on minimizing memory of the robots and as a result the time complexity of their exploration algorithms is very high. Fraigniaud et al. [20] shows that a robot with $\Theta(D \log \Delta)$ bits of memory can explore an anonymous graph, but may take time $O(\Delta^{D+1})$. Cohen et al. [12] considers the model where the nodes also have memory. Then with some initial preprocessing, they solve exploration with less memory bits, but the exploration takes $O(\Delta^{10}m)$ time. Further, the exploration algorithm of Cohen et al. with $O(1)$ bits memory at each node cannot solve dispersion immediately. Diks et al. [17] shows that exploration in a tree is possible with $O(\log^2 n)$ bits of memory. Ambühl et. al. [3] improved this memory bound to $O(\log n)$ bits. Dynia et al. [18] and Ortolf et al. [25] present optimal-time rooted tree exploration algorithms with k robots, but they assume unlimited memory of robots. Our paper focuses on the trade-off between running time and memory requirement to solve dispersion.

Another similar problem to dispersion is scattering or uniform deployment of k robots $k \leq n$ on a graph. In the scattering problem on a graph, k robots need to uniformly deploy over n nodes in the graph. Several papers studied the scattering problem on graphs; e.g., on rings [19,28] and on grids [7], but in different settings.

Most of the above algorithms for graph exploration or scattering on graphs are deterministic. To the best of our knowledge, our paper is the first presenting randomized solutions to dispersion and improve the previous results.

A slightly different way of looking at the dispersion problem is as load balancing in graphs. Load balancing requires nodes to distribute the load over nodes evenly. Here, if we consider robots as the load, then dispersion of robots is similar to load balancing, where the power to move load around the graph lies with the load as opposed to the nodes. Load balancing is a well explored problem, in particular in graphs [8,13,24,26,29]. Our model is closer to diffusion based load balancing [13,24,29] with discrete loads [8,26].

1.4   Organization of Paper

The rest of the paper is organized as follows. In Sect. 2, we introduce the technical preliminaries needed for our results. In Sect. 3, we present an important primitive, *Local-Leader-Election*, which we use extensively in our algorithms. In Sects. 4, 5, and 6, we present our algorithms to achieve dispersion on rooted rings, rooted trees, and rooted graphs respectively. In Sect. 7, we present a simple memory optimal algorithm to achieve dispersion on arbitrary graphs. We show how to extend our algorithms to handle dispersion with an arbitrary number of robots in Sect. 8. The lower bound on memory per robot is presented in Sect. 9. Finally, in Sect. 10, we present conclusions and some future work.

2   Technical Preliminaries

We consider a connected undirected graph of n nodes, m edges, diameter D, and maximum degree of any node Δ. The nodes are anonymous, i.e. they do not have unique labels. For every edge connected to a node, the node has a corresponding port number for the edge. The same edge may have different port number assigned to it at each of its attached nodes. For every node, there exists a total ordering on the port numbers from that node. A robot with x bits of memory has access to 2^x ports of that node. For a given node with y ports, if $2^x \geq y$, then the robot has access to all ports of the node. When $2^x < y$, the robot only has access to a subset of the ports, where the exact subset of ports is chosen arbitrarily by nature.[3] Thus, for a given node with Δ ports, any robot needs at least $\log \Delta$ bits of memory in order to access all ports.[4]

---

[3] The robot's memory restricts it to only use a subset of the available ports when determining which port to move through. Importantly, the robot does not know that there are more ports than it is seeing. Thus it cannot purposely choose which subset of ports to see. We call this lack of control "by nature".

[4] This does not necessarily give a $\Omega(\log \Delta)$ memory lower bound for dispersion. We discuss this further in the lower bound section (Sect. 9).

We assume a synchronous system, i.e. time progresses in rounds, and each robot knows when the current round ends and a new round starts, although robots may not know the round number. Each round proceeds as follows: (i) First, robots colocated at the node exchange messages with each other and perform local computation. (ii) Second, robots move through a port of current node and reach a new node. Robots may also choose to stay at the current node. Note that in step (i) of the round, we consider local computation and message exchange to be bounded, but free[5].

Robots are anonymous, i.e. they do not have unique labels. Each robot has a limited amount of memory for computation and to store information. The exact limit depends on the algorithm to be run and is explicitly given in each section of the paper. Each robot has access to a fair coin that be used to generate an infinite number of random bits. However, the number of random bits that can be stored and used for any purpose is limited by the robots memory. When a robot is present at a node, it can access the port numbers of that node, subject to memory restrictions as defined earlier. The robot can only view other robots colocated at the same node as it and cannot see anything beyond its current node (its "view" of the graph is limited). Robots do not know the value of n, m, D, or Δ. Note that our algorithms require do not require robots to know the actual values of D and Δ, but require the robots to have enough memory store either $O(\log D)$, $O(\log \Delta)$, or $O(\Delta)$ bits according to the algorithm in question. Thus robots may have an upper bound on those values but do not explicitly know those values.

We characterize the efficiency of solutions to dispersion along two metrics. First, how many bits of memory is each robot required to have. Second, what is the running time of the algorithm until dispersion is achieved. For all algorithms, save the algorithm in Sect. 7, robots execute the dispersion algorithm and then terminate within the running time we specify for the algorithm. For the algorithm in Sect. 7, we allow robots to be active indefinitely.

We now present several definitions of terms we use in the paper. We call a graph a *rooted graph* if all robots are initially placed at one node of that graph called its root. A similar definition applies for specific types of graphs such as rings and trees. We say that a robot *settles at a given node* if that robot chooses to stay at that node in the final dispersion configuration. We call a node with a robot that settles on it a *settled node*. The algorithm in Sect. 7 is based on random walks. A *simple random walk* in an undirected graph is defined as: in each step, the walk chooses a random adjacent edge from the current node and moves to that neighbor. The probability of choosing a random neighbor u from the current node v is $1/d(v)$, where $d(v)$ is the degree of v. The *cover time* of a random walk is defined to be the time required by the random walk to visit all the nodes in the graph. It is known that the cover time of any graph is bounded by $O(mn)$ [2]. We refer the reader to the survey [23] for details on random walks and cover time.

---

[5] This can be considered a realistic assumption when the time taken for a robot to move through an edge is significantly more than the time taken for either local message exchange or local computation.

We now formally define dispersion of k robots on an n node graph. Initially, k robots are arbitrarily placed on the graph. Dispersion asks that robots move around the graph to reach a configuration such that at most $\lceil k/n \rceil$ robots are present at any given node.

3   Local Leader Election

In this section, we describe a procedure which we use throughout the rest of the paper. The procedure allows any number of k robots co-located at an unsettled node to choose exactly one leader (robot) for the node within one round of an algorithm. Importantly, each robot only requires $O(1)$ bits of memory and access to a random number generator in order to execute the algorithm. We first describe the algorithm and prove our claims on it. We subsequently discuss the applications of the algorithm that immediately arise. In order to differentiate between different instances of communication occurring between robots at a node within the same round, we refer to each instance of communication among at most k robots as one sub-round and use that terminology while describing our algorithm. Note that, as per our model assumptions, any amount of communication is allowed to take place between robots within a single round of the system, so long as the amount of communication is bounded. As we shall see, our procedure satisfies that requirement with high probability.

3.1   Algorithm Local-Leader-Election

Each robot starts off as a candidate for leader. In every sub-round, a robot that is a candidate leader flips a fair coin. If heads, it broadcasts that it is alive to other robots. If tails (and at least one other robot broadcasts in that sub-round) it stops being a candidate for leader and doesn't broadcast anymore. If tails and no other robot broadcasts in that sub-round, it remains a candidate for leader. This process is repeated until exactly one robot broadcasts in a given sub-round. Then all robots know that that robot is the leader, and it is chosen as the robot which settles down at the given node. Subsequently, all other robots can then move to other nodes according to a given algorithm. Note that if it occurs that no robot broadcasts in a given sub-round, that sub-round is ignored and all robots that were still alive previously broadcast again.

Theorem 1. Local-Leader-Election *can be run by multiple robots, each having* $O(1)$ *bits of memory and co-located at a node, to select a common leader within one round of the system.*

Proof. We first show that the algorithm can be run by robots and completes within one round of the system. Then we argue about its memory complexity.

It is easy to that *Local-Leader-Election* takes $O(\log n)$ sub-rounds on expectation for a leader to be chosen. Applying a Chernoff bound, it is also easy to see that it takes $O(\log n)$ sub-rounds with high probability for termination. Thus the number of sub-rounds is bounded with high probability. Recall that any amount of bounded communication between co-located robots is allowed in one round of the system. Thus the algorithm successfully executes within one round of the system.

Each robot requires only $O(1)$ bits because a robot needs 1 bit to check if it's a candidate leader, 1 bit to check if it's the leader, and 2 bits to check if it heard $0, 1$, or more than 1 robot broadcast in a given sub-round.   □

3.2   Applications

Local-Leader-Election can be directly applied to past deterministic algorithms in [4] to replace the use of $O(\log n)$ bits to compare multiple robots to decide which one settles at a node. In addition, if the termination condition is relaxed, some of the resulting algorithms use dramatically less memory as a result. We list the improvements to algorithms in [4], as a result of these two modifications, below:

1. Algorithm *Path-Ring-Tree-LogN* achieves dispersion of n robots in $O(n)$ rounds on paths, trees, and rings when robots have $O(\log \Delta)$ bits of memory and do not terminate.
2. Algorithm *Rooted-Graph-LogN* achieves dispersion of n robots in $O(m)$ rounds on rooted graphs when robots have $O(\log \Delta)$ bits of memory and do not terminate.

4   Dispersion on Rooted Rings

In this section, we describe our algorithm to achieve dispersion of n robots on a rooted ring in $O(n)$ rounds when robots have $O(1)$ bits of memory. This does not contradict our lower bound because here Δ is a constant so $O(\log \Delta) = O(1)$. Recall that for a ring, any algorithm to achieve dispersion takes at least $\Omega(n)$ rounds because the diameter of the graph is $n/2$. Thus our algorithm is asymptotically time optimal.

4.1   Algorithm Rooted-Ring

Each robot performs a traversal of the ring in a deterministic manner until it becomes the leader of the node it is at, as chosen in *Local-Leader-Election*. Once it becomes the leader of that node, it settles down and terminates execution of the algorithm.

The traversal of the ring is done in the following manner. Initially have robots move through port 0. Subsequently, if the robot enters a node through port i and it does not become the leader, have it leave through port $i + 1 \mod 2$.

Theorem 2. *Algorithm* Rooted-Ring *can be run by* n *robots with* $O(1)$ *bits of memory each to ensure dispersion occurs in* $O(n)$ *rounds on rooted rings.*

Proof. It is easy to see that the entire ring is traversed by robots in $O(n)$ rounds. Further, each robot terminates as soon as it becomes a leader. Finally, each node has exactly one leader robot assigned to it. Thus dispersion is achieved in $O(n)$ rounds. The only memory requirement is a bit to remember which port the robot entered the node through and $O(1)$ bits to execute *Local-Leader-Election*. Thus each robot only requires $O(1)$ bits of memory.   □

5   Dispersion on Rooted Trees

In this section, we describe an algorithm to achieve dispersion of n robots on a rooted tree in $O(n)$ rounds when each robot has $O(\log \Delta)$ bits of memory.

5.1   Algorithm Rooted-Tree

The algorithm has two phases of execution. In the first phase, the algorithm has every robot perform a deterministic depth first search (DFS) in order to uniquely settle down at a node. In the second phase, the final robot to settle down then backtracks to the root of the tree and performs a second DFS to inform each robot to terminate execution.

In the first phase, each robot that does not settle down performs a DFS in the following manner. It remembers the port i that it entered the node through. It then leaves through port $(i + 1) \mod \delta$, where δ is the local degree of the node. Initially at the root, let the robots move through port 0 (since the robots did not initially enter the root through any node). At each empty node, a robot is chosen by *Local-Leader-Election* to settle down at it. This node remembers the port it entered the node through and we call the port the *parent pointer*.

In the second phase, the final robot to settle, x, changes its status to reflect that phase two has begun and backtracks to the root of the tree and then performs a second DFS until it finally settles down again and terminates. The robots it comes in contact with terminate when a special condition is met, as defined below. Consider the set of nodes on the path from the node where the last robot settled to the root and call it R. Let S represent the set of all nodes in the graph. For every node $u \in S \setminus R$, the robot at u terminates execution when x leaves u through u's parent pointer. For each node $u \in R$, we have the robot at u remember the port that x backtracked through to reach the root, i.e. the port x entered node u through as it backtracked, and we call that port the *pointer to final node*.[6] Once x passes through the pointer to final node of u, u terminates execution. Finally, when x reaches its empty node, it settles down and terminates execution.

Theorem 3. *Algorithm* Rooted-Tree *can be run by n robots with $O(\log \Delta)$ bits of memory each to ensure dispersion occurs in $O(n)$ rounds on rooted trees.*

---

[6] It is possible for the robot at u to differentiate x from just another robot executing phase one of the DFS because x has changed its status to reflect that the second phase has begun.

Proof. We first prove that the execution of *Rooted-Tree* results in dispersion being achieved and all robots terminating within $O(n)$ rounds of the start of the algorithm. Subsequently, we argue that each robot requires only $O(\log \Delta)$ bits of memory. □

It is clear that the first phase results in robots performing a DFS until an empty node is found to settle down in. Thus at the end of phase one, there is exactly one robot on each node. Note that we use notation from the algorithm. We now show that in phase two, due to the careful way we trigger termination of the algorithm in robots, all robots will terminate at the end of the DFS of x and x will end up back at the node it originally settled at. We first show this for robots at the nodes in the set $S \setminus R$ and then for the remaining robots.

Lemma 1. *For every node $u \in S \setminus R$, the robot in every node in the subtree rooted at u terminates execution before the robot in u terminates execution.*

Proof. Consider only the subtree rooted at u and let $S' \subseteq S \setminus R$ denote the set of nodes of the subtree including u. Let the maximum depth of any node in this subtree be d. u is at depth 0 in this subtree. Now, we prove by induction on the depths of nodes in the subtree that for any node $v \in S'$ at depth d', the following hypothesis holds. The robots in all descendant nodes of v have terminated before the robot in v terminates.

For a node at depth d, i.e. a leaf node, the hypothesis holds trivially because it has no descendants. Let the hypothesis hold true for all nodes at some depth d' in the subtree. We now prove that it holds for all nodes at depth $d' - 1$. Consider a node $w \in S'$ at depth $d' - 1$. Before moving through the parent pointer of w, x will have explored each child of w and moved through that child's robot's corresponding parent pointer. Thus the robot at each child is triggered to end termination of the algorithm before the robot at w is triggered to end execution. Thus the invariant holds true for w. □

For every node $u \in R$ excluding the final node that x settles down at, it is clear to see that once x passes through u's pointer to the final node, u will not be visited again. So the robot at u can terminate without a problem. And finally, once x reaches its empty node, it will also terminate execution. Thus, we see that for all nodes in S, after x completes its DFS in phase two, all robots at those nodes will terminate execution. The time taken to perform two DFS's on a tree and have x move to the root from a settled node is $O(n)$ rounds. Thus the algorithm successfully completes in $O(n)$ rounds.

Regarding the memory requirements of each robot, $O(\log \Delta)$ bits are required for a parent pointer and pointer to the final node. $O(1)$ bits are required to remember which phase the robot is in, to denote whether the robot is in the root of the tree, to denote whether a robot becomes the exploring robot x, and to perform *Local-Leader-Election*. So totally, each robot requires $O(\log \Delta)$ bits of memory. □

6   Dispersion on Rooted Graphs

In this section we describe two algorithms to achieve dispersion of n robots on a rooted graph in $O(m)$ rounds. One algorithm requires robots to have $O(\max\{\log \Delta, \log D\})$ bits of memory each while the other has a requirement of $O(\Delta)$ bits of memory.

6.1   Algorithm Rooted-Graph-LogDelta-LogD

This algorithm can be thought of as an extension to the algorithm *Rooted-Tree* found in Sect. 5.1. Similar to that algorithm, *Rooted-Graph-LogDelta-LogD* proceeds in two phases. In the first phase, robots again perform a deterministic DFS in order to find nodes to settle down at. However, the key difference between this algorithm and *Rooted-Tree* lies in how this algorithm deals with cycles in the graph. In the second phase, again the last settled robot goes to the root and performs a second DFS, triggering other robots to terminate execution of the algorithm in the process.

In the first phase, robots again perform a DFS by remembering the port i they entered the node through and subsequently leaving through port $(i+1) \mod \delta$, where δ is the local degree of the port. For the root node, robots initially move through port 0. At each empty node along the way, robots perform *Local-Leader-Election* to choose one robot to settle down at that node. If a robot is exploring and comes across a node with a robot already settled on it, the exploring robot backtracks to its previous node and tries the next port from that node. Finally, the last robot settles down at the last node, marking the end of phase one. Note that settled robots maintain a parent pointer which records the port through which they first entered the given node. For the root, the value of its parent pointer is null. Also, each robot (both settled and exploring) maintain a counter indicating its distance from the root with respect to the tree of nodes formed by the DFS.

In the second phase, the last robot to settle down, x, changes its state info to indicate that it is in phase two and makes its way to the root of the graph. From here it performs a DFS, similar to that done in phase one. However, in this DFS, cycles are detected when the robot moves from a node at distance ℓ from the root to a node at distance $< \ell$ from the root. In such a case, the robot backtracks to its previous node and proceeds through the next available port. Similar to *Rooted-Tree*, consider all nodes along the path from the root to the node that x finally settles at. Call this set R. Let every robot belonging to a node in R except for x maintain a pointer to the final node, indicating the port through which x must go to return to the node it must settle at. Once x moves through this port in the course of the DFS, all robots belonging to nodes in R save x itself terminate execution. Let the set of all nodes in the graph be S. For a robot belonging to a node $u \in S \setminus R$, once x passes from u to u's parent via u's parent pointer, the robot at u terminates execution of the algorithm. Finally, x completes the DFS, returns to the node it must settle at, and terminates execution.

Theorem 4. *Algorithm* Rooted-Graph-LogDelta-LogD *can be run by n robots with* $O(\max\{\log \Delta, \log D\})$ *bits of memory each to ensure dispersion occurs in* $O(m)$ *rounds on rooted graphs.*

Proof. This proof is very similar to the proof of Theorem 3. We first prove that all robots running *Rooted-Graph-LogDelta-LogD* achieve dispersion in $O(m)$ rounds. Then we argue that each robot requires $O(\max\{\log \Delta, \log D\})$ bits of memory to execute the algorithm.

To prove our claim on dispersion, we make use of the following Claim. We omit the proof as the Claim and its proof are very similar to Lemma 1 and its proof.   □

Claim. For every node $u \in S \setminus R$, the robot in every node in the subtree rooted at u terminates execution before the robot at u terminates execution.

Note that the proof of Claim 6.1 requires one extra argument in addition to the argument required in the proof of Lemma 1. We must show that x will immediately backtrack when a cycle is detected and not trigger the termination condition of a robot out of order of the DFS by accidentally further exploring through that robot's parent pointer. Our cycle checking strategy requires robots to maintain their distance from the root. x can easily identify that it is in a cycle if it moves from a node at distance ℓ to one at distance $< \ell$. Importantly, in a DFS, cross edges do not exist but only forward and back edges. This means that x moved to an ancestor of the node. x would not have yet moved through the parent pointer of the robot attached to the ancestor, and thus would not have triggered that robot to terminate execution. So, x can communicate with the robot at the ancestor, discover the distance of the ancestor from the root, discover that it is in a cycle, and backtrack immediately.

Thus, at the end of phase two, x will successfully complete execution of the DFS. At the end of the DFS, all robots would be triggered to terminate execution. Furthermore, we know that a DFS on a graph takes $O(m)$ rounds, so that is the execution time of the algorithm.

As to the memory requirement of robots. Each robot must use $O(\log \Delta)$ bits to store information about parent pointer and pointer to final node. Additionally, $O(\log D)$ bits are required to store information about the distance to root. Finally, $O(1)$ bits are needed to store information about which phase a robot is in, whether the robot is in the root or not, whether the robot is the exploring robot x or not, and to execute *Local-Leader-Election*. Therefore, each robot requires $O(\max\{\log \Delta, \log D\})$ bits of memory to execute the algorithm.   □

6.2   Algorithm Rooted-Graph-Delta

This algorithm is a variation of *Rooted-Graph-LogDelta-LogD* that provides a memory trade-off: instead of needing $O(\log D)$ bits of memory, this algorithm requires $O(\Delta)$ bits of memory. This algorithm again runs in two phases, with the goals of each phase being the same as that of the previous algorithm. We focus only on the variation between the two algorithms now.

In the previous algorithm, each settled robot was required to remember its distance from the root. In this algorithm, instead we require each settled robot to remember which of its ports lead to forward edges and which do not. This is done by maintaining a bit string of size at most Δ bits where each bit from LSB to MSB corresponds to one of the ports leading out of that node. Let us call it *list of forward ports*. Initially all bits except parent pointer's bit are set to one to indicate that all those ports possibly lead to forward edges. In the course of the DFS in phase one, if a robot uses a port from a node u and then realizes it is in a cycle, the robot will backtrack to u, inform the robot at u about the given port, and then move on with the DFS. The robot at u sets the corresponding bit in the list of forward ports to zero. Thus at the end of phase one, all settled robots have an accurate list of forward ports. Now, in phase two, when x performs its DFS, at a given node it only considers those ports whose corresponding bit in the list of forward ports is one.

Theorem 5. *Algorithm* Rooted-Graph-Delta *can be run by n robots with $O(\Delta)$ bits of memory each to ensure dispersion occurs in $O(m)$ rounds on rooted graphs.*

Proof (Proof Sketch). The proof of this theorem is identical to that of Theorem 4, so we omit details. However, we focus on two things: the proof that x does not end up in a cycle during the phase two DFS and the memory requirements of robots.

In phase two of the algorithm, x does not need to traverse an edge to discover if it is a back edge. Instead, at a given node u, it needs only rely on the list of forward ports maintained by the robot at u, which will be properly built in phase one of the algorithm. Therefore, the second DFS will be successful and all robots will terminate execution of the algorithm in $O(m)$ rounds.

Instead of using $O(\log D)$ bits of memory to remember the distance from root, each robot is required to maintain the list of forward ports, which is a bit string of size at most Δ. Thus each robot requires $O(\Delta)$ bits of memory.   □

7   Dispersion on Arbitrary Graphs (Without Termination)

In this section, we assume that the robots are initially arbitrarily located at nodes in the graph (i.e., not necessarily at a single node). We describe a simple randomized algorithm that can be run by robots to achieve dispersion and requires each robot to have $O(\log \Delta)$ bits of memory each. The algorithm is a Las Vegas style randomized algorithm in that the time until dispersion is achieved is variable and is bounded by the cover time of the graph with high probability.

7.1   Algorithm Arbitrary-Graph

The idea of the algorithm is that each robot performs a simple random walk on the graph in parallel until it finds an empty node that it can settle at. The algorithm is described below in more detail. Each robot performs a random walk on the graph in parallel. During the random walk, if a robot finds an empty node, it settles down in that node. If multiple robots are present at an empty node in the same round, they compute a local leader using the procedure *Local-Leader-Election* and the leader settles at that node. Each robot performs the random walk until it settles down. A settled robot stays active indefinitely since it must inform other exploring robots about the occupancy of the node.

Theorem 6. *Suppose n robots are placed arbitrarily over an n node graph. Then Algorithm* Arbitrary-Graph *solves dispersion with $O(\log \Delta)$ bits of memory in cover time of the graph with high probability. The robots are active indefinitely.*

Proof. First of all, a robot can perform a simple random walk with $O(\log \Delta)$ bits of memory as there are at most Δ adjacent edges at any node. Hence the robot can pick a random adjacent edge with $O(\log \Delta)$ bits of memory (i.e., it can generate a random number from 1 to Δ with $O(\log \Delta)$ bits of memory).

Consider a particular robot exploring the graph by performing a random walk. Since the random walk is independent of all the other robots' random walks, the robot visits all the nodes in the graph by the cover time of the graph with high probability (see the definition of the cover time in Sect. 2). Hence, with high probability, by at most the cover time of the graph, the robot settles at some node. Since every robot performs a random walk in parallel (until it settles down) and independently, every robot will settle down after the cover time of the graph with high probability. Hence dispersion is achieved in cover time of the graph with high probability. The cover time of a graph lies in the range between $\Omega(n \log n)$ and $O(mn)$ depending on the graph structure.

Whenever a robot settles at some node, the robot has to stay active until all the robots settle down. This is because the settled robot needs to inform other exploring robots that the node is already occupied; otherwise multiple robots may be settled down at a single node and dispersion will not be achieved. Since it takes cover time of the graph until all robots settle down, and the cover time of any graph is $\Omega(n \log n)$ [1,11], a robot cannot maintain a counter to count the rounds until cover time is achieved with only $O(\log \Delta)$ bits of memory and hence cannot terminate after the cover time. Thus, all the settled robots need to stay active for an indefinite number of rounds in order to achieve dispersion. □

Note that the random walk based exploration algorithm outperforms the $O(m)$ time algorithms in several graph classes. For example, consider regular expander graphs. The cover time of a regular expander graph is $\Theta(n \log n)$ [11]. However, in dense regular expander graphs, the number of edges is $\omega(n \log n)$ and it could be as high as $O(n^2)$ (e.g., a complete graph). In fact, the *Arbitrary-Graph* algorithm is asymptotically faster than deterministic algorithms with more memory from [4] in the graphs where $m = \omega(\text{cover-time})$. However, the algorithm is non-terminating. At the same time, *Arbitrary-Graph* requires robots to only have $O(\log \Delta)$ bits of memory. Moreover, since the random walks (corresponding to the robots) are independent of each other, the algorithm also works in an asynchronous system[7].

---

[7] It is required that *Local-Leader-Election* works without issue in that setting. This is the case.

8   Extending Algorithms to Arbitrary k

In this section, we describe how to extend our previous algorithms to work with an arbitrary number of robots. That is, we want to achieve dispersion of k robots on n nodes, for any positive integer value of k. There are two difficulties inherent in extending results to an arbitrary k, depending on whether $k < n$ or $k > n$. When $k < n$, if an algorithm relies on a certain condition to be met before robots terminate, we must ensure that this condition is still met even with less than n robots participating. When $k > n$, we must figure out how to have robots settle in stages, because we do not want to maintain a counter to allow $\lceil k/n \rceil$ robots to settle at each node because the memory requirement will be $O(\log k)$, which could be arbitrarily large.

Rooted-Ring works for any arbitrary k. At the end of a given round, one robot has settled and terminated execution of the algorithm and the remaining robots are colocated at the next node. It takes one round to settle one robot, and the ring is settled node by node. Thus, for k robots, we can achieve dispersion with *Rooted-Ring* in $O(k)$ rounds where each robot requires $O(\log \Delta)$ bits of memory.

Rooted-Tree works without any modifications when $k < n$. This is because the last robot to settle down initiates phase two of the algorithm. It is easy for a robot to detect this because no other robot will participate in *Local-Leader-Election* with it. Since only k nodes of the graph are explored in phase one or two, *Rooted-Tree* only takes $O(k)$ rounds to complete. When $k > n$, we have robots perform a second check to see if it is the last robot to settle. In a complete DFS traversal of a tree, a robot will reach the last node of the traversal when it has traversed the last port of each node from the root to a leaf node[8]. Thus, if a robot performing the DFS maintains a flag that indicates whether the robot traversed this path, we can detect the final node in the DFS traversal. Specifically, the robot sets the flag to true when at the root, and changes it to false if it traverses a port other than the last port of a given node. If the robot reaches a leaf node and the flag is true, and the robot is selected by *Local-Leader-Election* to settle down at the leaf node, then it knows that it is settling down at the last empty node in the tree. All other robots that got to this node but did not settle down go to the root node and wait. The last robot to settle down executes phase two of the algorithm as usual. Meanwhile, the robots waiting at the root node will execute a new iteration of the algorithm, i.e. start phase one, once the exploring robot in phase two passes through the final port of the root. This delay in start time guarantees that the exploring robot will trigger other robots in the tree to terminate execution in time for the robots executing phase one to re-populate the tree with settled robots. The above process is repeated $\lfloor k/n \rfloor$ times and takes $O(n)$ rounds for each repetition and with an additional repetition done by some $k - n * \lfloor k/n \rfloor$ robots that takes $O(k - n * \lfloor k/n \rfloor)$ rounds, for a total of $O(k)$ rounds to achieve dispersion of all robots. Thus, for any k, *Rooted-Tree* achieves

---

[8] Since ports are ordered, each robot can determine which port is last in the order. Furthermore, this ordering is unique to each node, so different robots will see the same ordering at each node.

dispersion in $O(k)$ rounds and requires robots to have $O(\log \Delta)$ bits of memory each.

Rooted-Graph-LogDelta-LogD and *Rooted-Graph-Delta* both work when $k < n$ without any modifications. When $k > n$, both work using the extension described in the previous paragraph on *Rooted-Tree*. However, since both algorithms take $O(m)$ rounds to settle n robots, the total running times are both $O(mk/n)$. However, an interesting change to the *Rooted-Graph-Delta* algorithm can ensure a running time of $O(m + k)$. The change is to add an extra bit to each robot that indicates if they are participating in an "even" or "odd" set of phases of the algorithm. Order the robots in sets of n robots referencing the set of phases in which they are settled and terminate execution of the algorithm. If the set is an odd (even) number set in this order, it is called odd (even) set. Now, for every set of phases a and b where a immediately precedes b in the order, order of phases is as follows. First, phase one of set a occurs, then phase one of set b, then phase two of set a, then phase two of set b.[9] A node can be identified as a possible soon to be empty node (in the sense that the settled robot terminates execution) if only a robot with a different phase (odd vs. even) robot is settled at it. Thus in phase one of set b, have each settled robot copy the values of the at most Δ bits of the previous set's robot at that node indicating which ports lead to back edges. Thus, only in phase one of the first set of nodes will the DFS take $O(m)$ rounds, and subsequent DFS's take only $O(n)$ rounds each for the remaining $O(k/n)$ sets of nodes (the algorithm is slightly tweaked so that future sets of robots take advantage of this info). Thus the running time becomes $O(m + k)$ rounds.

Arbitrary-Graph works without any modifications when $k < n$. When $k > n$, in general the algorithm will not work because robots never terminate execution of the algorithm. Thus, unlike the previous algorithms, we cannot have robots work in stages where the first set of robots settles down and terminates, then the next set, and so on. Instead, we would need a counter of $O(\lceil k/n \rceil)$ bits to count how many robots have already settled at a node and settle down (if chosen by *Local-Leader-Election*) if that counter is $< \lceil k/n \rceil$. This leads to the following special case where dispersion is possible. When $n < k \leq \Delta^c n$, where c is a positive constant, we can achieve dispersion using *Arbitrary-Graph*, modified with a counter as earlier described, in the cover time of the graph with high probability where each robot needs $O(\log \Delta)$ bits of memory.

9 Lower Bound on Memory

In [4], they showed that, assuming all robots have the same memory, a lower bound of $\Omega(\log n)$ bits is required for dispersion when considering deterministic algorithms. The bound resulted from an argument that robots needed enough bits to uniquely choose a robot from a set of robots at each node. However, as we

---

[9] For 3 sets of phases a, b, and c, sequence is: phase one of a, phase one of b, phase two of a, phase one of c, phase two of b, phase two of c. This sequence can be easily seen now for more sets.

see later in this paper, we are able to circumvent this with the use of randomness. Now we argue another lower bound in the presence of randomness.

Theorem 7. *Consider k robots trying to achieve dispersion on an n node graph. Assuming all robots have the same amount of memory, robots require $\Omega(\log \Delta)$ bits of memory each for any randomized algorithm to achieve dispersion on any graph.*

Proof. We first describe a situation where robots containing $o(\log \Delta)$ bits of memory will be unable to achieve dispersion. We then show that for all algorithms that attempt to achieve dispersion using $o(\log \Delta)$ bits of memory, we can arrive at this situation.

Consider any number of robots present at a given node with degree $O(\Delta)$. If each robot has $o(\log \Delta)$ bits of memory, it is impossible for any of them to individually access the entire list of possible ports to move through. Since the selection of ports by each of these robots is decided by nature, it may then occur that one particular port is never chosen in any of the subsets of ports. Let us focus on such a port.

Let this port lead to an edge which acts as a cut between the set of nodes with robots currently on them and the set of empty nodes. If $k < n$, additionally assume that the set of nodes with robots on them is of size $\leq k - 1$. Thus, the robots being unable to traverse that port prevents dispersion from being achieved. For any given algorithm, we can construct a graph such that there exists a node with associated cut edge satisfying the above description and all the robots start on nodes on the side of the cut with the node. Thus, for any algorithm, dispersion is impossible when robots have $o(\log \Delta)$ bits of memory each. □

10   Conclusion and Future Work

In this paper, we showed how to achieve dispersion on various types of graphs using less memory than required by other algorithms in the literature so far. Importantly, we showed how to leverage randomness in a novel way in the form of the *Local-Leader-Election* algorithm and utilize this primitive to reduce memory requirements. There are several interesting lines of research that result from this paper. We present two open problems of interest below.

Open Problem 1: All algorithms in our paper, save *Arbitrary-Graph*, work only for rooted versions of different types of graphs. The trade-off when implementing *Arbitrary-Graph* is that robots then must stay active indefinitely. Is it possible to develop algorithms for non-rooted versions of the graphs in question without requiring robots to stay active indefinitely?

Open Problem 2: Our algorithms for rooted graphs require robots to have possibly $\omega(\log \Delta)$ bits of memory each, depending on the values of Δ and D. Is it possible to develop algorithms with tighter upper bounds for rooted graphs?

References

1. Aldous, D.J.: Lower bounds for covering times for reversible Markov chains and random walks on graphs. J. Theor. Probab. **2**(1), 91–100 (1989)
2. Aleliunas, R., Karp, R.M., Lipton, R.J., Lovász, L., Rackoff, C.: Random walks, universal traversal sequences, and the complexity of maze problems. In: Proceedings of the 20th Annual Symposium on Foundations of Computer Science (FOCS), pp. 218–223 (1979)
3. Ambühl, C., Gąsieniec, L., Pelc, A., Radzik, T., Zhang, X.: Tree exploration with logarithmic memory. ACM Trans. Algorithms (TALG) **7**(2), 17 (2011)
4. Augustine, J., Moses Jr., W.K.: Dispersion of mobile robots: a study of memory-time trade-offs. In: Proceedings of the 19th International Conference on Distributed Computing and Networking, ICDCN 2018, Varanasi, India, 4–7 January 2018, pp. 1:1–1:10. ACM (2018)
5. Augustine, J., Moses Jr., W.K.: Dispersion of mobile robots: a study of memory-time trade-offs. CoRR abs/1707.05629 (v4) (2018). arxiv.org/abs/1707.05629
6. Baldoni, R., Bonnet, F., Milani, A., Raynal, M.: On the solvability of anonymous partial grids exploration by mobile robots. In: Baker, T.P., Bui, A., Tixeuil, S. (eds.) OPODIS 2008. LNCS, vol. 5401, pp. 428–445. Springer, Heidelberg (2008). https://doi.org/10.1007/978-3-540-92221-6_27
7. Barriere, L., Flocchini, P., Mesa-Barrameda, E., Santoro, N.: Uniform scattering of autonomous mobile robots in a grid. Int. J. Found. Comput. Sci. **22**(03), 679–697 (2011)
8. Bodlaender, H.L., van Leeuwen, J.: Distribution of Records on a Ring of Processors. Department of Computer Science, University of Utrecht (1986)
9. Brass, P., Cabrera-Mora, F., Gasparri, A., Xiao, J.: Multirobot tree and graph exploration. IEEE Trans. Rob. **27**(4), 707–717 (2011)
10. Brass, P., Vigan, I., Xu, N.: Improved analysis of a multirobot graph exploration strategy. In: 2014 13th International Conference on Control Automation Robotics & Vision (ICARCV), pp. 1906–1910. IEEE (2014)
11. Broder, A.Z., Karlin, A.R.: Bounds on the cover time. J. Theor. Probab. **2**(1), 101–120 (1989)
12. Cohen, R., Fraigniaud, P., Ilcinkas, D., Korman, A., Peleg, D.: Label-guided graph exploration by a finite automaton. ACM Trans. Algorithms (TALG) 4(4), 42 (2008)
13. Cybenko, G.: Dynamic load balancing for distributed memory multiprocessors. J. Parallel Distrib. Comput. **7**(2), 279–301 (1989)
14. Datta, A.K., Lamani, A., Larmore, L.L., Petit, F.: Enabling ring exploration with myopic oblivious robots. In: Parallel and Distributed Processing Symposium Workshop (IPDPSW), pp. 490–499. IEEE (2015)
15. Dereniowski, D., Disser, Y., Kosowski, A., Pająk, D., Uznański, P.: Fast collaborative graph exploration. Inf. Comput. **243**, 37–49 (2015)
16. Devismes, S., Lamani, A., Petit, F., Raymond, P., Tixeuil, S.: Optimal grid exploration by asynchronous oblivious robots. In: Richa, A.W., Scheideler, C. (eds.) SSS 2012. LNCS, vol. 7596, pp. 64–76. Springer, Heidelberg (2012). https://doi.org/10.1007/978-3-642-33536-5_7
17. Diks, K., Fraigniaud, P., Kranakis, E., Pelc, A.: Tree exploration with little memory. In: Proceedings of the Thirteenth Annual ACM-SIAM Symposium on Discrete Algorithms, pp. 588–597. Society for Industrial and Applied Mathematics (2002)

18. Dynia, M., Kutyłowski, J., auf der Heide, F.M., Schindelhauer, C.: Smart robot teams exploring sparse trees. In: Královič, R., Urzyczyn, P. (eds.) MFCS 2006. LNCS, vol. 4162, pp. 327–338. Springer, Heidelberg (2006). https://doi.org/10.1007/11821069_29
19. Elor, Y., Bruckstein, A.M.: Uniform multi-agent deployment on a ring. Theor. Comput. Sci. 412(8–10), 783–795 (2011)
20. Fraigniaud, P., Ilcinkas, D., Peer, G., Pelc, A., Peleg, D.: Graph exploration by a finite automaton. Theor. Comput. Sci. 345(2–3), 331–344 (2005)
21. Kshemkalyani, A.D., Ali, F.: Efficient dispersion of mobile robots on graphs. arXiv preprint arXiv:1805.12242 (2018)
22. Lamani, A., Potop-Butucaru, M.G., Tixeuil, S.: Optimal deterministic ring exploration with oblivious asynchronous robots. In: Patt-Shamir, B., Ekim, T. (eds.) SIROCCO 2010. LNCS, vol. 6058, pp. 183–196. Springer, Heidelberg (2010). https://doi.org/10.1007/978-3-642-13284-1_15
23. Lovász, L.: Random walks on graphs: a survey. In: Combinatorics, Paul Erdős is Eighty, vol. 2, pp. 1–46 (1993)
24. Muthukrishnan, S., Ghosh, B., Schultz, M.H.: First-and second-order diffusive methods for rapid, coarse, distributed load balancing. Theor. Comput. Syst. 31(4), 331–354 (1998)
25. Ortolf, C., Schindelhauer, C.: A recursive approach to multi-robot exploration of trees. In: Halldórsson, M.M. (ed.) SIROCCO 2014. LNCS, vol. 8576, pp. 343–354. Springer, Cham (2014). https://doi.org/10.1007/978-3-319-09620-9_26
26. Peleg, D., Van Gelder, A.: Packet distribution on a ring. J. Parallel Distrib. Comput. 6(3), 558–567 (1989)
27. Reingold, O.: Undirected connectivity in log-space. J. ACM 55(4), 17:1–17:24 (2008)
28. Shibata, M., Mega, T., Ooshita, F., Kakugawa, H., Masuzawa, T.: Uniform deployment of mobile agents in asynchronous rings. In: Proceedings of the 2016 ACM Symposium on Principles of Distributed Computing, pp. 415–424. ACM (2016)
29. Subramanian, R., Scherson, I.D.: An analysis of diffusive load-balancing. In: Proceedings of the Sixth Annual ACM Symposium on Parallel Algorithms and Architectures, pp. 220–225. ACM (1994)

Building Resource Auto-scaler with Functional-Link Neural Network and Adaptive Bacterial Foraging Optimization

Thieu Nguyen[1] , Binh Minh Nguyen[1]([⊠]) , and Giang Nguyen[2]

[1] School of Information and Communication Technology,
Hanoi University of Science and Technology, Hanoi, Vietnam
nguyenthieu2102@gmail.com, minhnb@soict.hust.edu.vn
[2] Institute of Informatics, Slovak Academy of Sciences, Bratislava, Slovakia
giang.ui@savba.sk

Abstract. In this paper, we present a novel intelligent proactive auto-scaling solution for cloud resource provisioning systems. The solution composes of an improvement variant of functional-link neural network and adaptive bacterial foraging optimization with life-cycle and social learning for proactive resource utilization forecasting as a part of our auto-scaler module. We also propose several mechanisms for processing simultaneously different resource metrics for the system. This enables our auto-scaler to explore hidden relationships between various metrics and thus help make more realistic for scaling decisions. In our system, a decision module is developed based on the cloud Service-Level Agreement (SLA) violation evaluation. We use Google trace dataset to evaluate the proposed solution well as the decision module introduced in this work. The gained experiment results demonstrate that our system is feasible to work in real situations with good performance.

Keywords: Proactive auto-scaling · Functional-link neural network ·
Adaptive bacterial foraging optimization ·
Multivariate time series data · Cloud computing · Google trace dataset

1 Introduction

Cloud technologies bring many benefits for both vendors and users. From the view of providers, they can maximize physical server utilization via time-sharing of using resources among multiple customers. Meanwhile, from the view of users, they do not need to care about infrastructure deployment costs, and only pay for what they used conforming to the pay-as-you-go model. Consequently, cloud hosting services are becoming more and more popular today.

One of the main outstanding features of cloud computing is the capability of elasticizing resources provided for applications. This mechanism thus supports

T. V. Gopal and J. Watada (Eds.): TAMC 2019, LNCS 11436, pp. 501–517, 2019.
https://doi.org/10.1007/978-3-030-14812-6_31

to increase the availability as well as Quality of Service (QoS) for those applications [4]. Usually, the resource elasticity function is offered to users through an auto-scaling service that operates based on two techniques covering threshold and prediction. While the first already has been used in clouds, the second technique is still an attractive research problem today because there are barriers while applying the prediction-based auto-scaling system to clouds in practice. Firstly, accuracy of the prediction models must reach a certain level while having the ability of processing multiple resource metrics at the same time to meet practical demands. Secondly, the prediction models must be simple to deploy and operate but keep the effectiveness in forecast. Thirdly, the prediction-based auto-scaling system also must have a scaling decision mechanism which ensures QoS defined in Service-Level Agreement (SLA) signed between cloud customers and vendors.

In this paper, we focus on developing a cloud prediction-based auto-scaler, which can simultaneously resolve the barriers presented above. In this way, our auto-scaler use a simple prediction model while still achieving the same level of accuracy and stability as compared with other complex methods. For the prediction module of the auto-scaler, we propose a novel variant of functional-linking neural network (FLNN) using adaptive bacterial foraging optimization with life-cycle and social learning (in short from here ABFOLS or Adaptive BFOLS) to train forecast model. Our prediction module also can simultaneously process multiple resources based on several data preprocessing mechanisms. To build scaling decision module for the auto-scaler, we exploit SLA-awareness to make decisions as well as evaluate these scaling actions. Our designed system is experimented and assessed with a real cluster trace dataset published by Google in 2011 [18], and [17]. The gained outcomes show that our auto-scaler archives good performance and can be applied to practice.

The structure of this paper is as follows. In Sect. 2, we classify and analyze existing studies to highlight our work contributions. Section 3 presents our cloud auto-scaler proposal (in short from here FLABL) with the intelligent core built based on the combination of FLNN and ABFOLS. The preprocessing raw data mechanisms, which helps the prediction module to exploit multiple data metrics is also described here. In the Sect. 4, we present tests and evaluations for the proposed auto-scaler to prove its effectiveness. The last Sect. 5 concludes and defines our future work directions.

2   Related Work

As mentioned in the previous Section, there are two cloud resources scaling mechanism types, namely reactive (i.e. systems will response to unpredictable resource consumption changes using usage thresholds) and proactive [7] (i.e. systems attempt to predict resource requirements to make scaling decisions in advance). According to that classification, we focus on analyzing related works in the proactive technique category.

Resource Consumption Prediction. In [3], several predictive time series models were proposed and demonstrated for cloud workload forecast problem such as ARMA (autoregressive-moving average), non-stationary, long memory, three families of seasonal, multiple input-output, intervention and multivariate ARMA models. As in [22], the authors used two datasets collected from the Intel Netbatch logs and Google cluster data center to compare and evaluate prediction models together, including first-order autoregressive, simple exponential smoothing, double exponential smoothing, ETS, Automated ARIMA, neural network (NN) autoregression. Although these studies investigate methods for resource usage prediction, there is still lack of investigation of applying of FLNN variance in combination with evolutionary optimization in that domain like in [13].

Functional-Link Neural Network (FLNN). Structurally, FLNN has only single neuron, which was proposed by Pao in [14] for pattern-recognition task. The network is quite simple but brings a good effective performance in the aspects of accuracy and training speed. Consequently, this learning model has been applied to many domains covering stock market prediction [10], and money exchange rate forecast [8]. In [5], the authors used FLNN for four datasets of wine sales. The gained results prove that FLNN is more efficient in comparison with random walk model and feed-forward neural network (FFNN). The authors of [19] used FLNN with Chebyshev and Legendre polynomials to predict impact on tall building structure under seismic loads. The experiments presented in that work shown FLNN yields better outcomes as compared with multi-layer neural network (MLNN) also in terms of accuracy and computation time. However, the main disadvantage of FLNN is the use of back-propagation (BP) mechanism with gradients descent to train the learning model ([5, 8, 10], and [19]).

Adaptive Bacterial Foraging Optimization (ABFOLS). Recently, a new evolutionary computing technique called Bacterial Foraging Optimization (BFO) has been proposed in the work [15]. It is inspired based on principle of bacterial movement (i.e. tumbling, swimming or repositioning) to food-seeking. Their behavior is achieved through a series of three processes on a population of simulated cells: "Chemotaxis", "Reproduction", and "Elimination-dispersal" [15]. BFO has been applied to several industry applications like PID controller tuning [6], power system [1], stock market [9]. Unfortunately, BFO has certain shortages e.g., the appropriate time and method for dispersal and reproduction must be carefully selected, otherwise the stability of population may be destroyed [23]. Therefore, the authors of [23] proposed an improvement for BFO including life-cycle of bacteria, social learning and adaptive search step length (ABFOLS), which offer significant improvements over the original version in complexity and competitive performances as compared with other algorithms (e.g. GA) on higher-dimensional problems. At present, there are no works that deal in optimizing FLNN with ABFOLS algorithm, especially in cloud auto-scaling issue.

Cloud Resources Provision Under SLA Conditions. In [16], the authors presented a job scheduling system using machine learning to predict workloads and allocate resources complying with SLA. The authors of [20] used SLA to estimate the amount of resources required for making scaling actions. They argued that their algorithm is able to reduce the number of SLA violations up to 95% and decrease resource requirements up to 33%. In [21], the authors proposed a proactive cloud scaling system that enables to estimate SLA violations based on multiple metric parameters. Although the works [16], and [20] already introduced SLA evaluation models, they operate based on single workload metric such as CPU usage, response time, or job execution deadline. Meanwhile, the work [21] allows processing multiple metrics, SLA violation still is evaluated based on resource usages (e.g. CPU, and memory usage). Conversely, in this study, we assess SLA to ensure QoS using the number of provided virtual machines (VMs). This is scaling unit, which cloud customers often care when making resources increase or decrease decisions.

In comparison with the existing works analyzed above, the differences and contributions of our work are as follows.

1. Proposing an improvement (called FLABL) for FLNN, in which the network is trained by ABFOLS instead of back-propagation mechanism.
2. Proposing an auto-scaler that uses FLABL, which can process multivariate metric data and predict resource consumptions.
3. Proposing an SLA violation evaluation module that operates based on VM number rather than resource metrics in order to make scaling decisions for the proposed auto-scaler.

3   Resource Auto-scaler Using Functional-Link Adaptive Bacterial Foraging Optimization Neural Network

The designs for our resource auto-scaling module are built based on underlying Cloud System. The module consists of 3 main phases, including Extraction, Learning and Scaling. The Scaling has two components namely Forecasting and Decision.

Raw resource monitoring data is collected from VMs in Cloud system. There are a lot of available monitoring services (e.g. CloudWatch, IBM cloud monitoring, and Rackspace Monitoring, Nagios, Prometheus and Zabbix and so forth) that can be used or deployed on cloud infrastructures for monitor problem. Based on the monitoring services, we gather diverse VM metrics such as CPU, memory utilization, disk IO, network IO, etc. The detail descriptions of Extraction, Learning and Scaling phases are described in the following subsections.

3.1   Extraction Phase

There are seven mechanisms deployed in Extraction phase to pre-process raw data and prepare normalized data for Scaling phase in our system. As shown in Fig. 1, those mechanisms cover:

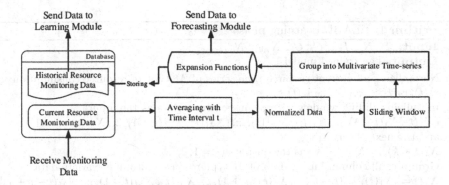

Fig. 1. FLABL data extraction process

1. Using collected raw monitoring data from the underlying Cloud System.
2. Transforming current raw data in the predefined period for model training into corresponding time series $r_i(t)(i = 1, 2, ..., M)$ with time interval ρ.
3. Averaging values of given metrics in the given interval of ρ for each point in time series $r_i(t)$.
4. Normalizing the time series data in the range of $[0, 1]$.
5. Transforming the normalized time series data to supervised data using sliding technique with window width k, which is the number of used values before the time t to predict a value at the time t.
6. The supervised data undergoes through an expansion function (e.g. Chebyshev or Power series) to enable the ability of catching nonlinear relationship between inputs and outputs.
7. Finally, the output data is put into a database in form of historical resources data, which is used to build prediction model in Learning phase. The data also is provided for Forecasting module in Scaling phase to predict the resource consumption.

3.2   Learning Phase

Generated data from Extraction phase is divided into three different sets, namely training, validating, and testing. While the first two sets are employed to learn and select parameters, the third set is used for validating trained prediction model. A novel method is proposed in our Learning phase based on FLNN and trained by ABFOLS algorithm to speed up the convergence and increase the prediction accuracy. Due to combinations, our learning method is called by Functional-Link Adaptive Bacterial Foraging Lifecycle Neural Network (called by FLABL).

The final trained model of Learning is used in Forecasting module that belongs to Scaling phase. Meanwhile, the real-time monitoring data collected from cloud VM after preprocessing will be use as Real-time monitoring resource usage before putting into our Final Model of Forecasting module. The gained predictive results then will be denormalized to be able to use in the next module.

Algorithm 1. FLABL Learning phase

Input: S, N_s, P_{ed}, C_s, C_e, N_{split}, N_{adapt}
Output: The trained model

1: Normalizing all current resource time series of M type of resource consumption:
   $r_1(t), r_2(t), ..., r_i(t), ..., r_M(t)$
2: Initializing sliding window with p consecutive points as the inputs
   $(X_1(t), X_1(t-1), ..., X_1(t-p+1)), ..., (X_M(t), X_M(t-1), ..., X_M(t-p+1))$
   and the next observation
   $X_1(t+k), ..., X_M(t+k)$ as the outputs $(t = 1, 2, ..., n)$
3: Grouping all column inputs data of M type resource into multivariate data
   $X(t) = [X_1(t), X_1(t-1), ..., X_1(t-p+1), ..., X_M(t), X_M(t-1), ..., X_M(t-p+1)]$
4: Applying expansion functions to multivariate data:

$$x(t) = [x_1^1(t), x_1^2(t), ..., x_1^5(t), x_1^1(t-1), x_1^2(t-1), ..., x_1^5(t-p+1), ...,$$
$$x_M^1(t), x_M^2(t), ..., x_M^5(t), x_M^1(t-1), x_M^2(t-1), ..., x_M^5(t-1), ..., \qquad (1)$$
$$x_M^1(t-p+1), x_M^2(t-p+1), .., x_M^5(t-p+1)]$$

5: Training the constructed FLNN (from step 1 to 4) by ABFOLS (from step 6 to 31)
6: Initialize population $Pop = \{Cell_1, ..., Cell_s\}$ with S cells (*bacterium*),
   where $Cell_i = (c_{i1}, ..., c_{id})$, $c_{ij} \in [-1, 1]$ with d-dimensions, $d = length(x(t) + 1)$
   Initialize $Nutrient(i) \leftarrow 0$ for all $Cell_i$
7: **while** (termination conditions are not met) **do**
8:     $S \leftarrow size(Pop)$; i $\leftarrow 0$
9:     **while** ($i < S$) **do**
10:        $i \leftarrow i + 1$; $fitLast \leftarrow fitCurrent(i)$
11:        Generate a tumble angle for $Cell_i$ by Eq. 4
12:        Update the position of $Cell_i$ by Eq. 3
13:        Recalculate $fitCurrent(i)$ and $Nutrient(i)$
14:        Update personal best ($Cell_{pBest}$) of ith cell and global best ($Cell_{gBest}$)
15:        **for** m = 1 to N_s **do**
16:            **if** $fitCurrent(i) < fitLast$ **then**
17:                $fitLast \leftarrow fitCurrent(i)$
18:                Run one step using Eq. 3
19:                Recalculate $fitCurrent(i)$ and $Nutrient(i)$
20:                Update $Cell_{pBest}$ and $Cell_{gBest}$
21:            **else**
22:                **break**
23:        **if** $Nutrient(i) > threshold_{split}$ (Eq. 6 is True) **then**
24:            Split $Cell_i$ into two *cell*
25:            **break**
26:        **if** $Nutrient(i) < threshold_{dead}$ (Eq. 7 is True) **then**
27:            Remove $Cell_i$ from Pop
28:            **break**
29:        **if** $Nutrient(i) < 0$ and $random(0, 1) < P_e$ **then**
30:            Move $Cell_i$ to a random position
31: Passing the **trained model** ($Cell_{gBest}$) to Forecasting module.
32: Repeating step 1 to step 30 in every time period T

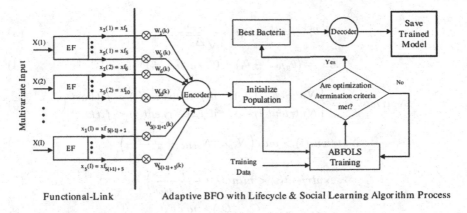

Functional-Link   Adaptive BFO with Lifecycle & Social Learning Algorithm Process

Fig. 2. FLABL learning phase with FLNN and ABFOLS

The details of our proposal FLABL are as follows: Used parameters in Algorithm 1 are explained in Table 1. While $Nutrient(i)$ is the nutrient of ith bacterium, $fitCurrent(i)$ is the current fitness value of ith bacterium, $fitLast$ is the fitness value of the last position of ith bacterium, $Cell_{pBest}$ is the past position of ith, bacterium which gained the best fitness so far. $threshold_{split}$ and $threshold_{dead}$ is corresponding to right side of Eqs. 6 and 7.

We use Mean Absolute Error (MAE) for bacterial fitness function according to the Eq. 2. In chemotactic steps, a bacterium will update its position based on the direction which created by combination of information of its personal best position and the population's global best position in Eqs. 3 and 4. The gain of nutrients process is updated when bacteria move, so if the new position is better than the last one (the fitness higher), it is regarded that the bacterium will gain nutrient from the environment and the nutrient is added by one. Otherwise, it loses nutrient in the searching process and its nutrient is reduced by one Eq. 5.

In an intelligence optimization algorithm, it is important to balance its exploration ability and exploitation ability. In the early stage, we should enhance the exploration ability to search all the areas. In the later stage of the algorithm, we should enhance the exploitation ability to search the good areas intensively. So in BFO, the step size which bacterium used to move play an important role. We use the decreasing step length based on bacterium's fitness (Eq. 8). In the early stage of BFOLS algorithm, larger step length provides the exploration ability. And at the later stage, small step length is used to make the algorithm turn to exploitation. Besides, we also deploy an adaptive search strategy, which change the step length of each bacteria based on their own nutrient, which calculated using (Eq. 9). The higher nutrient value, the bacterium's step length is shortened further. This is also in accordance with the food searching behaviors in natural. The higher nutrient value indicates that the bacterium is located in potential nutrient rich area with a larger probability. So, it is necessary to exploit the area carefully with smaller step length.

$$fit = MAE = \frac{\sum_{i=1}^{N} |forecast(i) - actual(i)|}{N} \tag{2}$$

$$\theta_i^{t+1} = \theta_i^t + C(i) * \frac{\Delta_i}{\sqrt{\Delta_i * \Delta_i^T}} \tag{3}$$

$$\Delta_i = (\theta_{gBest} - \theta_i) + (\theta_{i,pBest} - \theta_i) \tag{4}$$

$$Nutrient(i) = \begin{cases} Nutrient(i) + 1, & \text{if } fitCurrent(i) > fitLast \\ Nutrient(i) - 1, & \text{if } fitCurrent(i) < fitLast \end{cases} \tag{5}$$

$$Nutrient(i) > max\left(N_{split}, N_{split} + \frac{S^i - S}{N_{adapt}}\right) \tag{6}$$

$$Nutrient(i) < min\left(0, 0 + \frac{S^i - S}{N_{adapt}}\right) \tag{7}$$

$$C = C_s - \frac{(C_s - C_e) * nowFit}{totalFit} \tag{8}$$

$$C(i) = \begin{cases} \dfrac{C}{Nutrition(i)}, & \text{if } Nutrient(i) > 0 \\ C, & \text{if } Nutrient(i) \leq 0 \end{cases} \tag{9}$$

Table 1. FLABL parameters used for ABFOLS optimization

Name	Description
S	The number of cells (bacteria) in the population
d	d-dimension/problem size
N_s	Swimming length after which tumbling of bacteria will be undertaken in a chemotactic loop
P_{ed}	Probability for eliminate bacteria
C_s, C_e	The step size at the beginning and the end of chemotactic process
N_{split}, N_{adapt}	Parameters used to control and adjust the split criterion and dead criterion

3.3 Scaling Phase

The Scaling phase contains of Forecasting and Scaling components. The functionalities and collaboration of these two components are based on SLA-violation auto-scaling strategy as described in Algorithm 2.

Forecasting Module. Forecasting phase uses real-time monitoring data (is preprocessed by Preprocessing phase) as inputs for the trained model to predict new values (i.e. resource consumption) in advance. Then, the obtained outputs are de-normalized into the real values. In every time period T, learning algorithm is updated with new monitoring data.

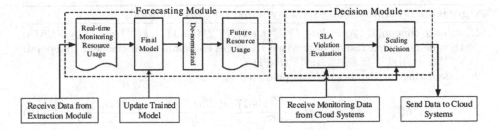

Fig. 3. FLABL scaling phase

Decision Module. Decision module is responsible for calculating the number of provided VMs according to the predictive resource consumption. In addition, we develop SLA Violation Evaluation component that operates based on the number of VMs allocated at previous time and with the VM numbers predicted in the future to decide the how many VMs will be provisioned as in Eqs. 10, 11, and 12. Finally, predictive scaling information will be sent to Resource Scaling Trigger in Cloud System to create appropriate actions. With resource usages predicted by Forecasting module, we assume that cloud system can provision unlimited amount of VMs like the systems presented in [12], which has the same hardware configuration. We also do not consider VM scheduling policies in our scaling strategy in this work.

In terms of QoS guarantee, the number of predicted VMs will be multiplied by s ($s \geq 1$). The larger s, the more VMs allocated to applications and this reduces the SLA violations. In this direction, the total number of SLA violations in an earlier period with length L is used to determine the increases (or decreases) of allocated VMs.

4   Experiments

To evaluate our proposed auto-scaler, we carry out two experiments, including:

1. Comparing prediction accuracy and the system run time between various neural network models with our FLABL, involving ANN, MLNN, traditional FLNN, FL-GANN and FL-BFONN using univariate and multivariate data.
2. Evaluating scaling performance of our proposed auto-scaler under SLA violation measurement.

4.1   Experimental Setup

Dataset. In our experiment, we use dataset gathered by Google from one of their production data center clusters. The log records operations of approximate 12000 servers for one month [17,18]. This log thus contains a lot of jobs submitted to the cluster. Each job is a collection of multiple tasks that are run simultaneously on many machines. Resource utilization of tasks is measured by several metrics

---

Algorithm 2. Auto-scaling strategy based on SLA-violations

1: At each timepoint t, calculate $F_1^{pred}(t+1), F_1^{pred}(t+1), .., F_1^{pred}(t+1)$ by FLABL
   then de-normalize results back to get resources consumption at the next timepoint:
   $r_1^{pred}(t+1), r_2^{pred}(t+1), ..., r_i^{pred}(t+1)$
2: Calculating the number of VMs predicted at time $(t+1)$ as

$$n_{VM}^{pred}(t+1) = max\left\{\frac{r_1^{pred}(t+1)}{C_1}, \frac{r_2^{pred}(t+1)}{C_2}, .., \frac{r_i^{pred}(t+1)}{C_i}\right\} \quad (10)$$

3: **for** $(z \leftarrow t - L + 1)$ to t calculate the number of VMs violations

$$n_{VM}^{violation}(t) = max\left\{0, n_{VM}^{actual}(t) - n_{VM}^{alloc}(t)\right\} \quad (11)$$

4: Calculating the number of allocated VMs

$$n_{VM}^{alloc}(t+1) = s * n_{VM}^{pred}(t+1) + \frac{1}{L}\sum_{z=t-L+1}^{t} n_{VM}^{violation}(z) \quad (12)$$

5: **if** $n_{VM}^{alloc}(t+1) > n_{VM}^{alloc}(t)$ **then**
6:     Making decision to instantiate $n_{VM}^{alloc}(t+1) - n_{VM}^{alloc}(t)$ VMs
7: **else**
8:     Making decision to destroy $n_{VM}^{alloc}(t) - n_{VM}^{alloc}(t+1)$ VMs
9: Repeating from step 1 to 8 at the next point of time.

---

such as CPU, and memory usage, disk I/O, and so on with more than 1233 million data items. To evaluate our predictive model as well as the auto-scaling strategy, we chose a long-running job with ID 6176858948, which consists of 60171 divergent tasks during the 20-day period.

We assume that the cloud system has enough VMs to process the chosen long-running job. VM capacity is equal to the minimum configuration of a machine in Google cluster with $C_{CPU} = 0.245$ and $C_{RAM} = 0.03$ (normalized values). We also assume that time required to instantiate a new VM is five minutes. Therefore, the resource usages in the incoming five minutes are predicted to make scaling decisions in advance. In this way, average time t is set by 5 min, forecast horizon $k = 1$ in all experiments. While, the collected data from 1st to 14th day is used to train the networks, data from 15th to 17th day employed to validate network's parameters, and data from 18th to 20th day is used to test the prediction performance. We select both CPU and memory metric types (visualization in Fig. 4) in the Google dataset as multivariate data for our proposed system. Before we group CPU and memory data to be multivariate data, we make sure the data range in the same by normalized step in the Extraction phrase. The prediction accuracy is assessed by Root Mean Square Error $RMSE = \sqrt{\frac{\sum_{i=1}^{N}(forecast(i)-actual(i))^2}{N}}$.

Test Models. Our tested ANN model is cofigured with three layers (one input, one hidden, and one output), MLNN model is configured with five layers (one

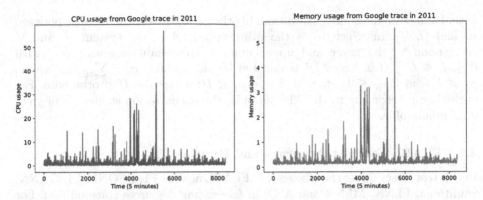

Fig. 4. CPU (left) and memory (right) from Google trace dataset (Google normalized data)

input, three hidden and one output). Meanwhile, FLNN, FL-GANN, FL-BFONN and FLABL have only one input and output layer with structure $(x, 1)$. The polynomial is set as functional-link for all models. Here, x is the sliding window value used in the Extraction phase in our system. Activation function used for our tested networks are Exponential Linear Unit (ELU) [2].

Based on previous research in [13], we use standard GA algorithm for FL-GANN model. The GA parameters are set as follows. The population size $p_s = 500$, the maximum number of generations $g_{max} = 700$, the probability of two individuals exchanging crossovers $p_c = 0.95$ and the probability of individual mutation $p_m = 0.025$. For BFO algorithm, the parameters are configured in the same way as works described in [15] which are set as follows: The length of the lifetime of the bacteria as measured by the number of chemotactic steps they take during their life $N_c = 50$, swimming length after which tumbling of bacteria will be undertaken in a chemotactic loop $N_s = 4$, maximum number of reproduction to be undertaken $N_{re} = 4$, the number of elimination-dispersal steps $N_{ed} = 8$, probability for eliminate bacteria $P_e = 0.25$, and $Step_{size} = 0.1$. Finally, for our ABFOLS algorithm, the parameter settings are similar to [23] work which are $N_s = 4$, $P_e = 0.25$ (same as BFO configurations), $N_{split} = 30$, $N_{adapt} = 5$, the step size at the beginning of process $C_s = 0.1(U_b - L_b)$, the step size at the end of the process $C_e = 0.00001(U_b - L_b)$, where U_b and L_b are referred to the upper and lower bound of the variables.

SLA Evaluation Mechanisms. We use SLAVTP (SLA Violation Time Percentage) proposed in [21] to evaluate our proposed VM calculation mechanism (formula 12) with $SLAVTP = \frac{T_{under-provisioning}}{T_{execution}}$, where $T_{under-provisioning}$ is the time when at least one allocation resource causes "under-provisioning" (i.e. lack of RAM or CPU core), and $T_{execution}$ is total time of the application running in cloud system.

To evaluate performance of our auto-scaling strategies, we use ADI (Auto-scaling Demand Index) measurement [11], which considers the difference between actual and desired resource utilization. In other words, the total distance between u_t and $[L, U]$. In which u_t is the utilization level of the system, L and U correspond to the lower and upper limits of reasonable resource use, with $0 \leq L \leq U \leq 1.0$. The ADI is denoted by the variable $\sigma = \sum_{t \in T} \sigma_t$ where, $\sigma_t = L - u_t$ if $u_t \leq L$; $\sigma_t = 0$ if $L < u_t < U$; $\sigma_t = u_t - U$ if otherwise. For each time t, according to the ADI formula, the optimization strategy will yield a minimum of σ.

4.2 FLABL Forecast Accuracy and Runtime

In this test, we evaluate the efficiency of FLABL against FL-BFONN, FL-GANN, traditional FLNN, MLNN and ANN in forecasting resources consumption. For each model, we use both univariate (single input metric) and multivariate (multiple input metrics) data. We also change sliding windows size k from 3 to 5 ($k = 3, 4, 5$) in the experiment to test effectiveness fluctuation. Our achieved outcomes are given in Table 2.

Table 2. RMSE comparison between FLABL and other models

Input type	Model	CPU			RAM		
		k = 3	k = 4	k = 5	k = 3	k = 4	k = 5
Univariate	ANN	0.4962	0.4952	0.5054	0.0344	0.0352	0.0354
	MLNN	0.4903	0.4930	0.4966	0.0345	0.0349	0.0355
	FLNN	0.5171	0.510	0.5253	0.038	0.0356	0.0373
	FL-GANN	0.4892	0.4762	0.4877	0.0389	0.0376	0.0375
	FL-BFONN	0.4777	0.4872	0.4798	0.0342	0.0431	0.035
	FLABL	**0.469**	**0.4726**	**0.4732**	**0.0335**	**0.0338**	**0.0338**
Multivariate	ANN	0.4884	0.4863	0.507	0.0343	0.0336	0.0348
	MLNN	0.4913	0.4814	0.5026	0.0359	0.0355	0.0356
	FLNN	0.4877	0.5179	0.5043	0.0367	0.0365	0.0366
	FL-GANN	0.49	0.491	0.488	0.037	0.0361	0.036
	FL-BFONN	0.4976	0.4762	0.4963	0.0334	0.0332	0.0337
	FLABL	**0.4678**	**0.4705**	**0.4793**	**0.0333**	0.0337	0.0344

According to the achieved results, there are some observations that can be made as follows. In almost all cases, RMSE accuracy of FLABL are smaller than ANN, MLNN, traditional FLNN, FL-GANN and FL-BFONN model with different sliding window values as well as input types. Concretely, for univariate data input, FLABL brings the best results as compared with others, also for multivariate data input except $k = 4$ and 5 in memory prediction case. However, the difference is trivial for the test scenarios (0.0337 (FLABL) compares with 0.0332 (FL-BFONN) with $k = 4$, and 0.0344 (FLABL) compares with 0.0337(FL-BFONN) with $k = 5$). This shows the advantage of FLABL in prediction in

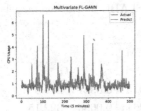
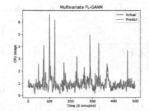
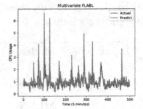

Fig. 5. CPU prediction outcomes of FL-GANN (first), FL-BFONN (second) and FLABL (third) with multivariate data, sliding window = 5

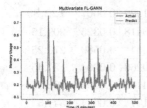
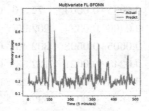
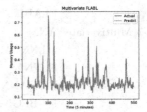

Fig. 6. Memory prediction outcomes of FL-GANN (first), FL-BFONN (second) and FLABL (third) with multivariate data, sliding window = 5

comparison with other models. Figures 5 and 6 illustrate the prediction outcomes of CPU and memory metrics in accuracy comparison tests among experimental models.

In this test, we compare the speed of those prediction models, which based on 3 factor includes: t_e is average time for 1 epoch (second/epoch), t_p is time to predict test data (second) and t_s is the total time of all system (preprocessing, training and testing - second). Because each model has different epoch configuration, so t_e should be choice instead of the total time of training process. Our speed comparison are given in Table 3.

4.3   Auto-scaling Strategy

In this test, we evaluate auto-scaling decisions made based on prediction results and SLA violation assessments as presented in Subsect. 3.3. Figure 7 shows the number of VMs calculated using the formula 12. An observation can be made from the obtained outcomes as follows. With $s < 1.3$, there are still some under-provision VMs as compared with resource requirements. For $s \geq 1.3$, VMs are allocated sufficiently for the demand usages. It's also shown in below test when we consider the lack and over-provision of resources.

In order to assess the lack or over-provision of resources, we use SLAVTP and ADI measurements. Concretely, Table 4 shows SLAVTP estimations of various models in VM allocation process with window size $p = 3$. In general, our proposed model gains smaller SLAVTP values than other models when the scaling coefficient is changed. For example, when $s = 2.0$, FLABL univariate model has

Table 3. System run time (second) comparison between FLABL and other models with sliding window = 5

Input type	Model	CPU			RAM		
		t_e	t_p	t_s	t_e	t_p	t_s
Univariate	ANN	0.0379	0.0319	198.4	0.044	0.0321	220.3
	MLNN	0.0583	0.0587	291.58	0.0644	0.0686	322.15
	FLNN	0.023	0.0003	114.83	0.0248	0.0003	124.03
	FL-GANN	0.3154	0.0004	220.79	0.2612	0.0004	182.89
	FL-BFONN	3.473	0.0006	2778.5	3.289	0.0014	2521.5
	FLABL	0.1095	0.0004	**71.16**	0.1229	0.0007	**122.89**
Multivariate	ANN	0.0406	0.0324	202.95	0.0418	0.0325	209.15
	MLNN	0.0605	0.0598	302.48	0.054	0.0589	270.25
	FLNN	0.0245	0.0021	122.53	0.0263	0.0004	131.45
	FL-GANN	0.3944	0.0004	276.15	0.4176	0.0004	292.35
	FL-BFONN	3.166	0.0007	2532.6	3.177	0.0034	2541.5
	FLABL	0.1708	0.0008	170.82	0.1419	0.0009	141.91

only 0.18% violation, while the multivariate model violates 0.12%. Especially, when $s = 2.5$ both of univariate and multivariate model reaches 0% violation. Of course, in these cases, the number of VMS provisioned is quite large.

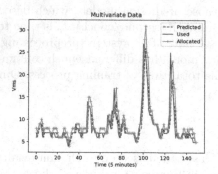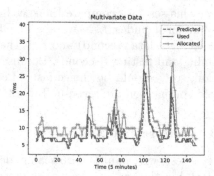

Fig. 7. The number of predicted, allocated, used VMs with sliding window = 3, adaptation length $L = 5$, and scaling coefficient $s = 1$ (left), $s = 1.3$ (right)

Based on the ADI measurement, Table 5 shows the ADI evaluation of different predictive models, with the desired utilization [60%, 80%] and window size $p = 3$. It is easy to observe that ADI of FLABL for both univariate and multivariate data types is smaller than others. Specifically, in the test, optimal ADI is at 3.7 when $s = 1.3$. The results demonstrate significant effect of our prediction as well

Table 4. Violation percentage in comparison among various models with adaptation length = 5, and sliding window = 3

Input type	Model	Scaling coefficient						
		s = 1.0	s = 1.3	s = 1.5	s = 1.7	s = 2.0	s = 2.2	s = 2.5
Univariate	ANN	10.3	1.87	1.02	0.54	0.18	0.12	0.06
	MLNN	10.24	1.81	0.96	0.6	0.18	0.12	0.06
	FLNN	9.28	1.81	0.96	0.54	0.24	0.24	0.12
	FL-GANN	11.33	1.69	0.9	0.48	0.24	0.18	0.06
	FL-BFONN	11.81	1.75	0.84	0.54	0.24	0.06	0
	FLABL	10.3	1.93	1.02	0.54	**0.18**	**0.06**	0
Multivariate	ANN	15.96	7.59	5.42	4.28	3.13	2.53	1.99
	MLNN	10.42	1.69	0.9	0.6	0.18	0.06	0.06
	FLNN	11.08	1.75	0.84	0.54	0.12	0.06	0
	FL-GANN	10.3	1.69	1.27	0.48	0.24	0.12	0.06
	FL-BFONN	9.7	1.81	0.84	0.6	0.18	0.12	0.06
	FLABL	**9.76**	**1.69**	0.9	0.54	**0.12**	**0.06**	0

Table 5. ADI in comparison among various models with adaptation length = 5, and sliding window = 3

Input type	Model	Scaling coefficient						
		s = 1.0	s = 1.3	s = 1.5	s = 1.7	s = 2.0	s = 2.2	s = 2.5
Univariate	ANN	154.01	15.41	37.88	112.63	227.11	293.53	372.4
	MLNN	176.59	3.81	4.87	78.94	205.75	270.58	353.31
	FLNN	184.6	4.41	11.29	82.46	207.8	277.2	357.88
	FL-GANN	176.67	3.86	15.9	89.18	214.8	283.6	363.81
	FL-BFONN	179.62	5.21	4.42	77.31	202.46	267.21	350.03
	FLABL	189.27	**3.7**	5.54	74.73	**199.9**	**267**	**349.5**
Multivariate	ANN	342.11	320.86	356.43	394.72	456.56	491.43	539.16
	MLNN	203.73	10.03	7.84	67.19	192.52	254.28	337.38
	FLNN	183.2	8.63	14.53	83.34	203.7	271.43	352.82
	FL-GANN	195.78	9.79	10.24	74.45	192.43	259.98	341.88
	FL-BFONN	219.81	40.24	36.47	93.23	197.92	260.28	340.58
	FLABL	192.27	**8.22**	**5.5**	72.16	**183.9**	259.61	342.76

as decision solutions. As mentioned above, when s is increased, SLA violation decreases but resource over-provision phenomenon occurs.

5   Conclusion and Future Work

In this paper, we presented our designs for a novel cloud proactive auto-scaler with complete modules from prediction to decision making. Thus, functional-link neural network is used for the forecast phase. To overcome back-propagation drawback, we integrate adaptive bacterial foraging life-cycle and social learning optimization with the artificial neural network. This improvement brings better accuracy for our proposed model. The auto-scaler designed in this paper also enables the capability of analyzing multiple monitoring metrics at the same time. This mechanism supports our system to be able to discover the implicit relationships among metrics types and thus help make scaling decisions more precisely. For the decision module, we proposed an efficient way to calculate the number of VMs that are provided for cloud-based applications using SLA violation measurement. We tested the auto-scaler with a real dataset generated by Google cluster. The obtained outcomes show that our system can work efficiently as compared with other methods and it can be applied to practice in clouds. For the future, we would like to implement the auto-scaler in private cloud middleware like Openstack or OpenNebula. Based on the infrastructures, we will test the proposal with applications under real conditions.

Acknowledgements. This research is supported by Vietnamese MOETs project "Research on developing software framework to integrate IoT gateways for fog computing deployed on multi-cloud environment" No. B2017-BKA-32, Slovak APVV-17-0619 "Urgent Computing for Exascale Data", and EU H2020-777536 EOSC-hub "Integrating and managing services for the European Open Science Cloud".

References

1. Ali, E., Abd-Elazim, S.: Bacteria foraging optimization algorithm based load frequency controller for interconnected power system. Int. J. Electr. Power Energy Syst. **33**(3), 633–638 (2011)
2. Clevert, D.A., Unterthiner, T., Hochreiter, S.: Fast and accurate deep network learning by exponential linear units (ELUS). arXiv preprint arXiv:1511.07289 (2015)
3. Hipel, K.W., McLeod, A.I.: Time Series Modelling of Water Resources and Environmental Systems, vol. 45. Elsevier, Amsterdam (1994)
4. Hluchý, L., Nguyen, G., Astaloš, J., Tran, V., Šipková, V., Nguyen, B.M.: Effective computation resilience in high performance and distributed environments. Comput. Inform. **35**(6), 1386–1415 (2017)
5. Khandelwal, I., Satija, U., Adhikari, R.: Forecasting seasonal time series with functional link artificial neural network. In: 2015 2nd International Conference on Signal Processing and Integrated Networks (SPIN), pp. 725–729. IEEE (2015)
6. Kim, D.H., Cho, J.H.: Adaptive tuning of PID controller for multivariable system using bacterial foraging based optimization. In: Szczepaniak, P.S., Kacprzyk, J., Niewiadomski, A. (eds.) AWIC 2005. LNCS (LNAI), vol. 3528, pp. 231–235. Springer, Heidelberg (2005). https://doi.org/10.1007/11495772_36
7. Lorido-Botrán, T., Miguel-Alonso, J., Lozano, J.A.: Auto-scaling techniques for elastic applications in cloud environments. Technical report EHU-KAT-IK-09 12, 2012, Department of Computer Architecture and Technology, University of Basque Country (2012)
8. Majhi, B., Rout, M., Majhi, R., Panda, G., Fleming, P.J.: New robust forecasting models for exchange rates prediction. Expert Syst. Appl. **39**(16), 12658–12670 (2012)

9. Majhi, R., Panda, G., Sahoo, G., Dash, P.K., Das, D.P.: Stock market prediction of S&P 500 and DJIA using bacterial foraging optimization technique. In: IEEE Congress on 2007 Evolutionary Computation, CEC 2007, pp. 2569–2575. IEEE (2007)

10. Majhi, R., Panda, G., Sahoo, G.: Development and performance evaluation of FLANN based model for forecasting of stock markets. Expert syst. Appl. **36**(3), 6800–6808 (2009)

11. Netto, M.A., Cardonha, C., Cunha, R.L., Assunçao, M.D.: Evaluating auto-scaling strategies for cloud computing environments. In: 2014 IEEE 22nd International Symposium on Modelling, Analysis & Simulation of Computer and Telecommunication Systems (MASCOTS), pp. 187–196. IEEE (2014)

12. Nguyen, B.M., Tran, D., Nguyen, G.: Enhancing service capability with multiple finite capacity server queues in cloud data centers. Clust. Comput. **19**(4), 1747–1767 (2016)

13. Nguyen, T., Tran, N., Nguyen, B.M., Nguyen, G.: A resource usage prediction system using functional-link and genetic algorithm neural network for multivariate cloud metrics. In: 2018 IEEE 11th Conference on Service-Oriented Computing and Applications (SOCA), pp. 49–56. IEEE (2018)

14. Pao, Y.H.: Adaptive Pattern Recognition and Neural Networks. Addison-Wesley Longman Publishing Co., Inc., Boston (1989)

15. Passino, K.M.: Biomimicry of bacterial foraging for distributed optimization and control. IEEE Control Syst. **22**(3), 52–67 (2002)

16. Reig, G., Alonso, J., Guitart, J.: Prediction of job resource requirements for deadline schedulers to manage high-level SLAs on the cloud. In: 2010 9th IEEE International Symposium on Network Computing and Applications (NCA), pp. 162–167. IEEE (2010)

17. Reiss, C., Tumanov, A., Ganger, G.R., Katz, R.H., Kozuch, M.A.: Heterogeneity and dynamicity of clouds at scale: Google trace analysis. In: Proceedings of the Third ACM Symposium on Cloud Computing, p. 7. ACM (2012)

18. Reiss, C., Wilkes, J., Hellerstein, J.L.: Google cluster-usage traces: format + schema. White Paper, pp. 1–14. Google Inc. (2011)

19. Sahoo, D.M., Chakraverty, S.: Functional link neural network learning for response prediction of tall shear buildings with respect to earthquake data. IEEE Trans. Syst. Man Cybern. Syst. **48**(1), 1–10 (2018)

20. Souza, A.A.D., Netto, M.A.: Using application data for sla-aware auto-scaling in cloud environments. In: 2015 IEEE 23rd International Symposium on Modeling, Analysis and Simulation of Computer and Telecommunication Systems (MASCOTS), pp. 252–255. IEEE (2015)

21. Tran, D., Tran, N., Nguyen, G., Nguyen, B.M.: A proactive cloud scaling model based on fuzzy time series and SLA awareness. Procedia Comput. Sci. **108**, 365–374 (2017)

22. Vazquez, C., Krishnan, R., John, E.: Time series forecasting of cloud data center workloads for dynamic resource provisioning. JoWUA **6**(3), 87–110 (2015)

23. Yan, X., Zhu, Y., Zhang, H., Chen, H., Niu, B.: An adaptive bacterial foraging optimization algorithm with lifecycle and social learning. Discrete Dyn. Nat. Soc. **2012**, 20 pp. (2012)

On the Enumeration of Bicriteria Temporal Paths

Petra Mutzel and Lutz Oettershagen[✉]

Department of Computer Science, TU Dortmund University, Dortmund, Germany
{petra.mutzel,lutz.oettershagen}@tu-dortmund.de

Abstract. We discuss the complexity of path enumeration in weighted temporal graphs. In a weighted temporal graph, each edge has an availability time, a traversal time and some cost. We introduce two bicriteria temporal min-cost path problems in which we are interested in the set of all efficient paths with low costs and short duration or early arrival times, respectively. Unfortunately, the number of efficient paths can be exponential in the size of the input. For the case of strictly positive edge costs, however, we are able to provide algorithms that enumerate the set of efficient paths with polynomial time delay and polynomial space. If we are only interested in the set of Pareto-optimal solutions (not in the paths themselves), then we show that in the case of nonnegative edge costs these sets can be found in polynomial time. In addition, for each Pareto-optimal solution, we are able to find an efficient path in polynomial time.

1 Introduction

A weighted temporal graph $G = (V, E)$ consists of a set of vertices and a set of temporal edges. Each temporal edge $e \in E$ is associated with an edge cost and is only available (for departure) at a specific integral point in time. Traversing an edge takes a specified amount of traversal time. We can imagine a temporal graph as a sequence of $T \in \mathbb{N}$ graphs $G_1, G_2, \ldots, G_T$ sharing the common set of vertices V; each graph G_i has its own set of edges E_i. Therefore, G can change its structure over a finite sequence of integral time steps.

Given a directed weighted temporal graph $G = (V, E)$, a source $s \in V$ and a target $z^1 \in V$, we want to find *earliest arrival* or *fastest* (s, z)-paths with *minimal costs*. A motivation can be found in typical queries in (public) transportation networks. Here, each vertex represents a bus stop, metro stop or a transfer point and each edge a connection between two such points. In this model, the availability time of an edge is the departure time of the bus or metro, the traversal time is the time a vehicle takes between the two transfer points, and the edge cost provides the ticket price. Two natural questions are the following: (1) Minimize the costs and the arrival times, or (2) minimize the costs and the total travel time.

---

[1] We use t to denote *time steps* and z for the target vertex.

© Springer Nature Switzerland AG 2019
T. V. Gopal and J. Watada (Eds.): TAMC 2019, LNCS 11436, pp. 518–535, 2019.
https://doi.org/10.1007/978-3-030-14812-6_32

In general, there is no path that minimizes both objectives simultaneously, and therefore we are interested in the set of all efficient paths. A path is called *efficient* if there is no other path that is strictly better in one of the criteria and at least as good concerning both criteria. In other words, a path is efficient iff its cost vector is Pareto-optimal.

We denote by MCFENUM and MCEAENUM the enumeration problems, in which the task is to enumerate the set of all efficient paths w.r.t. cost and duration or cost and arrival time, respectively. Unfortunately, there can be an exponential number of efficient paths. So we cannot expect to find polynomial time algorithms for the above mentioned enumeration problems. However, Johnson, Yannakakis, and Papadimitriou [7] have defined complexity classes for enumeration problems, where the time complexity is expressed not only in terms of the input size but also in the output size. We use the output complexity model to analyze the proposed enumeration problems and show that the problems belong to the class of polynomial time delay with polynomial space (**PSDelayP**). If we are only interested in the sets of Pareto-optimal solutions and not in all the paths themselves, we show that then the problems can be solved in polynomial time for nonnegative edge weights. In these cases we can also provide an associated path with each solution.

Contribution – In this paper we show the following:

1. MCFENUM and MCEAENUM are in **PSDelayP** for weighted temporal graphs with strictly positive edge costs.
2. In case of nonnegative edge costs, finding the Pareto-optimal set of cost vectors is possible in polynomial time (thus the set of Pareto-optimal solutions is polynomially bounded in the size of the input), and for each Pareto-optimal solution we can find an efficient path in polynomial time.
3. The decision versions that ask if a path is efficient, are in **P**. Deciding if there exists an efficient path with given cost and duration or arrival time is possible in polynomial time.

In the remainder of this section we discuss the related work. In Sect. 2 we provide all necessary preliminaries. Next, in Sect. 3 structural results are presented. These are the foundation of the algorithms for MCFENUM and MCEAENUM in the following Sects. 4 and 5, respectively. Finally, in Sect. 6 conclusions are drawn.

Related Work – Temporal graphs and related problems are discussed in several recent works. A general overview is provided, e.g., in [1,6,8]. Xuan et al. [12] discuss communication in dynamic and unstable networks. They introduce algorithms for finding a fastest path, an earliest arrival path and a path that uses the least number of edges. Wu et al. [11] further discuss the fastest, shortest and earliest arrival path problems in temporal graphs and introduce the latest-departure path problem. Their algorithm for calculating an earliest arrival (s, v)-path has time complexity $\mathcal{O}(n + m)$ and space complexity $\mathcal{O}(n)$, where n denotes the number of vertices and m the number of edges in the given temporal graph. Furthermore, they present an algorithm that finds a fastest (s, v)-path in

$\mathcal{O}(n + m \log c)$ time and $\mathcal{O}(\min\{n \cdot \mathcal{S}, n + m\})$ space. Here, $\mathcal{S}$ is the number of distinct availability times of edges leaving vertex s, and c the minimum of $\mathcal{S}$ and the maximal in-degree over all vertices of G. The algorithm uses a label setting approach to find a fastest (s, v)-path for each $v \in V$. However, Xuan et al. [12] and Wu et al. [11] do not consider weighted temporal graphs.

Hansen [5] introduces bicriteria path problems in static graphs, and provides an example for a family of graphs for which the number of efficient paths grows exponentially with the number of vertices. Meggido shows that deciding if there is an (s, z)-path that respects an upper bound on both objective functions is **NP**-complete (Meggido 1977, private communication with Garey and Johnson [3]). Martins [9] presents a label setting algorithm based on the well known Dijkstra algorithm for the bicriteria shortest path problem, that finds the set of all efficient (s, v)-paths for all $v \in V$. Ehrgott and Gandibleux [2] provide an overview of the work on bi- and multicriteria shortest path problems. Hamacher et al. [4] propose an algorithm for the bicriteria time-dependent shortest path problem in networks in which edges have time dependent costs and traversal times. The traversal time of an edge is given as a function of the time upon entering the edge. Moreover, each edge has a two-dimensional time dependent cost vector. Waiting at a vertex may be penalized by additional bicriteria time dependent costs. They propose a label setting algorithm that starts from the target vertex and finds the set of all efficient paths to each possible start vertex.

We are not aware of any work discussing the enumeration of efficient paths in weighted temporal graphs.

2   Preliminaries

A weighted temporal graph $G = (V, E)$ consists of a set V of $n \in \mathbb{N}$ vertices and a set E of $m \in \mathbb{N}$ weighted and directed temporal edges. A weighted and directed temporal edge $e = (u, v, t, \lambda, c) \in E$ consists of the starting vertex $u \in V$, the end vertex $v \in V$ (with $u \neq v$), availability time $t \in \mathbb{N}$, traversal time $\lambda \in \mathbb{N}$ and cost $c \in \mathbb{R}_{\geq 0}$. Each edge $e = (u, v, t, \lambda, c) \in E$ is only available for entering at its availability time t and traversing e takes λ time. For $T := \max\{t \mid (u, v, t, \lambda, c) \in E\}$, we can view G as a finite sequence $G_1, G_2, \ldots, G_T$ of static graphs over the common set of vertices V; each G_i with its own set of edges $E_i := \{(u, v, \lambda, c) \mid e = (u, v, i, \lambda, c) \in E\}$. Figure 1 shows an example. We denote the set of incoming (outgoing) temporal edges of a vertex $v \in V$ by $\delta^-(v)$ $(\delta^+(v))$. Note that in general for temporal graphs the number of edges is not bounded by a function in the number of vertices, i.e. we may have arbitrarily more temporal edges than vertices. In the following, we use a *stream representation* of temporal graphs. A temporal graph is given as a sequence of the m edges, which is ordered by the availability time of the edges in increasing order with ties being broken arbitrarily.

2.1   Temporal Path Problems

A *temporal* (u, v)-*walk* $P_{u,v}$ is a sequence $(e_1, \ldots, e_i = (v_i, v_{i+1}, t_i, \lambda_i, c_i), \ldots, e_k)$ of edges with $e_i \in E$ for $1 \leq i \leq k$, and with $v_1 = u$, $v_{k+1} = v$ and $t_i + \lambda_i \leq t_{i+1}$ for $1 \leq i < k$. If a temporal (u, v)-walk visits each $v \in V$ at most once, it is *simple* and we call it (u, v)-*path*. We denote by $s(P_{u,v}) := t_1$ the *starting time*, and by $a(P_{u,v}) := t_k + \lambda_k$ the *arrival time* of $P_{u,v}$. Furthermore, we define the *duration* as $d(P_{u,v}) := a(P_{u,v}) - s(P_{u,v})$. A path $P_{u,v}$ is *faster* than a path $Q_{u,v}$ if $d(P_{u,v}) < d(Q_{u,v})$. The *cost* of a path $P = (e_1, \ldots, e_k)$ is the sum of the edge cost, i.e. $c(P) := \sum_{i=1}^{k} c_i$. Finally, for a path $P = (e_1, \ldots, e_i, \ldots, e_k)$, we call $(e_1, \ldots, e_i)$ *prefix-path* and $(e_i, \ldots, e_k)$ *suffix-path* of P. In Fig. 1 the (s, z)-path $((s, b, 1, 1, 2), (b, z, 2, 1, 1))$ has arrival time 3, duration 2 and cost 3. The (s, z)-path $((s, z, 3, 1, 3))$ also has cost 3, but it has a later arrival time of 4 and is faster with a duration of only 1.

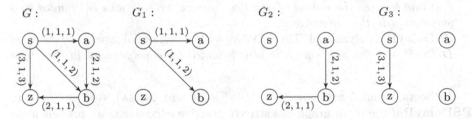

Fig. 1. Example for a weighted temporal graph G. Each edge label (t, λ, c) describes the time t when the edge is available, its traversal time λ and its cost c. For each time step $t \in \{1, 2, 3\}$, layer G_t is shown.

For the discussion of bicriteria path problems, we use the following definitions. Let $\mathcal{X}$ be the set of all feasible (s, z)-paths, and let $f(P)$ be the temporal value of P, i.e. either arrival time $f(P) := a(P)$ or duration $f(P) := d(P)$. We call a path $P \in \mathcal{X}$ *efficient* if there is no other path $Q \in \mathcal{X}$ with $c(Q) < c(P)$ and $f(Q) \leq f(P)$ or $c(Q) \leq c(P)$ and $f(Q) < f(P)$. We map each $P \in \mathcal{X}$ to a vector $(f(P), c(P))$ in the two-dimensional objective space which we denote by $\mathcal{Y}$. Complementary to efficiency in the decision space, we have the concept of *domination* in the objective space. We say $(f(P), c(P)) \in \mathcal{Y}$ *dominates* $(f(Q), c(Q)) \in \mathcal{Y}$ if either $c(P) < c(Q)$ and $f(P) \leq f(Q)$ or $c(P) \leq c(Q)$ and $f(P) < f(Q)$. We call $(f(P), c(P))$ *nondominated* if and only if P is efficient. We define the bicriteria enumeration problems MCFENUM and MCEAENUM as follows.

MIN-COST FASTEST PATHS ENUMERATION PROBLEM (MCFENUM)
Given: A weighted temporal graph $G = (V, E)$ and $s, z \in V$.
Task: Enumerate all and only (s, z)-paths that are efficient w.r.t. duration and costs.
MIN-COST EARLIEST ARRIVAL PATHS ENUMERATION PROBLEM (MCEAENUM)
Given: A weighted temporal graph $G = (V, E)$ and $s, z \in V$.

Task: Enumerate all and only (s, z)-paths that are efficient w.r.t. arrival time and costs.

We denote by MCF and MCEA the optimization versions, in which the task is to find a single efficient (s, z)-path.

2.2   Complexity Classes for Enumeration Problems

Bi- and multicriteria optimization problems are often not easily comparable using the traditional notion of worst-case complexity, due to their potentially exponential number of efficient solutions. We use the *output complexity* model as proposed by Johnson, Yannakakis, and Papadimitriou [7]. Here, the time complexity is stated as a function in the size of the input and the output.

Definition 1. *Let $\mathcal{E}$ be an enumeration problem. Then $\mathcal{E}$ is in*

1. **DelayP** (Polynomial Time Delay) *if the time delay until the output of the first and between the output of any two consecutive solutions is bounded by a polynomial in the input size.*
2. **PSDelayP** (Polynomial Time Delay with Polynomial Space) *if $\mathcal{E}$ is in DelayP and the used space is also bounded by a polynomial in the input size.*

In Sects. 4 and 5 we show that MCFENUM and MCEAENUM are both in **PSDelayP** if the input graph has strictly positive edge costs. We provide algorithms that enumerate the set of all efficient paths in polynomial time delay and use space bounded by a polynomial in the input size.

3   Structural Results

We show that it is possible to find an efficient (s, z)-path for the Min-Cost Earliest Arrival Path Problem (MCEA) in a graph G, if we are able to solve MCF. We use a transformed graph G', in which a new source vertex and a single edge is added. The reduction is from a search problem to another search problem. We show that it preserves the existence of solutions and we also provide a mapping between the solutions. This is also known as *Levin* reduction.

Lemma 1. *There is a Levin reduction from MCEA to MCF.*

Proof. Let $I = (G = (V, E), s, z)$ be an instance of MCEA. We construct the MCF instance $I' = (G' = (V \cup \{s'\}, E \cup \{e_0 = (s', s, 0, 0, 0)\}), s', z)$. Furthermore, let $\mathcal{X}_I$ be the sets of all (s, z)-paths for I, and $\mathcal{X}_{I'}$ be the sets of all (s', z)-paths for I'. We define $g : \mathcal{X}_I \rightarrow \mathcal{X}_{I'}$ as bijection that prepends edge e_0 to the paths in $\mathcal{X}_I$, i.e. $g((e_1, \ldots, e_k)) = (e_0, e_1, \ldots, e_k)$. We show that $P \in \mathcal{X}_I$ is efficient (for MCEA) iff $g(P) \in \mathcal{X}_{I'}$ is efficient (for MCF).

Let $P = (e_1, \ldots, e_k)$ be an efficient (s, z)-path in G w.r.t. costs and arrival time. Then $Q := g(P) = (e_0, e_1, \ldots, e_k)$ is an (s', z)-path in G' with $a(Q) = a(P)$ and $c(Q) = c(P)$. Now, assume Q is not efficient w.r.t. costs and duration in G'.

Then there is a path Q' with less costs and at most the duration of Q or with shorter duration and at most the same costs of Q. Path Q' also begins with edge e_0, and G contains a path P' that uses the same edges as Q' with exception of edge e_0. Then, at least one of the following two cases holds.

- Case $c(Q') \leq c(Q)$ and $d(Q') < d(Q)$: Since the costs of e_0 are 0, it follows that $c(P') = c(Q') \leq c(Q) = c(P)$. Because the paths start at time 0 and for each path $d(P) = a(P) - s(P)$, it follows $d(Q') = a(Q') = a(P') < a(P) = a(Q) = d(Q)$.
- Case $c(Q') < c(Q)$ and $d(Q') \leq d(Q)$: Analogously, here we have $c(P') = c(Q') < c(Q) = c(P)$. And $a(P') = a(Q') \leq a(Q) = a(P)$.

Either of these two cases leads to a contradiction to the assumption that P is efficient.

Let $Q = (e_0, e_1, \ldots, e_k)$ and assume it is an efficient (s', z)-path in G' w.r.t. to cost and duration. Then there exists an (s, z)-path $P = (e_1, \ldots, e_k)$ in G such that $g(P) = Q$, $a(Q) = a(P)$ and $c(Q) = c(P)$. Now, assume that P is not efficient. Then there is a path P' with less costs and not later arrival time than P or with earlier arrival time and at most the costs of P. In G' exists the path $Q' = g(P')$ that uses the same edges as P', and additionally the edge e_0 as prefix-path from s' to s. We have the cases $c(P') < c(P)$ and $a(P') \leq a(P)$ or $c(P') \leq c(P)$ and $a(P') < a(P)$. Again, either of them leads to a contradiction to the assumption that Q is efficient. □

Based on this result, we first present an algorithm for MCFENUM in Sect. 4 that we use in a modified version to solve MCEAENUM in Sect. 5. In the rest of this section, we focus on graphs with strictly positive edge costs.

Observation 1. *Let $G = (V, E)$ be a weighted temporal graph. If for all edges $e = (u, v, t, \lambda, c) \in E$ it holds that $c > 0$, then all efficient walks for MCEAENUM and MCFENUM are simple, i.e. are paths.*

Similar to the non-temporal static case, it would be possible to delete the edges of a cycle contained in the non-simple walk. We denote the two special cases for graphs with strictly positive edge costs by $(c{>}0)$-MCFENUM and $(c{>}0)$-MCEAENUM. Our enumeration algorithms use a label setting technique. A label $l = (b, a, c, p, v, r, \Pi)$ at vertex $v \in V$ corresponds to a (s, v)-path and consists of

- the starting time $b = s(P_{s,v})$,
- the arrival time $a = a(P_{s,v})$,
- the cost $c = c(P_{s,v})$,
- the predecessor label p,
- the current vertex v,
- the availability time r of the previous edge and
- a reference to a list of equivalent labels Π.

Moreover, each label is uniquely identifiable by an additional identifier and has a reference to the edge that lead to its creation (denoted by $l.edge$). The proposed algorithms process the edges in order of their availability time. When processing an edge $e = (u, v, t, \lambda, c_e)$, all paths that end at vertex u can be extended by pushing labels over edge e to vertex v. Pushing a label $l = (b, a, c, p, u, r, \Pi)$ over e means that we create a new label $l_{new} = (b, t + \lambda, c + c_e, l, v, t, \cdot)$ at vertex v.

If we would create and store a label for each efficient path, we possibly would need exponential space in the size of the input. The reason is that the number of efficient paths can be exponential in the size of the input.

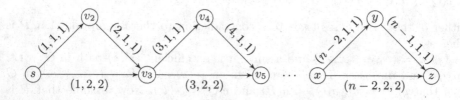

Fig. 2. Example for an exponential number of efficient paths for MCEAENUM and MCFENUM. All (s, v)-paths for $v \in V$ are efficient.

Figure 2 shows an example for MCFENUM and MCEAENUM, which is similar to the one provided by Hansen [5], but adapted to the weighted temporal case. G has m edges and $n = \frac{2m}{3} + 1$ vertices. There are two paths from s to v_3, four paths from s to v_5, eight paths from s to v_7 and so on. All (s, v)-paths for $v \in V$ are efficient. In total, there are $2^{\lfloor \frac{n}{2} \rfloor}$ efficient (s, z)-paths to be enumerated. However, the following lemma shows properties of the problems that help us to achieve polynomial time delay and a linear or polynomial space complexity. Let $\mathcal{Y}_A$ ($\mathcal{Y}_F$) denote the objective space for MCEAENUM (MCFENUM, respectively). Moreover, we define $\mathcal{S}$ to be the number of distinct availability times of edges leaving the source vertex s.

Lemma 2. *For* MCEAENUM*, the number of nondominated points in* $\mathcal{Y}_A$ *is in* $\mathcal{O}(m)$*. For* MCFENUM*, the number of nondominated points in* $\mathcal{Y}_F$ *is in* $\mathcal{O}(m^2)$*.*

Proof. Let $G = (V, E)$ be a weighted temporal graph and $s, z \in V$. First we show that the statement holds for MCEAENUM. The possibilities for different arrival times at vertex z is limited by the number of incoming edges at z. For each path $P_{s,z}$ we have

$$a(P_{s,z}) \in \{\alpha \mid \alpha := t_e + \lambda_e \text{ with } e = (u, z, t_e, \lambda_e, c) \in \delta^-(z)\}.$$

Consequently, there are at most $|\delta^-(z)| \in \mathcal{O}(m)$ different arrival times. For each arrival time a, there can only be one nondominated point $(a, c) \in \mathcal{Y}_A$ that has the minimum costs of c, and which represents exactly all efficient paths with arrival time a and costs c.

Now consider the case for McfEnum. The number of distinct availability times of edges leaving the source vertex S is bounded by $|\delta^+(s)| \in \mathcal{O}(m)$. Because the duration of any (s, z)-path P equals $a(P) - s(P)$, there are at most $S \cdot |\delta^-(z)| \in \mathcal{O}(m^2)$ different durations possible at vertex z. For each duration, there can only be one nondominated point $(d, c) \in \mathcal{Y}_F$ that has minimum costs c.   □

Note that for general bicriteria optimization (path) problems there can be an exponential number of nondominated points in the objective space. Skriver and Andersen [10] give an example for a family of graphs with an exponential number of nondominated points for a bicriteria path problem. The fact that in our case the number of nondominated points in the objective space is polynomially bounded, allows us to achieve polynomial time delay and space complexity for our algorithms. The idea is to consider equivalence classes of labels at each vertex, such that we only have to proceed with a single representative for each class. First, we define the following relations between labels.

Definition 2. *Let $l_1 = (b_1, a_1, c_1, p_1, v, r_1, \Pi_1)$ and $l_2 = (b_2, a_2, c_2, p_2, v, r_2, \Pi_2)$ be two labels at vertex v.*

1. *Label l_1 is equivalent to l_2 iff $c_1 = c_2$ and $b_1 = b_2$.*
2. *Label l_1 predominates l_2 if l_1 and l_2 are not equivalent, $b_1 \geq b_2$, $a_1 \leq a_2$ and $c_1 \leq c_2$ with at least one of the inequalities being strict.*
3. *Finally, label l_1 dominates l_2 if $a_1 - b_1 \leq a_2 - b_2$ and $c_1 \leq c_2$ with at least one of the inequalities being strict.*

For each class of equivalent labels, we have a representative l and a list Π_l that contains all equivalent labels to l. For each vertex $v \in V$, we have a set R_v that contains all representatives. The algorithms consist of two consecutive phases:

- *Phase 1* calculates the set of non-equivalent representatives R_v for every vertex $v \in V$ such that every label in R_v represents a set of equivalent paths from s to v. For each of the nonequivalent labels $l \in R_v$ we store the list Π_l that contains all labels equivalent to l.
- *Phase 2* recombines the sets of equivalent labels in a backtracking fashion, such that we are able to enumerate exactly all efficient (s, z)-paths without holding the paths in memory.

A label in Π_l at vertex v represents all (s, v)-paths that are extension of all paths represented by its predecessor, and $l \in R_v$ is a representative for all labels in Π_l. The representative itself is in Π_l and has minimum arrival time among all labels in Π_l.

We have to take into account that a prefix-path $P_{s,w}$ of an efficient (s, z)-path may not be an efficient (s, w)-path. Figure 3(a) shows an example for a weighted temporal graph with a non-optimal prefix-path. Consider the following paths:

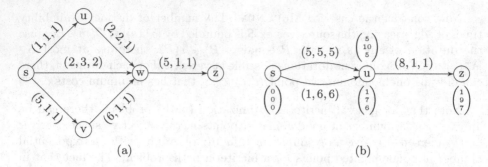

(a)   (b)

Fig. 3. (a) An example for non-efficient prefix-paths. (b) The vertices are annotated with labels that describe the starting time, arrival time and costs of the paths starting at s.

- $P_{s,z} = ((s, w, 2, 3, 2), (w, z, 5, 1, 1))$ with arrival time 6 and duration 4
- $P_{s,w}^1 = ((s, w, 2, 3, 2))$ with arrival time 5 and duration 3
- $P_{s,w}^2 = ((s, u, 1, 2, 1), (u, w, 2, 1, 1))$ with arrival time 3 and duration 3
- $P_{s,w}^3 = ((s, v, 5, 1, 1), (v, w, 6, 1, 1))$ with arrival time 7 and duration 2

All (s, z)-paths have cost 3 and all (s, w)-paths have cost 2. Path $P_{s,z}$ is efficient for MCFENUM and MCEAENUM. For MCEAENUM the prefix-path $P_{s,w}^1$ is not efficient, because $P_{s,w}^2$ arrives earlier. However, for MCFENUM the only efficient (s, w)-path is $P_{s,w}^3$. Consequently, we cannot discard a non-efficient path that possibly is a prefix-path of an efficient path. We use the predomination relation to remove labels that do not represent a prefix-path of an efficient path.

Lemma 3. Let $l_1 = (b_1, a_1, c_1, p_1, v, r_1, \Pi_1)$ and $l_2 = (b_2, a_2, c_2, p_2, v, r_2, \Pi_2)$ be two distinct labels at vertex $v \in V$. If l_1 predominates l_2, then l_2 cannot be a label representing a prefix-path of any efficient path.

Proof. There are two distinct paths $P_{s,v}^1$ and $P_{s,v}^2$ from s to v corresponding to l_1 and l_2. Due to the predomination of l_1 over l_2 it follows that $a_1 = a(P_{s,v}^1) \leq a(P_{s,v}^2) = a_2$, $b_1 = s(P_{s,v}^1) \geq s(P_{s,v}^2) = b_2$ and $c_1 = c(P_{s,v}^1) \leq c(P_{s,v}^2) = c_2$ with at least one of the later two relations being strict, due to the fact that the labels are not equivalent. Let $P_{s,w}$ be a path from s to some $w \in V$ such that $P_{s,v}^2$ is prefix-path of $P_{s,w}$, and assume that $P_{s,w}$ is efficient. Let $P_{s,w}'$ be the path where the prefix-path $P_{s,v}^2$ is replaced by $P_{s,v}^1$. This is possible because $a(P_{s,v}^1) \leq a(P_{s,v}^2)$. Now, since $s(P_{s,v}^1) \geq s(P_{s,v}^2)$ and $c(P_{s,v}^1) \leq c(P_{s,v}^2)$ with at least one of the inequalities being strict, it follows that $a(P_{s,w}') - s(P_{s,w}') \leq a(P_{s,w}) - s(P_{s,w})$ and $c(P_{s,w}') \leq c(P_{s,w})$ also with one of the inequalities being strict. Therefore, $P_{s,w}'$ dominates $P_{s,w}$, a contradiction to the assumption that $P_{s,w}$ is efficient. $\square$

Figure 3(b) shows an example for a non-predominated label of a prefix-path that we cannot discard. Although path $P_1 = ((s, u, 5, 5, 5))$ dominates path $P_2 = ((s, u, 1, 6, 6))$, we cannot discard P_2. The reason is that the arrival time of P_1 is later than the availability time of the only edge from u to z. Therefore, P_2 is the prefix-path of the only efficient path $((s, u, 1, 6, 6), (u, z, 8, 1, 1))$.

4   Min-Cost Fastest Path Enumeration Problem

In this section, we present the algorithm for MCFENUM. Algorithm 1 expects as input a weighted temporal graph with strictly positive edge costs in the edge stream representation, the source vertex $s \in V$ and the target vertex $z \in V$. First, we insert an initial label l_{init} into R_s and $\Pi_{l_{init}}$. Next, the algorithm processes successively the m edges in order of their availability time. For each edge $e = (u, v, t, \lambda, c)$, we first determine the set $S \subseteq R_u$ of labels with distinct starting times, minimal costs and an arrival time less or equal to t at vertex u (line 4). Next, we push each label in S over e. We check for predomination and equivalence with the other labels in R_v and discard all predominated labels. In case the new label is predominated, we discard it and continue with the next label in S. In case that the new label l_{new} is equivalent to a label $l = (a, c, p, v, t_e, \Pi) \in R_v$, we add l_{new} to Π. If l_{new} arrives earlier at v than the arrival time of l, we replace the representative l with l_{new} in R_v. If the new label is not predominated and not equivalent to any label in R_v, we insert l_{new} into R_v and $\Pi_{l_{new}}$. In this case, l_{new} is a new representative and we initialize $\Pi_{l_{new}}$ (which contains only l_{new} at this point). For the following discussion, we define the set of all labels at vertex $v \in V$ as $L_v := \bigcup_{l \in R_v} \Pi_l$.

Lemma 4. *Let $P_{s,v}$ be an efficient path and $P_{s,w}$ a prefix-path of $P_{s,v}$. At the end of Phase 1 of Algorithm 1, R_w contains a label representing $P_{s,w}$.*

Proof. We show that each prefix-path $P_0, P_1, \ldots, P_k$, with $P_0 = P_{s,s}$ and $P_k = P_{s,v}$ is represented by a label at the last vertex of each prefix-path by induction over the length h. Note that all prefix-paths have the same starting time $b = s(P_{s,v})$. For $h = 0$ we have $P_0 = P_{s,s}$ and since s does not have any incoming edges, the initial label $l_{init} = l_0$ representing P_0 is in L_s after Phase 1 finishes. Assume the hypothesis is true for $h = i-1$ and consider the case for $h = i$ and the prefix-path $P_i = P_{s,v_{i+1}} = (e_1, \ldots, e_i = (v_i, v_{i+1}, t_i, \lambda_i, c_i))$, which consists of the prefix-path $P_{i-1} = P_{s,v_i} = (e_1, \ldots, e_{i-1} = (v_{i-1}, v_i, t_{i-1}, \lambda_{i-1}, c_{i-1}))$ and edge $e_i = (v_i, v_{i+1}, t_i, \lambda_i, c_i)$. Due to the induction hypothesis, we conclude that L_{v_i} contains a label $l_{i-1} = (b, a_{i-1}, c_{i-1}, p_{i-1}, v_i, r_{i-1}, \Pi_{i-1})$ that represents P_{i-1}. Because P_{i-1} is a prefix-path of $P_{s,v}$ the representing label l_{i-1} must have the minimum cost in L_{v_i} under all labels with starting time b before edge e_i arrives. Else, it would have been predominated and replaced by a cheaper one (Lemma 3). The set S contains a label that represents l_{i-1}, because the representative of Π_{i-1} has an arrival time less or equal to a_{i-1}. Therefore, the algorithm pushes $l_{new} = (b, t_i + \lambda_i, c_{i-1} + c_i, l_{i-1}, v_{i+1}, t_i, \cdot)$ over edge e_i. If $R_{v_{i+1}}$ is empty the label l_{new} gets inserted into $R_{v_{i+1}}$ and $\Pi_{l_{new}}$. Otherwise we have to check for predomination and equivalence with every label $l' = (b', a', c', p', v_{i+1}, r', \Pi') \in R_{v_{i+1}}$. There are the following cases:

1. l_{new} predominates l': We can remove l' from $R_{v_{i+1}}$ because it will never be part of an efficient path (Lemma 3). The same is true for each label in Π' and therefore we delete Π'. However, we keep l_{new} and continue with the next label.

Algorithm 1. for McFEnum

Input: Graph G in edge stream representation, source $s \in V$ and target $z \in V$
Output: All efficient (s, z)-paths

    Phase 1
1: initialize R_v for each $v \in V$
2: insert label $l_{init} = (0, 0, 0, -, s, -, \Pi_{l_{init}})$ into R_s and $\Pi_{l_{init}}$
3: **for** each edge $e = (u, v, t_e, \lambda_e, c_e)$ **do**
4:     $S \leftarrow \{(b, a, c, p, v, r, \cdot) \in R_u \mid a \leq t_e,\ c \text{ minimal and distinct starting times } b\}$
5:     **for** each $l = (b, a, c, p, v, r, \cdot) \in S$ with $a \leq t_e$ **do**
6:         **if** $u = s$ **then**
7:             $l_{new} \leftarrow (t_e, t_e + \lambda_e, c_e, l, s, t_e, \cdot)$
8:         **else**
9:             $l_{new} \leftarrow (b, t_e + \lambda_e, c + c_e, l, u, t_e, \cdot)$
10:         **for** each $l' = (b', a', c', p', v, t', \Pi') \in R_v$ **do**
11:             **if** l_{new} predominates l' **then**
12:                 remove l' from R_v and delete Π'
13:             **else if** l' is equivalent to l_{new} **then**
14:                 insert l_{new} into Π'
15:                 set reference $\Pi \leftarrow \Pi'$
16:                 **if** $t_e + \lambda_e < a'$ **then**
17:                     replace l' in R_v by l_{new}
18:                 goto 5
19:             **else if** l' predominates l_{new} **then**
20:                 delete l_{new}
21:                 goto 5
22:         insert l_{new} into R_v and initialize $\Pi_{l_{new}}$ with l_{new}
    Phase 2
23: mark nondominated labels in R_z
24: **for** each marked label $l' = (b, a, c, r, z, p, \Pi) \in R_z$ **do**
25:     **for** each label $l \in \Pi$ with minimal arrival time **do**
26:         initialize empty path P
27:         call OutputPaths(l, P);

    Procedure for outputting paths
28: **procedure** OutputPaths(label $l = (b, a, c, p, cur, r, \Pi)$, path P)
29:     prepend edge $l.edge$ to P
30:     **if** l has predecessor $p = (b_p, a_p, c_p, p_p, v_p, r_p, \Pi_p)$ **then**
31:         **for** each label $l' = (b_{l'}, a_{l'}, c_{l'}, p_{l'}, v_{l'}, r_{l'}, \Pi_p)$ in Π_p **do**
32:             **if** $a_{l'} \leq r$ **then**
33:                 call OutputPaths(l', P, *visited*)
34:         **return**
35:     output path P

---

2. l_{new} and l' are equivalent: We add l_{new} to Π'. In this case we represent the path P_i by the representative of Π'. Consequently, the path is represented by a label in $L_{v_{i+1}}$.

If neither of these two cases apply for any label in $R_{v_{i+1}}$, we add l_{new} to $R_{v_{i+1}}$ and to $\Pi_{l_{new}}$. The case that a label l is not equivalent to l_{new} and predominates l_{new} cannot be for the following reason. If l predominates l_{new}, there is a path P' from s to v_{i+1} with less costs or later starting time (because to l and l_{new} are not equivalent) and a not later arrival time. Replacing the prefix-path P_i with P' in the path $P_{s,v}$ would lead to a (s,v)-path with less costs and/or shorter duration. This contradicts our assumption that $P_{s,v}$ is efficient. Therefore, after Phase 1 finished, the label l_{new} representing the prefix-path P_i is in L_{v_i}. It follows that if $P_{s,v} = (e_1, \ldots, e_k)$ is an efficient path, then after Phase 1 the set R_v contains a label representing $P_{s,v}$ (possibly, such that a label in R_v represents a list of equivalent labels, that contains the label representing $P_{s,v}$). $\qquad\square$

After all edges have been processed, the algorithm continues with Phase 2. First, the algorithm marks all nondominated labels in R_z. For each marked label l the algorithm iterates over the list of equivalent labels Π_l and calls the output procedure for each label in Π_l. We show that all and only efficient paths are enumerated.

Theorem 1. *Let $G = (V, E)$ be a weighted temporal graph with strictly positive edge costs and $s, z \in V$ an instance of MCFENUM. Algorithm 1 outputs exactly the set of all efficient (s, z)-paths.*

Proof. Lemma 4 implies that for each efficient path $P_{s,z}$ there is a corresponding representative label in R_z after Phase 1 is finished. Note that there might also be labels in L_z that do not represent efficient paths. First, we mark all non-dominated labels in R_z. For every marked representative $l' = (b, a, c, p, z, r, \Pi_{l'})$ in R_z we proceed by calling the output procedure for all labels $l \in \Pi_{l'}$ with minimal arrival time. Each such label l represents at least one efficient (s, z)-path and we call the output procedure with l and the empty path P. Let path $Q = (e_1 \ldots, e_k)$ be an efficient (s, z)-paths represented by l. We show that the output procedure successively constructs the suffix-paths $P_i = (e_{k-i+1}, \ldots, e_k)$ of Q for $i \in \{1, \ldots, k\}$ and finally outputs $Q = P_k = (e_1, \ldots, e_k)$.

We use induction over the length $i \geq 1$ of the suffix-path. For $i = 1$ the statement is true. $P_1 = (e_k)$ is constructed by the first instruction which prepends the last edge of Q to the initially empty path P. Now, assume the statement holds for $i = j - 1 < k$, i.e. the suffix-path P_{j-1} of Q with length $j - 1$ has been constructed, by calling the output procedure with P_{j-2} and label $l_{j-1} = (b, a, c, p, v_{k-j+2}, r, \Pi)$. The suffix-path $P_j = (e_{k-j+1}, \ldots, e_k)$ equals P_{j-1} with the additional edge $e_{k-j+1} = (v_{k-j+1}, v_{k-j+2}, t, \lambda, c)$ with v_{k-j+2} being the first vertex of P_{j-1}. The predecessor of l_{j-1} is label $p = (b, a_p, c_p, p_p, v_{k-j+1}, r_p, \Pi')$. We recursively call the output procedure for each label in the list of equivalent labels Π' and verify that the arrival time of each of these labels is less or equal to t. Due to Lemma 4, we particularly call the output procedure for the label that represents the beginning of the suffix-path P_j and which has an arrival time less than t. Consequently, there is a call of the output procedure that constructs P_j. If P_j does not have a predecessor, we arrived at vertex s and the algorithm outputs the found path $Q = P_k$.

We still have to show that only efficient paths are enumerated. In order to enumerate a non-efficient (s, z)-path Q', there has to be a label l_q in L_z for which the output procedure is called and which represents Q'. For Q' to be non-efficient there has to be at least one label l_d in L_z that dominates l_q. In line 23 the algorithm marks all nondominated labels in R_z. This implies that l_d and l_q have the same cost and starting times and that they are in the same list, let this list be Π_x for some label $x \in R_z$. Because l_q is dominated by l_d the arrival time of l_d is strictly earlier than the arrival time of l_q. However, we call the output procedure only for the labels in Π_x with the minimal arrival time. Consequently, it is impossible that the non-efficient path Q' is enumerated.

Finally, because all edge costs are strictly positive and due to Observation 1 only paths are enumerated.   □

Example: Figure 4 shows an example for Algorithm 1 at the end of Phase 1. The edges are numbered according to their position in the sequence of the edge stream. The representative labels at the vertices only show the starting time, arrival time and cost. The lists Π of equivalent labels are not shown. All of them contain only the representative, with exception of Π_l represented by label l in R_w. The list Π_l contains label $l = (3, 7, 4)^T$ representing path $((s, w, 3, 4, 4))$ and the equivalent label $(3, 8, 4)^T$ representing path $((s, u, 3, 3, 3), (u, w, 6, 1, 1))$. There are three efficient paths. Starting the output procedure from vertex z with the label $(7, 10, 6)^T$ yields path (e_5, e_8), and starting with label $(3, 9, 5)^T$ yields the two paths (e_1, e_4, e_7) and (e_2, e_7). Notice that label $(7, 9, 2)^T$ in R_w which dominates label $(3, 7, 4)^T$, is not part of an efficient (s, z)-path, due to its late arrival time.

Lemma 5. *Phase 1 of Algorithm 1 has a time complexity of $\mathcal{O}(S \cdot m^2)$.*

Proof. The outer loop iterates over m edges. For each edge $e = (u, v, t_e, \lambda_e, c_e)$ we have to find the set $S \subseteq R_u$ consisting of all labels with minimal cost, distinct starting times and arrival time less or equal to t_e (see line 4). This can be done in $\mathcal{O}(m)$ time. For each label in S we have to check for predominance or equivalence with each label in R_v in $\mathcal{O}(S \cdot m)$ total time. Since we have $|S| \leq S$, we get a total time of $\mathcal{O}(S \cdot m^2)$.   □

The following lemma shows that the number of labels is polynomially bounded in the size of the input.

Lemma 6. *The total number of labels generated and hold at the vertices in Algorithm 1 is less than or equal than $S \cdot m + 1$.*

Proof. We need one initial label l_{init}. For each incoming edge $e = (u, v, t, \lambda, c)$ in the edge stream we generate at most $|S| \leq S$ new labels which we push over e to vertex v. Therefore, we generate at most $S \cdot m + 1$ labels in total.   □

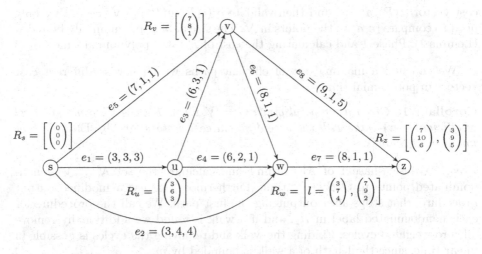

Fig. 4. Example for Algorithm 1. Each vertex is annotated with the representatives after Phase 1 finished.

Theorem 2. $(c>0)$-MCFENUM $\in$ *PSDelayP*.

Proof. Phase 1 takes only polynomial time in size of the input, i.e. number of edges (Lemma 5). In Phase 2 of Algorithm 1, we first find and mark all nondominated labels in R_z in $\mathcal{O}(m^2)$ time. For each nondominated label, we call the output procedure which visits at most $\mathcal{O}(m^2)$ labels and outputs at least one path. It follows that the time between outputting two consecutively processed paths is also bounded by $\mathcal{O}(m^2)$. Therefore, $(c>0)$-MCFENUM is in **DelayP**. The space complexity is dominated by the number of labels we have to manage throughout the algorithm. Due to Lemma 6, the number of labels is in $\mathcal{O}(m^2)$. Consequently, $(c>0)$-MCFENUM is in **PSDelayP**. □

The following problems are easy to decide, even if we allow zero weighted edges.

Theorem 3. *Given a weighted temporal graph* $G = (V, E)$, $s, z \in V$ *and*

1. *an (s, z)-path P, deciding if P is efficient for* MCFENUM, *or*
2. *$c \in \mathbb{R}_{\geq 0}$ and $d \in \mathbb{N}$, deciding if there exists a (s, z)-path P with $d(P) \leq d$ and $c(P) \leq c$ is possible in polynomial time.*

Proof. We use Phase 1 of Algorithm 1 and calculate the set $\mathcal{N} \subseteq \mathcal{Y}$ of nondominated points. Due to the possibility of edges with cost 0, there may be non-simple paths, i.e. walks, that have zero-weighted cycles. Nonetheless, Phase 1 terminates after processing the m edges. If there exists an efficient (s, z)-walk W with $(c(W), a(W))$, then there also exists a simple and efficient (s, z)-path Q with the same cost vector. Q is the same path as W but without the zero-weighted cycles. In order to decide if the given path P is efficient, we first calculate the cost vector $(c(P), a(P))$, and then validate if $(c(P), a(P)) \in \mathcal{N}$. For 2., we only need to compare (c, d) to the points in $\mathcal{N}$. The size of $\mathcal{N}$ is polynomially bounded (Lemma 2). Phase 1 and calculating the cost of P takes polynomial time.   □

We can find a maximal set of efficient paths with pairwise different cost vectors in polynomial time.

Corollary 1. *Given a temporal graph $G = (V, E)$ and $s, z \in V$, a maximal set of efficient (s, z)-paths with pairwise different cost vectors for MCFENUM can be found in $\mathcal{O}(\mathcal{S} \cdot m^2)$.*

Proof. We use Phase 1 of Algorithm 1 and calculate the set $\mathcal{N} \subseteq \mathcal{Y}$ of non-dominated points in $\mathcal{O}(\mathcal{S} \cdot m^2)$ time. Furthermore, we use a modified output procedure, that stops after outputting the first path. We call the procedure for each nondominated label in R_z, and if a walk is found we additionally remove all zero-weighted cycles. Finding the walk and removing the cycles is possible in linear time, since the length of a walk is bounded by m.   □

5   Min-Cost Earliest Arrival Path Enumeration Problem

Based on the reduction given in the beginning of Sect. 3, we modify Algorithm 1 to solve $(c>0)$-MCEAENUM. The modified algorithm only needs a linear amount of space and less time for Phase 1. Let $(G = (V, E), s, z)$ be the instance for $(c>0)$-MCEAENUM and $(G' = (V', E'), s', z)$ be the transformed instance in which all paths start at time 0 at the new source s'. Although edge $(s, s', 0, 0, 0)$ has costs 0, because s' has no incoming edges any efficient walk in the transformed instance is simple, i.e. a path. With all paths starting at 0, there are the following consequences for the earlier defined relations between labels. First, consider the equivalence from Definition 2 and let $l_1 = (0, a_1, c_1, p_1, v, r_1, \Pi_1)$ and $l_2 = (0, a_2, c_2, p_2, v, r_2, \Pi_2)$ be two labels at vertex v. Because the starting time of both labels is 0, the labels are equivalent if $c_1 = c_2$. It follows, that label l_1 predominates l_2 if $a_1 \leq a_2$ and $c_1 < c_2$, hence there is no distinction between domination and predomination.

Algorithm 2 shows a modified version of Algorithm 1, that sets all starting times to 0. In line 4 we only need to find a single label with the minimum costs, instead of the set S. At each vertex v we only have one representative l in R_v with minimal costs (w.r.t. the other labels in R_v), due to the equivalence of labels that have the same costs. In Phase 2 we do not need to explicitly find the nondominated labels in R_z. Because each label l in R_z has a unique cost value, we consider each represented class Π_l and call the output procedure with the labels that have the minimum arrival time in Π_l.

Theorem 4. *Algorithm 2 outputs exactly all efficient (s, z)-paths w.r.t. arrival time and costs.*

Proof. Lemma 4 implies that for each efficient path $P_{s,z}$ there is a corresponding representative label in R_z after Phase 1 is finished. For every representative

Algorithm 2. for MCEAENUM

Input: Graph G in edge stream representation, source $s \in V$ and target $z \in V$
Output: All efficient (s, z)-paths

    Phase 1
1: initialize R_v for each $v \in V$
2: insert label $l_{init} = (0, 0, 0, 0, -, s, \Pi_{l_{init}})$ into R_s and $\Pi_{l_{init}}$
3: **for** each edge $e = (u, v, t_e, \lambda_e, c_e)$ **do**
4:     $l \leftarrow (0, a, c, p, u, \cdot, \cdot) \in R_u$ with $a \leq t_e$ and c minimal
5:     $l_{new} \leftarrow (0, t_e + \lambda_e, c + c_e, l, v, t_e, \Pi)$
6:     **for** each $l' = (0, a', c', p', v, r', \Pi') \in R_v$ **do**
7:         **if** l_{new} dominates l' **then**
8:             remove l' from R_v and delete Π'
9:         **else if** l' dominates l_{new} **then**
10:            delete l_{new}
11:            **goto** line 4
12:         **else if** l' is equivalent to l_{new} **then**
13:            set reference $\Pi \leftarrow \Pi'$
14:            insert p_{new} into Π'
15:            **if** $t_e + \lambda_e < a'$ **then**
16:                replace l' in R_v by l_{new}
17:            **goto** line 4
18:     $\Pi \leftarrow \Pi_{l_{new}}$
19:     insert l_{new} into R_v and u into $\Pi_{l_{new}}$
    Phase 2
20: **for** each label $l' = (0, a, c, p, z, r, \Pi_{l'}) \in R_z$ **do**
21:     **for** each label $l \in \Pi_{l'}$ with minimal arrival time **do**
22:         initialize empty path P
23:         call OUTPUTPATHS(l, P)

$l' = (0, a, c, p, z, r, \Pi)$ in R_z, it holds by construction that all labels in $\Pi_{l'}$ have the same costs. Therefore, we only need to consider the nondominated labels in $\Pi_{l'}$ with minimal arrival time $a_{min} := \min\{a \mid l = (0, a, c, p, w, r, \Pi) \in \Pi_{l'}\}$. Hence, for each $l' \in R_z$ we call the output procedure for every label in $l \in \Pi_{l'}$ if l has minimal arrival time a_{min}. $\square$

Algorithm 2 uses a linear number of labels.

Lemma 7. *The total number of labels generated and hold at the vertices in Algorithm 2 is at most $m + 1$.*

Proof. We need one initial label l_{init} at the source vertex s. For each incoming edge $e = (u, v, t, \lambda, c)$ in the edge stream, in line 4 we choose the label l with minimal costs and arrival time at most t. We only push l and generate at most one new label l_{new} at vertex v. Therefore, we generate at most $m + 1$ labels in total. $\square$

Lemma 8. *Phase 1 of Algorithm 2 has a time complexity of $\mathcal{O}(m^2)$.*

Proof. The outer loop iterates over m edges. In each iteration we have to find the representative label $l \in R_u$ with minimum costs and arrival time $a \leq t_e$. This is possible in constant time, since we always keep the label with the earliest arrival time of each equivalence class as representative in R_u. Next we have to check the domination and equivalence between l_{new} and each label $l' \in R_v$. Each of the cases takes constant time, and there are $\mathcal{O}(m)$ labels in R_v. Altogether, a time complexity of $\mathcal{O}(m^2)$ follows.   □

Algorithm 2 lists all efficient paths in polynomial delay and uses only linear space.

Theorem 5. $(c>0)$-MCEAENUM $\in$ *PSDelayP*.   □

Using Algorithm 2, also the results of Theorem 3 and Corollary 1 can be adapted for the earliest arrival case.

Theorem 6. *Given a weighted temporal graph $G = (V, E)$, $s, z \in V$ and*

1. *an (s, z)-path P, deciding if P is efficient for MCEAENUM, or*
2. *$c \in \mathbb{R}_{\geq 0}$ and $a \in \mathbb{N}$, deciding if there exists a (s, z)-path P with $a(P) \leq a$ and $c(P) \leq c$ is possible in polynomial time.*   □

Corollary 2. *Given a temporal graph $G = (V, E)$ and $s, z \in V$, a maximal set of efficient (s, z)-paths with pairwise different cost vectors for MCEAENUM can be found in time $\mathcal{O}(m^2)$.*   □

6   Conclusion

We discussed the bicriteria optimization problems MIN-COST EARLIEST ARRIVAL PATHS ENUMERATION PROBLEM (MCEAENUM) and MIN-COST FASTEST PATHS ENUMERATION (MCFENUM). We have shown that enumerating the sets of all efficient paths with low costs and early arrival time or short duration is possible in polynomial time delay and linear or polynomial space if the input graph has strictly positive edge costs. In case of nonnegative edge costs, it is possible to determine a maximal set of efficient paths with pairwise different cost vectors in $\mathcal{O}(m^2)$ time or $\mathcal{O}(\mathcal{S} \cdot m^2)$ time, respectively, where $\mathcal{S}$ is the number of distinct availability times of edges leaving the source vertex s. We can find an efficient path for each nondominated point in polynomial time.

For the cases of zero-weighted or even negative edge weights, we cannot guarantee polynomial time delay for our algorithms to solve MCFENUM or MCEAENUM. However, the proposed algorithms can be used to determine all efficient (s, z)-walks in polynomial time delay. Because of each edge in a temporal graph can only be used for departure at a certain time, the number of different walks is finite and the algorithms terminate. So far, we are not aware of a way to ensure that only simple paths are enumerated without loosing the property that the delay between the output of two paths stays polynomially bounded.

References

1. Casteigts, A., Flocchini, P., Quattrociocchi, W., Santoro, N.: Time-varying graphs and dynamic networks. Int. J. Parallel Emergent Distrib. Syst. **27**(5), 387–408 (2012)
2. Ehrgott, M., Gandibleux, X.: A survey and annotated bibliography of multiobjective combinatorial optimization. OR-Spektrum **22**(4), 425–460 (2000)
3. Garey, M.R., Johnson, D.S.: Computers and Intractability: A Guide to the Theory of NP-Completeness. W. H. Freeman & Co., New York (1979)
4. Hamacher, H.W., Ruzika, S., Tjandra, S.A.: Algorithms for time-dependent bicriteria shortest path problems. Discrete Optim. **3**(3), 238–254 (2006)
5. Hansen, P.: Bicriterion path problems. In: Fandel, G., Gal, T. (eds.) Multiple Criteria Decision Making Theory and Application, pp. 109–127. Springer, Berlin (1980). https://doi.org/10.1007/978-3-642-48782-8_9
6. Holme, P., Saramäki, J.: Temporal networks. Phys. Rep. **519**(3), 97–125 (2012)
7. Johnson, D.S., Yannakakis, M., Papadimitriou, C.H.: On generating all maximal independent sets. Inf. Process. Lett. **27**(3), 119–123 (1988)
8. Kostakos, V.: Temporal graphs. Phys. A: Stat. Mech. Appl. **388**(6), 1007–1023 (2009)
9. Martins, E.Q.V.: On a multicriteria shortest path problem. Eur. J. Oper. Res. **16**(2), 236–245 (1984)
10. Skriver, A., Andersen, K.: A label correcting approach for solving bicriterion shortest-path problems. Comput. Oper. Res. **27**(6), 507–524 (2000)
11. Wu, H., Cheng, J., Huang, S., Ke, Y., Lu, Y., Xu, Y.: Path problems in temporal graphs. Proc. VLDB Endow. **7**(9), 721–732 (2014)
12. Xuan, B.B., Ferreira, A., Jarry, A.: Computing shortest, fastest, and foremost journeys in dynamic networks. Int. J. Found. Comput. Sci. **14**(02), 267–285 (2003)

An Output-Sensitive Algorithm
for the Minimization of 2-Dimensional
String Covers

Alexandru Popa[1,2(⊠)] and Andrei Tanasescu[3]

[1] University of Bucharest, Bucharest, Romania
alexandru.popa@fmi.unibuc.ro
[2] National Institute of Research and Development in Informatics,
Bucharest, Romania
[3] Politehnica University of Bucharest, Bucharest, Romania
andrei.tanasescu@mail.ru

Abstract. String covers are a powerful tool for analyzing the quasi-periodicity of 1-dimensional data and find applications in automata theory, computational biology, coding and the analysis of transactional data. A *cover* of a string T is a string C for which every letter of T lies within some occurrence of C. String covers have been generalized in many ways, leading to *k-covers*, *λ-covers*, *approximate covers* and were studied in different contexts such as *indeterminate strings*.

In this paper we generalize string covers to the context of 2-dimensional data, such as images. We show how they can be used for the extraction of textures from images and identification of primitive cells in lattice data. This has interesting applications in image compression, procedural terrain generation and crystallography.

1 Motivation

Redundancy is an ubiquitous phenomenon in engineering and computer science [18,19]. Periodicity is the most common and useful form of redundancy. Periodicity is a key phenomenon when analyzing physical data such as an analogue signal. Natural data is very redundant or repetitive and exhibits some patterns or regularities [11,23,24] which we may assert to be the intended information [21] within the data. Periodicity itself has been thoroughly studied in various fields such as signal processing [22], bioinformatics [8], dynamical systems [14] and control theory [6], each bringing its own insights.

However, natural data is imperfect. It is highly unlikely that natural data can ever be periodic. In fact, the data is *almost* or *quasi*-periodic [4]. This has been firstly studied over strings, the most general representation of digital data [17].

For example, assume that we want to send the word *aba* over a noisy channel as a digital signal where letters are modulated using *amplitude shift keying* [17]. Since the simple transmission is unlikely to yield the result due to the imperfect transmission channel, we add redundancy and thus send the word *aba* multiple

© Springer Nature Switzerland AG 2019
T. V. Gopal and J. Watada (Eds.): TAMC 2019, LNCS 11436, pp. 536–549, 2019.
https://doi.org/10.1007/978-3-030-14812-6_33

times. However, when errors occur, the received signal only partially retains its periodicity.

2 Our Results

In this paper we study the generalization of the String Cover operator on finite-dimensional images. First, throughout this paper, given two integers a and b, $a \leq b$, we define $\overline{a,b} = \{a, a+1, \ldots, b\}$. Then a 2-dimensional string (or an image) is function $I : \overline{0, H-1} \times \overline{0, W-1} \rightarrow \Sigma$, where Σ is the alphabet. Let $Mat_{H,W}$ be the set of all matrices with H rows and W columns. For a matrix M we define M_j^i to be the element from row i and column j.

Definition 1 (2-dimensional string cover). *A* cover *of a 2D image T is a 2D image C for which every element of T lies within some occurrence of C.*

In Sect. 3, we find two alternative ways of formalizing the 2D cover problem, by using masks and prove their equivalence. We then turn our attention towards the decision problem:

Problem 1 (Image Cover Decision). Given two images T and C, does one cover the other?

We give an $\mathcal{O}(WH)$ algorithm based upon Bird's 1977 [7] 2-dimensional matching algorithm. Then, using this algorithm we study the minimization problem (Sect. 4).

Problem 2 (Weak Minimal Image Cover). Given an image $T \in Mat_{H,W}(\Sigma)$ and an evaluation function $eval : \overline{1, h} \times \overline{1, w} \rightarrow \mathbb{R}$, where $h \leq H$ and $w \leq W$, which induces an order onto the covers, which is the cover $C \in Mat_{h, w}(\Sigma)$ of T minimal with respect to $eval(h, w)$?

We give an $\mathcal{O}(W^2 H^2) \Theta(eval)$ algorithm. Since the minimization problem is actually $\Omega(WH) \Theta(eval)$, we aim for a better algorithm. Using sorting of the input candidates according to $eval$ we obtain $\mathcal{O}(nWH) \Theta(eval)$ (the bound does not contain the time necessary for sorting), where the n-th entry in the vector sorted by $eval$ determines a cover of T. Note that to assume that the candidates are sorted is not in general realistic, so a more honest complexity bound is $\mathcal{O}(WH(n + \log(WH) \Theta(eval)))$. However, there is a very important optimization criterion where the sorting is very cheap, namely the *size* of an image.

Problem 3 (Strong Minimal Image Cover). Given an image $T \in Mat_{H,W}(\Sigma)$ which is the cover $C \in Mat_{h, w}(\Sigma)$ of T minimal with respect to its area (that is, wh), ℓ_1 norm (that is, $w + h$) and ℓ_∞ norm (that is, $\max(w, h)$)?

For this problem we augment the general minimization algorithm with a preprocessing routine, based on the optimal 1-dimensional Minimal String Cover algorithm [5], which reduces the number of candidate pairs that we have to check from $\Theta(WH)$ to $\mathcal{O}(1)$ on the average case, reducing the complexity to $\Theta(WH)$ on the average case and, particularly, $\mathcal{O}(W)$ in the worst-case for $H = 1$. We argue that the use of this routine never hinders performance and offers the same boost for the general case of an unknown *eval* function.

We conclude the article with a few very interesting applications of other generalizations of the Minimal String Cover Problem (Sects. 5 and 6) such as k-covers [12] and the Approximate String Cover Problem introduced by Amir et al. [2,3] to lattice unit-cell recognition from generic images, detection of the unit cells of some quasicrystals [25], extraction of the elementary set of tiles in a Wang Tiling, recognizing the minimal (quasi)periodic Wang Tile pattern in an image and the minimal modification required of an image for the existence of a non-trivial minimal (quasi)periodic Wang Tile pattern.

3   Image Covers

The simplest class of images is that of binary images, i.e. $\Sigma \cong \{0, 1\}$. Binary images can be thought of as sets over $\mathbb{Z}^2$, as follows: the set contains the position (i.e., row and column) of the elements of the binary image that have value 1.

Example 1. The set $\{(1, 2), (2, 2), (3, 3)\}$ corresponds to the image

$$\begin{bmatrix} 0 & 1 & 0 \\ 0 & 1 & 0 \\ 0 & 0 & 1 \end{bmatrix}.$$

Given a set S and an element x, the characteristic function of S, denoted by $\chi_S(x)$ has value 1 if $x \in S$, and 0 otherwise.

Definition 2. *A mask of an image T with respect to an image C is a binary image M which marks the first position of some occurrences of C in T.*

   Formally if $T \in \mathrm{Mat}_{H,W}(\Sigma)$ and $C \in \mathrm{Mat}_{h,w}(\Sigma)$ then $M \in \mathrm{Mat}_{H,W}(\{0, 1\})$ is a mask of T with respect to C if

$$\forall i \in \overline{1, H}, j \in \overline{1, W}, M_j^i = 1 \Rightarrow T_{j+x-1}^{i+y-1} = C_x^y , y \in \overline{1, h}, x \in \overline{1, w}.$$

By the correspondence between binary images and sets, there exists a maximal mask with respect to cardinality and it identifies all occurrences of an image in another.

Definition 3. *The maximal mask of an image T with respect to an image C is a binary image M^* which marks the first position of all occurrences of C in T.*

   Formally if $T \in \mathrm{Mat}_{H,W}(\Sigma)$ and $C \in \mathrm{Mat}_{h,w}(\Sigma)$ then $M^* \in \mathrm{Mat}_{H,W}(\{0, 1\})$ is the maximal mask of T with respect to C if

$$\forall i \in \overline{1, H}, j \in \overline{1, W}, M^{*i}_j = 1 \iff T_{j+x-1}^{i+y-1} = C_x^y , y \in \overline{1, h}, x \in \overline{1, w}.$$

Extrapolating from the definition of string covers, we can informally define a cover of an image. A *cover* of an image T is an image C for which every element of T lies within some occurrence of C. We can formalize this definition using masks. We introduce two equivalent definition candidates.

Definition 4 (Weak Image Covers). *If $T \in \mathrm{Mat}_{H,W}(\Sigma)$ and $C \in \mathrm{Mat}_{h,w}(\Sigma)$, then C covers T if there exists some mask M of T with respect to C such that:*

$$\forall Y \in \overline{1,H},\, X \in \overline{1,W}\; \exists i \in \overline{Y-h+1,Y},\, j \in \overline{X-w+1,X}\; M_j^i = 1.$$

Equivalently, we may define Image Covers with respect to the maximal mask:

Definition 5 (Strong Image Covers). *If $T \in \mathrm{Mat}_{H,W}(\Sigma)$ and $C \in \mathrm{Mat}_{h,w}(\Sigma)$, then C covers T if the maximal mask M^* of T with respect to C is such that:*

$$\forall Y \in \overline{1,H},\, X \in \overline{1,W}\; \exists i \in \overline{Y-h+1,Y},\, j \in \overline{X-w+1,X}\; M^{*\,i}_j = 1$$

By these definitions a cover $C \in \mathrm{Mat}_{h,w}(\Sigma)$ of an image $T \in \mathrm{Mat}_{H,W}(\Sigma)$ can be identified with the (h, w) pair.

The weak definition is a more natural extension of the definition of String Covers, while the strong definition provides us with a more clear understanding of the combinatorial properties of Image Covers. For example, the strong definition suggests that Image Covers are susceptible to dynamic programming, which we later use to obtain the minimal cover.

Theorem 1. *The weak and strong definitions are equivalent.*

Proof. Consider the set $\mathcal{S} = \overline{1,H} \times \overline{1,W}$. There exists a bijection between its power set, $\mathcal{P}(\mathcal{S})$, and the W-long, H-tall binary images $\mathrm{Mat}_{H,W}(\{0,1\})$ as explained at the beginning of the section. Formally, the bijection f is defined as

$$\mathcal{P}(\mathcal{S}) \ni S \leftrightarrow f(S) \in \mathrm{Mat}_{H,W}(\{0,1\}):\; f(S)_j^i = \chi_S((i,j))\,\forall i \in \overline{1,H},\, j \in \overline{1,W}.$$

However, the image of the Boolean algebra $(\mathcal{P}(\mathcal{S}),\, \cup, \cap, \bar{\cdot}, \emptyset, \mathcal{S})$ is thus by f onto $\mathrm{Mat}_{H,W}(\{0,1\})$. The new structure can be verified to be

$$(\mathrm{Mat}_{H,W}(\{0,1\}),\, \max, \min, \mathbf{M} \to 1 - \mathbf{M}, \mathbf{0}, \mathbf{1})$$

Thus the image of the inclusion order $\subseteq$ is the order $\leq$ and so, if there exists a mask M such that

$$\forall Y \in \overline{1,H},\, X \in \overline{1,W}\; \exists i \in \overline{Y-h+1,Y},\, j \in \overline{X-w+1,X}\; M_j^i = 1$$

then since $M \leq M^*$ we also have

$$\forall Y \in \overline{1,H},\, X \in \overline{1,W}\; \exists i \in \overline{Y-h+1,Y},\, j \in \overline{X-w+1,X}\; M^{*\,i}_j = 1$$

and vice versa: if M^* satisfies the later, then there exists at least one such mask M (precisely M^*) which satisfies the former. Thus, the two definitions are indeed equivalent. □

While from a formal standpoint the two definitions are equivalent, from a computational standpoint it is more convenient for us to work with the strong definition, since we do not have to consider all masks.

Lemma 1. *Given two images* $T \in \text{Mat}_{H,W}(\Sigma)$ *and* $C \in \text{Mat}_{h,w}(\Sigma)$ *the construction of the maximal mask of* T *with respect to* C *takes* $\Theta(WH)$ *time.*

Proof. Since the size of the output is WH we have the lower bound $\Omega(WH)$. We effectively only have to prove the upper bound of $\mathcal{O}(WH)$.

We begin by studying the case $H = 1$. In this case the maximal mask of T with respect to C consists of all occurrences of C in T. This can be found in linear time, for example using the Knuth-Morris-Pratt algorithm (KMP [15]), with a runtime of $\mathcal{O}(W + w)$, which is $\mathcal{O}(WH)$ since $H = h = 1$ and $w < W$.

For the case $H \neq 1$, we look for a two dimensional generalization of the Knuth-Morris-Pratt algorithm. One such generalization is Bird's algorithm [7] which uses KMP and a generalization of it due to Aho and Corasick [1] to find the rows and then columns where the pattern occurs.

The output of Bird's algorithm is the list of occurrences of C in T, i.e. the pairs (i, j) such that $M^{*i}_j = 1$. Consequently, we can recover M^* by taking $M^{*i}_j = stage(i + h - 1, j + w - 1)$. This yields the maximal mask in $\mathcal{O}(WH + wh) = \mathcal{O}(WH)$ time. $\qquad\square$

Theorem 2 (Image Cover Decision). *Given two images* $T \in \text{Mat}_{H,W}(\Sigma)$ *and* $C \in \text{Mat}_{h,w}(\Sigma)$ *checking if* C *is a cover of* T *takes* $\Theta(WH)$ *time (Algorithm 1).*

Proof. We can instantly disqualify images C having $h > H$ or $w > W$. Otherwise, since we must at least read T, the decision problem is at least $\Omega(HW)$. Thus, we prove only the upper bound, $\mathcal{O}(WH)$.

By Lemma 1 we compute M^* in $\mathcal{O}(WH)$ time. We now check if M^* "tiles up to" T, as per Definition 5. Thus we check that every (x, y) of T belongs to some occurrence of C, whose north-west corner is located at some point in $D(x, y)$, where

$$D(x, y) = \{(x - w + 1, y - h + 1) \le (x', y') \le (x, y) \,|\, M^{*y'}_{x'} = 1\}.$$

At this point we could simply walk through M^* and check that every location is indeed covered. However since there are up to $\mathcal{O}(WH)$ occurrences of C in T the naive approach takes $\mathcal{O}(W^2H^2)$ time.

For the rest of the proof we show that we can compute whether there exists some (x, y) for which $D(x, y) = \emptyset$ in $\mathcal{O}(WH)$. We call points for which $D(x, y) \neq 0$ admissible and points for which $D(x, y) = 0$ inadmissible. We say that the points $D(x, y)$ "support" the hypothesis that (x, y) is admissible.

Let $\le_{lex}$ be the lexicographical order and the function $N(x, y)$ be the closest (north-west corner of an) occurrence of C form (x, y), i.e.

$$N(x, y) = \arg\min_{\le_{lex}}\{(x - x', y - y') \,|\, (x', y') \in D(x, y)\},$$

for which, by definition, $N(x, y) = (\infty, \infty)$ if and only if $D(x, y) = \emptyset$.

Note that if the minimal support for the western neighbor of a point, $N(x-1, y)$, does not support it, then (x, y) is the only point that can support itself but not its northern neighbor, $(x, y-1)$, i.e.

$$N(x-1, y) \notin D(x, y) \Rightarrow M^{*y'}_{x'} = 0 \; \forall x' \in \overline{x-w+1, x-1}, \, y' \in \overline{y-h+1, y} \Rightarrow$$
$$\Rightarrow D(x, y) \subseteq D(x, y-1) \cup \{(x, y)\}.$$

Similarly, if the minimal support for the northern neighbor of a point, $N(x, y-1)$, does not support it, then (x, y) is the only point that can support itself but not its western neighbor, $(x-1, y)$, i.e.

$$N(x, y-1) \notin D(x, y) \Rightarrow M^{*y'}_{x} = 0 \; \forall y' \in \overline{y-h+1, y-1} \Rightarrow$$
$$\Rightarrow D(x, y) \subseteq D(x-1, y) \cup \{(x, y)\}.$$

By the above, if neither minimal support for the western and northern neighbors supports (x, y) then only (x, y) may support itself, i.e.

$$N(x, y-1) \notin D(x, y), \, N(x-1, y) \notin D(x, y) \Rightarrow$$
$$\Rightarrow M^{*y'}_{x'} = 0 \; \forall x' \in \overline{x-w+1, x}, \, y' \in \overline{y-h+1, y}, \, (x', y') \neq (x, y) \Rightarrow$$
$$\Rightarrow D(x, y) \subseteq \{(x, y)\}.$$

Moreover, if $(x_1, y_1) \leq_{lex} (x_2, y_2)$ we have

$$(x_1 - x_1^*, \, y_1 - y_1^*) \leq_{lex} (x_1 - x_1', \, y_1 - y_1') \Leftrightarrow$$
$$\Leftrightarrow (x_2 - x^*, \, y_2 - y^*) \leq_{lex} (x_2 - x', \, y_2 - y')$$

and thus, if (x', y') supports both (x, y) and one of its western or northern neighbors, but is not the minimal support of that neighbor, then it is not the minimal support of (x, y). We obtain the dynamic programming scheme

$$N(x, y) \in \{N(x-1, y), \, N(x, y-1), \, (x, y)\}.$$

This scheme can be implemented in $\mathcal{O}(WH)$ time as shown in Algorithm 1. We have proven that it correctly decides whether the maximal mask does indeed cover the entire image, i.e. C is a cover of T. We conclude that the complexity of the decision problem is indeed $\Theta(WH)$. □

4 Minimal Image Covers

Among the family of covers of an image T, our goal is to find a "minimal" one. To achieve this goal we have to define the optimization criterion. This criterion takes the form of an evaluation function: $eval : \overline{1, W} \times \overline{1, H} \to \mathbb{R}$.

Proposition 1. *Obtaining the minimal image cover C of T with respect to eval takes time $\mathcal{O}(W^2 H^2 + WH\Theta(eval))$.*

Algorithm 1. Image Cover Decision

```
1: procedure CHECK(T, w, h)
2:     Preprocess T (per Bird's algorithm)
3:     for x ∈ 1, H do
4:        for y ∈ 1, W do
5:           N (x, y) = (−∞, −∞)
6:           if x > 1 and (x − w + 1, y − h + 1) ≤ N (x − 1, y) then
7:              N (x, y) = N (x − 1, y)
8:           end if
9:           if y > 1 and (x − w + 1, y − h + 1) ≤ N (x, y − 1) then
10:              if (x, y) − N (x, y − 1) ≤_lex N (x, y) − N (x, y − 1) then
11:                 N (x, y) = N (x, y − 1)
12:              end if
13:           end if
14:           if stage (y, x) (per Bird's algorithm) then
15:              N (x, y) = (x, y)
16:           end if
17:           if N (x, y) = (−∞, −∞) then
18:              return Mismatch: (x, y)
19:           end if
20:        end for
21:     end for
22:     return Match
23: end procedure
```

Proof. A brute force approach checks all possible (w, h) pairs (which are $\Theta\left(WH\right)$) and uses the decision algorithm above. If a cover is found it is evaluated. This yields complexity $\mathcal{O}\left(W^2H^2 + WH\Theta\left(eval\right)\right)$.

Moreover, if *eval* is arbitrary all (w, h) pairs must be checked since $eval\left(.\right)$ can be unbounded (or some large finite value) for all (w, h) except (w^*, h^*) which shows the bound is tight. □

Proposition 2. *If the minimal C is the n-th candidate according to the order induced by eval, we can obtain C in $\mathcal{O}\left(nWH\right)$ if the input is already sorted according to this order.*

Proof. If the ordering induced by the *eval* function is known, we queue up the would-be covers in that order (by sorting for example). For instance, if the n-th candidate is the first cover encountered, the runtime of the minimization algorithm described above is $\mathcal{O}\left(WH\Theta(eval) + nWH\right)$. This can be achieved via sorting, yielding a complexity of

$$\mathcal{O}(WH\Theta(eval) + WH\log(WH) + nWH)$$

□

4.1   The Size Criteria

We now study minimality with respect to a natural criterion, namely the size, as given by the area, ℓ_1 norm and ℓ_∞ norm.

For the area, the evaluation function is defined as

$$\overline{1, W} \times \overline{1, H} \ni (w, h) \to eval\,(w, h) = wh \in \mathbb{R}.$$

Suppose we knew that $wh \leq w_0 h_0$. Then we have

$$h \in \overline{1, \min\left(\lfloor w_0 h_0 / w \rfloor, H\right)},$$

and thus (w, h) is one of the lattice points of the intersection of the rectangle $((1, 1), (1, H), (W, 1), (W, H))$ with the triangle $((1, 1), (1, w_0 h_0),$ $(w_0 h_0, 1))$.

These contain at most WH and $w_0^2 h_0^2 / 2$ lattice points respectively. Thus, the optimal (w, h) pair is found after at most $n \approx \mathcal{O}\left(\min\left(w_0^2 h_0^2, WH\right)\right)$ attempts which leads to an upper bound of $\mathcal{O}\left(\min\left(w_0^2 h_0^2, WH\right) WH\right)$.

For the ℓ_1 norm the evaluation function is

$$\overline{1, W} \times \overline{1, H} \ni (w, h) \to eval\,(w, h) = w + h \in \mathbb{R}.$$

Suppose we knew that $w + h \leq w_0 + h_0$. Then we have

$$h \in \overline{1, \min\left(w_0 + h_0 - w, H\right)},$$

and thus (w, h) is one of the lattice points of the intersection of the rectangle $((1, 1), (1, H), (W, 1), (W, H))$ with the triangle $((1, 1), (1, w_0 + h_0 - 1),$ $(w_0 + h_0 - 1, 1))$.

These contain at most WH and $(w_0 + h_0)^2 / 2$ lattice points respectively. Thus, the optimal (w, h) pair is found after at most $n \approx \mathcal{O}\left(\min\left(w_0^2 + h_0^2, WH\right)\right)$ attempts which leads to an upper bound of $\mathcal{O}\left(\min\left(w_0^2 + h_0^2, WH\right) WH\right)$.

For the ℓ_∞ norm the evaluation function is defined as

$$\overline{1, W} \times \overline{1, H} \ni (w, h) \to eval\,(w, h) = \max\,(w, h) \in \mathbb{R}.$$

Suppose we knew that $\max\,(w, h) \leq \max\,(w_0, h_0)$. Then we have

$$h \in \overline{1, \min\left(\max\,(w_0, h_0), H\right)},$$

and thus (w, h) is one of the lattice points of the intersection of the rectangle $((1, 1), (1, H), (W, 1), (W, H))$ with the square $((1, 1), (1, \max\,(w_0, h_0),),$ $(\max\,(w_0, h_0), 1), (\max\,(w_0, h_0), \max\,(w_0, h_0)))$.

These contain at most WH and $\max\,(w_0, h_0)^2$ lattice points respectively. Thus, the optimal (w, h) pair is found after at most $n \approx \mathcal{O}\left(\min\left(\max\,(w_0, h_0)^2, WH\right)\right)$ attempts which leads to an upper bound of $\mathcal{O}\left(\min\left(\max\,(w, h)^2, WH\right) WH\right)$.

Note that we never used w_0 or h_0 other than for the calculation of the algorithm runtime. Thus, these calculations remain valid even if we do not know anything about w_0 and h_0. Their value is automatically substituted for the width and height of the minimal cover.

4.2   Boosting Average Performance by Preprocessing

In many cases, we do not have to verify all candidates. For instance, if the candidate a (w, h) is a cover, then the first and the last w columns and h rows are image-covered by $T_{1,h}^{\overline{1,w}}$. Based on this criterion we construct a preprocessing routine.

Suppose we knew that $h \geq h_0$. This means that C is at least h_0-tall. Hence, $T_{1,h_0}^{\overline{1,W}}$ covers $T_{1,H}^{\overline{1,W}}$. Note that $M^{*i}_j = 0 \forall j \geq 2$ since there is not enough space to accomodate another tile horizontally. Consequently, $T_i^{\overline{1,w}}$ covers $T_i^{\overline{1,W}}$ for all $i \leq h_0$ or $i \geq H - h_0$. Thus if C_i are all the covers of $T_i^{\overline{1,W}}$ then

$$w \geq \min \left\{ |c| \mid c \in C_i, \forall i \in \overline{1, h_0} \cup \overline{H - h_0, H} \right\}$$

This bound can be calculated in $\mathcal{O}(W h_0)$. Since C is a cover of T if and only if C^T is a cover of T^T, if we knew that $w_C \geq w_0$ then $h_{C'} \geq w_0$ and hence we can similarly obtain a lower bound for $w_{C'} = h_C$ in $\mathcal{O}(H w_0)$.

Suppose we knew that $w \geq w_0$ and $h \geq h_0$. It takes $\mathcal{O}(W h_0 + w_0 H)$ to check that this first test does not already disprove the eligibility of (w_0, h_0). Notably, the covers of $T_i^{\overline{1,W}}$ and $T_{1,H}^i$ can be pre-computed (or cached) such that the cumulative preprocessing time is $\mathcal{O}(WH)$, which is essentially free.

Since we have established that this preprocessing is effectively free we can do it entirely *a priori*, i.e. obtain the transitive closure of the preprocessing function. Let S be the matrix of string covers returned by the optimal Minimal String Cover algorithm for each line and S' for columns, i.e. $S_j^i = 1$ if the first j characters on the i-th line cover the i-th line and $S'^i_j = 1$ if the first i characters on the j-th column cover the j-th column. The current preprocessing is equivalent to computing the Hadamard product of the matrices

$$S_1{}^i_j = \min \left(S_j^i, S_1{}^{i-1}_j \right)$$

$$S'_1{}^i_j = \min \left(S'^i_j, S'_1{}^i_{j-1} \right)$$

$$S^* = \min (S_1, S'_1) = S_1 \odot S'_1.$$

Notably, the number of elements that are not pruned is the number of non-zero elements of S^*. However

$$S_1{}^i_j = \prod_{i'=1}^{i} S_j^{i'}$$

$$S'_1{}^i_j = \prod_{j'=1}^{j} S'^i_{j'}$$

$$S^*{}^i_j = S_1{}^i_j S'^i_j = \prod_{i'=1}^{i} \prod_{j'=1}^{j} S_j^{i'} S'^i_{j'}$$

We now check the effectiveness of our preprocesing.

Proposition 3. *Computing the matrix S^* reduces the number of candidates that need to be checked to $\Theta(1)$ average time for arbitrary H and $\Theta(1)$ worst-case for $H = 1$.*

Proof. Assume that there is a p probability for any tile in S_j^i and S'^i_j to be 1, and even the additional condition that $S_j^{mi} \geq S_j^i$ for all m and assuming that there is no single-character line nor column. Then by the Euler approximation, the probability that S_{1j}^i be 1 is $p^{i/\log(i)}$, that S'^i_j be 1 is $p^{j/\log(j)}$ and thus the probability that S^{*i}_j be 1 is $p^{i/\log(i)+j/\log(j)}$. Thus the expected number of 1s, considering that $S_W^i = S_j^H = 1$, is

$$\left(1 + \sum_{i=2}^{H} p^{i/\log(i)}\right)\left(1 + \sum_{j=2}^{W} p^{j/\log(j)}\right) \leq \left(1 + \frac{p}{1-p}\right)^2 = \frac{1}{(1-p)^2}.$$

We conclude that there exists a solution that is linear on the average case, $\mathcal{O}(WH)$ and quadratic in the worst, with the output-sensitive complexity: $\mathcal{O}(whWH)$, but which reduces to $\mathcal{O}(W)$ for the 1-dimensional case.

5   A Connection with Lattices

A lattice [25] is an additive subgroup $\mathcal{L}$ of $\mathbb{R}^n$ isomorphic to $\mathbb{Z}^n$. By definition, it is infinite and yet it is generated by n elements. Consider the isomorphism $\phi : \mathbb{Z}^n \to \mathcal{L}$. The projection of the unit volume $\{0, 1\}^n$ through this isomorphism $\phi(\{0, 1\}^n)$ is called the primitive cell of the lattice and it can be tiled by translations to form the entire $\mathcal{L}$. Note that by isomorphism we have:

$$\phi\left(\sum_{i=1}^{n}\lambda_i \mathbf{e}_i\right) = \sum_{i=1}^{n}\lambda_i \phi(\mathbf{e}_i)$$

Moreover, if $\mathcal{L}$ is a lattice, R is a rotation and S is a scaling matrix i.e. $S_i^j = 0 \Leftrightarrow i \neq j$ then $SR\mathcal{L}$ is isomorphic to $\mathcal{L}$ and thus when classifying lattices we can assume that there exists some $\phi(\mathbf{e}_i) = \mathbf{e}_1$. Moreover since $\mathbb{Z}^n$ is isomorphic to itself by the maps $\mathbf{e}_i \to \mathbf{e}_{\sigma(i)}$ for any permutation $\sigma \in S_n$, we can assume that $\phi(\mathbf{e}_1) = \mathbf{e}_1$. Thus, all 2-dimensional lattices can be characterized by the relative phase and length of the second vector (Figs. 1, 2 and 3).

Given a volume in n-dimensional space and a lattice $\mathcal{L} \subseteq \mathbb{E}^n$, we can divide it according to the lattice i.e. given $\psi : \mathbb{Z}^n \to \mathcal{L}$ we have

$$\mathbb{E}_{\mathcal{L}}^n = \{C_1 = \overline{conv}(\{\phi(1 + \mathbf{v}) | \mathbf{v} \in \{0, 1\}^n\}) | 1 \in \mathbb{Z}^n\}$$

Note that the translation $\phi(1) \to \phi(1')$ maps C_1 to $C_{1'}$ and thus the volume of any two cells is the same for a given $\mathcal{L} \subseteq \mathbb{E}^n$. Thus we can define the quantity $vol(\mathcal{L}) = vol(C_0)$ to be the unit volume of a lattice $\mathcal{L}$.

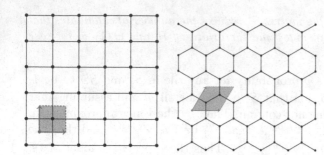

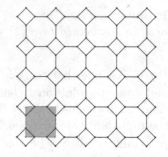

Fig. 1. A grid lattice

Fig. 2. A hexagonal row lattice

Fig. 3. A mixed tiling which is actually a grid lattice

Given some volumetric data $\mathbb{E}^n \supseteq \mathbf{V} \ni \mathbf{x} \to \psi(\mathbf{x}) \in \mathbb{R}$ we say that a lattice $\mathcal{L}$ is legal with respect to ψ if ψ is also translation invariant i.e.

$$\psi(\phi(\mathbf{x})) = \psi(\phi(\mathbf{x} + \mathbf{v})) \, \forall \mathbf{x} \in \mathbb{Z}, \, \mathbf{v} \in \{0, 1\}^n, \, \phi(\mathbf{x}) \in \mathbf{V}, \, \mathbf{x} + \mathbf{v} \in \mathbf{V}$$

Moreover, $\mathcal{L}$ is natural with respect to ψ if it is a legal lattice, minimal with respect to the unit volume.

We would like to obtain the unit cell of the natural lattice given *not the lattice points but instead a tiling of the unit cell that is cropped to a W-long, H-tall image that contains at least one copy of the unit cell* i.e. volumetric 2-dimensional data.

Once we have found an unit cell, any translation or rotation of it is still an unit cell which describes the same geometry and thus we have no interest in selecting any particular one. We accept any unit cell of any natural lattice.

Since a legal lattice is invariant to translations, we may always fix the origin of one unit cell on T_1^1. Since it is invariant to rotations we may always fix that one of its unit vectors is along the T^1 row. However, it may be that the other axis is not along the T_1 column, as is the case for hexagonal lattices. Moreover, it may be that our image does not end after an integer number of tiles, but instead a fractional one. In this case, the end fraction has to appear in the cover. We conclude that the shortest cover may never contain more than the volume of the box-cover of 4 unit tiles. In fact, it never contains 2 entire unit tiles on any side. Moreover, it will always contain at least one unit tile or a seed of it.

Note that this approach is especially interesting in the case of quasi-periodic crystals (which do not admit a Bravais lattice) [25]. This extends the k-covers problem [12] and asks for the k unit cells which have been used, for example in a Penrose tiling [9].

6　Applications in Computer Graphics

Consider the task of producing huge, unique maps for games, such as mazes or dungeons. Without procedural terrain generation this task is anything between infeasible and impossible, depending on the desired size and the available time and budget. Many games use Wang Tiles [10,16] to produce huge maps (an interesting example is the Infamous game produced by Sucker Punch). They have recently garnered around them a very large community.

Wang tiles are formal systems visually modeled by square tiles with colors on each side. Two Wang tiles may only be tiled along an edge if the colors match. The most popular problems concerning them were: whether a set of Wang tiles can cover the plane and whether this can be done in a periodic way [13].

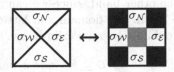

A Wang tile can also be represented as a 3-by-3 image. Two such images may be tiled together either along an edge or a corner. The formal system isomorphism is trivial: two 3-by-3 images may be tiled together on an edge if the respective colors on the Wang tiles match. This is very much like String Covers, except two such images may never be tiled one alongside another.

Consider the following problems:

Problem 4 (Minimal Wang Cover). Given a tiling of some Wang Tiles check if there exists a periodic pattern covering it.

Problem 5 (k-Wang Covers). Given a tiling of some Wang Tiles check if there exist k patterns which, when tiled together cover the image.

Problem 6 (Approximate Wang Cover). Given a tiling of some Wang Tiles find the minimal number of pixels to be changed for it to be covered by a single periodic pattern.

When given a 3-tall image the first two collapse to vectorial String Cover and vectorial k-Covers. For the last one, we must also impose that the black and gray pixels which we added ourselves are never corrupted. Thus we impose that the distance between two tilings is infinite if a black or gray pixel is corrupted. Hence it is equivalent to the Approximate String Cover of Amir et al. with an almost-Hamming metric.

Problem 7 (Generalization to pseudo-metrics). Given a compression palette $\Gamma \subseteq \Sigma$ and an algorithm that is consistent with respect to the colors it replaces i.e. $\mathcal{A} : \Sigma \to \Gamma$ and a tiling of some Wang Tiles, check if the solutions to the above problems change.

The last problem is not important from a computational perspective; in fact it is quite trivial, but it gives substance to the pseudo-metric variations of String Cover problems.

Given computationally efficient algorithms that solve Problem 7, there are several interesting applications in computer aided design (see e.g. [10]). One use of Wang tiles is procedure terrain generation in video games. If a player knows that the game he is playing uses Wang tiles, he can use an image cover algorithm to predict the next challenge. Another application is image compression: we can use these algorithms on images produced by designers in order to extract textures or motifs.

Consider a game with hexagonal tiles that wants to make use of Perlin noise [20]. It is unnatural that it be used purely, since the rectangular lattice is not actually legal. On the other hand, since we can obtain $\mathcal{L}$, we by default have a mapping $\phi^{-1} : \mathcal{L} \to \mathbb{Z}^n$. In this domain, our lattice is indeed rectangular. Thus, it is here that we should apply our Perlin noise.

Definition 6. *Given a lattice* $\mathbb{Z}^n \xrightarrow{\phi} \mathcal{L} \subseteq \mathbb{E}^n$ *and a noise-function appropriate for rectangular latices* $\mathcal{P} : \mathbb{Z}^n \to \mathbb{R}$, *we can lift it to* $\mathcal{L}$:

$$\mathcal{L} \ni \mathbf{x} \to \mathcal{P}_{\mathcal{L}}(\mathbf{x}) = \mathcal{P}\left(\phi^{-1}(\mathbf{x})\right) \in \mathbb{R}$$

Thus we can define the Perlin noise appropriate for a given Wang system. Note that the magnitude of Perlin noise is an input parameter. Thus, without changing the game or inducing unnatural patterns, as a game developer we can easily add a diversity grade for games using Wang tiles for terrain generation.

References

1. Aho, A.V., Corasick, M.J.: Efficient string matching: an aid to bibliographic search. Commun. ACM **18**(6), 333–340 (1975)
2. Amir, A., Levy, A., Lewenstein, M., Lubin, R., Porat, B.: Can we recover the cover? In: 28th Annual Symposium on Combinatorial Pattern Matching, CPM 2017, Warsaw, Poland, 4–6 July 2017, pp. 25:1–25:15 (2017)
3. Amir, A., Levy, A., Lubin, R., Porat, E.: Approximate cover of strings. In: 28th Annual Symposium on Combinatorial Pattern Matching, CPM 2017, Warsaw, Poland, 4–6 July 2017, pp. 26:1–26:14 (2017)
4. Apostolico, A., Breslauer, D.: Of periods, quasiperiods, repetitions and covers. In: Mycielski, J., Rozenberg, G., Salomaa, A. (eds.) Structures in Logic and Computer Science. LNCS, vol. 1261, pp. 236–248. Springer, Heidelberg (1997). https://doi.org/10.1007/3-540-63246-8_14
5. Apostolico, A., Farach, M., Iliopoulos, C.S.: Optimal superprimitivity testing for strings. Inf. Process. Lett. **39**(1), 17–20 (1991)
6. Bacciotti, A., Rosier, L.: Liapunov Functions and Stability in Control Theory. Springer, Heidelberg (2006). https://doi.org/10.1007/b139028
7. Bird, R.S.: Two dimensional pattern matching. Inf. Process. Lett. **6**(5), 168–170 (1977)
8. Brodzik, A.K.: Quaternionic periodicity transform: an algebraic solution to the tandem repeat detection problem. Bioinformatics **23**(6), 694–700 (2007)
9. Bursill, L., Lin, P.J.: Penrose tiling observed in a quasi-crystal. Nature **316**(6023), 50–51 (1985)

10. Derouet-Jourdan, A., Salvati, M., Jonchier, T.: Procedural Wang tile algorithm for stochastic wall patterns. CoRR, abs/1706.03950 (2017)
11. Havlin, S., et al.: Fractals in biology and medicine. Chaos Solitons Fractals **6**, 171–201 (1995). Complex Systems in Computational Physics
12. Iliopoulos, C., Smith, W.: An on-line algorithm of computing a minimum set of k-covers of a string. In: Proceedings of Ninth Australian Workshop on Combinatorial Algorithms (AWOCA), pp. 97–106 (1998)
13. Jeandel, E., Rao, M.: An aperiodic set of 11 Wang tiles. arXiv preprint arXiv:1506.06492 (2015)
14. Katok, A., Hasselblatt, B.: Introduction to the Modern Theory of Dynamical Systems, vol. 54. Cambridge University Press, Cambridge (1997)
15. Knuth, D.E., Morris Jr., J.H., Pratt, V.R.: Fast pattern matching in strings. SIAM J. Comput. **6**(2), 323–350 (1977)
16. Kopf, J., Cohen-Or, D., Deussen, O., Lischinski, D.: Recursive wang tiles for real-time blue noise. ACM Trans. Graph. **25**(3), 509–518 (2006)
17. Middlestead, R.: Digital Communications with Emphasis on Data Modems: Theory, Analysis, Design, Simulation, Testing, and Applications. Wiley, Hoboken (2017)
18. Ming, L., Vitányi, P.M.: Kolmogorov complexity and its applications. In: Algorithms and Complexity, pp. 187–254. Elsevier (1990)
19. Muchnik, A., Semenov, A., Ushakov, M.: Almost periodic sequences. Theor. Comput. Sci. **304**(1–3), 1–33 (2003)
20. Perlin, K.: Improving noise. ACM Trans. Graph. **21**(3), 681–682 (2002)
21. Searle, J.R., Kiefer, F., Bierwisch, M., et al.: Speech Act Theory and Pragmatics, vol. 10. Springer, Dordrecht (1980). https://doi.org/10.1007/978-94-009-8964-1
22. Sethares, W.A., Staley, T.W.: Periodicity transforms. IEEE Trans. Sig. Process. **47**(11), 2953–2964 (1999)
23. Timmermans, M., Heijmans, R., Daniels, H.: Cyclical patterns in risk indicators based on financial market infrastructure transaction data (2017)
24. Tychonoff, A.: Théorèmes d'unicité pour l'équation de la chaleur. Matematiceskij sbornik **42**(2), 199–216 (1935)
25. Wlodawer, A., Minor, W., Dauter, Z., Jaskolski, M.: Protein crystallography for aspiring crystallographers or how to avoid pitfalls and traps in macromolecular structure determination. FEBS J. **280**(22), 5705–5736 (2013)

Introducing Fluctuation into Increasing Order of Symmetric Uncertainty for Consistency-Based Feature Selection

Sho Shimamura[1] and Kouichi Hirata[2($\boxtimes$)]

[1] Graduate School of Computer Science and Systems Engineering,
Kyushu Institute of Technology, Kawazu 680-4, Iizuka 820-8502, Japan
`shimamura@dumbo.ai.kyutech.ac.jp`
[2] Department of Artificial Intelligence,
Kyushu Institute of Technology, Kawazu 680-4, Iizuka 820-8502, Japan
`hirata@dumbo.ai.kyutech.ac.jp`

Abstract. In order to select correlated and relevant features in a feature selection, several filter methods adopt a *symmetric uncertainty* as one of the feature ranking measures. In this paper, we introduce a *fluctuation* into the increasing order of the symmetric uncertainty for the consistency-based feature selection algorithms. Here, the fluctuation is an operation of transforming the sorted sequence of features to a new sequence of features. Then, we compare the selected features by the algorithms with a fluctuation with those without fluctuations.

Keywords: Fluctuation · Symmetric uncertainty ·
Consistency-based feature selection algorithm · Feature selection

1 Introduction

A *feature selection* [3] is well-known as one of the research fields of machine learning. In particular, a *consistency-based feature selection* [4,16] is the feature selection based on the filter evaluating the worth of a subset of features by the level of *consistency* in the class values. Hence, it is classified to *filter* methods [2, 3] in the feature selection.

On the other hand, a *symmetric uncertainty*, which is originally introduced by [8], is adopted as one of the feature ranking measures [5] in several filter methods in the feature selection. Note that the symmetric uncertainty is formulated as the normalizing values of the *information gain* [6], or, alternatively, *mutual information* [9], which are familiar measures for machine learning. The symmetric uncertainty can measure the relevancy between features based on the value computed from just a single feature.

The author would like to express thanks for support by Grant-in-Aid for Scientific Research 17H00762, 16H02870 and 16H01743 from the Ministry of Education, Culture, Sports, Science and Technology, Japan.

T. V. Gopal and J. Watada (Eds.): TAMC 2019, LNCS 11436, pp. 550–565, 2019.
https://doi.org/10.1007/978-3-030-14812-6_34

As the filter methods in the feature selection, Hall [5] has introduced a *correlation-based feature selection*, where one of the correlation is formulated by using the symmetric uncertainty. Yu and Liu [15] have developed the correlation-based feature selection as a *fast correlation-based filter* and combined it with several feature selection algorithms such as FOCUS [1] and ReliefF [7].

Furthermore, the consistency-based feature selection algorithms such as INTERACT [16], LCC [10,12,13] and CWC [11–13] first sort features by the increasing order of the symmetric uncertainty. For every feature selected in this order, if the set of features after eliminating the feature is still consistent, then the algorithms update the set of features as the eliminated set.

It is known that the symmetric uncertainty is a heuristic to measure correlation between features and succeeds to select correlated features empirically [5,15]. On the other hand, we cannot guarantee that such a correlation is always correct, for example, every pair of correlated features is always sorted as a pair of adjacent features in the increasing order of the symmetric uncertainty.

In order to verify such an effect in the consistency-based feature selection algorithms empirically, in this paper, we introduce a *fluctuation* into the increasing order of the symmetric uncertainty for the algorithms of LCC and CWC. Here, the fluctuation is an operation of transforming the sorted sequence of features to a new sequence of features. Then, we formulate the following four kinds of the fluctuations, that is, (1) *the sort of blocks*, (2) *the move of the initial segment*, (3) *the move of the initial segment before the selected features* and (4) *the collection of the selected features and the sort of features*.

The sort of blocks is an operation of transforming to a sequence obtained by sorting every block by the decreasing order after dividing some blocks to the sorted sequence of features. *The move of the initial segment* is an operation of transforming to a sequence obtained by moving the initial segment with a fixed ratio to the last of the sorted sequence of features.

The remained fluctuations assume to be selected features by applying the algorithms of LCC and CWC. Then, *the move of the initial segment before the selected features* is an operation of transforming to a sequence obtained by moving the initial segment whose symmetric uncertainty is smaller than that of every selected feature to the last of the sorted sequence of features. *The collection of the selected features and the sort of features* is an operation of transforming a sequence obtained by collecting the selected features, moving them to the initial of the sorted sequence of features and then sorting the collected features and the remained features by the increasing or the decreasing order, respectively.

Finally, we give experimental results to evaluate the influence of the fluctuation in the algorithms of LCC and CWC, by using artificial data, real data and nucleotide sequences. We generate artificial data by varying the number of features, the number of class labels, the number of instances and the number of feature values. As real data, we adopt DEXTER[1], DOROTHEA

---

[1] NIPS 2003 Workshop on Feature Extraction and Feature Selection Challenge. http://clopinet.com/isabelle/Projects/NIPS2003/#challenge.

(see footnote 1), GISETTE (see footnote 1), HIVA[2], NOVA (see footnote 2) and SYLVA (see footnote 2). As nucleotide sequences, we adopt 8 RNA segments of influenza A (H1N1) viruses provided from NCBI[3]. Then, by measuring the *ratio of different numbers of selected features*, the *similarity of selected features* and the *difference of accuracy* under fluctuations, we compare the selected features by the algorithms with a fluctuation with those without fluctuations.

2   Consistency-Based Feature Selection Algorithms

We call an $m \times (n + 1)$ matrix on $\mathbf{N}$ a *data set* and denote it by $D = [v_{ij}]$. Also we call every row $\boldsymbol{v}_i = [v_{i1}, \ldots, v_{in}, v_{i(n+1)}]$ in D an *instance* of D and the $(n + 1)$-th element $v_{i(n+1)}$ in $\boldsymbol{v}_i$ a *class label* of $\boldsymbol{v}_i$. We denote the set of all the class labels in D by C. In the following, we omit the subscript i. Then, we denote that $\boldsymbol{v}$ is an instance of D by $\boldsymbol{v} \in D$ and the class label of $\boldsymbol{v}$ by $\boldsymbol{v}_c$.

Let $F = \{1, \ldots, n\}$, which we call a *total feature set*, and $\boldsymbol{v} = \boldsymbol{v}_i \in D$ an instance. Then, we denote $[v_{i1}, \ldots, v_{in}]$ by $\boldsymbol{v}_F$. For a subset $X = \{j_1, \ldots, j_k\} \subseteq F$, which we call a *feature set*, we denote $[v_{ij_1}, \ldots, v_{ij_k}]$ by $\boldsymbol{v}_X$. For a data set D and a feature set $X \subseteq F$, we denote the data set consisting of the j-th column for every $j \in X \cup \{n + 1\}$, that is, the collection of rows $[\boldsymbol{v}_X, \boldsymbol{v}_c]$ for $\boldsymbol{v} \in D$ by D_X.

Let $\boldsymbol{v} \in D$ be an instance, $X \subseteq F$ a feature set such that $|X| = k$ and C the set of all class labels. Then, for $\boldsymbol{x} \in \mathbf{N}^k$ and $y \in C$, we denote the probability that $\boldsymbol{v}_X$ is $\boldsymbol{x}$ by $P(X = \boldsymbol{x})$ and the probability that $\boldsymbol{v}_c$ is y by $P(C = y)$. Then, the entropy $H(X)$ of X and the entropy $H(C)$ of C are defined as follows.

$$H(X) = - \sum_{\boldsymbol{x} \in \mathbf{N}^k} P(X = \boldsymbol{x}) \log P(X = \boldsymbol{x}),$$

$$H(C) = - \sum_{y \in C} P(C = y) \log P(C = y).$$

Also the conditional entropy $H(C|X)$ of C after observing X is defined as follows.

$$H(C|X) = - \sum_{\boldsymbol{x} \in \mathbf{N}^k} P(X = \boldsymbol{x}) \sum_{y \in C} P(C = y|X = \boldsymbol{x}) \log P(C = y|X = \boldsymbol{x})$$

$$= - \sum_{\boldsymbol{x} \in \mathbf{N}^k} \sum_{y \in C} P(X = \boldsymbol{x}, C = y) \log \frac{P(X = \boldsymbol{x}, C = y)}{P(X = \boldsymbol{x})}.$$

Hence, the *information gain* [6] $IG(C|X)$ (or the *mutual information* [9]) of C provided by X, that is, the amount by which the entropy of C decreases reflects additional information about C provided by X, is defined as follows.

---

[2] WCCI 2004 Performance Prediction Challenge.
   http://clopinet.com/isabelle/Projects/modelselect/datasets/.
[3] NCBI, National Center for Biotechnology Information.
   http://www.ncbi.gov/genome/FLU/.

$$IG(C|X) = H(C) - H(C|X) = H(X) - H(X|C).$$

The symmetry of the information gain is a desired property for a measure of correlations between features [5,15]. We can derive the following alternative formula (*cf.* [14]):

$$IG(C|X) = \sum_{x \in \mathbf{N}^k} \sum_{y \in C} P(X = x, C = y) \frac{P(X = x, C = y)}{P(X = x)P(C = y)}.$$

Furthermore, the *symmetric uncertainty* [8] $SU(C, X)$ is defined as follows.

$$SU(C, X) = \frac{2 \cdot IG(C|X)}{H(X) + H(C)}.$$

Note that we use the symmetric uncertainty as the form of $SU(C, \{i\})$ for every feature $i \in F$, in order to measure the relevancy between features in F by using the value computed from just a single feature i.

In this paper, we deal with two consistency measures, a *Bayesian risk* [10] and a *binary consistency* [11–13]. For $X \subseteq F$, the *Bayesian risk* $BR(X)$ of X is defined as follows:

$$BR(X) = 1 - \sum_{x \in \mathbf{N}^k} \max_{y \in C} P(X = x, C = y).$$

On the other hand, the *binary consistency* $BC(X)$ supports:

$$BC(X) = \begin{cases} 0, & \forall u, v \in D \ (u_X = v_X \Rightarrow u_c = v_c), \\ 1, & \text{otherwise.} \end{cases}$$

Let $\mu \in \{BR, BC\}$ be a consistency measure and δ a threshold ($0 \leq \delta < 1$). Then, we say that X is *consistent with respect to D under μ and δ* if $\mu(X) \leq \delta$; *inconsistent* otherwise. Note here that δ is not necessary when $\mu = BC$.

Let D be a data set, $X \subseteq F$ a feature set, $\mu \in \{BR, BC\}$ a consistency measure and δ a threshold. Then, we call the set of instances by eliminating all the inconsistent instances of D for X under μ and δ from D the set of *explanatory instances* of D for X and denote it by $e_{\mu,\delta}(D, X)$. It is obvious that X is consistent with $e_{\mu,\delta}(D, X)$ under μ and δ.

Then, as a general framework, we formulate *a consistency-based feature selection problem* as follows.

Definition 1. Let D be a data set, F a total feature set, μ a consistency measure and δ a threshold. Then, the *consistency-based feature selection problem* is to find a feature set $X \subseteq F$ such that $|X|$ is minimum when $|e_{\mu,\delta}(D, X)|$ is maximum.

In order to solve the consistency-based feature selection problem heuristically and efficiently, Shin *et al.* have introduced the algorithms LCC (Linear Consistency-Constrained) [10,12,13] and CWC (Combination of Weakest Components) [11–13] illustrated in Algorithm 1. Here, the procedure **sort** in lines

1 and 7 sorts a total feature set F as $\{i_1, \ldots i_n\}$ by the increasing order of the symmetric uncertainty $SU(C, \{i\})$ for every $i \in F$ and the procedure **denoise** in line 6 removes presumable noise examples from D.

Furthermore, we can regard that the algorithms LCC and CWC solve the relaxed version of the consistency-based feature selection problem with maximizing $\sum_{x \in X} |e_{\mu,\delta}(D, x)|$ with sorting by the symmetric uncertainty, instead of maximizing $|e_{\mu,\delta}(D, X)|$ directly.

---

procedure $\text{LCC}_\delta(D, F)$

    /* D: data set, F: total feature set, δ: threshold */

1   **sort** F as $\{i_1, \ldots i_n\}$ under the symmetric uncertainty;

2    $S \leftarrow \{i_1, \ldots, i_n\}$;

3   **for** $j = 1$ **to** n **do**

4   **if** $BR(S \setminus \{i_j\}) \leq \delta$ **then** $S \leftarrow S \setminus \{i_j\}$;

5   **output** S;

procedure $\text{CWC}(D, F)$

    /* D: data set, F: total feature set */

6   **denoise** D;

7   **sort** F as $\{i_1, \ldots i_n\}$ under the symmetric uncertainty;

8    $S \leftarrow \{i_1, \ldots, i_n\}$;

9   **for** $j = 1$ **to** n **do**

10   **if** $BC(S \setminus \{i_j\}) = 0$ **then** $S \leftarrow S \setminus \{i_j\}$;

11   **output** S;

Algorithm 1. LCC and CWC.

---

3   Fluctuation for Order of Symmetric Uncertainty

In this section, we introduce the fluctuations into the increasing order of the symmetric uncertainty. We assume that a total feature set F is sorted by the increasing order of the symmetric uncertainty, so F is regarded as a sequence.

Let $X, X_1, X_2 \subseteq F$ be the sets (regarded as the sequences) of features such that $X_1 \cap X_2 = \emptyset$. Then, X^i (*resp.*, X^d) denotes the sequence obtained by sorting X by the increasing (*resp.*, decreasing) order of the symmetric uncertainty. Also, $X_1 \circ X_2$ denotes the concatenation of X_1 to X_2, that is, X_2 occurs after X_1, and $F \setminus X$ denotes the sequence obtained by deleting X from F with preserving the order of F.

Then, we formulate the following four kinds of the fluctuations, that is, (1) *the sort of blocks*, (2) *the move of the initial segment*, (3) *the move of the initial segment before the selected features* and (4) *the collection of the selected features and the sort of features*, as the operations of transforming a total feature set F to a new sequence of features.

Definition 2 (Fluctuation). Let F be a total feature set $\{i_1, \ldots, i_n\}$, r a constant such that $0 < r \leq 1$ and $l = \lfloor 1/r \rfloor - 1$. Then, let B_k $(0 \leq k \leq l)$ be the set $\{i_{krn+1}, \ldots, i_{(k+1)rn}\}$ of features.

1. *The sort of blocks,* denoted by $Sort_r$, is an operation of transforming F to $B_0^d \circ \ldots \circ B_l^d$.
2. *The move of the initial segment,* denoted by $Move_r$, is an operation of transforming F to $(F \setminus B_0) \circ B_0$.

In the following, let $S \subseteq F$ be the selected features by applying the algorithm $A \in \{\text{CWC}, \text{LCC}\}$.

3. Let $i_j = \text{argmin}\{SU(C, \{i\}) \mid i \in S\}$ and B the set $\{i_1, \ldots, i_{j-1}\}$. Then, *the move of the initial segment before the selected features,* denoted by $Move_A$, is an operation of transforming F to $(F \setminus B) \circ B$.
4. *The collection of the selected features and the sort of features* is an operation of transforming F to one of $S^i \circ (F \setminus S)^i$, $S^i \circ (F \setminus S)^d$, $S^d \circ (F \setminus S)^i$ and $S^d \circ (F \setminus S)^d$. We denote them by $CSort_A^{ii}$, $CSort_A^{id}$, $CSort_A^{di}$ and $CSort_A^{dd}$, respectively.

Example 1. Let $F = \{1, 2, 3, 4, 5, 6, 7, 8, 9, 10\}$ be a total feature set sorted by the increasing order of the symmetric uncertainty. Also let $c = 0.2$. Furthermore, suppose that the algorithms CWC and LCC select the features $\{4, 5, 7\}$ and $\{5, 7, 8\}$, respectively. Then, we obtain the following fluctuations.

1. The fluctuation $Sort_{0.2}$ transforms F to $\{2, 1, 4, 3, 6, 5, 8, 7, 10, 9\}$.
2. The fluctuation $Move_{0.2}$ transforms F to $\{3, 4, 5, 6, 7, 8, 9, 10, 1, 2\}$.
3. The fluctuation $Move_{\text{LCC}}$ transforms F to $\{5, 6, 7, 8, 9, 10, 1, 2, 3, 4\}$ and the fluctuation $Move_{\text{CWC}}$ transforms F to $\{4, 5, 6, 7, 8, 9, 10, 1, 2, 3\}$.
4. For the fluctuations $CSort_{\text{LCC}}^{**}$ and $CSort_{\text{CWC}}^{**}$ $(*, \star \in \{i, d\})$:
   (a) The fluctuation $CSort_{\text{CWC}}^{ii}$ transforms F to $\{4, 5, 7, 1, 2, 3, 6, 8, 9, 10\}$ and the fluctuation $CSort_{\text{LCC}}^{ii}$ transforms F to $\{5, 7, 8, 1, 2, 3, 4, 6, 9, 10\}$.
   (b) The fluctuation $CSort_{\text{CWC}}^{id}$ transforms F to $\{4, 5, 7, 10, 9, 8, 6, 3, 2, 1\}$ and the fluctuation $CSort_{\text{LCC}}^{id}$ transforms F to $\{5, 7, 8, 10, 9, 6, 4, 3, 2, 1\}$.
   (c) The fluctuation $CSort_{\text{CWC}}^{di}$ transforms F to $\{7, 5, 4, 1, 2, 3, 6, 8, 9, 10\}$ and the fluctuation $CSort_{\text{LCC}}^{di}$ transforms F to $\{8, 7, 5, 1, 2, 3, 4, 6, 9, 10\}$.
   (d) The fluctuation $CSort_{\text{CWC}}^{dd}$ transforms F to $\{7, 5, 4, 10, 9, 8, 6, 3, 2, 1\}$ and the fluctuation $CSort_{\text{LCC}}^{dd}$ transforms F to $\{8, 7, 5, 10, 9, 6, 4, 3, 2, 1\}$.

4 Experimental Results

In this section, we evaluate the the influence from the fluctuation for the number of selected features and the accuracy of them, by using both artificial data and real data. In the following, we refer the number of features, the number of class labels, the number of instances and the number of different integers to n, c, m and d, respectively.

4.1 Data Set, Experimental Setting and Evaluation Method

For artificial data, we set $n = 1,000$, $c = 20$, $m = 800$ and $d = 4$ as initial parameters. First, we randomly generate 5 data sets with the initial parameters. Next, by selecting every parameter just once, we randomly generate 5 data sets such that the selected parameter is increased in five steps and the other parameters are fixed. Here, n is increased by $1,000$ from $2,000$ to $6,000$, c is by 10 from 30 to 70, m is by 100 from 900 to $1,300$ and d is by one from 5 to 9 in every step. As a result, we obtain 100 data sets. We refer the data set obtained by increasing a parameter n (*resp.*, c, m and d) and fixing the other parameters by AD_n (*resp.*, AD_c, AD_m and AD_d).

On the other hand, we adopt DEXTER (see footnote 1), DOROTHEA (see footnote 1), GISETTE (see footnote 1), HIVA (see footnote 2), NOVA (see footnote 2) and SYLVA (see footnote 2) as real data. Here, the parameters of n, c, m and d are summarized in Table 1.

Table 1. Summary of parameters in real data.

Data	n	c	m	d	Data	n	c	m	d
DEXTER	20,000	2	299	6,261	HIVA	1,617	2	3,845	2
DOROTHEA	100,000	2	800	2	NOVA	16,969	2	1,754	2
GISETTE	5,000	2	6,000	10	SYLVA	216	2	13,086	331

Furthermore, as nucleotide sequences, we adopt 8 RNA segments, consisting of PB2, PB1, PA, HA, NP, NA, MP and NS, of influenza A(H1N1) viruses provided from NCBI (see footnote 3). Since we deal with nucleotide sequences consisting of four nucleotides of A, U, G and C and an additional missing value, the parameter d is 5. The other parameters of n, c and m are summarized in Table 2. Here, the number (n) of features is the length of a nucleotide sequence and the class label is the name of the recorded country.

Table 2. Summary of parameters in nucleotide sequences.

Data	n	c	m	data	n	c	m
PB2	2,725	80	12,357	NP	1,636	86	12,690
PB1	2,650	84	12,251	NA	1,775	111	24,827
PA	3,817	81	12,263	MP	1,223	88	17,889
HA	2,358	126	31,908	NS	992	79	12,958

As an experimental setting, by increasing the value of r by 0.2 from 0.2 to 1 and 0.8, we deal with 5 fluctuations of $Sort_r$ and 4 fluctuations of $Move_r$, respectively. We fix a threshold δ in LCC for every data. Then, we deal with 2 fluctuations of $Move_{LCC}$ and $Move_{CWC}$, 4 fluctuations of $CSort_{CWC}^{ii}$, $CSort_{CWC}^{id}$,

$CSort_{\mathrm{CWC}}^{di}$ and $CSort_{\mathrm{CWC}}^{dd}$ for CWC and 4 fluctuations of $CSort_{\mathrm{LCC}}^{ii}$, $CSort_{\mathrm{LCC}}^{id}$, $CSort_{\mathrm{LCC}}^{di}$ and $CSort_{\mathrm{LCC}}^{dd}$ for LCC.

As an evaluation method, we evaluate the number of selected features, the similarity of selected features and the accuracy of selected features by LCC and CWC with or without fluctuations.

Let F be a a fluctuation and $A \in \{\mathrm{CWC}, \mathrm{LCC}\}$ an algorithm. Also we denote the set of features selected by A under the fluctuation F to sort features under the symmetric uncertainty (in lines 1 and 7 in Algorithm 1) by $SF(F, A)$. Here, we refer to ε as F if we adopt no fluctuations and we omit the subscripts of CWC and LCC in fluctuations. Furthermore, let $\alpha(F, A)$ be the accuracy obtained by applying 10-fold cross validations in LIBSVM[4] to $D_{SF(F,A)}$ (possibly $F = \varepsilon$).

Then, we introduce the following three measures to evaluate fluctuations without difference between data.

Definition 3 (Evaluation measures for fluctuations). Let F be a fluctuation and $A \in \{\mathrm{CWC}, \mathrm{LCC}\}$ an algorithm.

1. The *ratio of different numbers of selected features* under F by A, denoted by $diff\_num(F, A)$, is defined as follows:

$$diff\_num(F, A) = \frac{|SF(\varepsilon, A)| - |SF(F, A)|}{n},$$

where n is the cardinality of a total feature set.

2. The *similarity of selected features* under F by A, denoted by $sim(F, A)$, is defined as follows:

$$sim(F, A) = \frac{|SF(F, A) \cap SF(\varepsilon, A)|}{|SF(F, A) \cup SF(\varepsilon, A)|},$$

i.e., the Jaccard similarity coefficient between $SF(\varepsilon, A)$ and $SF(F, A)$.

3. The *difference of accuracy* under F by A, denoted by $diff\_acc(F, A)$, is defined as follows:

$$diff\_acc(F, A) = \alpha(F, A) - \alpha(\varepsilon, A).$$

The value of $diff\_num(F, A)$ is positive if the number of selected features by A with the fluctuation F is smaller than that without F. The value of $sim(F, A)$ is near to 1 (*resp.*, 0) if the set of selected features by A with the fluctuation F is similar (*resp.*, not similar) as that without F. The value of $diff\_acc(F, A)$ is positive if the accuracy of $D_{SF(F,A)}$ is larger than that of $D_{SF(\varepsilon,A)}$.

4.2   Results

Let $F \in \{Sort_r, Move_r, Move_A, CSort_A^{**}\}$, $A \in \{\mathrm{CWC}, \mathrm{LCC}\}$ and $*, \star \in \{i, d\}$. In the following, we discuss experimental results for artificial data, real data and

---

[4] LIBSVM: A Library for Support Vector Machine:
   https://www.csie.ntu.edu.tw/~cjlin/libsvm/.

Table 3. The running time (msec.) of the algorithms CWC and LCC without fluctuations (ε) and the maximum, the minimum and the average (ave.) running time of those with fluctuations for the real data.

DEXTER		DOROTHEA		GISETTE		HIVA		NOVA		SYLVA	
CWC	Time	CWC	Time	CWC	Time	CWC	Time	CWC	Time	CWC	Time
ε	272	ε	2,521	ε	3,611	ε	592	ε	279	ε	675
$Sort_1$	583	$Sort_{0.5}$	41,242	$Sort_{0.1}$	3,912	$Move_{0.4}$	738	$Sort_{0.5}$	1,092	$Move_{0.9}$	787
$Move_{0.3}$	125	$CSort^{id}$	2,180	$CSort^{ii}$	3,413	$Sort_{0.5}$	503	$CSort^{id}$	271	$CSort^{ii}$	589
ave.	295	ave.	16,303	ave.	3,636	ave.	569	ave.	590	ave.	656
DEXTER		DOROTHEA		GISETTE		HIVA		NOVA		SYLVA	
LCC	Time	LCC	Time	LCC	Time	LCC	Time	LCC	Time	LCC	Time
ε	326	ε	2,858	ε	4,282	ε	544	ε	362	ε	671
$Sort_{0.1}$	651	$Sort_{0.2}$	63,549	$Sort_{0.2}$	3,878	$Move_{0.3}$	743	$Sort_{0.1}$	1,484	$CSort^{dd}$	749
$CSort^{di}$	123	$CSort^{di}$	2,348	$CSort^{id}$	3,431	$CSort^{id}$	517	$CSort^{id}$	287	$Sort_{0.6}$	611
ave.	335	ave.	25,341	ave.	3,677	ave.	577	ave.	764	ave.	643

nucleotide sequences by using the algorithm CWC and those for real data by using the algorithm LCC. Note that the results for artificial data by LCC are same as those by CWC. Also, we set the threshold δ to 0.01 for DEXTER, 0.005 for DOROTHEA, GISETTE, HIVA and NOVA and 0.001 for SYLVA.

First of all, we give the results of the running time. Table 3 illustrates the running time of the algorithms CWC and LCC without fluctuations and the maximum, the minimum and the average running time of those with fluctuations for the real data of DEXTER, DOROTHEA, GISETTE, HIVA, NOVA and SYLVA.

Table 3 shows that, for the data of DOROTHEA, the running time of both algorithms with fluctuations tends to be much larger than that without fluctuations. Note that, there exist some fluctuations such that the running time of the algorithms with the fluctuation is smaller than that without fluctuations, for example, $CSort^{**}$ ($*, \star \in \{i, d\}$), shown as the minimum running time. On the other hand, for the data of GISETTE, the running time of the algorithm LCC with fluctuations is always smaller than that without fluctuations. The running time of both algorithms for the data of DOROTHEA and GISETTE is much larger than that of other data.

In considering the data with smaller running time, for the data of HIVA, the average running time of the algorithm CWC with fluctuations is smaller than that without fluctuations. Also, for the data of SYLVA, the average running time of both algorithms with fluctuations is smaller than that without fluctuations.

In the remainder of this section, by using the measures in Definition 3, we evaluate the selected features with fluctuations by comparing with those without fluctuations.

Figures 1, 2 and 3 illustrate the values of $diff\_num(F, \text{CWC})$ for artificial data of AD_n, AD_c, AD_m and AD_d, real data of DEXTER, DOROTHEA, GISETTE, HIVA, NOVA and SYLVA, and nucleotide sequences for 8 RNA segments of PB2, PB1, PA, HA, NP, NA, MP and NS, respectively, and the values of $diff\_num(F, \text{LCC})$ for real data.

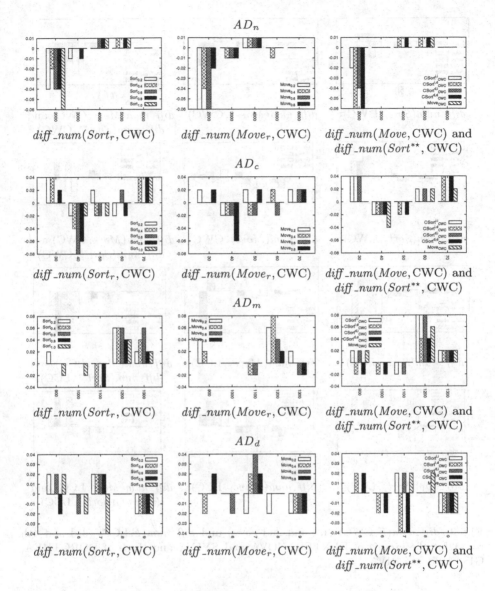

Fig. 1. $diff\_num(F, \mathrm{CWC})$ for AD_n, AD_c, AD_m and AD_d.

For artificial data, the x-axis in Fig. 1 denotes the value of the parameter n, c, m and d in AD_n, AD_c, AD_m and AD_n, respectively. Also, for real data, since the values of $diff\_num$ for DEXTER, NOVA and SYLVA are much larger than those of DOROTHEA, GISETTE and HIVA, we group the former and the latter in Fig. 2.

Figures 1, 2 and 3 show that the values of $diff\_num$ for artificial data is much smaller than those for both real data and nucleotide sequences. On the

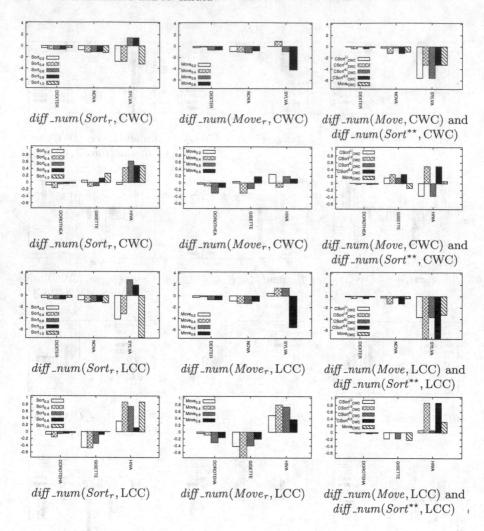

Fig. 2. $diff\_num(F, \text{CWC})$ (1st and 2nd rows) and $diff\_num(F, \text{LCC})$ (3rd and 4th rows) for the groups of DEXTER, NOVA and SYLVA and those of DOROTHEA, GISETTE and HIVA.

other hand, while the values of $diff\_num$ for nucleotide sequences are almost negative in Fig. 3, those for artificial data and real data are either positive or negative in Figs. 1 and 2.

Figure 1 shows that, for every threshold r in fluctuations $Sort_r$ and $Move_r$, there exist many cases such that the values of $diff\_num$ are just positive or just negative. For AD_n, $diff\_num$ tends to be increasing when n is increasing. For AD_c, $diff\_num$ has the maximum value when $c = 30$ (the minimum value of c) and 70 (the maximum value of c). For AD_d, $diff\_num$ is always negative when $d = 9$ (the maximum value of d).

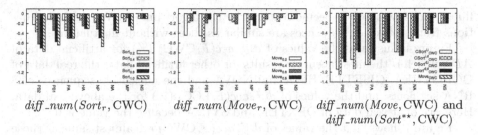

$$diff\_num(Sort_r, \text{CWC}) \qquad diff\_num(Move_r, \text{CWC}) \qquad diff\_num(Move, \text{CWC}) \text{ and}$$
$$diff\_num(Sort^{**}, \text{CWC})$$

Fig. 3. $diff\_num(F, \text{CWC})$ for nucleotide sequences of PB2, PB1, PA, HA, NP, NA, MP and NS.

Figure 2 shows that, for the algorithm CWC, in many cases for the data of HIVA, introducing fluctuations implies the positive values of $diff\_num$ (but the value is less than 1) and the values of $diff\_num$ for the data of DEXTER and DOROTHEA are always negative. On the other hand, for the algorithm LCC, the values of $diff\_num$ for the data of DEXTER, DOROTHEA and GISETTE are always negative.

Furthermore, for the data of SYLVA, the fluctuations of $Sort_{0.6}$ and $Sort_{0.8}$ give the maximum and the second maximum values and $Move_{0.4}$ gives the third maximum value of $diff\_num$ in the real data, by using both CWC and LCC. On the other hand, the fluctuations $Move_A$ and $CSort_A^{**}$ for $A \in \{\text{CWC}, \text{LCC}\}$ gives negative values of $diff\_num(F, A)$ and one of them gives the minimum value in the real data.

Table 4 illustrates the distribution of the values of $sim(F, A)$ for every fluctuation. Here, the number of artificial data, real data and nucleotide sequences is 100, 6 and 8, respectively. Note that $sim(Sort_r, A)$ and $sim(Move_r, A)$ are same even if r is varied, so we represent them as one column.

Table 4 shows that the selected features by the algorithms of CWC and LCC with fluctuations from artificial and real data are not similar as those without

Table 4. The distribution of the values of $sim(F, A)$.

Artificial data				Real data				Real data			Nucleotide sequences					
CWC				CWC				LCC			CWC					
F	0	0.1	0.2	0.3	F	0	0.1	0.2	F	0	0.1	F	0.6	0.7	0.8	0.9	1
$Sort_r$	0	87	13	0	$Sort_r$	2	4	0	$Sort_r$	4	2	$Sort_r$	0	0	3	2	3
$Move_r$	0	27	72	1	$Move_r$	2	2	2	$Move_r$	3	3	$Move_r$	0	0	0	2	6
$Move$	17	72	11	0	$Move$	4	2	0	$Move$	6	0	$Move$	0	0	0	0	8
$CSort^{ii}$	17	72	11	0	$CSort^{ii}$	4	2	0	$CSort^{ii}$	6	0	$CSort^{ii}$	1	2	0	3	2
$CSort^{id}$	14	69	17	0	$CSort^{id}$	4	2	0	$CSort^{id}$	6	0	$CSort^{id}$	2	1	0	3	2
$CSort^{di}$	17	71	12	0	$CSort^{di}$	4	2	0	$CSort^{di}$	6	0	$CSort^{di}$	1	2	0	3	2
$CSort^{dd}$	14	69	17	0	$CSort^{dd}$	4	2	0	$CSort^{dd}$	6	0	$CSort^{dd}$	1	2	0	3	2

fluctuations, while the selected features by the algorithm of CWC with fluctuations from nucleotide sequences are similar as those without fluctuations.

Figure 4 illustrates the values of $diff\_acc(F, \text{CWC})$ (%) for artificial data of AD_c, which is the most accurate results for other artificial data, the real data of DOROTHEA, GISETTE, HIVA and NOVA and the nucleotide sequences of 8 RNA segments, and the values of $diff\_acc(F, \text{LCC})$ (%) for the above real data. Here, we omit real data of DEXTER and SYLVA because the value is 0.

Figure 4 shows that the values of $diff\_acc(F, \text{CWC})$ are almost same as those of $diff\_acc(F, \text{LCC})$ for every fluctuation F, which are always negative. On the other hand, for the artificial data of AD_c, $diff\_acc$ tends to be positive and large if c is small ($c = 30, 40$) but negative otherwise ($c = 50, 60, 70$). For the nucleotide sequences, the values of $diff\_acc$ for PB2 and NA tend to be positive but those for PB1, PA, HA and NP are always negative.

4.3   Discussion

For the real data of SYLVA, the values of $diff\_acc(F, A)$ is always 0 for every fluctuation F but there exists an r ($0.2 \leq r \leq 1$) such that $diff\_num(Sort_r, A)$ and $diff\_num(Move_r, A)$ are positive in Fig. 2. Also, for the real data HIVA, the values of $diff\_acc(F, A)$ is negative but near to 0 (that is, about -0.03%) for every fluctuation F in Fig. 4 but almost values of $diff\_num(F, A)$ are positive in Fig. 2. In both cases, $sim(F, A)$ is near to 0 in Table 4.

These results imply the existence of a fluctuation such that the accuracy of a data set constructed from the selected features by the algorithm with it is same or similar as that without it, the number of selected features with it is smaller than that without it, and the selected features with it is different from that without it, which we call the successful case (1).

For the nucleotide sequences of PB2 and NA, $diff\_acc(F, \text{CWC})$ is positive in Fig. 4, $diff\_num(F, \text{CWC})$ is negative in Fig. 3 and $sim(F, \text{CWC})$ is near to 1 in Table 4 for every $F \in \{Sort_r, Move_{\text{CWC}}, CSort^{**}_{\text{CWC}}\}$.

This result implies the existence of a fluctuation such that the accuracy of a data set constructed from the selected features by the algorithm with it is larger than that without it, the number of selected features with it is larger than that without it, and the selected features with it is similar as that without it, which we call the successful case (2).

Table 5 illustrates the values of $|SF(F, A)|$, $\alpha(F, A)$ and $sim(F, A)$, by comparing them without fluctuations, for the real data of SYLVA and HIVA and the nucleotide sequences of PB2 and NA. Here, we represent the case that $|SF(F, A)|$ is minimum for SYLVA and HIVA and the case that $\alpha(F, A)$ is maximum for PB2 and NA.

For the real data of SYLVA and HIVA, Table 5 shows the successful case (1) whose number of features is smaller and whose accuracy is slightly smaller than the case without fluctuations and whose features are different from the case for the algorithm CWC. On the other hand, for the algorithm LCC, while the number of features is much smaller than the case without fluctuations and for the algorithm CWC, the accuracy is much smaller.

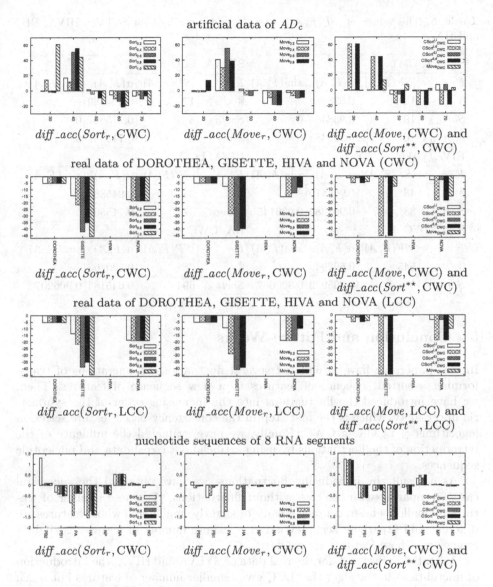

artificial data of AD_c

$diff\_acc(Sort_r, \mathrm{CWC})$ $diff\_acc(Move_r, \mathrm{CWC})$ $diff\_acc(Move, \mathrm{CWC})$ and
$diff\_acc(Sort^{**}, \mathrm{CWC})$

real data of DOROTHEA, GISETTE, HIVA and NOVA (CWC)

$diff\_acc(Sort_r, \mathrm{CWC})$ $diff\_acc(Move_r, \mathrm{CWC})$ $diff\_acc(Move, \mathrm{CWC})$ and
$diff\_acc(Sort^{**}, \mathrm{CWC})$

real data of DOROTHEA, GISETTE, HIVA and NOVA (LCC)

$diff\_acc(Sort_r, \mathrm{LCC})$ $diff\_acc(Move_r, \mathrm{LCC})$ $diff\_acc(Move, \mathrm{LCC})$ and
$diff\_acc(Sort^{**}, \mathrm{LCC})$

nucleotide sequences of 8 RNA segments

$diff\_acc(Sort_r, \mathrm{CWC})$ $diff\_acc(Move_r, \mathrm{CWC})$ $diff\_acc(Move, \mathrm{CWC})$ and
$diff\_acc(Sort^{**}, \mathrm{CWC})$

Fig. 4. $diff\_acc(F, \mathrm{CWC})$ for artificial data, real data and nucleotide sequences and $diff\_acc(F, \mathrm{LCC})$ for real data.

For the nucleotide sequences of PB2 and HA, Table 5 shows the successful case (2) whose accuracy is larger and whose number of features is slightly larger than the case without fluctuations and whose features are similar as the case.

Table 5. The values of $|SF(F,A)|$, $\alpha(F,A)$ and $sim(F,A)$ for SYLVA, HIVA, PB2 and HA.

SYLVA, CWC				SYLVA, LCC			
F	$|SF(F,A)|$	$\alpha(F,A)$	$sim(F,A)$	F	$|SF(F,A)|$	$\alpha(F,A)$	$sim(F,A)$
$\emptyset$	14	0.938484	–	$\emptyset$	17	0.904789	–
$Sort_{0.5}$	10	0.938484	0	$Sort_{0.5}$	8	0.715507	0
$Move_{0.3}$	10	0.938484	0	$Move_{0.3}$	8	0.715507	0
HIVA, CWC				HIVA, LCC			
F	$|SF(F,A)|$	$\alpha(F,A)$	$sim(F,A)$	F	$|SF(F,A)|$	$\alpha(F,A)$	$sim(F,A)$
$\emptyset$	48	0.965670	–	$\emptyset$	54	0.945883	–
$Sort_{0.6}$	38	0.964889	0.036145	$Move_{0.3}$	37	0.715507	0
PB2, CWC				HA, CWC			
F	$|SF(F,A)|$	$\alpha(F,A)$	$sim(F,A)$	F	$|SF(F,A)|$	$\alpha(F,A)$	$sim(F,A)$
$\emptyset$	545	0.592943	–	$\emptyset$	688	0.639863	–
$Sort_{0.2}$	551	0.605972	0.826667	$Sort_{0.8}$	694	0.645184	0.906207

5   Conclusion and Future Works

In this paper, we have formulated several fluctuations as operations of transforming a sorted sequence of features to a new sequence of features. Then, we have introduced the fluctuations into the increasing order of the symmetric uncertainty, which is the first step in the consistency-based feature selection algorithms of LCC and CWC. Finally, we have evaluated the influence of the introduction of the fluctuations by using artificial data, real data and nucleotide sequences.

As a result, by introducing fluctuations, we have obtained the successful cases, comparing with the case without fluctuations, (1) whose number of features is smaller, whose accuracy is same or slightly smaller and whose features are different and (2) whose accuracy is larger, whose number of features is slightly larger and whose features are similar.

As stated in Sect. 4.3, for the real data of SYLVA and HIVA, the introduction of fluctuations in the algorithm LCC gives smaller number of features but much smaller accuracy than that in the algorithm CWC. Also, in comparison of the real data with the nucleotide sequences of PB2 and HA and additionally AD_c in Fig. 4, larger number of classes tends to imply smaller accuracy. Then, it is a future work to analyze which of parameters influences the results introducing fluctuations.

It remains open how to select fluctuations and whether a more appropriate fluctuation exists, in order to select better features. It is also a future work to introduce a measure to evaluate the selected features instead of accuracy.

References

1. Almuallim, H., Dietterich, T.G.: Learning boolean concepts in the presence of many irrelevant features. Artif. Intell. **69**, 279–305 (1994)
2. Arauzo-Azofra, A., Benitez, J.M., Castro, J.L.: Consistency measures for feature selection. J. Intell. Inf. Syst. **30**, 273–292 (2008)
3. Bolón-Canedo, V., Sánchez-Maroño, N., Alonso-Betanozos, A.: Feature Selection for High-Dimensional Data. Springer, Heidelberg (2015). https://doi.org/10.1007/978-3-319-21858-8
4. Dash, M., Liu, H.: Consistency-based search in feature selection. Artif. Intell. **151**, 155–176 (2003)
5. Hall, M.A.: Correlation-based feature selection for machine learning, Ph.D thesis, Waikato University (1998)
6. Quinlan, J.R.: Induction of decision trees. Mach. Learn. **1**, 81–106 (1986)
7. Kononenko, I.: Estimating attributes: analysis and extensions of RELIEF. In: Bergadano, F., De Raedt, L. (eds.) ECML 1994. LNCS, vol. 784, pp. 171–182. Springer, Heidelberg (1994). https://doi.org/10.1007/3-540-57868-4_57
8. Press, W.H., Flannery, B.P., Teukolsky, S.A., Vetterling, W.T.: Numerical recipes in C. Cambridge University Press, Cambridge (1988)
9. Shannon, C.E., Weaver, W.: The Mathematical Theory of Communication. University of Illinois Press, Champaign (1948)
10. Shin, K., Xu, X.M.: Consistency-based feature selection. In: Velásquez, J.D., Ríos, S.A., Howlett, R.J., Jain, L.C. (eds.) KES 2009. LNCS (LNAI), vol. 5711, pp. 342–350. Springer, Heidelberg (2009). https://doi.org/10.1007/978-3-642-04595-0_42
11. Shin, K., Fernanndes, D., Miyazaki, S.: Consistency measures for feature selection: a formal definition, relative sensitivity comparison, and a fast algorithm. In: Proceedings IJCAI 2011, pp. 1491–1497. AAAI/IJCAI Press (2011)
12. Shin, K., Kuboyama, T., Hashimoto, T., Shepard, D.: Super-CWC and super-LCC: Super fast feature selection algorithms. In: Proceedings BigData 2015, pp. 1–7. IEEE (2015)
13. Shin, K., Miyazaki, S.: A fast and accurate feature selection algorithm based on binary consistency measures. Comput. Intell. **32**, 646–666 (2016)
14. Webb, A.R., Kopsey, K.D.: Statistical Pattern Recognition, 3rd edn. Wiley, Hoboken (2011)
15. Yu, L., Liu, H.: Efficient feature selection via analysis of relevance and redundancy. J. Mach. Learn. Res. (JMLR) **5**, 1205–1224 (2004)
16. Zhao, Z., Liu, H.: Searching for interacting features. In: Proceedings IJCAI 2007, pp. 1156–1161. AAAI/IJCAI Press (2007)

Card-Based Cryptography
with Invisible Ink

Kazumasa Shinagawa[1,2]([⊠])

[1] Tokyo Institute of Technology, Meguro-ku, Japan
[2] National Institute of Advanced Industrial Science and Technology,
Koto-ku, Japan
shinagawakazumasa@gmail.com

Abstract. It is known that secure computation can be done by using a
deck of physical cards; *card-based cryptography* makes people understand
the correctness and security of secure computation, even for people who
are not familiar with mathematics. In this paper, we propose a new
type of cards, *layered polygon cards*, based on the use of *invisible ink*.
A deck of cards with invisible ink naturally hides the contents of cards
and allows to open some pieces of contents, which we referred to it as
partial opening. Based on them, we construct novel protocols for various
interesting functions such as carry of addition, equality, and greater-than.

1 Introduction

Secure computation enables a set of parties each having inputs to jointly compute
a predetermined function of their inputs without revealing their inputs beyond
the output. Card-based cryptography (ex. [3,4,10]) is secure computation that
can be done by using a deck of physical cards, instead of computer devices. This
makes people understand the correctness and security of secure computation,
even for people who are not familiar with mathematics. Indeed, it is applied to
educational situations; some universities (e.g., Cornell University [7], University
of Waterloo [2], and Tohoku University [8]) adopt card-based cryptography as a
teaching material for beginner students.

While most of all existing works [1–3,5,6,9–12,15] are mainly focused on
binary computation only, a lot of secure computation that arises in everyday
and classroom situations needs to take multi-valued inputs. For instance, secure
computation of the average score, which takes a number of scores and outputs the
average of them, is such a canonical example. In order to compute multi-valued
functions efficiently, Shinagawa et al. [14] proposed a deck of *regular polygon
cards*, whose shape is a regular n-sided polygon for the base number n. They
proposed a two-card addition protocol that outputs $x + y \bmod n$ given two cards
having $x, y \in \{0, 1, \cdots, n-1\}$.

Does a deck of regular polygon cards realize sufficiently efficient secure computation for multi-valued functions? Up until now, there exist efficient protocols
only for a very restrictive class of functions such as addition and subtraction,
however, it requires a large number of cards for computing a function in the outside of the class (in general, it requires $O(n^k)$ cards for k inputs). Unfortunately,

© Springer Nature Switzerland AG 2019
T. V. Gopal and J. Watada (Eds.): TAMC 2019, LNCS 11436, pp. 566–577, 2019.
https://doi.org/10.1007/978-3-030-14812-6_35

Table 1. Comparison between our protocols and previous protocols: "RPC", and "LPC" denote regular polygon cards and layered polygon cards, respectively.

	Type of cards	Number of cards	Number of shuffles
∘ Addition and subtraction			
Shinagawa et al. [14]	RPC	2	1
Ours	LPC	2	1
∘ Carry: the predicate "$x + y \geq n$"			
Shinagawa et al. [14]	RPC	$n^2 + n + 2$	2
Ours	LPC	2	5
∘ Equality with zero: the predicate "$x = 0$"			
Shinagawa et al. [14]	RPC	$2n + 1$	1
Ours	LPC	2	4
∘ Equality: the predicate "$x = y$"			
Shinagawa et al. [14]	RPC	$n^2 + n + 2$	2
Ours	LPC	2	6
∘ Greater than: the predicate "$x \geq y$"			
Shinagawa et al. [14]	RPC	$n^2 + n + 2$	2
Ours	LPC	2	5

there are no efficient protocols even for very simple functions such as *addition with carry*, where given two integers $x, y \in \{0, 1, \cdots, n - 1\}$, it outputs a carry of addition, the predicate "$x + y \geq n$". To compute a carry of addition efficiently is one of the open problems in this area. In this paper, we solve it by designing a new type of cards.

1.1 Our Contribution

In this paper, we apply *invisible ink* to the area of card-based cryptography for the first time. The use of invisible ink makes it easier to design a new types of cards and allows a *partial opening*, which partially reveals the contents of cards. Then, we design a new type of cards, *layered polygon cards*, and an encoding rule that is compatible with partial opening: it reveals either a value part $x \in \{0, 1, \cdots, n - 1\}$ of information or a sign part $s \in \{0, 1\}$ of information. In Sect. 3.3, we propose a *conversion protocol* that takes a card with a sign $s \in \{0, 1\}$ and outputs a card with a value $s \in \{0, 1\}$. Using the conversion protocol, we construct an efficient protocol for computing a carry of addition, the predicate "$x + y \geq n$". This can be also applied to other interesting predicates (equality with zero "$x = 0$", equality "$x = y$", and greater than "$x \geq y$"). Table 1 shows a comparison between our protocols and the previous protocols [14] with regular polygon cards (RPC). Somewhat surprisingly, our protocols with layered polygon cards (LPC) for these predicates requires only two cards while RPC-based protocols needs $\mathsf{poly}(n)$ number of cards for the same predicates.

1.2 Related Work

Binary Cards. A deck of binary cards is the first proposed type of card, where the front sides are either ♣ or ♡, and the back sides are the same pattern ?. In order to encode a binary value, the following encoding rule is used:

$$\boxed{\clubsuit}\,\boxed{\heartsuit} = 0, \qquad \boxed{\heartsuit}\,\boxed{\clubsuit} = 1.$$

Regular Polygon Cards. Shinagawa et al. [14] proposed a deck of regular n-sided polygon cards, $n \geq 3$, whose shape is a regular n-sided polygon, in order to efficiently compute a function over $\mathbb{Z}/n\mathbb{Z} = \{0,1,2,\cdots,n-1\}$. Its front side has an arrow and back side has nothing. For example, a regular 4-sided polygon card has a square shape and an arrow in its front side. In order to encode a value in $\mathbb{Z}/4\mathbb{Z}$, the following encoding rule is used:

$$\boxed{\rightarrow} = 0, \quad \boxed{\uparrow} = 1, \quad \boxed{\leftarrow} = 2, \quad \boxed{\downarrow} = 3.$$

Triangle Cards. Shinagawa and Mizuki [13] proposed a deck of triangle cards, whose shape is a regular triangle. Its front and back sides are indistinguishable while a regular 3-sided polygon card has distinguishable front/back sides.

$$\triangle = 0, \quad \triangle = 1, \quad \triangle = 2.$$

In order to hide the contents of the cards, both faces of a card can be hidden by placing seals as follows:

$$\triangle.$$

We note that the use of invisible ink makes it easier to make triangle cards. Indeed, our work is inspired by triangle cards.

1.3 Notation

For any integer $n > 0$, we use "$[n]$" to denote the set $\mathbb{Z}/n\mathbb{Z} = \{0,1,\cdots,n-1\}$. We use "$\mathsf{p}(\cdot)$" to denote predicates, which is a function outputting either 0 (as false symbol) or 1 (as true symbol). For example, $\mathsf{p}(x = y)$ is a function that takes x, y as input and outputs 1 if $x = y$ and 0 otherwise. Throughout of this paper, we use *radian* to denote the angle of rotation.

2 Layered Polygon Cards

In this section, we propose a new type of cards, *layered polygon cards*, using invisible ink. Thanks to the invisibility of invisible ink, it naturally hides the contents of cards and allows a *partial opening*. In Sect. 2.1, we newly design a deck of layered polygon cards using invisible ink. In Sects. 2.2 and 2.3, we introduce partial openings and shuffles, which are the most important operations in our protocols. In Sect. 2.4, we remark on complexity measure.

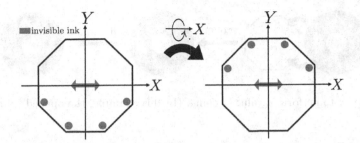

Fig. 1. A layered polygon card ($n = 4$): the left (right) card turns into the right (left) card by rotating π rad with the X-axis.

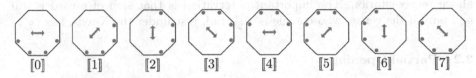

Fig. 2. The encoding for layered polygon card when $n = 4$

2.1 Layered Polygon Cards

Invisible Ink. Invisible ink is used for writing, which is invisible but can be made visible with illuminating a black light. It can be used for *steganography*, which hides the existence of plain text while *cryptography* hides the contents of plain text. Our idea is to use invisible ink for designing cards. Card with invisible ink has nice properties: (1) it naturally hides the contents of cards, (2) it allows to do "partial opening" (see Sect. 2.2), and (3) thanks to partial opening, it is meaningful to put lots of information on a single card.

Layered Polygon Cards. For the base number $n \geq 2$, a layered polygon card has a regular $2n$-sided polygon. Figure 1 shows an example of a layered polygon card when $n = 4$. Each vertex under the X-axis has a symbol "•" written by invisible ink on both of front and back sides, and the center of the card has a single "⇔" written by invisible ink. Thus, by rotating π rad with the X-axis, the left (resp. right) card (having all "•"s under the X-axis) turns into the right (resp. left) card (having all "•"s over the X-axis). Note that although this rotation, which flips with the X-axis, seems equivalent to the rotation π rad around the center point in the plane, they are not the same in general.

Encoding Rule. In order to encode a value, we use the *angle of rotation*. We denote a card with an angle of $\frac{X \cdot \pi}{n}$ rad by $[\![X]\!]$ as in Fig. 2. We usually pick X from an element of $[2n] = \{0, 1, \cdots, 2n - 1\}$ but sometimes from outside of it. For instance, a card $[\![\frac{1}{2}]\!]$ has an angle of $\frac{\pi}{2n}$. It is easily observed that for a layered polygon card $[\![X]\!]$, add $A \in [2n]$ (resp. subtract A) to $[\![X]\!]$ is done with rotating it by $\frac{A \cdot \pi}{n}$ rad (resp. $-\frac{A \cdot \pi}{n}$ rad), without revealing X. We can observe that by rotating π rad with the X-axis, a card $[\![X]\!]$ turns into $[\![n - X]\!]$, especially, $[\![0]\!]$ turns into $[\![n]\!]$. We will use these properties in our construction.

Fig. 3. How to perform a value opening. (In this example, the opened value is 1.)

Canonical Representation. For a card $[\![X]\!]$, $X \in [2n]$, we sometimes represent it as $[\![x + sn]\!]$, where $x \in [n]$ and $s \in \{0, 1\}$. The number x is called a *value* of the card and the bit s is called a *sign* of the card. Hereafter, we mainly use the above representation. The important observation is that sign of a card is 0 if and only if the vertex whose angle is $-\frac{\pi}{2n}$ rad, just under the X-axis, has "•".

2.2   Partial Opening

Opening is an important operation, which reveals contents of cards. While the standard opening reveals all pieces of information in a card, our opening reveals some pieces of information (*partial opening*). In particular, we define two types of opening, *value opening* and *sign opening*: for a layered polygon card $[\![x + sn]\!]$, the former reveals the value $x \in [n]$ and the latter reveals the sign $s \in \{0, 1\}$.

Value Opening. Recall that a value of a card is encoded at the inside area of the card by angle of the arrow. Figure 3 shows an example of how to proceeds a value opening. It is done by using a doughnut-shaped cover that hides the outside area of the card. Then, by illuminating with black light, the arrow is popped up; then, we obtain the value, i.e., if the angle of the arrow is $\frac{v \cdot \pi}{n}$ rad, the opened value is $v \in [n]$.

Sign Opening. Recall that a sign of a card is encoded at the vertex whose angle is $-\frac{\pi}{2n}$ rad. Figure 4 shows an example of how to proceed a sign opening. It is done by using a wrench-shaped cover that hides the other area of the card as follows: Then, by illuminating with black light, we can obtain the sign, i.e., if "•" is popped up, the sign is 0, otherwise, the sign is 1.

2.3   Shuffle

Shuffle is as important operation as opening in card-based cryptography. It is a probabilistic transformation from a sequence of cards to a sequence of cards. The important request for achieving security is that nobody can guess which possible sequence is chosen by the shuffle. In this paper, we use four types of shuffles, which are implementable physically with helping objects like clips, rubber bands, and wooden boards.

Rotation Shuffle. It takes a number of cards $([\![X_1]\!], [\![X_2]\!], \cdots, [\![X_k]\!])$ as input and outputs $([\![X_1 + R]\!], [\![X_2 + R]\!], \cdots, [\![X_k + R]\!])$, where $R \in [2n]$ is a uniformly random number and hidden from all parties. In order to implement a rotation

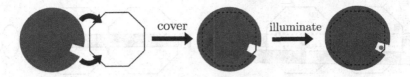

Fig. 4. How to perform a sign opening. (In this example, the opened value is 0.)

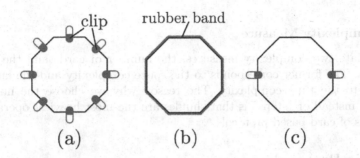

Fig. 5. Implementations of a rotation shuffle, (a) and (b), and a two-sided rotation shuffle (c), using clips and a rubber band.

shuffle, all cards are stacked and fixed either by $2n$ clips as (a) in Fig. 5 or by a rubber band as (b) in Fig. 5; and then, it is randomly rotated like a frisbee or a roulette until nobody can guess the angle of rotation.

Two-Sided Rotation Shuffle. It takes a number of cards ($[\![X_1]\!], [\![X_2]\!], \cdots , [\![X_k]\!]$) as input and outputs ($[\![X_1+rn]\!], [\![X_2+rn]\!], \cdots , [\![X_k+rn]\!]$), where $r \in \{0, 1\}$ is a uniformly random bit and hidden from all parties. In order to implement a two-sided rotation shuffle, all cards are stacked and fixed by 2 clips as (c) in Fig. 5; and then, it is randomly rotated like a frisbee or a roulette until nobody can guess the angle of rotation.

Flipping Shuffle. It takes a number of cards ($[\![X_1]\!], [\![X_2]\!], \cdots , [\![X_k]\!]$) as input and outputs ($[\![(-1)^r X_1 + rn]\!], [\![(-1)^r X_2 + rn]\!], \cdots , [\![(-1)^r X_n + rn]\!]$), where $r \in \{0, 1\}$ is a uniformly random bit and hidden from all parties. It is used in the conversion protocol (Sect. 3.3). In order to implement a flipping shuffle, all cards are placed in a single line and fixed by two wooden boards as in Fig. 6; and then, it is randomly rotated with the X-axis until nobody can guess which sequence is chosen.

Perfect Randomization. It takes a card $\lfloor X \rfloor$ as input and outputs a random card $[\![R]\!]$, where $R \in [2n]$ is a uniformly random number and hidden from all parties. In order to implement a perfect randomization, just throwing it in the air like a coin tossing.

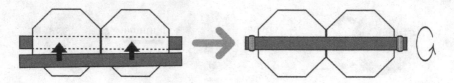

Fig. 6. An implementation of a flipping shuffle using two wooden boards.

2.4   Complexity Measure

We associate two complexity measures, the number of cards and the number of shuffles. The former corresponds to the space complexity and the latter corresponds to the time complexity. The reason why we choose the number of "shuffles" instead of "steps" is that shuffles are the most heaviest operations in executions of card-based protocols.

3   Basic Protocols

In this section, we design secure computation protocols for basic functionality. In Sect. 3.1, we adopt the addition/subtraction protocols of regular polygon cards [14] to layered polygon cards. In Sect. 3.2, we construct a sign normalization protocol, which enforces a sign of the input card to 0 without revealing the original sign. In Sect. 3.3, we construct a conversion protocol, which converts a sign of the input into a value of the output.

3.1   Addition Protocol

It takes two cards $[\![X]\!]$ and $[\![Y]\!]$, $X, Y \in [2n]$, as input and outputs a card $[\![X+Y]\!]$ without revealing the inputs. For simplicity, we use double representation instead of canonical representation. It proceeds as follows:

1. Place two cards in a single line as $([\![X]\!], [\![Y]\!])$.
2. Flip the second card with the X-axis; then, we obtain a new sequence:

$$([\![X]\!], [\![-Y]\!]).$$

3. Apply a rotation shuffle to them; then, for a random value $R \in [2n]$, we obtain a new sequence:

$$([\![X + R]\!], [\![-Y + R]\!]).$$

4. Flip the second card with the X-axis; then, we obtain a new sequence:

$$([\![X + R]\!], [\![Y - R]\!]).$$

5. Open the second card by value opening and sign opening; and then, rotate the first card by $\frac{V \cdot \pi}{2n}$ rad, where $V = Y - R$ is the opened value. The current sequence is:

$$([\![X + Y]\!], [\![Y - R]\!]).$$

6. Apply a perfect randomization to the second card, open it by value opening and sign opening, and rotate it so as to be $[\![0]\!]$; then, we obtains the result sequence:

$$([\![X+Y]\!], [\![0]\!]).$$

Subtraction Protocol. The above addition protocol can be converted into a subtraction protocol with two modifications: (1) skip Step 2 and (2) rotate the first card by $-\frac{V \cdot \pi}{2n}$ instead of $\frac{V \cdot \pi}{2n}$. Then, it outputs $([\![X-Y]\!], [\![0]\!])$.

Complexity. It requires two cards and a single shuffle.

3.2 Sign Normalization Protocol

Sign normalization protocol converts a card with unknown sign into a card with a fixed sign without revealing the sign. Specifically, it takes a single card $[\![x+sn]\!]$, $x \in [n]$ and $s \in \{0, 1\}$, as input and outputs a card $[\![x]\!]$ without revealing x and s. It proceeds as follows:

1. Apply a two-sided rotation shuffle to the input card; then, we obtain a new card $[\![x + (s+r)n]\!]$ for a uniformly random value $r \in \{0, 1\}$.
2. Open the sign of the card; and then, rotate it by π rad if the sign is 1. Then, we obtain the result card $[\![x]\!]$.

Complexity. It requires a single card and a single shuffle.

3.3 Conversion Protocol

It plays an important role to securely compute interesting predicates in Sect. 4. We use two types of conversion: positive conversion and negative conversion. Both of them takes a single card $[\![x+sn]\!]$ as input and outputs a card $[\![s]\!]$ if it is positive and $[\![1-s]\!]$ if it is negative. Let type $\in \{\mathsf{positive}, \mathsf{negative}\}$ be a conversion type of the protocol. It proceeds as follows:

1. Place two cards as follows:

$$([\![x+sn]\!], [\![0]\!]).$$

2. Apply a two-sided rotation shuffle to them; then, for $r \in \{0, 1\}$, we obtain a new sequence:

$$([\![x + (s+r)n]\!], [\![rn]\!]).$$

3. Open the sign of the first card; if it is 1, then rotate the second card by π rad. Then, we obtain a new sequence:

$$([\![x + (s+r)n]\!], [\![sn]\!]).$$

4. Randomize the first card; and then, make a card $[\![0]\!]$ by applying opening and rotation. Then, we obtain a new sequence $([\![0]\!], [\![sn]\!])$.

5. Rotate the first card by $-\frac{\pi}{2n}$ if type = positive and by $\frac{\pi}{2n}$ if type = negative; then, we obtain a new sequence:

$$\begin{cases} [\![-(1/2)]\!], [\![sn]\!]) & \text{if type = positive} \\ [\![1/2]\!], [\![sn]\!]) & \text{if type = negative.} \end{cases} \tag{1}$$

6. Apply a flipping shuffle to them; then, we obtain a new sequence:

$$\begin{cases} ([\![(-1)^{1+t}(1/2) + tn]\!], [\![(s+t)n]\!]) & \text{if type = positive} \\ ([\![(-1)^{t}(1/2) + tn]\!], [\![(s+t)n]\!]) & \text{if type = negative.} \end{cases}$$

7. Open the sign of the second card; if it is 1, then flip the first card with the X-axis and rotate the second card by π rad. Then, we obtain a new sequence:

$$\begin{cases} ([\![(-1)^{1+s}(1/2) + sn]\!], [\![0]\!]) & \text{if type = positive} \\ ([\![(-1)^{s}(1/2) + sn]\!], [\![0]\!]) & \text{if type = negative.} \end{cases}$$

8. Rotate the first card by $\frac{\pi}{2n}$ if type = positive and by $-\frac{\pi}{2n}$ if type = negative; then, we obtain a new sequence:

$$\begin{cases} ([\![s + sn]\!], [\![0]\!]) & \text{if type = positive} \\ ([\![1 - s + sn]\!], [\![0]\!]) & \text{if type = negative.} \end{cases}$$

9. Apply the sign normalization protocol in Sect. 3.2 to the first card; then, we obtain the result sequence:

$$\begin{cases} ([\![s]\!], [\![0]\!]) & \text{if type = positive} \\ ([\![1 - s]\!], [\![0]\!]) & \text{if type = negative.} \end{cases}$$

Complexity. It requires two cards and four shuffles (in Steps 2, 4, 6, and 9).

Optimization. Step 9 can be skipped if we do not care about the sign of the output card. Then, the number of shuffles is reduced to three.

4   Application: Protocols for Interesting Predicates

In this section, we design secure computation protocols for interesting predicates. In Sect. 4.1, we construct a carry protocol for the predicate "$x + y \geq n$". This is used for addition with carry: for two integers $x, y \in [n]$, addition with carry outputs a tuple of integers $(x + y \geq n), x + y \bmod n$. In Sects. 4.2 and 4.3, we construct equality protocols for the predicates of $\mathsf{p}(x = 0)$ and $\mathsf{p}(x = y)$, respectively. In Sect. 4.4, we construct a greater-than protocol for the predicate $\mathsf{p}(x \geq y)$. We note that there are no efficient solutions for them by using regular polygon cards.

4.1   Carry Protocol

It takes two cards $[\![x]\!]$, $[\![y]\!]$ as input and outputs a card $[\![\mathsf{p}(x+y \geq n)]\!]$. It proceeds as follows:

1. Place two cards as follows:
$$([\![x]\!], [\![y]\!]).$$

2. Apply the addition protocol in Sect. 3.1 to the sequence; then, we obtain a new sequence:
$$([\![x+y]\!], [\![0]\!]).$$

3. Apply the positive conversion protocol in Sect. 3.3; and then, we obtain the result sequence:
$$([\![\mathsf{p}(x+y \geq n)]\!], [\![0]\!]).$$

   (Note that the sign of $[\![x+y]\!]$ is 1 only if $x+y \geq n$.)

Complexity. It requires two cards and five shuffles.

4.2   Equality with Zero Protocol

It takes a single card $[\![x]\!]$ as input and outputs a card $[\![\mathsf{p}(x = 0)]\!]$. It proceeds as follows:

1. Place two cards as follows:
$$([\![x]\!], [\![0]\!]).$$

2. Flip the first card; and then, rotate it by $-\pi$ rad. The current sequence is:
$$([\![-x]\!], [\![0]\!]).$$

3. Apply the negative conversion protocol in Sect. 3.3; and then, we obtain the result sequence:
$$([\![\mathsf{p}(x = 0)]\!], [\![0]\!]).$$

   (Note that the sign of $[\![-x]\!]$ is 0 only if $x = 0$.)

Complexity. It requires two cards and four shuffles.

4.3   Equality Protocol

It takes two cards $[\![x]\!]$, $[\![y]\!]$ as input and outputs a card $[\![\mathsf{p}(x = y)]\!]$. It proceeds as follows:

1. Place two cards as follows:
$$([\![x]\!], [\![y]\!]).$$

2. Apply the subtraction protocol in Sect. 3.1 to the sequence; then, we obtain a new sequence:

$$([\![x - y]\!], [\![0]\!]).$$

3. Apply the sign normalization protocol in Sect. 3.2 to the first card; then, we obtain a new sequence:

$$([\![|x - y|]\!], [\![0]\!]).$$

(This step is for the equality with zero protocol that takes a card with sign 0 as input.)

4. Apply the equality with zero protocol in Sect. 4.2; then, we obtain the result sequence:

$$([\![\mathsf{p}(|x - y| = 0)]\!], [\![0]\!]).$$

Complexity. It requires two cards and six shuffles.

4.4   Greater-Than Protocol

It takes two cards $[\![x]\!]$, $[\![y]\!]$ as input and outputs a card $[\![\mathsf{p}(x \geq y)]\!]$. It proceeds as follows:

1. Place two cards as follows:

$$([\![x]\!], [\![y]\!]).$$

2. Apply the subtraction protocol in Sect. 3.1 to the sequence; then, we obtain a new sequence:

$$([\![x - y]\!], [\![0]\!]).$$

3. Apply the negative conversion protocol in Sect. 3.3; and then, we obtain the result sequence:

$$([\![\mathsf{p}(x \geq y)]\!], [\![0]\!]).$$

(Note that the sign of $[\![x - y]\!]$ is 0 only if $x \geq y$.)

Complexity. It requires two cards and five shuffles.

5   Conclusion and Future Work

In this paper, we designed a new type of cards, layered polygon cards, with invisible ink, and constructed efficient protocols for various interesting predicates. We believe that the use of invisible ink makes it easier to design a new type of cards that enable to construct efficient secure computation protocols. An interesting research direction is to find such a new type of cards and objects, e.g., polyhedron.

Acknowledgments. This work was supported in part by JSPS KAKENHI Grant Number 17J01169.

References

1. Abe, Y., Hayashi, Y., Mizuki, T., Sone, H.: Five-card AND protocol in committed format using only practical shuffles. In: Proceedings of the 5th ACM on ASIA Public-Key Cryptography Workshop, APKC@AsiaCCS, Incheon, Republic of Korea, 4 June 2018, pp. 3–8. ACM (2018)
2. Cheung, E., Hawthorne, C., Lee, P.: CS 758 project: secure computation with playing cards (2013). https://csclub.uwaterloo.ca/~cdchawth/files/papers/secure_playing_cards.pdf
3. Crépeau, C., Kilian, J.: Discreet solitary games. In: Stinson, D.R. (ed.) CRYPTO 1993. LNCS, vol. 773, pp. 319–330. Springer, Heidelberg (1994). https://doi.org/10.1007/3-540-48329-2_27
4. den Boer, B.: More efficient match-making and satisfiability *the five card trick*. In: Quisquater, J.-J., Vandewalle, J. (eds.) EUROCRYPT 1989. LNCS, vol. 434, pp. 208–217. Springer, Heidelberg (1990). https://doi.org/10.1007/3-540-46885-4_23
5. Kastner, J., et al.: The minimum number of cards in practical card-based protocols. In: Takagi, T., Peyrin, T. (eds.) ASIACRYPT 2017, Part III. LNCS, vol. 10626, pp. 126–155. Springer, Cham (2017). https://doi.org/10.1007/978-3-319-70700-6_5
6. Koch, A., Walzer, S., Härtel, K.: Card-Based Cryptographic Protocols Using a Minimal Number of Cards. In: Iwata, T., Cheon, J.H. (eds.) ASIACRYPT 2015, Part I. LNCS, vol. 9452, pp. 783–807. Springer, Heidelberg (2015). https://doi.org/10.1007/978-3-662-48797-6_32
7. Marcedone, A., Wen, Z., Shi, E.: Secure dating with four or fewer cards. Cryptology ePrint Archive, Report 2015/1031 (2015). https://eprint.iacr.org/2015/1031
8. Mizuki, T.: Applications of card-based cryptography to education. IEICE Tech. Rep. **116**(289), 13–17 (2016)
9. Mizuki, T., Kumamoto, M., Sone, H.: The five-card trick can be done with four cards. In: Wang, X., Sako, K. (eds.) ASIACRYPT 2012. LNCS, vol. 7658, pp. 598–606. Springer, Heidelberg (2012). https://doi.org/10.1007/978-3-642-34961-4_36
10. Mizuki, T., Sone, H.: Six-card secure AND and four-card secure XOR. In: Deng, X., Hopcroft, J.E., Xue, J. (eds.) FAW 2009. LNCS, vol. 5598, pp. 358–369. Springer, Heidelberg (2009). https://doi.org/10.1007/978-3-642-02270-8_36
11. Mizuki, T., Uchiike, F., Sone, H.: Securely computing XOR with 10 cards. Australas. J. Comb. **36**, 279–293 (2006)
12. Niemi, V., Renvall, A.: Secure multiparty computations without computers. Theor. Comput. Sci. **191**(1–2), 173–183 (1998)
13. Shinagawa, K., Mizuki, T.: Card-based protocols using triangle cards. In: 9th International Conference on Fun with Algorithms, FUN 2018, La Maddalena, Italy, 13–15 June 2018. LIPIcs, vol. 100, pp. 31:1–31:13 (2018)
14. Shinagawa, K., et al.: Card-based protocols using regular polygon cards. IEICE Trans. **100-A**(9), 1900–1909 (2017)
15. Stiglic, A.: Computations with a deck of cards. Theor. Comput. Sci. **259**(1–2), 671–678 (2001)

Read-Once Certification of Linear Infeasibility in UTVPI Constraints

K. Subramani[✉] and Piotr Wojciechowski

Lane Department of Computer Science and Electrical Engineering,
West Virginia University, Morgantown, WV, USA
k.subramani@mail.wvu.edu, pwojciec@mix.wvu.edu

Abstract. In this paper, we discuss the design and analysis of a polynomial time algorithm for a problem associated with a linearly infeasible system of Unit Two Variable Per Inequality (UTVPI) constraints, viz., the read-once refutation (ROR) problem. Recall that a UTVPI constraint is a linear relationship of the form: $a_i \cdot x_i + a_j \cdot x_j \leq b_{ij}$, where $a_i, a_j \in \{0, 1, -1\}$. A conjunction of such constraints is called a UTVPI constraint system (UCS) and can be represented in matrix form as: $\mathbf{A} \cdot \mathbf{x} \leq \mathbf{c}$. These constraint find applications in a host of domains including but not limited to operations research and program verification. For the linear system $\mathbf{A} \cdot \mathbf{x} \leq \mathbf{b}$, a refutation is a collection of m variables $\mathbf{y} = [y_1, y_2, \ldots, y_m]^T \in \mathbb{R}^m_+$, such that $\mathbf{y} \cdot \mathbf{A} = \mathbf{0}, \mathbf{y} \cdot \mathbf{b} < 0$. Such a refutation is guaranteed to exist for any infeasible linear program, as per Farkas' lemma. The refutation is said to be read-once, if each $y_i \in \{0, 1\}$. Read-once refutations are **incomplete** in that their existence is not guaranteed for infeasible linear programs, in general. Indeed they are not complete, even for UCSs. Hence, the question of whether an arbitrary UCS has an ROR is both interesting and non-trivial. In this paper, we reduce this problem to the problem of computing a minimum weight perfect matching (MWPM) in an undirected graph. This results in an algorithm that runs in time polynomial in the size of the input UCS.

1 Introduction

This paper is concerned with the design and analysis of a polynomial time algorithm for the problem of checking if a linearly infeasible system of UTVPI constraints has a read-once refutation (ROR). A linear relationship of the form: $a_i \cdot x_i + a_j \cdot x_j \leq b_{ij}$ is called a UTVPI constraint, if $a_i, a_j \in \{0, 1, -1\}$. A conjunction of such constraints is called a UTVPI constraint system (UCS). Observe that

K. Subramani—This research was supported in part by the Air Force Research Laboratory Information Directorate, through the Air Force Office of Scientific Research Summer Faculty Fellowship Program and the Information Institute®, contract numbers FA8750-16-3-6003 and FA9550-15-F-0001.

P. Wojciechowski—This research was made possible by NASA WV EPSCoR grant #NNX15AK74A.

© Springer Nature Switzerland AG 2019
T. V. Gopal and J. Watada (Eds.): TAMC 2019, LNCS 11436, pp. 578–593, 2019.
https://doi.org/10.1007/978-3-030-14812-6_36

a UCS is a specialized linear program and hence can be represented in matrix form as: $\mathbf{U}$: $\mathbf{A} \cdot \mathbf{x} \le \mathbf{b}$. Accordingly, as per Farkas' lemma, if $\mathbf{U}$ is empty, then there must exist a non-negative m-vector $\mathbf{y}$, such that $\mathbf{y} \cdot \mathbf{A} = \mathbf{0}$, $\mathbf{y} \cdot \mathbf{b} < 0$. This vector $\mathbf{y}$ is called a refutation of $\mathbf{U}$, since it serves as a "certificate" for the emptiness of $\mathbf{U}$. We are concerned with a specialized refutation called *read-once refutation*. The refutation $\mathbf{y}$ is said to be read-once, if each $y_i \in \{0, 1\}$.

Read-once refutations are the simplest form of refutations, since they correspond to summing a subset of the constraints of $\mathbf{U}$ to derive a contradiction. It follows that users would prefer to receive such refutations of infeasibility. Unfortunately, read-once refutations are **incomplete**, in that there exist infeasible linear programs which do not have read-once refutations. Indeed, such is the case for UTVPI constraints systems (see Sect. 2). Consequently, the question of whether an arbitrary UCS has a read-once refutation is an interesting one. In this paper, we reduce this problem to the problem of finding a minimum weight perfect matching (MWPM) in an undirected graph. This results in an algorithm that runs in time polynomial in the size of the input UCS.

UTVPI constraints find applications in a number of domains such as program verification [18] and abstract interpretation [19]. Providing easily checkable, short certificates of infeasibility is therefore a practically significant endeavor.

The rest of the paper is organized as follows: Sect. 2 formally describes the ROR problem in UTVPI constraints. An enumeration of our contributions is also provided in this section. The motivation for our work and related approaches in the literature are described in Sect. 3. In Sect. 4, we detail the new polynomial time algorithm for the ROR existence problem. We conclude in Sect. 5 by summarizing our contributions and identifying avenues for future research.

2 Statement of Problems

In this section, we formally specify the problems under consideration and define the terms that will be used in the rest of the paper.

System (1) denotes a system of linear inequalities (or linear program).

$$\mathbf{A} \cdot \mathbf{x} \le \mathbf{b} \tag{1}$$

We assume without loss of generality that $\mathbf{A}$ has dimensions $m \times n$ and that $\mathbf{b}$ is an integral m-vector.

Definition 1. *A constraint of the form $a_i \cdot x_i < b_i$ is called an absolute constraint if $a_i \in \{1, -1\}$.*

Definition 2. *A constraint of the form $a_i \cdot x_i + a_j \cdot x_j \le b_{ij}$ is called a difference constraint, if $a_i, a_j \in \{1, -1\}$ and $a_i = -a_j$.*

A conjunction of difference constraints and absolute constraints is called a difference constraint system (DCS).

Definition 3. *A constraint of the form $a_i \cdot x_i + a_j \cdot x_j \leq b_{ij}$ is called a Unit Two Variable per Constraint (UTVPI), if $a_i, a_j \in \{0, 1, -1\}$, and a_i and a_j are not both 0.*

A conjunction of UTVPI constraints is called a UTVPI constraint system (UCS).

In the above definitions, b_{ij} is called the defining constant of the constraint. For the constraint systems studied in this paper, we require that $b_{ij} \in \mathbb{Z}$.

For instance, $x_1 \leq 5$ is an absolute constraint, $x_1 - x_2 \leq 4$ is a difference constraint, and $x_1 + x_2 \leq 4$ is a UTVPI constraints. Note that difference constraints are also UTVPI constraints.

In a UTVPI constraint, each variable x_i can appear as either x_i or $-x_i$. These are referred to as literals.

We are interested in certificates of infeasibility; in particular, we are interested in resolution refutations. In linear programs (systems of linear inequalities), we use the following rule, which plays the role that resolution does in clausal formulas:

$$\frac{\sum_{i=1}^{n} a_i \cdot x_i \leq b_1 \qquad \wedge \qquad \sum_{i=1}^{n} a_i' \cdot x_i \leq b_2}{\sum_{i=1}^{n} (a_i + a_i') \cdot x_i \leq b_1 + b_2} \qquad (2)$$

We refer to Rule (2) as the *Addition rule*. It is easy to see that Rule (2) is sound in that any assignment satisfying the hypotheses **must** satisfy the consequent. Furthermore, the rule is **complete** in that if System (1) is unsatisfiable, then repeated application of Rule (2) will result in a contradiction of the form: $0 \leq -a$, $a > 0$. The completeness of the Addition rule was established by Farkas [7], in a lemma that is famously known as Farkas' Lemma for systems of linear inequalities [22]:

Lemma 1. *Let $\mathbf{A} \cdot \mathbf{x} \leq \mathbf{b}$ denote a system of m linear constraints over n variables.*

Then, either $\exists \mathbf{x}\ \mathbf{A} \cdot \mathbf{x} \leq \mathbf{b}$ or (mutually exclusively), $\exists \mathbf{y} \in \mathbb{R}_+^m\ \mathbf{y} \cdot \mathbf{A} = \mathbf{0}$, $\mathbf{y} \cdot \mathbf{b} < 0$.

The above lemma along with the fact that linear programs must have basic feasible solutions establishes that the linear programming problem is in the complexity class **NP** $\cap$ **coNP**. Farkas' lemma is one of several lemmata that consider pairs of linear systems in which exactly one element of the pair is feasible. These lemmata are collectively referred to as "Theorems of the Alternative" [20]. The $\mathbf{y}$ variables are called the Farkas' variables corresponding to the system $\mathbf{A} \cdot \mathbf{x} \leq \mathbf{b}$ and they serve as a witness that certifies the infeasibility of this system. In general, the Farkas variables can assume any real value for a given constraint system. In this paper, we are interested only in cases where the Farkas' variables are restricted, as discussed below (see Sect. 2.1).

In case of UTVPI constraints, Rule (2) can be restricted to the following rule:

$$\frac{a_i \cdot x_i + a_j \cdot x_j \leq b_{ij} \qquad \wedge \qquad -a_j \cdot x_j + a_k \cdot x_k \leq b_{jk}}{a_i \cdot x_i + a_k \cdot x_k \leq b_{ij} + b_{jk}} \qquad (3)$$

Rule (3) is known as the *transitive inference rule*. Although it is a restricted version of the addition rule, it is both sound and complete for linear feasibility in UTPVI constraint systems [18]. This rule will be used in the reductions in Sect. 4.

It is well-known that a system of difference constraints can be represented as a constraint network (directed weighted graph) [3]. It follows from Farkas' lemma that the system of difference constraints is infeasible if and only if the corresponding constraint network contains a negative cost cycle. Similar constraint networks have been developed for UTVPI constraints [18,19,21,26]. These constraint networks have been used in the design of graph-based algorithms for linear and integer feasibility testing in UTVPI constraints.

2.1 The Read-Once Refutation (ROR) Problem

Definition 4. *A refutation is said to be read-once, if each input constraint is used at most once in the derivation of a contradiction.*

We are interested in the problem of determining if a UCS has a read-once refutation or not. Accordingly, we have:

> The **Read-once Refutation (ROR)** problem: Given a UCS **U**, does **U** have a read-once refutation.

Example 1. Note that not every system of constraints has a read-once refutation. Consider the following UCS:

$$l_1 : x_1 - x_2 \leq -3 \quad l_2 : -x_1 + x_4 \leq 1 \quad l_3 : -x_1 - x_4 \leq 1 \tag{4}$$
$$l_4 : x_2 + x_3 \leq 1 \quad l_5 : \quad x_2 - x_3 \leq 1$$

First observe that l_1 is the only constraint with a negative defining constant; hence, it must be included in *any* refutation.

In order to eliminate x_1, we must include both l_2 and l_3 in the refutation; otherwise, x_4 is not eliminated. Similarly, to eliminate $-x_2$, we must include both l_4 and l_5; otherwise, x_3 is not eliminated.

It follows that all the constraints in System (4) must be included in any refutation. However, the sum of the five constraints yields the constraint l_6 : $-x_1 + x_2 \leq 1$. The only way to arrive at a contradiction is to include l_1 a second time. In other words, System (4) lacks an ROR.

On the other hand, every infeasible UTVPI system has a refutation in which each constraint is used at most twice [26].

Given an unsatisfiable linear system, specified as in System (1), the ROR problem is to determine if there exists a read-once refutation. In other words, we wish to find a subset of constraints from System (1) whose sum produces a contradiction of the form $0 < -a, a > 0$.

Using Farkas' Lemma, the ROR problem is easily modeled as the following integer program:

$$\exists \mathbf{y} \; \mathbf{y} \cdot \mathbf{A} = \mathbf{0} \tag{5}$$
$$\mathbf{y} \cdot \mathbf{b} \leq -1$$
$$\mathbf{y} \in \{0,1\}^m$$

Proposition 1. *Let R be a read-once refutation of a UCS* **U**. *If we add the constraints in R, we get a contradiction of the form:* $0 \leq b$, $b < 0$.

Proof. Follows immediately from System (5).   □

The principal contribution of this paper is designing a polynomial time algorithm for the ROR problem in UTVPI constraint systems.

2.2   The Minimum Weight Perfect Matching (MWPM) Problem

In this subsection, we briefly discuss the problem of finding the minimum weight perfect matching (MWPM) in an undirected, weighted graph. This digression is necessitated by the fact that Sect. 4 involves reduction to the MWPM problem.

Let $\mathbf{G} = \langle \mathbf{V}, \mathbf{E}, \mathbf{c} \rangle$ denote an undirected graph, with vertex set $\mathbf{V}$, edge set $\mathbf{E}$ and edge cost function $\mathbf{c}$. Let $n = |\mathbf{V}|$ and let $m = |\mathbf{E}|$. A *matching* is any collection of vertex-disjoint edges. A *perfect matching* is a matching in which each vertex $v \in \mathbf{V}$ is matched. Without loss of generality, we assume that n is even, since $\mathbf{G}$ cannot have a perfect matching, otherwise.

The MWPM problem is one of the classical problems in combinatorial optimization [17]. Over the years, there has been a steady stream of papers documenting improvements in algorithms for this problem [5,6,8].

The fastest combinatorial algorithm for the MWPM problem runs in time $O(m \cdot n + n^2 \cdot \log n)$ [9]. It is this bound that we shall be using in our paper.

3   Motivation and Related Work

The problem of finding short refutations is one of the principal problems in proof complexity [2]. Research proceeds along the lines of finding lower bounds on the lengths of refutations for propositional tautologies (contradictions) in proof systems of increasing complexity, with a view towards separating the complexity class **NP** from the class **coNP** [27]. Resolution is one of the weakest proof systems, but even in this proof system it was difficult to obtain exponential lower bounds on the length of proofs. The first non-trivial lower bound on the length of resolution proofs is due to Haken [11], who showed that any resolution proof for the pigeonhole principle required exponentially many steps. In [13], it was shown that the problem of finding the shortest resolution proofs in arbitrary 3CNF formulas is **NP-complete**. A stronger result was obtained in [1]; they showed that the problem of finding the shortest resolution proof in Horn formulas is not linearly approximable, unless **P** = **NP**. This result is interesting because it is easy to see that every unsatisfiable Horn formula has a resolution refutation that is quadratic in the number of clauses.

On the read-once refutation side, [14] showed that the problem of checking if an arbitrary CNF has an ROR is **NP-complete**. This result was strengthened in [16], which showed that the problem of checking whether a CNF formula has a read-once unit resolution refutation is **NP-complete**. In [15], it was shown that the problem of checking if a 2CNF formula has a read-once refutation is **NP-complete**. In this paper, we examine this problem on continuous (as opposed to discrete) variables.

As we can see, much of the work in finding short refutations focused on discrete domains (CNF formulas). In a departure from existing work, [24] considered difference constraint systems from the perspective of determining the optimal length resolution refutations. That paper shows that short refutations exist for difference constraints and also shows that the optimal length refutation can be determined in polynomial time. The algorithm therein is based on dynamic programming and runs in time $O(n^3 \cdot \log n)$ on DCSs with n variables. In [25], a different dynamic program was used to achieve a time of $O(m \cdot n \cdot k)$, where m is the number of constraints and k is the length of the shortest refutation. It is worth noting that in DCSs, linear and integer feasibility coincide and therefore, the departure is not strict. Furthermore, as pointed out in [24], every minimal refutation (i.e., a refutation without redundant constraints) is necessarily read-once, since every minimal refutation corresponds to a simple negative cost cycle in the corresponding constraint network [3]. In this paper though, we consider the problem of read-once refutations in UCSs. Unlike DCSs, linear feasibility does not imply integer feasibility in UCSs [26]. UTVPI constraints occur in a number of problem domains including but not limited to program verification [18], abstract interpretation [4,19], real-time scheduling [10] and operations research [12].

To the best of our knowledge, this is the first paper to study read-once refutations for linear infeasibility in UCSs.

4   The ROR Problem in UTVPI Constraints

In this section, we show that read-once refutations in systems of UTVPI constraints (if they exist) can be found in polynomial time.

4.1   Construction

We convert the UCS $\mathbf{U} : \mathbf{A} \cdot \mathbf{x} \leq \mathbf{b}$ into an undirected graph $\mathbf{G}' = \langle \mathbf{V}', \mathbf{E}', \mathbf{c}' \rangle$ as follows:

1. For each variable x_i in $\mathbf{U}$, add the vertices x_i^+, $x_i'^+$, x_i^-, and $x_i'^-$ to $\mathbf{V}'$. Additionally, add the edges $x_i^- \overset{0}{\rule{1.2em}{0.4pt}} x_i^+$ and $x_i'^- \overset{0}{\rule{1.2em}{0.4pt}} x_i'^+$ to $\mathbf{E}'$.

2. Add the vertices x_0^+ and x_0^- to $\mathbf{V}'$. Additionally, add the edge $x_0^- \overset{0}{\rule{1.2em}{0.4pt}} x_0^+$ to $\mathbf{E}'$.

3. For each constraint l_k of **U**, add the vertices l_k and l'_k to **V'** and the edge $l_k \xrightarrow{\;0\;} l'_k$ to **E'**. Additionally:

(a) If l_k is $x_i + x_j \le b_k$, add the edges $x_i^+ \xrightarrow{\frac{b_k}{2}} l_k$, $x_i'^+ \xrightarrow{\frac{b_k}{2}} l_k$, $x_j^+ \xrightarrow{\frac{b_k}{2}} l'_k$, and $x_j'^+ \xrightarrow{\frac{b_k}{2}} l'_k$ to **E'**.

(b) If l_k is $x_i - x_j \le b_k$, add the edges $x_i^+ \xrightarrow{\frac{b_k}{2}} l_k$, $x_i'^+ \xrightarrow{\frac{b_k}{2}} l_k$, $x_j^- \xrightarrow{\frac{b_k}{2}} l'_k$, and $x_j'^- \xrightarrow{\frac{b_k}{2}} l'_k$ to **E'**.

(c) If l_k is $-x_i + x_j \le b_k$, add the edges $x_i^- \xrightarrow{\frac{b_k}{2}} l_k$, $x_i'^- \xrightarrow{\frac{b_k}{2}} l_k$, $x_j^+ \xrightarrow{\frac{b_k}{2}} l'_k$, and $x_j'^+ \xrightarrow{\frac{b_k}{2}} l'_k$ to **E'**.

(d) If l_k is $-x_i - x_j \le b_k$, add the edges $x_i^- \xrightarrow{\frac{b_k}{2}} l_k$, $x_i'^- \xrightarrow{\frac{b_k}{2}} l_k$, $x_j^- \xrightarrow{\frac{b_k}{2}} l'_k$, and $x_j'^- \xrightarrow{\frac{b_k}{2}} l'_k$ to **E'**.

(e) If l_k is $x_i \le b_k$, add the edges $x_i^+ \xrightarrow{\frac{b_k}{2}} l_k$, $x_i'^+ \xrightarrow{\frac{b_k}{2}} l_k$, $x_0^+ \xrightarrow{\frac{b_k}{2}} l'_k$, and $x_0^- \xrightarrow{\frac{b_k}{2}} l'_k$ to **E'**.

(f) If l_k is $-x_i \le b_k$, add the edges $x_i^- \xrightarrow{\frac{b_k}{2}} l_k$, $x_i'^- \xrightarrow{\frac{b_k}{2}} l_k$, $x_0^+ \xrightarrow{\frac{b_k}{2}} l'_k$, and $x_0^- \xrightarrow{\frac{b_k}{2}} l'_k$ to **E'**.

In this construction, each variable is represented by a pair of 0-weight edges. As we shall see, this permits each vertex can be used twice by a read-once refutation. However, we still only have one 0-weight edge for each constraint, which prevents a read-once refutation from reusing edges.

Note that if **U** has m constraints over n variables, then **G'** has $(4 \cdot n + 2 \cdot m + 2)$ vertices and $(2 \cdot n + 5 \cdot m + 1)$ edges. In other words, **G'** has $O(m + n)$ vertices and $O(m + n)$ edges.

Example 2. Let us consider the UCS represented by System (6).

$$\begin{array}{lll} l_1: & x_1 + x_2 \le -2 \quad l_2: & x_1 + x_3 \le -2 \quad l_3: -x_1 - x_4 \le -2 \\ l_4: -x_1 - x_5 \le -2 & l_5: -x_2 - x_3 \le 2 & l_6: \quad x_4 + x_5 \le 2 \end{array} \tag{6}$$

The undirected graph corresponding to UCS (6) is shown in Fig. 1.

The minimum weight perfect matching in this graph is $x_1^+ \xrightarrow{-1} l_2$, $l'_2 \xrightarrow{-1} x_3^+$, $x_3^- \xrightarrow{1} l'_5$, $l_5 \xrightarrow{1} x_2^-$, $x_2^+ \xrightarrow{-1} l'_1$, $l_1 \xrightarrow{-1} x_1'^+$, $x_1'^- \xrightarrow{-1} l_3$, $l'_3 \xrightarrow{-1} x_4^-$, $x_4^+ \xrightarrow{1} l_6$, $l'_6 \xrightarrow{1} x_5^+$, $x_5^- \xrightarrow{-1} l'_4$, $l_4 \xrightarrow{-1} x_1^-$, $x_2'^+ \xrightarrow{0} x_2'^-$, $x_3'^+ \xrightarrow{0} x_3'^-$, $x_4'^+ \xrightarrow{0} x_4'^-$, and $x_5'^+ \xrightarrow{0} x_5'^-$. This matching has weight -4 and corresponds to the read-once refutation obtained by summing all six constraints (see Theorem 1).

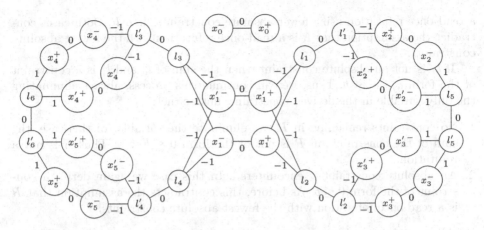

Fig. 1. Undirected graph

4.2   Correctness

We now proceed to argue the correctness of the above construction.

We first establish some structural properties of certain read-once refutations in a UCS (see Lemmas 2 and 4). These properties will be used in Theorem 1.

Let R be a read-once refutation of $\mathbf{U}$ which uses the fewest number of constraints, i.e., a shortest read-once refutation. Let x_i be a variable used by R. Let $|R|_i$ be the number of constraints in R that use the literal x_i. Since R is a refutation, $|R|_i$ is also the number of constraints that use the literal $-x_i$.

We will first show that we can assume without loss of generality, that R has zero or two absolute constraints.

Lemma 2. *Let $\mathbf{U} : \mathbf{A} \cdot \mathbf{x} \leq \mathbf{b}$ denote a UCS. If $\mathbf{U}$ has a read-once refutation using absolute constraints, then $\mathbf{U}$ has a read-once refutation using zero or two absolute constraints.*

Proof. Let R be a read-once refutation of $\mathbf{U}$ with the minimum number of absolute constraints, and let $|R|_a$ represent the number of absolute constraints in R.

If R has an odd number of absolute constraints, then the total number of literals in R would also be odd. Thus, summing the constraints in R would not result in a constraint of the form $0 \leq b$ where $b < 0$, which is a requirement of a read-once refutation (see Proposition 1). Thus, R must have an even number of absolute constraints.

Assume that $|R|_a > 2$. Let $l_0 : a_i \cdot x_i \leq b_0$ be an absolute constraint in R. Since R is a refutation, there must be a constraint l_1 with the term $-a_i \cdot x_i$.

If l_1 is an absolute constraint, then the sum of l_0 and l_1 is a constraint of the form $0 \leq b$. If $b < 0$, then the constraints l_0 and l_1 form a refutation using fewer absolute constraints than R. If $b \geq 0$, then the remaining constraints form a read-once refutation using fewer absolute constraints than R. Both cases contradict the assumption that R is a read-once refutation with the fewest absolute constraints.

If l_1 is not an absolute constraint, then the sum of l_0 and l_1 is a constraint of the form $a_j \cdot x_j \leq b$. Thus, we can continue this process, always eliminating the only variable in the derived constraint, until either:

1. No constraints remaining in R can eliminate the variable. In this case, the sum of the constraints in R is not of the form $0 \leq b < 0$. Thus R is not a refutation.
2. An absolute constraint is encountered. In this case we again derive a constraint of the form $0 \leq b$. As before, this contradicts the assumption that R is a read-once refutation with the fewest absolute constraints.

All possible cases lead to a contradiction, thus we must have that $|R|_a \leq 2$. Since $|R|_a$ is even, this means that $|R|_a \in \{0, 2\}$. □

Assume that $|R|_i \geq 3$. By Lemma 2, we can assume without loss of generality that R has at most 2 absolute constraints. Thus R must have a non-absolute constraint that uses x_i. Let $l_0 : x_i + a_j \cdot x_j \leq b_{ij}$ be one such non-absolute constraint in R. l_0 is called the *current constraint*. In what follows, we will proceed through a sequence of stages eliminating the non-x_i variable in the current constraint, until eventually x_i itself is eliminated or a contradiction results.

Algorithm 4.1 represents our approach.

Lemma 3. PROCESS-REFUTATION(R, l_0) *cannot return any value.*

Proof. Recall that R is a shortest read-once refutation of $\mathbf{U}$ and that $|R|_i \geq 3$. Assume that PROCESS-REFUTATION(R, l_0) returns a value of u. We will show that every value of u results in a contradiction.

1. $u = -1$ - In this case, there is a non-x_i literal that cannot be canceled by any of the the remaining constraints in R'. Thus, the sum of the constraints in R is not of the form $0 \leq b < 0$. Thus R is not a read-once refutation.
2. $u = -2$ - In this case, $\sum_{l_h \in sum} l_h$ is a constraint of the form $0 \leq b$. Note that the literal x_i is used only once by the constraints in sum. Thus, $sum \subset R$. If $b < 0$, then the constraints in sum form a read-once refutation of $\mathbf{U}$. If $b \geq 0$, then the constraints in $R \setminus sum$ form a read-once refutation of $\mathbf{U}$. Thus R is not the shortest read-once refutation of $\mathbf{U}$.
3. $u = -3$ - In this case, $\sum_{l_h \in sum} l_h$ is a constraint of the form $-x_i - x_i \leq b$ and $\sum_{l_h \in sum'} l_h$ is a constraint of the form $x_i + x_i \leq b'$, we have that $\sum_{l_h \in sum \cup sum'} l_h$ is a constraint of the form $0 \leq b + b'$.
   Note that sum contains no constraints that use x_i and that sum' contains two constraints that use x_i. Thus, $sum \cup sum' \subset R$. If $b < 0$, then the constraints in $sum \cup sum'$ form a read-once refutation of $\mathbf{U}$. If $b \geq 0$, then the constraints in $R \setminus (sum \cup sum')$ form a read-once refutation of $\mathbf{U}$. Thus R is not the shortest read-once refutation of $\mathbf{U}$.

```
1: procedure PROCESS-REFUTATION(R, l₀)
2:     sum := {l₀}.                                    ▷ The set of constraints processed thus far.
3:     R' := R \ {l₀}.                                 ▷ The set of constraints not yet processed.
4:     numᵢ := 1.
5:     while (R' ≠ ∅) do
6:         Let lₛ ∈ R' be a constraint which cancels the non-xᵢ variable in Σₗₕ∈ₛᵤₘ lₕ.
7:         if (lₛ does not exist) then
8:             return(−1).
9:         end if
10:        R' := R' \ {lₛ}.
11:        sum := sum ∪ {lₛ}.
12:        if (lₛ is an absolute constraint) then
13:            Let lₛ be the remaining absolute constraint in R.
14:            R' := R' \ {lₛ}.
15:            sum := sum ∪ {lₛ}.
16:        end if
17:        if (lₛ uses the literal −xᵢ) then
18:            if (numᵢ = 1) then
19:                return(−2).
20:            else
21:                return(−3).
22:            end if
23:        end if
24:        if (lₛ uses the literal xᵢ) then
25:            if (numᵢ = 1) then
26:                numᵢ := 2.
27:                Let lₛ ∈ R be a non-absolute constraint with the literal −xᵢ.
28:                R' := R' \ {lₛ}.
29:                sum' := sum.
30:                sum := {lₛ}.
31:            else
32:                return(−2).
33:            end if
34:        end if
35:    end while
36:    return(−4).
37: end procedure
```

Algorithm 4.1. Shortest refutation property

---

4. $u = -4$ - In this case, we have processed every constraint in R. Thus, we have processed every constraint in R that uses the literal x_i. Let $l_1, l_2 \in R$ be the first two such constraints processed. Since $|R|_i \geq 3$, these constraints are guaranteed to exist.

When processing l_1, we have that $num_i = 1$. Thus, when constraint l_1 triggers the **if** statement on line 24, Algorithm 4.1 sets num_i to 2. This meant that, when constraint l_2 triggers the **if** statement on line 24, Algorithm 4.1 returns -2. Thus $u = -2$. □

Lemma 4. *Let* $\mathbf{U} : \mathbf{A} \cdot \mathbf{x} \leq \mathbf{b}$ *denote an infeasible UCS. If* $\mathbf{U}$ *has a read-once refutation, then it has a read-once refutation, in which each literal is used at most twice.*

Proof. Let R be the shortest read-once refutation of $\mathbf{U}$. Assume that $|R|_i \geq 3$ for some literal x_i. Recall that R must have a non-absolute constraint l_0 of the form $x_i - a_j \cdot x_j \leq b_{ij}$.

From Lemma 3, PROCESS-REFUTATION(R, L_0) cannot return any value. However, R is finite. Thus, PROCESS-REFUTATION(R, L_0) must eventually halt and return a value. This is a contradiction. Thus, we must have that $|R|_i \leq 2$ for every literal x_i.   □

Theorem 1. *Let* $\mathbf{U} : \mathbf{A} \cdot \mathbf{x} \leq \mathbf{b}$ *denote a UCS and let* $\mathbf{G}' = \langle \mathbf{V}', \mathbf{E}', \mathbf{c}' \rangle$ *denote the graph constructed from* $\mathbf{U}$, *as described in Subsect. 4.1.*

Then, $\mathbf{U}$ *has a read-once refutation if and only if* $\mathbf{G}'$ *has a negative weight perfect matching.*

Proof. First assume that $\mathbf{U}$ has a read-once refutation R. As argued in Lemma 4, we can assume that each literal in $\mathbf{U}$ occurs in R at most twice.

We construct a negative weight perfect matching P of $\mathbf{G}'$ as follows:

1. For each variable x_i in $\mathbf{U}$:
   (a) If R does not use the literal x_i, add the edges $x_i^+ \xrightarrow{0} x_i^-$ and $x_i'^+ \xrightarrow{0} x_i'^-$ to P.
   (b) If R uses the literal x_i only once, add the edge $x_i'^+ \xrightarrow{0} x_i'^-$ to P.
   (c) If R uses the literal x_i twice, do not add any edges to P.
2. Assume that the constraints in $\mathbf{U}$ are assigned an arbitrary order. For constraint l_k in $\mathbf{U}$:
   (a) If $l_k \notin R$, add the edge $l_k \xrightarrow{0} l_k'$ to P.
   (b) If $l_k \in R$ is of the form $a_i \cdot x_i + a_j \cdot x_j \leq b_k$ such that $a_i, a_j \neq 0$:

   i. If l_k is the first edge to use the literal x_i, add the edge $x_i^+ \xrightarrow{\frac{b_k}{2}} l_k$ to P. If it is the second, add the edge $x_i'^+ \xrightarrow{\frac{b_k}{2}} l_k$.

   ii. If l_k is the first edge to use the literal $-x_i$, add the edge $x_i^- \xrightarrow{\frac{b_k}{2}} l_k$ to P. If it is the second, add the edge $x_i'^- \xrightarrow{\frac{b_k}{2}} l_k$.

   iii. If l_k is the first edge to use the literal x_j, add the edge $x_j^+ \xrightarrow{\frac{b_k}{2}} l_k'$ to P. If it is the second, add the edge $x_j'^+ \xrightarrow{\frac{b_k}{2}} l_k'$.

   iv. If l_k is the first edge to use the literal $-x_j$, add the edge $x_j^- \xrightarrow{\frac{b_k}{2}} l_k'$ to P. If it is the second, add the edge $x_j'^+ \xrightarrow{\frac{b_k}{2}} l_k'$.
   (c) If $l_k \in R$ is of the form $a_i \cdot x_i \leq b_k$:

    i. If l_k is the first edge to use the literal x_i, add the edge $x_i^+ \xrightarrow{\frac{b_k}{2}} l_k$ to P. If it is the second, add the edge $x_i'^+ \xrightarrow{\frac{b_k}{2}} l_k$.

    ii. If l_k is the first edge to use the literal $-x_i$, add the edge $x_i^- \xrightarrow{\frac{b_k}{2}} l_k$ to P. If it is the second, add the edge $x_i'^- \xrightarrow{\frac{b_k}{2}} l_k$.

    iii. If l_k is the first absolute constraint, then add the edge $x_0^+ \xrightarrow{\frac{b_k}{2}} l_k'$ to P.

    iv. If l_k is the second absolute constraint, then add the edge $x_0^- \xrightarrow{\frac{b_k}{2}} l_k'$ to P. From Lemma 2, we can assume without loss of generality that there are at most two absolute constraints in R.

(d) If R has no absolute constraints, then add the edge $x_0^+ \xrightarrow{0} x_0^-$ to P.

We now make the following observations:

1. For each variable x_i:
   (a) If the literal x_i is not used by R, then the verticies x_i^+, x_i^-, $x_i'^+$, and $x_i'^-$ are the end points of the edges $x_i^+ \xrightarrow{0} x_i^-$ and $x_i'^+ \xrightarrow{0} x_i'^-$ which are the only such edges in P.
   (b) If the literal x_i is used in only one constraint $l_k \in R$, then the literal x_i must also be used by only one constraint $l_{k'} \in R$. Otherwise R would not be a read-once refutation. Thus the vertices x_i^+, x_i^-, $x_i'^+$, and $x_i'^-$ are the end points of the edges $x_i'^+ \xrightarrow{0} x_i'^-$, $x_i^+ \xrightarrow{\frac{b_k}{2}} l_k$, and $x_i^- \xrightarrow{\frac{b_{k'}}{2}} l_{k'}$ which are the only such edges in P.
   (c) If the literal x_i is used two constraint $l_{k_1}, l_{k_2} \in R$, then the literal x_i must also be used by two constraints $l_{k_1'}, l_{k_2'} \in R$. Otherwise R would not be a read-once refutation. Thus the vertices x_i^+, x_i^-, $x_i'^+$, and $x_i'^-$ are the end points of the edges $x_i^+ \xrightarrow{\frac{b_{k_1}}{2}} l_{k_1}$, $x_i^- \xrightarrow{\frac{b_{k_1'}}{2}} l_{k_1'}$, $x_i'^+ \xrightarrow{\frac{b_{k_2}}{2}} l_{k_2}$, $x_i'^- \xrightarrow{\frac{b_{k_2'}}{2}} l_{k_2'}$ which are the only such edges in P.
2. For each constraint l_k:
   (a) If $l_k \in R$, then, by construction, P contains two edges of weight $\frac{b_k}{2}$ one with end point l_k and one with endpoint l_k'. Thus both l_k and l_k' are the endpoints of exactly one edge in P.
   (b) If $l_k \notin R$, then, by construction, P contains the edge $l_k \xrightarrow{0} l_k'$ and none of the weight $\frac{b_k}{2}$ edges. Thus both l_k and l_k' are the endpoints of exactly one edge in P.
3. By Lemma 2, we can assume without loss of generality that R contains 0 or 2 absolute constraints. Thus:
   (a) If R contains no absolute constraints, then the vertices x_0^+ and x_0^- are the endpoints of the edge $x_0^+ \xrightarrow{0} x_0^-$ which is the only such edge in P.

(b) If R contains two absolute constraints l_{k_1} and l_{k_2}, then the vertices x_0^+ and x_0^- are the endpoints of the edges $x_0^+ \xrightarrow{\frac{b_{k_1}}{2}} l'_{k_1}$ and $x_0^- \xrightarrow{\frac{b_{k_2}}{2}} l'_{k_2}$ which are the only such edges in P.

Thus, every vertex in $\mathbf{G}'$ is an endpoint of exactly one edge in P. It follows that P is a perfect matching. Additionally, for each constraint $l_k \in R$, P has two edges of weight $\frac{b_k}{2}$ and all other edge in P have weight 0. Thus,

$$\sum_{e \in P} \mathbf{c}'(e) = \sum_{l_k \in R} \left(\frac{b_k}{2} + \frac{b_k}{2} \right) = \sum_{l_k \in R} b_k < 0.$$

This means that P has negative weight.

Now assume that $\mathbf{G}'$ has a negative weight perfect matching P. We construct a read-once refutation R as follows:

For each constraint l_j in $\mathbf{U}$, if P does not use the edge $l_j \xrightarrow{0} l'_j$, then add the constraint l_j to R.

P is a perfect matching. Thus, for each variable x_i, we have the following:

1. If a constraint $l_j \in R$ uses the literal x_i, then either:

   (a) One of the edges $l_j \xrightarrow{\frac{b_j}{2}} x_i^+$ or $l'_j \xrightarrow{\frac{b_j}{2}} x_i^+$ is in P. Thus, the edge $x_i^+ \xrightarrow{0} x_i^-$ is not in P. This means that for some k, one of the edges $l_k \xrightarrow{\frac{b_k}{2}} x_i^-$ or $l'_k \xrightarrow{\frac{b_k}{2}} x_i^-$ is in P. Thus, the literal x_i is used by the constraint $l_k \in R$.

   (b) One of the edges $l_j \xrightarrow{\frac{b_j}{2}} x_i'^+$ or $l'_j \xrightarrow{\frac{b_j}{2}} x_i'^+$ is in P. Thus, the edge $x_i'^+ \xrightarrow{0} x_i'^-$ is not in P. This means that for some k, one of the edges $l_k \xrightarrow{\frac{b_k}{2}} x_i'^-$ or $l'_k \xrightarrow{\frac{b_k}{2}} x_i'^-$ is in P. Thus, the literal x_i is used by the constraint $l_k \in R$.

   Note that if there are two constraints in R that use the literal x_i, then each one corresponds to a constraint in R that uses the literal $-x_i$.

2. Similarly, if one or two constraints in R use the literal $-x_i$, then the same number of constraints in R use the literal x_i.

Thus, the literal x_i appears in the same number of constraints in R as the literal $-x_i$.

If a constraint $l_k \in R$, then the edge $l_k \xrightarrow{0} l'_k$ is not in P. Thus, P contains two edges of weight $\frac{b_k}{2}$ one with end point l_k and one with endpoint l'_k. Conversely, if the constraint $l_k \notin R$, then the edge $l_k \xrightarrow{0} l'_k$ is in P. It follows that none of the edges of weight $\frac{b_j}{2}$ from l_j and l'_j are in P. Thus, for each constraint $l_k \in R$, P has two edges of weight $\frac{b_k}{2}$ and all other edge in P have weight 0.

Thus, summing the constraints in R yields

$$0 \leq \sum_{l_j \in R} b_j = \sum_{l_j \in R} \left(\frac{b_j}{2} + \frac{b_j}{2} \right) = \sum_{e \in P} \mathbf{c}'(e) < 0.$$

By construction, each constraint appears at most once in R. Thus, R is a read-once refutation of $\mathbf{U}$. $\square$

4.3 Resource Analysis

To construct $\mathbf{G}'$, we need to process each variable and each constraint in $\mathbf{U}$. Each variable and each constraint can be processed in constant time. Thus the reduction can be performed in $O(m + n)$ time. The minimum weight perfect matching of $\mathbf{G}'$ can be found in $O(|\mathbf{E}'| \cdot |\mathbf{V}'| + |\mathbf{V}'|^2 \cdot \log |\mathbf{V}'|)$ time [9]. As discussed in Subsect. 4.1, $\mathbf{G}'$ has $O(m + n)$ vertices and $O(m + n)$ edges. Thus, by using this reduction, the ROR problem for UTVPI constraints can be solved in $O((m + n)^2 \cdot \log(m + n))$ time.

5 Conclusion

In this paper, we investigated the problems of checking whether a UTVPI constraint system has a read-once refutation. Unlike difference constraints, a UCS may not have an ROR. Therefore, a sophisticated search is called for. We reduced the ROR problem in UTVPI constraints to the problem of finding a minimum weight perfect matching (MWPM) in an undirected graph. One of the advantages of our reduction-based algorithm is that we can leverage improvements in algorithms for the MWPM problem to improvements in our algorithm for the ROR problem [5]. UTVPI constraints are an important class of linear constraints that find applications in abstract interpretation and program verification. It follows that certificates of infeasibility for UTVPI constraint systems are of enormous practical significance. Read-once certificates are particularly useful in applications, since they are "short" by definition.

An important extension of this research is to implement in our algorithm within the framework of an SMT solver such as Yices [23].

References

1. Alekhnovich, M., Buss, S., Moran, S., Pitassi, T.: Minimum propositional proof length is NP-hard to linearly approximate. In: Brim, L., Gruska, J., Zlatuška, J. (eds.) MFCS 1998. LNCS, vol. 1450, pp. 176–184. Springer, Heidelberg (1998). https://doi.org/10.1007/BFb0055766
2. Beame, P., Pitassi, T.: Propositional proof complexity: past, present, future. Bull. EATCS **65**, 66–89 (1998)
3. Cormen, T.H., Leiserson, C.E., Rivest, R.L., Stein, C.: Introduction to Algorithms. MIT Press, Cambridge (2001)

4. Cousot, P., Cousot, R.: Abstract interpretation: a unified lattice model for static analysis of programs by construction or approximation of fixpoints. In: POPL, pp. 238–252 (1977)
5. Duan, R., Pettie, S., Su, H.-H.: Scaling algorithms for weighted matching in general graphs. ACM Trans. Algorithms **14**(1), 8:1–8:35 (2018)
6. Edmonds, J.: An introduction to matching. In: Mimeographed Notes. Engineering Summer Conference, University of Michigan, Ann Arbor, MI (1967)
7. Farkas, G.: Über die Theorie der Einfachen Ungleichungen. Journal für die Reine und Angewandte Mathematik **124**(124), 1–27 (1902)
8. Gabow, H.N.: An efficient implementation of Edmonds' algorithm for maximum matching on graphs. J. ACM **23**(2), 221–234 (1976)
9. Gabow, H.N.: Data structures for weighted matching and nearest common ancestors with linking. In: Johnson, D. (ed.) Proceedings of the 1st Annual ACM-SIAM Symposium on Discrete Algorithms (SODA 1990), San Francisco, CA, USA, pp. 434–443. SIAM, January 1990
10. Gerber, R., Pugh, W., Saksena, M.: Parametric dispatching of hard real-time tasks. IEEE Trans. Comput. **44**(3), 471–479 (1995)
11. Haken, A.: The intractability of resolution. Theor. Comput. Sci. **39**(2–3), 297–308 (1985)
12. Hochbaum, D.S., (Seffi) Naor, J.: Simple and fast algorithms for linear and integer programs with two variables per inequality. SIAM J. Comput. **23**(6), 1179–1192 (1994)
13. Iwama, K.: Complexity of finding short resolution proofs. In: Prívara, I., Ružička, P. (eds.) MFCS 1997. LNCS, vol. 1295, pp. 309–318. Springer, Heidelberg (1997). https://doi.org/10.1007/BFb0029974
14. Iwama, K., Miyano, E.: Intractability of read-once resolution. In: Proceedings of the 10th Annual Conference on Structure in Complexity Theory (SCTC 1995), Los Alamitos, CA, USA, pp. 29–36. IEEE Computer Society Press, June 1995
15. Büning, H.K., Wojciechowski, P.J., Subramani, K.: Finding read-once resolution refutations in systems of 2CNF clauses. Theor. Comput. Sci. **729**, 42–56 (2018)
16. Büning, H.K., Zhao, X.: The complexity of read-once resolution. Ann. Math. Artif. Intell. **36**(4), 419–435 (2002)
17. Korte, B., Vygen, J.: Combinatorial Optimization. Algorithms and Combinatorics, vol. 21. Springer, New York (2010). https://doi.org/10.1007/978-3-642-24488-9
18. Lahiri, S.K., Musuvathi, M.: An efficient decision procedure for UTVPI constraints. In: Gramlich, B. (ed.) FroCoS 2005. LNCS (LNAI), vol. 3717, pp. 168–183. Springer, Heidelberg (2005). https://doi.org/10.1007/11559306_9
19. Miné, A.: The octagon abstract domain. High.-Order Symb. Comput. **19**(1), 31–100 (2006)
20. Nemhauser, G.L., Wolsey, L.A.: Integer and Combinatorial Optimization. Wiley, New York (1999)
21. Revesz, P.Z.: Tightened transitive closure of integer addition constraints. In: Symposium on Abstraction, Reformulation, and Approximation (SARA), pp. 136–142 (2009)
22. Schrijver, A.: Theory of Linear and Integer Programming. Wiley, New York (1987)
23. SRI International. Yices: An SMT solver. http://yices.csl.sri.com/
24. Subramani, K.: Optimal length resolution refutations of difference constraint systems. J. Autom. Reason. (JAR) **43**(2), 121–137 (2009)

25. Subramani, K., Williamson, M., Gu, X.: Improved algorithms for optimal length resolution refutation in difference constraint systems. Formal Aspects Comput. **25**(2), 319–341 (2013)
26. Subramani, K., Wojciechowski, P.J.: A combinatorial certifying algorithm for linear feasibility in UTVPI constraints. Algorithmica **78**(1), 166–208 (2017)
27. Urquhart, A.: The complexity of propositional proofs. Bull. Symb. Logic **1**(4), 425–467 (1995)

Generalizations of Weighted Matroid Congestion Games: Pure Nash Equilibrium, Sensitivity Analysis, and Discrete Convex Function

Kenjiro Takazawa$^{(\boxtimes)}$ (iD)

Hosei University, Tokyo 184-8584, Japan
takazawa@hosei.ac.jp
http://ds.ws.hosei.ac.jp/index.html

Abstract. Congestion games provide a model of human's behavior of choosing an optimal strategy while avoiding congestion. In the past decade, matroid congestion games have been actively studied and their good properties have been revealed. In most of the previous work, the cost functions are assumed to be univariate or bivariate. In this paper, we discuss generalizations of matroid congestion games in which the cost functions are n-variate, where n is the number of players. First, we prove the existence of pure Nash equilibria in matroid congestion games with monotone cost functions, which extends that for weighted matroid congestion games by Ackermann, Röglin, and Vöcking (2009). Second, we prove the existence of pure Nash equilibria in matroid resource buying games with submodular cost functions, which extends that for matroid resource buying games with marginally nonincreasing cost functions by Harks and Peis (2014). Finally, motivated from polymatroid congestion games with M$^\natural$-convex cost functions, we conduct sensitivity analysis for separable M$^\natural$-convex optimization, which extends that for separable convex optimization over base polyhedra by Harks, Klimm, and Peis (2018).

Keywords: Pure Nash equilibrium · Matroid congestion game ·
Monotone set function · Resource buying game ·
Submodular function · Polymatroid congestion game ·
Sensitivity analysis · M$^\natural$-convex function

1 Introduction

Congestion games, introduced by Rosenthal [19], provides a model of human's behavior of choosing an optimal strategy while avoiding congestion. A congestion game consists of a set N of players and a set E of resources. Each player $i \in N$ has a family of resource subsets $\mathcal{B}_i \subseteq 2^E$, which represents the possible strategies of i. Each resource $e \in E$ is associated with a nondecreasing cost function $c_e \colon \mathbb{R} \to \mathbb{R}$. The cost function means that, if the number of players using e is x_e, then each of

© Springer Nature Switzerland AG 2019
T. V. Gopal and J. Watada (Eds.): TAMC 2019, LNCS 11436, pp. 594–614, 2019.
https://doi.org/10.1007/978-3-030-14812-6_37

those x_e players should pay $c_e(x_e)$ to use e. Thus, if a player i chooses a strategy $B_i \in \mathcal{B}_i$, then the total cost paid by i is equal to $\sum_{e \in B_i} c_e(x_e)$.

Rosenthal [19] proved that every congestion game is a potential game, and thus has a pure Nash equilibrium. Up to the present date, a number of generalized or focused models of congestion games have been considered, and pure Nash equilibria of those models have been intensively analyzed.

Network congestion games (or *routing games*) form a typical class of congestion games, in which the resources are the edges in a network and the strategies of a player i are the routes (or flows) between two nodes s_i and t_i. Examples of recent work on network congestion games appear in [3,4,12].

There is another class of congestion games showing successful results, *matroid congestion games*, in which the strategies of each player are the bases of a matroid. Ackermann, Röglin, and Vöcking [2] proved that every weighted matroid congestion game, in which the cost c_e for using a resource $e \in E$ is determined by the sum of the weights of the players using e, and every matroid congestion games with player-specific cost functions have pure Nash equilibria. Since weighted congestion games and congestion games with player-specific cost functions do not necessarily have pure Nash equilibria, this result demonstrates a good structure of matroid congestion games. Recently, Harks, Klimm, and Peis [8] discussed *polymatroid congestion games*, in which the strategies of each player are the integer points in a base polyhedron of a polymatroid. In their model, the cost functions are player-specific and bivariate. That is, for a player $i \in N$ and resource $e \in E$, the cost function $c_{i,e} \colon \mathbb{R}^2 \to \mathbb{R}$ has two variables $x_{i,e}$ and $x_{-i,e} = \sum_{j \in N \setminus \{i\}} x_{j,e}$, where $x_{j,e}$ represents the multiplicity of the usage of e by a player $j \in N$. Harks, Klimm, and Peis [8] proved that polymatroid congestion games have pure Nash equilibria if the cost functions are *discrete convex* and *regular*. The proof is based on a sensitivity analysis for separable discrete convex optimization over base polyhedra: the deviation of an optimal strategy of a player i when the strategies of the other players change by a unit vector. Special classes of this model include matroid congestion games with player-specific costs [2] and singleton integer-splittable congestion games by Tran-Thanh et al. [21]. Other results on matroid congestion games appear, e.g., in [1,11].

A variant of congestion games is *resource buying games*, introduced by Harks and Peis [9,10]. A distinctive feature of resource buying games is that, if a resource e is used by x_e players, then it is required that the sum of the payment by those x_e players for using e is at least $c_e(x_e)$. This is in contrast to the usual models of congestion games in which every player using e should pay $c_e(x_e)$ for using e. Harks and Peis [9,10] proved that matroid resource buying games have pure Nash equilibria if the cost functions are marginally nonincreasing, and that resource buying games have pure Nash equilibria if the cost functions are marginally nondecreasing.

Now the purpose of this paper is to provide a new direction of generalizing matroid congestion games: matroid congestion games with n-variate cost functions, where n denotes the number of players. As mentioned above, in most of matroid congestion games, the cost functions are univariate (x_e) or bivariate

($x_{i,e}$ and $x_{-i,e}$). We generalize this point so that each cost function has variables $x_{1,e}, \ldots, x_{n,e}$: a nondecreasing cost function is generalized to a *monotone set function*; a marginally nonincreasing cost function is generalized to a *submodular function*; and a bivariate discrete convex function is generalized to an $M^\natural$-*convex function*. These generalizations enable us to establish more elaborate models in which the cost for using a resource e is not determined by the sum of the weights of the players using e, but by the set of players using e.

For this purpose, we often make use of the theory of *discrete convex analysis* [15]. The importance of $M^\natural$-convex and $L^\natural$-convex functions in game theory has been well appreciated [16]. For instance, one relation between network congestion games and M-convex function minimization is revealed by Fujishige et al. [6].

Our contribution consists of three extensions of the aforementioned results. First, we discuss matroid congestion games in which the cost functions are monotone set functions. This class generalizes that of weighted matroid congestion games. By applying the arguments for weighted matroid congestion games by Ackermann, Röglin, and Vöcking [2], we prove that every matroid congestion game with monotone cost functions has a pure Nash equilibrium.

Second, we consider matroid resource buying games in which the cost functions are monotone and submodular. This class generalizes that of matroid resource buying games with marginally nonincreasing cost functions. By extending the proof by Harks and Peis [9,10], we show that every matroid resource buying game with monotone submodular cost functions has a pure Nash equilibrium.

It is a kind of interest that submodular functions together with matroidal domain have a good structure. In the context of discrete convex analysis [15], polymatroids form the domain of $M^\natural$-convex functions ($M^\natural$-*convex set*), while submodular functions form a special class of $L^\natural$-*convex functions*. It is a basic fact in discrete convex analysis that $M^\natural$-convex functions and $L^\natural$-convex functions have a duality relation. Thus, it is of interest that $L^\natural$-convex functions with domain defined from $M^\natural$-convex sets have a good property.

Finally, for a polymatroid congestion game with n-variate $M^\natural$-convex cost functions, we analyze the deviation of an optimal strategy of a player when the strategies of the other players change by a unit vector. As Harks, Klimm, and Peis [8] proved the existence of pure Nash equilibria in polymatroid congestion games with bivariate cost functions based on the sensitivity analysis for convex optimization over base polyhedra, we anticipate that our sensitivity result will find an application in the analysis of polymatroid congestion games with $M^\natural$-convex cost functions.

We remark that our sensitivity result differs from that for minimization of a potential function over polymatroids by Fujishige et al. [7]. Fujishige et al. [7] discuss the deviation of the optimal value, whereas we consider the deviation of the optimal solution. Another related topic is proximity results for M-convex optimization [13,18], which deal with the behavior of minimizers of an M-convex function when the function is extended to a continuous function.

The rest of the paper is organized as follows. In Sect. 2, we review basic facts on matroids, submodular functions, and M$^\natural$-convex functions. In Sects. 3.1 and 3.2, we prove the existence of pure Nash equilibria in matroid congestion games with monotone cost functions, and in matroid resource buying games with monotone submodular cost functions, respectively. Finally, in Sect. 4, we show the sensitivity result for separable M$^\natural$-convex minimization.

2 Preliminaries

In this section, we review definitions and some basic facts on matroids, submodular functions, and M$^\natural$-convex functions, which will be used in the following sections. For more details, the readers are referred to [5,15,20].

2.1 Matroids and Submodular Functions

In this paper, we define matroids by their base families. Let E be a finite set and $\mathcal{B} \subseteq 2^E$. Now $(E, \mathcal{B})$ is called a *matroid* if it satisfies the following axioms:

1. $\mathcal{B} \neq \emptyset$; and
2. if $B, B' \in \mathcal{B}$ and $e \in B \setminus B'$, there exists $f \in B' \setminus B$ with $(B \setminus \{e\}) \cup \{f\} \in \mathcal{B}$.

Let $w \colon E \to \mathbb{R}$ be a weight function. For $B \subseteq E$, let $w(B) = \sum_{e \in B} w(e)$. In this paper, we are mainly concerned with *minimum-weight bases*, that is, a base $B^* \in \mathcal{B}$ such that $w(B^*) \leq w(B)$ for each $B \in \mathcal{B}$. For minimization, matroid bases have a good structure: the global optimality is assured by some local optimality, which is described below and will often appear in this paper.

Lemma 1 (see, e.g., [15,20]). *Let $(E, \mathcal{B})$ be a matroid and $w \colon E \to \mathbb{R}$ be a weight function. Then, for a base $B^* \in \mathcal{B}$, it holds that $w(B^*) \leq w(B)$ for each $B \in \mathcal{B}$ if and only if $w(B^*) \leq w((B^* \setminus \{e\}) \cup \{f\})$ for every $e \in B^*$ and $f \in E \setminus B^*$ such that $(B^* \setminus \{e\} \cup \{f\} \in \mathcal{B}$.*

Let N be a finite set. A set function $c \colon 2^N \to \mathbb{R}$ is called *submodular* if

$$c(X \cup \{i\}) - c(X) \geq c(Y \cup \{i\}) - c(Y) \quad \text{if } X \subseteq Y \subseteq N \text{ and } i \in N \setminus Y.$$

It is recognized that submodular functions provide a good model of economies of scale. For instance, submodular functions generalize marginally nonincreasing function. A function $\phi \colon \mathbb{Z} \to \mathbb{R}$ is called *marginally nonincreasing* if

$$\phi(x + \delta) - \phi(x) \geq \phi(y + \delta) - \phi(y) \quad (x, y, \delta \in \mathbb{Z}, x \leq y).$$

It is straightforward to see that a submodular function $c \colon 2^N \to \mathbb{R}$ such that $c(X) = c(X')$ if $|X| = |X'|$ is essentially a marginally nonincreasing function.

2.2   M-convex Functions and M♮-convex Functions

Let N be a finite set. For $i \in N$, define a vector $\chi^i \in \{0,1\}^N$ by

$$\chi^i_j = \begin{cases} 1 & (j = i), \\ 0 & (j \in N \setminus \{i\}). \end{cases}$$

For a vector $\boldsymbol{x} \in \mathbb{Z}^N$, define $\mathrm{supp}^+(\boldsymbol{x}), \mathrm{supp}^-(\boldsymbol{x}) \subseteq N$ respectively by

$$\mathrm{supp}^+(\boldsymbol{x}) = \{i \in N \mid x_i > 0\}, \quad \mathrm{supp}^-(\boldsymbol{x}) = \{i \in N \mid x_i < 0\}.$$

Let $\overline{\mathbb{R}}$ denote $\mathbb{R} \cup \{+\infty\}$. For a function $c \colon \mathbb{Z}^N \to \overline{\mathbb{R}}$, its *effective domain* $\mathrm{dom}\, c$ is defined as $\{\boldsymbol{x} \mid c(\boldsymbol{x}) < +\infty\}$. Now a function $c \colon \mathbb{Z}^N \to \overline{\mathbb{R}}$ with $\mathrm{dom}\, c \neq \emptyset$ is an M♮-*convex function* [15,17] if it satisfies the following exchange axiom:

For $\boldsymbol{x}, \boldsymbol{y} \in \mathrm{dom}\, c$ and $i \in \mathrm{supp}^+(\boldsymbol{x} - \boldsymbol{y})$, it holds that

$$c(\boldsymbol{x}) + c(\boldsymbol{y}) \geq c(\boldsymbol{x} - \chi^i) + c(\boldsymbol{y} + \chi^i)$$

or there exists $j \in \mathrm{supp}^-(\boldsymbol{x} - \boldsymbol{y})$ such that

$$c(\boldsymbol{x}) + c(\boldsymbol{y}) \geq c(\boldsymbol{x} - \chi^i + \chi^j) + c(\boldsymbol{y} + \chi^i - \chi^j).$$

A set $\mathcal{B} \subseteq \mathbb{Z}^N$ is called an M♮-*convex set* if its indicator function $\delta_{\mathcal{B}} \colon \mathbb{Z}^N \to \overline{\mathbb{R}}$ defined by

$$\delta_{\mathcal{B}}(\boldsymbol{x}) = \begin{cases} 0 & (\boldsymbol{x} \in \mathcal{B}), \\ +\infty & (\boldsymbol{x} \in \mathbb{Z}^N \setminus \mathcal{B}) \end{cases}$$

is an M♮-convex function.

We remark that, if $|N| = 1$, i.e., c is a univariate function, M♮-convexity coincides with discrete convexity in [8], which is defined as

$$c(x) - c(x - 1) \leq c(x + 1) - c(x) \quad \text{for every } x \in \mathbb{Z}. \tag{1}$$

Also, the following property, which is an immediate consequence of the exchange axiom, is also useful: if $\boldsymbol{y} = \boldsymbol{x} + \chi^i$ for some $i \in N$, then

$$c(\boldsymbol{x}) - c(\boldsymbol{x} - \chi^j) \leq c(\boldsymbol{y}) - c(\boldsymbol{y} - \chi^j) \quad \text{for each } j \in N. \tag{2}$$

The class of M♮-convex functions are closed under the following operations.

Lemma 2 (see [15]). *Let $c \colon \mathbb{Z}^N \to \overline{\mathbb{R}}$ be an M♮-convex function.*

1. *For $\boldsymbol{a} \in \mathbb{Z}^N$, $c(\boldsymbol{x} - \boldsymbol{a})$ is an M♮-convex function.*
2. *For $N' \subseteq N$, the function $c_{N'} \colon \mathbb{Z}^{N'} \to \overline{\mathbb{R}}$ defined by*

$$c_{N'}(\boldsymbol{x}') = c(\boldsymbol{x}', \boldsymbol{0}_{N \setminus N'}) \quad (\boldsymbol{x}' \in \mathbb{Z}^{N'})$$

is an M♮-convex function, provided $\mathrm{dom}\, c_{N'} \neq \emptyset$

3. *The direct sum of two $M^{\natural}$-convex functions is an $M^{\natural}$-convex function.*

We next define M-convex functions, which form a special, but essentially equivalent class of $M^{\natural}$-convex functions. A function $c \colon \mathbb{Z}^N \to \overline{\mathbb{Z}}$ with $\operatorname{dom} c \neq \emptyset$ is called an M-*convex function* [14,15] if it satisfies the following exchange axiom:

For $\boldsymbol{x}, \boldsymbol{y} \in \operatorname{dom} c$ and $i \in \operatorname{supp}^{+}(\boldsymbol{x} - \boldsymbol{y})$, there exists $j \in \operatorname{supp}^{-}(\boldsymbol{x} - \boldsymbol{y})$ such that

$$c(\boldsymbol{x}) + c(\boldsymbol{y}) \geq c(\boldsymbol{x} - \chi^{i} + \chi^{j}) + c(\boldsymbol{y} + \chi^{i} - \chi^{j}).$$

A set $\mathcal{B} \subseteq \mathbb{Z}^N$ is called an M-*convex set* if its indicator function $\delta_{\mathcal{B}} \colon \mathbb{Z}^N \to \overline{\mathbb{R}}$ is an M-convex function. We remark that an M-convex set is a generalization of the base family of a matroid, and consists of the integer points in a base polyhedron of a polymatroid.

For M-convex functions, the following optimality criterion, which extends Lemma 1, is established.

Lemma 3 ([14,15]). *Let $c \colon \mathbb{Z}^N \to \overline{\mathbb{R}}$ be an M-convex function. A vector $\boldsymbol{x} \in \mathbb{Z}^N$ is a minimizer of c if and only if*

$$c(\boldsymbol{x}) \leq c(\boldsymbol{x} - \chi^{k} + \chi^{l}) \quad \text{for each } k, l \in N.$$

3 Generalizations of Weighted Matroid Congestion Games

In this section, we prove the existence of pure Nash equilibria in matroid congestion games with monotone cost functions (Sect. 3.1), and in matroid resource buying games with monotone submodular cost functions (Sect. 3.2).

3.1 Matroid Congestion Games with Monotone Cost Functions

Our model of a matroid congestion game is represented by a tuple $(N, E, (\mathcal{B}_i)_{i \in N}, (c_e)_{e \in E})$. Here, N denotes the set of players, and E denotes the set of resources. Let n denote the number of players and let $N = \{1, \ldots, n\}$. Each player $i \in N$ has a family of resource subsets $\mathcal{B}_i \subseteq 2^E$, called *configurations*. In matroid congestion games, $\mathcal{B}_i$ is the base family of some matroid on E for each $i \in N$. Finally, each resource $e \in E$ has a nonnegative cost function $c_e \colon 2^N \to \mathbb{R}_{+}$. We assume that c_e is monotone and normalized for each resource $e \in E$, i.e., $c_e(X) \leq c_e(Y)$ if $X \subseteq Y$ and $c_e(\emptyset) = 0$.

A *state* is a collection $B = (B_1, \ldots, B_n)$ of configurations of all players. Let B be a state. For a player $i \in N$, let B_{-i} denote a collection of the configurations of the other players under the state B, that is, $B_{-i} = (B_1, \ldots, B_{i-1}, B_{i+1}, \ldots, B_n)$. For a resource $e \in E$, let $N_e(B) \subseteq N$ denote the set of players using e, that is, $N_e(B) = \{i \in N \mid e \in B_i\}$. The cost for using a resource $e \in E$ is described as $c_e(N_e(B))$. Thus, the cost paid by a player $i \in N$ under a state

B is $\sum_{e \in B_i} c_e(N_e(B))$. Now a state B is called a *pure Nash equilibrium* if no player has an incentive to change the configuration to have less cost, i.e.,

$$\sum_{e \in B_i} c_e(N_e(B)) \leq \sum_{e \in B_i'} c_e(N_e((B_i', B_{-i}))) \quad \text{for each } B_i' \in \mathcal{B}_i.$$

It is straightforward to see that the above model of matroid congestion games generalizes weighted matroid congestion games [2], which is defined as follows. In weighted matroid congestion games, each player $i \in N$ has a positive weight $w_i \in \mathbb{Z}_+$, and each resource $e \in E$ is associated with a univariate cost function $c_e : \mathbb{Z}_+ \to \mathbb{R}_+$. The cost for using e in a state B is defined by $c_e(w(N_e(B)))$.

Ackermann, Röglin, and Vöcking [2] proved that every weighted matroid congestion game has a pure Nash equilibrium. They also showed that a pure Nash equilibrium is attained by a finite number of *locally best responses*. Under a state B, a *locally best response* of a player i is to choose a configuration $B_i^* = (B_i \setminus \{e^*\}) \cup \{f^*\}$ for some $e^* \in B_i$ and $f^* \in E \setminus B_i$ satisfying

$$\sum_{e \in B_i^*} c_e(N_e(B_i^*, B_{-i})) \leq \sum_{e \in B_i'} c_e(N_e(B_i', B_{-i}))$$

for each $B_i' \in \mathcal{B}_i$ with $B_i' = (B_i \setminus \{e'\}) \cup \{f'\}$ for some $e' \in B_i$ and $f' \in E \setminus B_i$. Note that B_i^* may not be a global minimizer, i.e., a configuration B_i° minimizing $\sum_{e \in B_i} c_e(N_e(B_i, B_{-i}))$ among all strategies $B_i \in \mathcal{B}_i$.

The following theorem extends these results to matroid congestion games with monotone cost functions. The proof below is a straightforward extension of that for weighted matroid congestion games [2].

Theorem 1. *Every matroid congestion game with monotone cost functions has a pure Nash equilibrium. Moreover, a pure Nash equilibrium is attained after a finite number of locally best responses.*

Proof. Let B be a state. For each resource $e \in E$, associate a pair $\gamma_e(B) = (c_e(N_e(B)), |N_e(B)|)$. For two resources $e, f \in E$, we define that $\gamma_e(B) \leq \gamma_f(B)$ if either $c_e(N_e(B)) < c_f(N_f(B))$ or $c_e(N_e(B)) = c_f(N_f(B))$ and $|N_f(B)| \leq |N_f(B)|$ holds. Now let $\Gamma(B)$ be a sequence of $\gamma_e(B)$ for every $e \in E$ in nondecreasing order. For two states B, B', we denote the lexicographic order of $\Gamma(B)$ and $\Gamma(B')$ by $<_{\text{lex}}$. That is, for $\Gamma(B) = (\gamma_{e_1}(B), \ldots, \gamma_{e_m}(B))$ and $\Gamma(B') = (\gamma_{f_1}(B'), \ldots, \gamma_{f_m}(B'))$, we denote $\Gamma(B) <_{\text{lex}} \Gamma(B')$ if there exists an integer l with $1 \leq l \leq m$ such that $\gamma_{e_k}(B) = \gamma_{f_k}(B')$ for every $k < l$ and $\gamma_{e_l}(B) < \gamma_{f_l}(B')$.

If a state B is not a pure Nash equilibrium, there exists a player $i \in N$ who can decrease the cost when the other players' configurations B_{-i} is fixed. If i chooses a configuration $B_i' \in \mathcal{B}_i$, the cost paid by i is

$$\sum_{e \in B_i'} c_e(N_e(B_{-i}) \cup \{i\}).$$

By defining a weight function $w\colon E \to \mathbb{R}$ by

$$w(e) = c_e(N_e(B_{-i}) \cup \{i\}) \quad (e \in E),$$

the cost paid by i is described as $w(B_i')$. We remark that the weight function w is independent of the configuration of i, and thus minimizing the cost paid by i amounts to choosing a minimum-weight base in $\mathcal{B}_i$ with respect to w. This means that the current configuration B_i is not a minimum-weight base, and it follows from Lemma 1 that there exists resources $e \in B_i$ and $f \in N \setminus B_i$ such that $B_i^* = (B_i \setminus \{e^*\}) \cup \{f^*\} \in \mathcal{B}_i$ satisfies $w(B_i^*) < w(B_i)$, i.e.,

$$c_{f^*}(N_{f^*}(B_{-i}) \cup \{i\}) < c_{e^*}(N_{e^*}(B_{-i}) \cup \{i\}). \tag{3}$$

Note that such B_i^* can be obtained by a locally best response.

Now consider a state $B^* = (B_i^*, B_{-i})$. In what follows, we will show that $\Gamma(B^*) <_{\text{lex}} \Gamma(B)$. Since the number of the states is finite, this strict inequality implies that after a finite number of iteration of locally best responses we reach a state in which no player can improve her cost, a pure Nash equilibrium.

First, it is immediate from (3) that

$$c_{f^*}(N_{f^*}(B^*)) = c_{f^*}(N_{f^*}(B_{-i}) \cup \{i\}) < c_{e^*}(N_{e^*}(B_{-i}) \cup \{i\}) = c_{e^*}(N_{e^*}(B)).$$

Next, by the monotonicity of c_e, we have that

$$c_{e^*}(N_{e^*}(B^*)) = c_{e^*}(N_{e^*}(B) \setminus \{i\}) \leq c_{e^*}(N_{e^*}(B)).$$

Finally, it is obvious that

$$|N_{e^*}(B^*)| = |N_{e^*}(B)| - 1 < |N_{e^*}(B)|.$$

The above three inequalities imply that $\gamma_{f^*}(B^*) < \gamma_{e^*}(B)$ and $\gamma_{e^*}(B^*) < \gamma_{e^*}(B)$. Since $\gamma_e(B^*) = \gamma_e(B)$ for each $e \in E \setminus \{e^*, f^*\}$, we conclude that $\Gamma(B^*) <_{\text{lex}} \Gamma(B)$, which completes the proof. □

3.2 Matroid Resource Buying Games with Submodular Cost Functions

A *matroid resource buying game* is represented by a tuple $(N, E, (\mathcal{B}_i)_{i \in N}, (c_e)_{e \in E})$, which is the same as that in Sect. 3.1. Recall that each c_e is monotone and normalized. A distinguishing feature of resource buying games is that the players using a resource $e \in E$ should buy the resource e by cooperatively paying the cost for e, which is formally described in the following way.

Let $p_{i,e} \in \mathbb{R}_+$ be the payment of a player i for a resource e. The vector $(p_{i,e})_{i \in N, e \in E} \in \mathbb{R}_+^{N \times E}$ is called the *payment vector*. For a payment vector p and a player i, let p_{-i} denote a restriction of p to $\mathbb{R}_+^{(N \setminus \{i\}) \times E}$, i.e., $p_{-i} = (p_{j,e})_{j \in N \setminus \{i\}, e \in E}$. A *strategy profile* is a tuple (B, p) of a state $B = (B_1, \ldots, B_n)$

and a payment vector $p \in \mathbb{R}_+^{N \times E}$. Under a strategy profile (B, p), a resource $e \in E$ is *bought* if

$$\sum_{i \in N} p_{i,e} \geq c_e(N_e(B)),$$

and the *private cost* π_i of a player $i \in N$ is defined by

$$\pi_i(B, p) = \begin{cases} \sum_{e \in E} p_{i,e} & \text{if every resource in } B_i \text{ is bought,} \\ +\infty & \text{otherwise.} \end{cases}$$

Now a strategy profile (B, p) is a *pure Nash equilibrium* if, for each player $i \in N$, it holds that

$$\pi_i(B, p) \leq \pi_i(B', p')$$

for an arbitrary strategy profile (B', p') satisfying $B'_{-i} = B_{-i}$ and $p'_{-i} = p_{-i}$. That is, a pure Nash equilibrium is a strategy profile under which every player $i \in N$ cannot strictly decrease her private cost π_i if the configurations and payments of the other players do not change.

The above model generalizes resource buying games with univariate cost functions introduced by Harks and Peis [9,10]. In their model, each player $i \in N$ is associated with a weight $w_i \in \mathbb{Z}_+$, and the cost for a resource e under a state B is determined as $c_e(w(N_e(B)))$ by a univariate function $c_e \colon \mathbb{Z}_+ \to \mathbb{R}_+$ for each $e \in E$. Harks and Peis [9,10] proved that matroid resource buying games have pure Nash equilibria if each cost function c_e is marginally nonincreasing.

Now we prove that every matroid resource buying game with monotone submodular cost functions has a pure Nash equilibrium. Indeed, the arguments by Harks and Peis [9,10] can be extended to our model straightforwardly, by replacing the marginally nonincreasing property of the cost functions by submodularity. Below we exhibit the extended argument, with some refinement of propositions in [9,10].

First, it is straightforward to see that the following condition is necessary for a strategy profile (B, p) to be a pure Nash equilibrium:

$$p_{i,e} = 0 \qquad \begin{array}{l} \text{for each } i \in N \text{ and } e \in E \setminus B_i, \\ \text{equivalently, for each } e \in E \text{ and } i \in N \setminus N_e(B), \end{array} \tag{4}$$

$$\sum_{i \in N} p_{i,e} = c_e(N_e(B)) \quad \text{for each resource } e \in E. \tag{5}$$

Here we make this condition sufficient as well, by adding one more constraint.

Lemma 4. *In a matroid resource buying game with monotone cost functions, a strategy profile (B, p) is a pure Nash equilibrium if and only if it satisfies* (4), (5), *and*

$$p_{i,e} \leq c_f(N_f(B) \cup \{i\}) - c_f(N_f(B))$$
$$(i \in N, e \in B_i, f \in N \setminus B_i, (B_i \setminus \{e\}) \cup \{f\} \in \mathcal{B}_i). \tag{6}$$

Proof. We first show necessity. As mentioned above, (4) and (5) are obviously necessary for a pure Nash equilibrium (B, p). Suppose to the contrary that (6) does not hold for $i \in N$, $e \in B_i$, and $f \in N \setminus B_i$ satisfying $(B_i \setminus \{e\}) \cup \{f\} \in \mathcal{B}_i$, that is,

$$p_{i,e} > c_f(N_f(B) \cup \{i\}) - c_f(N_f(B)). \tag{7}$$

Then, the player i can change the configuration and payments in the following manner:

$$B_i' := (B_i \setminus \{e\}) \cup \{f\}, \qquad p_{i,e}' := 0, \qquad p_{i,f}' := c_f(N_f(B) \cup \{i\}) - c_f(N_f(B)).$$

After the change, the resource f remains bought, as well as the other resources in B_i'. Thus, the private cost π_i changes by $-p_{i,e} + (c_f(N_f(B) \cup \{i\}) - c_f(N_f(B)))$, which is negative by (7). This contradicts to the fact that (B, p) is a pure Nash equilibrium.

We next prove sufficiency. Take a player $i \in N$, and fix B_{-i} and p_{-i}. We show that (B_i, p_i) yields the minimum private cost π_i for i if (4), (5), and (6) hold.

If i chooses a configuration $B_i' \in \mathcal{B}_i$, then her payments $p_{i,e}'$ achieving the minimum private cost $\pi_i(B_i')$ is

$$p_{i,e}' = \begin{cases} \max \left\{ c_e(N_e(B_{-i}) \cup \{i\}) - \displaystyle\sum_{j \in N \setminus \{i\}} p_{j,e}, \ 0 \right\} & (e \in B_i'), \\ 0 & (e \in E \setminus B_i'). \end{cases}$$

Here, if $e \in B_i$, then

$$c_e(N_e(B_{-i}) \cup \{i\}) - \sum_{j \in N \setminus \{i\}} p_{j,e} = c_e(N_e(B)) - \sum_{j \in N \setminus \{i\}} p_{j,e} = p_{i,e} \geq 0,$$

where the last equality follows from (5). If $e \notin B_i$, then

$$c_e(N_e(B_{-i}) \cup \{i\}) - \sum_{j \in N \setminus \{i\}} p_{j,e} = c_e(N_e(B) \cup \{i\}) - \sum_{j \in N \setminus \{i\}} p_{j,e} \geq 0,$$

where the inequality follows from the monotonicity of c_e and (5). We thus obtain

$$p_{i,e}' = \begin{cases} c_e(N_e(B_{-i}) \cup \{i\}) - \displaystyle\sum_{j \in N \setminus \{i\}} p_{j,e} & (e \in B_i'), \\ 0 & (e \in E \setminus B_i'), \end{cases}$$

$$\pi_i(B_i') = \sum_{e \in B_i'} \left(c_e(N_e(B_{-i}) \cup \{i\}) - \sum_{j \in N \setminus \{i\}} p_{j,e} \right).$$

Note that (B_i, p_i) satisfies the above condition by (4) and (5). Thus, it suffices to prove that B_i attains the minimum private cost $\pi_i(B_i')$ among the configurations $B_i' \in \mathcal{B}_i$.

Now define a weight function $w\colon E \to \mathbb{R}$ by

$$w(e) = c_e(N_e(B_{-i}) \cup \{i\}) - \sum_{j \in N \setminus \{i\}} p_{j,e} \quad (e \in E).$$

Note that w is independent of the configuration of i. Thus, a configuration $B_i \in \mathcal{B}_i$ attaining the minimum private cost π_i is a minimum-weight base with respect to w. It then follows from Lemma 1 that

$$w(B_i) \le w((B_i \setminus \{e\}) \cup \{f\}) \tag{8}$$

for every $e \in B_i$ and $f \in E \setminus B_i$ such that $(B_i \setminus \{e\}) \cup \{f\} \in \mathcal{B}_i$. It is derived from (5) that (8) is equivalent to

$$w(e) \le w(f)$$
$$\iff c_e(N_e(B_{-i}) \cup \{i\}) - \sum_{j \in N \setminus \{i\}} p_{j,e} \le c_f(N_f(B_{-i}) \cup \{i\}) - \sum_{j \in N \setminus \{i\}} p_{j,f}$$
$$\iff c_e(N_e(B)) - \sum_{j \in N \setminus \{i\}} p_{j,e} \le c_f(N_f(B) \cup \{i\}) - \sum_{j \in N \setminus \{i\}} p_{j,f}$$
$$\iff p_{i,e} \le c_f(N_f(B) \cup \{i\}) - c_f(N_f(B)).$$

Since (6) holds, we have proved that B_i is a minimum-weight base in $\mathcal{B}_i$ with respect to w, and thus (B_i, p_i) yields the minimum private cost π_i. $\quad\square$

Let $B = (B_1, \ldots, B_n)$ be a state. For $i \in N$, define $F_i \subseteq E$ by

$$F_i = \{e \in E \mid e \in B_i' \text{ for each } B_i' \in \mathcal{B}_i\}.$$

It is obvious that $F_i \subseteq B_i$ for each $i \in N$. It is also straightforward to see that, for each $e \in B_i \setminus F_i$, there exists $f \in E \setminus B_i$ such that $(B_i \setminus \{e\}) \cup \{f\} \in \mathcal{B}_i$. For a player $i \in N$ and a resource $e \in B_i$, define $\Delta_{i,e}(B) \in \overline{\mathbb{R}}$ by

$$\Delta_{i,e}(B) =$$
$$\begin{cases} \min\{c_f(N_f(B) \cup \{i\}) - c_f(N_f(B)) \mid f \in N \setminus B_i, (B_i \setminus \{e\}) \cup \{f\} \in \mathcal{B}_i\} & (e \in B_i \setminus F_i), \\ +\infty & (e \in F_i). \end{cases}$$

Theorem 2. *In a matroid resource buying game with monotone cost functions, a state $B = (B_1, \ldots, B_n)$ has a payment vector $p \in \mathbb{R}^{N \times E}$ such that (B, p) is a pure Nash equilibrium if and only if*

$$c_e(N_e(B)) \le \sum_{i \in N_e(B)} \Delta_{i,e}(B) \quad \text{for each resource } e \in E. \tag{9}$$

Proof. We first show necessity. If (B, p) is a pure Nash equilibrium, then it follows from Lemma 4 that

$$c_e(N_e(B)) = \sum_{i \in N_e(B)} p_{i,e} \le \sum_{i \in N_e(B)} \Delta_{i,e}(B),$$

where the equality follows from (4) and (5), and the inequality follows from (6) and the definition of $\Delta_{i,e}(B)$.

We next prove sufficiency. Define $p \in \mathbb{R}^{N \times E}$ in the following manner. If $e \in F_{i^*}$ for some player $i^* \in N$, define

$$p_{i,e} = \begin{cases} c_e(N_e(B)) & (i = i^*), \\ 0 & (i \in N \setminus \{i^*\}). \end{cases}$$

Otherwise, define

$$p_{i,e} = \begin{cases} \dfrac{\Delta_{i,e}(B)}{\displaystyle\sum_{j \in N_e(B)} \Delta_{j,e}(B)} c_e(N_e(B)) & (i \in N_e(B)), \\ 0 & (i \in N \setminus N_e(B)). \end{cases}$$

Then, it is not difficult to see that (4) and (5) hold. Moreover, it follows from (9) that $p_{i,e} \le \Delta_{i,e}(B)$, and thus (6) holds. Therefore, by Lemma 4, we conclude that (B, p) is a pure Nash equilibrium. $\qquad\square$

Theorem 3. *In a matroid resource buying game with monotone submodular cost functions, a state $B = (B_1, \ldots, B_n)$ minimizing $\sum_{e \in E} c_e(N_e(B))$ has a payment vector $p \in \mathbb{R}^{N \times E}$ such that (B, p) is a pure Nash equilibrium.*

Proof. Let B be a state minimizing $\sum_{e \in E} c_e(N_e(B))$. By Theorem 2, it suffices to show (9). Suppose to the contrary that there exists a resource $e^* \in E$ satisfying

$$c_{e^*}(N_{e^*}(B)) > \sum_{i \in N_{e^*}(B)} \Delta_{i,e^*}(B). \tag{10}$$

Without loss of generality, let $N_{e^*}(B) = \{1, \ldots, k\}$. For each $i \in \{1, \ldots, k\}$, it is derived from (10) that $e \notin F_i$, and denote the resource attaining $\Delta_{i,e^*}(B)$ by f_i^* and let $\hat{B}_i = (B_i \setminus \{e^*\}) \cup \{f_i^*\}$, that is,

$$\Delta_{i,e^*}(B) = c_{f_i^*}(N_{f_i^*}(B) \cup \{i\}) - c_{f_i^*}(N_{f_i^*}(B)) \quad \text{and}$$
$$\hat{B}_i = (B_i \setminus \{e^*\}) \cup \{f_i^*\} \in \mathcal{B}_i.$$

Then, define states $B^{(0)}, B^{(1)}, \ldots B^{(k)}$ by

$$B^{(0)} = B,$$
$$B^{(i)} = \left(\hat{B}_i, B^{(i-1)}_{-i}\right) = (\hat{B}_1, \ldots, \hat{B}_i, B_{i+1}, \ldots, B_k) \quad (i = 1, \ldots, k).$$

Now for every $i = 1, \ldots, k$, it is straightforward to see that $N_{f_i^*}(B^{(i)}) \supseteq N_{f_i^*}(B)$, and hence

$$c_{f_i^*}(N_{f_i^*}(B^{(i)})) - c_{f_i^*}(N_{f_i^*}(B^{(i-1)})) = c_{f_i^*}(N_{f_i^*}(B^{(i-1)} \cup \{i\})) - c_{f_i^*}(N_{f_i^*}(B^{(i-1)}))$$
$$\le c_{f_i^*}(N_{f_i^*}(B \cup \{i\})) - c_{f_i^*}(N_{f_i^*}(B))$$
$$= \Delta_{i,e^*}(B),$$

where the inequality follows from the submodularity of $c_{f_i^*}$. Thus,

$$\sum_{e \in E} c_e(N_e(B^{(k)})) - \sum_{e \in E} c_e(N_e(B))$$

$$= \sum_{i=1}^{k} \left(\sum_{e \in E} c_e(N_e(B^{(i)})) - \sum_{e \in E} c_e(N_e(B^{(i-1)})) \right)$$

$$= \sum_{i=1}^{k} \left(\left(c_{f_i^*}(N_{f_i^*}(B^{(i)})) + c_{e^*}(N_{e^*}(B^{(i)})) \right) - \left(c_{f_i^*}(N_{f_i^*}(B^{(i-1)})) + c_{e^*}(N_{e^*}(B^{(i-1)})) \right) \right)$$

$$\leq c_{e^*}(N_{e^*}(B^{(k)})) - c_{e^*}(N_{e^*}(B^{(0)})) + \sum_{i=1}^{k} \Delta_{i,e^*}(B)$$

$$= -c_{e^*}(N_{e^*}(B)) + \sum_{i=1}^{k} \Delta_{i,e^*}(B)$$

$$< 0.$$

This contradicts that the state B minimizes $\sum_{e \in E} c_e(N_e(B))$. Therefore, we have shown that (9) holds, and thus by Theorem 2 we conclude that B has a payment vector p such that (B, p) is a pure Nash equilibrium.   □

4 Sensitivity Analysis for M$^\natural$-convex Minimization

In this section, we present sensitivity analysis for an optimization problem arising from the following generalization of weighted matroid congestion games [2] and polymatroid congestion games [8].

Again, let N denote the set of players, and E denote the set of resources. In this generalization, there are two main differences from the models in the previous section. The first difference is that a configuration of a player $i \in N$ is described by a nonnegative integer vector $\boldsymbol{x}_i \in \mathbb{Z}_+^E$, and accordingly a state is represented by a vector $\boldsymbol{x} \in \mathbb{Z}_+^{N \times E}$. This means that the usage of a resource e by each player i is not identical, and its multiplicity is represented by a nonnegative integer $x_{i,e}$. Thus, the models in the previous section amounts to special cases of this model where the configurations belong to $\{0,1\}^E$.

The second difference is that the cost functions are player-specific: the costs are represented by functions $c_{i,e} \colon \mathbb{Z}_+^N \to \mathbb{R}_+$ for each $i \in N$ and $e \in E$. For a state $\boldsymbol{x} \in \mathbb{Z}_+^{N \times E}$ and a resource $e \in E$, let $\boldsymbol{x}_e$ be the restriction of the vector $\boldsymbol{x} \in \mathbb{Z}_+^{N \times E}$ to $\mathbb{Z}_+^{N \times \{e\}}$. Then, the cost paid by $i \in N$ for using e is $c_{i,e}(\boldsymbol{x}_e)$, and the total cost paid by i is $\sum_{e \in E} c_{i,e}(\boldsymbol{x}_e)$.

Now, as a generalization of matroid congestion games, each player $i \in N$ is associated with an M$^\natural$-convex set $\mathcal{B}_i \subseteq \mathbb{Z}_+^E$ and a demand $d_i \in \mathbb{Z}_+$. These impose that a configuration of a player i should belong to $\mathcal{B}_i$, and the number of resources used by i, including the multiplicities, is equal to d_i. Thus, the set of configurations $\mathcal{B}_i(d_i) \subseteq \mathbb{Z}_+^E$ is described as

$$\mathcal{B}_i(d_i) = \{\boldsymbol{x}_i \in \mathbb{Z}_+^E \mid \boldsymbol{x}_i \in \mathcal{B}_i, \boldsymbol{x}_i(E) = d_i\}.$$

We remark that $\mathcal{B}_i(d_i)$ is an M-convex set (or the set of the integer points in a base polyhedron of a polymatroid), and thus a generalization of the base family of a matroid.

For the ease of notation, we often abbreviate d_i as d, when the player i is clear from the context. Given the configurations $\boldsymbol{x}_{-i} \in \mathbb{Z}_+^{(N\setminus\{i\})\times E}$ of the other players, a player i determines her configuration as an optimal solution for the following minimization problem $P(\boldsymbol{x}_{-i}, d)$ in variable $\boldsymbol{x}_i \in \mathbb{Z}_+^E$:

$$\text{Minimize} \quad \sum_{e\in E} c_{i,e}(x_{i,e}, \boldsymbol{x}_{-i,e})$$
$$\text{subject to} \quad \boldsymbol{x}_i \in \mathcal{B}_i(d).$$

From now on, we assume that the cost function $c_{i,e}$ is $M^\natural$-convex for each $i \in N$ and $e \in E$, and present a sensitivity analysis for $P(\boldsymbol{x}_{-i}, d)$. That is, we analyze the deviation of an optimal solution $\boldsymbol{x}_i^*$ when the parameters $\boldsymbol{x}_{-i}$ and d change by a unit vector.

First, the deviation of $\boldsymbol{x}_i^*$ in the change of d is immediately derived from the theory of $M^\natural$-convex minimization. It follows from Assertion 3 in Lemma 2 that $\sum_{e\in E} c_{i,e}(\boldsymbol{x}_e)$ is an $M^\natural$-convex function in variable $\boldsymbol{x} \in \mathbb{Z}^{N\times E}$. Then, by fixing the variable in $\mathbb{Z}^{(N\setminus\{i\})\times E}$ to be $\boldsymbol{x}_{-i}$, we obtain $\sum_{e\in E} c_{i,e}(x_{i,e}, \boldsymbol{x}_{-i,e})$ in variable $\boldsymbol{x}_i \in \mathbb{Z}^E$, which is $M^\natural$-convex by Assertions 1 and 2 in Lemma 2. Thus, the following theorem immediately follows from [17, Theorem 3.2].

Theorem 4. *Define* $\underline{d}, \overline{d} \in \mathbb{Z}_+$ *by* $\underline{d} = \min\{x_i(E) \mid \boldsymbol{x}_i \in \mathcal{B}_i\}$ *and* $\overline{d} = \max\{x_i(E) \mid \boldsymbol{x}_i \in \mathcal{B}_i\}$, *respectively. Let* $\boldsymbol{x}_{-i} \in \mathcal{B}_1 \times \cdots \mathcal{B}_{i-1} \times \mathcal{B}_{i+1} \times \cdots \mathcal{B}_n$, $d \in \mathbb{Z}_+$ *satisfy* $\underline{d} \leq d \leq \overline{d}$, *and* $\boldsymbol{x}_i^* \in \mathbb{Z}_+^E$ *be an optimal solution for* $P(\boldsymbol{x}_{-i}, d)$.

1. *If* $d < \overline{d}$, *there exists* $e \in E$ *such that* $\boldsymbol{x}_i^* + \chi^e$ *is an optimal solution for* $P(\boldsymbol{x}_{-i}, d+1)$.
2. *If* $d > \underline{d}$, *there exists* $e \in E$ *such that* $\boldsymbol{x}_i^* - \chi^e$ *is an optimal solution for* $P(\boldsymbol{x}_{-i}, d-1)$.

In the following, we discuss the deviation of $\boldsymbol{x}_i^*$ when $\boldsymbol{x}_{-i}$ changes by $\chi^{j,e}$ for some $j \in N\setminus\{i\}$ and $e \in E$. For this analysis, we assume that each cost function $c_{i,e}$ is *i-regular*, defined as follows.

Definition 1 (*i-regular function*). *For* $i \in N$, *a function* $c: \mathbb{Z}_+^N \to \mathbb{Z}$ *is an* i-regular function *if*

$$c(x_i + 2, \boldsymbol{x}_{-i}) - c(x_i + 1, \boldsymbol{x}_{-i}) \geq c(x_i + 1, \boldsymbol{x}_{-i} + \chi^j) - c(x_i, \boldsymbol{x}_{-i} + \chi^j)$$

for every $\boldsymbol{x} \in \mathbb{Z}_+^N$ *and every* $j \in N\setminus\{i\}$.

The fact that the cost function $c_{i,e}$ is *i*-regular means that the forward difference $c_{i,e}(x_{i,e}+1, \boldsymbol{x}_{-i,e}) - c_{i,e}(x_{i,e}, \boldsymbol{x}_{-i,e})$ becomes larger in the case when the usage $x_{i,e}$ of i increases by one, than the case when the usage $x_{j,e}$ of another player j increases by one. We remark that i-regularity does not imply $M^\natural$-convexity, and vice versa.

We also remark that i-regular functions generalize *regular functions*[8], which are defined for bivariate functions. Harks, Klimm, and Peis [8] presented a sensitivity result for the case where the functions $c_{i,e}$ are bivariate, discrete convex, and regular. Now our main result in this section is a sensitivity analysis extended to the case where the functions $c_{i,e}$ are n-variate, $M^\natural$-convex, and i-regular.

Theorem 5. *Let $i \in N$, $\boldsymbol{x}_{-i} \in \mathbb{Z}_+^{(N \setminus \{i\}) \times E}$, and $d \in \mathbb{Z}_+$. Suppose that $\mathrm{P}(\boldsymbol{x}_{-i}, d)$ is feasible and let $\boldsymbol{x}_i^* \in \mathbb{Z}_+^E$ be its optimal solution. Then, for each $j \in N \setminus \{i\}$ and $e \in E$, the problem $\mathrm{P}(\boldsymbol{x}_{-i} + \chi^{j,e}, d)$ is feasible and there exists $f \in E$ such that $\boldsymbol{x}_i^* - \chi^e + \chi^f$ is its optimal solution.*

Proof. The feasibility of $\mathrm{P}(\boldsymbol{x}_{-i} + \chi^{j,e}, d)$ is clear: both of the feasible regions of $\mathrm{P}(\boldsymbol{x}_{-i}, d)$ and $\mathrm{P}(\boldsymbol{x}_{-i} + \chi^{j,e}, d)$ are $\mathcal{B}_i(d)$, and hence $\mathrm{P}(\boldsymbol{x}_{-i} + \chi^{j,e}, d)$ is feasible when $\mathrm{P}(\boldsymbol{x}_{-i}, d)$ is feasible.

We then prove that $\mathrm{P}(\boldsymbol{x}_{-i} + \chi^{j,e}, d)$ has an optimal solution $\boldsymbol{x}_i^* + \chi^e - \chi^f$ for some $f \in E$. We will denote $\sum_{e \in E} c_{i,e}(\boldsymbol{x}_e)$ by $c_i(\boldsymbol{x})$. Let $f \in E$ satisfy

$$c_i(\boldsymbol{x}_i^* - \chi^e + \chi^f, \boldsymbol{x}_{-i} + \chi^{j,e}) \leq c_i(\boldsymbol{x}_i^* - \chi^e + \chi^{e'}, \boldsymbol{x}_{-i} + \chi^{j,e}) \quad (\forall e' \in E), \quad (11)$$

and let

$$\boldsymbol{x}_i' = \boldsymbol{x}_i^* - \chi^e + \chi^f.$$

Note that there always exists $f \in E$ such that $\boldsymbol{x}_i' \in \mathcal{B}_i(d)$, since $\boldsymbol{x}_i^* - \chi^e + \chi^e = \boldsymbol{x}_i^* \in \mathcal{B}_i(d)$. Our goal is to prove that $\boldsymbol{x}_i'$ is an optimal solution for $\mathrm{P}(\boldsymbol{x}_{-i} + \chi^{j,e}, d)$.

Since the restriction of an $M^\natural$-convex function on $\mathbb{Z}^E$ to a hyperplane $\sum_{e \in E} \boldsymbol{x}_i(e) = d$ is an M-convex function, we can apply Lemma 3 to show the optimality of $\boldsymbol{x}_i'$. Let $k, l \in E$ and

$$\boldsymbol{x}_i'' = \boldsymbol{x}_i' - \chi^k + \chi^l.$$

We will prove that

$$c_i(\boldsymbol{x}_i', \boldsymbol{x}_{-i} + \chi^{j,e}) \leq c_i(\boldsymbol{x}_i'', \boldsymbol{x}_{-i} + \chi^{j,e}) \quad \text{for every } k, l \in E. \quad (12)$$

Hereafter we assume $k \neq l$, since $k = l$ directly implies (12).

We have two cases: $e \neq f$ (Case 1); and $e = f$, i.e., $\boldsymbol{x}_i' = \boldsymbol{x}_i^*$ (Case 2). We first discuss Case 1.

Case 1.1 ($k = e$ and $l = f$). Suppose that $k = e$ and $l = f$, i.e., $\boldsymbol{x}_i'' = \boldsymbol{x}_i' - \chi^e + \chi^f$. Let $\boldsymbol{x}_i^\circ = \boldsymbol{x}_i' - \chi^e = \boldsymbol{x}_i'' - \chi^f$. To prove (12), it suffices to show that

$$c_i(\boldsymbol{x}_i', \boldsymbol{x}_{-i} + \chi^{j,e}) - c_i(\boldsymbol{x}_i^\circ, \boldsymbol{x}_{-i} + \chi^{j,e})$$
$$\leq c_i(\boldsymbol{x}_i'', \boldsymbol{x}_{-i} + \chi^{j,e}) - c_i(\boldsymbol{x}_i^\circ, \boldsymbol{x}_{-i} + \chi^{j,e}). \quad (13)$$

The LHS of (13) is equal to

$$c_i(\boldsymbol{x}_i', \boldsymbol{x}_{-i} + \chi^{j,e}) - c_i(\boldsymbol{x}_i' - \chi^e, \boldsymbol{x}_{-i} + \chi^{j,e})$$
$$= c_{i,e}(x_i'(e), (\boldsymbol{x}_{-i} + \chi^{j,e})_e) - c_{i,e}(x_i'(e) - 1, (\boldsymbol{x}_{-i} + \chi^{j,e})_e)$$
$$= c_{i,e}(x_i^*(e) - 1, \boldsymbol{x}_{-i,e} + \chi^j) - c_{i,e}(x_i^*(e) - 2, \boldsymbol{x}_{-i,e} + \chi^j). \quad (14)$$

The RHS of (13) is equal to

$$c_i(\boldsymbol{x}_i'', \boldsymbol{x}_{-i} + \chi^{j,e}) - c_i(\boldsymbol{x}_i'' - \chi^f, \boldsymbol{x}_{-i} + \chi^{j,e})$$
$$= c_{i,f}(x_{i,f}'', (\boldsymbol{x}_{-i} + \chi^{j,e})_f) - c_{i,l}(x_{i,f}'' - 1, (\boldsymbol{x}_{-i} + \chi^{j,e})_f)$$
$$= c_{i,f}(x_{i,f}' + 1, \boldsymbol{x}_{-i,f}) - c_{i,f}(x_{i,f}', \boldsymbol{x}_{-i,f}). \tag{15}$$

Let $\boldsymbol{x}_i^\bullet = \boldsymbol{x}_i^* - \chi^e = \boldsymbol{x}_i' - \chi^f$. Since $\boldsymbol{x}^*$ is an optimal solution for $\mathrm{P}(\boldsymbol{x}_{-i}, d)$, it follows that

$$c_i(\boldsymbol{x}_i^*, \boldsymbol{x}_{-i}) - c_i(\boldsymbol{x}_i^\bullet, \boldsymbol{x}_{-i}) \le c_i(\boldsymbol{x}_i', \boldsymbol{x}_{-i}) - c_i(\boldsymbol{x}_i^\bullet, \boldsymbol{x}_{-i})$$
$$\Longleftrightarrow c_{i,e}(x_{i,e}^*, \boldsymbol{x}_{-i,e}) - c_{i,e}(x_{i,e}^* - 1, \boldsymbol{x}_{-i,e})$$
$$\le c_{i,f}(x_{i,f}', \boldsymbol{x}_{-i,f}) - c_{i,f}(x_{i,f}' - 1, \boldsymbol{x}_{-i,f}). \tag{16}$$

By Assertions 1 and 2 in Lemma 2, we have that $c_{i,f}(\cdot, \boldsymbol{x}_{-i,f})$ is a univariate $\mathrm{M}^\natural$-convex function in variable $x_{i,f} \in \mathbb{Z}$. Then the convexity (1) of $c_{i,f}(\cdot, \boldsymbol{x}_{-i,f})$ implies that (15) is at least the RHS of (16). On the other hand, it follows from the i-regularity of $c_{i,e}$ that the LHS of (16) is at least (14). We thus obtain (14) $\le$ (15), which implies (13).

Case 1.2 ($k = e$ and $l \in E \setminus \{e, f\}$). Suppose that $k = e$ and $l \in E \setminus \{e, f\}$. In this case $\boldsymbol{x}_i'' = \boldsymbol{x}_i' - \chi^e + \chi^l$.

Again let $\boldsymbol{x}_i^\circ = \boldsymbol{x}_i' - \chi^e = \boldsymbol{x}_i'' - \chi^l$ and $\boldsymbol{x}_i^\bullet = \boldsymbol{x}_i^* - \chi^e = \boldsymbol{x}_i' - \chi^f$. Note that $\boldsymbol{x}_i^\bullet = \boldsymbol{x}_i'' + \chi^e - \chi^f - \chi^l$. To prove (12), it suffices to show that

$$c_i(\boldsymbol{x}_i', \boldsymbol{x}_{-i} + \chi^{j,e}) - c_i(\boldsymbol{x}_i^\circ, \boldsymbol{x}_{-i} + \chi^{j,e})$$
$$\le c_i(\boldsymbol{x}_i'', \boldsymbol{x}_{-i} + \chi^{j,e}) - c_i(\boldsymbol{x}_i^\circ, \boldsymbol{x}_{-i} + \chi^{j,e}). \tag{17}$$

For the LHS of (17), it holds that

$$c_i(\boldsymbol{x}_i', \boldsymbol{x}_{-i} + \chi^{j,e}) - c_i(\boldsymbol{x}_i^\circ, \boldsymbol{x}_{-i} + \chi^{j,e})$$
$$= c_{i,e}(x_{i,e}', \boldsymbol{x}_{-i,e} + \chi^{j,e}) - c_{i,e}(x_{i,e}' - 1, \boldsymbol{x}_{-i,e} + \chi^{j,e})$$
$$\le c_{i,e}(x_{i,e}' + 1, \boldsymbol{x}_{-i,e}) - c_{i,e}(x_{i,e}', \boldsymbol{x}_{-i,e})$$
$$= c_{i,e}(x_{i,e}^*, \boldsymbol{x}_{-i,e}) - c_{i,e}(x_{i,e}^* - 1, \boldsymbol{x}_{-i,e}), \tag{18}$$

where the inequality follows from the i-regularity of $c_{i,e}$. Since $\boldsymbol{x}^*$ is an optimal solution for $\mathrm{P}(\boldsymbol{x}_{-i}, d)$, it follows that

$$c_i(\boldsymbol{x}_i^*, \boldsymbol{x}_{-i}) - c_i(\boldsymbol{x}_i^\bullet, \boldsymbol{x}_{-i}) \le c_i(\boldsymbol{x}_i', \boldsymbol{x}_{-i}) - c_i(\boldsymbol{x}_i^\bullet, \boldsymbol{x}_{-i})$$
$$\Longleftrightarrow c_{i,e}(x_{i,e}^*, \boldsymbol{x}_{-i,e}) - c_{i,e}(x_{i,e}^* - 1, \boldsymbol{x}_{-i,e})$$
$$\le c_{i,f}(x_{i,f}', \boldsymbol{x}_{-i,f}) - c_{i,f}(x_{i,f}' - 1, \boldsymbol{x}_{-i,f}). \tag{19}$$

The RHS of (17) is equal to

$$c_{i,l}(x_{i,l}'', \boldsymbol{x}_{-i,l}) - c_{i,l}(x_{i,l}'' - 1, \boldsymbol{x}_{-i,l}) = c_{i,l}(x_{i,l}^\bullet + 1, \boldsymbol{x}_{-i,l}) - c_{i,f}(x_{i,l}^\bullet, \boldsymbol{x}_{-i,l}). \tag{20}$$

By the choice (11) of f, we have that

$$c_i(\boldsymbol{x}_i', \boldsymbol{x}_{-i} + \chi^{j,e}) \leq c_i(\boldsymbol{x}_i^* - \chi^e + \chi^l, \boldsymbol{x}_{-i} + \chi^{j,e})$$
$$\Longleftrightarrow c_i(\boldsymbol{x}_i', \boldsymbol{x}_{-i} + \chi^{j,e}) - c_i(\boldsymbol{x}_i^{\bullet}, \boldsymbol{x}_{-i} + \chi^{j,e})$$
$$\leq c_i(\boldsymbol{x}_i^{\bullet} + \chi^l, \boldsymbol{x}_{-i} + \chi^{j,e}) - c_i(\boldsymbol{x}_i^{\bullet}, \boldsymbol{x}_{-i} + \chi^{j,e})$$
$$\Longleftrightarrow c_{i,f}(x_{i,f}', \boldsymbol{x}_{-i,f}) - c_{i,f}(x_{i,f}' - 1, \boldsymbol{x}_{-i,f})$$
$$\leq c_{i,l}(x_{i,l}^{\bullet} + 1, \boldsymbol{x}_{-i,l}) - c_{i,l}(x_{i,l}^{\bullet}, \boldsymbol{x}_{-i,l}) \tag{21}$$

From (18)–(21), we obtain (17).

Case 1.3 ($k = f$ **and** $l \in E \setminus \{f\}$). Suppose that $k = e$ and $l \in E \setminus \{f\}$. In this case, we have that $\boldsymbol{x}_i'' = \boldsymbol{x}_i^* - \chi^e + \chi^l$, and thus (12) follows from (11).

Case 1.4 ($k \in E \setminus \{e, f\}$ **and** $l = e$). Suppose that $k \in E \setminus \{e, f\}$ and $l = e$. In this case $\boldsymbol{x}_i'' = \boldsymbol{x}_i^* - \chi^k + \chi^f$. Let $\boldsymbol{x}^{\circ} = \boldsymbol{x}_i^* - \chi^k = \boldsymbol{x}_i'' - \chi^f$ and $\boldsymbol{x}_i^{\bullet} = \boldsymbol{x}_i^* - \chi^e = \boldsymbol{x}_i' - \chi^f$. To prove (12), it suffices to show that

$$c_{i,k}(x_{i,k}', (\boldsymbol{x}_{-i} + \chi^{j,e})_k) - c_{i,k}(x_{i,k}' - 1, (\boldsymbol{x}_{-i} + \chi^{j,e})_k)$$
$$\leq c_{i,e}(x_{i,e}'', (\boldsymbol{x}_{-i} + \chi^{j,e})_e) - c_{i,e}(x_{i,e}'' - 1, (\boldsymbol{x}_{-i} + \chi^{j,e})_e). \tag{22}$$

Firstly, the LHS of (22) is equal to

$$c_{i,k}(x_{i,k}', \boldsymbol{x}_{-i,k}) - c_{i,k}(x_{i,k}' - 1, \boldsymbol{x}_{-i,k})$$
$$= c_{i,k}(x_{i,k}^*, \boldsymbol{x}_{-i,k}) - c_{i,k}(x_{i,k}^{\circ}, \boldsymbol{x}_{-i,k}). \tag{23}$$

Secondly, since $\boldsymbol{x}^*$ is an optimal solution for $P(\boldsymbol{x}_{-i}, d)$, it follows that

$$c_i(\boldsymbol{x}_i^*, \boldsymbol{x}_{-i}) - c_i(\boldsymbol{x}_i^{\circ}, \boldsymbol{x}_{-i}) \leq c_i(\boldsymbol{x}_i'', \boldsymbol{x}_{-i}) - c_i(\boldsymbol{x}_i^{\circ}, \boldsymbol{x}_{-i})$$
$$\Longleftrightarrow c_{i,k}(x_{i,k}^*, \boldsymbol{x}_{-i,k}) - c_{i,k}(x_{i,k}^{\circ}, \boldsymbol{x}_{-i,k}) \leq c_{i,f}(x_{i,f}'', \boldsymbol{x}_{-i,f}) - c_{i,f}(x_{i,f}^{\circ}, \boldsymbol{x}_{-i,f})$$
$$\Longleftrightarrow c_{i,k}(x_{i,k}^*, \boldsymbol{x}_{-i,k}) - c_{i,k}(x_{i,k}^{\circ}, \boldsymbol{x}_{-i,k})$$
$$\leq c_{i,f}(x_{i,f}', \boldsymbol{x}_{-i,f}) - c_{i,f}(x_{i,f}^{\bullet}, \boldsymbol{x}_{-i,f}). \tag{24}$$

Finally, it follows from (11) that

$$c_i(x_i', \boldsymbol{x}_{-i} + \chi^{j,e}) \leq c_i(x_i^*, \boldsymbol{x}_{-i} + \chi^{j,e})$$
$$\Longleftrightarrow c_i(x_i', \boldsymbol{x}_{-i} + \chi^{j,e}) - c_i(x_i^{\bullet}, \boldsymbol{x}_{-i} + \chi^{j,e})$$
$$\leq c_i(x_i^*, \boldsymbol{x}_{-i} + \chi^{j,e}) - c_i(x_i^{\bullet}, \boldsymbol{x}_{-i} + \chi^{j,e})$$
$$\Longleftrightarrow c_{i,f}(x_{i,f}', \boldsymbol{x}_{-i,f}) - c_{i,f}(x_{i,f}^{\bullet}, \boldsymbol{x}_{-i,f})$$
$$\leq c_{i,e}(x_{i,e}^*, \boldsymbol{x}_{-i,e} + \chi^j) - c_{i,e}(x_{i,e}^{\bullet}, \boldsymbol{x}_{-i} + \chi^j)$$
$$\Longleftrightarrow c_{i,f}(x_{i,f}', \boldsymbol{x}_{-i,f}) - c_{i,f}(x_{i,f}^{\bullet}, \boldsymbol{x}_{-i,f})$$
$$\leq c_{i,e}(x_{i,e}'', \boldsymbol{x}_{-i,e} + \chi^j) - c_{i,e}(x_{i,e}'' - 1, \boldsymbol{x}_{-i} + \chi^j). \tag{25}$$

Therefore, from (23)–(25), we obtain (22).

Case 1.5 ($k \in E \setminus \{e, f\}$ **and** $l = f$). In this case, it holds that $x_i'' = x_i^* - \chi^e - \chi^k + 2 \cdot \chi^f$. To prove (12), it suffices to prove

$$
\begin{aligned}
&c_{i,k}(x_{i,k}', (x_{-i} + \chi^{j,e})_k) + c_{i,f}(x_{i,f}', (x_{-i} + \chi^{j,e})_f) \\
&\le c_{i,k}(x_{i,k}'', (x_{-i} + \chi^{j,e})_k) + c_{i,f}(x_{i,f}'', (x_{-i} + \chi^{j,e})_f) \\
&\Longleftrightarrow c_{i,k}(x_{i,k}', x_{-i,k}) - c_{i,k}(x_{i,k}' - 1, x_{-i,k}) \\
&\qquad\qquad \le c_{i,f}(x_{i,f}'', x_{-i,f}) - c_{i,f}(x_{i,f}'' - 1, x_{-i,f}).
\end{aligned}
\tag{26}
$$

Let $x_i^\circ = x_i^* - \chi^k + \chi^f = x_i'' - \chi^f + \chi^e$ and $x_i^\bullet = x_i^* - \chi^k = x_i^\circ - \chi^f$. Since x^* is an optimal solution for $\mathrm{P}(x_{-i}, d)$, we have that

$$
\begin{aligned}
&c_i(x_i^*, x_{-i}) - c_i(x_i^\bullet, x_{-i}) \le c_i(x_i^\circ, x_{-i}) - c_i(x_i^\bullet, x_{-i}) \\
&\Longleftrightarrow c_{i,k}(x_{i,k}^*, x_{-i,k}) - c_{i,k}(x_{i,k}^* - 1, x_{-i,k}) \\
&\qquad\qquad \le c_{i,f}(x_{i,f}^\circ, x_{-i,f}) - c_{i,f}(x_{i,f}^\circ - 1, x_{-i,f}).
\end{aligned}
\tag{27}
$$

By the M$^\natural$-convexity (1) of $c_{i,f}(\cdot, x_{-i,f})$, it holds that

$$
\begin{aligned}
&c_{i,f}(x_{i,f}^\circ, x_{-i,f}) - c_{i,f}(x_{i,f}^\circ - 1, x_{-i,f}) \\
&= c_{i,f}(x_{i,f}'' - 1, x_{-i,f}) - c_{i,f}(x_{i,f}'' - 2, x_{-i,f}) \\
&\le c_{i,f}(x_{i,f}'', x_{-i,f}) - c_{i,f}(x_{i,f}'' - 1, x_{-i,f}).
\end{aligned}
\tag{28}
$$

Now (26) follows from (27) and (28).

Case 1.6 ($k, l \in E \setminus \{e, f\}$). In this case $x_i'' = x_i^* - \chi^e - \chi^k + \chi^f + \chi^l$. Observe that

$$
\begin{aligned}
c_i(x_i', x_{-i} + \chi^{j,e}) - c_i(x_i', x_{-i}) &= c_{i,e}(x_{i,e}', x_{-i,e} + \chi^{j,e}) - c_{i,e}(x_{i,e}', x_{-i,e}) \\
&= c_{i,e}(x_{i,e}'', x_{-i,e} + \chi^{j,e}) - c_{i,e}(x_{i,e}'', x_{-i,e}) \\
&= c_i(x_i'', x_{-i} + \chi^{j,e}) - c_i(x_i'', x_{-i}),
\end{aligned}
$$

and thus (12) is equivalent to

$$
c_i(x_i', x_{-i}) \le c_i(x_i'', x_{-i}).
\tag{29}
$$

Let $x_i^\circ = x_i^* - \chi^e + \chi^l = x_i' - \chi^f + \chi^l$. Now consider an M-convex function $c_i(\cdot, x_{-i})$ with variable x_i on the hyperplane $x_i(E) = d$. By applying the exchange axiom for M-convex functions in which x, y and i are replaced with x_i^*, x_i'' and e, we have that at least one of the following inequalities holds:

$$
\begin{aligned}
c_i(x_i^*, x_{-i}) + c_i(x_i'', x_{-i}) &\ge c_i(x_i^* - \chi^e + \chi^f, x_{-i}) + c_i(x_i'' + \chi^e - \chi^f, x_{-i}) \\
&= c_i(x_i', x_{-i}) + c_i(x_i'' + \chi^e - \chi^f, x_{-i})
\end{aligned}
\tag{30}
$$

or

$$
\begin{aligned}
c_i(x_i^*, x_{-i}) + c_i(x_i'', x_{-i}) &\ge c_i(x_i^* - \chi^e + \chi^l, x_{-i}) + c_i(x_i'' + \chi^e - \chi^l, x_{-i}) \\
&= c_i(x_i^\circ, x_{-i}) + c_i(x_i'' + \chi^e - \chi^l, x_{-i}).
\end{aligned}
\tag{31}
$$

If (30) holds, then (29) follows from the optimality of x_i^* for $P(x_{-i}, d)$:
$c_i(x_i^*, x_{-i}) \le c_i(x_i'' + \chi^e - \chi^f, x_{-i})$.

Suppose that (31) holds. We have

$$c_i(x_i', x_{-i}) \le c_i(x_i^\circ, x_{-i}), \tag{32}$$

because

$$
\begin{aligned}
c_i(x_i', x_{-i} + \chi^{j,e}) - c_i(x_i', x_{-i}) &= c_{i,e}(x_{i,e}', x_{-i,e} + \chi^{j,e}) - c_{i,e}(x_{i,e}', x_{-i,e}) \\
&= c_{i,e}(x_{i,e}^\circ, x_{-i,e} + \chi^{j,e}) - c_{i,e}(x_{i,e}^\circ, x_{-i,e}) \\
&= c_i(x_i^\circ, x_{-i} + \chi^{j,e}) - c_i(x_i^\circ, x_{-i}),
\end{aligned}
$$

and $c_i(x_i', x_{-i} + \chi^{j,e}) \le c_i(x_i^\circ, x_{-i} + \chi^{j,e})$ by (11). Meanwhile, from (31) and the optimality of x_i^* for $P(x_{-i}, d)$, i.e., $c_i(x_i^*, x_{-i}) \le c_i(x_i'' + \chi^e - \chi^l, x_{-i})$, we obtain

$$c_i(x_i^\circ, x_{-i}) \le c_i(x_i'', x_{-i}). \tag{33}$$

By (32) and (33), we conclude that (29) holds.

We next deal with Case 2: $e = f$, i.e., $x_i' = x_i^*$.

Case 2.1 ($k = e$ and $l \in E \setminus \{e\}$). In this case, (12) directly follows from (11).

Case 2.2 ($k = E \setminus \{e\}$ and $l = e$). We have that

$$
\begin{aligned}
c_i(x_i'', x_{-i} + \chi^{j,e}) - c_i(x_i'', x_{-i}) &= c_{i,e}(x_{i,e}'', x_{-i,e} + \chi^{j,e}) - c_{i,e}(x_{i,e}'', x_{-i,e}) \\
&= c_{i,e}(x_{i,e}^* + 1, x_{-i,e} + \chi^{j,e}) - c_{i,e}(x_{i,e}^* + 1, x_{-i,e}) \\
&\ge c_{i,e}(x_{i,e}^*, x_{-i,e} + \chi^{j,e}) - c_{i,e}(x_{i,e}^*, x_{-i,e}) \\
&= c_i(x_i^*, x_{-i} + \chi^{j,e}) - c_i(x_i^*, x_{-i}),
\end{aligned}
$$

where the inequality follows from the $M^\natural$-convexity (2) of $c_{i,e}$. Then, $c_i(x_i^*, x_{-i}) \le c_i(x_i'', x_{-i})$ implies $c_i(x_i^*, x_{-i} + \chi^{j,e}) \le c_i(x_i'', x_{-i} + \chi^{j,e})$, and thus (12) holds.

Case 2.3 ($k, l \in E \setminus \{e\}$). We have that

$$
\begin{aligned}
c_i(x_i'', x_{-i} + \chi^{j,e}) - c_i(x_i'', x_{-i}) &= c_{i,e}(x_{i,e}'', x_{-i,e} + \chi^{j,e}) - c_{i,e}(x_{i,e}'', x_{-i,e}) \\
&= c_{i,e}(x_{i,e}^*, x_{-i,e} + \chi^{j,e}) - c_{i,e}(x_{i,e}^*, x_{-i,e}) \\
&= c_i(x_i^*, x_{-i} + \chi^{j,e}) - c_i(x_i^*, x_{-i}),
\end{aligned}
$$

Then, $c_i(x_i^*, x_{-i}) \le c_i(x_i'', x_{-i})$ implies $c_i(x_i^*, x_{-i} + \chi^{j,e}) \le c_i(x_i'', x_{-i} + \chi^{j,e})$, and thus (12) holds. $\qquad\square$

We remark that the following form of Theorem 5 is also true:

For each $j \in N \setminus \{i\}$ and $e \in E$, $P(x_{-i} - \chi^{j,e}, d)$ is feasible and there exists $f \in E$ such that $x_i^* + \chi^e - \chi^f$ is its optimal solution.

This implies the following corollary.

Corollary 1. *Let x_i^* be an optimal solution for $P(x_{-i}, d)$, and x'_{-i} satisfy $\|x'_{-i} - x_{-i}\|_1 = 1$. Then, there exists an optimal solution x'_i for $P(x_{-i}, d)$ such that $\|x'_i - x_i^*\|_1 \leq 2$.*

By Theorem 4 and Corollary 1, we obtain the following sensitivity theorem.

Theorem 6. *Let $d, d' \in \mathbb{Z}_+$ and $x_{-i}, x'_{-i} \in \mathbb{Z}_+^{N \setminus \{i\}}$ satisfy that $P(x_{-i}, d)$ and $P(x'_{-i}, d')$ are feasible. Then, for an optimal solution x_i^* for $P(x_{-i}, d)$, there exists an optimal solution x°_{-i} for $P(x'_{-i}, d')$ such that $\|x_i^{\circ} - x_i^*\|_1 \leq 2\|x'_{-i} - x_{-i}\|_1 + |d' - d|$.*

Acknowledgements. This work is partially supported by JSPS KAKENHI Grant Numbers JP16K16012, JP26280001, JP26280004, Japan.

References

1. Ackermann, H., Röglin, H., Vöcking, B.: On the impact of combinatorial structure on congestion games. J. ACM **55**(6), 25:1–25:22 (2008). https://doi.org/10.1145/1455248.1455249
2. Ackermann, H., Röglin, H., Vöcking, B.: Pure Nash equilibria in player-specific and weighted congestion games. Theor. Comput. Sci. **410**(17), 1552–1563 (2009). https://doi.org/10.1016/j.tcs.2008.12.035
3. Bhaskar, U., Fleischer, L., Hoy, D., Huang, C.-C.: On the uniqueness of equilibrium in atomic splittable routing games. Math. Oper. Res. **40**(3), 634–654 (2015). https://doi.org/10.1287/moor.2014.0688
4. Cominetti, R., Correa, J.R., Moses, N.E.S.: The impact of oligopolistic competition in networks. Oper. Res. **57**(6), 1421–1437 (2009). https://doi.org/10.1287/opre.1080.0653
5. Fujishige, S.: Submodular Functions and Optimization. Annals of Discrete Mathematics, vol. 58, 2nd edn. Elsevier, Amsterdam (2005)
6. Fujishige, S., Goemans, M.X., Harks, T., Peis, B., Zenklusen, R.: Congestion games viewed from M-convexity. Oper. Res. Lett. **43**(3), 329–333 (2015). https://doi.org/10.1016/j.orl.2015.04.002
7. Fujishige, S., Goemans, M.X., Harks, T., Peis, B., Zenklusen, R.: Matroids are immune to Braess' paradox. Math. Oper. Res. **42**(3), 745–761 (2017). https://doi.org/10.1287/moor.2016.0825
8. Harks, T., Klimm, M., Peis, B.: Sensitivity analysis for convex separable optimization over integral polymatroids. SIAM J. Optim. **28**, 2222–2245 (2018). https://doi.org/10.1137/16M1107450
9. Harks, T., Peis, B.: Resource buying games. Algorithmica **70**(3), 493–512 (2014). https://doi.org/10.1007/s00453-014-9876-6
10. Harks, T., Peis, B.: Resource buying games. In: Schulz, A.S., Skutella, M., Stiller, S., Wagner, D. (eds.) Gems of Combinatorial Optimization and Graph Algorithms, pp. 103–111. Springer, Cham (2015). https://doi.org/10.1007/978-3-319-24971-1_10
11. Harks, T., Timmermans, V.: Uniqueness of equilibria in atomic splittable polymatroid congestion games. J. Comb. Optim. **36**(3), 812–830 (2018). https://doi.org/10.1007/s10878-017-0166-5

12. Huang, C.-C.: Collusion in atomic splittable routing games. Theory Comput. Syst. **52**(4), 763–801 (2013). https://doi.org/10.1007/s00224-012-9421-4
13. Moriguchi, S., Shioura, A., Tsuchimura, N.: M-convex function minimization by continuous relaxation approach: proximity theorem and algorithm. SIAM J. Optim. **21**(3), 633–668 (2011). https://doi.org/10.1137/080736156
14. Murota, K.: Convexity and Steinitz's exchange property. Adv. Math. **125**, 272–331 (1996). https://doi.org/10.1006/aima.1996.0084
15. Murota, K.: Discrete Convex Analysis. Society for Industrial and Applied Mathematics, Philadelphia (2003)
16. Murota, K.: Discrete convex analysis: a tool for economics and game theory. J. Mech. Inst. Des. **1**, 151–273 (2016). https://doi.org/10.22574/jmid.2016.12.005
17. Murota, K., Shioura, A.: M-convex function on generalized polymatroid. Math. Oper. Res. **24**, 95–105 (1999). https://doi.org/10.1287/moor.24.1.95
18. Murota, K., Tamura, A.: Proximity theorems of discrete convex functions. Math. Program. **99**(3), 539–562 (2004). https://doi.org/10.1007/s10107-003-0466-7
19. Rosenthal, R.W.: A class of games possessing pure-strategy Nash equilibria. Int. J. Game Theory **2**, 65–67 (1973). https://doi.org/10.1007/BF01737559
20. Schrijver, A.: Combinatorial Optimization - Polyhedra and Eciency. Springer, Heidelberg (2003)
21. Tran-Thanh, L., Polukarov, M., Chapman, A., Rogers, A., Jennings, N.R.: On the existence of pure strategy Nash equilibria in integer–splittable weighted congestion games. In: Persiano, G. (ed.) SAGT 2011. LNCS, vol. 6982, pp. 236–253. Springer, Heidelberg (2011). https://doi.org/10.1007/978-3-642-24829-0_22

The Complexity of Synthesis for 43 Boolean Petri Net Types

Ronny Tredup$^{(\boxtimes)}$ and Christian Rosenke

Institut für Informatik, Theoretische Informatik, Universität Rostock,
Albert-Einstein-Straße 22, 18059 Rostock, Germany
ronny.tredup@uni-rostock.de

Abstract. Synthesis for a type of Petri nets is the problem of finding, for a given transition system A, a Petri net N of this type having a state graph that is isomorphic to A, if such a net exists. This paper studies the computational complexity of synthesis for 43 boolean types of Petri nets. It turns out that for 36 of these types synthesis can be done in polynomial time while for the other seven it is NP-hard.

1 Introduction

Synthesis for a type of Petri nets τ, that is, the task of finding a Petri net N of type τ that implements a given transition system A, was originally invented for the type of elementary net systems by Ehrenfeucht and Rozenberg [8]. As recently presented by Badouel, Bernardinello and Darondeau [4], synthesis has also been studied for many other types of Petri nets.

Synthesis for types of Petri nets yields implementations which are correct by design and can be used to extract concurrency and distributability data from sequential specifications like transition systems or languages [5]. It is applied in the field of process mining to reconstruct a model from its execution traces [1]. Also, it has applications for the synthesis of speed independent circuits [7].

This paper deals with the computational complexity aspect of synthesis. Research in this area has been conducted for several different Petri net types. In [2], it has been shown that the problem can be solved in polynomial time for the type of bounded P/T-nets. This is also true for pure bounded P/T-nets [4]. We like to point out that, although both of these types are bounded, that is, there exists a bound $b \in \mathbb{N}^+$ on the number of tokens that places will ever receive, these bounds are rather implicit than chosen *a priori*.

In contrast, for the type of pure 1-bounded P/T-nets the limit $b = 1$ is chosen *a priori*. Interestingly, [3] shows here that synthesis is NP-hard. Results from [17,19] reveal that this remains true even for remarkable input restrictions. Hence, the kind of bound in the considered Petri net type has a noticeable impact on the complexity of synthesis. Then again, Schmitt [14] extends pure 1-bounded P/T-nets by the so-called swap-interaction which results in the type of flip-flop nets. Here, synthesis becomes tractable, again. Thus, the interactions admitted by the Petri net type influence the complexity, too.

© Springer Nature Switzerland AG 2019
T. V. Gopal and J. Watada (Eds.): TAMC 2019, LNCS 11436, pp. 615–634, 2019.
https://doi.org/10.1007/978-3-030-14812-6_38

The types of pure 1-bounded P/T-nets and flip-flop nets both belong to the family of *boolean* types of Petri nets [4]. The Petri net types of this family stand out by their a priori bound $b = 1$ on the number of tokens per place and they are distinguished by the set of admitted interactions. Until now, research has explicitly defined seven boolean net types. Beyond the two types mentioned above, there are contextual nets [11], event/condition nets [16], inhibitor nets [13], set nets [10] and trace nets [6]. Our elaborate case study [18] shows that synthesis for the latter five and 71 further boolean types is NP-hard. The corresponding presentation, however, is extensive and not in the scope of this paper.

In this paper, we study the computational complexity of synthesis for 43 further boolean types of Petri nets. This is a considerable step towards a full characterization of the computational complexity of synthesis for all possible 256 boolean Petri net types. For 36 of the 43 types, synthesis can be done in polynomial time: Firstly, we present a new polynomial time algorithm that works for 16 of these 36 types. Secondly, we show that synthesis for another 16 types is solvable by a generalization of Schmitt's algorithm [14]. Finally, we argue for the remaining four types that their synthesis is a rather trivial problem.

For seven of the 43 types, we demonstrate that synthesis is NP-hard by turning to the decision version of the synthesis problem, also known as *feasibility*. Here, we ask for a given transition system A whether there exists a Petri net of the particular type, with a state graph isomorphic to A. To prove the NP-completeness of feasibility, we use the well known equivalent formulation of the problem by the so-called *event state separation property* (ESSP) and *state separation property* (SSP). As a matter of fact, a transition system A is feasible with respect to a Petri net type if and only if it has the type related ESSP *and* SSP [4].

2   Preliminaries

This section provides short formal definitions and preliminary notions used in the paper and applies them for a running example. A *transition system* (TS, for short) $A = (S, E, \delta)$ consists of finite disjoint sets S of states and E of events and a partial transition function $\delta : S \times E \to S$. Usually, we think of A as an edge-labeled directed graph with node set S where every triple $\delta(s, e) = s'$ is interpreted as an e-labeled edge $s \xrightarrow{e} s'$ called *transition*. We say that an event e *occurs* at state s if $\delta(s, e) = s'$ for some state s' and abbreviate this with $s \xrightarrow{e}$. This notation is extended to words in E^* by allowing $s \xrightarrow{\varepsilon} s$ on the empty word ε for all $s \in S$ and by inductively leading back $s \xrightarrow{w} s'$ to the satisfaction of $s \xrightarrow{w'} s''$ and $s'' \xrightarrow{e} s'$ for any word $w = w'e$ with $e \in E, w' \in E^*$ and states $s, s', s'' \in S$. By $s \xrightarrow{w}$, we state for $s, \in S, w \in E^*$ the existence of a state $s' \in S$ such that $s \xrightarrow{w} s'$. An *initialized* TS $A = (S, E, \delta, s_0)$ is a TS with an initial state $s_0 \in S$ *reaching* all states, that is, there is a word $w \in E^*$ with $s_0 \xrightarrow{w} s$ for all states $s \in S$. If not explicitly stated otherwise, we assume all TSs in the sequel of this paper to be initialized. We consistently refer to the components of a TS A by S_A, E_A, δ_A, and optionally $s_{0,A}$.

The following notion of *type of nets* has been developed in [4]. It allows us to uniformly capture all 43 boolean Petri net types in one general scheme. This means, every boolean Petri net type can be seen as an instantiation of this scheme. In this spirit, we describe a *type of nets* τ as a (non-initialized) TS $\tau = (S_\tau, E_\tau, \delta_\tau)$. A Petri net $N = (P, T, f, M_0)$ of type τ (τ-net, for short) is given by finite and disjoint sets P of places and T of transitions, an initial marking $M_0 : P \longrightarrow S_\tau$, and a flow function $f : P \times T \to E_\tau$. The meaning of a τ-net is its dynamic behavior realized by cascades of firing transitions. In particular, a transition $t \in T$ can fire in a marking $M : P \longrightarrow S_\tau$ and thereby produces the marking $M' : P \longrightarrow S_\tau$ if for all $p \in P$ the transition $M(p) \xrightarrow{f(p,t)} M'(p)$ is present in τ. The firing of t and the corresponding marking transfer is shortly denoted by $M \xrightarrow{t} M'$. This notation is again extended to words $\sigma \in T^*$ and, based on that, the set of all reachable markings of N is defined by $RS(N) = \{M \mid \exists \sigma \in T^* : M_0 \xrightarrow{\sigma} M\}$. Thus, given a τ-net $N = (P, T, f, M_0)$, its behavior is captured by the transition system $A_N = (RS(N), T, \delta, M_0)$, called the state graph of N, where for every reachable marking M of N and transition $t \in T$ with $M \xrightarrow{t} M'$ the transition function δ of A_N is defined by $\delta(M, t) = M'$.

In this paper, we deal with *boolean* Petri net types τ characterized by $S_\tau = \{0, 1\}$. For boolean type of nets τ, fixing any $e \in E_\tau$, thus, reduces δ_τ to the partial function $e : \{0, 1\} \to \{0, 1\}$ with $e(s) = \delta_\tau(s, e)$. There exist only nine functions of this signature including the entirely undefined function $\perp$. The table below lists the other eight possible functions nop, inp, out, set, res, swap, used, and free:

s	$\mathsf{nop}(s)$	$\mathsf{inp}(s)$	$\mathsf{out}(s)$	$\mathsf{set}(s)$	$\mathsf{res}(s)$	$\mathsf{swap}(s)$	$\mathsf{used}(s)$	$\mathsf{free}(s)$
0	0		1	1	0	1		0
1	1	0		1	0	0	1	

Consequently, as two different events of E_τ must represent different partial functions, the event set of any *boolean* type of Petri nets has to be a subset of these nine events. In fact, we would also like to rule out $\perp \in E_\tau$. The reason is that every τ-net N having a place $p \in P_N$ and a transition $t \in T_N$ such that $f_N(p, t) = \perp$ would make transition t unable to ever fire and, thus, irrelevant for the behavior of N. Therefore, t could be discarded from N making $\perp$ useless in the first place.

Altogether, this justifies to define every boolean Petri net type τ by arranging E_τ as a subset of the interactions $I = \{\mathsf{nop}, \mathsf{inp}, \mathsf{out}, \mathsf{set}, \mathsf{res}, \mathsf{swap}, \mathsf{used}, \mathsf{free}\}$. Figure 1 depicts a visualization for every element of I. Notice that there have to be exactly 256 boolean types of nets, one for every subset of I. In the following, we often refer to a type of nets τ simply via E_τ.

The subsequent notion of τ-regions allows us, on the one hand, to define the type related ESSP, respectively SSP and, on the other hand, to discover in which way we are able to obtain a τ-net N for a given TS A if it exists. Figure 2 shows an example for all introduced notions.

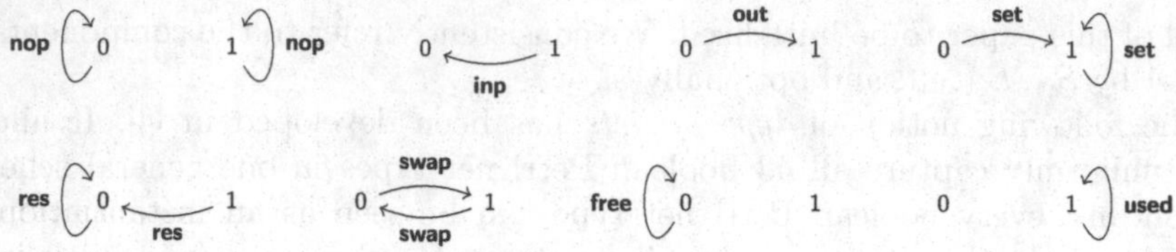

Fig. 1. The interactions I as transitions in a type of nets.

If τ is a type of nets then a τ-region (sup, sig) of a TS A is a pair of mappings (sup, sig), where $sup : S_A \longrightarrow S_\tau$ and $sig : E_A \longrightarrow E_\tau$, such that, for every transition $s \xrightarrow{e} s'$ of A, the transition $sup(s) \xrightarrow{sig(e)} sup(s')$ is present in τ. As this paper studies boolean types of nets τ where $S_\tau = \{0, 1\}$, every τ-region (sup, sig) and state $s \in S_A$ implies $sup(s) \in \{0, 1\}$. For the sake of simplicity, we also write $s \in sup$ ($s \notin sup$) if $sup(s) = 1$ ($sup(s) = 0$) and $S \subseteq sup$ for a subset $S \subseteq S_A$ that satisfies $sup(s) = 1$ for all $s \in S$.

Any pair of distinct states $s, s' \in S_A$ defines an *SSP atom* (s, s'). The atom is said to be τ-solvable if there is a τ-region (sup, sig) of A such that $sup(s) \neq sup(s')$. Any event $e \in E_A$ and state $s \in S_A$ define an *ESSP atom* (e, s) if e does not occur at s, that is $\neg s \xrightarrow{e}$. This atom is said to be τ-solvable if there is a τ-region (sup, sig) of A such that $\neg sup(s) \xrightarrow{sig(e)}$. Any τ-region solving an ESSP, respectively an SSP, atom (x, y) is called a *witness* for the τ-solvability of (x, y). If (x, y) is a τ-solvable ESSP, respectively SSP, atom then x and y are said to be τ-separable. If all ESSP, respectively all SSP, atoms of A are τ-solvable then A is said to have the τ-ESSP, respectively the τ-SSP. We define a TS A to be τ-feasible if it has the τ-ESSP and the τ-SSP.

It is noteworthy, that, by definition, a TS A has at most $|S_A|^2$ SSP atoms and at most $|S_A| \cdot |E_A|$ ESSP atoms. Hence, a valid non-deterministic guess containing a τ-solving region for every atom is verifiable in polynomial time. This puts τ-feasibility into NP for every of the 256 boolean types of nets.

The following fact is well known from [4]: A set $\mathcal{R}$ of τ-regions of A contains a witness for all ESSP and SSP atoms if and only if the so-called *synthesized τ-net* $N_A^{\mathcal{R}} = (\mathcal{R}, E_A, f, M_0)$ with flow function $f((sup, sig), e) = sig(e)$ and initial marking $M_0((sup, sig)) = sup(s_{0,A})$ for all $(sup, sig) \in \mathcal{R}, e \in E_A$ has a state graph that is isomorphic to A. Notice that the regions of A in $\mathcal{R}$ become places and the events of E_τ become transitions of $N_A^{\mathcal{R}}$. Hence, for a τ-feasible TS A where $\mathcal{R}$ is known we can easily synthesize a net N with state graph isomorphic to A by constructing $N_A^{\mathcal{R}}$.

Types of nets τ and $\tilde{\tau}$ have an isomorphism ϕ if $s \xrightarrow{i} s'$ is a transition of τ if and only if $\phi(s) \xrightarrow{\phi(i)} \phi(s')$ is one of $\tilde{\tau}$. This paper benefits from isomorphisms using the following lemma:

Lemma 1 (Without proof). *If τ and $\tilde{\tau}$ are isomorphic types of nets then a TS A has the (E)SSP for τ if and only if A has the (E)SSP for $\tilde{\tau}$.*

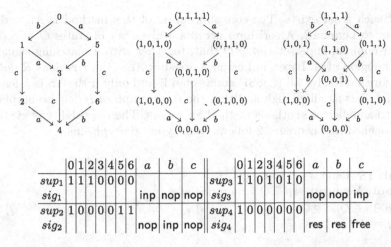

	0	1	2	3	4	5	6	a	b	c			0	1	2	3	4	5	6	a	b	c
sup_1	1	1	1	0	0	0	0				sup_3	1	1	0	1	0	1	0			
sig_1								inp	nop	nop	sig_3								nop	nop	inp
sup_2	1	0	0	0	0	1	1				sup_4	1	0	0	0	0	0	0			
sig_2								nop	inp	nop	sig_4								res	res	free

Fig. 2. Upper left: our running example TS A borrowed from [4]. Bottom: τ_1-regions $\mathcal{R}_1 = \{R_1, R_2, R_3, R_4\}$ for type of nets $E_{\tau_1} = \{nop, inp, res, free\}$ where $R_i = (sup_i, sig_i)$ for $i \in \{1, \ldots, 4\}$. At the same time, the set of all possible non-trivial τ_2-regions $\mathcal{R}_2 = \{R_1, R_2, R_3\}$ for $E_{\tau_2} = \{nop, inp\}$. While $\mathcal{R}_1$ τ_1-solves all ESSP and SSP atoms of A, $\mathcal{R}_2$ fails to τ_2-solve atom $(c, 0)$. Upper middle: the state graph $A_{N_A^{\mathcal{R}_1}}$ of the synthesized τ_1-net $N_A^{\mathcal{R}_1} = (\mathcal{R}_1, \{a, b, c\}, f_1, \mathcal{R}_1)$ with flow function $f_1(R_i, e) = sig_i(e)$ for $i \in \{1, \ldots, 4\}, e \in \{a, b, c\}$. Every state in $A_{N_A^{\mathcal{R}_1}}$ is a marking of $N_A^{\mathcal{R}_1}$, thus, a subset of the regions R_1, R_2, R_3, R_4 denoted as a 4-tuple. As expected, $A_{N_A^{\mathcal{R}_1}}$ is isomorphic to A. Upper right: the state graph $A_{N_A^{\mathcal{R}_2}}$ of the synthesized τ_2-net $N_A^{\mathcal{R}_2} = (\mathcal{R}_2, \{a, b, c\}, f_2, \mathcal{R}_2)$ with flow function $f_2(R_i, e) = sig_i(e)$ for $i \in \{1, 2, 3\}, e \in \{a, b, c\}$. By A's lack of τ_2-ESSP there is no τ_2-net with isomorphic state graph and, thus, $A_{N_A^{\mathcal{R}_2}}$ is not isomorphic to A, either.

3 Polynomial Time Net Synthesis

Theorem 1. *There is a polynomial time algorithm, that, on input TS A, synthesizes a τ-net N with state graph isomorphic to A or rejects A if N does not exist, for every τ where*

1. $\tau = \{nop, res\} \cup \omega$ with $\omega \subseteq \{inp, used, free\}$,
2. $\tau = \{nop, set\} \cup \omega$ with $\omega \subseteq \{out, used, free\}$,
3. $\tau = \{nop, swap\} \cup \omega$ with $\omega \subseteq \{inp, out, used, free\}$,
4. $\tau = \{nop\} \cup \omega$ with $\omega \subseteq \{used, free\}$.

The remainder of this section is dedicated to prove every item of Theorem 1.

3.1 Proof of Theorems 1.1 and 1.2

As a first step, we introduce an algorithm for τ as defined in Theorem 1.1 to compute a corresponding τ-region set $\mathcal{R}$ solving all (E)SSP atoms of a given

TS A, if such a set exists. The core subroutine of this method is Algorithm 1. According to Lemma 2, Algorithm 1 accepts a given set of states $Q \subseteq S_A$ and returns a minimal superset $sup \supseteq Q$ that, together with a matching signature, forms a τ-region of A. Later, in Lemma 3, we show that the τ-regions $\mathcal{R}$ derived from Algorithm 1 solve all (E)SSP atoms of A if and only if the TS is τ-feasible. Using that $\mathcal{R}$ is small enough and that the state graph of $N_A^{\mathcal{R}}$ is isomorphic to A leads to an efficient synthesis method for τ-nets. The tractability of synthesis for τ as defined in Theorem 1.2 follows from type isomorphisms.

Input: TS A and set of states $Q \subseteq S_A$
Output: A support $sup \supseteq Q$ for a region of A.
while $\exists\, s \in Q, s' \notin Q, e \in E_A : (s' \xrightarrow{e} s) \vee (s \xrightarrow{e} s' \wedge z \xrightarrow{e} z'$ for $z, z' \in Q)$ **do**
$\quad |\quad Q = Q \cup \{s'\};$
end
return $sup = Q;$

Algorithm 1. Given $Q \subseteq S_A$, the algorithm minimally extends Q to the support of a τ-region (sup, sig) of A for all types of nets $\tau = \{\mathsf{nop}, \mathsf{res}\} \cup \omega$ extended with $\omega \subseteq \{\mathsf{inp}, \mathsf{used}, \mathsf{free}\}$.

Lemma 2. *If $\tau = \{\mathsf{nop}, \mathsf{res}\} \cup \omega$ is a type of nets with $\omega \subseteq \{\mathsf{inp}, \mathsf{used}, \mathsf{free}\}$ and A is a TS and $Q \subseteq S_A$ then the result sup of Algorithm 1 started on Q forms a τ-region (sup, sig) of A with*

$$
sig(e) = \begin{cases}
used, & \text{if } used \in \tau \text{ and } \{s, s' \mid s \xrightarrow{e} s'\} \subseteq sup, \\
free, & \text{if } free \in \tau \text{ and } \{s, s' \mid s \xrightarrow{e} s'\} \cap sup = \emptyset, \\
inp, & \text{if } inp \in \tau \text{ and for all } s \xrightarrow{e} s' : s \in sup, s' \notin sup, \\
res, & \text{if } inp \notin \tau \text{ and for all } s \xrightarrow{e} s' : s \in sup, s' \notin sup, \\
res, & \text{if for at least one but not all } s \xrightarrow{e} s' : s \in sup, s' \notin sup, \\
nop, & \text{otherwise,}
\end{cases}
$$

for all $e \in E_A$. Moreover, for all τ-regions (sup', sig') of A with $Q \subseteq sup'$ it is true that even $sup \subseteq sup'$. Algorithm 1 terminates after $\mathcal{O}(|E_A| \cdot |S_A|^5)$ time.

Proof. That the algorithm terminates is trivial as every iteration extends Q. This is possible for at most $|S_A|$ times. After termination, sup obviously contains input Q. Moreover, there are no events $e \in E_A$ participating in a transition $s' \xrightarrow{e} s$ with $sup(s) = 1, sup(s') = 0$. On the other hand, if there is a transition $s \xrightarrow{e} s'$ with $s \in sup, s' \notin sup$ then no other transition $z \xrightarrow{e} z'$ satisfies the condition $z, z' \in sup$. Hence, that case implies for all transition $s \xrightarrow{e} s'$ that $\delta_\tau(sup(s), \mathsf{res}) = sup(s')$ is present in τ. If, additionally, $s \in sup, s' \notin sup$ for *every* transition $z \xrightarrow{e} z'$ and inp is available then $\delta_\tau(sup(s), \text{inp}) = sup(s')$ is always defined in τ for all transition $s \xrightarrow{e} s'$. Otherwise, if all transitions of $s \xrightarrow{e} s'$ satisfy that $s, s' \in sup$ or $s, s' \notin sup$ then $\delta_\tau(sup(s), \text{nop}) = sup(s')$ is present in τ. If e's transitions $s \xrightarrow{e} s'$ consistently satisfy $s, s' \in sup$, respectively $s, s' \notin sup$, then even $\delta_\tau(sup(s), \text{used}) = sup(s')$, respectively $\delta_\tau(sup(s), \text{free}) = sup(s')$, is defined in τ, given that used, respectively free, belongs to E_τ. Hence, (sup, sig) is a τ-region of A in every case.

Now let (sup', sig') be any τ-region of A with $Q \subseteq sup'$. We show inductively that the set Q_i that results from i while-iterations of Algorithm 1 fulfills $Q_i \subseteq sup'$. For a start, $Q_0 = Q \subseteq sup'$. Assume that $Q_i \subseteq sup'$ and $Q_{i+1} \not\subseteq sup'$ and let $\{s'\} = Q_{i+1} \backslash Q_i$ which, thus, fulfills $s' \notin sup'$. As s' is added to Q_{i+1}, there are $s \in Q_i \subseteq sup'$ and $e \in E_A$ such that either $s' \xrightarrow{e} s$ or $s \xrightarrow{e} s'$ and $z \xrightarrow{e} z'$ with $z, z' \in Q_i \subseteq sig'$. But then $sig'(e) \notin \{\text{nop}, \text{res}, \text{inp}, \text{used}, \text{free}\}$, a contradiction. When the while loop terminates after n iterations, then sup becomes Q_n and, thus, fulfills $sup = Q_n \subseteq sup'$.

As there are at most $|S_A|$ while-iterations and as checking the condition takes $\mathcal{O}(|E_A| \cdot |S_A|^4)$ time, Algorithm 1 runs in $\mathcal{O}(|E_A| \cdot |S_A|^5)$ time. $\square$

Lemma 2 proposes a way to reliably produce τ-regions for the net types of interest. We argue that they are either sufficient to solve all (E)SSP atoms or otherwise, that A is not τ-feasible.

Lemma 3. *If $\tau = \{\text{nop}, \text{res}\} \cup \omega$ with $\omega \subseteq \{\text{inp}, \text{used}, \text{free}\}$ and A is a TS then an ESSP atom (e, s) of A is τ-solvable if and only if*

1. *inp $\in \tau$ and the region (sup, sig) returned by Algorithm 1 on input $Q = \{z \mid z \xrightarrow{e}\}$ satisfies $sig(e) = \text{inp}$ and $sup(s) = 0$, or*
2. *used $\in \tau$ and the region (sup, sig) returned by Algorithm 1 on input $Q = \{z, z' \mid z \xrightarrow{e} z'\}$ satisfies $sig(e) = \text{used}$ and $sup(s) = 0$, or*
3. *free $\in \tau$ and the region (sup, sig) returned by Algorithm 1 on input $Q = \{s\}$ satisfies $sig(e) = \text{free}$.*

Two states $s, s' \in S_A$ are τ-separable if and only if the region (sup, sig) returned by Algorithm 1 for $Q = \{s\}$ fulfills $sup(s') = 0$ or $sup(s) = 0$ for $Q = \{s'\}$.

Proof. The if-direction for an ESSP atom (e, s) is trivial, as (e, s) is solved by the pair (sup, sig) from Algorithm 1, a τ-region according to Lemma 2.

For the only-if-direction let (sup', sig') be a τ-region that solves (e, s). This implies that $\delta_\tau(sup'(s), sig'(e))$ is not defined and, hence, $sig'(e) \notin \{\text{nop}, \text{res}\}$. Let $X = \{x \mid x \xrightarrow{e}\}$, $Y = \{y \mid \xrightarrow{e} y\}$, and $Z = \{z, z' \mid z \xrightarrow{e} z'\}$. If $sig'(e) = \text{inp}$ then $sup'(s) = 0$ and $sup' \cap Y = \emptyset$. Otherwise, if $sig'(e) = \text{used}$ then $sup'(s) = 0$ and $Z \subseteq sup'$. Eventually, if $sig'(e) = \text{free}$ then $sup'(s) = 1$ and $Z \cap sup' = \emptyset$.

If $sig'(e) = \text{inp}$ then we define $Q = X$. If $sig'(e) = \text{used}$ then we take $Q = Z$ and if $sig'(e) = \text{free}$ then we define $Q = \{s\}$. Let sup be the result of Algorithm 1 on input Q. The returned support satisfies $Q \subseteq sup \subseteq sup'$. Moreover, if $Q = X$ then $sup(s) = 0$, $sup \cap Y = \emptyset$ and $sig(e) = \text{inp}$ and if $Q = Z$ then $sup(s) = 0$,

$Z \subseteq sup$, $sig(e) =$ used and if $Q = \{s\}$ then $sup(s) = 1$, $Z \cap sup = \emptyset$, $sig(e) =$ free. Consequently, (sup, sig) τ-solves (e, s).

The if-direction for the SSP atom (s, s) is trivial, again, as the τ-region of Algorithm 1 separates s, s'. For the only-if-direction let s, s' be separated by a τ-region (sup', sig') where, without loss of generality, $sup'(s) = 1$ and $sup'(s') = 0$. The result (sup, sig) of Algorithm 1 on $Q = \{s\}$ is a τ-region by Lemma 2 that fulfills $Q \subseteq sup \subseteq sup'$. Hence, $sup(s) = 1$ and $sup(s') = 0$, too.   □

By the required versatility of the regions from Algorithm 1, we can now assemble a full synthesis algorithm for the examined types of τ-nets.

Corollary 1. *If $\tau = \{nop, res\} \cup \omega$ is a type of nets with $\omega \subseteq \{inp, used, free\}$ or $\tau = \{nop, set\} \cup \omega'$ with $\omega' \subseteq \{out, used, free\}$ then a given TS A can be synthesized into a τ-net N with state graph A_N isomorphic to A, respectively rejected if N does not exist, in $\mathcal{O}(|E_A| \cdot |S_A|^6 \cdot \max\{|E_A|, |S_A|\})$ time.*

Proof. Let $S = S_A$ and $E = E_A$ and, for a start, let $\tau = \{nop, res\} \cup \omega$ with $\omega \subseteq \{inp, used, free\}$. The idea is to firstly produce a τ-region set $\mathcal{R}$ that solves all (E)SSP atoms of A. If we cannot find $\mathcal{R}$, then we reject A. There are $\mathcal{O}(|E| \cdot |S|)$ ESSP atoms (e, s). Depending on the availability of inp, used, free in τ, we have to test the τ-solvability of (e, s) by up to three calls of Algorithm 1 with inputs $Q_{inp} = \{z \mid z \xrightarrow{e} \}$, $Q_{used} = \{z, z' \mid z \xrightarrow{e} z'\}$, and $Q_{free} = \{s\}$. In every case, the method's running time of $\mathcal{O}(|E| \cdot |S|^5)$ heavily dominates the time for the creation of the input. If for all ESSP atoms at least one of the available tests succeeds, then we have picked up enough regions to τ-solve all ESSP atoms. This takes $\mathcal{O}(|E|^2 \cdot |S|^6)$ time. Otherwise, Lemma 3 allows us to reject A. Notice that, in case of $\tau = \{nop, res\}$ there must not be any ESSP atoms. The reason is that nop and res cannot be used to τ-solve ESSP atoms. In this case, we reject A if there is an event $e \in E$ and a state $s \in S$ with $\neg(s \xrightarrow{e})$.

Next, there are $\mathcal{O}(|S|^2)$ SSP atoms (s, s'). By Lemma 3, we have to call Algorithm 1 with $Q_s = \{s\}$ and $Q_{s'} = \{s'\}$ to decide the τ-separability of s, s'. After $\mathcal{O}(|E| \cdot |S|^7)$ time, either $\mathcal{R}$ τ-solves all SSP atoms or we can reject A.

Hence, using $\mathcal{O}(|E| \cdot |S|^6 \cdot \max\{|E|, |S|\})$ time in total, we decide the τ-feasibility of A and, in the positive case, get $\mathcal{R}$. Computing $N_A^{\mathcal{R}}$ consumes $\mathcal{O}(|\mathcal{R}| \cdot |E|) = \mathcal{O}(|E| \cdot |S| \cdot \max\{|E|, |S|\})$ time, which is dominated by the previous costs.

If $\tau = \{nop, set\} \cup \omega'$ with $\omega' \subseteq \{out, used, free\}$ our approach is to synthesize a net N' for the isomorphic type τ' that replaces set with res, out with inp, used with free, and free with used. In order to obtain a τ-net N, we simply revert the interaction replacement in the flow function $f_{N'}$. Obviously, A_N is isomorphic to $A_{N'}$, which is isomorphic to A.   □

For an example of our method, we use our the TS A in the upper left of Fig. 2 and demonstrate how to synthesize a τ_1-net with isomorphic state graph for the type of nets $E_{\tau_1} = \{nop, inp, res, free\}$. We first solve all ESSP atoms and start with $(a, 3)$. As inp $\in E_\tau$, Lemma 3 tells us to start Algorithm 1 on $Q_{inp}^a = \{0, 1, 2\}$. But Q_{inp}^a does not satisfy any condition of the while-loop and

Algorithm 1 immediately terminates and returns $sup = \{0,1,2\}$. According to Lemma 2, this results in the region R_1 presented in the bottom of Fig. 2. As all other ESSP atoms $(a,\cdot)$ with event a are automatically τ_1-solved by R_1, we do not bother Algorithm 1 for them. Secondly, for $(b,1)$ we start the algorithm on input $Q_{\mathsf{inp}}^b = \{0,5,6\}$ and, again without further iterations, obtain R_2 from Fig. 2, which also τ_1-solves all atoms $(b,\cdot)$ involving b. For the third step, we solve $(c,2)$ by running Algorithm 1 on $Q_{\mathsf{inp}}^c = \{1,3,5\}$. Since $1 \in Q_{\mathsf{inp}}^c, 0 \notin Q_{\mathsf{inp}}^c, b \in E_A$ and $\delta_A(0,b) = 1$ satisfies the while-condition, Q_{inp}^c is updated and becomes $Q = \{0,1,3,5\}$, which does not satisfy the the while-conditions, anymore. Thus, we obtain R_3 from Fig. 2 which τ_1-solves (c,s) with $s \in \{2,4,6\}$ and only leaves $(c,0)$. We next try input $Q_{\mathsf{free}}^c = \{0\}$. This does not satisfy any while-condition and immediately leads to R_4 from Fig. 2. Altogether, $R_1,\ldots,R_4$ are sufficient to τ_1-solve all (E)SSP atoms of A. Hence, as discussed in the proof of Corollary 1, the synthesized net $N_A^{\mathcal{R}}$, defined by $\mathcal{R} = \{R_1,\ldots,R_4\}$, has a state graph isomorphic to A, namely $A_{N_A^{\mathcal{R}_1}}$ depicted in the upper middle of Fig. 2.

3.2 Proof of Theorem 1.3

The types covered in this subsection are relatives of flip-flop nets. Therefore, we attack them by a generalization of Schmitt's algorithm [14], which was originally invented for the type $\tau = \{\mathsf{nop}, \mathsf{inp}, \mathsf{out}, \mathsf{swap}\}$ of flip-flop nets. The fundamental idea is to reduce ESSP and SSP to systems of linear equations over the additive group $\mathbb{Z}_2$ of integers modulo 2. To this end, let τ' be a type of nets as defined in Theorem 1.3. Moreover, in order to rule out trivial cases, we assume A to be a TS with $|S_A| \geq 2$. Also, we enumerate the events and let $E_A = \{e_1,\ldots,e_n\}$. The equations presented in the following have to be considered modulo 2.

Before we start, we have to introduce some further definitions that, eventually, allow us to interpret regions as solutions of equation systems over $\mathbb{Z}_2$. We start by reformulating τ' into a corresponding $\mathbb{Z}_2$-interpretation τ as follows: The type of nets $\tau = (\{0,1\}, E_\tau, \delta_\tau)$ is defined by $E_\tau = (E_{\tau'} \setminus \{\mathsf{nop}, \mathsf{swap}\}) \cup \{0,1\}$ and, for all $s \in \{0,1\}$, we define $\delta_\tau(s,e) = \delta_{\tau'}(s,e)$ for all $e \in E_{\tau'} \setminus \{\mathsf{nop}, \mathsf{swap}\}$ and $\delta_\tau(s,1) = \delta_{\tau'}(s,\mathsf{swap})$ and $\delta_\tau(s,0) = \delta_{\tau'}(s,\mathsf{nop})$. Observe, that τ' and τ are isomorphic. In fact, Fig. 3 demonstrates that the 0-labeled transitions, respectively 1-labeled transitions, of τ mimic nop and swap of τ'. Hence, by Lemma 1, we can analyze τ instead of τ'.

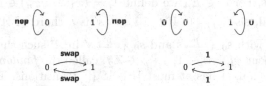

Fig. 3. Upper left: nop-labeled transitions of τ'. Upper right: 0-labeled transitions of τ mimicking nop. Lower left: swap-labeled transitions of τ'. Lower right: 1-labeled transitions of τ mimicking swap.

Next, the notion of abstract regions allows us to translate the τ-solvability of ESSP, respectively SSP, atoms into the solvability of linear equation systems: A τ-region (sup, abs) of A is called an *abstract τ-region* if $abs : E_A \longrightarrow \{0,1\}$. We call abs an *abstract* signature. One easily verifies that $sup, abs : \{0,1\} \longrightarrow \{0,1\}$ define an abstract τ-region if and only if $sup(s') = sup(s) + abs(e)$ for every transition $s \xrightarrow{e} s'$ of A. Using induction, we obtain that this is equivalent to the condition that for every word $p \in E_A^*$ with $s_{0,A} \xrightarrow{p} s$ the so-called *path equation* holds: $sup(s) = sup(s_{0,A}) + abs(e_1') + \cdots + abs(e_m')$ where $p = e_1' \ldots e_m'$. We say that $\psi_p = (\#_{e_1}^p, \ldots, \#_{e_n}^p) \in \mathbb{Z}_2^n$ is the Parikh-vector of p, counting the number of occurrences $\#_{e_i}^p$ of every event $e_i \in E_A$ in the word p modulo 2. Then we identify the abstract signature abs with the element $abs = (abs(e_1), \ldots, abs(e_n)) \in \mathbb{Z}_2^n$. For two elements $v, w \in \mathbb{Z}_2^n$, we define $v \cdot w = v_1 w_1 + \cdots + v_n w_n$. Hence, considering p and abs as elements of $\mathbb{Z}_2^n$ allows us to reformulate the path equation to $sup(s) = sup(s_{0,A}) + \psi_p \cdot abs$. In particular, if p, p' are two different words with $s_{0,A} \xrightarrow{p} s$ and $s_{0,A} \xrightarrow{p'} s$ then $\psi_p \cdot abs = \psi_{p'} \cdot abs$. This makes the support sup fully determined by $sup(s_{0,A})$ and abs. By the validity of the path equation, every abstract signature abs implies two different abstract τ-regions of A, one for $sup(s_{0,A}) = 1$ and one for $sup(s_{0,A}) = 0$.

We proceed by showing how the notion of abstract regions translates the τ-solvability for every (E)SSP atom (x, y) of A into the solvability of a system of linear equations $M_a \cdot abs = c_a$ with $M_a \in \mathbb{Z}_2^{m \times n}, c_a \in \mathbb{Z}_2^m$ by an abstract signature $abs \in \mathbb{Z}_2^n$. In particular, the system $M_a \cdot abs = c_a$ implements a basic and an extended part. The basic part is built from equations describing the properties of abstract signatures abs by their possible solutions. The extended part depends on (x, y) and includes additional equations to make sure that solutions abs actually τ-solve the atom.

The equations of the basic part are build on any spanning tree A' of A. More precisely, a spanning tree A' is a sub-transition system $A' = (S_A, E_A, \delta_{A'}, s_{0,A})$ of A with the same states and events but a restricted transition function $\delta_{A'}$ such that, firstly, $\delta_{A'}(s, e) = s'$ entails $\delta_A(s, e) = s$ and, secondly, for every $s \in S_{A'}$ there is *exactly* one word $p \in E_A^*$ with $s_{0,A} \xrightarrow{p} s$ in A'. In other words, the underlying undirected graph of A' is a directed labeled tree in the common graph-theoretical sense. A transition $s \xrightarrow{e} s'$ of A that is not in A' is called a *chord* (of A'). The chords of A' are exactly the edges that would introduce cycles into the graph underlying A'. This gives raise to the following notion of fundamental cycles. For $e_i \in E_A$ we define $1_i = (x_1, \ldots, x_n) \in \mathbb{Z}_2^n$, where every $x_j = 1$ if $j = i$ and, else $x_j = 0$. If $t = s \xrightarrow{e_i} s'$ is a chord of A' then there are *unique* words p, p' with $s_{0,A} \xrightarrow{p} s$ and $s_{0,A} \xrightarrow{p'} s'$ in A' such that t corresponds to the unique element $\psi_t = \psi_p + 1_i + \psi_{p'} \in \mathbb{Z}_2^n$, called the *fundamental cycle* of t.

As δ_A is a function, A has at most $|E| \cdot |S_A|^2$ transitions. This makes computing a spanning tree A' doable in $\mathcal{O}(|E| \cdot |S_A|^3)$ time [15]. Then A' contains $|S_A| - 1$ transitions and, thus, has at most $|E| \cdot |S_A|^2 - |S_A| + 1 \leq |E| \cdot |S_A|^2 - 1$

chords (taking $|S_A| \geq 2$ into consideration). The next lemma shows how to use the chords to generate abstract signatures:

Lemma 4. *If A' is a spanning tree of a TS A with chords $t_1, \ldots, t_k$ then $abs \in \mathbb{Z}_2^n$ is an abstract signature of A if and only if $\psi_{t_i} \cdot abs = 0$ for all $i \in \{1, \ldots, k\}$. Two different spanning trees A'_1 and A'_2 of A provide equivalent systems of equations.*

Proof. We start with the first statement. *If*: Let $abs \in \mathbb{Z}_2^n$ such that $\psi_{t_i} \cdot abs = 0$ for all $i \in \{1, \ldots, k\}$ and $sup(s_{0,A}) \in \{0,1\}$. For every $s \in S_A$ there is a unique word p with $s_{0,A} \xrightarrow{p} s$ in A'. Defining $sup(s) = sup(s_{0,A}) + \psi_p \cdot abs$ we inductively obtain that every transition $s \xrightarrow{e} s'$ of A' satisfies $sup(s') = sup(s) + abs(e)$. It remains to prove that this definition can be made consistent with the remaining transitions of A, that is, the chords of A'. Let $t = s \xrightarrow{e_i} s'$ be a chord of A' and let p, p' be the unique words with $s_{0,A} \xrightarrow{p} s$ and $s_{0,A} \xrightarrow{p'} s'$ in A'. By $sup(s) = sup(s_{0,A}) + \psi_p \cdot abs$ and $sup(s') = sup(s_{0,A}) + \psi_{p'} \cdot abs$ we have that $0 = \psi_t \cdot abs \iff 0 = (\psi_{p'} + 1_i + \psi_p) \cdot abs \iff 0 = \psi_{p'} \cdot abs + abs(e) + \psi_p \cdot abs \iff \psi_{p'} \cdot abs = abs(e) + \psi_p \cdot abs \iff sup(s_{0,A}) + \psi_{p'} \cdot abs = sup(s_{0,A}) + \psi_p \cdot abs + abs(e) \iff sup(s') = sup(s) + abs(e)$ where $0 = \psi_t \cdot abs$ is true by assumption. Hence, abs is an abstract signature of A and the proof also describes how to get a corresponding abstract region (sup, abs) for A.

Only-if: If abs is an abstract region of A then we have $sup(s') = sup(s) + abs(e)$ for every transition in A. Hence, if $t = s \xrightarrow{e} s'$ is a chord of the spanning tree A' then working backwards through the equalities above proves $\psi_t \cdot abs = 0$.

The second statement is implied by the first: If A'_1 and A'_2 are two spanning trees of A with fundamental cycles $\psi_{t_1}^{A'_1}, \ldots, \psi_{t_k}^{A'_1}$ and $\psi_{t'_1}^{A'_2}, \ldots, \psi_{t'_k}^{A'_2}$, respectively, then we have for $abs \in \mathbb{Z}_2^n$ that $\psi_{t_i}^{A'_1} \cdot abs = 0, i \in \{1, \ldots, k\}$ if and only if abs is an abstract signature of A if and only if $\psi_{t'_i}^{A'_2} \cdot abs = 0, i \in \{1, \ldots, k\}$. $\square$

Justified by Lemma 4, we let A' be any fixed spanning tree of A with chords $t_1, \ldots, t_k$. For $s \in S_A$ we abridge the Parikh-vector ψ_p of the unique word $p \in E_A^*$ with $s_{0,A} \xrightarrow{p} s$ by ψ_s. The next lemma, borrowed from [9], is the crucial ingredient of our polynomial time estimations:

Lemma 5 ([9]). *If $M \in \mathbb{Z}_2^{k \times n}$ and $c \in \mathbb{Z}_2^k$ then deciding if there is an element $x \in \mathbb{Z}_2^n$ such that $Mx = c$ is doable in $\mathcal{O}(nk\, max\{n,k\})$ time.*

Given an (E)SSP atom of A, the following two lemmas clarify how to obtain the *extended* part of $M_a \cdot abs = c_a$ and, altogether, show how τ-feasibility, τ-ESSP and τ-SSP become decidable in polynomial time.

Lemma 6. *An SSP atom (s, s') of A is τ-solvable if and only if there is an abstract signature abs of A with $\psi_s \cdot abs \neq \psi_{s'} \cdot abs$. Deciding whether A has the τ-SSP can be done in $\mathcal{O}(|E_A|^3 \cdot |S_A|^6)$ time.*

Proof. If: Setting $sup(s_{0,A}) = 0$ implies a τ-region (sup, abs) solving (s, s').

Only-if: For a τ-region (sup, sig) separating s, s' and every transition $z \xrightarrow{e} z'$ of A we define $abs(e) = x \in \mathbb{Z}_2$ if and only if $sup(z) \xrightarrow{x} sup(z') \in \tau$. Consequently, $sup(s_{0,A}) + \psi_s \cdot abs = sup(s) \neq sup(s') = sup(s_{0,A}) + \psi_{s'} \cdot abs$ implies $\psi_s \cdot abs \neq \psi_{s'} \cdot abs$.

The basic part of $M_a \cdot abs = c_a$ provides at most $|E| \cdot |S_A|^2 - 1$ equations, for (s, s') and we add one more equation to extend M_a, namely $(\psi_s - \psi_{s'}) \cdot abs = 1$. Thus, we have to solve a linear systems with at most $|E_A| \cdot |S_A|^2$ equations with $|E_A|$ unknown. By $|E_A||S_A|^2 \geq |E_A|$ and Lemma 5, this is doable in $\mathcal{O}(|E_A|^3 \cdot |S_A|^4)$ time. As at most $|S_A|^2$ different SSP atoms exist, we can decide τ-SSP in $\mathcal{O}(|E_A|^3 \cdot |S_A|^6)$ time. □

Lemma 7. *A ESSP atom (e, s) of A is τ-solvable if and only if there is an abstract signature abs of A satisfying $(\psi_{s'} - \psi_{s''}) \cdot abs = 0$ and $(\psi_{s'} - \psi_s) \cdot abs = 1$ for all pairs of states s', s'' where e occurs (that is, $s' \xrightarrow{e}$ and $s'' \xrightarrow{e}$, and one of the following conditions is true:*

1. *$E_\tau \cap \{inp, out\} \neq \emptyset$ and $abs(e) = 1$,*
2. *$E_\tau \cap \{used, free\} \neq \emptyset$ and $abs(e) = 0$.*

Deciding if A has the τ-ESSP is doable in $\mathcal{O}(|E_A|^4 \cdot |S_A|^5)$ time.

Proof. Let $s_1 \xrightarrow{e} s'_1, \ldots, s_m \xrightarrow{e} s'_m$ be the transitions of A labeled with event e.

If: If $inp \in E_\tau$ and $abs(e) = 1$ or if $used \in E_\tau$ and $abs(e) = 0$ then we define $sup(s_{0,A}) = 0$ if $\psi_{s_1} \cdot abs = 1$ and, otherwise, if $\psi_{s_1} \cdot abs = 0$ then $sup(s_{0,A}) = 1$. By $(\psi_{s_i} - \psi_{s_j}) \cdot abs = 0$ and $(\psi_{s_1} - \psi_s) \cdot abs = 1$, this makes sure that $sup(s_i) = sup(s_{0,A}) + \psi_{s_i} \cdot abs = 1 \neq sup(s) = sup(s_{0,A}) + \psi_s \cdot abs$, for $i, j \in \{0, \ldots, m\}$. The latter inequality implies $sup(s) = 0$. By $abs(e) = 1$, respectively by $abs(e) = 0$, we have that $\delta_\tau(sup(s_i), inp) = sup(s'_i)$, respectively $\delta_\tau(sup(s_i), used) = sup(s'_i)$, is present in τ. Hence, we obtain a τ-region (sup, sig) τ-solving (e, s) by defining $sup(s_{0,A})$ as described above and setting for $e' \in E_A$: $sig(e') = inp$, respectively $sig(e') = used$, if $e' = e$ and $sig(e') = abs(e')$, otherwise.

If $out \in E_\tau$ and $abs(e) = 1$ or if $free \in E_\tau$ and $abs(e) = 0$ then we work complementary to the previous case and define $sup(s_{0,A}) = 0$ if $\psi_{s_1} \cdot abs = 0$ and $sup(s_{0,A}) = 1$ if $\psi_{s_1} \cdot abs = 1$. Similarly to the discussion above, this yields a solution τ-region (sup, sig) with $sig(e') = out$, respectively $sig(e') = free$, if $e' = e$ and $sig(e') = abs(e')$, otherwise, for all $e' \in E_A$.

Only-if: Defining for $z \xrightarrow{e'} z' \in A$ that $abs(e') = x \in \{0, 1\}$ if and only if $sup(z) \xrightarrow{x} sup(z')$ is present in τ yields an abstract signature of A. As e occurs at $s_1, \ldots, s_m$, we obtain $sup(s_0) + \psi_{s_i} \cdot abs = sup(s_0) + \psi_{s_j} \cdot abs$ for all $i, j \in \{1, \ldots, m\}$. This implies the equation $(\psi_{s_i} - \psi_{s_j}) \cdot abs = 0$. By $sup(s_1) \neq sup(s)$, we have $(\psi_{s_1} - \psi_s) \cdot abs = 1$.

To estimate the given computational complexity we observe the following: The basic part of the system defined by the chords of A' has at most $|E_A| \cdot |S_A|^2$ equations. Moreover, e occurs at most at $|S_A| - 1$ states, which brings at most

$|S_A|^2$ additional equations that ensure that the source states of e all have the same support. One further equation is to added for the condition $(\psi_{s_1} - \psi_s) \cdot abs = 1$. Hence, the system has at most $2 \cdot |E_A| \cdot |S_A|^2$ equations. For a single atom (s, e), we have to solve at most 2 systems of equations, namely one for each fixed value $abs(e) \in \{0, 1\}$. By Lemma 5, this takes $\mathcal{O}(|E_A|^3 \cdot |S_A|^4)$ time. Since, we have at most $|S_A| \cdot |E_A|$ ESSP atoms, deciding if A has the τ-ESSP can be done in $\mathcal{O}(|E_A|^4 \cdot |S_A|^5)$ time. $\square$

Corollary 2. *If τ' is a type of nets as defined in Theorem 1.3 then a τ'-net N with a state graph isomorphic to a given TS A, if it exists, can be computed in $\mathcal{O}(|E_A \cdot |^3 |S_A|^5 \cdot \max\{|E_A|, |S_A|\})$ time.*

Proof. If τ is the $\mathbb{Z}_2$-interpretation of τ' and if $\mathcal{R}$ is a set of τ-regions containing a witness for all (E)SSP atoms of A then, by Lemma 1, $\mathcal{R}$ corresponds directly to a set $\mathcal{R}'$ of τ'-regions proving the τ'-feasibility of A. The τ'-net $N_A^{\mathcal{R}'}$ has a state graph isomorphic to A and, from here, the corollary is a direct implication of Lemmas 1, 6, and 7. $\square$

3.3 Proof of Theorem 1.4

It remains to prove the trivial synthesis for the types of nets from Theorem 1.4. A related polynomial time algorithm is established in the following lemma.

Lemma 8. *If $\tau = \{nop\} \cup \omega$ is a type of nets with $\omega \subseteq \{used, free\}$ then τ-feasibility of a given TS A can be decided in $\mathcal{O}(1)$ time. A τ-feasible TS A can be synthesized into a τ-net N with state graph A_N isomorphic to A in $\mathcal{O}(|E_A|)$ time.*

Proof. Let $s_0 \xrightarrow{e} s_1$ be any transition of A with $s_0 \neq s_1$. To separate these states, we require a τ-region (sup, sig) with $sup(s_0) \neq sup(s_1)$. However, $sig(e) \in \tau$ implies that $sup(s_0) = sup(s_1)$. Hence, s_0 and s_1 are not τ-separable. As all states have to be reachable, this implies that input TSs with more than one state cannot be τ-feasible and, thus, are discarded after a constant time check.

As $A = (\{s\}, E, \{s \xrightarrow{e} s \mid e \in E\}, s)$ is the only single-state TS with a given event set E, the net $N = (\{p\}, E_A, \{(p, e, nop) \mid e \in E_A\}, \{p\})$, which has a state graph isomorphic to A, can be written to the output in $\mathcal{O}(|E_A|)$ time. $\square$

4 NP-Hard Net Synthesis

Second corner stone of this paper is the following Theorem revealing 7 boolean types of nets with an NP-complete feasibility problem.

Theorem 2. *Deciding for input TSs A if there is a τ-net N with state graph isomorphic to A is NP-complete if*

1. *$\tau = \{nop, inp, free\}$ or $\tau = \{nop, inp, used, free\}$,*
2. *$\tau = \{nop, out, used\}$ or $\tau = \{nop, out, used, free\}$, and*
3. *$\tau = \{nop, set, res\} \cup \omega$ with non-empty $\omega \subseteq \{used, free\}$.*

The remainder of this section proves Theorem 2. In particular, we present polynomial time reductions of the NP-complete cubic monotone one-in-three 3-SAT problem [12] to the τ-feasibility of net types τ from Theorem 2.1 and Theorem 2.3. Our reductions make sure for every TS A_φ^τ, being constructed from a given cubic monotone 3-CNF φ, that the τ-ESSP implies the τ-SSP for the considered type of nets τ. Hence, with respect to the created TSs, τ-feasibility and τ-ESSP are the same problem. This justifies the reduction from monotone one-in-three 3-SAT to the respective τ-ESSP rather than to actual τ-feasibility. The proof for Theorem 2.1 and Lemma 1 imply hardness for Theorem 2.2 as the respective types are isomorphic.

Input to our reductions is a monotone set $\varphi = \{C_0, \ldots, C_{m-1}\}$ of negation free 3-clauses $C_i = \{X_{i,0}, X_{i,1}, X_{i,2}\}$ where every variable occurs in exactly three clauses. Finding a one-in-three model for φ, that is, a subset $M \subseteq V(\varphi)$ of the variables in φ that covers every clause exactly once, $|M \cap C_i| = 1, i \in \{0, \ldots, m-1\}$, is known to be NP-complete from [12]. For every considered type of nets τ, the result is a TS A_φ^τ which is τ-feasible if and only if φ has a one-in-three model. The idea is always to install a *key event* k and a *key state* q in A_φ^τ such that the so called *key* atom (k, q) is τ-solvable by a key τ-region (sup, sig) if and only if M exists. More precisely, we use the variables $V(\varphi)$ as events in A_φ^τ and their key signature sig tells us how to find M and vice versa. The other ESSP atoms, and even more all SSP atoms, are secondary as they become τ-solvable as soon as the key atom (k, q) is τ-solved.

This idea is put into practice by creating six directed labeled paths per clause $C_i = \{X_{i,0}, X_{i,1}, X_{i,2}\}$ that commonly start at state $t_{i,0}$, terminate at $t_{i,5}$ and consist of three transitions permuting the events $X_{i,0}, X_{i,1}, X_{i,2}$. The TS A_φ^τ fulfills the following conditions:

1. For every $i \in \{0, \ldots, m-1\}$ and every permutation (α, β, γ) of $\{0, 1, 2\}$ there is a path $t_{i,0} \xrightarrow{X_{i,\alpha}} t \xrightarrow{X_{i,\beta}} t' \xrightarrow{X_{i,\gamma}} t_{i,5}$ in A_φ^τ.
2. If (sup, sig) is a key τ-region of A_φ^τ, that is, one that solves (k, q), then $sup(t_{0,0}) = \cdots = sup(t_{m-1,0}) \neq sup(t_{0,5}) = \cdots = sup(t_{m-1,5})$.
3. If $i \in \{0, \ldots, m-1\}$ and (sup, sig) is a key τ-region of A_φ^τ then *exactly one* of $sig(X_{i,\alpha}), sig(X_{i,\beta}), sig(X_{i,\gamma})$ is different from nop. Hence, the signature tells us how to build $M = \{X \in V(\varphi) \mid sig(X) \neq \mathsf{nop}\}$, a one-in-three model of φ.
4. If φ has a one-in-three model then (k, q) is τ-solvable.
5. If (k, q) is τ-solvable, then A_φ^τ has the τ-ESSP and the τ-SSP.

Clearly, having these conditions proves that there is a one-in-three model for φ if and only if A_φ^τ has the τ-ESSP and the τ-SSP.

Next, we define how A_φ^τ is constructed from φ. See Fig. 4 for a visualization of the following concepts. Firstly, we call A_φ the basic TS with states $S = \{s_0, s_1, q\} \cup \{t_{i,0}, \ldots, t_{i,8} \mid 0 \leq i \leq m-1\}$ and events $E = \{k, h\} \cup \{h_i, r_i \mid 0 \leq i \leq m-1\} \cup V(\varphi)$. To omit a lengthy and complex definition of the transitions in A_φ, we use Fig. 4, which depicts $\delta(s, e)$ with solid black edges for all states $s \in \{s_0, s_1, q, t_{i,0}, \ldots, t_{i,8}\}$ and all events $e \in \{k, h, h_i, r_i, X_{i,0}, X_{i,1}, X_{i,2}\}$. If $\tau = \{\mathsf{nop}, \mathsf{res}, \mathsf{free}\}$ or $\tau = \{\mathsf{nop}, \mathsf{res}, \mathsf{used}, \mathsf{free}\}$ then we simply use the basic TS, that is, $A_\varphi^\tau = A_\varphi$.

If τ is one of $\{\mathsf{set}, \mathsf{res}, \mathsf{used}\}$, $\{\mathsf{set}, \mathsf{res}, \mathsf{free}\}$, or $\{\mathsf{set}, \mathsf{res}, \mathsf{used}, \mathsf{free}\}$, the construction of A_φ^τ is more complex. We first require the extended TS A_φ^+ with extended states $S_{A_\varphi^+} = S_{A_\varphi} \cup \{m_0, \ldots, m_4\} \cup \{p_{i,0}, \ldots, p_{i,3} \mid 0 \le i < m\}$ and extended events $E_{A_\varphi^+} = E_{A_\varphi} \cup \{a, c, u, v\} \cup \{a_i, b_i, x_i \mid 0 \le i < m\}$. The transitions of A_φ^+ are also an extension in the way that $\delta_{A_\varphi^+}(s, e) = \delta_{A_\varphi}(s, e)$ for basic states $s \in S_{A_\varphi}$ and basic events $e \in E_{A_\varphi}$ where $\delta_{A_\varphi}(s, e)$ is defined. Thus, the solid black arcs in Fig. 4 illustrate one part of the extended transitions. Aside from this, the solid brown arcs present the remaining transition function $\delta_{A_\varphi^+}(s, e)$ for all $s \in \{s_0, t_{i,0}, \ldots, t_{i,8}, m_0, \ldots, m_4, p_{i,0}, \ldots, p_{i,3}\}$ and all $e \in \{k, h, h_i, a, c, u, v, a_i, b_i, x_{i,0}, x_{i,1}, x_{i,2}\}$.

While still being depictable, A_φ^+ is not yet a complete TS A_φ^τ for the type of nets defined by Theorem 2.3. This requires the *loop-enhancement* $A_\varphi^\times$ of A_φ^+ on the same states $S_{A_\varphi^\times} = S_{A_\varphi^+}$ and events $E_{A_\varphi^\times} = E_{A_\varphi^+}$ but with loop-enhanced transitions, that is, for all $s, s' \in S_{A_\varphi^+}$ and $e \in E_{A_\varphi^+}$ where $\delta_{A_\varphi^+}(s, e) = s'$ we have $\delta_{A_\varphi^\times}(s, e) = s'$ and $\delta_{A_\varphi^\times}(s', e) = s'$. Now, $A_\varphi^\tau = A_\varphi^\times$. However, for understanding it is mostly better to deal with A_φ^+ instead of the complicated $A_\varphi^\times$. Therefore, Fig. 4 desists from showing all the loops.

At this point, we are ready to provide the main piece of our proof. The next lemma shows the equivalence between the one-in-three satisfiability of φ and the τ-solvability of (k, q):

Lemma 9. *If $\tau \in \{\{\mathsf{nop}, \mathsf{inp}, \mathsf{free}\}, \{\mathsf{nop}, \mathsf{inp}, \mathsf{used}, \mathsf{free}\}, \{\mathsf{nop}, \mathsf{set}, \mathsf{res}\} \cup \omega\}$, where $\omega \in \{\mathsf{used}, \mathsf{free}\}$, then the key atom (k, q) is τ-solvable in A_φ^τ if and only if φ is one-in-three satisfiable.*

Proof. Only-if: Let (sup, sig) be a τ-region solving (k, q) in A_φ^τ. We show for every clause C_i that there is exactly one variable event $X \in \{X_{i,0}, X_{i,1}, X_{i,2}\}$ with $sig(X) \in \{\mathsf{inp}, \mathsf{used}, \mathsf{free}\}$ while the other two have nop-signature. Consequently, the set $M = \{X \in V(\varphi) \mid sig(X) \ne \mathsf{nop}\}$ will be a one-in-three model of φ.

For a start, $\tau = \{\mathsf{nop}, \mathsf{inp}, \mathsf{free}\}$ or $\tau = \{\mathsf{nop}, \mathsf{inp}, \mathsf{used}, \mathsf{free}\}$. As (k, q) is τ-solved, assume first that $sig(k) = \mathsf{free}$ and $sup(q) = 1$. By $sig(k) = \mathsf{free}$, it is $sup(s_0) = sup(s_1) = 0$ which means that $sig(h) \notin \tau$. Hence, $sig(k) \in \{\mathsf{inp}, \mathsf{used}\}$ and $sup(q) = 0$. This implies $sup(t_{i,0}) = 1$ and $sig(h_i) \in \{\mathsf{nop}, \mathsf{free}\}$ and, thus, $sup(t_{i,5}) = 0$. By this, we get $sig(X_{i,0}), sig(X_{i,1}), sig(X_{i,2}) \in \{\mathsf{nop}, \mathsf{inp}, \mathsf{free}\}$. In A_φ^τ, there is a path $t_{i,0} \xrightarrow{X_{i,\alpha}} t_1 \xrightarrow{X_{i,\beta}} t_2 \xrightarrow{X_{i,\gamma}} t_{i,5}$ for every permutation (α, β, γ) of $\{0, 1, 2\}$. As all variable events of the clause occur exactly once on each of these six paths, there has to be exactly one $X \in \{X_{i,0}, X_{i,1}, X_{i,2}\}$ with $sig(X) = \mathsf{inp}$. Moreover, for every $j \in \{0, 1, 2\}$ there are both, a path that starts with $X_{i,j}$ and another that ends on $X_{i,j}$. Consequently, for $Y \in \{X_{i,0}, X_{i,1}, X_{i,2}\} \setminus X$ there

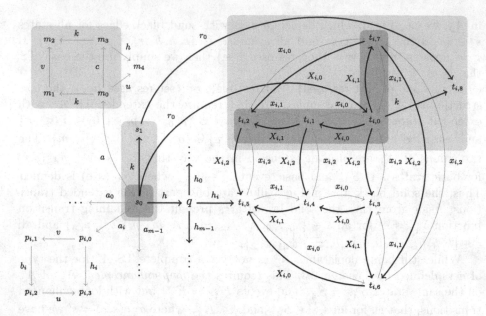

Fig. 4. The black arcs in isolation illustrate A_φ. Aside from the static center on states s_0, s_1, q, TS A_φ contains a compartment of states $t_{i,0}, \ldots, t_{i,8}$ for every clause C_i of φ. Together with the brown arcs, we get A_φ^+ which adds one compartment on $m_0, \ldots, m_4$ and one on $p_{i,0}, \ldots, p_{i,3}$ for every clause C_i. Restricted to the presented parts of A_φ, respectively A_φ^+, the blue areas, respectively red areas, mark the support of the τ-region that solves (k, q) defined in Lemma 9. (Color figure online)

are transitions $s \xrightarrow{Y} s'$ and $z \xrightarrow{Y} z'$ with $sup(s) = sup(s') \neq sup(z) = sup(z')$ implying $sig(Y) = \mathsf{nop}$.

Next, let $\tau = \{\mathsf{nop}, \mathsf{set}, \mathsf{res}\} \cup \omega$ with non-empty $\omega \in \{\mathsf{used}, \mathsf{free}\}$. Assume $sig(k) = \mathsf{used}$ and $sup(q) = 0$, which implies $sup(s_0) = sup(m_0) = sup(t_{i,0}) = 1$ and $sig(h) = \mathsf{res}$ and, thus, $sup(m_4) = 0$. We immediately get $sig(u) = \mathsf{res}$ implying $sup(p_{0,3}) = 0$ and $sig(h_i) \in \{\mathsf{nop}, \mathsf{res}, \mathsf{free}\}$ and, thus, $sup(t_{i,5}) = 0$. For every $j \in \{0, 1, 2\}$, we have $sig(X_{i,j}) \in \{\mathsf{nop}, \mathsf{res}, \mathsf{free}\}$ by $\xrightarrow{X_{i,j}} t_{i,5}$. Accordingly, $\xrightarrow{x_{i,j}} t_{i,0}$ leads to $sig(x_{i,j}) \in \{\mathsf{nop}, \mathsf{set}, \mathsf{used}\}$. As $sup(t_{i,0}) \neq sup(t_{i,5})$, there must be at least one event in $X_{i,0}, X_{i,1}, X_{i,2}$, respectively $x_{i,0}, x_{i,1}, x_{i,2}$, with signature not in $\{\mathsf{nop}, \mathsf{free}\}$, respectively $\{\mathsf{nop}, \mathsf{used}\}$. We show that there must be exactly one j with $sig(X_{i,j}) = \mathsf{res}$ and $sig(x_{i,j}) = \mathsf{set}$, while the other four events have nop-signature. For example, if $sig(X_{i,2}) = \mathsf{res}$ then $sup(t_{i,3}) = 0$ and, thus, $sig(x_{i,0}) = sig(x_{i,1}) = \mathsf{nop}$ and $sig(x_{i,2}) = \mathsf{set}$. This implies $sup(t_{i,1}) = sup(t_{i,2}) = 1, sup(t_{i,4}) = 0$ and, consequently, $sig(X_{i,0}) = sig(X_{i,1}) = \mathsf{nop}$. Notice that none of the events can get used or free. A similar explanation works for $j \in \{0, 1\}$.

The case $sig(e) = \mathsf{free}$ and $sup(q) = 1$, follows by the same argumentation by inverting the support and interchanging set with res and used with free.

If: Let $M \subseteq V(\varphi)$ be a one-in-three model of φ and, for a start, let $\tau = \{\mathsf{nop}, \mathsf{inp}, \mathsf{free}\}$ or $\tau = \{\mathsf{nop}, \mathsf{inp}, \mathsf{used}, \mathsf{free}\}$. We define a τ-region (sup, sig) of A_φ^τ that solves (k, q) by $sup = \{s_0\} \cup \{s \mid s\overset{X}{\longrightarrow} : X \in M\}$ and $sig(k) = \mathsf{inp}$, $sig(X) = \mathsf{inp}$ for all $X \in M$ and $sig(e) = \mathsf{nop}$ for all other events e in $E_{A_\varphi^\tau}$.

If τ contains, beside $\{\mathsf{nop}, \mathsf{set}, \mathsf{res}\}$, the interaction used then we let $M' = \{x_i \mid X_i \in M\}$ and create a τ-region (sup', sig') of A_φ^τ to solve (k, q) by $sup' = sup \cup \{s_1, t_{0,8}, \ldots, t_{m-1,8}, m_0, \ldots, m_3\}$ and $sig'(k) = \mathsf{used}$, $sig'(X) = \mathsf{res}$ for all $X \in M$, $sig'(x) = \mathsf{set}$ for all $x \in M'$ and $sig'(e) = \mathsf{nop}$ for all other events e in $E_{A_\varphi^\tau}$. If used is not in τ then free is and we get a τ-region by inverting the support and interchanging set with res and used with free.   □

Using Lemma 1, we have also covered the type of nets defined by Theorem 2.2. Using the same lemma, we can finish our proof for Theorem 2 by the following lemma:

Lemma 10. *If* $\tau = \{\mathsf{nop}, \mathsf{inp}, \mathsf{free}\}$ *or* $\tau = \{\mathsf{nop}, \mathsf{inp}, \mathsf{used}, \mathsf{free}\}$ *or, otherwise, if* $\tau = \{\mathsf{nop}, \mathsf{set}, \mathsf{res}\} \cup \omega$ *with non-empty* $\omega \in \{\mathsf{used}, \mathsf{free}\}$ *and the key atom* (k, q) *is* τ-*solvable in* A_φ^τ *then* A_φ^τ *has the* τ-*ESSP and the* τ-*SSP.*

Proof. We present for each event $e \in E_{A_\varphi^\tau}$, respectively states $s, s' \in S_{A_\varphi^\tau}$, a corresponding set of τ-regions which τ-solves every valid atom (e, s), respectively (s, s'), in A_φ^τ.

Assume that $\tau = \{\mathsf{nop}, \mathsf{inp}, \mathsf{free}\}$ or $\tau = \{\mathsf{nop}, \mathsf{inp}, \mathsf{used}, \mathsf{free}\}$. For brevity, we use the following scheme to define $sig(e)$ based on a given support sup: If $sup(s) = 1$ and $sup(s') = 0$ for all $s\overset{e}{\longrightarrow}s'$ then $sig(e) = \mathsf{inp}$, if $sup(s) = sup(s') = 0$ for all $s\overset{e}{\longrightarrow}s'$ then $sig(e) = \mathsf{free}$ and, otherwise, $sig(e) = \mathsf{nop}$. Using this, we can define a region simply by defining sup.

The τ-solvability of (k, q) and (h, s_1) already follows from Lemma 9. Furthermore, if (sup, sig) τ-solves (k, q) then the τ-region $sup \cup \{q\}$ solves (r_i, s_1) for $i \in \{0, \ldots, m - 1\}$. For $X \in V(\varphi)$ let $sup_X = \{s \mid s\overset{k}{\longrightarrow}\} \cup \{s \mid s\overset{X}{\longrightarrow}\} \cup \{q\} \cup \{t_{n,0}, \ldots, t_{n,7} \mid 0 \le n < m, X \notin C_n\}$. The τ-region sup_X solves (X, s) for all $s \in S_X = \{t_{n,1}, \ldots, t_{n,8} \mid 0 \le n < m, X \in C_n\}$. Moreover, $sup_X' = S_{A_\varphi^\tau} \setminus S_X$ τ-solves (X, s) for the states $s \in sup_X'$, where the signature of X is free. The τ-regions $sup_{X_0}, \ldots, sup_{X_m}$ also complete the τ-solvability of (k, s) for every state s in question as for every $t \in \{t_{i,1}, \ldots, t_{i,8}\}$ there is $X \in \{X_{i,0}, X_{i,1}, X_{i,2}\}$ with $\neg(t\overset{X}{\longrightarrow})$. The τ-regions $\{s_0, s_1\}$ and $\{s_0, s_1, q\}$ complete the τ-solvability of (c, s) for every $c \subset \{h, r_0, \ldots, r_{m-1}\}$ and every remaining state in question.

The τ-solution of (k, q) separates s_0 and s_1 and the region $\{s_0, s_1\}$ separates s_0, s_1 from all the other states. The τ-region $\{s_0, s_1, s_2\}$ completes the τ-separation of q. The τ-regions $sup_{X_{i,0}}, sup_{X_{i,1}}, sup_{X_{i,2}}$ complete the τ-separation of $t_{i,0}, \ldots, t_{i,8}$.

Now assume $\tau = \{\mathsf{nop}, \mathsf{set}, \mathsf{res}\} \cup \omega$ with non-empty $\omega \in \{\mathsf{used}, \mathsf{free}\}$. For the τ-ESSP of $A_\varphi^\times$, it is sufficient to prove the τ-solvability of (e, s) for $e \in E_{A_\varphi^+}$ and

$s \in S_{A_\varphi^+}$ where $\neg s \xrightarrow{e}$ and $\neg \xrightarrow{e} s$. By definition, if $\xrightarrow{e} s$ in A_φ^+ then $s \xrightarrow{e} s$ in $A_\varphi^\times$ and (e, s) is not a valid ESSP atom. We proceed like in the previous case and use the following scheme to define a signature $sig(e)$ for given support sup: If $sup(s) = sup(s') = 1$ for all $s \xrightarrow{e} s'$ then $sig(e) = $ used, if $sup(s) = 1$ and $sup(s') = 0$ for all $s \xrightarrow{e} s'$ then $sig(e) = $ res, if $sup(s) = 0$ and $sup(s') = 1$ then $sig(e) = $ set and, otherwise, $sig(e) = $ nop.

The τ-solution of (k, s) where $s \in \{q, m_4, t_{0,5}, t_{1,5}, \dots, t_{m-1,5}, p_{0,0}, \dots, p_{m-1,3}\}$ is done by the key region of Lemma 9. For $X \in V(\varphi)$ let $sup_X = \{s, s' \mid s \xrightarrow{k} s'\} \cup \{s \mid s \xrightarrow{X}\} \cup \{q, m_4\} \cup \{t_{n,0}, \dots, t_{n,7} \mid 0 \le n < m, X \notin C_n\}$. The τ-regions $sup_{X_0}, \dots, sup_{X_{m-1}}$ solve every remaining ESSP atom $(k, \cdot)$ in A_φ^τ because for every state $t \in \{t_{i,1}, \dots, t_{i,7}, t_{i,8}\}$ there is an event $X \in \{X_{i,0}, X_{i,1}, X_{i,2}\}$ such that $\neg(t \xrightarrow{X})$. Moreover, they prove (r_i, s) to be τ-solvable for all $i \in \{0, \dots, m - 1\}$ and all states $s \in S_{A_\varphi^\tau}$ except for $s \in \{m_0, \dots, m_3\}$.

For $X \in \{X_{i,0}, X_{i,1}, X_{i,2}\}$ $(x \in \{x_{i,0}, x_{i,1}, x_{i,2}\})$ and $t \in \{t_{i,0}, \dots, t_{i,7}\}$ with $\neg t \xrightarrow{X}$ $(\neg t \xrightarrow{x})$ we have $\xrightarrow{X} t$ $(\xrightarrow{x} t)$. Hence, for $X \in V(\varphi)$ the τ-regions $sup_{X,x}^1 = \{t_{n,0}, \dots, t_{n,7} \mid 0 \le n < m, X \in C_n\} \cup \{p_{n,0}, \dots, p_{n,3} \mid X \xrightarrow{} s \xleftarrow{h_n}\}$ and $sup_{X,x}^2 = S_{A_\varphi^\tau} \setminus \{m_0, \dots, m_4, p_{0,0}, \dots, p_{m-1,3}\}$ τ-solve every valid atom $(X, \cdot)$ and every valid atom $(x, \cdot)$ of its corresponding event x in A_φ^τ. Moreover, $sup_{X,x}^2$ τ-solves (r_i, s) for $i \in \{0, \dots, m - 1\}$ and $s \in \{m_0, \dots, m_4\}$ τ-solving all atoms $(r_i, \cdot)$, too. The τ-regions $sup_h^1 = S_{A_\varphi^\tau} \setminus (\{m_0, m_1, m_2\} \cup \{s \mid s \xrightarrow{k} s\})$ and $sup_h^2 = \{s_0, s_1, q, m_0, \dots, m_4, \}$ solve every atom $(h, \cdot)$. The τ-region $sup_{b_n}^1 = S_{A_\varphi^\tau} \setminus \{m_3, m_4, p_{0,0}, p_{0,3}, \dots, p_{m-1,0}, p_{m-1,3}\}$ and the τ-region $sup_{b_n}^1 = \{p_{n,0}, \dots, p_{n,3}\}$ solve every $(b_n, \cdot)$. Furthermore, the τ-region $\{s_0, p_{n,0}\}$ solves every $(a_n, \cdot)$. The τ-regions $sup_u^1 = \{s_0, q, m_0, m_4\} \cup \{t_{n,0}, \dots, t_{n,7}, p_{i,2}, p_{n,3} \mid 0 \le n < m\}$ and $sup_u^2 = \{p_{n,0}, \dots, p_{n,3} \mid 0 \le n < m\}$ settle that every possible atom $(u, \cdot)$ is τ-solvable. The τ-regions $sup_v^1 = \{m_1, m_2\} \cup \{s \mid s \xrightarrow{k} s\} \cup \{p_{i,0}, \dots, p_{n,3} \mid 0 \le n < m\}$ and $sup_v^2 = \{m_0, \dots, m_4\} \cup \{p_{n,0}, \dots, p_{n,1} \mid 0 \le n < m\}$ prove all atoms $(v, \cdot)$ to be τ-solvable. Finally, the τ-region $sup_c = \{m_0, m_3\}$, respectively $sup_a = \{s_0, m_0\}$, τ-solves every atom $(c, \cdot)$, respectively every atom $(a, \cdot)$.

It is easy to see that the state separating τ-regions defined above for the types of Theorem 2.1 can be used here for the types of Theorem 2.3 to separate the same states simply by replacing inp by res. Moreover, the states S_{A_φ} are clearly separable from the states $S_{A_\varphi^\times} \setminus S_{A_\varphi}$ as $S_{A_\varphi^\times}$ is itself a τ-support. The remaining SSP-atoms are τ-solved by $\{m_0\}, \{m_0, m_1\}, \{m_0, m_3\}, \{m_0, m_3, m_4\}, \{m_0, m_3, m_4\}$ and by $\{p_{i,0}\}, \{p_{i,0}, p_{i,1}\}, \{p_{i,0}, p_{i,1}, p_{i,2}\}$ for $i \in \{0, \dots, m - 1\}$. If used is not available, then free is and we can modify all τ-regions accordingly by inverting the support and interchanging set with res and used with free.   □

5   Conclusion

In this paper, we take the first step towards a full characterization of the computational complexity of synthesis for all 256 boolean types of nets. Beyond the only two previously known cases of elementary net systems [3] and flip-flop nets [14], we present the synthesis complexity for 42 *new* classes. It turns out that for 36 of these types synthesis can be done in polynomial time while it is NP-hard for the remaining 7 types.

It is noteworthy that, in particular the polynomial time results of Theorem 1 may also be exploitable for the synthesis of other types of nets discussed in literature and practice. For example, the type τ' of trace nets is defined by $\tau' = \{\mathsf{nop}, \mathsf{res}, \mathsf{inp}, \mathsf{set}, \mathsf{out}, \mathsf{used}, \mathsf{free}\}$ [4,6] and we already know from another analyses that τ'-synthesis is NP-hard [18]. Nevertheless, by Theorem 1.1, synthesis for the type $\tau = \{\mathsf{nop}, \mathsf{res}, \mathsf{inp}, \mathsf{used}, \mathsf{free}\}$ is doable in polynomial time and, moreover, we have that $E_\tau \subset E_{\tau'}$. Consequently, every τ-region of a TS A is a τ'-region, too. Hence, to solve the τ'-synthesis for A, one may previously use Algorithm 1 as part of a preprocessing method to solve as many of A's atoms as possible by τ-regions. This might significantly simplify the synthesis process or, as to be seen for our running example in Fig. 2, even solve τ'-synthesis completely.

References

1. van der Aalst, W.M.P.: Process Mining - Discovery, Conformance and Enhancement of Business Processes. Springer, Heidelberg (2011). https://doi.org/10.1007/978-3-642-19345-3

2. Badouel, E., Bernardinello, L., Darondeau, P.: Polynomial algorithms for the synthesis of bounded nets. In: Mosses, P.D., Nielsen, M., Schwartzbach, M.I. (eds.) CAAP 1995. LNCS, vol. 915, pp. 364–378. Springer, Heidelberg (1995). https://doi.org/10.1007/3-540-59293-8_207

3. Badouel, E., Bernardinello, L., Darondeau, P.: The synthesis problem for elementary net systems is NP-complete. Theor. Comput. Sci. **186**(1–2), 107–134 (1997). https://doi.org/10.1016/S0304-3975(96)00219-8

4. Badouel, E., Bernardinello, L., Darondeau, P.: Petri Net Synthesis. Texts in Theoretical Computer Science. An EATCS Series. Springer, Heidelberg (2015). https://doi.org/10.1007/978-3-662-47967-4

5. Badouel, É., Caillaud, B., Darondeau, P.: Distributing finite automata through Petri net synthesis. Formal Asp. Comput. **13**(6), 447–470 (2002). https://doi.org/10.1007/s001650200022

6. Badouel, E., Darondeau, P.: Trace nets and process automata. Acta Informatica **32**(7), 647–679 (1995). https://doi.org/10.1007/BF01186645

7. Cortadella, J., Kishinevsky, M., Kondratyev, A., Lavagno, L., Yakovlev, A.: A region-based theory for state assignment in speed-independent circuits. IEEE Trans. Comput.-Aided Des. Integr. Circuits Syst. **16**(8), 793–812 (1997). https://doi.org/10.1109/43.644602

8. Ehrenfeucht, A., Rozenberg, G.: Partial (set) 2-structures. Part I: basic notions and the representation problem. Acta Informatica **27**(4), 315–342 (1990). https://doi.org/10.1007/BF00264611

9. Goldmann, M., Russell, A.: The complexity of solving equations over finite groups. Inf. Comput. **178**(1), 253–262 (2002). https://doi.org/10.1006/inco.2002.3173

10. Kleijn, J., Koutny, M., Pietkiewicz-Koutny, M., Rozenberg, G.: Step semantics of Boolean nets. Acta Informatica **50**(1), 15–39 (2013). https://doi.org/10.1007/s00236-012-0170-2

11. Montanari, U., Rossi, F.: Contextual nets. Acta Informatica **32**(6), 545–596 (1995). https://doi.org/10.1007/BF01178907
12. Moore, C., Robson, J.M.: Hard tiling problems with simple tiles. Discret. Comput. Geom. **26**(4), 573–590 (2001). https://doi.org/10.1007/s00454-001-0047-6
13. Pietkiewicz-Koutny, M.: Transition systems of elementary net systems with inhibitor arcs. In: Azéma, P., Balbo, G. (eds.) ICATPN 1997. LNCS, vol. 1248, pp. 310–327. Springer, Heidelberg (1997). https://doi.org/10.1007/3-540-63139-9_43
14. Schmitt, V.: Flip-flop nets. In: Puech, C., Reischuk, R. (eds.) STACS 1996. LNCS, vol. 1046, pp. 515–528. Springer, Heidelberg (1996). https://doi.org/10.1007/3-540-60922-9_42
15. Tarjan, R.E.: Finding optimum branchings. Networks **7**(1), 25–35 (1977). https://doi.org/10.1002/net.3230070103
16. Thiagarajan, P.S.: Elementary net systems. In: Brauer, W., Reisig, W., Rozenberg, G. (eds.) Petri Nets: Central Models and Their Properties. LNCS, vol. 254, pp. 26–59. Springer, Heidelberg (1986). https://doi.org/10.1007/BFb0046835
17. Tredup, R., Rosenke, C.: Narrowing down the hardness barrier of synthesizing elementary net systems. In: Schewe, S., Zhang, L. (eds.) 29th International Conference on Concurrency Theory, CONCUR 2018. LIPIcs, Beijing, China, 4–7 September 2018, vol. 118, pp. 16:1–16:15. Schloss Dagstuhl - Leibniz-Zentrum fuer Informatik (2018). https://doi.org/10.4230/LIPIcs.CONCUR.2018.16
18. Tredup, R., Rosenke, C.: Towards completely characterizing the complexity of Boolean nets synthesis. CoRR abs/1806.03703 (2018). http://arxiv.org/abs/1806.03703
19. Tredup, R., Rosenke, C., Wolf, K.: Elementary net synthesis remains NP-complete even for extremely simple inputs. In: Khomenko, V., Roux, O.H. (eds.) PETRI NETS 2018. LNCS, vol. 10877, pp. 40–59. Springer, Cham (2018). https://doi.org/10.1007/978-3-319-91268-4_3

Space Lower Bounds
for Graph Stream Problems

Paritosh Verma[✉]

Birla Institute of Technology & Science, Pilani, Pilani, India
paritoshverma97@gmail.com

Abstract. This work concerns with proving space lower bounds for graph problems in streaming model. It is known that the single source shortest path problem in streaming model requires $\Omega(n)$ space, where $|V| = n$. In the first part of the paper we try find whether the same lower bound hold for a similar problem defined on trees. We prove lower bounds for single and multi pass version of the problem.

We then apply the ideas used in above lower bound results to prove space lower bounds (single and multipass) for other graph problems like finding min s-t cut, detecting negative weight cycle and finding whether two nodes lie in the same strongly connected component.

1 Introduction

Streaming model is a computation model in which data arrives in the form of a stream. Unlike the RAM model, random access of input is not allowed in this model. This study concerns with space lower bounds for streaming problems which take graphs as input. Graph streaming algorithms find their use in applications where the size of input graph is too large to be stored in a single machine or where the data naturally arrives in an order for example, network packets arriving in a router. They can also be used when the input graph is dynamic. Studying graph streaming algorithms also yields insights into complexity of stream computation [3].

Communication complexity is a concept from information theory that is used as a tool in many space lower bound results for problems in streaming model [1,2,5]. Such lower bound proofs rely on capturing the underlying information exchange/communication happening during the stream computation.

Communication problems like Index problem, set disjointness problem, pointer chasing have been used to produce many known lower bounds for streaming problems [1,2,5]. In this study we use known communication complexity lower bound results to prove space lower bounds for graph streaming problems. For some of our lower bound results, we use a technique that does not rely on communication complexity.

In Sect. 2, we define all the communication problems and their known lower bound results that are used in this work. Most of our multipass space lower

---

This work was done as a part of author's undergradute thesis.

© Springer Nature Switzerland AG 2019
T. V. Gopal and J. Watada (Eds.): TAMC 2019, LNCS 11436, pp. 635–646, 2019.
https://doi.org/10.1007/978-3-030-14812-6_39

bound proofs use the communication complexity lower bound results proved in [2].

In Sect. 3, we prove single and multi pass space lower bounds for the problem of finding depth of a node in a tree being streamed. The motivation to study this problem is that it is a simpler version of the general shortest path problem in graphs and study of this problem can yield insights into the latter problem. We prove a $\Omega(n.\log n)$ lower bound for the single pass version of the problem which also applies to the shortest path problem. To the best of author's knowledge, such a bound is not known for the shortest path problem in streaming model.

In Sects. 4 and 5, we use the ideas used in the above proofs to obtain single and multi pass lower bounds for the problem of finding min s-t cuts in a graph and detecting negative weight cycles.

2   Preliminaries

2.1   Streaming Model and Communication Complexity

In the streaming model the data is presented in the form of a stream i.e. data arrives in an order and random access on the input is not permitted. The space available for the algorithm is also limited. This model is useful in modelling scenarios in big data processing and cloud computing.

A p-pass (or multi pass) streaming algorithm refers to an algorithm to which input is streamed p times (or many number of times). If the input is streamed only once, the streaming algorithm is called single pass.

Communication complexity is a concept from Information theory that has been used to prove space lower bounds for many problems in streaming model.

Communication complexity of a function f is defined as the worst case communication (over all inputs) required by the best communication protocol for the following communication problem - the input of the function f is divided amongst different entities which can communicate only through a channel and the goal is to compute the value of the function f using minimum communication [1,5]. Next subsection describes a specific communication complexity model.

Yao's Communication Model. This model consists of two players Alice and Bob (can be n players in the general case). Alice and bob can communicate to each other via a channel. Alice has a binary string $x \in \{0,1\}^n$ and Bob has a binary string $y \in \{0,1\}^n$ such that both the players are unaware of the other person's string.

Both of them are interested in computing the value $f(x,y)$ where f is function of strings of both the players [5]. Both of them can apriori agree on a communication protocol which they will use in order to compute the value $f(x,y)$. A trivial protocol could be that Alice sends her input x to Bob via the communication channel and Bob upon receiving x computes $f(x,y)$ which he then passes on to Alice. Both the players Alice and Bob are assumed to be computationally unbounded.

Communication complexity of a function $f(x, y)$, $CC(f)$ is the minimum amount of bits required to be transferred through the channel by any communication protocol (for computing $f(x, y)$) in the worst case. The communication complexity of a function is in general difficult to compute because the first quantifier in it's definition is over all possible protocols.

$$CC(f) = \min_{\forall \text{ protocols } P} \max_{\forall x, y} (\text{bits communicated to compute } f(x, y) \text{ using } P)$$

Most space lower bound proof involving communication complexity use the following idea:

- For a given streaming problem S identify a underlying communication problem C i.e. a communication problem that can be reduced to the given streaming problem. Which means that using an algorithm A for S one should be able to construct a protocol for the communication problem C.
- Prove a communication complexity lower bound for the communication problem C.
- Translate the communication complexity lower bound for C to space lower bound for streaming problem P. This step is based on the construction of the reduction.

The above idea is central to most space lower bounds results for streaming algorithms [1,2,4–6]. The following communication problems and their known communication complexity lower bounds are used in this work.

2.2 Index Problem

In this problem there are two players Alice and Bob. Alice has an array A, $A \in \{0, 1\}^n$ and Bob has $i \in [n]$ (where $[n]$ represents $\{0, 1\}^n$). Both Alice and Bob are not aware of the other player's input and one way communication from Alice to Bob is allowed through a channel. However, Bob is not allowed to communicate to Alice. Bob wants to find out the value stored at the ith position of Alice's string, A_i. It is known that the one way communication complexity of the index problem is $\Omega(n)$ [5].

Pointer and Set Chasing. A (p, r)-communication problem is defined as a communication problem which consists of p players $P_1, P_2 \ldots P_p$. Players communicate for r rounds and are constrained to speak in the order $P_1 \to P_2 \to \ldots$ to $P_p \to P_1 \to P_2$ and so on. In the last round the player P_p has to output the required value to be computed.

If $f : [n] \to 2^{[n]}$ be a function mapping the set $[n] = \{1, 2, 3 \ldots n\}$ to $2^{[n]}$ (power set of $[n]$), then a function $f' : 2^{[n]} \to 2^{[n]}$ can be defined using f as follows [2]:

$$f'(S) = \bigcup_{i \in s} f(i)$$

Pointer Chasing. Pointer chasing problem $PC_{n,p}$ for positive integers n and p, is defined as a $(p, p-1)$-communication problem where $\forall i \in [p]$ player P_i has a function $f_i : [n] \to [n]$. They are interested in computing the value $f_1(f_2(f_3(\ldots f_p(1))))$ [2,4].

Theorem 1. *Any randomized communication protocol that solves the pointer chasing problem $PC_{n,p}$ with error probability at most $1/10$ must require at least $\Omega(n/p^4 - p^2 \log(n))$ bits of communication [4].*

Set Chasing Intersection Problem. The set chasing problem $SC_{n,p}$ is defined similarly, as a $(p, p-1)$-communication problem where the ith player P_i has a function $f_i : [n] \to 2^{[n]}$ and they are interested in computing the value $f'_1(f'_2(f'_3 \cdots f'_p(\{1\})))$.

For pointer chasing and set chasing problem become easy once the number of rounds are increased from $p-1$ to p or if the order of communication is inverted. The set chasing intersection problem $INTERSECT(SC_{n,p})$ is a $(2p, p-1)$-communication problem in which the first p players have one instance of the set chasing problem and the other p players have another instance of the set chasing problem. In all the $p-1$ rounds they can communicate only in the order $P_1 \to P_2 \to P_3 \to P_4 \cdots \to P_{2p} \to P_1$ and so on. Finally the goal is to check whether $f'_1(f'_2(f'_3 \cdots f'_p(\{1\}))) \cap f'_{p+1}(f'_{p+2}(f'_{p+3} \cdots f'_{2p}(\{1\}))) = \phi?$, in other words they are interested in knowing whether the output of the two set chasing instances intersect or not [2].

The following communication complexity lower bound on the set intersection problem has been proved.

Theorem 2. *For some positive constant p such that $1 < p \leq \log(n)/(\log(\log(n)))$, any randomized communication protocol that solves the $INTERSECT(SC_{n,p})$ problem with a probability greater than $9/10$ requires $\Omega\big(n^{1+1/2(p+1)}/(p^{16} \cdot \log^{3/2} n)\big)$ [2].*

Using this result, space lower bounds for multi pass streaming algorithm for the problems like finding perfect matching, checking if there exists a directed path between two vertices has been proved in [2].

3   Finding Depth of a Node in a Tree

The problem that is considered here is is the following - Let T be a rooted tree whose root is denoted by a known symbol r and u is some node in the tree. Given a stream σ consisting of the node u followed by edges of the tree T, the problem is to compute the depth of the node u in T.

This problem is a simpler version of the general problem of finding distance between two nodes in a graph.

3.1 Multipass Lower Bound

Using the communication complexity lower bound of the pointer chasing problem we prove the following multi pass space lower bound for the mentioned problem.

Claim 1. *Any randomized p-pass streaming algorithm that computes the depth of a given node in a rooted tree with error probability at most $1/10$ must require at least $\Omega((n/p^7) - \log(n/p))$ space.*

Proof. Given a p-pass streaming algorithm A for finding the depth of a node in a rooted tree that uses s bits of space we design a communication protocol that solves the pointer chasing problem $PC_{n,p+1}$ using $s.\Theta(p^2)$ bits of communication (through channel between players).

The construction of the communication protocol is based on the idea of visualizing the computation performed in the pointer chasing problem as a graph. The function f_i of each player P_i, $\forall i \in [p + 1]$ can be visualized as bipartite graphs. As a result, the composition of the functions f_i can be viewed as side by side concatenation of these bipartite graphs as shown in Fig. 1. In this view, the goal of the pointer chasing problem is to find the node to which the bold edges emerging from the node n_1 lead. The node to which the bold edges lead corresponds to the value $f_1(f_2(f_3(\ldots f_{p+1}(1))))$.

Given the algorithm A in each of the p rounds all the $p+1$ players will stream the edges of the bipartite graph (corresponding to their function f_i) to A and then they will pass the memory transcript of A to the next player according to the communication order defined in pointer chasing problem. At the end of each round player P_{p+1} will stream the edges shown in blue (see Fig. 1) in addition to the edges corresponding to f_{p+1}. All the edges that are streamed to A can be viewed as a tree rooted at node N (see Lemma 1 below). This means that after p round of communication, the value of $f_1(f_2(f_3(\ldots f_{p+1}(1))))$ can be found out by finding the depth of the node n_1 (see Fig. 3) in the computation graph/tree. Let $d(v)$ denote the depth of a node v and x be the node corresponding to $.f_1(f_2(f_3(\ldots f_{p+1}(1))))$. Then, $d(x) = d(n_1) - 1 - p$. The depth of node n_1 is the output algorithm A gives at the end of the last round. Hence by knowing the value of $d(n_1)$ we can determined $d(x)$ thereby determining the node x itself. This is because each of the rightmost node (Fig. 1) has different depth with respect to root N.

The communication complexity of this protocol is $s.\Theta(p^2)$ as the size of the memory transcript is s and order of the total messages sent between players is $\Theta(p^2)$ ($\Theta(p)$ messages in p rounds). According to the communication complexity lower bound of the pointer chasing problem

$$s.\Theta(p^2) = \Omega\left(\frac{(n/p)}{p^4} - p^2 \log(n/p)\right)$$

$$s = \Omega\left(n/p^7 - \log(n/p)\right)$$

Where n is the total number of nodes in the computation graph and n/p represents the domain of functions f_i.

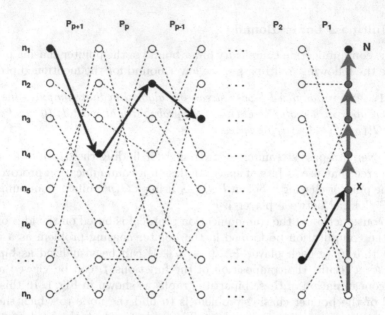

Fig. 1. Computation graph G for $PC_{n,p+1}$, with blue edges added for node depth lower bound (Color figure online)

Lemma 1. *The computation graph G for the pointer chasing problem is a tree.*

Proof. Let's assume that G has a cycle $\mathcal{C}$. Each edge of the cycle C can be classified into either a blue edge of an edge corresponding to some function f_i, $\forall i \in [p+1]$. Also, the cycle cannot be completely composed of blue edges. Due to this, we can find the minimum i, such that edge corresponding to f_i is in $\mathcal{C}$. Call this i'. Since $\mathcal{C}$ is a cycle, there will be two edges corresponding to function $f_{i'}$, which is a contradiction since $f_{i'}$ is a function. Hence no cycle can exist.

This result also shows that any one pass streaming algorithm must require at least $\Omega(n)$ space to compute the depth of a given node in a tree being streamed (for $p = 1$). In the next section we prove a stronger $\Omega(n.\log(n))$ lower bound for the single pass version of the problem.

3.2 Stronger Bound for the Single Pass

In this section we prove a stronger space lower bound for the single pass version of the problem. The given lower bound proof does not follow the standard reduction procedure that is used to prove most space lower bound results in streaming model.

Claim 2. *Any one pass streaming algorithm that computes the depth of a given node in a rooted tree must require $\Omega(n.log(n))$ space.*

Proof. To establish this result we first choose a particular input instance and then we prove a lower bound on the space required by the best algorithm for that instance, this value according to yao's minimax principle serves as a space lower bound for the problem.

The input instance we consider is the instance in which the node v (whose depth is to be calculated) is a leaf node and the edge connecting v to its parent in the tree $(v, p(v))$ arrives as the last edge of the stream. The idea here is that the essential information required to compute the depth of the node v i.e. the location of the node v in the tree is deferred in the input stream.

Let A be some algorithm that computes the depth of a given node. Now consider the memory transcript $\mathcal{M}$ of the algorithm A when it has processed all the edges except the last edge $(v, p(v))$ of the constructed input instance. Now we claim that $|\mathcal{M}| = \Omega(n.\log(n))$, proving this is sufficient to prove the claim.

To prove $|\mathcal{M}| = \Omega(n.\log(n))$ we argue that given the memory transcript $\mathcal{M}$ and the algorithm A one can recover the depths of all nodes in $V \setminus \{v\}$ using $2|\mathcal{M}|$ space. Given $\mathcal{M}$ and A one can run the algorithm $n - 1$ times, each time continuing the computation on $\mathcal{M}$ and streaming the edge $(v, u) \; \forall u \in V \setminus \{v\}$. From all of these runs we can recover the depths of all the nodes in the set $V \setminus \{v\}$, as depth of node u is one less than depth of node v and $u \in V \setminus \{v\}$. To complete the proof we show that the depth information of all the nodes of a tree requires $\Theta(n.\log(n))$ bits to store which implies that $|\mathcal{M}| = \Omega(n.\log(n))$. Let $\mathcal{D}$ denote the number of different functions $d : V \to \mathbb{Z}^+$ such that $d(v)$ represents the depth of the node $u \in V \setminus \{v\}$.

Let $\mathcal{D}_i$ denote the number of different functions $d : V \to \mathbb{Z}^+$ such that $d(v)$ represents the depth of the node and $\max_{v \in V} d(v) = i$ and $S(n,k)$ is stirling number of second kind.

$$|\mathcal{M}| = \Omega(\log_2(\mathcal{D}))$$

$$|\mathcal{M}| = \Omega(\log_2(\sum_{i \in [n-2]} \mathcal{D}_i))$$

$$|\mathcal{M}| = \Omega(\log_2(\sum_{i \in [n-2]} n.S(n-1,i)))$$

$$|\mathcal{M}| = \Omega(\log_2(n.\sum_{i \in [n-2]} i^{n-i}))$$

$$|\mathcal{M}| = \Omega(\log_2(n.(n/2)^{n/2}))$$

$$|\mathcal{M}| = \Omega(n.\log_2(n))$$

Corollary 1. *Any streaming one pass algorithm that computes the distance between two nodes in a graph requires at least $\Omega(n.\log_2(n))$ space.*

This above corollary is applicable even for the shortest path problem, that is any streaming algorithm that finds the distance between two given vertices of a graph must require $\Omega(n.\log n)$ space.

Using the same idea of deferring the essential input in the stream, we prove a space lower bound for the problem of min s-t cut in weighted graph.

4   Min $s - t$ Cut Problem

To prove this lower bound we use the following known lower bound for the unweighted min cut problem.

4.1   Single Pass Lower Bound

Theorem 3. *Any one pass streaming algorithm that computes the min cut of a unweighted graph requires at least $\Omega(n^2)$ memory [6].*

The input for this problem is a weighted graph stream i.e. the individual tokens of the stream are weighted edges of the form $((u,v), w)$ where w is the weight of the edge (u, v).

Claim 3. *Any one pass streaming algorithm that computes the min s-t cut value for given pair of nodes s and t in a weighted graph must require at least $\Omega(n^2)$ space.*

Proof. Let A be any streaming algorithm for the weighted min s-t cut problem and the space required by the algorithm be s. Now we use this algorithm A to construct a streaming algorithm A_s for solving the unweighted min cut problem that uses $\Theta(s)$ space. Since A_s requires $\Theta(s)$ space, using Theorem 3 implies that $s = \Omega(n^2)$, which completes the proof.

Let $G = (V, E)$ be the graph whose min cut is to be calculated, then we construct another weighted graph $G' = (V \cup \{u, v\}, E \cup \{(u, x), (v, y)\})$, for $x, y \in V$ such that weight of the edges (u, x) and (v, y) is n and every other edge weighs one. These weights ensure that the edges (u, x) and (v, y) never lie in the min u-v cut of G'. This implies that min u-v cut of graph G' is same as the min x-y cut of graph G, G being an unweighted graph.

Now to compute the min cut for G, the instance G' is streamed to the algorithm A multiple times, one for each $y \in V \setminus \{x\}$, for a fixed value of x. The graph G' is streamed in such a way that the edge (v, y) arrives as the last edge of the stream. The min cut value of graph G is calculated by taking the min value of the all the min u-v cut values calculated for the graph G' (same as min x-y cut for G).

All the $n - 1$ min cut values (corresponding to $y \in V \setminus \{x\}$) that are required to be computed can be computed using $2s$ space as the memory transcript $\mathcal{M}$ of the algorithm A - after it has processed all the edges $E \cup \{(u, x)\}$, can be reused to compute all the min u-v (x-y) cuts $\forall y \in V \setminus \{x\}$.

To calculate the value of min x-y' cut for some $y' \in V \setminus \{x\}$, the memory transcript $\mathcal{M}$ is used along with the algorithm A and the edge (y', v) is streamed to A, then the value computed by the algorithm A is the min u-v cut for G' (or min x-y' cut for G). The memory transcript can be copied and used similarly $n - 1$ times for each $y' \in V \setminus \{x\}$, using $2s$ space.

This leads to a $2s$ space streaming algorithm that computes min cut of a unweighted graph being streamed, it implies that $s = \Omega(n^2)$. Which means that for all possible algorithms A, the space required must be $\Omega(n^2)$.

In the next section, using the known communication complexity lower bound for the set chasing intersection problem we prove a multi pass space lower bound for the unweighted min cut problem.

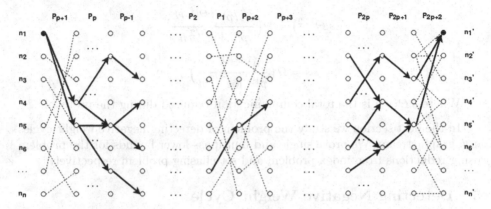

Fig. 2. Computation graph for $INTERSECT(SC_{n,p+1})$, directions are used to indicated paths emerging and leading to nodes n_1 and n'_1. All the edges corresponding to functions f'_i are not shown for clarity.

4.2 Multi Pass Lower Bound

Claim 4. *Any p pass streaming algorithm that computes the min s-t cut of a unweighted graph must require $\Omega\left(n^{1+1/2(p+1)}/(p^{19} \cdot \log^{3/2} n)\right)$ bits of space.*

Proof. Suppose there exists such an algorithm A that uses at most s bits of memory. Then we can use this algorithm to design a protocol for the set chasing intersection problem $INTERSECT(SC_{n,p+1})$ that uses $s.\Theta(p^2)$ bits of communication.

The computation performed in the set chasing intersection problem can be visualized as a computation graph, shown in Fig. 2. For all the $2(p+1)$ players their functions f'_i can be viewed as bipartite graphs.

We use the algorithm A to compute the min n_1-n'_1 cut of the computation graph G (shown in Fig. 2). This is done as follows. Each player P_i uses the algorithm A and streams the edges corresponding to their function f_i into the algorithm (as the individual functions can be visualized as a bipartite graph) and passes the memory transcript to the next player according to the communication order constraint. In this view, any path going from node n_1 to n'_1 would represent an element in the set $f'_1(f'_2(f'_3 \ldots f'_{p+1}(\{1\}))) \cap f'_{p+2}(f'_{p+3}(f'_{p+4} \cdots f'_{2p+2}(\{1\})))$. Now we can claim that by knowing the value of min n_1-n'_1 cut we can find out whether the two instances of set chasing intersect. This is because if the two instance of set chasing do not intersect then the size of min n_1-n'_1 cut would be zero as there is not path from the vertex n_1 to n'_1. If on the other hand the outputs of the two set chasing instances intersect then the min cut value would be greater than zero as the intersection would yield a path from n_1 to n'_1. Hence by checking whether the min cut value is zero or not one can solve the set chasing intersection problem. Using the communication complexity lower bound:

$$s.\Theta(p^2) = \Omega\left(\frac{(n/p)^{1+\Theta(1/p)}}{p^{16}.\log^{3/2} n}\right)$$

$$s = \Omega\left(\frac{n^{1+\Theta(1/p)}}{p^{19}.\log^{3/2} n}\right)$$

Where $s.\Theta(p^2)$ is the total communication required during the protocol.

In the next section we study the problem of detecting negative weight cycles in a graph stream. We prove single and multipass lower bounds for the problem using reductions from index problem and set chasing problem respectively.

5 Detecting Negative Weight Cycle

5.1 Single Pass Lower Bound

Using the known communication complexity lower bound for the index problem we first prove a single pass space lower bound for detecting negative weight cycle.

Claim 5. *Any streaming algorithm that can detect the presence of a negative weight cycle in a weighted graph stream must use at least $\Omega(n^2)$ space.*

Proof. Let C be a streaming algorithm that detects the presence of a negative weight cycle using s space. Then C can be used to design a communication protocol for the index protocol as follows.

Let (A, i) be an instance of index problem in which A is of the size $\Theta(n^2)$, then the binary string A can be interpreted as a graph on n vertices. Alice can stream the edges of this graph to the algorithm C associating with each edge a weight of positive one unit. Then she can send the memory transcript obtained (of size s) to Bob.

Let (a, b) be the edge corresponding to Bob's input i. After receiving the memory transcript Bob streams the edges (a, v) and (b, v) to the streaming algorithm associating with each edge a weight of negative one.

Bob can now find the ith bit of Alice's string by knowing whether a negative weight cycle is present or not. This is due to the fact that if edge (a, b) is present in the graph, the vertices a, b and v form a cycle of weight negative one. If the edge (a, b) is not present in the graph then the minimum length of the cycle containing the negative weight edges is 4 which means that it's weight will be non negative, hence negative weight cycle will not exist. Negative weight cycle exists if and only if edge (a, b) is present in the graph i.e. when $A_i = 1$.

This leads to a s bit one way communication protocol that solves the index problem. It means that according to the lower bound result of Index problem, $s = \Omega(n^2)$.

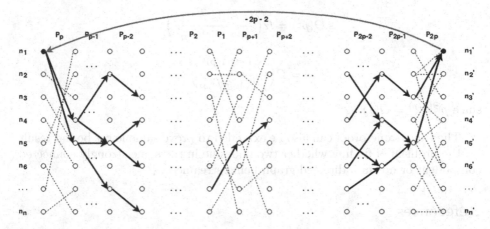

Fig. 3. Computation graph for $INTERSECT(SC_{n,p+1})$, blue edge is added by the player P_{2p} in every round. (Color figure online)

5.2 Multi Pass Lower Bound

A multi pass lower bound for the same problem can be proved using the communication complexity lower bound for the set chasing intersection problem.

Claim 6. *Any p pass streaming algorithm that detects the presence of a negative weight cycle in a weighted graph being streamed must require at least $\Omega\big(n^{1+1/2(p+1)}/(p^{19}.\log^{3/2} n)\big)$ bits of space.*

Proof. Suppose there exists a p-pass streaming algorithm A for the problem of detecting a negative weight cycle which uses at most s bits of space. Then such an algorithm can be used to design a protocol for the set chasing intersection problem $INTERSECT(SC_{n,p+1})$ as follows:

In each of the p rounds all the $2(p+1)$ players stream edges the edges corresponding to their function (as shown in Fig. 3) and pass the resultant memory transcript of the algorithm to the next player. The weight of one is assigned to every edge being streamed. The player $P_{2(p+1)}$ also adds the edge shown in blue (see Fig. 3) having weight $-2(p+1) - 1$ in every round.

After the completion of p rounds, the set chasing problem can be solved by asking whether the graph that is streamed to algorithm A has a negative weight cycle or not. This is because there exists a negative weight cycle in the graph if and only if the output sets intersect in the set chasing problem. As if the output set does not intersect, there is no path from node n_1 to n_1' which means no negative weight cycle. If the output sets intersect, we have a cycle of weigh -1, consisting of the blue edge (weight $-2(p+1) - 1$) and a path of $2(p+1)$ edges (from n_1 to n_1') of weight 1 each.

Since the size of the memory transcript is s bits and in each of the p rounds the memory transcript is transferred $\Theta(p)$ times, the total communication done by the protocol is $s.\Theta(p^2)$ which according to the lower bound on communication complexity of the set chasing intersection problem means that:

$$s.\Theta(p^2) = \Omega\Big(\frac{(n/p)^{1+\Theta(1/p)}}{p^{16}.\log^{3/2} n}\Big)$$

$$s = \Omega\Big(\frac{n^{1+\Theta(1/p)}}{p^{19}.\log^{3/2} n}\Big)$$

since $p^{\Theta(1/p)} = O(1)$.

The exact same proof can also be extended to prove same lower bound result for the problem of finding whether two nodes lie in the same strongly connected component or not in a directed graph being streamed.

References

1. Chakrabarti, A.: Data Stream Algorithms, Lecture Notes, Fall 2011, October 2014
2. Guruswami, V., Onak, K.: Superlinear lower bounds for multipass graph processing. Algorithmica **76**(3), 654–683 (2016). https://doi.org/10.1007/s00453-016-0138-7
3. McGregor, A.: Graph stream algorithms: a survey. SIGMOD Rec. **43**(1), 9–20 (2014). https://doi.org/10.1145/2627692.2627694
4. Nisan, N., Wigderson, A.: Rounds in communication complexity revisited. SIAM J. Comput. **22**(1), 211–219 (1993). https://doi.org/10.1137/0222016
5. Roughgarden, T.: Communication complexity for algorithm designers. Found. Trends Theor. Comput. Sci. **11**(3–4), 217–404 (2016). https://doi.org/10.1561/0400000076
6. Zelke, M.: Intractability of min- and max-cut in streaming graphs. Inf. Process. Lett. **111**(3), 145–150 (2011). https://doi.org/10.1016/j.ipl.2010.10.017

Bounded Jump and the High/Low Hierarchy

Guohua Wu[1] and Huishan Wu[2(✉)]

[1] Division of Mathematical Sciences, School of Physical and Mathematical Sciences,
Nanyang Technological University, Singapore, Singapore
guohua@ntu.edu.sg
[2] School of Information Science,
Beijing Language and Culture University, Beijing, China
huishanwu@blcu.edu.cn

Abstract. The notion of bounded-jump operator, $A^\dagger$, was proposed by Anderson and Csima in paper [1], where they tried to find an appropriate jump operator on weak-truth-table (*wtt* for short) degrees. For a set A, the bounded-jump of A is defined as the set $A^\dagger = \{e \in \mathbb{N} : \exists i \leq e[\varphi_i(e) \downarrow$ & $\Phi_e^{A\upharpoonright\varphi_i(e)}(e) \downarrow]\}$. In [1], Anderson and Csima pointed out this bounded-jump operator † behaves likes Turing jump $'$, like (1) $\emptyset^\dagger$ and $\emptyset'$ are 1-equivalent, (2) for any set A, $A <_{wtt} A^\dagger$, and (3) for any sets A, B, if $A \leq_{wtt} B$, then $A^\dagger \leq_{wtt} B^\dagger$. A set A is bounded-low, if $A^\dagger \leq_{wtt} \emptyset^\dagger$, and a set $B \leq_{wtt} \emptyset^\dagger$ is bounded-high if $\emptyset^{\dagger\dagger} \leq_{wtt} B^\dagger$. Anderson, Csima and Lange constructed in [2] a high bounded-low set and a low bounded-high set, showing that the bounded jump and Turing jump can behave very different. In this paper, we will answer several questions raised by Anderson, Csima and Lange in their paper [2] and show that:
(1) there is a bounded-low c.e. set which is low, but not superlow;
(2) $0'$ contains a bounded-low c.e. set;
(3) there are bounded-low c.e. sets which are high, but not superhigh;
(4) there are bounded-high sets which are high, but not superhigh.
In particular, we will develop new pseudo-jump inversion theorems via bounded-low sets and bounded-high sets respectively.

Keywords: Bounded jump · High/low hierarchy ·
Bounded high/low sets · Pseudo-jump inversion

1 Introduction

The notion of bounded-jump operator, $A^\dagger$, was proposed by Anderson and Csima in paper [1], where they tried to find an appropriate jump operator on weak-truth-table (*wtt* for short) degrees. For a set A, the bounded-jump of A is defined

Guohua Wu is partially supported by M4020333 (MOE2016-T2-1-083(S)), M4011672 (RG32/16) and M4011274 (RG29/14) from Ministry of Education of Singapore. Huishan Wu is supported by Science Foundation of Beijing Language and Culture University (by "the Fundamental Research Funds for the Central Universities") (Grant No. 18YBB19).

T. V. Gopal and J. Watada (Eds.): TAMC 2019, LNCS 11436, pp. 647–658, 2019.
https://doi.org/10.1007/978-3-030-14812-6_40

as the set $A^\dagger = \{e \in \mathbb{N} : \exists n \le e[\varphi_n(e) \downarrow \ \& \ \Phi_e^{A\restriction\varphi_n(e)}(e) \downarrow]\}$, where $A\restriction_z = \{x \in A : x \le z\}$. Note that $A^\dagger \le_T A \oplus \emptyset'$. In [1], Anderson and Csima pointed out that this bounded-jump operator † behaves like the Turing jump $'$ does, i.e., (1) $\emptyset^\dagger$ and $\emptyset'$ are 1-equivalent, (2) for any set A, $A <_{wtt} A^\dagger$, and (3) for any sets A, B, if $A \le_{wtt} B$, then $A^\dagger \le_{wtt} B^\dagger$. (2) implies that $A^\dagger \equiv_T A$ when $A \ge_T \emptyset'$. In the same paper, Anderson and Csima proved an analogue of Shoenfield's jump inversion theorem for the bounded-jump operator †.

Say that a set A is *bounded-low* if $A^\dagger \le_{wtt} \emptyset^\dagger$, and *bounded-high* if $A \le_{wtt} \emptyset^\dagger$ and $\emptyset^{\dagger\dagger} \le_{wtt} A^\dagger$. It is easy to see that A is bounded-low if and only if $A^\dagger$ is ω-c.e. Thus, all superlow sets are bounded-low, because for all superlow sets A, $A^\dagger \le_1 A' \le_{tt} \emptyset' \equiv_1 \emptyset^\dagger$. Also note that $\emptyset' \equiv_1 \emptyset^\dagger$ is bounded-high, because $\emptyset^{\dagger\dagger} \le_{wtt} (\emptyset')^\dagger$.

Anderson, Csima and Lange considered in [2] the interaction between the bounded-jump operator and the Turing jump, and constructed a high bounded-low set, and asked *whether bounded-lowness and superlowness agree with each other on low c.e. sets*. In this paper, we will give a negative answer to this question by showing the existence of bounded-low c.e. sets, which are low, but not superlow.

Theorem 1. *There are low c.e. sets which are bounded-low but not superlow.*

After showing that superlow sets are always bounded-low, Anderson, Csima and Lange [2] asked *whether superhigh sets are also bounded-high*. Our second result in this paper will provide a negative answer to this question. That is, we show the existence of bounded-low c.e. sets which are superhigh.

Theorem 2. $\mathbf{0}'$ *contains a bounded-low c.e. set.*

Note that $\emptyset'$ is a bounded-high set, and it is well-known that $\emptyset'$ is also superhigh. In [2], Anderson, Csima and Lange further asked whether there are bounded-high sets which are high but not superhigh. Similarly, based on Theorem 2, we can also ask whether there are bounded-low sets which are high but not superhigh.

One way to obtain high, but not superhigh c.e. sets is through the pseudo-jump inversion theorem of Jockusch and Shore [3], which was used by Mohrherr in her paper [4]. Let W_e be a c.e. operator, the corresponding pseudo-jump operator V_e is defined as $V_e^A = A \oplus W_e^A$ for a given set A. The pseudo-jump inversion theorem of Jockusch and Shore says that for any c.e. operator W_e, there is a c.e. set C such that $V_e^C \equiv_T \emptyset'$.

The procedure of obtaining high but not superhigh c.e. sets are as follows:

(1) Through the construction of low but not superlow c.e. sets, one can obtain a c.e. operator W such that for any set A, $A \oplus W^A$ is low over A, but not superlow over A, i.e., $(A \oplus W^A)' \le_T A'$, but $(A \oplus W^A)' \not\le_{tt} A'$;

(2) By applying the pseudo-jump inversion theorem to this operator W, there exists a c.e. set C such that $C \oplus W^C \equiv_T \emptyset'$.

Then $\emptyset'' \equiv_1 (C \oplus W^C)' \leq_T C'$ (hence, C is high), and $\emptyset'' \equiv_1 (C \oplus W^C)' \nleq_{tt} C'$ (hence, C is not superhigh).

In this paper, we will provide two strengthened versions of Jockusch and Shore's pseudo-jump inversion theorem.

Theorem 3. *For any c.e. operator W,*

(1) there is a bounded-low c.e. set C such that $C \oplus W^C \equiv_T \emptyset'$;
(2) there is a bounded-high set C such that $C \oplus W^C \equiv_T \emptyset'$.

We have the following consequence immediately:

Corollary 1. *There are bounded-low c.e. sets which are high but not superhigh, and there are bounded-high sets which are high but not superhigh.*

Proof. Let W be the c.e. operator such that for any set A, $A \oplus W^A$ is low but not superlow over A, then any C satisfying $C \oplus W^C \equiv_T \emptyset'$ is high but not superhigh. By Theorem 3, there are bounded-low c.e. set C_1 and bounded-high set C_2 such that $C_1 \oplus W^{C_1} \equiv_T \emptyset'$ and $C_2 \oplus W^{C_2} \equiv_T \emptyset'$. Thus C_1 is a bounded-low c.e. set which is high but not superhigh, while C_2 is a bounded-high set which is high but not superhigh.

The organization of this paper is as follows. We will give a proof of Theorem 1 in Sect. 2, and a proof of Theorem 2 in Sect. 3. We will provide basic idea for proving Theorem 3, two strengthened versions of Jockusch and Shore's pseudo-jump inversion theorem, in Sect. 4.

2   A Bounded-Low c.e. Set which is Low, But not Superlow

To prove Theorem 1, we will construct a low c.e. set A such that $A^\dagger$ is ω-c.c. (so A is bounded-low), and A' is not ω-c.e. (so A is not superlow). One approach of constructing a set low, but not superlow, is to apply Sacks splitting theorem, to split $\emptyset'$ into two low c.e. sets A and B. Both A and B cannot be superlow, as if A is superlow, then $\emptyset'$ is wtt-reducible to B, by a theorem of Bickford and Mills [5]. In [5], Bickford and Mills also pointed out that there are superlow c.e. sets A and B such that $A \oplus B$ computes $\emptyset'$. Most of the results in [5] can be found in Nies' book [6]. One feature of Sacks splitting is that there is no effective bound on the number of injuries from a strategy with higher priority, and hence we cannot have a recursive bound on the number of changes of $A'(e)$ in advance. It also turns out to be an obstacle for us to obtain a bounded-low c.e. set which is low, but not superlow, by using Sacks splitting. Instead, we will construct a low, but not superlow, c.e. set directly.

2.1   Requirements and Strategies

By the definition of $A^\dagger$, for each e, $A^\dagger(e)$ has the following computable approximations at stage s:

$$A^\dagger(e)[s] := \begin{cases} 1, \text{ if } \exists n \leq e[\varphi_n(e)[s] \downarrow \ \& \ \ \Phi_e^{A\restriction\varphi_n(e)}(e)[s] \downarrow]; \\ 0, \text{ otherwise.} \end{cases}$$

To make $A^\dagger$ ω-c.e., we ensure that the number of changes of $A^\dagger$ at e is bounded by $2(e+1)^2$. That is, $|\{s \in \mathbb{N} : A^\dagger(e)[s] \neq A^\dagger(e)[s+1]\}| \leq 2(e+1)^2$.

We will construct a c.e. set A meeting the following requirements:

$\mathcal{L}_e$: If there are infinitely many stages s such that $\Phi_e^A(e)[s] \downarrow$, then $\Phi_e^A(e) \downarrow$.
$\mathcal{R}_e$: $|\{s \in \mathbb{N} : A^\dagger(e)[s] \neq A^\dagger(e)[s+1]\}| \leq 2(e+1)^2$.
$\mathcal{P}_{\langle i,j \rangle}$: If φ_i and φ_j^2 are both total, then there is some x such that either

$$A'(x) \neq \lim_{t \to \infty} \varphi_j^2(x, t)$$

or

$$|\{t \in \mathbb{N} : \varphi_j^2(x, t) \neq \varphi_j^2(x, t+1)\}| \geq \varphi_i(x).$$

Here $\{\varphi_i : i \in \mathbb{N}\}$ and $\{\varphi_j^2 : j \in \mathbb{N}\}$ are standard enumerations of all partial computable functions in one variable and two variables respectively, $\{\Phi_e : e \in \mathbb{N}\}$ is a standard enumeration of all partial computable functionals, and $\langle \cdot, \cdot \rangle$ is an effective bijection between $\mathbb{N}^2$ and $\mathbb{N}$.

We assign the priority of requirements in our finite injury construction as

$$\mathcal{L}_0 \prec \mathcal{R}_0 \prec \mathcal{P}_0 \prec \mathcal{L}_1 \prec \mathcal{R}_1 \prec \mathcal{P}_1 \prec \cdots \prec \mathcal{L}_e \prec \mathcal{R}_e \prec \mathcal{P}_e \prec \cdots$$

If all the $\mathcal{L}$-requirements are satisfied, then A is low. If all the $\mathcal{R}$-requirements are satisfied, then the whole construction will ensure that $A^\dagger$ is ω-c.e., which ensures that A is bounded-low. If all the $\mathcal{P}$-requirements are satisfied, then A' is not ω-c.e., and hence A is not superlow.

The strategy of satisfying one $\mathcal{L}_e$ is standard, i.e., we will set restraints to preserve the desired computations we have seen and this strategy can only be injured by $\mathcal{P}$-strategies with higher priority, which will happen at most finitely many times. Thus, after a stage large enough, after which no $\mathcal{P}$-strategies with higher priority act, if the strategy sets restraints to preserve computations, these computations will be preserved forever, and $\mathcal{L}_e$ is satisfied.

The strategy for satisfying one $\mathcal{R}_e$ is also to set restraints to preserve computations. At each stage s, when we see that $\varphi_n(e)$ converges, where $n \leq e$, we set a restraint to protect $A \restriction \varphi_n(e)$. Note that $\mathcal{P}$-strategy with higher priority can injure $\mathcal{R}_e$ by enumerating small numbers x into A, but as $\varphi_n(e)$ will be fixed, and hence, once $\mathcal{P}$ enumerates x into A, $\mathcal{P}$ selects a new number x' say, bigger than $\varphi_n(e)$, and the further enumeration of x' will not injure $\mathcal{R}_e$, as this enumeration will not affect the computation $\Phi_e^{A\restriction\varphi_n(e)}(e)$, if $\Phi_e^{A\restriction\varphi_n(e)}(e)$ converges. Note that there are at most $e+1$ many such restraints $A \restriction \varphi_n(e)$, and each $\mathcal{P}$-strategy with higher priority can enumerate numbers less than $\varphi_n(e)$ at most once, the total number of such enumerations is at most $(e + 1)^2$, as one such an enumeration entails at most two changes of $A^\dagger(e)$, the number of changes of $A^\dagger(e)$ is bounded by $2(e + 1)^2$.

We now describe how to satisfy a $\mathcal{P}_{\langle i,j \rangle}$-requirement. As in [6] for the construction of nonsuperlow sets, we will apply the recursion theorem for the construction. That is, our construction will be uniform in parameters, r say, and in the r-th construction, we will build a partial computable functional Γ_r, a c.e. set of axioms $\langle \sigma, n, m \rangle$, where $\langle \sigma, n, m \rangle$ is enumerated into Γ_r at stage s, $\Gamma_r^A(n)[s]$ is defined as m with use $A_s \upharpoonright_{\gamma_r^A(n)[s]} = \sigma$. From Γ_r, if $\Gamma_r = W_r$, we can have a computable function p_r such that

$$\forall X \forall x [\Gamma_r^X(x) \simeq \Phi_{p_r(x)}^X(p_r(x))].$$

This will ensure that $\Gamma_r^A(x) \downarrow \Longleftrightarrow p_r(x) \in A'$, and hence, by controlling Γ_r^A, we can defeat φ_j^2 being an approximation of A' with number of changes bounded by φ_i. More specifically, when $\varphi_j^2(p_r(x), \cdot)$ is a current approximation of $A'(p_r(x))$, then either it has a current limit 1 which means that $p_r(x)$ goes into A' or it has a current limit 0 which means that $p_r(x)$ moves out of A'. In the former case, we will start convergence modules and force $p_r(x)$ to leave out A' by enumerating $\gamma_r^X(x)$ into A provided that it is defined currently, while in the latter case, we will start divergence modules and force $p_r(x)$ to enumerate into A' by defining $\Gamma_r^X(x)$ provided that it is undefined currently.

Convergence modules and divergence modules actually switch to each other, and $\varphi_j^2(p_r(x), \cdot)$ will change at least once after each switch. So to make $\varphi_j^2(p_r(x), \cdot)$ change at least $\varphi_i(p_r(x))$ times, we can keep $\varphi_i(p_r(x)) + 1$ many convergence modules as well as $\varphi_i(p_r(x)) + 1$ many divergence modules. During the following $\mathcal{P}_{\langle i,j \rangle}$-strategy, we use $C(n)$ and $D(n)$ to denote the n-th convergence module and the n-th divergence module respectively. We also define a number $\ell(n)$ with $\ell(-1) = 0$ such that $\varphi_j^2(p_r(x), \ell(n-1)) \neq \varphi_j^2(p_r(x), \ell(n))$. Intuitively, $\ell(n)$ records the n-th change of $\varphi_j^2(p_r(x), \cdot)$ in the sense that when $\varphi_j^2(p_r(x), \cdot)$ is total, then

$$|\{q < \ell(n) : \varphi_j^2(p_r(x), q) \neq \varphi_j^2(p_r(x), q + 1)\}| \geq n.$$

Since the construction is uniform in r, there is a computable function g such that $\Gamma_r = W_{g(r)}$. By Recursion Theorem, there is a r_0 such that $W_{g(r_0)} = W_{r_0}$, i.e., $\Gamma_{r_0} = W_{r_0}$. In the following, we assume we are in the r_0-th construction, and fix the computable function p_{r_0} in advance.

A $\mathcal{P}_{\langle i,j \rangle}$-strategy proceeds as follows:

1. Choose a fresh witness x, let $z = p_{r_0}(x)$.
2. Wait for $\varphi_i(z) \downarrow$ at some stage s.
   (*If we wait at (2) forever, then $\varphi_i(z) \uparrow$, and φ_i is partial and $\mathcal{P}_{\langle i,j \rangle}$ is satisfied.*)
3. Set $N = \varphi_i(z) + 1$, and run the following modules $C(n)$ and $D(n)$ for $n \leq N$. We start with $C(0)$ in which case $\varphi_j^2(z, 0) \downarrow = 1$ or $D(0)$ in which case $\varphi_j^2(z, 0) \downarrow = 0$, and let $\ell(-1) = 0$. We will end when reaching $C(N)$ or $D(N)$,

because this shows that $\varphi_i(z)$ cannot bound the number of changes of $\varphi_j^2(z,q)$, showing that $\mathcal{P}_{\langle i,j \rangle}$ is satisfied.

$C(n)$: There is a $q_1 > \ell(n-1)$ such that φ_j^2 converges (at stage s) to 1 on all (z,q) with $\ell(n-1) < q \leq q_1$, and for those q_2 with $q_1 < q_2 \leq s$, if $\varphi_j^2(z,q_2)$ converges at stage s, then $\varphi_j^2(z,q_2) = 1$.

If such a q_1 does not exist, then do nothing.

- If this is because of the fact that $\varphi_j^2(z, \ell(n-1)+1)$ never converges, then φ_j^2 is partial and $\mathcal{P}_{\langle i,j \rangle}$ is satisfied.
- If it is because of the existence of q_2 with $\varphi_j^2(z,q_2) = 0$, then wait for a bigger stage at which $\varphi_j^2(z,q)$ converges for all $q \leq q_2$, leave $C(n)$ with no actions and switch to $D(n+1)$ with $\ell(n) = q_2$.
  (*Again, if $\varphi_j^2(z,q)$ never converges for some $q < q_2$, then φ_j^2 is partial and $\mathcal{P}_{\langle i,j \rangle}$ is satisfied.*)

That is, at stage s, we guess that $\varphi_j^2(z,q)$ has limit 1, and we should not keep such a guess if we see $\varphi_j^2(z,q)$ converges to 0 after $\ell(n-1)$.

Action for $C(n)$: If $\Gamma_{r_0}^A(x)$ has definition at stage $s-1$, then enumerate $\gamma_{r_0}(x)$ into A, to undefine $\Gamma_{r_0}^A(x)$. Otherwise, do nothing.

- Wait for a stage $t > s$ such that $\Phi_z^A(z)[t]$ diverges.
  (*For r_0-th construction, such a stage t exists, because otherwise, we would have $\Gamma_{r_0}^A(x)$ diverges, but $\Phi_z^A(z)$ converges.*)
- Wait for a bigger stage at which we see that $\varphi_j^2(z,q_3)$ converges to 0, where $q_3 \geq q_1$, and $\varphi_j^2(z,q)$ converges for all $q \leq q_3$. Let $\ell(n) = q_3$ and switch to $D(n+1)$.
  (*Again, if $\varphi_j^2(z,q)$ never converges for some $q < q_3$, then φ_j^2 is partial and $\mathcal{P}_{\langle i,j \rangle}$ is satisfied. Or, if $\varphi_j^2(z,q)$ converges to 1 for all $q \geq q_1$, then $\lim_{x \to \infty} \varphi_j^2(z,q) = 1$, while $z \notin A'$, meaning that $\lim_{x \to \infty} \varphi_j^2(z,q)$ is not the characteristic function of A', and $\mathcal{P}_{\langle i,j \rangle}$ is satisfied.*)

$D(n)$: There is a $q_1 > \ell(n-1)$ such that φ_j^2 converges (at stage s) to 0 on all (z,q) with $\ell(n-1) < q \leq q_1$, and for those q_2 with $q_1 < q_2 \leq s$, if $\varphi_j^2(z,q_2)$ converges at stage s, then $\varphi_j^2(z,q_2) = 0$.

If there is no such a q_1, then do nothing.

- If this is because of the fact that $\varphi_j^2(z, \ell(n-1)+1)$ never converges, then φ_j^2 is partial and $\mathcal{P}_{\langle i,j \rangle}$ is satisfied.
- If it is because of the existence of q_2 with $\varphi_j^2(z,q_2) = 1$, then wait for a bigger stage at which $\varphi_j^2(z,q)$ converges for all $q \leq q_2$, leave $D(n)$ with no actions and switch to $C(n+1)$ with $\ell(n) = q_2$. (*Again, if $\varphi_j^2(z,q)$ never converges for some $q < q_2$, then φ_j^2 is partial and $\mathcal{P}_{\langle i,j \rangle}$ is satisfied.*)

That is, at stage s, we guess that $\varphi_j^2(z,q)$ has limit 0, and we should not keep such a guess if we see $\varphi_j^2(z,q)$ converges to 1 after $\ell(n-1)$.

Action for $D(n)$: If $\Gamma_{r_0}^A(x)$ has no definition at stage $s-1$, define $\Gamma_{r_0}^A(x)$ with use $\gamma_{r_0}(x)$ big.

- Wait for a stage $t > s$ such that $\Phi_z^A(z)[t]$ converges.
  (*For the r_0-th construction, such a stage t exists, because otherwise, we would have $\Gamma_{r_0}^A(x)$ converges, but $\Phi_z^A(z)$ diverges.*)
- Wait for a bigger stage at which we see that $\varphi_j^2(z, q_3)$ converges to 1, where $q_3 \geq q_1$, and $\varphi_j^2(z, q)$ converges for all $q \leq q_3$. Let $\ell(n) = q_3$ and switch to $C(n + 1)$.
  (*Again, if $\varphi_j^2(z, q)$ never converges for some $q < q_3$, then φ_j^2 is partial and $\mathcal{P}_{\langle i,j \rangle}$ is satisfied. Or, if $\varphi_j^2(z, q)$ converges to 0 for all $q \geq q_1$, then $\lim_{x \to \infty} \varphi_j^2(z, q) = 0$, while $z \in A'$, meaning that $\lim_{q \to \infty} \varphi_j^2(z, q)$ is not the characteristic function of A', and $\mathcal{P}_{\langle i,j \rangle}$ is satisfied.*)

Thus, a $\mathcal{P}_{\langle i,j \rangle}$-strategy either waits at some $C(n)$ or $D(n)$ for some $n < N$, or reaches $C(N)$ or $D(N)$, both of which will show that $\mathcal{P}_{\langle i,j \rangle}$ is satisfied, as explained above. Thus, in the construction, a $\mathcal{P}_{\langle i,j \rangle}$-strategy can injure those strategies with lower priority at most finitely often.

2.2   Construction and Verification

Construction of A.

Stage 0: Let $A_0 = \emptyset$, and initialize all strategies.
Stage $s > 0$: Find a requirement with the highest priority, Q say, among

$$\mathcal{L}_0 \prec \mathcal{R}_0 \prec \mathcal{P}_0 \prec \cdots \mathcal{L}_s \prec \mathcal{R}_s \prec \mathcal{P}_s$$

that requires attention at stage s, and act accordingly.

Say that $\mathcal{L}_e$ *requires attention* at stage s if $\Phi_e^A(e)[s-1] \uparrow$ and $\Phi_e^A(e)[s] \downarrow$.
Say $\mathcal{R}_e$ *requires attention* at stage s if there is a $n \leq e \leq s$ such that $\varphi_n(e) \downarrow$ at stage s and $\mathcal{R}_e$ has not set A-restraint $\varphi_n(e)$ yet.

For both cases, we act as follows:
Action: Set A-restraint as s, and initialize all strategies with lower priority.

Say that $\mathcal{P}_{\langle i,j \rangle}$ *requires attention* at stage s if one of the following applies:

1. $\mathcal{P}_{\langle i,j \rangle}$ has no witness currently.

   Action: Pick a fresh number x as a witness for $\mathcal{P}_{\langle i,j \rangle}$.
2. x is selected, $p_{r_0}(x)$ converges to z, $\varphi_i(z)[s] \downarrow$ and N is not defined yet.

   Action: Definite N as $\varphi_i(z) + 1$.
3. N is defined, $\mathcal{P}_{\langle i,j \rangle}$ is not in $C(n)$ or $D(n)$ for any $n \leq N$, and $\varphi_j^2(z, 0)[s]$ converges.

   Action: Let $\mathcal{P}_{\langle i,j \rangle}$ start by running $C(0)$, if $\varphi_j^2(z, 0)[s] \downarrow = 1$, or $D(0)$, if $\varphi_j^2(z, 0)[s] \downarrow = 0$. Here we let $\ell(-1) = 0$.
4. $\mathcal{P}_{\langle i,j \rangle}$ switches from $C(n)$ to $D(n + 1)$, where $n + 1 < N$.

   Action: If $\Gamma^A(x)$ has no definition at stage $s - 1$, then define $\Gamma^A(x)[s] = 0$ with use $\gamma(x)[s]$ larger than all numbers used before.

5. $\mathcal{P}_{\langle i,j \rangle}$ switches from $D(n)$ to $C(n+1)$, where $n+1 < N$.

   Action: If $\Gamma^A(x)$ has definition at stage $s-1$, then enumerate $\gamma(x)$ into A, undefining $\Gamma^A(x)$.

6. $\mathcal{P}_{\langle i,j \rangle}$ switches from $C(N-1)$ to $D(N)$ or from $D(N-1)$ to $C(N)$.

   Action: Declare that $\varphi_i(z)$ cannot bound the number of changes of $\varphi_j^2(z,q)$, and that $\mathcal{P}_{\langle i,j \rangle}$ is satisfied.

   If $\mathcal{P}_{\langle i,j \rangle}$ acts as above, then all strategies with lower priority are initialized. This ends the construction of A at stage s.

Lemma 1. *For each requirement, Q say,*

(1) Q can be initialized at most finitely often;
(2) Q acts at most finitely often and is satisfied;
(3) The restraint set by Q is finite, and Q can initialize all strategies with lower priority finitely often.

Proof. We will show that for each e, (1)–(3) are true for all $\mathcal{L}_e, \mathcal{R}_e, \mathcal{P}_{\langle i,j \rangle}$, where $\langle i,j \rangle = e$. We prove it by induction.

When $e = 0$, $\mathcal{L}_0$ can never be initialized, due to its highest priority. (1) holds. $\mathcal{L}_0$ acts and sets a restraint only when $\Phi_0^A(0)$ converges, and once we set such a restraint, $\Phi_0^A(0)$ converges at any further stage, and (2) holds. In the construction, $\mathcal{L}_0$ sets a restraint only when $\Phi_0^A(0)$ converges, and after it acts, it will never act again. (3) holds.

Note that $\mathcal{R}_0$ can be initialized by $\mathcal{L}_0$ at most once. (1) holds for $\mathcal{R}_0$. Note that $\mathcal{R}_0$ acts only when $\varphi_0(0)$ converges, and after this, if $\Phi_0^{A \restriction \varphi_0(0)}(0)$ converges, then it will converge forever, as no $\mathcal{P}$ can enumerate numbers less than $\varphi_0(0)$ into A, so if $\Phi_0^{A \restriction \varphi_0(0)}(0)$ converges again, then it converges forever, which means that $\mathcal{R}_0$ is satisfied. (2) holds. The restraint set by $\mathcal{R}_0$ is $\varphi_0(0)$, if it converges, and after it acts, it will never act again. (3) holds.

For $\mathcal{P}_{\langle 0,0 \rangle}$, it can be initialized by $\mathcal{L}_0$ and $\mathcal{R}_0$, so $\mathcal{P}_{\langle 0,0 \rangle}$ can be initialized by $\mathcal{L}_0$ and $\mathcal{R}_0$ at most three times. (1) holds. Let s_0 be the least stage after which $\mathcal{P}_{\langle 0,0 \rangle}$ can not be initialized again, and suppose that $\mathcal{P}_{\langle 0,0 \rangle}$ selects a witness x as a big number at stage $s_1 > s_0$. Then x cannot be canceled later. Then after stage s_1, $\mathcal{P}_{\langle 0,0 \rangle}$ acts only when

1. it selects N; or
2. it starts module $C(0)$ or $D(0)$; or
3. it switches from $C(n-1)$ to $D(n)$, or from $D(n-1)$ to $C(n)$; or
4. it reaches $C(N)$ or $D(N)$,

which can happen at most $N+2$ many times. If it reaches $C(N)$ or $D(N)$, then $\varphi_0(z)$ cannot be a bound of the number of changes of $\varphi_0^2(z,q)$, and $\mathcal{P}_{\langle 0,0 \rangle}$ is satisfied. If it never selects N, then $\varphi_0(z)$ never converges (recall that p_{r_0} is assumed to be total, by the relativized s_n^m-theorem). Here $z = p_{r_0}(x)$. If it never starts module $C(0)$ or $D(0)$, then $\varphi_0^2(z,0)$ diverges. If it stays at some module $C(n)$ or $D(n)$ forever, then either $\varphi_0^2(z,q)$ diverges for some q, or we will have $\lim_{q\to\infty} \varphi_0^2(z,q) \neq A'(z)$. $\mathcal{P}_{\langle 0,0\rangle}$ is satisfied for all cases. (2) holds for $\mathcal{P}_{\langle 0,0\rangle}$. Thus, after a stage s_3 large enough, $\mathcal{P}_{\langle 0,0\rangle}$ will have no actions, and hence the restraint set by $\mathcal{P}_{\langle 0,0\rangle}$ is finite, and it will not initialize strategies with lower priority. (3) holds.

For $e > 0$, we assume that the lemma is true for $e' < e$. The proof that the lemma is true for e is very similar to the proof for the basic case, i.e. $e = 0$, and we will assume that after a stage s big enough, no strategies $\mathcal{L}_{e'}$, $\mathcal{R}_{e'}$, and $\mathcal{P}_{e'}$ can act. Then we can show that after stage s, $\mathcal{L}_e$, $\mathcal{R}_e$, and $\mathcal{P}_e$ will behave exactly like $\mathcal{L}_0$, $\mathcal{R}_0$, and $\mathcal{P}_0$ above, and hence the lemma is true for e.

This completes the proof of Lemma 1.

By Lemma 1, A is low but not superlow. Lemma 2 below shows that A is bounded-low.

Lemma 2. $A^\dagger$ *is ω-c.e., and hence A is bounded-low.*

Proof. To show that $A^\dagger$ is ω-c.e., we need to show that there is a computable function bounding the changes of e in $A^\dagger$ for all e.

Fix e, and $n \leq e$. Then in the construction, whenever we see $\varphi_n(e)$ converges, at stage s say, $\mathcal{R}_e$ acts and initializes all strategies with lower priority. Thus, no strategy with lower priority can put a number less that $\varphi_n(e)$ into A later. So if $\Phi_e^{A\restriction\varphi_n(e)}$ converges later (e enters $A^\dagger$), this computation can only be changed by $\mathcal{P}_{e'}$-strategies (where $e' < e$, which have higher priority) by enumerating $\gamma_{r_0}^A(y)[t]$ into A. After this, if $\gamma_{r_0}^A(y)$ is defined later, then it will be defined as a number bigger than $\varphi_n(e)$. Thus, because of this, $\Phi_e^{A\restriction\varphi_n(e)}$ is injured at most $e + 1$ many times, then $A^\dagger(e)$ can change at most $2(e + 1)$ many times, for this particular n. Thus, in total, $A^\dagger(e)$ can change at most $2(e+1)^2$ many times, and $A^\dagger$ is ω-c.e.

This completes the proof of Theorem 1.

3   A Bounded-Low Set with Turing Degree $0'$

In this section, we give a sketchy proof of Theorem 2. That is, we will construct a bounded-low c.e. set A Turing computing $\emptyset'$. We have seen how to construct a bounded-low c.e. set in the proof of Theorem 1, so to prove Theorem 2, we only need to show how to make A Turing complete. We will construct a partial computable functional Γ such that $\emptyset' = \Gamma^A$. Of course, the construction of Γ will be consistent with the $\mathcal{R}$-strategies of making A bounded-low.

For the construction of Γ, we will make sure that for each e, $\Gamma^A(e)$ is defined and computes $\emptyset'(e)$ correctly. The idea of constructing Γ is quite standard. That is, in the construction, at stage s, we find the least x with $\Gamma^A(x)$ not defined and define $\Gamma^A(x)[s] = \emptyset'(x)[s]$ with use $\gamma(x)$ big, like s. If later, at stage $t > s$, x enters $\emptyset'$, then at this stage, we enumerate $\gamma(x)$ into A, undefining $\Gamma^A(x)$. In general, the construction of Γ satisfies the following rules:

1. For any x_1, x_2 with $x_1 < x_2$, if $\Gamma^A(x_2)$ has definition at stage s, then $\Gamma^A(x_1)$ also has definition at this stage, with uses $\gamma(x_1) < \gamma(x_2)$.
2. For any x_1, x_2 with $x_1 < x_2$, if $\Gamma^A(x_1)$ is undefined at stage s, then $\Gamma^A(x_2)$ is also undefined at this stage, due to a number less than $\gamma(x_1)$ enters A.
3. For any x, if $\Gamma^A(x)$ has definition at stage s_1 and s_2, then $\gamma(x)[s_1] \leq \gamma(x)[s_2]$, and if $\gamma(x)[s_1] < \gamma(x)[s_2]$, $\Gamma^A(x)$ must have been undefined between these two stages.
4. For any x, $\Gamma^A(x)$ can be undefined at most finitely often during the construction.
5. For any x, $\Gamma^A(x) = \emptyset'(x)$.

We now show how to modify an $\mathcal{R}_e$-strategy in the proof of Theorem 1 so that it is consistent with the rules above. Recall that the $\mathcal{R}_e$-strategy can act finitely many times, and each time when it acts, because of $\varphi_i(e) \downarrow$ for some $i \leq e$, it initializes all strategies with lower priority. To make it consistent with the construction of Γ, it needs to have an additional action: enumerate $\gamma(e)[s]$ into A. This enumeration undefines $\Gamma^A(e)$, so when it is defined again, the use will be defined as a big number, which is bigger than the value $\varphi_i(e)$. Thus, for a fixed e, the membership of e in $A^\dagger$ can be changed by $\mathcal{R}_{e'}$-strategies with $e' \leq e$, i.e., by further enumerations of the γ-uses, or by rectifying $\Gamma^A(x) = 1$ when some $x <\, e$ enumerating into $\emptyset'$, then $A^\dagger(e)$ changes at most $2(e+1)^2$ many times, which guarantees that $A^\dagger$ is ω-c.e.

We will skip the construction and verification parts, due to the fact that the construction is an easy finite injury argument.

4   Basic Idea of Proving Theorem 3

Fix a c.e. operator W. To prove the first part of Theorem 3, we need to build a bounded-low c.e. set C (i.e., $C^\dagger$ is ω-c.e.) such that $C \oplus W^C \equiv_T \emptyset'$. As C will be c.e., $C^\dagger$ has the following natural computable approximations:

$$f_{C^\dagger}(x, s) := \begin{cases} 1, \text{ if } \exists n \leq x[\varphi_n(x)[s] \downarrow \ \& \ \Phi_x^{C \upharpoonright \varphi_n(x)}(x)[s] \downarrow]; \\ 0, \text{ otherwise.} \end{cases}$$

Besides the requirements for the standard pseudo-jump inversion theorem, C will also satisfy the following negative requirements:

$$\mathcal{N}_e : |\{s : f_{C^\dagger}(e, s) \neq f_{C^\dagger}(e, s+1)\}| \leq 2(e+1)^2 + 1,$$

and we will call these requirements as bounded-low requirements.

We will build functional Γ such that $\Gamma^{C \oplus W^C} = \emptyset'$. If e enters into $\emptyset'$, we will rectify $\Gamma^{C \oplus W^C}(e) = 1$ by enumerating use $\gamma^{C \oplus W^C}(e)$ into C. To make $W^C \leq_T \emptyset'$ (i.e., W^C is Δ_2^0), as usual, we will preserve $e \in W^C$ by setting a C-restraint $\psi^C(e)$ if $\Psi^C(e) \downarrow$. As in the standard pseudo-jump inversion theorem, the definition of $\Gamma^{C \oplus W^C}(e)$ is compatible with preserving $W^C(e)$. For the bounded-low requirement $\mathcal{N}_e$, in order to preserve $C^\dagger(e)$, at each stage s, we set a C-restraint by

$$r^{\mathcal{N}}(e,s) = \max\{\varphi_n(e)[s] : n \le e \ \& \ \varphi_n(e)[s] \downarrow\}.$$

If $\Gamma^{C \oplus W^C}(e)[s] \downarrow$ with use $\gamma^{C \oplus W^C}(e)[s] \le r^{\mathcal{N}}(e,s)$, $\mathcal{N}_e$ will put $\gamma^{C \oplus W^C}(e)[s]$ into C, thus making $\Gamma^{C \oplus W^C}(e)$ undefined, and $\Gamma^{C \oplus W^C}(e)$ will be defined with new use $> r^{\mathcal{N}}(e,s)$. Since $r^{\mathcal{N}}(e,s)$ can move up at most $e+1$ times, $\mathcal{N}_e$ will enumerate at most $e+1$ many such uses.

Now computations of the form $\Phi_e^{C\restriction \varphi_n(e)}(e)[s] \downarrow$ (of course, $\varphi_n(e)[s] \downarrow$) for some $n \le e$ are injured at stage $s+1$ only if there is some $i \le e$ such that $\gamma^{C \oplus W^C}(i)[s] \le \varphi_n(e)$ is enumerated into C_{s+1}. If $\Gamma^{C \oplus W^C}(i)$ is not defined currently, it will be defined with a use $> \varphi_n(e)$. So the enumeration of $\gamma^{C \oplus W^C}(i)$ at a later stage will not injure a new computation of the form $\Phi_e^{C\restriction \varphi_n(e)}(e)$, which implies that for fixed $i \le e$ and $n \le e$, the enumeration of $\gamma^{C \oplus W^C}(i)$-uses can injure $\Phi_e^{C\restriction \varphi_n(e)}(e)$ at most once. Thus, for a fixed n, $\Phi_e^{C\restriction \varphi_n(e)}(e)$ is injured at most $e+1$ times, as a consequence, the approximations of $C^\dagger(e)$ can be injured at most $(e+1)^2$ many times. Note that each injury like this entails at most two changes of $f_{C^\dagger}(e,s)$, and hence $|\{s : f_{C^\dagger}(e,s) \ne f_{C^\dagger}(e,s+1)\}| \le 2(e+1)^2+1$. Hence, $C^\dagger$ is ω-c.e., i.e., C is bounded-low.

For the second part, we need to build a bounded-high set C (i.e., $\emptyset^{\dagger\dagger} \le_{wtt} C^\dagger$) such that C is ω-c.e., and $C \oplus W^C \equiv_T \emptyset'$. C will be constructed to satisfy the following requirements:

$\mathcal{P}_e$: $\emptyset'(e) = \Gamma^{C \oplus W^C}(e)$, where Γ is a partial computable functional built by us.

$\mathcal{Q}_e$: $\emptyset^{\dagger\dagger}(e) = \Delta^{C^\dagger \restriction g(e)}(e)$, where Δ is a partial computable functional built by us

and g is a computable function such that for all e, the use $\delta^{C^\dagger}(e) \le g(e)$.

$\mathcal{R}_e$: If there are infinitely many stages s with $e \in W^C[s]$, then $e \in W^C$.

The construction of C is a combination of the bounded-high strategy developed by Anderson, Csima and Lange in [2] and the proof of Jockusch and Shore's pseudo-jump inversion theorem.

For the $\mathcal{P}_e$-strategy, when e enters $\emptyset'$, we rectify $\Gamma^{C \oplus W^C}(e) = 1$ by enumerating the use $\gamma^{C \oplus W^C}(e)$ into C. To meet $\mathcal{R}_e$, we preserve $C \restriction_{\psi^C(e)[s]}$ if e enters W^C at stage s, i.e., $\Psi^C(e)[s] \downarrow$ and $\Psi^C(e)[s-1] \uparrow$. Set a C-restraint

$$r(e,s) = \max\{\psi^C(e)[s'] : s' \le s \wedge \Psi^C(e)[s'] \downarrow\}.$$

When $r(e,s)$ increases to a higher level, as in pseudo-jump inversion theorem, $\Gamma^{C \oplus W^C}(j) \uparrow$ for all $j \ge e$; we also undefine $\Delta^{C^\dagger \restriction g(e)}(e)$-computations by enumerating or exacting numbers into $C^\dagger$ (we apply recursion theorem here). So $\mathcal{R}_e$ can only be injured by $\mathcal{P}_j, \mathcal{Q}_j$ with $j < e$. Moreover, due to the feature of $\mathcal{P}_j, \mathcal{Q}_j$-strategies, the number of injuries to $\mathcal{R}_e$ is recursively bounded.

Now consider the $\mathcal{Q}_e$-strategy. $\emptyset^{\dagger\dagger}$ is ω^2-c.e. under the standard notation for ω^2. Then there is a partial computable function $\chi : \omega \times \omega^2 \to \{0,1\}$ such that for each e, there is a least ordinal $\beta_e < \omega^2$ such that $\chi(e,\beta_e) \downarrow = \emptyset^{\dagger\dagger}(e)$, and we also have computable approximations for β_e and $\chi(e,\beta_e)$. That is,

- for each e, there is a first stage s_e and a least ordinal $\beta_{e,s_e} = \omega \cdot i_{s_e} + j_{s_e} < \omega^2$ such that $\chi(e, \beta_{e,s_e})[s_e] \downarrow$; and

- for each e and stage $s \geq s_e$, there is a least ordinal $\beta_{e,s} = \omega \cdot i_{e,s} + j_{e,s}$ such that $\chi(e, \beta_{e,s})[s] \downarrow$.

For convenience, we also choose χ such that:

- for each e and stage $s \geq s_e$, if $\beta_{e,s} \neq \beta_{e,s+1}$, then $\chi(e, \beta_{e,s+1}) \neq \chi(e, \beta_{e,s})$. *In this case, $\beta_{e,s+1} < \beta_{e,s}$, that is, either $i_{e,s+1} < i_{e,s}$, or $i_{e,s+1} = i_{e,s}$ with $j_{e,s+1} < j_{e,s}$.*

As χ is partial computable, i_{s_e} and j_{s_e} are computable functions on e; $i_{e,s}$ and $j_{e,s}$ are computable functions on e, s.

Now $\beta_e = \lim_{s \to \infty} \beta_{e,s}$, and $\emptyset^{\dagger\dagger}(e) = \lim_{s \to \infty} \chi(e, \beta_{e,s})$. Based on the approximation $\emptyset^{\dagger\dagger}(e)[s] = \chi(e, \beta_{e,s})$, we define $\Delta^{C^\dagger \restriction g(e)}(e)[s'] = \chi(e, \beta_{e,s'})$. If there is a least stage $s > s'$ such that $\beta_{e,s} < \beta_{e,s-1}$ (i.e., $i_{e,s+1} < i_{e,s}$, or $i_{e,s+1} = i_{e,s}$) or $\mathcal{Q}_e$ is injured by higher $\mathcal{R}$-strategies at stage s, then we need to rectify $\Delta^{C^\dagger}(e)$ at stage s by enumerating numbers into $C^\dagger$ or extracting numbers out of $C^\dagger$. We will ensure that the number that $\Delta^{C^\dagger}(e)$ will be undefined is recursively bounded, and then the use $\delta^{C^\dagger}(e)$ can be bounded by some computable function g. This shows that $\emptyset^{\dagger\dagger} \leq_{wtt} C^\dagger$, and hence, C is bounded-high.

References

1. Anderson, B., Csima, B.: A bounded jump for the bounded turing degrees. Notre Dame J. Formal Logic **55**, 245–264 (2014)
2. Anderson, B., Csima, B., Lange, K.: Bounded low and high sets. Arch. Math. Logic **56**, 523–539 (2017)
3. Jockusch Jr., C.G., Shore, R.A.: Pseudo-jump operators. I: the r.e. case. Trans. Am. Math. Soc. **275**, 599–609 (1983)
4. Mohrherr, J.: A refinement of low_n and $high_n$ for r.e. degrees. Z. Math. Logik Grundlag. Math. **32**, 5–12 (1986)
5. Bickford, M., Mills, C.: Lowness properties of r.e. sets (1982, typewritten unpublished manuscript)
6. Nies, A.: Computability and Randomness. Oxford University Press, Inc., New York (2009)

Supportive Oracles for Parameterized Polynomial-Time Sub-Linear-Space Computations in Relation to L, NL, and P

Tomoyuki Yamakami[✉]

Faculty of Engineering, University of Fukui, 3-9-1 Bunkyo, Fukui 910-8507, Japan
TomoyukiYamakami@gmail.com

Abstract. We focus our attention onto polynomial-time sub-linear-space computation for decision problems, which are parameterized by size parameters $m(x)$, where the informal term "sub linear" means a function of the form $m(x)^\varepsilon \cdot polylog(|x|)$ on input instances x for a certain absolute constant $\varepsilon \in (0, 1)$ and a certain polylogarithmic function $polylog(n)$. The parameterized complexity class PsubLIN consists of all parameterized decision problems solvable simultaneously in polynomial time using sub-linear space. This complexity class is associated with the linear space hypothesis. There is no known inclusion relationships between PsubLIN and para-NL, where the prefix "para-" indicates the natural parameterization of a given complexity class. Toward circumstantial evidences for the inclusions and separations of the associated complexity classes, we seek their relativizations. However, the standard relativization of Turing machines is known to violate the relationships of L $\subseteq$ NL = co-NL $\subseteq$ DSPACE$[O(\log^2 n)] \cap$ P. We instead consider special oracles, called NL-supportive oracles, which guarantee these relationships in the corresponding relativized worlds. This paper vigorously constructs such NL-supportive oracles that generate relativized worlds where, for example, para-L $\neq$ para-NL $\nsubseteq$ PsubLIN and para-L $\neq$ para-NL $\subseteq$ PsubLIN.

Keywords: Supportive oracle · Parameterized decision problem · Relativization · Sub-linear space computation · Log-space computation

1 Prelude: Quick Overview

1.1 Size Parameters and Parameterized Decision Problems

Among decision problems computable in polynomial time, nondeterministic logarithmic-space (or log-space, for short) computable problems are of special interest, partly because these problems contain practical problems, such as the *directed s-t connectivity problem* (DSTCON) and the *2-CNF Boolean formula*

© Springer Nature Switzerland AG 2019
T. V. Gopal and J. Watada (Eds.): TAMC 2019, LNCS 11436, pp. 659–673, 2019.
https://doi.org/10.1007/978-3-030-14812-6_41

satisfiability problem (2SAT). These problems form a complexity class known as NL (nondeterministic log-space class).

For many problems, their computational complexity have been discussed, from a more practical aspect, according to the "size" of particular items of each given instance. As a concrete example of such a "size", let us consider an efficient algorithm of Barnes, Buss, Ruzzo, and Schieber [2] that solves DSTCON on input graphs of n vertices and m edges simultaneously using $(m+n)^{O(1)}$ time and $n^{1-c/\sqrt{\log n}}$ space for an appropriately chosen constant $c > 0$. In this case, the number of vertices and the number of edges in a directed graph G are treated as the "size" or the "basis unit" of measuring the computational complexity of DSTCON. For an input CNF Boolean formula, in contrast, we can take the number of variables and the number of clauses as the "size" of the Boolean formula. In a more general fashion, we denote the "size" of instance x by $m(x)$ and we call this function m a *size parameter* of a decision problem L. A decision problem L together with a size parameter m naturally forms a *parameterized decision problem* and we use a special notation (L, m) to describe it.

Throughout this paper, we intend to study the properties of parameterized decision problems and their collections. Such collections are distinctively called *parameterized complexity classes*. To distinguish such parameterized complexity classes from standard binary-size complexity classes, we often append the term "para-" as in para-NL and para-L, which are respectively the parameterizations of NL and L (see Sect. 2 for their formal definitions).

The aforementioned parameterized decision problems (DSTCON, m_{ver}) and (2SAT, m_{vbl}), where $m_{ver}(\cdot)$ and $m_{vbl}(\cdot)$ respectively indicate the number of vertices and the number of variables, fall into para-NL [11]. It is, however, unclear whether we can improve the performance of the aforementioned algorithm of Barnes et al. to run using only $O(m_{ver}(x)^\varepsilon \ell(|x|))$ space for a certain absolute constant $\varepsilon \in [0, 1)$ and a certain polylogarithmic (or polylog, for short) function ℓ. Given a size parameter $m(\cdot)$, the informal term "sub linear" generally refers to a function of the form $m(x)^\varepsilon \cdot \ell(|x|)$ for a certain constant $\varepsilon \in [0, 1)$ and a certain polylog function ℓ. We denote by PsubLIN the collection of all parameterized decision problems solved by deterministic Turing machines running simultaneously in $(|x|m(x))^{O(1)}$ (polynomial) time using $O(m(x)^\varepsilon polylog(|x|))$ (sublinear) space [11]. It follows that para-L $\subseteq$ PsubLIN $\subseteq$ para-P. Various sub-linear reducibilities were further studied in [10] in association with PsubLIN. The *linear space hypothesis* (LSH), which is a practical working hypothesis proposed in [11], asserts that $(2SAT_3, m_{vbl})$ cannot belong to PsubLIN, where $2SAT_3$ is a variant of 2SAT with an extra restriction that every 2-CNF Boolean formula given as an instance must have each variable appearing at most 3 times in the form of literals. We do not know whether LSH is true or even para-NL $\not\subseteq$ PsubLIN. A characterization of LSH was given in [12] in connection to state complexity of finite automata.

1.2   Relativizations of L, NL, P, and PsubLIN

The current knowledge seems not good enough to determine the exact complexity of PsubLIN in comparison with para-L, para-NL, and para-P. This fact makes us look for *relativizations* of these classes by way of forcing underlying Turing machines to make queries to appropriately chosen oracles. The notion of relativization in computational complexity theory dates back to an early work of Baker, Gill, and Solovay [1], who constructed various relativized worlds in which numerous inclusion and separation relationships among P, NP, and co-NP are possible. Generally speaking, relativization is a methodology by which we can argue that a certain mathematical property holds or does not hold in the presence of external information source, called an *oracle*. The use of an oracle A creates a desired relativized world where certain desired conditions, such as $P^A \neq NP^A$ together with $NP^A = \text{co-NP}^A$, hold. In a similar vein, we want to discuss the possibility/impossibility of the inclusion of para-NL in PsubLIN by considering relativized worlds where various conflicting relationships between para-NL and PsubLIN hold.

Unlike P and NP, it has been known that there is a glitch in defining the relativization of NL. Ladner and Lynch [5] first considered relativization of NL in a way similar to that of Baker, Gill, and Solovay [1]. Despite our knowledge regarding basic relationships among NL, co-NL, and P, this relativization leads to the existence of oracles A and B such that $NL^A \not\subseteq P^A$ and $NL^B \neq \text{co-NL}^B$ although co-NL = NL $\subseteq$ P holds in the unrelativized world. Quite different from time-bounded oracle Turing machines, there have been several suggested models for space-bounded oracle Turing machines. Ladner-Lynch relativization does not always guarantee both relationships $NL^A \subseteq P^A$ and $NL^B = \text{co-NL}^B$ since certain oracles A, B refute those relations. This looks like contradicting the fact that, in the un-relativised world, L $\subseteq$ NL $\subseteq$ P, NL = co-NL, and NL $\subseteq$ LOG^2SPACE hold [3,6,8,9], where $\text{LOG}^2\text{SPACE} = \text{DSPACE}[O(\log^2 n)]$.

Another, more restrictive relativization model was proposed in 1984 by Ruzzo, Simon, and Tompa [7], where an oracle machine behaves deterministically while writing a query word on its query tape. More precisely, after a query tape becomes blank, if the oracle machine starts writing the first symbol of a query word, then the machine must make deterministic moves until the query word is completed and an oracle is called. After the oracle answers, the query tape is automatically erased to be blank again and a tape head instantly jumps back to the initial tape cell. This restrictive model can guarantee the inclusions $L^A \subseteq NL^A \subseteq P^A$ for any oracle A; however, it is too restrictive because it leads to the conclusion that L = NL iff $L^A = NL^A$ for any oracle A [4].

In this paper, we expect our relativization of an underlying machine to guarantee that all log-space nondeterministic oracle Turing machines can be simulated by polynomial-time deterministic Turing machines, yielding three relationships that para-NL^A $\subseteq$ para-P^A, para-NL^A = co-para-NL^A, and para-$NL^A \subseteq$ para-$\text{LOG}^2\text{SPACE}^A$, where para-$\text{LOG}^2\text{SPACE}$ is the parameterization of LOG^2SPACE and co-para-NL is the collection of all (L, m) for which $(\overline{L}, m)$ belongs to para-NL.

In spite of an amount of criticism, relativization remains an important research subject to pursue. Returning to parameterized complexity classes, nonetheless, it is possible to consider "conditional" relativizations that support the aforementioned three relations. To distinguish such relativization from the ones we have discussed so far, we need a new type of relativization, which we will explain in the next subsection. In Sect. 5, we will return to a discussion on the usefulness (and the vindication) of this new relativization.

1.3   Main Contributions

We use the notation PsubLIN^A to denote the collection of all parameterized decision problems solvable by oracle Turing machines with adaptive access to oracle A simultaneously using $(|x|m(x))^{O(1)}$ time and $O(m(x)^\varepsilon \ell(|x|))$ space on all inputs x for a certain constant $\varepsilon \in [0, 1)$ and a certain polylog function ℓ.

A key concept of our subsequent discussion is "supportive oracles". Instead of restricting the way of accessing oracles (such as non-adaptive queries and limited number of queries), we use the most natural query mechanism but we force the oracles to support certain known inclusion relationships among complexity classes. In this way, an oracle A is said to be NL-*supportive* if the following three relations hold: (1) para-$\text{L}^A \subseteq$ para-$\text{NL}^A \subseteq$ para-P^A, (2) para-$\text{NL}^A =$ co-para-NL^A, and (3) para-$\text{NL}^A \subseteq$ para-$\text{LOG}^2\text{SPACE}^A$. Note that Condition (1) is always satisfied for any oracle A.

We first claim the existence of recursive NL-supportive oracles generating various relativized worlds where three classes para-L, PsubLIN, and para-P have specific computational power. Notice that para-$\text{L}^A \subseteq \text{PsubLIN}^A \subseteq$ para-P^A holds for all oracles A.

Theorem 1. *There exist recursive* NL-*supportive oracles A, B, C, and D satisfying the following conditions.*

1. para-$\text{L}^A = \text{PsubLIN}^A =$ para-P^A.
2. para-$\text{L}^B \subsetneq \text{PsubLIN}^B \subsetneq$ para-P^B.
3. para-$\text{L}^C = \text{PsubLIN}^C \subsetneq$ para-P^C.
4. para-$\text{L}^D \subsetneq \text{PsubLIN}^D =$ para-P^D.

The difficulty in proving each claim in Theorem 1 lies in the fact that we need to (i) deal with the fluctuations of the values of size parameters of parameterized decision problems (notice that the standard binary length of inputs is monotonically increasing) and to (ii) satisfy three or four conditions simultaneously for parameterized complexity classes by avoiding any conflict occurring during the construction of the desired oracles.

We can prove other relationships among para-L, para-NL, and PsubLIN. Concerning a question of whether or not para-NL $\subseteq$ PsubLIN, we can present four different relativized worlds in which para-NL $\subseteq$ PsubLIN and para-NL $\not\subseteq$ PsubLIN separately hold between para-NL and PsubLIN in relation to para-L.

Theorem 2. *There exist recursive* NL-*supportive oracles A, B, C, and D satisfying the following conditions.*

1. para-L^A = para-NL^A = PsubLIN^A.
2. para-L^B = para-NL^B $\subsetneq$ PsubLIN^B.
3. para-L^C ≠ para-NL^C $\not\subseteq$ PsubLIN^C.
4. para-L^D ≠ para-NL^D $\subseteq$ PsubLIN^D.

The relationships given in Theorems 1 and 2 suggest that any relativizable proof is not sufficient to separate para-L, para-NL, PsubLIN, and para-P.

2   Preliminaries

We briefly explain basic terminology necessary for the later sections. We use $\mathbb{N}$ to denote the set of all *natural numbers* (i.e., nonnegative integers) and we set $\mathbb{N}^+ = \mathbb{N} - \{0\}$. Given a number $n \in \mathbb{N}^+$, $[n]$ expresses the set $\{1, 2, \ldots, n\}$. In this paper, all *polynomials* have nonnegative integer coefficients and all *logarithms* are taken to the base 2. We define $\log^* n$ as follows. First, we set $\log^{(0)} n = n$ and $\log^{(i+1)} n = \log(\log^{(i)} n)$ for each index $i \in \mathbb{N}$. Finally, we set $\log^* n$ to be the minimal number $k \in \mathbb{N}$ satisfying $\log^{(k)} n \leq 1$.

An *alphabet* Σ is a nonempty finite set and a *string over* Σ is a finite sequence of elements of Σ. A *language over* Σ is a subset of Σ^*. We freely identify a decision problem with its associated language over Σ. The *length* (or *size*) of a string x is the total number of symbols in x and is denoted $|x|$. We write $\overline{L}$ for the set $\Sigma^* - L$ when Σ is clear from the context. A function $f : \Sigma^* \to \Sigma^*$ (resp., $f : \Sigma^* \to \mathbb{N}$) is *polynomially bounded* if there exists a polynomial p satisfying $|f(x)| \leq p(|x|)$ (resp., $f(x) \leq p(|x|)$) for all $x \in \Sigma^*$.

We use *deterministic Turing machines* (DTMs) and *nondeterministic Turing machines* (NTMs), each of which has a read-only input tape and a rewritable work tape. If necessary, we also attach a write-only[1] output tape. All tapes have the left endmarker ¢ and stretch to the right. Additionally, an input tape has the right endmarker \$. When a DTM begins with an initial state and, whenever it enters a halting state (either an accepting state or a rejecting state), it halts. We say that a DTM M *accepts* (resp., *rejects*) input x if M starts with x written on an input tape (surrounded by the two endmarkers) and eventually enters an accepting (resp., a rejecting) state. Similarly, an NTM *accepts* x if there exists a series of nondeterministic choices that lead the NTM to an accepting state. Otherwise, the NTM *rejects* x. A machine M is said to *recognize* a language L if, for all $x \in L$, M accepts x, and for all $x \in \Sigma^* - L$, M rejects x.

The notation L (resp., NL) refers to the class of all languages recognized by DTMs (resp., NTMs) using space $O(\log n)$, where "n" is a symbolic input size. Moreover, P stands for the class of languages recognized by DTMs in time $n^{O(1)}$. It is known that L $\subseteq$ NL = co-NL $\subseteq$ LOG^{2}SPACE $\cap$ P [3,6,8,9].

An *oracle* is an external device that provides useful information to an underlying Turing machine, which is known as an *oracle Turing machine*. In this paper, oracles are simply languages over a certain alphabet. An oracle Turing

---

[1] A tape is *write only* if a tape head must move to the right whenever it writes any non-blank symbol.

machine M is equipped with an extra *query tape*, on which the machine writes a query word, say, w and enters a query state q_{query} that triggers an oracle query. We demand that any query tape should be *write only* because, otherwise, the query tape can be used as an extra work tape composed of polynomially many tape cells. Triggered by an oracle query, an oracle X responds by modifying the machine's inner state from q_{query} to either q_{yes} or q_{no}, depending on $w \in X$ or $w \notin X$, respectively. Simultaneously, the query tape becomes empty and its tape head is returned to ϕ. Given an oracle Turing machine M and an oracle A, the notation $L(M, A)$ expresses the set of all strings accepted by M relative to A.

A *size parameter* is a function from Σ^* to $\mathbb{N}^+$ for a certain alphabet Σ. A *log-space size parameter* $m : \Sigma^* \to \mathbb{N}$ is a size parameter for which there exists a DTM M equipped with a write-only output tape that takes a string $x \in \Sigma^*$ and produces $1^{m(x)}$ on the output tape using $O(\log |x|)$ space. As a special size parameter, we write "$||$" to denote the size parameter m defined by $m(x) = |x|$ for any x. The notation LSP indicates the set of all log-space size parameters. Given a size parameter m and any index $n \in \mathbb{N}$, we set $\Sigma_n = \{x \in \Sigma^* \mid m(x) = n\}$. Note that $\Sigma_i \cap \Sigma_j = \emptyset$ for any distinct pair $i, j \in \mathbb{N}$ and that $\Sigma^* = \bigcup_{n \in \mathbb{N}} \Sigma_n$. A pair (L, m) with a decision problem (equivalently, a language) L and a size parameter m is called a *parameterized decision problem* and any collection of parameterized decision problems is called a *parameterized complexity class*. We informally use the term "parameterization" for underlying decision problems and complexity classes if we supplement size parameters to their instances.

As noted in Sect. 1.1, the prefix "para-" is used to distinguish parameterized complexity classes from standard complexity classes. With this convention, for two functions s and t, the notation para-DTIME, SPACE$(t(|x|, m(x)),$ $s(|x|, m(x)))$, where "$m(x)$" in "$\log m(x)$" indicates a symbolic size parameter m with a symbolic input x, denotes the collection of all parameterized decision problems with log-space size parameters, each (L, m) of which is solved (or recognized) by a certain DTM M in time $O(t(|x|, m(x)))$ using space $O(s(|x|, m(x)))$. Its nondeterministic variant is denoted by para-NTIME, SPACE$(t(|x|, m(x)),$ $s(|x|, m(x)))$. We set para-NL to be $\bigcup_{c \in \mathbb{N}}$ para-NTIME, SPACE$((|x|m(x))^c,$ $\log |x|m(x))$ and para-L to be $\bigcup_{c \in \mathbb{N}}$ DTIME, SPACE$((|x|m(x))^c, \log |x|m(x))$. Moreover, we set para-LOG^{2}SPACE to be para-DTIME, SPACE$(|x|^{\log |x|}$ $m(x)^{\log m(x)}, \log^2 |x| + \log^2 m(x))$. When we take $m(x) = |x|$, those parameterized complexity classes coincide with the corresponding "standard" complexity classes. In addition, we define PsubLIN as $\bigcup_{c, k \in \mathbb{N}, \varepsilon \in [0,1)}$ DTIME, SPACE $((|x|m(x))^c, m(x)^\varepsilon \log^k |x|)$. Given a parameterized complexity class para-$\mathcal{C}$, its complement class co-para-$\mathcal{C}$ is composed of all parameterized decision problems (L, m) for which $(\overline{L}, m)$ belongs to para-$\mathcal{C}$. The relativization of para-NL with an oracle A is denoted by para-NL^A and is obtained by replacing underlying Turing machines for para-NL with oracle Turing machines. In a similar fashion, we define para-L^A, para-P^A, and PsubLIN^A.

Formally, we introduce the notion of NL-supportive oracles.

Definition 3. *An oracle A is said to be* NL-supportive *if the following three conditions hold: (1)* para-L^A $\subseteq$ para-NL^A $\subseteq$ para-P^A, *(2)* para-NL^A = co-para-NL^A, *and (3)* para-NL^A $\subseteq$ para-LOG^{2}SPACE^A.

Although Condition (1) holds for all oracles A, we include it for a clarity reason.

In the subsequence sections, we will provide the proofs of Theorems 1 and 2.

3 Proofs of Theorem 1

We will give necessary proofs that verify our main theorems. We begin with proving Theorem 1.

Our goal is to construct oracles A, B, C, and D that satisfy Theorem 1(1)–(4). Here, we start with the first claim of Theorem 1.

Proof of (1). Note that, if para-L^A = para-P^A, then A is NL-supportive because we obtain para-NL^A = co-para-NL^A = para-L^A $\subseteq$ para-LOG^{2}SPACE. Let A be any P-complete problem (via log-space many-one reductions). We first claim that para-P^A $\subseteq$ para-L^A. Since $A \in$ P, we obtain para-P^A $\subseteq$ para-P. Since A is P-complete, it follows that para-P $\subseteq$ para-L^A, as requested. □

In the proofs of Theorem 1(2)–(4) that will follow shortly, we need several effective enumerations of pairs consisting of machines and size parameters. First, let $\{(M_i, m_i)\}_{i \in \mathbb{N}^+}$ be an effective enumeration of all such pairs satisfying that, for each index $i \in \mathbb{N}^+$, m_i is in LSP and M_i is a DTM running in time at most $(|x| m_i(x))^{c_i} + c_i$ using space at most $m_i(x)^{\varepsilon_i} \log^{k_i} |x| + c_i$ on all inputs x and for all oracles for appropriately chosen constants $k_i, c_i > 0$ and $\varepsilon_i \in [0, 1)$. Moreover, let $\{(D_i, m_i)\}_{i \in \mathbb{N}^+}$ denote an effective enumeration of pairs for which each m_i belongs to LSP and each D_i is a DTM running in time at most $(|x| m_i(x))^{a_i} + a_i$ using space at most $a_i \log |x| m_i(x) + a_i$ on all inputs x and for all oracles for a certain constant $a_i > 0$. Next, we use an effective enumeration $\{(P_i, m_i)\}_{i \in \mathbb{N}^+}$, where each m_i is in LSP and each DTMs P_i runs in time at most $(|x| m_i(x))^{b_i} + b_i$ on all inputs x and for all oracles, where b_i is an absolute positive constant. We also assume an effective enumeration $\{(N_i, m_i)\}_{i \in \mathbb{N}^+}$ such that, for each $i \in \mathbb{N}^+$, m_i is in LSP and N_i is an NTM running in time at most $(|x| m_i(x))^{e_i} + e_i$ using space at most $e_i \log |x| m_i(x) + e_i$ on all inputs x and for all oracles for a certain absolute constant $e_i > 0$. For notational simplicity, we write $\overline{M}$ to express an oracle machine obtained from M by exchanging between Q_{acc} and Q_{rej}. With this notation, each $\overline{N}_i$ relative to oracle A together with m_i induces a parameterized decision problem in co-para-NL^A.

Since every log-space size parameter is computed by a certain log-space DTM equipped with an output tape, we can enumerate all log-space size parameters by listing all such DTMs as $(K_1, K_2, \ldots)$. For each index $i \in \mathbb{N}^+$, we write m_j for the size parameter computed by K_j as long as it is obvious from the context. Since m_j is polynomially bounded, it is possible to assume that $m_j(x) \leq |x|^{g_j} + g_j$ for all j and x, where g_j is an absolute positive constant.

Our construction of the desired oracles will proceed by stages. To prepare such stages, for a given finite set $\Theta \subseteq \mathbb{N}^+$, let us define an index set $\Lambda = \{(n,l) \mid l \in \Theta, n \in \mathbb{N}^+\} \cup \{0\}$ together with an appropriate effective enumeration of all elements in Λ defined by the following linear order $<$ on Λ: (1) $0 < t$ holds for all $t \in \Lambda - \{0\}$ and (2) $(n',l') > (n,l)$ iff either $n' > n$ or $n' = n \wedge l' > l$. Given a number $n \in \mathbb{N}^+$, let $S_n = \{(x,i) \mid x \in \Sigma^n, i \in [\log^* n]\}$ with $\Sigma = \{0,1\}$. Note that $|S_n| = 2^n \log^* n$. We also define a linear ordering $<$ on S_n with respect to $\{e_i\}_{i \in \mathbb{N}^+}$ in the following way: letting $k_{x,i} = (|x|m_i(x))^{e_i} + e_i$, $(x,i) < (y,j)$ iff one of the following conditions hold: $k_{x,i} < k_{y,j}$, $k_{x,i} = k_{y,j} \wedge i < j$, and $k_{x,i} = k_{y,j} \wedge i = j \wedge x < y$ (lexicographically). According to this ordering $<$, we choose all elements of S_n one by one in the increasing order.

Proof of (2). We wish to construct an oracle D that meets the following four conditions: (i) $\mathrm{PsubLIN}^B \not\subseteq \mathrm{para\text{-}L}^B$, (ii) $\mathrm{para\text{-}P}^B \not\subseteq \mathrm{PsubLIN}^B$, (iii) $\mathrm{para\text{-}NL}^B \subseteq \mathrm{para\text{-}LOG^2SPACE}^B$, and (iv) $\mathrm{co\text{-}para\text{-}NL}^B \subseteq \mathrm{para\text{-}NL}^B$. These conditions obviously ensure the desired claim of $\mathrm{para\text{-}L}^B \subsetneqq \mathrm{PsubLIN}^B \subsetneqq \mathrm{para\text{-}P}^B$.

We want to introduce two example languages for (i) and (ii). First, we set $y_j = 10^{r_x - j}1^j$ and $u_j = B(101^i x \# 0 y_j)$ for all $j \in [r_x]$, where $r_x = \lceil\sqrt{|x|}\rceil$. We write u for $u_1 u_2 \cdots u_{r_x}$ and set $L_1^B = \{101^i x \mid 101^i x \# 1u \in B\}$. It is not difficult to show that $(L_1^B, \|)$ belongs to $\mathrm{PsubLIN}^B$ for any oracle B by running the following algorithm: first produce all words $101^i x \# 0 y_j$ one by one, query them to obtain u from B, remember all answer bits u_j, and finally query $101^i x \# 1u$. Similarly, we define $y_j' = 10^{|x|-j}1^j$ and $u_j = B(1^2 01^i x \# 0 y_j')$ for each $j \in [\|x\|]$ and set $L_2^B = \{1^2 01^i x \mid 1^2 01^i x \# 1u' \in B\}$, where $u' = u_1 u_2 \cdots u_n$. Note that $(L_2^B, \|) \in \mathrm{para\text{-}P}^B$ for any oracle B. Through our oracle construction, we will define two sequences $\{n_{s_1}\}_{s_1 \in \mathbb{N}^+}$ and $\{n'_{s_2}\}_{s_2 \in \mathbb{N}^+}$. For readability, we set $k_{x,i} = (|x|m_i(x))^{a_i} + a_i$, $k'_{x,i} = (|x|m_i(x))^{c_i} + c_i$, and $k''_{x,i} = (|x|m_i(x))^{e_i} + e_i$ for any $x \in \Sigma^*$ and $i \in \mathbb{N}^+$.

In this proof, we set $\Theta = \{1,2,3\}$ and define Λ as stated before. For each $t \in \Lambda$, we want to construct two sets B_t and R_t. At Stage 0, we set $B_0 = R_0 = \emptyset$ and $n_0 = n'_0 = 1$. Moreover, we set two counters s_1 and s_2 to 1. In what follows, we deal with Stage $t = (n,l) \in \Lambda$ and the values of s_1 and s_2. By induction hypothesis, we assume that, for all $t' < t$, $B_{t'}$ and $R_{t'}$ have been already defined. Moreover, we assume that, for all $e_1 < s_1$ and $e_2 < s_2$, n_{e_1} and n'_{e_2} have been appropriately defined. For simplicity, let $B' = \bigcup_{t' < t} B_{t'}$ and $R' = \bigcup_{t' < t} R_{t'}$.

During the construction process of B, the value of $k_{x,i}$ may fluctuate, depending on (x,i), and this fact may make many words reserved, leaving no room for inserting extra strings to define B in (c)–(d). To avoid such a situation, we will use S_n and its linear ordering $<$ with respect to $\{c_i\}_{i \in \mathbb{N}^+}$. For our convenience, let $Z_{x,n}^{(3)} = \{1^3 01^i x \# 0^{k''_{x,i}} z \mid |z| = \lceil 2 \log \log n \rceil\}$ and $Z_{x,n}^{(4)} = \{1^4 01^i x \# z \mid |z| = k''_{x,i}\}$.

(a) Case $l = 1$. Our target is Condition (i). Consider the size parameter $m(x) = |x|$. If $n < \max\{n_{s_1}, n'_{s_2}\}$, then we skip this case and move to Case $l = 2$. Now, let us assume otherwise. We try to find a room for diagonalization by avoiding all the reserved words defined in the previous stages. For this purpose, we first check whether there exist a number $\tilde{n} \in \mathbb{N}^+$ and a string $x \in \Sigma^{\tilde{n}}$

satisfying that (*) $\max_{i \in [\log^* n]}\{4n^{2a_i g_i} + a_i\} < \tilde{n} \leq 2^n$ and, for any $j \in \{3,4\}$, $|R' \cap Z_{x,\tilde{n}}^{(j)}| + |\tilde{x}|^{2a_i} + a_i + r_x + 1 < \tilde{n}^{\log \tilde{n}}$, where $\tilde{x} = 101^i x$. The latter condition is to make enough room for L_1^B as well as the constructions in (c)–(d). If (*) is not satisfied for all $\tilde{n}$ and x, then we skip this case. Assuming that (*) is satisfied for certain $\tilde{n}$ and x, we fix such a pair $(\tilde{n}, x)$ for the subsequent argument.

Take the machine D_i and the input $\tilde{x} = 101^i x$. Recall that D_i runs in time at most $|\tilde{x}|^{2a_i} + a_i$ using space at most $2a_i \log|\tilde{x}| + a_i$. Let $\tilde{R}_t = \{101^i x \# 0 y_j \mid j \in [r_x]\}$. Although D_i may possibly query all words of the form $101^i x \# 0 y_j$ for $j \in [r_x]$, it cannot remember all oracle answers u_j, because the work tape space of D_i is smaller than r_x. From this fact and also by (*), there is a set $B_t \subseteq \tilde{R}_t \cup \{101^i x \# 1u \mid u = u_1 \cdots u_{r_x}, u_j = B'(101^i x \# 0 y_j), j \in [r_x]\} - R'$ for which $D_i^{B' \cup B_t}(\tilde{x}) \neq L_1^{B' \cup B_t}(\tilde{x})$. We define R_t to include $\tilde{R}_t \cup B_t$, all queried words of D_i on $\tilde{x}$ relative to $B' \cup B_t$, and $101^i x \# 1u$, where $u = u_1 \cdots u_{r_x}$ and $u_j = B'(101^i x \# 0 y_j)$ for all $j \in [r_x]$. Note that $|R' \cap R_t| < \tilde{n}^{\log \tilde{n}}$. Before leaving this case, we set n_{s_1+1} to be $\tilde{n}$ and then increment the counter from s_1 to $s_1 + 1$.

(b) Case $l = 2$. Hereafter, we try to satisfy Condition (ii). For this purpose, let us assume that B' and R' have been updated. We will make an argument similar to (a) using L_2^B instead of L_1^B. We skip this case and move to Case $l = 3$ if $n < \max\{n_{s_1}, n'_{s_2}\}$. Otherwise, we check if there are a number $\tilde{n}$ and a string $x \in \Sigma^{\tilde{n}}$ satisfying that (*) $\max_{i \in [\log^* n]}\{4n^{2c_i g_i} + c_i\} < \tilde{n} \leq 2^n$ and $|R' \cap Z_{x,\tilde{n}}^{(j)}| + |\tilde{x}|^{2c_i} + c_i \tilde{n} + 1 < \tilde{n}^{\log \tilde{n}}$ for any index $j \in \{3,4\}$, where $\tilde{x} = 1^2 01^i x$. If no pair $(\tilde{n}, x)$ satisfies (*), then we skip this case as well. Next, we assume (*) for certain $\tilde{n}$ and x. We consider the machine M_i and feed the input $\tilde{x} = 1^2 01^i x$ to M_i. We then choose a set $B_t \subseteq \tilde{R}_t \cup \{101^i x \# 1u \mid u = u_1 \cdots u_{\tilde{n}}, u_j = B'(1^2 01^i x \# 0 y_j), j \subset [\tilde{n}]\} - R'$, where $y_j = 1^{\tilde{n}-j} 0^j$, satisfying $M_i^{R' \cup B_t}(\tilde{x}) \neq L_2^{B' \cup B_t}(\tilde{x})$. Note that M_i cannot remember all values u_j for any $j \in [\tilde{n}]$ using its work tape because the work tape space is bounded by $|\tilde{x}|^{\varepsilon_i} \log^{k_i}|\tilde{x}| + c_i < \tilde{n}$. Let R_t be composed of $\tilde{R}_t$, all queried words of M_i on $\tilde{x}$ relative to $B' \cup B_t$, and $1^2 01^i x \# 1u$, where $u = u_1 \cdots u_{\tilde{n}}$ and $u_j = B'(1^2 01^i x \# 0 y_j)$ for all $j \in [\tilde{n}]$. Finally, we set $n'_{s_2+1} = \tilde{n}$ and increment the counter from s_2 to $s_2 + 1$.

(c) Case $l = 3$. We target Condition (iii). Consider S_n and its linear ordering $<$ with respect to $\{e_i\}_{i \in \mathbb{N}^+}$. We inductively choose all pairs (x, i) in S_n one by one in the increasing order. For each element (x, i), let us consider N_i with x and B'. For convenience, we write $B_{t,<(x,i)}$ to denote the union of all $B_{t,(y,j)}$ for any $(y, j) \in S_n$ with $(y, j) < (x, i)$. Similarly, we write $R_{t,<(x,i)}$. In this case, we need to find an appropriate query word deterministically to simulate N_i on x. Since $m_i(x) \leq |x|^{g_i} + g_i$, it follows that $k''_{x,i} = (|x| m_i(x))^{e_i} + e_i \leq 4|x|^{e_i g_i} + e_i$. We set $W_{x,i} = \{1^3 01^i x \# 0^{k''_{x,i}} y \mid |y| = \lceil 2 \log \log |x| \rceil\}$.

Here, we define a new machine H_i as follows. On input w, query all words of the form $1^3 01^i w \# 0^{k''_{w,i}} y$ for any y of length $\lceil 2 \log \log |w| \rceil$. Note that the number of different y's is $2^{\lceil 2 \log \log |w| \rceil}$, which is at most $2 \log^2 |w|$. Collect all answers from an oracle. Let u be the sequence of oracle answers in order. Finally, make a query of $1^3 01^i w \# 1^{k''_{w,i}} u$. If the oracle answers YES, accept w; otherwise, reject w. Take strings of the form $1^3 01^i x \# 0^{k''_{x,i}} y$ in $W_{x,i}$ so that, for the string u

obtained from $B' \cup B_{t,<(x,i)}$, $1^301^ix\#1^{k''_{x,i}}u$ is not in $R' \cup R_{t,<(x,i)}$. We include all those strings into $B_{t,(x,i)}$. It follows that $x \in L(N_i, B' \cup B_{t,<(x,i)} \cup B_{t,(x,i)})$ iff $x \in L(H_i, B' \cup B_{t,<(x,i)} \cup B_{t,(x,i)})$. Next, we define $R_{t,(x,i)}$ to include all queried strings of N_i, $\{1^301^ix\#0^{k''_{x,i}}y \mid |y| = \lceil 2\log\log|x|\rceil\}$, and $1^301^ix\#1^{k''_{x,i}}u$, where u is the word determined by the query answers. In the end, we set $B_t = \bigcup_{(x,i)\in S_n} B_{t,(x,i)}$ and $R_t = \bigcup_{(x,i)\in S_n} R_{t,(x,i)}$.

(d) Case $l = 4$. We aim at Condition (iv). Consider S_n and its linear ordering $<$ with respect to $\{c_i\}_{i\in\mathbb{N}^+}$. Choose all pairs $(x,i) \in S_n$ one by one and define two sets $B_{t,(x,i)}$ and $R_{t,(x,i)}$. We run $\overline{N}_i$ on the input x with the oracle C'. Note that $\overline{N}_i$ runs in time at most $k''_{x,i}$ using space at most $c_i\log|x|m_i(x) + c_i$ for all oracles. Note that $k''_{x,i} \leq 4|x|^{2c_ig_i} + c_i$ since $m_i(x) \leq |x|^{g_i} + g_i$. Let us consider the set $V_t = \{1^401^ix\#z \mid |z| = k''_{x,i}\}$. Note that the runtime bound of N_i makes it impossible for $\overline{N}_i$ to query any string in V_t. A new machine G_i is defined to work as follows: on input w, nondeterministically generate $1^401^iw\#z$ for all strings $z \in \Sigma^{k''_{w,i}}$ and query it. If an oracle answers YES, then accept w; otherwise, reject w. If (x,i) is the smallest element in S_n, then we set $B_{t,<(x,i)} = R_{t,<(x,i)} = \emptyset$; otherwise, we define $R_{t,<(x,i)} = \bigcup_{(y,j)<(x,i)} R_{t,(y,j)}$.

We define $B_{t,(x,i)}$ as follows: if $\overline{N}_i$ accepts x relative to B', then we set $B_{t,(x,i)} = \{1^401^ix\#z_{x,i}\}$, where $z_{x,i} = \min\{z \in \Sigma^{k''_{x,i}} \mid 1^401^ix\#z \notin R' \cup R_{t,<(x,i)}\}$; otherwise, we set $C_{t,(x,i)} = \emptyset$. We define $R_{t,(x,i)} = R' \cup \{w \mid w \text{ is queried by } \overline{N}_i \text{ on } x\}$. It follows by the definition that $x \in L(\overline{N}_i, B' \cup B_{t,<(x,i)} \cup B_{t,(x,i)})$ iff $x \notin L(G_i, B' \cup B_{t,<(x,i)} \cup B_{t,(x,i)})$. Before leaving this case, we set $B_t = \bigcup_{(x,i)\in S_n} B_{t,(x,i)}$ and $R_t = \bigcup_{(x,i)\in S_n} R_{t,(x,i)}$.

Finally, we define $B = \bigcup_{t\in\Lambda} B_t$. By the construction of B, Conditions (i)–(iv) are all satisfied. $\square$

The third claim of Theorem 1 is proven below.

Proof of (3). To verify the target claim (3), it suffices for us to construct a set C for which (i) para-$P^C \nsubseteq PsubLIN^C$, (ii) para-$NL^C \subseteq$ para-L^C, and (iii) $PsubLIN^C \subseteq$ para-L^C.

Similar to the proof of (2), we prepare Λ, S_n, C_t, and R_t with a counter s, starting at $s = 1$. Let us assume that we reach Stage $t = (n,l) \in \Lambda$ and the counter has advanced to s. Initially, we set $C' = \bigcup_{t'<t} C_{t'}$ and $R' = \bigcup_{t'<t} R_{t'}$. Case $l = 1$, which targets Condition (i), is similar to that in the proof of (2). In what follows, we discuss only Cases $l \in \{2,3\}$. For simplicity, we set $k_{x,i} = (|x|m_i(x))^{e_i} + e_i$ and $k'_{x,i} = (|x|m_i(x))^{c_i} + c_i$ for any x and i.

(a) Case $l = 2$. We aim at fulfilling Condition (ii). After treating Case $l = 1$, we assume that C' and R' have been properly updated. Using a linear ordering $<$ on S_n with respect to $\{e_i\}_{i\in\mathbb{N}^+}$, we choose all pairs (x,i) in S_n one by one in the increasing order. Consider the computation of N_i on the input x in time at most $k_{x,i}$ using space at most $e_i\log|x|m_i(x) + e_i$. Let $R_{t,(x,i)}$ denote the set of all query words of N_i on x relative to $C' \cup C_{t,<(x,i)}$. Here, we introduce a new DTM E_i that works as follows. On input w, we make a query of the form $1^201^iw\#0^{k_{w,i}}$ and accepts (resp., rejects) w if its oracle answer is YES (resp.,

NO). Define $C_{t,(x,i)} = \{1^2 01^i x \# 0^{k_{x,i}}\}$ if N_i accepts x relative to $B' \cup C_{t,<(x,i)}$, and $C_{t,(x,i)} = \emptyset$ otherwise. Since N_i cannot query $1^2 0^i x \# 0^{k_{x,i}}$, it follows that $x \in L(N_i, B' \cup C_{t,<(x,i)} \cup C_{t,(x,i)})$ iff $x \in L(E_i, C' \cup C_{t,<(x,i)} \cup C_{t,(x,i)})$.

(b) Case $l = 3$. Our goal is to meet Condition (iii). Here, we use a linear ordering $<$ on S_n with respect to $\{c_i\}_{i\in\mathbb{N}^+}$. Similarly to (a), we assume that B' and R' have been properly updated after Case $l = 2$. Consider M_i and pick up all pairs $(x,i) \in S_n$ one by one in the increasing order according to $<$. Let us assume that we have already defined $B_{t,<(x,i)}$ and $R_{t,<(x,i)}$. Let $R_{t,(x,i)}$ be composed of all query words of M_i on x relative to $C' \cup C_{t,<(x,i)}$. Since M_i runs in time at most $k'_{x,i}$, we define $C_{t,(x,i)} = \{1^3 01^i x \# 0^{k'_{x,i}}\}$ if M_i accepts x relative to $C' \cup C_{t,<(x,i)}$, and $C_{t,(x,i)} = \emptyset$ otherwise. Consider a new DTM F_i defined as follows. On input w, compute $k'_{w,i}$, query $1^3 1^i w \# 0^{k'_{w,i}}$, accept (resp., reject) w if the oracle answers YES (resp., NO). We then obtain a relationship that $x \in L(M_i, C' \cup C_{t,<(x,i)} \cup C_{t,(x,i)})$ iff $x \in L(F_i, C' \cup C_{t,<(x,i)} \cup C_{t,(x,i)})$. Notice that F_i uses only space $O(\log k'_{x,i})$.

Finally, we set $C = \bigcup_{t\in\Lambda} C_t$. Clearly, Conditions (i)–(iii) are satisfied by this oracle C. $\qquad\square$

Next, we want to prove Theorem 1(4).

Proof of (4). Our goal of this proof is to construct an oracle D such that (i) $\mathrm{PsubLIN}^D \nsubseteq \mathrm{para\text{-}L}^D$, (ii) $\mathrm{para\text{-}P}^D \subseteq \mathrm{PsubLIN}^D$, (iii) $\mathrm{para\text{-}NL}^D \subseteq \mathrm{para\text{-}LOG^2SPACE}^D$, and (iv) $\mathrm{co\text{-}para\text{-}NL}^D \subseteq \mathrm{para\text{-}NL}^D$. A basic idea of constructing D is similar to the one used in Theorem 1(2). Cases $l \in \{1,3,4\}$, which target Conditions (i) and (iii)–(iv), are similar to (2). From Condition (i), it follows that $\mathrm{para\text{-}P}^D \nsubseteq \mathrm{para\text{-}L}^D$. At Stage $t = (l,i)$, assume that $D' = \bigcup_{t'<t} D_{t'}$ and $R' = \bigcup_{t'<t} R_{t'}$ have been already defined. Let $r_x = \lceil \sqrt{|x|} \rceil$.

(a) Case $l = 2$. This case is meant to satisfy Condition (ii). We consider S_n together with a linear ordering $<$ with respect to $\{b_i\}_{i\in\mathbb{N}^+}$, as defined before. Take all pairs (x,i) one by one. Let $k_{x,i} = (|x|m_i(x))^{b_i} + b_i$ and consider the machine P_i, which runs on any input x in time at most $k_{x,i}$. To simulate P_i, we use the following DTM H_i. On input w, H_i makes queries of the form $1^4 01^i w \# 0^{k_{w,i}} y_j$ with $y_j = 10^{r_w - j} 1^j$ for all $j \in [r_w]$ and collect their oracle answers $u_j = D'(1^4 01^i w \# 0^{k_{w,i}} y_j)$. Letting $u = u_1 u_2 \cdots u_{r_w}$, H_i then queries the word $1^4 01^i w \# 1^{k_{w,i}} u$ to an oracle. If the oracle answers YES, then we accept w; otherwise, we reject w. This machine H_i is indeed an oracle PsubLIN-machine.

If x is the first string in Σ^n, then we set $D_{t,<(x,i)} = R_{t,<(x,i)} = \emptyset$. If x is not the first element, then we set $D_{t,<(x,i)}$ as $\bigcup_{y<(x,i)} D_{t,y}$ and set $R_{t,<(x,i)}$ as $\bigcup_{y<(x,i)} R_{t,y}$. Let $\tilde{R}_{t,x,i} = \{1^4 01^i x \# 1^{k_{x,i}} y_j \mid j \in [r_x]\}$. Since there is enough room for $D_{t,(x,i)}$ by Case $l = 1$, we can choose a set $D_{t,(x,i)} \subseteq \tilde{R}_{t,x,i} \cup \{1^4 01^i x \# 1^{k_{x,i}} u\} - R' \cup R_{t,<(x,i)}$ so that $x \in L(P_i, D' \cup D_{t,<(x,i)} \cup D_{t,(x,i)})$ iff $x \in L(H_i, D' \cup D_{t,<(x,i)} \cup D_{t,(x,i)})$. Finally, we define R_t to include $\tilde{R}_{t,x,i} \cup D_{t,(x,i)}$ and all query words of P_i as well as H_i on x relative to $D' \cup D_{t,<(x,i)}$ for all pairs $(x,i) \in S_n$.

It is not difficult to show that Condition (iv) is satisfied. $\qquad\square$

Combining all the proofs for (1)–(4), we now complete the proof of Theorem 1.

4   Proof of Theorem 2

We will verify Theorem 2. In Theorem 1, we utilize the fact that the inclusion relationship of para-$L^A \subseteq$ PsubLIN$^A \subseteq$ para-P^A holds for any oracle A. Unlike this case, we cannot expect a similar inclusion relationships for para-L^A, para-NL^A, and PsubLIN^A. Since the proof of Theorem 2 requires effective enumerations of various oracle machines, we need to recall from Sect. 3 the effective enumerations $\{(M_i, m_i)\}_{i \in \mathbb{N}^+}$, $\{(D_i, m_i)\}_{i \in \mathbb{N}^+}$, $\{(N_i, m_i)\}_{i \in \mathbb{N}^+}$, and $(K_1, K_2, \ldots)$. Moreover, we recall two index sets Λ with Θ and S_n with its linear ordering $<$ from Sect. 3.

Proof of (1)–(2). (1) This follows directly from Theorem 1(1).

(2) We require the following two conditions: (i) PsubLIN$^B \not\subseteq$ para-L^B and (ii) para-NL$^B \subseteq$ para-L^B. Note that Condition (ii) implies that para-NL^B = co-para-NL^C = para-$L^B \subseteq$ para-LOG^{2}SPACE^B. Condition (i) can be dealt with in a way similar to the proof of Theorem 1(2). Condition (ii) is also similar to the proof of Theorem 1(3).   □

Next, we prove the third claim of Theorem 2.

Proof of (3). Since para-$L^C \subseteq$ PsubLIN^C holds for any C, para-NL$^C \not\subseteq$ PsubLIN^C implies that para-$L^C \neq$ para-NL^C. Hence, we demand that the desired oracle C should satisfy that (i) para-NL$^C \not\subseteq$ PsubLIN^C, (ii) para-NL$^C \subseteq$ para-LOG^{2}SPACE^C, and (iii) co-para-NL$^C \subseteq$ para-NL^C. In this proof, we use an example language $L^C = \{101^i x \mid \exists z \in \Sigma^{|x|} [101^i x \# z \in C]\}$ for an oracle C. Note that $(L^C, \|) \in$ para-NL^C for any C.

In what follows, we want to construct the desired oracle C by stages. For the construction of C, we use $\Theta = \{1, 2, 3\}$ and define an index set Λ as done before. At each stage, we want to define C_t and also define a set R_t of *reserved words*. We will define a series $\{n_s\}_{s \in \mathbb{N}^+}$ of numbers by stages.

At Stage 0, we set $n_0 = 0$, $C_0 = R_0 = \emptyset$ and $n_0 = 1$. We also prepare a counter s, starting at $s = 1$. Let us consider Stage $t = (n, l)$ in Λ with a counter s. We assume that, for all elements t in Λ satisfying $t < (n, l)$, the sets C_t and R_t have been already defined. Moreover, we have already defined all n_e for $e < s$. For brevity, let $C' = \bigcup_{t < (n, l)} C_t$ and $R' = \bigcup_{t < (n, l)} R_t$. We will describe how to define C_t and R_t depending on the values of l and n_s. Cases $l \in \{2, 3\}$, which target Conditions (ii)–(iii), are similar to Theorem 1(2). Here, we explain only Case $l = 1$. Assume that s and n_s have been already defined.

(a) Case $l = 1$. In this case, we will target Condition (i). Take a simple size parameter m defined by $m(x) = |x|$ for all x. Whenever $n < n_s$, we skip this case. Next, we assume that $n = n_s$. Let $V_{x,n} = \{101^i x \# z \mid z \in \Sigma^n\}$. Letting $k_{x,i} = (|x| m_i(x))^{e_i} + e_i$, we define $Z_{x,n}^{(2)} = \{1^2 01^i x \# 0^{k_{x,i}} z \mid |z| = \lceil 2 \log \log n \rceil\}$ and $Z_{x,n}^{(3)} = \{1^3 01^i x \# z \mid |z| = k_{x,i}\}$. Check whether there exist a number $\tilde{n}$ and a string $x \in \Sigma^{\tilde{n}}$ such that (*) $\max_{i \in [\log^* n]}\{4n^{2c_i g_i} + c_i\} < \tilde{n} \leq 2^{n_s}$ and, for any

$j \in \{2,3\}$, $|R' \cap Z_{x,i}^{(j)}| + |\tilde{x}|^{2e_i} + e_i < \tilde{n}^{\log \tilde{n}}$, where $\tilde{x} = 101^i x$. If (*) is not satisfied for all $\tilde{n}$ and x, then we skip this case. Hereafter, we assume that (*) holds for certain $\tilde{n}$ and $x \in \Sigma^{\tilde{n}}$. take such a pair $(x, \tilde{n})$. We consider M_i, which on input $\tilde{x}$ runs in time at most $|\tilde{x}|^{2e_i} + e_i$ using space at most $|\tilde{x}|^{e_i} \log^{k_i} |x| + e_i$. Note that M_i cannot query all strings in $V_{x,\tilde{n}} - R'$. Our goal is to construct C_t (as-well as R_t) such that $\tilde{x} \in L(M_i, C' \cup C_t)$ iff $\tilde{x} \notin L^{C' \cup C_t}$, where $\tilde{x} = 101^i x$. For this purpose, we define $\tilde{R}_t$ to be the set of all queried words of M_i on $\tilde{x}$ relative to C'. We also define $C_t = \{101^i x \# z_{x,\tilde{n}}\}$ with $z_{x,\tilde{n}} = \min\{z \in \Sigma^{\tilde{n}} \mid 101^s x \# z \notin R' \cup \tilde{R}_t\}$ if M_i rejects $\tilde{x}$, and $C_t = \emptyset$ otherwise. Define $R_t = \tilde{R}_t \cup C_t$. Before leaving this case, we define n_{s+1} to be $\tilde{n}$ and then increment the counter from s to $s+1$.

By the construction of C, Conditions (i)–(iii) are all satisfied.   □

To close the proof of Theorem 2, we will verify the fourth claim of the theorem.

Proof of (4). Hereafter, we want to show that, for a certain recursive oracle D, (i) para-NL$^D \nsubseteq$ para-L^D, (ii) para-NL$^D \subseteq$ para-LOG^{2}SPACE^D, (iii) co-para-NL$^D \subseteq$ para-NL^D, and (iv) para-NL$^D \subseteq$ PsubLIN^D.

Stage by stage, we will construct the desired oracle D. We use the index set Λ. Cases $l \in \{2,3\}$, which respectively correspond to Conditions (ii)–(iii), are similar to the ones in the proof of Theorem 1(2). Hence, we will target Conditions (i) and (iv). We will define $\{n_s\}_{s \in \mathbb{N}^+}$, $\{D_t\}_{t \in \Lambda}$, and $\{R_t\}_{t \in \Lambda}$ by stages. At Stage $t = (l, i)$, we assume that all stages $t' < (n, l)$ have already been processed. Let $D' = \bigcup_{t' < t} D_{t'}$ and $R' = \bigcup_{t' < t} R_{t'}$. Let $k_{x,i} = (|x| m_i(x))^{e_i} + e_i$. We use $L^D = \{101^i x \mid \exists z \in \Sigma^{|x|} [101^i x \# z \in D]\}$ as an example language. We define $Z_{x,n}^{(2)} = \{1^2 01^i x \# 0^{k_{x,i}} z \mid |z| = \lceil 2 \log \log n \rceil\}$, $Z_{x,n}^{(3)} = \{1^3 01^i x \# z \mid |z| = k_{x,i}\}$, $Z_{x,n}^{(4)} = \{1^4 01^i x \# k_{x,i} y_j \mid y_j = 10^{r_x - j} 1^j, j \in [r_x]\}$, where $r_x = \lceil \sqrt{|x|} \rceil$.

(a) Case $l = 1$. This case corresponds to Condition (i) and it can be handled in a way similar to (a) of the proof of (3). Let s be the value of a counter. If $n < n_s$ holds, then we skip this case and move to Case $l = 2$. Let us assume that $n = n_s$. Check if there is a pair of $\tilde{n}$ and $x \in \Sigma^{\tilde{n}}$ satisfying that (*) $\max_{i \in \lceil \log^* n \rceil} \{4 n^{c_i g_i} + c_i\} < \tilde{n} \le 2^{n_s}$ and, for any index $j \in \{2,3,4\}$, $|R' \cap Z_{x,\tilde{n}}^{(j)}| + |\tilde{x}|^{2a_i} + a_i < 2^{\sqrt{\tilde{n}}}$. Let $V_{x,\tilde{n}} = \{101^i x \# z \mid z \in \Sigma^{\tilde{n}}\}$. If (*) is not satisfied for all pairs $(\tilde{n}, x)$, then we skip this case and advance to Case $l = 2$. Next, we assume that (*) holds for a certain pair $(\tilde{n}, x)$. Fix such a pair $(x, \tilde{n})$. Let us consider L^D and M_i running on the input $\tilde{x}$. Let $\tilde{R}_t$ be composed of all query words of D_i on $\tilde{x}$ relative to D'. It follows that $|\tilde{R}_t| \le |\tilde{x}|^{2a_i} + a_i$ because of the runtime of M_i. We choose the lexicographically smallest string $z_{x,i}$ in $\Sigma^{|\tilde{x}|}$ satisfying $101^i x \# z \notin R' \cup \tilde{R}_t$ and define $D_t = \{101^i x \# z_{x,i}\}$ if D_i rejects $\tilde{x}$, and $D_t = \emptyset$ otherwise. Finally, we set $R_t = \tilde{R}_t \cup V_{x,\tilde{n}}$. We also set $n_{s+1} = \tilde{n}$ and update the counter from s to $s+1$.

(b) Case $l = 4$. We aim at Condition (iv). Assume that D' and R' have been already updated. Recall an oracle PsubLIN-machine H_i from the proof of Theorem 1(4). Here, we use S_n and its linear ordering $<$ with respect to $\{e_i\}_{i \in \mathbb{N}^+}$. We pick up all pairs (x, i) in S_n one by one in order and assume that $D_{t, <(x,i)}$ and $R_{t, <(x,i)}$ have been defined. The machine H_i in the proof of Theorem 1(4) makes queries of the form $1^4 01^i x \# 0^{k_{x,i}} y_j$ with $y_j = 10^{r_x - j} 1^j$ for all $j \in [r_x]$, where $r_x = \lceil \sqrt{|x|} \rceil$. Collect their oracle answers $u_j = D'(1^4 01^i x \# 0 y_j)$. Finally, query the word $1^4 01^i x \# 1^{k_{x,i}} u$, where $u = u_1 u_2 \cdots u_{r_x}$. If its oracle answer is YES, then accept x; otherwise, reject x. We define $D_{t,(x,i)}$ as follows. If N_i rejects x relative to $D' \cup D_{t,<(x,i)}$, then we choose a set $D_{t,(x,i)} \subseteq \{1^4 01^i x \# 0^{k_{x,i}} y_j \mid j \in [r_x]\} \cup \{1^4 01^i x \# 1^{k_{x,i}} u\} - R' \cup R_{t,<(x,i)}$, where $u = u_1 u_2 \cdots u_{r_x}$ and $u_j = D'(1^4 01^i x \# 0^{k_{x,i}} y_j)$ for all $j \in [r_x]$, such that $x \in L(N_i, D' \cup D_{t,<(x,i)} \cup D_{t,(x,i)})$ iff $x \in L(H_i, D' \cup D_{t,<(x,i)} \cup D_{t,(x,i)})$. Otherwise, we set $D_{t,(x,i)} = \emptyset$. Since $|R' \cap \{1^4 01^i x \# z \mid z \in \Sigma^*\}| \leq |\tilde{x}|^{2a_i} + a_i < 2^{\sqrt{n}}$ by (a), $D_{t,(x,i)}$ must exist. Before leaving this case, we set $R_{t,(x,i)}$ to be the set of all query words of N_i and H_i on x.

Therefore, Conditions (i)–(iv) are satisfied.   □

5   A Brief Discussion on Supportive Oracles

We have introduced the notion of "NL-supportive oracle" to guarantee the known inclusion relationships associated with para-NL and thus make the relativization of para-NL keep its validity and meaningfulness of providing good information on the structural similarities and differences between para-NL and other parameterized complexity classes. With this notion, we have been able to demonstrate the existence of various relativized worlds in which different inclusion and separation relationships occur among four parameterized complexity classes: para-L, para-NL, PsubLIN, and para-P.

The usefulness of "supportive oracles" can be justified by the following argument. Take a quick look at a longstanding open problem: the P =?NP problem. There are known recursive oracles A and B for which $P^A = NP^A$ and $P^B \neq NP^B$ [1]. Once either P = NP or P ≠ NP is proven in the unrelativized world, the currently known relativization methodology produces a contradicting result against either P = NP or P ≠ NP. Therefore, we no longer use the current relativization of P and NP for a further study on relativized worlds associated with P and NP. However, if we consider "NP-supportive oracles", which supports the correct relativized relationship of either $P^A = NP^A$ or $P^A \neq NP^A$, depending on either P = NP or P ≠ NP, then we can avoid any conflicting relativized world and it thus remains worth investigating the relativization of complexity classes in relation to P and NP.

We strongly hope that the notion of supportive oracle will prove its importance in computational complexity.

References

1. Baker, T., Gill, J., Solovay, R.: Relativizations of the P=?NP question. SIAM J. Comput. **4**, 431–442 (1975)
2. Barnes, G., Buss, J.F., Ruzzo, W.L., Schieber, B.: A sublinear space, polynomial time algorithm for directed s-t connectivity. SIAM J. Comput. **27**, 1273–1282 (1998)
3. Immerman, N.: Nondeterministic space is closed under complement. SIAM J. Comput. **17**, 935–938 (1988)
4. Kirsig, B., Lange, K.J.: Separation with the Ruzzo, Simon, and Tompa relativization implies DSPACE[$\log n$] $\neq$ NSPACE[$\log n$]. Inf. Process. Lett. **25**, 13–15 (1987)
5. Ladner, R.E., Lynch, N.A.: Relativization of questions about log space computability. Math. Syst. Theory **10**, 19–32 (1976)
6. Reingold, O.: Undirected connectivity in log-space. J. ACM **55** (2008). Article 17 (24 pages)
7. Ruzzo, W.L., Simon, J., Tompa, M.: Space-bounded hierarchies and probabilistic computations. J. Comput. Syst. Sci. **28**, 216–230 (1984)
8. Savitch, W.J.: Relationships between nondeterministic and deterministic tape complexities. J. Comput. Syst. Sci. **4**, 177–192 (1970)
9. Szelepcsényi, R.: The method of forced enumeration for nondeterministic automata. Acta Informatica **26**, 279–284 (1988)
10. Yamakami, T.: Parameterized graph connectivity and polynomial-time sub-linear-space short reductions. In: Hague, M., Potapov, I. (eds.) RP 2017. LNCS, vol. 10506, pp. 176–191. Springer, Cham (2017). https://doi.org/10.1007/978-3-319-67089-8_13
11. Yamakami, T.: The 2CNF Boolean formula satisfiability problem and the linear space hypothesis. In: Proceedings of MFCS 2017. LIPIcs, vol. 83, pp. 62:1–62:14. Schloss Dagstuhl - Leibniz-Zentrum fuer Informatik (2017). A complete version is found at arXiv:1709.10453
12. Yamakami, T.: State complexity characterizations of parameterized degree-bounded graph connectivity, sub-linear space computation, and the linear space hypothesis. In: Konstantinidis, S., Pighizzini, G. (eds.) DCFS 2018. LNCS, vol. 10952, pp. 237–249. Springer, Cham (2018). https://doi.org/10.1007/978-3-319-94631-3_20. A complete and corrected version is found at arXiv:1811.06336

Dynamic Average Value-at-Risk Allocation on Worst Scenarios in Asset Management

Yuji Yoshida[✉] and Satoru Kumamoto

Faculty of Economics and Business Administration, University of Kitakyushu,
4-2-1 Kitagata, Kokuraminami, Kitakyushu 802-8577, Japan
{yoshida,kumamoto}@kitakyu-u.ac.jp

Abstract. A dynamic portfolio optimization model with average value-at-risks is discussed for drastic declines of asset prices. Analytical solutions for the optimization at each time are obtained by mathematical programming. By dynamic programming, an optimality equation for optimal average value-at-risks over time is derived. The optimal portfolios and the corresponding average value-at-risks are given as solutions of the optimality equation. A numerical example is given to understand the solutions and the results.

1 Introduction

In financial management, portfolio allocation is important technique to hedge risks and it is useful to make asset management stable. *Markowitz's mean-variance model* in classical portfolio theory is studied by many researchers, and the variance, i.e., volatility, is strongly related to risks in portfolio allocation [6,8,10,11]. Recently, *value-at-risk (VaR)* is used widely in financial management to estimate the risk of worst-scenarios. VaR is a risk-sensitive criterion based on percentiles, and it is one of the standard criteria in practical asset management [5,7,13,14]. VaR detects drastic declines of asset prices and it is useful to get rid of bad scenarios in investment, however it does not have coherency. *Coherent risk measures* have been studied to improve the criterion of risks with worst scenarios [1], and several improved risk measures based on value-at-risks are proposed: For example, conditional value-at-risk, expected shortfall, entropic value-at-risk, exponential type spectral measure [3,4,9,12]. *Average value-at-risk (AVaR)*, which is defined by VaR, is a coherent risk measure and it has good properties [1].

In this paper, we deal with a dynamic portfolio optimization problem with AVaR as a risk measure. To discuss worst-scenarios, at each time drastic declines of asset prices are estimated using AVaR and their total estimation, which is different from [13,14,16], is defined from worst values of AVaRs over time. Portfolio optimization at each time is solved by mathematical programming, and we obtain analytical solutions. Introducing AVaR based on conditional probability, dynamic optimization is discussed by dynamic programming, and we derive

© Springer Nature Switzerland AG 2019
T. V. Gopal and J. Watada (Eds.): TAMC 2019, LNCS 11436, pp. 674–683, 2019.
https://doi.org/10.1007/978-3-030-14812-6_42

the optimality equation and the optimal solution for the proposed problem. A numerical example is given to understand the obtained results.

2 A Dynamic Portfolio Model

Let $\mathbf{R} = (-\infty, \infty)$ and let $(\Omega, \mathcal{M}, P)$ be a probability space, where $\mathcal{M}$ is a σ-field and P is a non-atomic probability on a sample space Ω. Let $\mathcal{X}$ be the set of all integrable $\mathcal{M}$-adapted real-valued random variables X on Ω with a continuous distribution function $x \mapsto F_X(x) = P(X < x)$ for which there exists a non-empty open interval I such that $F_X(\cdot) : I \to (0,1)$ is strictly increasing and onto. Then there exists a strictly increasing and continuous inverse function $F_X^{-1} : (0,1) \to I$. *Value-at-risk (VaR)* at a positive probability p is given by the percentile of the distribution function F_X, i.e. $\mathrm{VaR}_p(X) = \sup\{x \in I \mid F_X(x) \leq p\}$ for $p \in (0,1)$ and $\mathrm{VaR}_1(X) = \sup I$. Hence we have $\mathrm{VaR}_p(X) = F_X^{-1}(p)$ for $p \in (0,1)$. Then *average value-at-risk (AVaR)* at a positive probability p is given by

$$\mathrm{AVaR}_p(X) = \frac{1}{p} \int_0^p \mathrm{VaR}_q(X)\, dq. \tag{1}$$

The following lemma is easily checked from [1,15].

Lemma 1. *Let a probability $p \in (0,1]$ and let random variables $X, Y \in \mathcal{X}$. Then the average value-at-risk AVaR_p has the following properties:*

 (i) *If $X \leq Y$, then $\mathrm{AVaR}_p(X) \leq \mathrm{AVaR}_p(Y)$. (monotonicity)*
 (ii) *$\mathrm{AVaR}_p(cX) = c\,\mathrm{AVaR}_p(X)$ for $c > 0$. (positive homogeneity)*
 (iii) *$\mathrm{AVaR}_p(X + c) = \mathrm{AVaR}_p(X) + c$ for $c \in \mathbf{R}$. (translation invariance)*
 (iv) *$\mathrm{AVaR}_p(X + Y) \geq \mathrm{AVaR}_p(X) + \mathrm{AVaR}_p(Y)$. (super-additivity)*

It is known that $-\mathrm{AVaR}_p$ is a *coherent risk measure* but $-\mathrm{VaR}_p$ is not coherent [1] because VaR_p does not have super-additivity in Lemma 1. To discuss the dynamics, we introduce AVaR based on conditional expectations. Let $\mathcal{G}$ be a sub-σ-field of $\mathcal{M}$ and let $E(\cdot \mid \mathcal{G})$ be the conditional expectation. Define a map $x \mapsto F_X(x \mid \mathcal{G}) = P(X < x \mid \mathcal{G}) = E(1_{\{X<x\}} \mid \mathcal{G})$, where 1_Γ denotes the characteristic function of a $\mathcal{M}$-measurable set Γ. Then we define VaR of $X(\in \mathcal{X})$ under condition $\mathcal{G}$ by $\mathrm{VaR}_p(X \mid \mathcal{G}) = \sup\{x \in I \mid F_X(x \mid \mathcal{G}) \leq p\}$ for a probability $p \in (0,1)$ and $\mathrm{VaR}_1(X \mid \mathcal{G}) = \sup I$. Further *AVaR under condition* $\mathcal{G}$ is also given by $\mathrm{AVaR}_p(X \mid \mathcal{G}) = \frac{1}{p} \int_0^p \mathrm{VaR}_q(X \mid \mathcal{G})\, dq$ for $p \in (0,1]$, and then we can easily check $\mathrm{VaR}_p(X \mid \mathcal{G})$ and $\mathrm{AVaR}_p(X \mid \mathcal{G})$ are $\mathcal{G}$-measurable random variables. Here the following lemma holds from [14].

Lemma 2. *Let $p \in (0,1]$ and let $\mathcal{G}$ be a sub-σ-field generated by Z and assume Y and $\mathcal{G}$ are independent. Then $\mathrm{AVaR}_p(\cdot \mid \mathcal{G})$ has the following properties:*

 (i) *$\mathrm{AVaR}_p(Y \mid \mathcal{G}) = \mathrm{AVaR}_p(Y)$.*
 (ii) *$\mathrm{AVaR}_p(Z \mid \mathcal{G}) = Z$.*
 (iii) *$\mathrm{AVaR}_p(ZX \mid \mathcal{G}) = Z\mathrm{AVaR}_p(X \mid \mathcal{G})$ if $Z \geq 0$.*
 (iv) *$\mathrm{AVaR}_p(X + Z \mid \mathcal{G}) = \mathrm{AVaR}_p(X \mid \mathcal{G}) + Z$.*

3 A Dynamic AVaR Allocation for Worst Scenarios

We deal with a portfolio optimization model with n risky assets and an expiration date T, where n and T are positive integers. For an asset $i = 1, 2, \cdots, n$, an *asset price process* $\{S_t^i\}_{t=0}^T$ is given by the *rate of return* $R_t^i (\in \mathcal{X})$ which satisfies $1 + R_t^i \geq 0$ and

$$S_t^i = S_{t-1}^i (1 + R_t^i) \tag{2}$$

for $t = 1, 2, \cdots, T$. Let $\mathcal{M}_t$ be a σ-field generated by random variables S_s^i ($s = 1, 2, \cdots, t$; $i = 1, 2, \cdots, n$) for $t \geq 1$ and let $\mathcal{M}_0 = \{\emptyset, \Omega\}$. We assume R_t^i is independent of the past information $\mathcal{M}_{t-1}$ for $t = 1, 2, \ldots, T$ and $i = 1, 2, \cdots, n$. Let a set of vectors $\mathcal{W} = \{(w^1, w^2, \ldots, w^n) \in \mathbf{R}^n \mid \sum_{i=1}^n w^i = 1$ and $w^i \geq 0$ $(i = 1, 2, \cdots, n)\}$. *Trading strategies* are given by *portfolio weight vectors* $(w^1, w^2, \cdots, w^n) \in \mathcal{W}$. Let time $t = 1, 2, \ldots, T$. Then the rate of return with a portfolio $(w_t^1, w_t^2, \cdots, w_t^n) \in \mathcal{W}$ is given by

$$R_t = \sum_{i=1}^n w_t^i R_t^i, \tag{3}$$

and we give the asset price with portfolio $(w_t^1, w_t^2, \cdots, w_t^n) \in \mathcal{W}$ as follows

$$S_t = S_{t-1} \sum_{i=1}^n w_t^i (1 + R_t^i) = S_{t-1}(1 + R_t), \tag{4}$$

where the initial asset price is given by $S_0 = 1$ for simplicity. This paper discusses drastic declines of asset prices in stock markets. Let a portfolio $(w_t^1, w_t^2, \cdots, w_t^n) \in \mathcal{W}$. The theoretical *bankruptcy* at time t occurs on scenarios ω satisfying $S_{t-1}(\omega) > 0$ and $S_t(\omega) \leq 0$, i.e. it follows $1 + R_t(\omega) \leq 0$ from (4). Similarly, for a constant $\delta \in [0, 1]$, a set of sample paths $\{\omega \in \Omega \mid 1 + R_t(\omega) \leq 1 - \delta\} = \{\omega \in \Omega \mid R_t(\omega) \leq -\delta\}$ is the scenarios where the asset price S_t will fall to lower than $100(1-\delta)$ % of the current price S_{t-1}, i.e. the rate is $100\,\delta$ %-falling. The parameter δ is called *the rate of falling*. Then the probability of falling is

$$p = P(R_t \leq -\delta). \tag{5}$$

For example, p denotes the probability of the falling below par value if '$\delta = 0$' and it indicates the probability of the bankruptcy if '$\delta = 1$'. From (5), for a probability p, the rate of falling is

$$\delta = -\mathrm{VaR}_p(R_t) \tag{6}$$

since P is non-atomic. This paper discusses the minimization of the rate of falling (6), i.e. the maximization of AVaR. In this paper, we deal with a case where $\mathrm{VaR}_p(R_t)$ in (6) has the following representation.

$$\begin{aligned}(\mathrm{VaR}) = \ &(\text{the mean}) - (\text{a positive constant } \kappa(p)) \\ &\times (\text{the standard deviation}),\end{aligned} \tag{7}$$

where the positive constant $\kappa(p)$ is given corresponding to probability p. Equation (7) holds if the rates of return R_t^i have normal distributions [2,7]. Estimating the total risks over time, we discuss the following dynamic portfolio problem regarding AVaR under information $\{\mathcal{M}_{t-1}\}_{t=1}^T$. Let a discount rate β be a positive constant.

Problem 1. Maximize the total AVaR

$$\bigwedge_{t=1}^T \beta^{t-1} E(\text{AVaR}_p(S_t \mid \mathcal{M}_{t-1})) \tag{8}$$

with portfolio weights $(w_t^1, w_t^2, \cdots, w_t^n) \in \mathcal{W}$ $(t = 1, 2, \cdots, T)$, where $\bigwedge_{t=1}^T = \min_{t=1}^T$.

This term implies the total risk of worst scenarios which occur on the transition from time $t-1$ to time t. In (4), portfolio weights $(w_t^1, w_t^2, \cdots, w_t^n) \in \mathcal{W}$ are decided sequentially and predictably, and then $1 + R_t = 1 + \sum_{i=1}^n w_t^i R_t^i$ is nonnegative and independent of the past information $\mathcal{M}_{t-1}$ for $t = 1, 2, \cdots .T$. By applying (4) and Lemma 2 to (8), we can easily check (8) follows

$$\bigwedge_{t=1}^T \beta^{t-1} \prod_{s=1}^{t-1}(1 + E(R_s)) \cdot (1 + \text{AVaR}_p(R_t)). \tag{9}$$

Because $1 + E(R_t) = 1 + \sum_{i=1}^n w_t^i E(R_t^i) \geq 0$ for $t = 1, 2, \cdots, T$, by dynamic programming we obtain the following equations from (9).

Theorem 1. *Let $\{v_t\}$ be a sequence defined inductively by the following backward optimality equations:*

$$v_t = \max_{(w_t^1, \cdots, w_t^n) \in \mathcal{W}} \min \left\{ 1 + \text{AVaR}_p \left(\sum_{i=1}^n w_t^i R_t^i \right), \right.$$
$$\left. \left(1 + \sum_{i=1}^n w_t^i E(R_t^i) \right) \beta v_{t+1} \right\} \tag{10}$$

for $t = 1, 2, \cdots, T-1$, and

$$v_T = \max_{(w_T^1, \cdots, w_T^n) \in \mathcal{W}} \left(1 + \text{AVaR}_p \left(\sum_{i=1}^n w_T^i R_T^i \right) \right). \tag{11}$$

Then v_1 is the optimal total AVaR in Problem 1.

4   Optimal Portfolios for AVaR at Each Time

We estimate the rate of return with portfolios based on fundamental results in [13,14], which have investigated quite different criteria from Problem 1. Let the mean and the covariance of the rate of return R_t^i respectively by $\mu_t^i = E(R_t^i)$ and $\sigma_t^{ij} = E((R_t^i - \mu_t^i)(R_t^j - \mu_t^j))$ for $i, j = 1, 2, \cdots, n$. Let vectors, matrix and real numbers

$$\mu_t = \begin{bmatrix} \mu_t^1 \\ \mu_t^2 \\ \vdots \\ \mu_t^n \end{bmatrix}, \Sigma_t = \begin{bmatrix} \sigma_t^{11} & \sigma_t^{12} & \cdots & \sigma_t^{1n} \\ \sigma_t^{21} & \sigma_t^{22} & \cdots & \sigma_t^{2n} \\ \vdots & \vdots & \ddots & \vdots \\ \sigma_t^{n1} & \sigma_t^{n2} & \cdots & \sigma_t^{nn} \end{bmatrix}, \mathbf{1} = \begin{bmatrix} 1 \\ 1 \\ \vdots \\ 1 \end{bmatrix},$$

$A_t = \mathbf{1}^\mathrm{T} \Sigma_t^{-1} \mathbf{1}, B_t = \mathbf{1}^\mathrm{T} \Sigma_t^{-1} \mu_t, C_t = \mu_t^\mathrm{T} \Sigma_t^{-1} \mu_t$ and $\Delta_t = A_t C_t - B_t^2$, where T denotes the transpose of a vector. We assume the determinant of variance-covariance matrix Σ_t is not zero and then there exists its inverse positive definite matrix Σ_t^{-1} and we have $A_t > 0$ and $\Delta_t > 0$. This assumption is natural since they can be realized easily by taking care of the combinations of assets. For a portfolio $(w_t^1, w_t^2, \cdots, w_t^n) \in \mathcal{W}$, the expectation and the variance of the rate of return $R_t = \sum_{i=1}^n w_t^i R_t^i$ with the portfolio are calculated respectively as follows: $E(R_t) = \sum_{i=1}^n w_t^i E(R_t^i) = \sum_{i=1}^n w_t^i \mu_t^i$ and $E((R_t - E(R_t))^2) = \sum_{i=1}^n \sum_{j=1}^n w_t^i w_t^j \sigma_t^{ij}$. Therefore, from (1) and (7), AVaR of the rate of return R_t is evaluated as

$$\mathrm{AVaR}_p(R_t) = \sum_{i=1}^n w_t^i \mu_t^i - \kappa \sqrt{\sum_{i=1}^n \sum_{j=1}^n w_t^i w_t^j \sigma_t^{ij}} \tag{12}$$

for $p \in (0, 1]$ with a positive constant $\kappa = \frac{1}{p} \int_0^p \kappa(q)\, dq$. Now step by step we discuss a portfolio problem to minimize the risk values. For a constant $\gamma \in \mathbf{R}$, first we deal with the following quadratic program.

Problem 2. Minimize the variance

$$\sum_{i=1}^n \sum_{j=1}^n w^i w^j \sigma_t^{ij}$$

with respect to portfolios $(w_t^1, w_t^2, \cdots, w_t^n) \in \mathcal{W}$ under a condition $\sum_{i=1}^n w_t^i \mu_t^i = \gamma$.

Problem 2 has the following solutions [14].

Lemma 3. *The minimum variance for Problem 2 is*

$$\frac{A_t \gamma^2 - 2B_t \gamma + C_t}{\Delta_t}$$

with an optimal portfolio

$$w_t = \xi_t \Sigma_t^{-1} \mathbf{1} + \eta_t \Sigma_t^{-1} \mu_t,$$

where $\xi_t = \frac{C_t - B_t \gamma}{\Delta_t}$ and $\eta_t = \frac{A_t \gamma - B_t}{\Delta_t}$.

Next we discuss the following risk-sensitive portfolio problem.

Problem 3. Maximize the average value-at-risks of the rate of return

$$\text{AVaR}_p(R_t) = \sum_{i=1}^{n} w_t^i \mu_t^i - \kappa \sqrt{\sum_{i=1}^{n} \sum_{j=1}^{n} w_t^i w_t^j \sigma_t^{ij}}$$

with respect to portfolios $(w_t^1, w_t^2, \cdots, w_t^n) \in \mathcal{W}$ under a condition $\sum_{i=1}^{n} w_t^i \mu_t^i = \gamma$.

Then the following lemma holds from Lemma 3.

Lemma 4. *Let κ satisfy $\kappa^2 > \Delta_t / A_t$. The optimal average value-at-risk for Problem 3 is*

$$\sup_{w_t \in \mathcal{W}: \sum_{i=1}^{n} w_t^i \mu_t^i = \gamma} \text{AVaR}_p(R_t) = \gamma - \kappa \sqrt{\frac{A_t \gamma^2 - 2B_t \gamma + C_t}{\Delta_t}}. \tag{13}$$

Hence we can easily check the following lemma for (13) [13,14].

Lemma 5. *If κ satisfies $\kappa^2 > \Delta_t / A_t$, then a function*

$$\gamma(\in \mathbf{R}) \mapsto \gamma - \kappa \sqrt{\frac{A_t \gamma^2 - 2B_t \gamma + C_t}{\Delta_t}}$$

is concave and it has the maximum

$$\frac{B_t - \sqrt{A_t \kappa^2 - \Delta_t}}{A_t}$$

at

$$\gamma = \frac{B_t}{A_t} + \frac{\Delta_t}{A_t \sqrt{A_t \kappa^2 - \Delta_t}}.$$

Finally we discuss the following problem.

Problem 4. Maximize the average value-at-risks of the rate of return

$$\text{AVaR}_p(R_t) = \sum_{i=1}^{n} w_t^i \mu_t^i - \kappa \sqrt{\sum_{i=1}^{n} \sum_{j=1}^{n} w_t^i w_t^j \sigma_t^{ij}}$$

with portfolio weights $(w_t^1, w_t^2, \cdots, w_t^n) \in \mathcal{W}$.

Since we have

$$\sup_{w \in \mathcal{W}} (12) = \sup_{\gamma} (\inf_{w \in \mathcal{W}: \sum_{i=1}^{n} w_t^i \mu_t^i = \gamma} (12)) = \sup_{\gamma} (13),$$

by Lemmas 3, 4 and 5 we arrive at the following analytical solutions for Problem 4.

Lemma 6. *Let κ satisfy $\kappa^2 > \Delta_t/A_t$. Then the optimal average value-at-risk for Problem 4 is*

$$\frac{B_t - \sqrt{A_t\kappa^2 - \Delta_t}}{A_t} \tag{14}$$

at the expected rate of return

$$\gamma_t^\circ = \frac{B_t}{A_t} + \frac{\Delta_t}{A_t\sqrt{A_t\kappa^2 - \Delta_t}}. \tag{15}$$

The corresponding optimal portfolio is given by

$$w_t^\circ = \xi_t^\circ \Sigma_t^{-1}\mathbf{1} + \eta_t^\circ \Sigma_t^{-1}\mu_t, \tag{16}$$

where $\xi_t^\circ = \frac{C_t - B_t\gamma_t^\circ}{\Delta_t}$ and $\eta_t^\circ = \frac{A_t\gamma_t^\circ - B_t}{\Delta_t}$.

5 Dynamic Optimal Portfolio Optimization with AVaRs

From (12), Theorem 1 is written as the followings.

Theorem 2. *Let $\{v_t\}$ be a sequence defined inductively by the following backward optimality equations:*

$$
\begin{aligned}
v_t = \max_{(w_t^1,\cdots,w_t^n)\in\mathcal{W}} \min\Bigg\{ &1 + \sum_{i=1}^n w_t^i\mu_t^i \\
&- \kappa\sqrt{\sum_{i=1}^n\sum_{j=1}^n w_t^i w_t^j \sigma_t^{ij}},\ \left(1 + \sum_{i=1}^n w_t^i\mu_t^i\right)\beta v_{t+1}\Bigg\}
\end{aligned}
\tag{17}
$$

for $t = 1, 2, \ldots, T-1$, and

$$v_T = \max_{(w_T^1,\cdots,w_T^n)\in\mathcal{W}}\left\{1 + \sum_{i=1}^n w_T^i\mu_T^i - \kappa\sqrt{\sum_{i=1}^n\sum_{j=1}^n w_T^i w_T^j \sigma_T^{ij}}\right\}. \tag{18}$$

Then v_1 is the optimal total AVaR in Problem 1.

In Theorem 2, we have the optimal solution for (18) from Lemma 6 and we can obtain the optimal solution for (17) with a condition $\sum_{i=1}^n w_t^i\mu_t^i = \gamma$ in a similar way to Lemma 5. Then Theorem 2 is represented as follows.

Theorem 3. *Assume $A_t > 0$, $B_t > 0$, $\Delta_t > 0$ and $\kappa^2 > \Delta_t/A_t$ for $t = 1, 2, \cdots, T$. Let $\{\gamma_t^*\}$ and $\{v_t\}$ be sequences defined inductively by the following backward optimality equations:*

$$
\gamma_t^* = \begin{cases}
\dfrac{B_t}{A_t} + \dfrac{\Delta_t}{A_t\sqrt{A_t\kappa^2 - \Delta_t}} \\
\quad\text{if } A_t + 2B_t + C_t \le \dfrac{\kappa^2}{(1-\beta v_{t+1})^2} \\[2mm]
\displaystyle\max_{l=1,2}\dfrac{B_t + \Gamma_t + (-1)^l\sqrt{(A_t + 2B_t + C_t)\Gamma_t - \Delta_t}}{A_t - \Gamma_t} \\
\quad\text{if } A_t + 2B_t + C_t > \dfrac{\kappa^2}{(1-\beta v_{t+1})^2}\ \text{ and } A_t \ne \Gamma_t \\[2mm]
\dfrac{C_t - A_t}{2(A_t + B_t)} \\
\quad\text{if } A_t + 2B_t + C_t > \dfrac{\kappa^2}{(1-\beta v_{t+1})^2}\ \text{ and } A_t = \Gamma_t,
\end{cases}
\tag{19}
$$

$$v_t = \begin{cases} \frac{A_t + B_t - \sqrt{A_t \kappa^2 - \Delta_t}}{A_t} \\ \quad\quad if \ \ A_t + 2B_t + C_t \leq \frac{\kappa^2}{(1-\beta v_{t+1})^2} \\ (1 + \gamma_t^*) \ \beta v_{t+1} \\ \quad\quad otherwise \end{cases} \tag{20}$$

for $t = 1, 2, \cdots, T-1$, and

$$v_T = 1 + \frac{B_T - \sqrt{A_T \kappa^2 - \Delta_T}}{A_T}, \tag{21}$$

where $\Gamma_t = \frac{\Delta_t(1-\beta v_{t+1})^2}{\kappa^2}$ for $t = 1, 2, \cdots, T$. Then v_1 is the optimal AVaR for Problem 1.

Corollary 1. *The optimal portfolio in Theorem 3 is given by*

$$w_t^* = \xi_t^* \Sigma_t^{-1} 1 + \eta_t^* \Sigma_t^{-1} \mu_t \tag{22}$$

for $t = 1, 2, \cdots, T$, where γ_t^* is the expected rate of return (19), $\xi_t^* = \frac{C_t - B_t \gamma_t^*}{\Delta_t}$ and $\eta_t^* = \frac{A_t \gamma_t^* - B_t}{\Delta_t}$.

Table 1. Expected returns μ_t and variance-covariance matrix Σ_t

μ_t^i		σ_t^{ij}	$j=1$	$j=2$	$j=3$	$j=4$
$i=1$	0.07	$i=1$	0.35	0.06	0.08	-0.07
$i=2$	0.06	$i=2$	0.06	0.37	-0.09	0.08
$i=3$	0.09	$i=3$	0.08	-0.09	0.34	-0.05
$i=4$	0.08	$i=4$	-0.07	0.08	-0.05	0.38

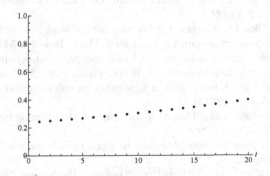

Fig. 1. Dynamic minimum average value-at-risks v_t

6   A Numerical Example

In this section, we give a numerical example for the previous sections. We assume R_t^i obeys the following distributions, and then the constant $\kappa = \frac{1}{p} \int_0^p \kappa(q)\, dq$ in (12) is given by $\kappa(p) = \Phi^{-1}(p)$ for $p \in (0, 1)$, where Φ is the cumulative function of the standard normal distribution function

$$\Phi(x) = \frac{1}{\sqrt{2\pi}} \int_{-\infty}^{x} e^{-\frac{z^2}{2}}\, dz \tag{23}$$

for $x \in \mathbf{R}$. We discuss a case of less than 3% part of the normal distribution, i.e. $p = 0.03$, and then $\kappa = 2.26807$. We deal with 4 assets, i.e. $n = 4$, and we give a vector of expected rates of return $\mu_t = [\mu_t^i]$ and a variance-covariance matrix $\Sigma = [\sigma_t^{ij}]$ by Table 1. At time t the constants A_t, B_t, C_t, Δ_t are calculated as follows: $A_t = \mathbf{1}^\mathsf{T} \Sigma_t^{-1} \mathbf{1} = 11.2675$, $B_t = \mathbf{1}^\mathsf{T} \Sigma_t^{-1} \mu_t = 0.863449$, $C_t = \mu_t^\mathsf{T} \Sigma_t^{-1} \mu_t = 0.0674479$ and $\Delta_t = A_t C_t - B_t^2 = 0.0144207$. By Lemma 6, we easily obtain the optimal portfolio $w^\circ = (0.193092, 0.222821, 0.318765, 0.265323)$ for Problem 4, and the corresponding average value-at-risk (14) $= -0.598968$, which implies average rate of falling is 59.8968%, and the expected rate of return $\gamma_t^\circ = 0.0768003$.

Next we investigate dynamic solutions for Problem 1. Let a discount rate $\beta = 0.95$, and let a terminal time $T = 20$. In dynamic case, v_t is the minimum regarding average value-at-risk between current time t and the terminal time T. By Theorem 3 we observe dynamic movement of $\{v_t\}$ in Fig. 1 and we get $v_1 = 0.243751$ for Problem 1.

Acknowledgments. This research is supported from JSPS KAKENHI Grant Number JP 16K05282.

References

1. Artzner, P., Delbaen, F., Eber, J.-M., Heath, D.: Coherent measures of risk. Math. Finance **9**, 203–228 (1999)
2. El Chaoui, L., Oks, M., Oustry, F.: Worst-case value at risk and robust portfolio optimization: a conic programming approach. Oper. Res. **51**, 543–556 (2003)
3. Cotter, J., Dowd, K.: Extreme spectral risk measures: an application to futures clearinghouse margin requirements. J. Bank. Finance **30**, 3469–3485 (2006)
4. Javidi, A.A.: Entropic value-at-risk: a new coherent risk measure. J. Optim. Theory Appl. **155**, 1105–1123 (2012)
5. Jorion, P.: Value at Risk: The New Benchmark for Managing Financial Risk. McGraw-Hill, New York (2006)
6. Markowitz, H.: Mean-Variance Analysis in Portfolio Choice and Capital Markets. Blackwell, Oxford (1990)
7. Meucci, A.: Risk and Asset Allocation. Springer, Heidelberg (2005). https://doi.org/10.1007/978-3-540-27904-4
8. Pliska, S.R.: Introduction to Mathematical Finance: Discrete-Time Models. Blackwell Publisher, New York (1997)

9. Rockafellar, R.T., Uryasev, S.: Optimization of conditional value-at-risk. J. Risk **2**, 21–41 (2000)
10. Ross, S.M.: An Introduction to Mathematical Finance. Cambridge University Press, Cambridge (1999)
11. Steinbach, M.C.: Markowitz revisited: mean-variance model in financial portfolio analysis. SIAM Rev. **43**, 31–85 (2001)
12. Tasche, D.: Expected shortfall and beyond. J. Bank. Finance **26**, 1519–1533 (2002)
13. Yoshida, Y.: A dynamic value-at-risk portfolio model. In: Torra, V., Narakawa, Y., Yin, J., Long, J. (eds.) MDAI 2011. LNCS (LNAI), vol. 6820, pp. 43–54. Springer, Heidelberg (2011). https://doi.org/10.1007/978-3-642-22589-5_6
14. Yoshida, Y.: A dynamic risk allocation of value-at-risks with portfolios. J. Adv. Comput. Intell. Intell. Inform. **16**, 800–806 (2012)
15. Yoshida, Y.: An ordered weighted average with a truncation weight on intervals. In: Torra, V., Narukawa, Y., López, B., Villaret, M. (eds.) MDAI 2012. LNCS (LNAI), vol. 7647, pp. 45–55. Springer, Heidelberg (2012). https://doi.org/10.1007/978-3-642-34620-0_6
16. Yoshida, Y.: An optimal process for average value-at-risk portfolios in financial management. In: Ntalianis, K., Croitoru, A. (eds.) APSAC 2017. LNEE, vol. 428, pp. 101–107. Springer, Cham (2018). https://doi.org/10.1007/978-3-319-53934-8_12

First-Order vs. Second-Order Encodings for LTL$_f$-to-Automata Translation

Shufang Zhu[1], Geguang Pu[1($\boxtimes$)], and Moshe Y. Vardi[2]

[1] East China Normal University, Shanghai, China
ggpu@sei.ecnu.edu.cn
[2] Rice University, Houston, TX, USA

Abstract. Translating formulas of Linear Temporal Logic (LTL) over finite traces, or LTL$_f$, to symbolic Deterministic Finite Automata (DFA) plays an important role not only in LTL$_f$ synthesis, but also in synthesis for Safety LTL formulas. The translation is enabled by using MONA, a powerful tool for symbolic, BDD-based, DFA construction from logic specifications. Recent works used a first-order encoding of LTL$_f$ formulas to translate LTL$_f$ to First Order Logic (FOL), which is then fed to MONA to get the symbolic DFA. This encoding was shown to perform well, but other encodings have not been studied. Specifically, the natural question of whether second-order encoding, which has significantly simpler quantificational structure, can outperform first-order encoding remained open.

In this paper we address this challenge and study second-order encodings for LTL$_f$ formulas. We first introduce a specific MSO encoding that captures the semantics of LTL$_f$ in a natural way and prove its correctness. We then explore is a *Compact* MSO encoding, which benefits from automata-theoretic minimization, thus suggesting a possible practical advantage. To that end, we propose a formalization of symbolic DFA in second-order logic, thus developing a novel connection between BDDs and MSO. We then show by empirical evaluations that the first-order encoding does perform better than both second-order encodings. The conclusion is that first-order encoding is a better choice than second-order encoding in LTL$_f$-to-Automata translation.

1 Introduction

Synthesis from temporal specifications [23] is a fundamental problem in Artificial Intelligence and Computer Science [8]. A popular specification is Linear Temporal Logic (LTL) [24]. The standard approach to solving LTL synthesis requires, however, determinization of automata on *infinite* words and solving *parity games*, both challenging algorithmic problems [17]. Thus a major barrier of temporal synthesis has been algorithmic difficulty. One approach to combating this difficulty is to focus on using fragments of LTL, such as the GR(1) fragment, for which temporal synthesis has lower computational complexity [1].

A new logic for temporal synthesis, called LTL$_f$, was proposed recently in [6,8]. The focus there is not on limiting the syntax of LTL, but on interpreting it semantically on *finite* traces, rather than *infinite* traces as in [24].

© Springer Nature Switzerland AG 2019
T. V. Gopal and J. Watada (Eds.): TAMC 2019, LNCS 11436, pp. 684–705, 2019.
https://doi.org/10.1007/978-3-030-14812-6_43

Such interpretation allows the executions being arbitrarily long, but not infinite, and is adequate for finite-horizon planning problems. While limiting the semantics to finite traces does not change the computational complexity of temporal synthesis (2EXPTIME), the algorithms for LTL$_f$ are much simpler. The reason is that those algorithms require determinization of automata on *finite* words (rather than *infinite* words), and solving *reachability* games (rather than *parity* games) [8]. Another application, as shown in [30], is that temporal synthesis of *Safety* LTL formulas, a syntactic fragment of LTL expressing *safety properties*, can be reduced to reasoning about finite words (see also [18,19]). This approach has been implemented in [31] for LTL$_f$ synthesis and in [30] for synthesis of *Safety* LTL formulas, and has been shown to outperform existing temporal-synthesis tools such as Acacia+ [2].

The key algorithmic building block in these approaches is a translation of LTL$_f$ to *symbolic* Deterministic Finite Automata (DFA) [30,31]. In fact, translating LTL$_f$ formula to DFA has other algorithmic applications as well. For example, in dealing with safety properties, which are arguably the most used temporal specifications in real-world systems [18]. As shown in [28], model checking of safety properties can benefit from using deterministic rather than nondeterminisic automata. Moreover, in runtime verification for safety properties, we need to generate monitors, a type of which are, in essence, deterministic automata [29]. In [28,29], the translation to deterministic automata is explicit, but symbolic DFAs can be useful also in model checking and monitor generation, because they can be much more compact than explicit DFAs, cf. [31].

The method used in [30,31] for the translation of LTL$_f$ to symbolic DFA used an encoding of LTL$_f$ to First-Order Logic (FOL) that captures directly the semantics of temporal connectives, and MONA [13], a powerful tool, for symbolic DFA construction from logical specifications. This approach was shown to outperform explicit tools such as SPOT [12], but encodings other than the first-order one have not yet been studied. This leads us here to study second-order translations of LTL$_f$, where we use Monadic Second Order (MSO) logic of one successor over finite words (called M2L-STR in [16]). Indeed, one possible advantage of using MSO is the simpler quantificational structure that the second-order encoding requires, which is a sequence of existential monadic second-order quantifiers followed by a single universal first-order quantifier. Moreover, instead of the syntax-driven translation of first-order encoding of LTL$_f$ to FOL, the second-order encoding employs a semantics-driven translation, which allows more space for optimization. The natural question arises whether second-order encoding outperforms first-order encoding.

To answer this question, we study here second-order encodings of LTL$_f$ formulas. We start by introducing a specific second-order encoding called MSO encoding that relies on having a second-order variable for each temporal operator appearing in the LTL$_f$ formula and proving the correctness. Such MSO encoding captures the semantics of LTL$_f$ in a natural way and is linear in the size of the formula. We then introduce a so called *Compact* MSO encoding, which captures the tight connection between LTL$_f$ and DFAs. We leverage the fact that while the translation from LTL$_f$ to DFA is doubly exponential [18], there is an exponential translation from *Past* LTL$_f$ to DFA (a consequence of [5,6]). Given an LTL$_f$ formula ϕ, we first construct a DFA that accepts exactly the reverse language satisfying *models*(ϕ) via *Past* LTL$_f$. We then encode this DFA using second-order logic and "invert" it to get a second-order formulation for the original LTL$_f$ formula. Applying this approach directly, however, would yield an MSO formula with an exponential (in terms of the original LTL$_f$ formula) number of quantified monadic predicates. To get a more compact formulation we can benefit from the fact that the DFA obtained by MONA from the *Past* LTL$_f$ formula is symbolic, expressed by binary decision diagrams (BDDs) [14]. We show how we can obtain a Compact MSO encoding directly from these BDDs. In addition, we present in this paper the first evaluation of the spectrum of encodings for LTL$_f$-to-automata from first-order to second-order.

To perform an empirical evaluation of the comparison between first-order encoding and second-order encoding of LTL$_f$, we first provide a broad investigation of different optimizations of both encodings. Due to the syntax-driven translation of FOL encoding, there is limit potential for optimization such that we are only able to apply different normal forms to LTL$_f$ formulas, which are Boolean Normal Form (BNF) and Negation Normal Form (NNF). The semantics-driven translation of second-order encoding, however, enables more potential for optimization than the FOL encoding. In particular, we study the following optimizations introduced in [21,22]: in the variable form, where a *Lean* encoding introduces fewer variables than the standard *Full* encoding; and in the constraint form, where a *Sloppy* encoding allows less tight constraints than the standard *Fussy* encoding. The main result of our empirical evaluations is the superiority of the first-order encoding as a way to get MONA to generate a symbolic DFA, which answers the question of whether second-order outperforms first-order for LTL$_f$-to-automata translation.

The paper is organized as follows. In Sect. 2 we provide preliminaries and notations. Section 3 introduces MSO encoding and proves the correctness. Section 4 describes a more compact second-order encoding, called Compact MSO encoding and proves the correctness. Empirical evaluation results of different encodings and different optimizations are presented in Sect. 5. Finally, Sect. 6 offers concluding remarks.

2   Preliminaries

2.1   LTL$_f$ Basics

Linear Temporal Logic over *finite traces* (LTL$_f$) has the same syntax as LTL [6]. Given a set $\mathcal{P}$ of propositions, the syntax of LTL$_f$ formulas is as follows:

$$\phi ::= \top \mid \bot \mid p \mid \neg\phi \mid \phi_1 \wedge \phi_2 \mid X\phi \mid \phi_1 U \phi_2$$

where $p \in \mathcal{P}$. We use $\top$ and $\bot$ to denote *true* and *false* respectively. X (Next) and U (Until) are temporal operators, whose dual operators are N (Weak Next) and

R (Release) respectively, defined as $N\phi \equiv \neg X\neg\phi$ and $\phi_1 R\phi_2 \equiv \neg(\neg\phi_1 U\neg\phi_2)$. The abbreviations (Eventually) $F\phi \equiv \top U\phi$ and (Globally) $G\phi \equiv \bot R\phi$ are defined as usual. Finally, we have standard boolean abbreviations, such as $\lor$ (or) and $\rightarrow$ (implies).

Elements $p \in \mathcal{P}$ are *atoms*. A literal l can be an atom or the negation of an atom. A *trace* $\rho = \rho[0], \rho[1], \ldots$ is a sequence of propositional assignments, where $\rho[x] \in 2^{\mathcal{P}}$ ($x \geq 0$) is the x-th point of ρ. Intuitively, $\rho[x]$ is the set of propositions that are *true* at instant x. Additionally, $|\rho|$ represents the length of ρ. The trace ρ is an *infinite* trace if $|\rho| = \infty$ and $\rho \in (2^{\mathcal{P}})^{\omega}$; otherwise ρ is *finite*, and $\rho \in (2^{\mathcal{P}})^*$. LTL$_f$ formulas are interpreted over finite traces. Given a finite trace ρ and an LTL$_f$ formula ϕ, we inductively define when ϕ is *true* for ρ at point x ($0 \leq x < |\rho|$), written $\rho, x \models \phi$, as follows:

- $\rho, x \models \top$ and $\rho, x \not\models \bot$;
- $\rho, x \models p$ iff $p \in \rho[x]$;
- $\rho, x \models \neg\phi$ iff $\rho, x \not\models \phi$;
- $\rho, x \models \phi_1 \land \phi_2$, iff $\rho, x \models \phi_1$ and $\rho, x \models \phi_2$;
- $\rho, x \models X\phi$, iff $x + 1 < |\rho|$ and $\rho, x + 1 \models \phi$;
- $\rho, x \models \phi_1 U\phi_2$, iff there exists y such that $x \leq y < |\rho|$ and $\rho, y \models \phi_2$, and for all z, $x \leq z < y$, we have $\rho, z \models \phi_1$.

An LTL$_f$ formula ϕ is *true* in ρ, denoted by $\rho \models \phi$, when $\rho, 0 \models \phi$. Every LTL$_f$ formula can be written in Boolean Normal Form (BNF) or Negation Normal Form (NNF) [27]. BNF rewrites the input formula using only $\neg$, $\land$, $\lor$, X, and U. NNF pushes negations inwards, introducing the dual temporal operators N and R, until negation is applied only to atoms.

2.2 Symbolic DFA and MONA

We start by defining the concept of *symbolic automaton* [31], where a boolean formula is used to represent the transition function of a Deterministic Finite Automaton (DFA). A symbolic deterministic finite automaton (Symbolic DFA) $\mathcal{F} = (\mathcal{P}, \mathcal{X}, X_0, \eta, f)$ corresponding to an explicit DFA $\mathcal{D} = (2^{\mathcal{P}}, S, s_0, \delta, F)$ is defined as follows:

- $\mathcal{P}$ is the set of atoms;
- $\mathcal{X}$ is a set of state variables where $|\mathcal{X}| = \lceil \log_2 |S| \rceil$;
- $X_0 \in 2^{\mathcal{X}}$ is the initial state corresponding to s_0;
- $\eta : 2^{\mathcal{X}} \times 2^{\mathcal{P}} \rightarrow 2^{\mathcal{X}}$ is a boolean transition function corresponding to δ;
- f is the acceptance condition expressed as a boolean formula over $\mathcal{X}$ such that f is satisfied by an assignment X iff X corresponds to a final state $s \in F$.

We can represent the symbolic transition function η by an indexed family $\eta_q : 2^{\mathcal{X}} \times 2^{\mathcal{P}} \rightarrow \{0, 1\}$ for $x_q \in \mathcal{X}$, which means that η_q can be represented by a binary decision diagram (BDD) [14] over $\mathcal{X} \cup \mathcal{P}$. Therefore, the symbolic DFA can be represented by a sequence of BDDs, each of which corresponding to a state variable.

The MONA tool [13] is an efficient implementation for translating FOL and MSO formulas over finite words into minimized symbolic deterministic automata. MONA represents symbolic deterministic automata by means of *Shared Multi-terminal BDDs* (ShMTBDDs) [3,20]. The symbolic LTL$_f$ synthesis framework of [31] requires standard BDD representation by means of symbolic DFAs as defined above. The transformation from ShMTBDD to BDD is described in [31].

2.3   FOL Encoding of LTL$_f$

First Order Logic (FOL) encoding of LTL$_f$ translates LTL$_f$ into FOL over finite linear order with monadic predicates. In this paper, we utilize the FOL encoding proposed in [6]. We first restrict our interest to *monadic structure*. Consider a finite trace $\rho = \rho[0]\rho[1]\cdots\rho[e]$, the corresponding *monadic structure* $\mathcal{I}_\rho = (\Delta^\mathcal{I}, <, \cdot^\mathcal{I})$ describes ρ as follows. $\Delta^\mathcal{I} = \{0, 1, 2, \cdots, last\}$, where $last = e$ indicating the last point along the trace. The linear order $<$ is defined over $\Delta^\mathcal{I}$ in the standard way [16]. The notation $\cdot^\mathcal{I}$ indicates the set of monadic predicates that describe the atoms of $\mathcal{P}$, where the interpretation of each $p \in \mathcal{P}$ is $Q_p = \{x \,:\, p \in \rho[x]\}$. Intuitively, Q_p is interpreted as the set of positions where p is true in ρ. In the translation below, $fol(\theta, x)$, where θ is an LTL$_f$ formula and x is a variable, is an FOL formula asserting the truth of θ at point x of the linear order. The translation uses the successor function $+1$, and the variable $last$ that represents the maximal point in the linear order.

- $fol(p, x) = (Q_p(x))$
- $fol(\neg\phi, x) = (\neg fol(\phi, x))$
- $fol(\phi_1 \wedge \phi_2, x) = (fol(\phi_1, x) \wedge fol(\phi_2, x))$
- $fol(\phi_1 \vee \phi_2, x) = (fol(\phi_1, x) \vee fol(\phi_2, x))$
- $fol(X\phi, x) = ((\exists y)((y = x + 1) \wedge fol(\phi, y)))$
- $fol(N\phi, x) = ((x = last) \vee ((\exists y)((y = x + 1) \wedge fol(\phi, y))))$
- $fol(\phi_1 U \phi_2, x) = ((\exists y)((x \leq y \leq last) \wedge fol(\phi_2, y) \wedge (\forall z)((x \leq z < y) \rightarrow fol(\phi_1, z))))$
- $fol(\phi_1 R \phi_2, x) = (((\exists y)((x \leq y \leq last) \wedge fol(\phi_1, y) \wedge (\forall z)((x \leq z \leq y) \rightarrow fol(\phi_2, z)))) \vee ((\forall z)((x \leq z \leq last) \rightarrow fol(\phi_2, z))))$

For FOL variables, MONA provides a built-in operator $+1$ for successor computation. Moreover, we can use built-in procedures in MONA to represent the variable $last$. Given a finite trace ρ, we denote the corresponding finite linear ordered FOL interpretation of ρ by $\mathcal{I}_\rho$. The following theorem guarantees the correctness of FOL encoding of LTL$_f$.

Theorem 1 ([15])**.** *Let ϕ be an LTL$_f$ formula and ρ be a finite trace. Then $\rho \models \phi$ iff $\mathcal{I}_\rho \models fol(\phi, 0)$.*

3   MSO Encoding

First-order encoding was shown to perform well in the context of LTL$_f$-to-automata translation [30], but other encodings have not been studied. Specifically, the natural question of whether second-order (MSO) outperforms first-order in the same context remained open. MSO is an extension of FOL that allows quantification over monadic predicates [16]. By applying a semantics-driven translation to LTL$_f$, we obtain an MSO encoding that has significantly simpler quantificational structure. This encoding essentially captures in MSO the standard encoding of temporal connectives, cf. [4]. Intuitively speaking, MSO encoding deals with LTL$_f$ formula by interpreting every operator with corresponding subformulas following the semantics of the operator. We now present MSO encoding that translates LTL$_f$ formula ϕ to MSO, which is then fed to MONA to produce a symbolic DFA.

For an LTL$_f$ formula ϕ over a set $\mathcal{P}$ of atoms, let $cl(\phi)$ denote the set of subformulas of ϕ. We define atomic formulas as atoms $p \in \mathcal{P}$. For every subformula in $cl(\phi)$ we introduce monadic predicate symbols as follows: for each atomic subformula $p \in \mathcal{P}$, we have a monadic predicate symbol Q_p; for each non-atomic subformula $\theta_i \in \{\theta_1, \ldots, \theta_m\}$, we have Q_{θ_i}. Intuitively speaking, each monadic predicate indicates the positions where the corresponding subformula is true along the linear order.

Let mso(ϕ) be the translation function that given an LTL$_f$ formula ϕ returns a corresponding MSO formula asserting the truth of ϕ at position 0. We define mso(ϕ) as following: mso(ϕ) = $(\exists Q_{\theta_1}) \cdots (\exists Q_{\theta_m})(Q_\phi(0) \wedge (\forall x)(\bigwedge_{i=1}^m \mathsf{t}(\theta_i, x))$, where x indicates the position along the finite linear order. Here $\mathsf{t}(\theta_i, x)$ asserts that the truth of every non-atomic subformula θ_i of ϕ at position x relies on the truth of corresponding subformulas at x such that following the semantics of LTL$_f$. Therefore, $\mathsf{t}(\theta_i, x)$ is defined as follows:

- If $\theta_i = (\neg\theta_j)$, then $\mathsf{t}(\theta_i, x) = (Q_{\theta_i}(x) \leftrightarrow \neg Q_{\theta_j}(x))$
- If $\theta_i = (\theta_j \wedge \theta_k)$, then $\mathsf{t}(\theta_i, x) = (Q_{\theta_i}(x) \leftrightarrow (Q_{\theta_j}(x) \wedge Q_{\theta_k}(x)))$
- If $\theta_i = (\theta_j \vee \theta_k)$, then $\mathsf{t}(\theta_i, x) = (Q_{\theta_i}(x) \leftrightarrow (Q_{\theta_j}(x) \vee Q_{\theta_k}(x)))$
- If $\theta_i = (X\theta_j)$, then $\mathsf{t}(\theta_i, x) = (Q_{\theta_i}(x) \leftrightarrow ((x \neq last) \wedge Q_{\theta_j}(x + 1)))$
- If $\theta_i = (N\theta_j)$, then $\mathsf{t}(\theta_i, x) = (Q_{\theta_i}(x) \leftrightarrow ((x = last) \vee Q_{\theta_j}(x + 1)))$
- If $\theta_i = (\theta_j U \theta_k)$, then $\mathsf{t}(\theta_i, x) = (Q_{\theta_i}(x) \leftrightarrow (Q_{\theta_k}(x) \vee ((x \neq last) \wedge Q_{\theta_j}(x) \wedge Q_{\theta_i}(x + 1))))$
- If $\theta_i = (\theta_j R \theta_k)$, then $\mathsf{t}(\theta_i, x) = (Q_{\theta_i}(x) \leftrightarrow (Q_{\theta_k}(x) \wedge ((x = last) \vee Q_{\theta_j}(x) \vee Q_{\theta_i}(x + 1))))$

Consider a finite trace ρ, the corresponding interpretation $\mathcal{I}_\rho$ of ρ is defined as in Sect. 2.3. The following theorem asserts the correctness of the MSO encoding.

Theorem 2. *Let ϕ be an LTL$_f$ formula, ρ be a finite trace. Then $\rho \models \phi$ iff $\mathcal{I}_\rho \models$ mso(ϕ).*

Proof. If ϕ is a propositional atom p, then mso(ϕ) = $Q_p(0)$. It is true that $\rho \models \phi$ iff $\mathcal{I}_\rho \models$ mso(ϕ). If ϕ is an nonatomic formula, we prove this theorem in two directions.

Suppose first that ρ satisfies ϕ. We expand the monadic structure $\mathcal{I}_\rho$ with interpretations for the existentially quantified monadic predicate symbols by setting Q_{θ_i}, the interpretation of subformula θ_i in $\mathcal{I}_\rho$, as the set collecting all points of ρ satisfying θ_i, that is $Q_{\theta_i} = \{x \ : \ \rho, x \models \theta_i\}$. We also have $Q_p = \{x \ : \ \rho, x \models p\}$ and denote the expanded structure by $\mathcal{I}_\rho^{mso}$. By assumption, $Q_\phi(0)$ holds in $\mathcal{I}_\rho^{mso}$. It remains to prove that $\mathcal{I}_\rho^{mso} \models \forall x.\mathsf{t}(\theta_i, x)$, for each nonatomic subformula $\theta_i \in cl(\phi)$, which we prove via structural induction over θ_i.

- If $\theta_i = (\neg\theta_j)$, then $\mathsf{t}(\theta_i, x) = (Q_{\theta_i}(x) \leftrightarrow (\neg Q_{\theta_j}(x)))$. This holds, since $Q_{(\neg\theta_j)} = \{x \ : \ \rho, x \not\models \theta_j\}$ and $Q_{\theta_j} = \{x \ : \ \rho, x \models \theta_j\}$.
- If $\theta_i = (\theta_j \wedge \theta_k)$, then $\mathsf{t}(\theta_i, x) = (Q_{\theta_i}(x) \leftrightarrow (Q_{\theta_j}(x) \wedge Q_{\theta_k}(x)))$. This holds, since $Q_{(\theta_j \wedge \theta_k)} = \{x \ : \ \rho, x \models \theta_j \text{ and } \rho, x \models \theta_k\}$, $Q_{\theta_j} = \{x \ : \ \rho, x \models \theta_j\}$ and $Q_{\theta_k} = \{x \ : \ \rho, x \models \theta_k\}$.
- If $\theta_i = (\theta_j \vee \theta_k)$, then $\mathsf{t}(\theta_i, x) = (Q_{\theta_i}(x) \leftrightarrow (Q_{\theta_j}(x) \vee Q_{\theta_k}(x)))$. This holds, since $Q_{(\theta_j \vee \theta_k)} = \{x \ : \ \rho, x \models \theta_j \text{ or } \rho, x \models \theta_k\}$, $Q_{\theta_j} = \{x \ : \ \rho, x \models \theta_j\}$ and $Q_{\theta_k} = \{x \ : \ \rho, x \models \theta_k\}$.
- If $\theta_i = (X\theta_j)$, then $\mathsf{t}(\theta_i, x) = (Q_{\theta_i}(x) \leftrightarrow ((x \neq last) \wedge Q_{\theta_j}(x+1)))$. This holds, since $Q_{(X\theta_j)} = \{x \ : \ \rho, x \models (X\theta_j)\} = \{x \ : \ x \neq last \text{ and } \rho, x+1 \models \theta_j\}$, and $Q_{\theta_j} = \{x \ : \ \rho, x \models \theta_j\}$.
- If $\theta_i = (N\theta_j)$, then $\mathsf{t}(\theta_i, x) = (Q_{\theta_i}(x) \leftrightarrow ((x = last) \vee Q_{\theta_j}(x+1)))$. This holds, since $Q_{(N\theta_j)} = \{x \ : \ \rho, x \models (N\theta_j)\} = \{x \ : \ x = last \text{ or } \rho, x+1 \models \theta_j\}$, and $Q_{\theta_j} = \{x \ : \ \rho, x \models \theta_j\}$.
- If $\theta_i = (\theta_j U \theta_k)$, then $\mathsf{t}(\theta_i, x) = (Q_{\theta_i}(x) \leftrightarrow (Q_{\theta_k}(x) \vee ((x \neq last) \wedge Q_{\theta_j}(x) \wedge Q_{\theta_i}(x+1))))$. This holds, since $Q_{(\theta_j U \theta_k)} = \{x \ : \ \rho, x \models \theta_j U \theta_k\} = \{x \ : \ \rho, x \models \theta_k \text{ or } x \neq last \text{ with } \rho, x \models \theta_j \text{ also } \rho, x+1 \models \theta_i\}$, $Q_{\theta_j} = \{x \ : \ \rho, x \models \theta_j\}$, and $Q_{\theta_k} = \{x \ : \ \rho, x \models \theta_k\}$;
- If $\theta_i = (\theta_j R \theta_k)$, then $\mathsf{t}(\theta_i, x) = (Q_{\theta_i}(x) \leftrightarrow (Q_{\theta_k}(x) \wedge ((x = last) \vee Q_{\theta_j}(x) \vee Q_{\theta_i}(x+1))))$. This holds, since $Q_{(\theta_j R \theta_k)} = \{x \ : \ \rho, x \models \theta_j R \theta_k\} = \{x \ : \ \rho, x \models \theta_k \text{ with } x = last \text{ or } \rho, x \models \theta_j \text{ or } \rho, x+1 \models \theta_i\}$, $Q_{\theta_j} = \{x \ : \ \rho, x \models \theta_j\}$, and $Q_{\theta_k} = \{x \ : \ \rho, x \models \theta_k\}$.

Assume now that $\mathcal{I}_\rho \models \mathsf{mso}(\phi)$. This means that there is an expansion of $\mathcal{I}_\rho$ with monadic interpretations Q_{θ_i} for each nonatomic subformula $\theta_i \in cl(\phi)$ such that this expanded structure $\mathcal{I}_\rho^{mso} \models (Q_\phi(0) \wedge ((\forall x) \bigwedge_{i=1}^{m} \mathsf{t}(\theta_i, x)))$. We now prove by induction on ϕ that if $x \in Q_\phi$, then $\rho, x \models \phi$ such that $Q_\phi(0)$ indicates that $\rho, 0 \models \phi$.

- If $\phi = (\neg\theta_j)$, then $\mathsf{t}(\phi, x) = (Q_\phi(x) \leftrightarrow (x \notin Q_{\theta_j}))$. Since $\mathsf{t}(\phi)$ holds at every point x of $\mathcal{I}_\rho^{mso}$, it holds that $x \in Q_\phi$ iff $x \notin Q_{\theta_j}$. It follows by induction that $\rho, x \not\models \theta_j$. Thus, $\rho, x \models \phi$.
- If $\phi = (\theta_j \wedge \theta_k)$, then $\mathsf{t}(\phi, x) = (Q_\phi(x) \leftrightarrow (Q_{\theta_j}(x) \wedge Q_{\theta_k}(x)))$. Since $\mathsf{t}(\phi)$ holds at every point x of $\mathcal{I}_\rho^{mso}$, it follows that $x \in Q_\phi$ iff $x \in Q_{\theta_j}$ and $x \in Q_{\theta_k}$. It follows by induction that $\rho, x \models \theta_j$ and $\rho, x \models \theta_k$. Thus, $\rho, x \models \phi$.
- If $\phi = (\theta_j \vee \theta_k)$, then $\mathsf{t}(\phi, x) = (Q_\phi(x) \leftrightarrow (Q_{\theta_j}(x) \vee Q_{\theta_k}(x)))$. Since $\mathsf{t}(\phi)$ holds at every point x of $\mathcal{I}_\rho^{mso}$, it follows that $x \in Q_\phi$ iff $x \in Q_{\theta_j}$ or $x \in Q_{\theta_k}$. It follows by induction that $\rho, x \models \theta_j$ or $\rho, x \models \theta_k$. Thus, $\rho, x \models \phi$.
- If $\phi = (X\theta_j)$, then $\mathsf{t}(\phi, x) = (Q_\phi(x) \leftrightarrow ((x \neq last) \wedge Q_{\theta_j}(x+1)))$. Since $t(\phi)$ holds at every point x of $\mathcal{I}_\rho^{mso}$, it follows that $x \in Q_\phi$ iff $x \neq last$ and $x + 1 \in Q_{\theta_j}$. It follows by induction that $x \neq last$ and $\rho, x \models \theta_j$. Thus, $\rho, x \models \phi$.

- If $\phi = (N\theta_j)$, then $\mathsf{t}(\phi, x) = (Q_\phi(x) \leftrightarrow ((x = last) \vee Q_{\theta_j}(x + 1)))$. Since $\mathsf{t}(\phi)$ holds at every point x of $\mathcal{I}_\rho^{mso}$, it follows that $x \in Q_\phi$ iff $x = last$ or $x + 1 \in Q_{\theta_j}$. It follows by induction that $x = last$ or $\rho, x \models \theta_j$. Thus, $\rho, x \models \phi$.

- If $\phi = (\theta_j U \theta_k)$, then $\mathsf{t}(\phi, x) = (Q_\phi(x) \leftrightarrow (Q_{\theta_k}(x) \vee ((x \neq last) \wedge Q_{\theta_j}(x) \wedge Q_\phi(x+1))))$. Since $\mathsf{t}(\phi)$ holds at every point x of $\mathcal{I}_\rho^{mso}$, it follows that $x \in Q_\phi$ iff $x \in Q_{\theta_k}$ or $x \neq last$ with $x \in Q_{\theta_j}$ also $x + 1 \in Q_\phi$. Thus, $\rho, x \models \phi$.

- If $\phi = (\theta_j R \theta_k)$, then $\mathsf{t}(\phi, x) = (Q_\phi(x) \leftrightarrow (Q_{\theta_k}(x) \wedge ((x = last) \vee Q_{\theta_j}(x) \vee Q_\phi(x+1))))$. Since $\mathsf{t}(\phi)$ holds at every point x of $\mathcal{I}_\rho^{mso}$, it follows that $x \in Q_\phi$ iff $x \in Q_{\theta_k}$ with $x = last$ or $x \in Q_{\theta_j}$ or $x + 1 \in Q_\phi$. Thus, $\rho, x \models \phi$. $\square$

4   Compact MSO Encoding

The MSO encoding described in Sect. 3 is closely related to the translation of LTL$_f$ to alternating automata [6], with each automaton state corresponding to a monadic predicate. The construction, however, is subject only to syntactic minimization. Can we optimize this encoding using *automata-theoretic minimization?* In fact, MONA itself applies automata-theoretic minimization. Can we use MONA to produce a more efficient encoding for MONA?

The key observation is that MONA can produce a compact symbolic representation of a non-deterministic automaton (NFA) representing a given LTL$_f$ formula, and we can use this symbolic NFA to create a more compact MSO encoding for LTL$_f$. This is based on the observation that while the translation from LTL$_f$ to DFA is 2-EXP [18], the translation from *past* LTL$_f$ to DFA is 1-EXP, as explained below. We proceed as follows: (1) Reverse a given LTL$_f$ formula ϕ to *Past* LTL$_f$ formula ϕ^R; (2) Use MONA to construct the DFA of ϕ^R, the reverse of which is an NFA, that accepts exactly the reverse language of the words satisfying $models(\phi)$; (3) Express this symbolic DFA in second-order logic and "invert" it to get $\mathcal{D}_\phi$, the corresponding DFA of ϕ.

The crux of this approach, which follows from [5,6], is that the DFA corresponding to the reverse language of an LTL$_f$ formula ϕ of length n has only 2^n states. The reverse of this latter DFA is an NFA for ϕ. We now elaborate on these steps.

4.1   LTL$_f$ to PLTL$_f$

Past Linear Temporal Logic over finite traces, i.e. PLTL$_f$, has the same syntax as PLTL over infinite traces introduced in [24]. Given a set of propositions $\mathcal{P}$, the grammar of PLTL$_f$ is given by:

$$\psi ::= \top \mid \bot \mid p \mid \neg\psi \mid \psi_1 \wedge \psi_2 \mid Y\psi \mid \psi_1 S\psi_2$$

Given a finite trace ρ and a PLTL$_f$ formula ψ, we inductively define when ψ is *true* for ρ at step x $(0 \leq x < |\rho|)$, written by $\rho, x \models \psi$, as follows:

- $\rho, x \models \top$ and $\rho, x \not\models \bot$;
- $\rho, x \models p$ iff $p \in \rho[x]$;
- $\rho, x \models \neg\psi$ iff $\rho, x \not\models \psi$;
- $\rho, x \models \psi_1 \wedge \psi_2$, iff $\rho, x \models \psi_1$ and $\rho, x \models \psi_2$;
- $\rho, x \models Y\psi$, iff $x - 1 \geq 0$ and $\rho, x - 1 \models \psi$;
- $\rho, x \models \psi_1 S\psi_2$, iff there exists y such that $0 \leq y \leq x$ and $\rho, y \models \psi_2$, and for all z, $y < z \leq x$, we have $\rho, z \models \psi_1$.

A PLTL$_f$ formula ψ is *true* in ρ, denoted by $\rho \models \psi$, if and only if $\rho, |\rho| - 1 \models \psi$. To reverse an LTL$_f$ formula ϕ, we replace each temporal operator in ϕ with the corresponding *past* operator of PLTL$_f$ thus getting ϕ^R. X(Next) and U(Until) correspond to Y(Before) and S(Since) respectively.

We define $\rho^R = \rho[|\rho| - 1], \rho[|\rho| - 2], \ldots, \rho[1], \rho[0]$ to be the reverse of ρ. Moreover, given language $\mathcal{L}$, we denote the reverse of $\mathcal{L}$ by $\mathcal{L}^R$ such that $\mathcal{L}^R$ collects all reversed sequences in $\mathcal{L}$. Formally speaking, $\mathcal{L}^R = \{\rho^R : \rho \in \mathcal{L}\}$. The following theorem shows that PLTL$_f$ formula ϕ^R accepts exactly the reverse language satisfying ϕ.

Theorem 3. *Let $\mathcal{L}(\phi)$ be the language of LTL$_f$ formula ϕ and $\mathcal{L}^R(\phi)$ be the reverse language, then $\mathcal{L}(\phi^R) = \mathcal{L}^R(\phi)$.*

Proof. $\mathcal{L}(\phi^R) = \mathcal{L}^R(\phi)$ iff for an arbitrary sequence $\rho \in \mathcal{L}(\phi)$ such that $\rho \models \phi$, it is true that $\rho^R \models \phi^R$. We prove the theorem by the induction over the structure of ϕ. *last* is used to denote the last instance such that $last = |\rho| - 1$.

- Basically, if $\phi = p$ is an atom, then $\phi^R = p$, $\rho \models \phi$ iff $p \in \rho[0]$ such that $p \in \rho^R[last]$. Therefore, $\rho^R \models \phi^R$;
- If $\phi = \neg\phi_1$, then $\phi^R = \neg\phi_1^R$, $\rho \models \neg\phi_1$ iff $\rho \not\models \phi_1$, such that by induction hypothesis $\rho^R \not\models \phi_1^R$ holds, therefore $\rho^R \models \phi^R$ is true;
- If $\phi = \phi_1 \wedge \phi_2$, then $\phi^R = \phi_1^R \wedge \phi_2^R$, $\rho \models \phi$ iff ρ satisfies both ϕ_1 and ϕ_2. By induction hypothesis $\rho^R \models \phi_1^R$ and $\rho^R \models \phi_2^R$ hold, therefore $\rho^R \models \phi^R$ is true;
- If $\phi = X\phi_1$, $\phi^R = Y\phi_1^R$, $\rho \models \phi$ iff suffix ρ' is sequence $\rho[1], \rho[2], \ldots, \rho[last]$ and $\rho' \models \phi_1$. By induction hypothesis, $\rho'^R \models \phi_1^R$ holds, in which case $\rho^R, last - 1 \models \phi_1^R$ is true, therefore $\rho^R \models \phi^R$ holds.
- If $\phi = \phi_1 U\phi_2$, $\rho \models \phi$ iff there exists y such that y $(0 \leq y \leq last)$, suffix $\rho' = \rho[y], \rho[y + 1], \ldots, \rho[last]$ satisfies ϕ_2. Also for all z such that z $(0 \leq z < y)$, $\rho'' = \rho[z], \rho[z + 1], \ldots, \rho[last]$ satisfies ϕ_1. By induction hypothesis, $\rho'^R \models \phi_1^R$ and $\rho''^R \models \phi_2^R$ hold, therefore we have $\rho^R, last - y \models \phi_2^R$ and $\forall z.last - y < z \leq last, \rho^R, z \models \phi_1^R$ hold such that $\rho^R \models \phi^R$. The proof is done. $\square$

4.2 PLTL$_f$ to DFA

The DFA construction from PLTL$_f$ formulas relies on MONA as well. Given PLTL$_f$ formula ψ, we are able to translate ψ to FOL formula as input of MONA, which returns the DFA. For PLTL$_f$ formula ψ over $\mathcal{P}$, we construct the corresponding FOL formula with respect to point x by a function $\mathsf{fol_p}(\psi, x)$ asserting the truth of ψ at x. Detailed translation of PLTL$_f$ to FOL is defined below. The translation uses the predecessor function -1, and the predicate $last$ referring to the last point along the finite trace.

- $\mathsf{fol_p}(p, x) = (Q_p(x))$
- $\mathsf{fol_p}(\neg\psi, x) = (\neg\mathsf{fol_p}(\psi, x))$
- $\mathsf{fol_p}(\psi_1 \wedge \psi_2, x) = (\mathsf{fol_p}(\psi_1, x) \wedge \mathsf{fol_p}(\psi_2, x))$
- $\mathsf{fol_p}(Y\psi, x) = ((\exists y)((y = x - 1) \wedge (y \geq 0) \wedge \mathsf{fol_p}(\psi, y)))$
- $\mathsf{fol_p}(\psi_1 S \psi_2, x) = ((\exists y)((0 \leq y \leq x) \wedge \mathsf{fol_p}(\psi_2, y) \wedge (\forall z)((y < z \leq x) \rightarrow \mathsf{fol_p}(\psi_1, z))))$

Consider a finite trace ρ, the corresponding interpretation $\mathcal{I}_\rho$ is defined as in Sect. 2.3. The following theorem guarantees the correctness of the above translation.

Theorem 4. [15] *Let ψ be a PLTL$_f$ formula, ρ be a finite trace. Then $\rho \models \psi$ iff $\mathcal{I}_\rho \models \mathsf{fol_p}(\psi, last)$, where $last = |\rho| - 1$.*

Proof. We prove the theorem by the induction over the structure of ψ.

- Basically, if $\psi = p$ is an atom, $\rho \models \psi$ iff $p \in \rho[last]$. By the definition of $\mathcal{I}$, we have that $last \in Q_p$. Therefore, $\rho \models \psi$ iff $\mathcal{I}_\rho \models \mathsf{fol_p}(p, last)$ holds;
- If $\psi = \neg\psi$, $\rho \models \neg\psi$ iff $\rho \not\models \psi$. By induction hypothesis it is true that $\mathcal{I}_\rho \not\models \mathsf{fol_p}(\psi, last)$, therefore $\mathcal{I}_\rho \models \mathsf{fol_p}(\neg\psi, last)$ holds;
- If $\psi = \psi_1 \wedge \psi_2$, $\rho \models \psi$ iff ρ satisfies both ψ_1 and ψ_2. By induction hypothesis, it is true that $\mathcal{I}_\rho \models \mathsf{fol_p}(\psi_1, last)$ and $\mathcal{I}_\rho \models \mathsf{fol_p}(\psi_2, last)$. Therefore $\mathcal{I}_\rho \models \mathsf{fol_p}(\psi_1, last) \wedge \mathsf{fol_p}(\psi_2, last)$ holds;
- If $\psi = Y\psi_1$, $\rho \models \psi$ iff prefix $\rho' = \rho[0], \rho[1], \ldots, \rho[last - 1]$ of ρ satisfies $\rho' \models \psi_1$. Let $\mathcal{I}'_\rho$ be the corresponding interpretation of ρ', thus for every atom $p \in \mathcal{P}$, $x \in Q'_p$ iff $x \in Q_p$ where Q'_p is the corresponding monadic predicate of p in $\mathcal{I}'_\rho$. By induction hypothesis it is true that $\mathcal{I}'_\rho \models \mathsf{fol_p}(\psi_1, last - 1)$, therefore $\mathcal{I}_\rho \models \mathsf{fol_p}(Y\psi_1, last)$ holds.
- If $\psi = \psi_1 S \psi_2$, $\rho \models \psi$ iff there exists y such that $0 \leq y \leq last$ and prefix $\rho' = \rho[0], \rho[1], \ldots, \rho[y]$ of ρ satisfies ψ_2 and for all z such that $y < z \leq last$, $\rho'' = \rho[0], \rho[1], \ldots, \rho[z]$ satisfies ψ_1. Let $\mathcal{I}'_\rho$ and $\mathcal{I}''_\rho$ be the corresponding interpretations of ρ' and ρ''. Thus for every atom $p \in \mathcal{P}$ it is true that $x \in Q'_p$ iff $x \in Q_p$, $x \in Q''_p$ iff $x \in Q_p$, where Q'_p and Q''_p correspond to the monadic predicates of p in $\mathcal{I}'_\rho$ and $\mathcal{I}''_\rho$ respectively. By induction hypothesis it is true that $\mathcal{I}'_\rho \models \mathsf{fol_p}(\psi_2, last - y)$ and $\mathcal{I}''_\rho \models \mathsf{fol_p}(\psi_1, last - z)$ hold, therefore $\mathcal{I}_\rho \models \mathsf{fol_p}(\psi_1 S \psi_2, last)$. $\qquad\square$

4.3   Reversing DFA via Second-Order Logic

For simplification, from now we use ψ to denote the corresponding PLTL$_f$ formula ϕ^R of LTL$_f$ formula ϕ. We first describe how BDDs represent a symbolic DFA. Then we introduce the Compact MSO encoding that inverts the DFA by formulating such BDD representation into a second-order formula. The connection between BDD representation and second-order encoding is novel, to the best of our knowledge.

As defined in Sect. 2.2, given a symbolic DFA $\mathcal{F}_\psi = (\mathcal{P}, \mathcal{X}, X_0, \eta, f)$ represented by a sequence $\mathcal{B} = \langle B_0, B_1, \ldots, B_{k-1}\rangle$ of BDDs, where there are k variables in $\mathcal{X}$, a run of such DFA on a word $\rho = \rho[0], \rho[1], \ldots, \rho[e-1]$ involves a sequence of states $\xi = X_0, X_1, \ldots, X_e$ of length $e + 1$. For the moment if we omit the last state reached on an input of length e, we have a sequence of states $\xi' = X_0, X_1, \ldots, X_{e-1}$ of length e. Thus we can think of the run ξ' as a labeling of the positions of the word with states, which is $(\rho[0], X_0), (\rho[1], X_1), \ldots, (\rho[e-1], X_{e-1})$. At each position with given word and state, the transition moving forward involves a computation over every B_q $(0 \leq q \leq k-1)$. To perform such computation, take the high branch in every node labeled by variable $v \in \{\mathcal{X} \cup \mathcal{P}\}$ if v is assigned 1 and the low branch otherwise.

The goal here is to write a formula $\mathsf{Rev}(\mathcal{F}_\psi)$ such that there is an accepting run over $\mathcal{F}_\psi$ of a given word ρ iff ρ^R is accepted by $\mathsf{Rev}(\mathcal{F}_\psi)$. To do this, we introduce one second-order variable V_q for each $x_q \in \mathcal{X}$ with $0 \leq q \leq k-1$, and one second-order variable N_α for every nonterminal node α in BDDs, u nonterminal nodes in total. The V_q variables collect the positions where x_q holds, and the N_α variables indicate the positions where the node α is visited, when computing the transition. To collect all transitions moving towards accepting states, we have BDD $B'_f = f(\eta(\mathcal{X}, \mathcal{P}))$.

Here are some notations. Let α be a nonterminal node, c be a terminal node in B_q such that $c \in \{0, 1\}$ and $d \in \{0, 1\}$ be the value of v. For nonterminal node α, we define:

$$\mathsf{Pre}(\alpha) = \{(\beta, v, d) \ : \ \text{there is an edge from } \beta \text{ to } \alpha \text{ labelled by } v = d\}$$
$$\mathsf{Post}(\alpha) = \{(\beta, v, d) \ : \ \text{there is an edge from } \alpha \text{ to } \beta \text{ labelled by } v = d\}$$

For every terminal node c in BDD B_q, we define:

$$\mathsf{PreT}(B_q, c) = \{(\beta, v, d) \ : \ \text{there is an edge from } \beta \text{ to } c \text{ labelled by } v = d \text{ in BDD } B_q\}$$

Also, we use $\in^d$ to denote $\in$ when $d = 1$ and $\notin$ when $d = 0$. For each BDD B_q, $\mathsf{root}(B_q)$ indicates the root node of B_q.

We use these notations to encode the following statements:

(1) At the **last** position, state X_0 should hold since ξ' is being inverted and X_0 is the starting point;

$$\mathsf{Rinit} = (x = last) \rightarrow \left(\bigwedge\nolimits_{0 \leq q \leq k-1, X_0(x_q)=d} x \in^d V_q\right);$$

(2) At position x, if the current computation is at nonterminal node α labeled by v, then (2.a) the current computation must come from a predecessor labeled by v' following the value of v', and (2.b) the next step is moving to the corresponding successor following the value of v;

$$\text{node} = \bigwedge_{1 \leq \alpha \leq u} (\ \text{PreCon} \wedge \text{PostCon}\ ); \text{ where}$$

$$\text{PreCon} = \left(x \in N_\alpha \rightarrow (\ \bigvee_{(\beta, v', d) \in \text{Pre}(\alpha)} [x \in N_\beta \wedge x \in^d v']) \right)$$

$$\text{PostCon} = \left(\bigwedge_{(\beta, v, d) \in \text{Post}(\alpha)} [x \in N_\alpha \wedge x \in^d v \rightarrow x \in N_\beta] \right).$$

(3) At position x such that $\mathbf{x} > \mathbf{0}$, if the current computation node α moves to a terminal node c of B_q, then the value of x_q at position $\mathbf{x\text{-}1}$ is given by the value of c. Such computations of all $B_q (0 \leq q \leq k-1)$ finish one transition;

$$\text{Rterminal} = \bigwedge_{0 \leq q \leq k-1} \left(\bigwedge_{(\beta, v, d) \in \text{PreT}(B_q, c)} [(x > 0 \wedge x \in N_\beta \wedge x \in^d v) \rightarrow (x - 1 \in^c V_q)] \right);$$

(4) At the **first** position, the current computation on B'_f has to surely move to terminal 1, therefore terminating the running trace of ξ.

$$\text{Racc} = (x = 0) \rightarrow \left(\bigvee_{(\beta, v, d) \in \text{PreT}(B'_f, 1)} [x \in N_\beta \wedge x \in^d v] \right).$$

To get all computations over BDDs start from the root at each position, we have

$$\text{roots} = \bigwedge_{0 \leq x \leq last} \bigwedge_{0 \leq q \leq k-1} x \in \text{root}(B_q)$$

$\text{Rev}(\mathcal{F}_\psi)$ has to take a conjunction of all requirements above such that

$$\text{Rev}(\mathcal{F}_\psi) = (\exists V_0)(\exists V_1) \ldots (\exists V_{k-1})(\exists N_1)(\exists N_2) \ldots (\exists N_u)(\forall x)(\text{Rinit} \wedge \text{node}$$
$$\wedge \text{Rterminal} \wedge \text{Racc} \wedge \text{roots}).$$

Therefore, let $\text{Cmso}(\phi)$ be the translation function that given an LTL$_f$ formula ϕ returns a corresponding second-order formula applying the Compact MSO encoding, we define $\text{Cmso}(\phi) = \text{Rev}(\mathcal{F}_\psi)$ asserting the truth of ϕ at position 0, where ψ is the corresponding PLTL$_f$ formula of ϕ, and $\mathcal{F}_\psi$ is the symbolic DFA of ψ. The following theorem asserts the correctness of the Compact MSO encoding.

Theorem 5. *The models of formula* $\text{Cmso}(\phi)$ *are exactly the words satisfying* ϕ.

Proof. We first have that $\mathcal{L}(\phi) = \mathcal{L}^R(\psi) = \mathcal{L}^R(\mathcal{F}_\psi)$ holds since ψ is the corresponding PLTL$_f$ formula of ϕ and $\mathcal{F}_\psi$ collects exactly the words satisfying ψ. Moreover, $\mathcal{L}(\text{Rev}(\mathcal{F}_\psi)) = \mathcal{L}^R(\mathcal{F}_\psi)$ is true following the construction rules of $\text{Rev}(\mathcal{F}_\psi)$ described above and $\text{Cmso}(\phi) = \text{Rev}(\mathcal{F}_\psi)$. Therefore, $\mathcal{L}(\phi) = \mathcal{L}(\text{Cmso}(\phi))$ holds, in which case the models of formula $\text{Cmso}(\phi)$ are exactly the words satisfying ϕ.   □

Notice that the size of $\text{Cmso}(\phi)$ is in linear on the size of the BDDs, which lowers the logical complexity comparing to the MSO encoding in Sect. 3. Moreover, in the Compact MSO encoding, the number of existential second-order symbols for state variables are nevertheless possibly less than that in MSO encoding, but new second-order symbols for nonterminal BDD nodes are introduced. BDDs provide a compact representation, in which redundant nodes are reduced. Such advantages allow Compact MSO encoding to use as few second-order symbols for BDD nodes as possible.

5 Experimental Evaluation

We implemented proposed second-order encodings in different parsers for LTL$_f$ formulas using C++. Each parser is able to generate a second-order formula corresponding to the input LTL$_f$ formula, which is then fed to MONA [13] for subsequent symbolic DFA construction. Moreover, we employed *Syft*'s [31] code to translate LTL$_f$ formula into first-order logic (FOL), which adopts the first-order encoding described in Sect. 2.3.

Benchmarks. We conducted the comparison of first-order encoding with second-order encoding in the context of LTL$_f$-to-DFA, thus only satisfiable but not valid formulas are interesting. Therefore, we first ran an LTL$_f$ satisfiability checker on LTL$_f$ formulas and their negations to filter the valid or unsatisfiable formulas. We collected 5690 formulas, which consist of two classes of benchmarks: 765 LTL$_f$-specific benchmarks, of which 700 are scalable LTL$_f$ pattern formulas from [10] and 65 are randomly conjuncted common LTL$_f$ formulas from [7,11,25] ; and 4925 LTL-as-LTL$_f$ formulas from [26,27], since LTL formulas share the same syntax as LTL$_f$.

Experimental Setup. To explore the comparison between first-order and second-order for LTL$_f$-to-DFA translation, we ran each formula for every encoding on a node within a high performance cluster. These nodes contain 12 processor cores at 2.2 GHz each with 8 GB of RAM per core. Time out was set to be 1000 s. Cases that cannot generate the DFA within 1000 s generally fail even if the time limit is extended, since in these cases, MONA typically cannot handle the large BDD.

5.1 Optimizations of Second-Order Encoding

Before diving into the optimizations of second-order encoding, we first study the potential optimization space of the first-order encoding that translates LTL$_f$ to

FOL. Due to the syntax-driven translation of FOL encoding, we are only able to apply different normal forms, Boolean Norma Form (BNF) and Negation Normal Form (NNF). We compared the impact on performance of FOL encoding with two LTL$_f$ normal forms. It turns out that the normal form does not have a measurable impact on the performance of the first-order encoding. Since FOL-BNF encoding performs slightly better than FOL-NNF, the best FOL encoding refers to FOL-BNF.

To explore the potential optimization space of the second-order encodings proposed in this paper, we hope to conduct experiments with different optimizations. We name second-order encoding with different optimizations *variations*. We first show optimizations of the MSO encoding described in Sect. 3, then describe variations of the Compact MSO encoding shown in Sect. 4 in the following.

The basic MSO encoding defined in Sect. 3 translates LTL$_f$ to MSO in a natural way, in the sense that introducing a second-order predicate for each non-atomic subformula and employing the $\leftrightarrow$ constraint. Inspired by [22,27], we define in this section several optimizations to simplify such encoding thus benefiting symbolic DFA construction. These variations indicating different optimizations are combinations of three independent components: (1) the Normal Form (choose between BNF or NNF); (2) the Constraint Form (choose between *Fussy* or *Sloppy*); (3) the Variable Form (choose between *Full* or *Lean*). In each component one can choose either of two options to make. Thus for example, the variation described in Sect. 3 is BNF-*Fussy-Full*. Note that BNF-*Sloppy* are incompatible, as described below, and so there are $2^3 - 2 = 6$ viable combinations of the three components above. We next describe the variations in details.

Constraint Form. We call the translation described in Sect. 3 the *Fussy* variation, in which we translate ϕ to MSO formula $\mathsf{mso}(\phi)$ by employing an *iff* constraint (see Sect. 3). For example:

$$\mathsf{t}(\theta_i, x) = (Q_{\theta_i}(x) \leftrightarrow (Q_{\theta_j}(x) \wedge Q_{\theta_k}(x))) \text{ if } \theta_i = (\theta_j \wedge \theta_k) \tag{1}$$

We now introduce *Sloppy* variation, inspired by [27], which allows less tight constraints that still hold correctness guarantees thus may speed up the symbolic DFA construction. To better reason the incompatible combination BNF-*Sloppy*, we specify the description for different normal forms, NNF and BNF separately.

For LTL$_f$ formulas in NNF, the *Sloppy* variation requires only a single implication constraint $\rightarrow$. Specifically the *Sloppy* variation $\mathsf{mso}_s(\phi)$ for NNF returns MSO formula $(\exists Q_{\theta_1}) \cdots (\exists Q_{\theta_m}) (Q_\phi(0) \wedge (\forall x)(\bigwedge_{i=1}^{m} \mathsf{t}_s(\theta_i, x)))$, where $\mathsf{t}_s(\theta_i)$ is defined just like $\mathsf{t}(\theta_i)$, replacing the $\leftrightarrow$ by $\rightarrow$. For example translation (1) under the *Sloppy* translation for NNF is $\mathsf{t}_s(\theta_i, x) = (Q_{\theta_i}(x) \rightarrow (Q_{\theta_j}(x) \wedge Q_{\theta_k}(x)))$.

The *Sloppy* variation cannot be applied to LTL$_f$ formulas in BNF since the $\leftrightarrow$ constraint defined in function $\mathsf{t}(\theta_i)$ is needed only to handle negation correctly. BNF requires a general handling of negation. For LTL$_f$ formulas in NNF, negation is applied only to atomic formulas such that handled implicitly by the base case $\rho, x \models p \leftrightarrow \rho, x \nvDash \neg p$. Therefore, translating LTL$_f$ formulas in NNF does not require the $\leftrightarrow$ constraint. For example, consider LTL$_f$ formula $\phi = \neg Fa$ (in BNF), where a is an atom. The corresponding BNF-*Sloppy* variation gives MSO formula $(\exists Q_{\neg Fa})(\exists Q_{Fa})(Q_{\neg Fa}(0) \wedge ((\forall x)((Q_{\neg Fa}(x) \rightarrow \neg Q_{Fa}(x)) \wedge (Q_{Fa}(x) \rightarrow (Q_a(x) \vee ((x \neq last) \wedge Q_{Fa}(x+1))))))))$ via $\mathsf{mso}_s(\phi)$. Consider finite trace $\rho = (a=0), (a=1)$, $\rho \models \phi$ iff $\rho \models \mathsf{mso}_s(\phi)$ does not hold since $\rho \not\models \neg Fa$. This happens because $\neg Fa$ requires $(Q_{\neg Fa}(x) \leftrightarrow \neg Q_{Fa}(x))$ as Fa is an non-atomic subformula. Therefore, *Sloppy* variation can only be applied to LTL$_f$ formulas in NNF.

The following theorem asserts the correctness of the *Sloppy* variation.

Theorem 6. *Let ϕ be an* LTL$_f$ *formula in* NNF *and ρ be a finite trance. Then $\rho \models \phi$ iff $\mathcal{I}_\rho \models \mathsf{mso}_s(\phi)$.*

The proof here is analogous to that of Theorem 2. The crux here is that the $\leftrightarrow$ in $\mathsf{t}(\theta_i)$ is needed only to handle negation correctly. *Sloppy* encoding, however, is applied only to LTL$_f$ formulas in NNF, so negation can be applied only to atomic propositions, which is handled by the base case ($\neg Q_p(x)$).

Variable Form. In all the variations of the MSO encoding we can get above, we introduced a monadic predicate for each non-atomic subformula in $cl(\phi)$, this is the *Full* variation. We now introduce *Lean* variation, a new variable form, aiming at decreasing the number of quantified monadic predicates. Fewer quantifiers on monadic predicates could benefit symbolic DFA construction a lot since quantifier elimination in MONA takes heavy cost. The key idea of *Lean* variation is introducing monadic predicates only for atomic subformulas and non-atomic subformulas of the form $\phi_j U \theta_k$ or $\phi_j R \theta_k$ (named as U- or R-subformula respectively).

For non-atomic subformulas that are not U- or R- subformulas, we can construct *second-order terms* using already defined monadic predicates to capture the semantics of them. Function $\mathsf{lean}(\theta_i)$ is defined to get such second-order terms. Intuitively speaking, $\mathsf{lean}(\theta_i)$ indicates the same positions where θ_i is true as Q_{θ_i} does, instead of having Q_{θ_i} explicitly. We use built-in second-order operators in MONA to simplify the definition of $\mathsf{lean}(\theta_i)$. ALIVE is defined using built-in procedures in MONA to collect all instances along the finite trace. MONA also allows to apply set union, intersection, and difference for second-order terms, as well as the -1 operation (which shifts a monadic predicate backwards by one position). $\mathsf{lean}(\theta_i)$ is defined over the structure of θ_i as following:

- If $\theta_i = (\neg \theta_j)$, then $\mathsf{lean}(\theta_i) = (\mathsf{ALIVE} \backslash \mathsf{lean}(\theta_j))$
- If $\theta_i = (\theta_j \wedge \theta_k)$, then $\mathsf{lean}(\theta_i) = (\mathsf{lean}(\theta_j) \text{ inter } \mathsf{lean}(\theta_k))$
- If $\theta_i = (\theta_j \vee \theta_k)$, then $\mathsf{lean}(\theta_i) = (\mathsf{lean}(\theta_j) \text{ union } \mathsf{lean}(\theta_k))$
- If $\theta_i = (X\theta_j)$, then $\mathsf{lean}(\theta_i) = ((\mathsf{lean}(\theta_j) - 1) \backslash \{last\})$
- If $\theta_i = (N\theta_j)$, then $\mathsf{lean}(\theta_i) = ((\mathsf{lean}(\theta_j) - 1) \text{ union } \{last\})$
- If $\theta_i = (\theta_j U \theta_k)$ or $\theta_i = (\theta_j R \theta_k)$, then $\mathsf{lean}(\theta_i) = Q_{\theta_a}$, where Q_{θ_a} is the corresponding monadic predicate.

The following lemma ensures that $\mathsf{lean}(\theta_i)$ keeps the interpretation of each non-atomic subformula $\theta_i \in cl(\phi)$.

Lemma 1. *Let ϕ be an* LTL$_f$ *formula, ρ be a finite trace. Then $\rho, x \models \theta_i$ iff $\mathsf{lean}(\theta_i)(x)$ holds, where x is the position in ρ.*

Proof. Suppose first that $\rho, x \models \theta_i$. We prove this inductively on the structure of θ_i.

- If $\theta_i = \neg\theta_j$, then $\mathsf{lean}(\theta_i) = (\mathsf{ALIVE}\backslash\mathsf{lean}(\theta_j))$. $\mathsf{lean}(\theta_i)(x)$ holds since $\mathsf{lean}(\theta_i) = \{x : x \notin \mathsf{lean}(\theta_j)\}$ and $\mathsf{lean}(\theta_j) = \{x : \rho, x \models \theta_j\}$.
- If $\theta_i = \theta_j \wedge \theta_k$, then $\mathsf{lean}(\theta_i) = (\mathsf{lean}(\theta_j) \text{ inter } \mathsf{lean}(\theta_k))$. $\mathsf{lean}(\theta_i)(x)$ holds since $\mathsf{lean}(\theta_i) = \{x : x \in \mathsf{lean}(\theta_j) \text{ and } x \in \mathsf{lean}(\theta_k)\}$, $\mathsf{lean}(\theta_j) = \{x : \rho, x \models \theta_j\}$ and $\mathsf{lean}(\theta_k) = \{x : \rho, x \models \theta_k\}$.
- If $\theta_i = \theta_j \vee \theta_k$, then $\mathsf{lean}(\theta_i) = (\mathsf{lean}(\theta_j) \text{ union } \mathsf{lean}(\theta_k))$. $\mathsf{lean}(\theta_i)(x)$ holds since $\mathsf{lean}(\theta_i) = \{x : x \in \mathsf{lean}(\theta_j) \text{ or } x \in \mathsf{lean}(\theta_k)\}$, $\mathsf{lean}(\theta_j) = \{x : \rho, x \models \theta_j\}$ and $\mathsf{lean}(\theta_k) = \{x : \rho, x \models \theta_k\}$.
- If $\theta_i = X\theta_j$, then $\mathsf{lean}(\theta_i) = ((\mathsf{lean}(\theta_j) - 1)\backslash\{last\})$. $\mathsf{lean}(\theta_i)(x)$ holds since $\mathsf{lean}(\theta_i) = \{x : x \neq last \text{ and } x + 1 \in \mathsf{lean}(\theta_j)\}$, $\mathsf{lean}(\theta_j) = \{x : \rho, x \models \theta_j\}$.
- If $\theta_i = N\theta_j$, then $\mathsf{lean}(\theta_i) = ((\mathsf{lean}(\theta_j) - 1) \text{ union } \{last\})$. $\mathsf{lean}(\theta_i)(x)$ holds since $\mathsf{lean}(\theta_i) = \{x : x = last \text{ or } x+1 \in \mathsf{lean}(\theta_j)\}$, $\mathsf{lean}(\theta_j) = \{x : \rho, x \models \theta_j\}$.
- If $\theta_i = \theta_j U\theta_k$ or $\theta_i = \theta_j R\theta_k$, then $\mathsf{lean}(\theta_i) = Q_{\theta_a}$. $\mathsf{lean}(\theta_i)(x)$ holds since $\mathsf{lean}(\theta_i) = Q_{\theta_a} = \{x : \rho, x \models \theta_a\}$, where Q_{θ_a} is the corresponding second-order predicate for formula θ_i.

Assume now that $\mathcal{I}_\rho^{mso} \models \mathsf{lean}(\theta_i)(x)$ with given interpretations of second-order predicates. We now prove $\rho, x \models \theta_i$ by induction over the structure on θ_i.

- If $\theta_i = \neg\theta_j$, then $\mathsf{lean}(\theta_i) = (\mathsf{ALIVE}\backslash\mathsf{lean}(\theta_j))$. Since $\mathsf{lean}(\theta_i)(x)$ holds, we also have that $x \in \mathsf{lean}(\theta_i)$ iff $x \notin \mathsf{lean}(\theta_j)$. It follows by induction that $\rho, x \models \theta_i$.
- If $\theta_i = \theta_j \wedge \theta_k$, then $\mathsf{lean}(\theta_i) = (\mathsf{lean}(\theta_j) \text{ inter } \mathsf{lean}(\theta_k))$. Since $\mathsf{lean}(\theta_i)(x)$ holds, we also have that $x \in \mathsf{lean}(\theta_i)$ iff $x \in \mathsf{lean}(\theta_j)$ and $x \in \mathsf{lean}(\theta_k)$. It follows by induction that $\rho, x \models \theta_i$.
- If $\theta_i = \theta_j \vee \theta_k$, then $\mathsf{lean}(\theta_i) = (\mathsf{lean}(\theta_j) \text{ union } \mathsf{lean}(\theta_k))$. Since $\mathsf{lean}(\theta_i)(x)$ holds, we also have that $x \in \mathsf{lean}(\theta_i)$ iff $x \in \mathsf{lean}(\theta_j)$ or $x \in \mathsf{lean}(\theta_k)$. It follows by induction that $\rho, x \models \theta_i$.
- If $\theta_i = X\theta_j$, then $\mathsf{lean}(\theta_i) = ((\mathsf{lean}(\theta_j) - 1)\backslash\{last\})$. Since $\mathsf{lean}(\theta_i)(x)$ holds, we also have that $x \neq last$ and $x + 1 \in \mathsf{lean}(\theta_j)$. It follows by induction that $\rho, x \models \theta_i$.
- If $\theta_i = N\theta_j$, then $\mathsf{lean}(\theta_i) = ((\mathsf{lean}(\theta_j) - 1) \text{ union } \{last\})$. Since $\mathsf{lean}(\theta_i)(x)$ holds, we also have that $x = last$ or $x + 1 \in \mathsf{lean}(\theta_j)$. It follows by induction that $\rho, x \models \theta_i$.
- If $\theta_i = \theta_j U\theta_k$ or $\theta_i = \theta_j R\theta_k$, then $\mathsf{lean}(\theta_i) = Q_{\theta_a}$, where Q_{θ_a} is the corresponding second-order predicate. It follows by induction that $\rho, x \models \theta_i$.

$\square$

Finally, we define *Lean* variation based on function $\mathsf{lean}(\phi)$. *Lean* variation $\mathsf{mso}_\lambda(\phi)$ returns MSO formula $(\exists Q_{\theta_1})\ldots(\exists Q_{\theta_n})$ $(\mathsf{lean}(\phi)(0) \wedge ((\forall x)(\bigwedge_{a=1}^n \mathsf{t}_\lambda(\theta_a, x))))$, where n is the number of U- and R- subformulas $\theta_a \in cl(\phi)$, and $\mathsf{t}_\lambda(\theta_a, x)$ is defined as follows: if $\theta_a = (\theta_j U\theta_k)$, then $\mathsf{t}_\lambda(\theta_a, x) = (Q_{\theta_a}(x) \leftrightarrow (\mathsf{lean}(\theta_k)(x) \vee ((x \neq last) \wedge \mathsf{lean}(\theta_j)(x) \wedge Q_{\theta_a}(x+1))))$; if $\theta_a = (\theta_j R\theta_k)$, then $\mathsf{t}_\lambda(\theta_a, x) = (Q_{\theta_a}(x) \leftrightarrow (\mathsf{lean}(\theta_k)(x) \wedge ((x = last) \vee \mathsf{lean}(\theta_j)(x) \vee Q_{\theta_a}(x+1))))$. The following theorem guarantees the correctness of *Lean* variation.

Theorem 7. *Let ϕ be an* LTL$_f$ *formula, ρ be a finite trace. Then $\rho \models \phi$ iff* $\mathcal{I}_\rho \models \mathsf{mso}_\lambda(\phi)$.

Proof. If ϕ is a propositional atom p, then $\mathsf{mso}_\lambda(\phi) = Q_p(0)$. It is true that $\rho \models \phi$ iff $\mathcal{I}_\rho \models \mathsf{mso}_\lambda(\phi)$. If ϕ is an nonatomic formula, we prove this theorem in two directions.

Suppose first that ρ satisfies ϕ. We expand the monadic structure $\mathcal{I}_\rho$ with interpretations for $Q_{\theta_1}, Q_{\theta_2}, \ldots, Q_{\theta_n}$ by setting $Q_{\theta_a} = \{x : \rho, x \models \theta_a\}$. Let the expanded structure be $\mathcal{I}_\rho^{mso}$. By assumption, $\mathsf{lean}(\phi)(0)$ holds in $\mathcal{I}_\rho^{mso}$. It remains to prove that $\mathcal{I}_\rho^{mso} \models (\forall x)(\bigwedge_{a=1}^{n} \mathsf{t}_\lambda(\theta_a, x))$, for each U or R subformula $\theta_a \in cl(\phi)$.

- If $\theta_a = (\theta_j U \theta_k)$, then $\mathsf{t}_\lambda(\theta_a, x) = (\mathsf{lean}(\theta_a)(x) \leftrightarrow (\mathsf{lean}(\theta_k)(x) \vee ((x \neq last) \wedge \mathsf{lean}(\theta_j)(x) \wedge \mathsf{lean}(\theta_a)(x+1))))$. This holds, since $\mathsf{lean}((\theta_j U \theta_k)) = \{x : \rho, x \models \theta_j U \theta_k\} = \{x : \rho, x \models \theta_k$ or $x \neq last$ with $\rho, x \models \theta_j$ also $\rho, x+1 \models \theta_a\}$, $\mathsf{lean}(\theta_j) = \{x : \rho, x \models \theta_j\}$, and $\mathsf{lean}(\theta_k) = \{x : \rho, x \models \theta_k\}$ with Lemma 1;
- If $\theta_a = (\theta_j R \theta_k)$, then $\mathsf{t}_\lambda(\theta_a, x) = (\mathsf{lean}(\theta_a)(x) \leftrightarrow (\mathsf{lean}(\theta_k)(x) \wedge ((x = last) \vee \mathsf{lean}(\theta_j)(x) \vee \mathsf{lean}(\theta_a)(x+1))))$. This holds, since $\mathsf{lean}((\theta_j R \theta_k)) = \{x : \rho, x \models \theta_j R \theta_k\} = \{x : \rho, x \models \theta_k$ with $x = last$ or $\rho, x \models \theta_j$ or $\rho, x+1 \models \theta_a\}$, $\mathsf{lean}(\theta_j) = \{x : \rho, x \models \theta_j\}$, and $\mathsf{lean}(\theta_k) = \{x : \rho, x \models \theta_k\}$ with Lemma 1.

Assume now that $\mathcal{I}_\rho \models \mathsf{mso}_\lambda(\phi)$. This means that there is an expansion of $\mathcal{I}_\rho$ with monadic interpretations Q_{θ_a} for each element θ_a of U or R subformulas in $cl(\phi)$ such that this expanded structure $\mathcal{I}_\rho^{mso} \models (\mathsf{lean}(\phi)(0)) \wedge ((\forall x)(\bigwedge_{a=1}^{n} \mathsf{t}_\lambda(\theta_a, x)))$. If ϕ is not an R or U subformula, then it has been proven by Lemma 1 that if $x \in \mathsf{lean}(\phi)$, then $\rho, x \models \phi$. We now prove by induction on ϕ that if $x \in Q_{\theta_a}$, then $\rho, x \models \phi$. Since $\mathcal{I}_\rho^{mso} \models (\mathsf{lean}(\phi)(0))$, it follows that $\rho, 0 \models \phi$.

- If $\phi = (\theta_j U \theta_k)$, then $\mathsf{t}_\lambda(\phi, x) = (\mathsf{lean}(\phi)(x) \leftrightarrow (\mathsf{lean}(\theta_k)(x) \vee ((x \neq last) \wedge \mathsf{lean}(\theta_j)(x) \wedge \mathsf{lean}(\phi)(x+1))))$. Since $\mathsf{t}_\lambda(\phi)$ holds at every point x of $\mathcal{I}_\rho^{mso}$, it follows that $x \in \mathsf{lean}(\phi)$ iff $x \in \mathsf{lean}(\theta_k)$ or $x \neq last$ with $x \in \mathsf{lean}(\theta_j)$ also $x + 1 \in \mathsf{lean}(\phi)$. Moreover, $\mathsf{lean}(\phi) = Q_{\theta_a}$, where Q_{θ_a} is the corresponding second-order predicate. Thus, by induction hypothesis $\rho, x \models \phi$.
- If $\phi = (\theta_j R \theta_k)$, then $\mathsf{t}_\lambda(\phi, x) = (\mathsf{lean}(\phi)(x) \leftrightarrow (\mathsf{lean}(\theta_k)(x) \wedge ((x = last) \vee \mathsf{lean}(\theta_j)(x) \vee \mathsf{lean}(\phi)(x+1))))$. Since $\mathsf{t}_\lambda(\phi)$ holds at every point x of $\mathcal{I}_\rho^{mso}$, it follows that $x \in \mathsf{lean}(\phi)$ iff $x \in \mathsf{lean}(\theta_k)$ with $x = last$ or $x \in \mathsf{lean}(\theta_j)$ or $x + 1 \in \mathsf{lean}(\phi)$. Moreover, $\mathsf{lean}(\phi) = Q_{\theta_a}$, where Q_{θ_a} is the corresponding second-order predicate. Thus, by induction hypothesis $\rho, x \models \phi$. $\qquad\square$

Having defined different variations of the MSO encoding, we now provide variations of the Compact MSO encoding described in Sect. 4.

Sloppy Formulation. The formulation described in Sect. 4 strictly tracks the computation over each BDD B_q, which we refer to *Fussy* formulation. That is, for each nonterminal node α, both the forward computation and previous computation must be tracked. This causes a high logical complexity in the formulation.

An alteration to diminish the logical complexity is to utilize a *Sloppy Formulation*, analogous to the *Sloppy* variation described above, that only tracks the forward computation. Since the previous computations are not tracked, none of the computations leading to terminal node 0 of the BDD B'_f enable an accepting condition.

To define the accepting condition of *Sloppy Formulation*, we have

$$\mathsf{Racc_s} = \left(\bigvee_{(\beta,v,d)\in\mathsf{PreT}(B'_f,0)} [x \in N_\beta \wedge x \in^d v] \right) \rightarrow (x \neq 0).$$

Moreover, $\mathsf{node_s}$ only requires $\mathsf{PostCon}$ of node. Therefore, we have

$$\mathsf{node_s} = \bigwedge_{1\leq\alpha\leq u} \mathsf{PostCon}.$$

The second-order formula $\mathsf{Rev_s}(\mathcal{F}_\psi)$ of *Sloppy Formulation* is defined as following:

$$\mathsf{Rev_s}(\mathcal{F}_\psi) = (\exists V_0)\ldots(\exists V_{k-1})(\exists N_1)\ldots(\exists N_u)(\forall x)(\mathsf{Rinit} \wedge \mathsf{node_s}$$
$$\wedge\mathsf{Rterminal} \wedge \mathsf{Racc_s} \wedge \mathsf{roots}),$$

where Rinit, $\mathsf{Rterminal}$ and roots are defined as in Sect. 4. Therefore, let $\mathsf{Cmso_s}(\phi)$ be the *Sloppy Formulation* of the Compact MSO encoding, we define $\mathsf{Cmso_s}(\phi) = \mathsf{Rev_s}(\mathcal{F}_\psi)$ asserting the truth of ϕ at position 0, where ψ is the corresponding PLTL$_f$ formula of ϕ, and $\mathcal{F}_\psi$ is the symbolic DFA of ψ. The following theorem asserts the correctness of the *Sloppy Formulation*.

Theorem 8. *The models of formula* $\mathsf{Cmso_s}(\phi)$ *are exactly the words satisfying* ϕ.

The proof here is analogous to that of the *Fussy Formulation*, where the crux is that we define the computation trace on a BDD as a sequence of sets of BDD nodes, instead of just a specific sequence of BDD nodes, see the definition of $\mathsf{node_s}$. Such definition still keeps unambiguous formulation of the symbolic DFA since we have stronger constraints on the accepting condition, as shown in the definition of $\mathsf{Racc_s}$.

5.2 Experimental Results

Having presented different optimizations, we now have 6 variations of the MSO encoding corresponding to specific optimizations, which are BNF-*Fussy-Full*, BNF-*Fussy-Lean*, NNF-*Fussy-Full*, NNF-*Fussy-Full*, NNF-*Sloppy-Full* and NNF-*Sloppy-Lean*. Moreover, we have two variations of the Compact MSO encoding, which are *Fussy* and *Sloppy*. The experiments were divided into two parts and resulted in two major findings. First we explored the benefits of the various optimizations of MSO encoding and showed that the most effective one is that of *Lean*. Second, we aimed to answer the question whether second-order outperforms first-order in the context of LTL$_f$-to-automata translation. To do so, we compared the best performing MSO encoding and Compact MSO encoding against the FOL encoding and showed the superiority of first-order.

Correctness. The correctness of the implementation of different encodings was evaluated by comparing the DFAs in terms of the number of states and transitions generated from each encoding. No inconsistencies were discovered.

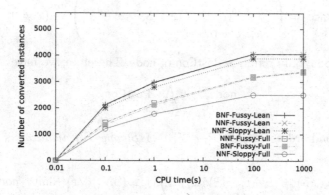

Fig. 1. Comparison over 6 variations of MSO encoding

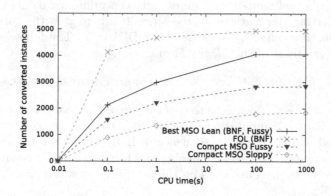

Fig. 2. Overall comparison of FOL, MSO and Compact MSO encodings

Lean Constraint Form Is More Effective in* MSO *Encodings. Fig. 1 presents the number of converted instances of each variation of MSO encoding, where the upper three are all for *Lean* variations and the lower ones are for *Full* variations. The choice of BNF vs NNF did not have a major impact, and neither did the choice of *Fussy* vs *Sloppy*. The one optimization that was particularly effective was that of *Lean* variation. The best-performing MSO encoding was BNF-*Fussy-Lean*. While in the Compact MSO encoding, the *Fussy* variation highly outperforms that of *Sloppy*, as shown in Fig. 2.

First-Order Logic Dominates Second-Order Logic for LTL$_f$ *-to-Automata Translation.* As presented in Fig. 2, FOL encoding shows its superiority over second-order encodings performance-wise, which are MSO encoding and Compact MSO encoding. Thus, the use of second-order logic, even under sophisticated optimization, did not prove its value in terms of performance. This suggests that nevertheless second-order encoding indicates a much simpler quantificational structure which theoretically leads to more potential space to optimize, it would be useful to have first-order as a better way in the context of LTL$_f$-to-automata translation in practice.

6   Concluding Remarks

In this paper, we revisited the translation from LTL$_f$ to automata and presented new second-order encodings, MSO encoding and Compact MSO encoding with various optimizations. Instead of the syntax-driven translation in FOL encoding, MSO encoding provides a semantics-driven translation. Moreover, MSO encoding allows a significantly simpler quantificational structure, which requires only a block of existential second-order quantifiers, followed by a single universal first-order quantifier, while FOL encoding involves an arbitrary alternation of quantifiers. The Compact MSO encoding simplifies further the syntax of the encoding, by introducing more second-order variables. Nevertheless, empirical evaluation showed that first-order encoding, in general, outperforms the second-order encodings. This finding suggests first-order encoding as a better way for LTL$_f$-to-automata translation.

To obtain a better understanding of the performance of second-order encoding of LTL$_f$, we looked more into MONA. An interesting observation is that MONA is an "aggressive minimizer": after each quantifier elimination, MONA re-minimizes the DFA under construction. Thus, the fact that the second-order encoding starts with a block of existential second-order quantifiers offers no computational advantage, as MONA eliminates the second-order quantifiers one by one, performing computationally heavy minimization after each quantifier. Therefore, a possible improvement to MONA would enable it to eliminate a whole *block* of quantifiers of the same type (existential or universal) in one operation, involving only one minimization. Currently, the quantifier-elimination strategy of one quantifier at a time is deeply hardwired in MONA, so the suggested improvement would require a major rewrite of the tool. We conjecture that, with such an extension of MONA, the second-order encodings would have a better performance, but this is left to future work.

Beyond the unrealized possibility of performance gained via second-order encodings, another motivation for studying such encodings is their greater expressivity. The fact that LTL$_f$ is equivalent to FOL [15] shows limited expressiveness of LTL$_f$. For this reason it is advocated in [6] to use *Linear Dynamic Logic* (LDL$_f$) to specify ongoing behavior. LDL$_f$ is expressively equivalent to MSO, which is more expressive than FOL. Thus, automata-theoretic reasoning for LDL$_f$, for example, reactive synthesis [8], cannot be done via first-order encoding

704 S. Zhu et al.

and requires second-order encoding. Similarly, synthesis of LTL_f with incomplete information requires the usage of second-order encoding [9]. We leave this too to future research.

Acknowledgments. Work supported in part by China HGJ Project No. 2017ZX0103 8102-002, NSFC Projects No. 61572197, No. 61632005 and No. 61532019, NSF grants IIS-1527668, IIS-1830549, and by NSF Expeditions in Computing project "ExCAPE: Expeditions in Computer Augmented Program Engineering". Special thanks to Jeffrey M. Dudek and Dror Fried for useful discussions.

References

1. Bloem, R., Galler, S.J., Jobstmann, B., Piterman, N., Pnueli, A., Weiglhofer, M.: Interactive pesentation: automatic hardware synthesis from specifications: a case study. In: DATE, pp. 1188–1193 (2007)
2. Bohy, A., Bruyère, V., Filiot, E., Jin, N., Raskin, J.-F.: Acacia+, a tool for LTL synthesis. In: Madhusudan, P., Seshia, S.A. (eds.) CAV 2012. LNCS, vol. 7358, pp. 652–657. Springer, Heidelberg (2012). https://doi.org/10.1007/978-3-642-31424-7_45
3. Bryant, R.E.: Symbolic Boolean manipulation with ordered binary-decision diagrams. ACM Comput. Surv. **24**(3), 293–318 (1992)
4. Burch, J., Clarke, E., McMillan, K., Dill, D., Hwang, L.: Symbolic model checking: 10^{20} states and beyond. Inf. Comput. **98**(2), 142–170 (1992)
5. Chandra, A., Kozen, D., Stockmeyer, L.: Alternation. J. ACM **28**(1), 114–133 (1981)
6. De Giacomo, G., Vardi, M.Y.: Linear temporal logic and linear dynamic logic on finite traces. In: IJCAI, pp. 854–860 (2013)
7. De Giacomo, G., De Masellis, R., Montali, M.: Reasoning on LTL on finite traces: insensitivity to infiniteness. In: AAAI, pp. 1027–1033 (2014)
8. De Giacomo, G., Vardi, M.Y.: Synthesis for LTL and LDL on finite traces. In: IJCAI, pp. 1558–1564 (2015)
9. De Giacomo, G., Vardi, M.Y.: LTL_f and LDL_f synthesis under partial observability. In: IJCAI, pp. 1044–1050 (2016)
10. Di Ciccio, C., Maggi, F.M., Mendling, J.: Efficient discovery of target-branched declare constraints. Inf. Syst. **56**, 258–283 (2016)
11. Ciccio, C.D., Mecella, M.: On the discovery of declarative control flows for artful processes. ACM Trans. Manag. Inf. Syst. **5**(4), 24:1–24:37 (2015)
12. Duret-Lutz, A., Lewkowicz, A., Fauchille, A., Michaud, T., Renault, E., Xu, L.: Spot 2.0 – a framework for LTL and ω-automata manipulation. In: ATVA, pp. 122–129 (2016)
13. Henriksen, J.G., et al.: Mona: monadic second-order logic in practice. In: Brinksma, E., Cleaveland, W.R., Larsen, K.G., Margaria, T., Steffen, B. (eds.) TACAS 1995. LNCS, vol. 1019, pp. 89–110. Springer, Heidelberg (1995). https://doi.org/10.1007/3-540-60630-0_5
14. Akers Jr., S.B.: Binary decision diagrams. IEEE Trans. Comput. **27**(6), 509–516 (1978)
15. Kamp, J.: Tense logic and the theory of order. Ph.D. thesis, UCLA (1968)
16. Klarlund, N., Møller, A., Schwartzbach, M.I.: MONA implementation secrets. In: Yu, S., Păun, A. (eds.) CIAA 2000. LNCS, vol. 2088, pp. 182–194. Springer, Heidelberg (2001). https://doi.org/10.1007/3-540-44674-5_15

17. Kupferman, O., Vardi, M.Y.: Safraless decision procedures. In: FOCS, pp. 531–540 (2005)
18. Kupferman, O., Vardi, M.Y.: Model checking of safety properties. Formal Methods Syst. Des. **19**(3), 291–314 (2001)
19. Lichtenstein, O., Pnueli, A., Zuck, L.: The glory of the past. In: Parikh, R. (ed.) Logic of Programs 1985. LNCS, vol. 193, pp. 196–218. Springer, Heidelberg (1985). https://doi.org/10.1007/3-540-15648-8_16
20. Biehl, M., Klarlund, N., Rauhe, T.: Mona: decidable arithmetic in practice. In: Jonsson, B., Parrow, J. (eds.) FTRTFT 1996. LNCS, vol. 1135, pp. 459–462. Springer, Heidelberg (1996). https://doi.org/10.1007/3-540-61648-9_56
21. Pan, G., Sattler, U., Vardi, M.Y.: BDD-based decision procedures for $\mathcal{K}$. In: Voronkov, A. (ed.) CADE 2002. LNCS (LNAI), vol. 2392, pp. 16–30. Springer, Heidelberg (2002). https://doi.org/10.1007/3-540-45620-1_2
22. Pan, G., Vardi, M.Y.: Optimizing a BDD-based modal solver. In: Baader, F. (ed.) CADE 2003. LNCS (LNAI), vol. 2741, pp. 75–89. Springer, Heidelberg (2003). https://doi.org/10.1007/978-3-540-45085-6_7
23. Pnueli, A., Rosner, R.: On the synthesis of a reactive module. In: POPL, pp. 179–190 (1989)
24. Pnueli, A.: The temporal logic of programs. In: FOCS, pp. 46–57 (1977)
25. Prescher, J., Di Ciccio, C., Mendling, J.: From declarative processes to imperative models. In: SIMPDA 2014, pp. 162–173 (2014)
26. Rozier, K.Y., Vardi, M.Y.: LTL satisfiability checking. In: Bošnački, D., Edelkamp, S. (eds.) SPIN 2007. LNCS, vol. 4595, pp. 149–167. Springer, Heidelberg (2007). https://doi.org/10.1007/978-3-540-73370-6_11
27. Rozier, K.Y., Vardi, M.Y.: A multi-encoding approach for LTL symbolic satisfiability checking. In: Butler, M., Schulte, W. (eds.) FM 2011. LNCS, vol. 6664, pp. 417–431. Springer, Heidelberg (2011). https://doi.org/10.1007/978-3-642-21437-0_31
28. Rozier, K.Y., Vardi, M.Y.: Deterministic compilation of temporal safety properties in explicit state model checking. In: Biere, A., Nahir, A., Vos, T. (eds.) HVC 2012. LNCS, vol. 7857, pp. 243–259. Springer, Heidelberg (2013). https://doi.org/10.1007/978-3-642-39611-3_23
29. Tabakov, D., Rozier, K.Y., Vardi, M.Y.: Optimized temporal monitors for SystemC. Formal Methods Syst. Des. **41**(3), 236–268 (2012)
30. Zhu, S., Tabajara, L.M., Li, J., Pu, G., Vardi, M.Y.: A symbolic approach to safety LTL synthesis. Hardware and Software: Verification and Testing. LNCS, vol. 10629, pp. 147–162. Springer, Cham (2017). https://doi.org/10.1007/978-3-319-70389-3_10
31. Zhu, S., Tabajara, L.M., Li, J., Pu, G., Vardi, M.Y.: Symbolic LTL$_f$ synthesis. In: IJCAI, pp. 1362–1369 (2017)

Author Index

Printed in the United States
By Bookmasters